21st Century Nanoscience –

21st Century Nanoscience – A Handbook: Nanophysics Sourcebook (Volume One)

21st Century Nanoscience – A Handbook: Design Strategies for Synthesis and Fabrication (Volume Two)

21st Century Nanoscience – A Handbook: Advanced Analytic Methods and Instrumentation (Volume Three)

21st Century Nanoscience – A Handbook: Low-Dimensional Materials and Morphologies (Volume Four)

21st Century Nanoscience – A Handbook: Exotic Nanostructures and Quantum Systems (Volume Five)

21st Century Nanoscience – A Handbook: Nanophotonics, Nanoelectronics, and Nanoplasmonics (Volume Six)

21st Century Nanoscience – A Handbook: Bioinspired Systems and Methods (Volume Seven)

21st Century Nanoscience – A Handbook: Nanopharmaceuticals, Nanomedicine, and Food Nanoscience (Volume Eight)

21st Century Nanoscience – A Handbook: Industrial Applications (Volume Nine)

21st Century Nanoscience – A Handbook: Public Policy, Education, and Global Trends (Volume Ten)

21st Century Nanoscience – A Handbook

Nanophotonics, Nanoelectronics, and Nanoplasmonics (Volume Six)

Edited by

Klaus D. Sattler

CRC Press
Taylor & Francis Group
Boca Raton London New York

CRC Press is an imprint of the
Taylor & Francis Group, an **informa** business

MATLAB® is a trademark of The MathWorks, Inc. and is used with permission. The MathWorks does not warrant the accuracy of the text or exercises in this book. This book's use or discussion of MATLAB® software or related products does not constitute endorsement or sponsorship by The MathWorks of a particular pedagogical approach or particular use of the MATLAB® software.

CRC Press
Taylor & Francis Group
6000 Broken Sound Parkway NW, Suite 300
Boca Raton, FL 33487-2742

First issued in paperback 2022

© 2021 by Taylor & Francis Group, LLC
CRC Press is an imprint of Taylor & Francis Group, an Informa business

No claim to original U.S. Government works

ISBN-13: 978-0-815-35641-7 (hbk)
ISBN-13: 978-1-03-233589-6 (pbk)
DOI: 10.1201/9780429351617

Library of Congress Cataloging-in-Publication Data

Names: Sattler, Klaus D., editor.
Title: 21st century nanoscience : a handbook / edited by Klaus D. Sattler.
Description: Boca Raton, Florida : CRC Press, [2020] | Includes bibliographical references and index. | Contents: volume 1. Nanophysics sourcebook—volume 2. Design strategies for synthesis and fabrication—volume 3. Advanced analytic methods and instrumentation—volume 5. Exotic nanostructures and quantum systems—volume 6. Nanophotonics, nanoelectronics, and nanoplasmonics—volume 7. Bioinspired systems and methods. | Summary: "This 21st Century Nanoscience Handbook will be the most comprehensive, up-to-date large reference work for the field of nanoscience. Handbook of Nanophysics, by the same editor, published in the fall of 2010, was embraced as the first comprehensive reference to consider both fundamental and applied aspects of nanophysics. This follow-up project has been conceived as a necessary expansion and full update that considers the significant advances made in the field since 2010. It goes well beyond the physics as warranted by recent developments in the field"—Provided by publisher.
Identifiers: LCCN 2019024160 (print) | LCCN 2019024161 (ebook) | ISBN 9780815384434 (v. 1 ; hardback) | ISBN 9780815392330 (v. 2 ; hardback) | ISBN 9780815384731 (v. 3 ; hardback) | ISBN 9780815355281 (v. 4 ; hardback) | ISBN 9780815356264 (v. 5 ; hardback) | ISBN 9780815356417 (v. 6 ; hardback) | ISBN 9780815357032 (v. 7 ; hardback) | ISBN 9780815357070 (v. 8 ; hardback) | ISBN 9780815357087 (v. 9 ; hardback) | ISBN 9780815357094 (v. 10 ; hardback) | ISBN 9780367333003 (v. 1 ; ebook) | ISBN 9780367341558 (v. 2 ; ebook) | ISBN 9780429340420 (v. 3 ; ebook) | ISBN 9780429347290 (v. 4 ; ebook) | ISBN 9780429347313 (v. 5 ; ebook) | ISBN 9780429351617 (v. 6 ; ebook) | ISBN 9780429351525 (v. 7 ; ebook) | ISBN 9780429351587 (v. 8 ; ebook) | ISBN 9780429351594 (v. 9 ; ebook) | ISBN 9780429351631 (v. 10 ; ebook)
Subjects: LCSH: Nanoscience—Handbooks, manuals, etc.
Classification: LCC QC176.8.N35 A22 2020 (print) | LCC QC176.8.N35 (ebook) | DDC 500—dc23
LC record available at https://lccn.loc.gov/2019024160
LC ebook record available at https://lccn.loc.gov/2019024161

Visit the Taylor & Francis Web site at
http://www.taylorandfrancis.com

and the CRC Press Web site at
http://www.crcpress.com

Contents

Editor vii
Contributors ix

1   Nanophotonic Devices Based on Low-Voltage Emission of 2D Electron Gas
    Hong Koo Kim, Myungji Kim, Daud Emon, Siyang Liu, Yu Shi, and Haiming Guo 1-1
2   Polarized Nano-Optics *Sophie Brasselet* 2-1
3   Optical Properties of Semiconductor Nanostructures *Andres Cantarero* 3-1
4   Hybrid Phase-Change Nanophotonic Circuits *Johannes Feldmann and Wolfram Pernice* . . 4-1
5   X-Ray Nanophotonics Based on Planar X-Ray Waveguide-Resonator *V. K. Egorov
    and E. V. Egorov* 5-1
6   Optical Tweezers *Jing Liu and Zhi-Yuan Li* 6-1
7   Metal Nanostructures with Plasmonically Enhanced Raman and Photoluminescence
    Signals *Jeong-Eun Park, Minho Kim, Jiwoong Son, Chungyeon Lee, Sung Min Ko,
    and Jwa-Min Nam* 7-1
8   GaN Nanoflowers: Growth to Optoelectronic Device *Neha Aggarwal,
    Shibin Krishna, and Govind Gupta* 8-1
9   The Future of Nanoelectronics *Simon Deleonibus* 9-1
10 Semimetal Electronics: Quantum Confinement and Surface Chemistry as Design Tools
    *Alfonso Sanchez-Soares, Christian König, Conor O'Donnell, Jean-Pierre Colinge, and
    James C. Greer* 10-1
11 Neuromorphic Nanoelectronics *Jessamyn A. Fairfield* 11-1
12 Single Electronics: Modeling and Simulation Techniques *B. S. Pês, E. Oroski,
    and J. G. Guimarães* 12-1
13 Single Electron Transport and Possible Quantum Computing in 2D Materials
    K. L. Chiu 13-1
14 Assembly of Plasmonic Nanoparticles *Eric H. Hill, Christoph Hanske, Cyrille Hamon,
    and Yuebing Zheng* 14-1
15 Coulomb Blockade Plasmonic Switch *Ali Khademi, Dao Xiang, and Reuven Gordon* 15-1
16 Amplification of Surface Plasmons *Gerard Tatel and M. S. Wartak* 16-1
17 Magneto-plasmonics in Purely Ferromagnetic Subwavelength Arrays *S. D. Pappas and
    E. Th. Papaioannou* 17-1
18 Cathodoluminescence of Nanoplasmonics *Naoki Yamamoto and Takumi Sannomiya* 18-1
19 Biosensing under Surface Plasmon Resonance Conditions *Boris Snopok* 19-1
20 Plasmonic Optical Antenna and Its Enhancement to Infrared Photodetectors
    Lin Li, Neda Mojaverian, Guiru Gu, and Xuejun Lu 20-1

Index I-1

Editor

Klaus D. Sattler pursued his undergraduate and master's courses at the University of Karlsruhe in Germany. He earned his PhD under the guidance of Professors G. Busch and H.C. Siegmann at the Swiss Federal Institute of Technology (ETH) in Zurich. For three years he was a Heisenberg fellow at the University of California, Berkeley, where he initiated the first studies with a scanning tunneling microscope of atomic clusters on surfaces. Dr. Sattler accepted a position as professor of physics at the University of Hawaii, Honolulu, in 1988. In 1994, his group produced the first carbon nanocones. His current work focuses on novel nanomaterials and solar photocatalysis with nanoparticles for the purification of water. He is the editor of the sister references, *Carbon Nanomaterials Sourcebook* (2016) and *Silicon Nanomaterials Sourcebook* (2017), as well as *Fundamentals of Picoscience* (2014). Among his many other accomplishments, Dr. Sattler was awarded the prestigious Walter Schottky Prize from the German Physical Society in 1983. At the University of Hawaii, he teaches courses in general physics, solid-state physics, and quantum mechanics.

Contributors

Neha Aggarwal
Advanced Materials & Devices
    Division
CSIR-National Physical Laboratory
New Delhi, India

Sophie Brasselet
Institut Fresnel
Aix Marseille University, CNRS,
    Centrale Marseille
Marseille, France

Andres Cantarero
Molecular Science Institute
University of Valencia
Valencia, Spain

K. L. Chiu
Department of Physics
National Sun Yat-sen University
Kaohsiung, Taiwan

Jean-Pierre Colinge
CEA, LETI
Grenoble, France

Simon Deleonibus
Université Grenoble Alpes
CEA, LETI
Grenoble, France

E. V. Egorov
IMT RAS
Chernogolovka, Russia
and
Department of Information Security
Financial University under the
    Government of Russian Federation
Moscow, Russia

V. K. Egorov
IMT RAS
Chernogolovka, Russia

Daud Emon
Department of Electrical and
    Computer Engineering
and
Petersen Institute of NanoScience and
    Engineering
University of Pittsburgh
Pittsburgh, Pennsylvania

Jessamyn A. Fairfield
School of Physics
National University of Ireland Galway
Galway, Ireland

Johannes Feldmann
Institute of Physics
WWU Münster
Münster, Germany

Reuven Gordon
Department Electrical and Computer
    Engineering
University of Victoria
Victoria, British Columbia, Canada

James C. Greer
University of Nottingham Ningbo
China
Ningbo, China

Guiru Gu
Department of Physics
Stonehill College
Easton, Massachusetts

J. G. Guimarães
UFSC
Blumenau, Brazil

Haiming Guo
Department of Electrical and
    Computer Engineering
and
Petersen Institute of NanoScience and
    Engineering
University of Pittsburgh
Pittsburgh, Pennsylvania

Govind Gupta
Advanced Materials & Devices
    Division
CSIR-National Physical Laboratory
New Delhi, India

Cyrille Hamon
Laboratoire de Physique des Solides,
    CNRS UMR8502
University of Paris-Sud
Orsay Cedex, France

Christoph Hanske
BioNanoPlasmonics Laboratory
CIC biomaGUNE
San Sebastián, Spain

Eric H. Hill
Department of Mechanical
    Engineering
Texas Materials Institute
The University of Texas at Austin
Austin, Texas

Ali Khademi
Department Electrical and Computer
    Engineering
University of Victoria
Victoria, British Columbia, Canada

Hong Koo Kim
Department of Electrical and
    Computer Engineering
and
Petersen Institute of NanoScience and
    Engineering
University of Pittsburgh
Pittsburgh, Pennsylvania

Minho Kim
Department of Chemistry
Seoul National University
Seoul, South Korea

Myungji Kim
Department of Electrical and
    Computer Engineering
and
Petersen Institute of NanoScience and
Engineering
University of Pittsburgh
Pittsburgh, Pennsylvania

Sung Min Ko
Department of Chemistry
Seoul National University
Seoul, South Korea

Christian König
Tyndall National Institute
Cork, Ireland

Shibin Krishna
Advanced Materials & Devices
    Division
CSIR-National Physical Laboratory
New Delhi, India

Chungyeon Lee
Department of Chemistry
Seoul National University
Seoul, South Korea

Lin Li
Department of Electrical and
    Computer Engineering
University of Massachusetts Lowell
Lowell, Massachusetts

Zhi-Yuan Li
School of Physics and Optoelectronics
South China University of Technology
Guangzhou, China

Jing Liu
Institute of Laser and Intelligent
    Manufacturing Technology
South-Central University for
    Nationalities
Wuhan, China

Siyang Liu
Department of Electrical and
    Computer Engineering
and
Petersen Institute of NanoScience and
    Engineering
University of Pittsburgh
Pittsburgh, Pennsylvania

Xuejun Lu
Department of Electrical and
    Computer Engineering
University of Massachusetts Lowell
Lowell, Massachusetts

Neda Mojaverian
Department of Electrical and
    Computer Engineering
University of Massachusetts Lowell
Lowell, Massachusetts

Jwa-Min Nam
Department of Chemistry
Seoul National University
Seoul, South Korea

Conor O'Donnell
Tyndall National Institute
Cork, Ireland

E. Oroski
UTFPR
Curitiba, Brazil

B. S. Pês
IFPR
Campo Largo, Brazil

E. Th. Papaioannou
Fachbereich Physik and
    Forschungszentrum OPTIMAS
Technische Universität Kaiserslautern
Kaiserslautern, Germany
and
Martin-Luther Universität Halle
    Wittenberg
Halle, Germany

S. D. Pappas
Fachbereich Physik and
    Forschungszentrum OPTIMAS
Technische Universität Kaiserslautern
Kaiserslautern, Germany

Jeong-Eun Park
Department of Chemistry
Seoul National University
Seoul, South Korea

Wolfram Pernice
Institute of Physics
WWU Münster
Münster, Germany

Alfonso Sanchez-Soares
EOLAS Designs
Cork, Ireland

Takumi Sannomiya
School of Materials and Chemical
    Technologies
Tokyo Institute of Technology
Yokohama, Japan

Yu Shi
Department of Electrical and
    Computer Engineering
and
Petersen Institute of NanoScience and
    Engineering
University of Pittsburgh
Pittsburgh, Pennsylvania

Boris Snopok
V. E. Lashkaryov Institute of
    Semiconductor Physics
National Academy of Sciences of
    Ukraine
Kyiv, Ukraine

Jiwoong Son
Department of Chemistry
Seoul National University
Seoul, South Korea

Gerard Tatel
Department of Physics and Computer
    Science
Wilfrid Laurier University
Waterloo, Ontario, Canada

M. S. Wartak
Department of Physics and Computer
    Science
Wilfrid Laurier University
Waterloo, Ontario, Canada

Dao Xiang
Department Electrical and Computer
    Engineering
University of Victoria
Victoria, British Columbia, Canada

Naoki Yamamoto
School of Materials and Chemical
    Technologies
Tokyo Institute of Technology
Yokohama, Japan

Yuebing Zheng
Department of Mechanical
    Engineering
Texas Materials Institute
The University of Texas at Austin
Austin, Texas

Nanophotonic Devices Based on Low-Voltage Emission of 2D Electron Gas

Hong Koo Kim, Myungji Kim,
Daud Emon, Siyang Liu, Yu Shi,
and Haiming Guo
University of Pittsburgh

1.1 Introduction 1-1
1.2 Theory: Low-Voltage Emission of 2DEG 1-2
Analysis of a Graphene/SiO₂/Si (GOS) Capacitor Structure under Dark,
Inversion Bias
1.3 UV-Enhanced Photodetection in a GOS Structure with a Void
Channel 1-4
Emission of 2DEG in MOS • Photocurrent Response of GOS with a Void
Channel • Emission and Transport of Photocarrier 2DEG • Spectral
Dependence of Photocurrent Response
1.4 Nanoscale Quantum-Dot Organic Light-Emitting
Diode (QD-OLED) 1-11
Device Fabrication: QD-OLED with Edge Injection • Edge Emission from
QD-OLED • Nanoparticle-Patterned Etch Mask for Nanohole Fabrication •
Nano-QD-OLED Fabrication and Characterization
Acknowledgments 1-16
References 1-16

1.1 Introduction

Electrons (or holes) can be harbored at an interface of heterogeneous materials, forming quasi-two-dimensional electron systems (2DESs) (Ando et al. 1982). Being confined in a potential well at the interface, charge carriers can freely travel along the in-plane direction but not in the normal direction and therefore are called two-dimensional electron gas (2DEG) or two-dimensional hole gas (2DHG). In a strict sense, the 2DES is not two-dimensional because wave functions spread to both sides of interface. In most materials systems, the spatial extent of lateral spreading ranges from 1–2 to ~10 nm, and quantum confinement effects can be readily observed in the tightly confined case.

The 2DES phenomenon has been observed in a variety of heterogeneous materials interfaces, broadly termed heterostructures. The best known examples are carriers confined to the vicinity of junctions between insulators and semiconductors, such as metal-oxide-semiconductor (MOS) structure, between layers of different semiconductors, such as heterojunctions of III–V, II–VI or III–VI compounds, and intercalated graphite. Electrons can also be confined to the surface of a material. Free electrons, for example, can float and move along on the surface of liquid helium: some of the earlier work in 2DEG utilized the carrier confinement effect of an image-charge potential formed on helium surface (Sommer 1964). More recently, it has been found that solid insulators, such as topological insulators, can

also support conductive surface electronic states [topological insulators].

In most of these 2D systems, the carrier concentration can be varied by altering surface conditions or by applying electric field normal to the surface, thereby opening up a new avenue for numerous device applications. The most successful example is the silicon MOS field-effect transistor (MOSFET), where a superb natural insulating oxide makes a high-quality interface with Si. The in-plane conductance of 2DES confined at the interface is modulated by vertical electric field applied across a MOS capacitor structure with gate bias voltage (Figure 1.1). The first proposal of this kind of field-effect devices dates back to the 1930s (Lilienfeld's and Heil's patents). Since Shockley and Pearson's study (Shockley & Pearson 1948), this device concept had been developed into MOSFET, which began to be employed for amplification and logic functions in the 1960s. Since then, MOSFET has been serving as a basic building block in microelectronics.

In the case of compound semiconductor heterostructures, the GaAs/AlGaAs system is one of the most extensively studied ones. This is because the material system provides a good match of their lattice constants and thermal expansion coefficients between GaAs and AlGaAs allowing epitaxial growth with high interface quality. The quasi-triangular potential well at the interface can confine electrons and can lead to quantum effects similar to those observed in MOS structures (Figure 1.1). This compound semiconductor

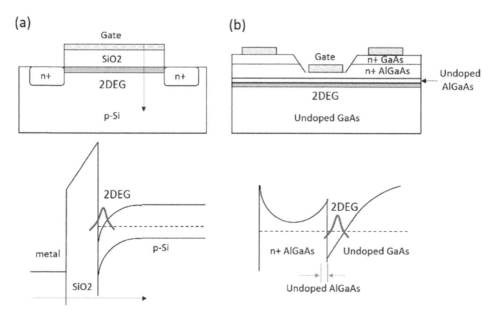

FIGURE 1.1 2DEG formed in semiconductor FET devices. (a) MOS. (b) High-electron-mobility transistor (HEMT).

heterostructure had been further developed into HEMTs: in this device concept, the 2DEG confined to a potential well exhibits greater mobility than those in MOSFET, since the former device utilizes an undoped channel thereby mitigating the deleterious effect of ionized impurity scattering (Sze 1990). The HEMT devices based on III–V compound materials (such as GaN or InP) are employed for high-speed electronic circuits operating at RF/microwave frequencies.

A different class of heterostructures that can host 2DEGs are oxides, e.g., ZnO/ZnMgO and LaAlO$_3$/SrTiO$_3$ (Kozuka et al. 2011, Ohtomo & Hwang 2004). The interface between LaAlO$_3$ and SrTiO$_3$ exhibits interesting properties that cannot be found in its constituent materials. Individually, LaAlO$_3$ and SrTiO$_3$ are nonmagnetic insulators, yet their interfaces can exhibit electrical conductivity and large magnetoresistance (Brinkman et al. 2007). In particular, it is notable that the LaAlO$_3$/SrTiO$_3$ interface can accommodate a very high level of 2DEG concentration, surpassing 10^{14} cm^{-2} level.

Recently, atomically thin solid materials have been developed, such as graphene and transition metal dichalcogenide (TMDC), where electrons are confined to atomic layer thickness (Novoselov et al. 2005, Zhang et al. 2005, Wang et al. 2012). The 2DES in graphene can be tuned to either 2DEG or 2DHG by applying transverse electric field (i.e., gating) or by performing chemical doping. Aside from being in a variety of practical device applications, 2D systems allow many novel physical phenomena such as quantum Hall effect in 2DEG (von Klitzing 1980) and Hofstadter's butterfly, a fractal structure in the energy versus magnetic field plot (Dean et al. 2013).

While the in-plane transport of 2DES along interface is the standard configuration commonly adopted in device applications, we have explored alternative channel media/configurations in a MOS structure by involving

low-voltage emission of 2DEG and a subsequent, ultrafast energy-efficient transport of electrons in a nanoscale vacuum (void) channel formed in SiO$_2$/Si substrate. We recently reported low-voltage ($<$3 V) operation of a vacuum FET (Srisonphan 2012): a 2DEG is extracted out at low bias ($\sim$1 V) and is injected into a nanovoid channel that is vertically etched into a MOS structure. Coulombic repulsion of electrons at channel edge is found to reduce the energy barrier for electron emission, leading to a high-current-density emission at low bias. In this chapter, we review two device applications that utilize this low-voltage emission phenomenon of 2DEG for Si-based nanophotonic devices.

1.2   Theory: Low-Voltage Emission of 2DEG

A MOS with 10–20 nm oxide thickness can easily harbor high-concentration (10^{12}–10^{13} cm^{-2}) electron (or hole) gas at relatively low bias ($<$5 V). Consider a MOS structure with a lateral extent that is finite, for example, terminated by cleaved edges or by vertically etched surface (Figure 1.2). The electric field near the edge is determined by an integral of Coulombic interactions with 2DEG. Considering the small thickness ($\sim$2 nm) of 2DEG, a strong shape-effect is expected in the in-plane direction near the edge: see the field and potential distributions plotted as a function of distance from the edge. The horizontal field sharply rises for nearer to the edge with surface field exceeding the critical level ($\sim$1 GV/m). From the potential distribution, it is also evident that the strong surface field near the edge has the effect of reducing energy barrier width. For example, 10^{13} cm^{-2} 2DEG concentration induces $\sim$1 GV/m surface field, which results in a narrow potential barrier (2-nm barrier width at 2-eV barrier height) enabling 2DEG emission into air at much reduced bias.

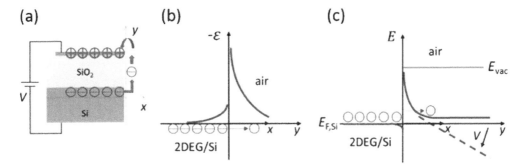

FIGURE 1.2 Low-voltage emission of 2DEG. (a) Schematic of a MOS structure with a cleaved edge. (b) Field distribution near the edge, plotted as a function of distance from the surface. (c) Potential distribution showing a significant reduction of energy barrier (height and width).

In graphene-based electronics/photonics research, a graphene-oxide-semiconductor (GOS) structure is commonly used in exploring device phenomena (Novoselov et al. 2012, Koppens et al. 2014, Kim & Kim 2015). Similar to the case of a MOS capacitor, a GOS structure can harbor two different types (inversion or accumulation) of 2D electron system at the oxide/Si interface, depending on the polarity of gate bias applied across the oxide layer. A switching control of this 2DES by oxide field is an essential function of GOS-based field-effect devices. Below we calculate the 2DEG density in a GOS structure under inversion bias.

1.2.1 Analysis of a Graphene/SiO$_2$/Si (GOS) Capacitor Structure under Dark, Inversion Bias

The bias voltage (V) applied to the graphene/SiO$_2$/Si structure drops across mainly three places: the flat band voltage (V_{FB}), across the band bending region in semiconductor (φ_s), and across the oxide layer (V_{ox}).

$$V = V_{FB} + \varphi_S + V_{ox} \tag{1.1}$$

Here the flat band voltage (V_{FB}) refers to the work function difference of graphene (ϕ_{graphene}) and semiconductor (ϕ_{Si}), expressed as $V_{FB} = \phi_{\text{graphene}} - \phi_{Si}$.

The work function of intrinsic (undoped) graphene is ~4.56 eV (Yu et al. 2009, Yan et al. 2012). The Fermi level of graphene (work function), however, varies depending on the carrier (electron or hole) concentration n_s, and the Fermi-level shift (referring to the Dirac point) can be characterized as

$$\Delta E_F = \hbar |v_F| \sqrt{\pi n_s} \tag{1.2}$$

where v_F is the Fermi velocity, 1.1×10^8 cm/s.

Considering the dependence of flat band voltage on the Fermi-level shift in graphene, Eq. (1.1) is adjusted as follows.

$$V = [(\phi_{\text{graphene}} \pm \Delta E_F) - \phi_{Si}] + \varphi_S + V_{ox} \tag{1.3}$$

Here the positive sign is for p-type graphene, and the negative sign is for the n-type case.

The space charge density (Q_s) in the semiconductor side can be determined by solving the Poisson equation, and is expressed as follows (Sze 1981).

$$Q_s = -\varepsilon_s E_s = -\frac{\sqrt{2}\varepsilon_s}{\beta L_D} \left[\left(e^{-\beta\varphi_s} + \beta\varphi_s - 1 \right) \right.$$
$$\left. + \frac{n_{p0}}{p_{po}} \left(e^{\beta\varphi_s} - \beta\varphi_s - 1 \right) \right]^{\frac{1}{2}} \text{ for p-Si} \tag{1.4}$$

where ε_s is the permittivity of semiconductor and E_s is the electric field at the interface with the oxide layer. φ_s is the band bending at the semiconductor/oxide interface called the surface potential. $\beta = q/kT$, and L_D is the extrinsic Debye length for holes, given as $L_D = \sqrt{\frac{\varepsilon_s}{q p_{po} \beta}}$. n_{po} and p_{po} are the equilibrium densities of electrons and holes, respectively.

The voltage drop across the oxide layer (V_{ox}) is related to the space charge in Si (Q_s) and oxide capacitance ($C_{ox} = \varepsilon_{ox}/d$) as follows.

$$V_{ox} = |Q_s| / C_{ox} \tag{1.5}$$

Across the oxide layer, the same amount of charge n_s ($= Q_s/q$) (of opposite polarity) appears in the graphene side.

Solving the above Eqs. (1.2)–(1.5) simultaneously, the space charge density in Si, Q_s can be calculated as a function of applied voltage V.

Figure 1.3 shows the surface potential in Si, φ_s, space charge density in Si, Q_s, depletion width (W), and the graphene work function (φ_g), which were calculated for bias voltage V up to 5 V reverse bias for p-Si substrate. In this GOS capacitor structure, the 2DEG density induced in the graphene layer is equal to the space charge density in Si, and at 5 V, it reaches ~5 × 10^{12} cm^{-2} level.

In this calculation, the following numbers were assumed: the work function of graphene, 4.56 eV; electron affinity of Si, 4.15 eV; electron affinity of SiO$_2$, 0.95 eV; dielectric constant of SiO$_2$, 3.9; dielectric constant of Si, 11.8 (Lide 2006); $N_A = 1.32 \times 10^{15}$ cm^{-3}; and SiO$_2$ thickness, 20 nm.

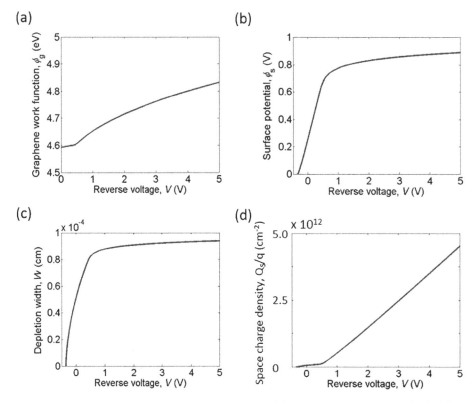

FIGURE 1.3 Analysis of a GOS capacitor structure under reverse bias. (a) Graphene work function (φ_g). (b) Surface potential (φ_s). (c) Depletion width (W). (d) Space charge density in Si (Q_s). Calculated as a function of reverse bias voltage (V) applied to a p-Si GOS in darkness.

1.3   UV-Enhanced Photodetection in a GOS Structure with a Void Channel

Detection of light is an essential function of silicon photonics. Silicon photodiodes are widely employed in sensing and imaging applications in the visible to near-infrared (NIR) range. Attaining high sensitivity in the ultraviolet (UV) range, however, remains an intrinsic challenge. This is primarily due to the fact that UV photons are absorbed near the surface of Si (e.g., within 10 nm absorption depth at 325 nm wavelength), and therefore most photocarriers are lost via surface recombination. In order to mitigate the problem, a shallow diffused layer or a charge-induced inversion layer is commonly introduced, which induces depletion field near the surface thereby sweeping photogenerated minority carriers away from defect-prone surface (Hansen 1978, Korde & Geist 1987). The performances of photodiodes with the adapted structures are still affected by presence of defects in the diffused or surface charge region, showing a limited degree of improvement (e.g., UV responsivity of ~0.2 A/W).

We have investigated photodetection properties of a graphene/SiO$_2$/Si structure with a void channel and demonstrated a UV-enhanced performance. Here, the GOS capacitor structure is designed to develop a depletion region directly controlled by bias voltage, where photocarrier separation occurs, and to have a potential well at a well-passivated Si surface, where photogenerated minority carriers are confined forming a 2D inversion channel. The photoinduced 2DEG is extracted out to air and transported to a graphene-covered void channel. The photocurrent response of this GOS structure demonstrates high responsivity over a broad spectral range, especially showing significant enhancement in UV (e.g., 0.43 A/W at 325 nm).

In this work, graphene is employed to serve two purposes: (i) the part placed on SiO$_2$ is utilized as a transparent electrode, which provides depletion bias while allowing direct illumination of light into Si substrate, and (ii) the part suspended on top of a void channel serves as a collector electrode for electrons that are transported out of plane from the 2DEG layer.

Compared to the graphene/Si Schottky-diode-like structure reported in literature (Li et al. 2010, Chen et al. 2011, Miao et al. 2012, An et al. 2013, Shin et al. 2015), this GOS capacitor structure provides better confinement of photocarriers in a potential well at SiO$_2$/Si, which is well passivated by thermally grown SiO$_2$ and therefore less prone to surface recombination. The perpendicular configuration of photocarrier separation and transport directions (i.e., carrier separation along the vertical depletion field, followed by a transport along the horizontal 2DEG layer) in Si of GOS structure also allows wider (thicker) depletion region formation, which would benefit photocarrier generation/separation without compromising the carrier transport and collection efficiency.

1.3.1 Emission of 2DEG in MOS

In extracting photocarriers out of Si, we exploit the phenomenon that a 2DEG induced at SiO_2/Si of a MOS structure can easily emit into air (void channel) at low bias voltage (<5 V) (Srisonphan et al. 2012, 2014). A MOS wafer of Al/SiO_2 (23 nm)/n-Si structure was cleaved into two pieces: the cleaved facet provides a vertical transport channel on the air side, which connects the 2DEG layer at SiO_2/Si and the Al electrode on top (Figure 1.4a). Note that the channel length of the vertical transport in air is determined by the oxide layer thickness (23 nm) of MOS and is designed to be smaller than the mean-free path of electrons in air (~60 nm). This is to ensure that emitted electrons travel scattering free in the ambient (air) channel, as if in a vacuum.

The current-versus-voltage (I–V) measurement of the cleaved MOS clearly demonstrates a rectifying characteristic. Under forward accumulation bias, the MOS current reveals $V^{3/2}$-voltage dependence, indicative of Child–Langmuir's space-charge-limited current in scattering-free vacuum environment (Child 1911, Langmuir 1913). The rise of current at relatively low voltage (<~1 V) indicates a significant reduction of energy barrier at the 2DEG/air interface. Besides the relatively high electric field (that is, ~1 V/23 nm = 4×10^5 V/cm) applied to this MOS capacitor, the Coulombic repulsion among electrons in 2DEG lowers the energy barrier at the cleaved edge as discussed above. Below we demonstrate this barrier lowering effect in both MOS and GOS structures and elucidate underlying mechanisms.

Electrons residing inside the bulk Si of a MOS structure are confined by energy barriers at the surface (with air) and at the interface (with SiO_2), typically with a vacuum barrier greater than the SiO_2 barrier. Electrons, therefore, cannot be easily emitted into air, unless the energy barrier is significantly lowered for thermionic or field emission (Langmuir & Kingdon 1923, Fowler & Nordheim 1928, Lau et al. 1994, Han et al. 2012) or for photoelectric emission (Fowler 1931, DuBridge 1932). For the thickness of SiO_2 layer (23 nm) employed in this work, the tunneling through the oxide barrier is expected to remain completely negligible. The leakage current through the oxide layer is measured to be of ~10 pA order as tested prior to cleaving (Figure 1.4b, bottom, dashed). This oxide leakage is several orders of magnitude smaller than the channel current observed with a cleaved MOS (~100 μA at 1 V), and this confirms negligible contribution by tunneling through oxide.

The situation can be very different for the electrons confined in a 2DEG layer. First, let's imagine an infinite extension of 2DEG at SiO_2/Si, that is, the case prior to cleaving of MOS wafer. The overall charge neutrality condition is maintained between the 2DEG in Si and the positive charges induced in the metal side (Ando et al. 1982, Mead 1961). Along the in-plane direction of 2DEG, Coulombic repulsion among electrons is canceled out due to the symmetry of electron distribution. Now consider a MOS structure whose lateral extent is finite, i.e., the 2DEG layer is terminated by a cleaved edge or a vertically etched trench structure. Electrons at the channel edge will then experience net repulsive force from neighboring electrons inside the 2DEG layer. In the case that the charge neutrality is maintained by relatively remote charges (e.g., opposite polarity charges induced across the oxide layer of MOS capacitor), strong in-plane Coulombic repulsion is expected to be dominant in the local area around the edge of 2DEG, and this can significantly alter the electrostatic potential there as discussed above (Figure 1.2) (Han & Ihm 2000, Zheng et al. 2004, Mayer 2005). For the case of the MOS capacitor structure being discussed here, for example, a 2DEG density of ~10^{12} cm^{-2} is obtainable at ~1 V accumulation bias (Srisonphan et al. 2012). The average spacing between electrons in the 2DEG is then estimated to be ~10 nm, smaller than the oxide thickness (23 nm). This will ensure that the in-plane interaction of 2DEG electrons becomes stronger than the dipole charge interaction across the oxide layer.

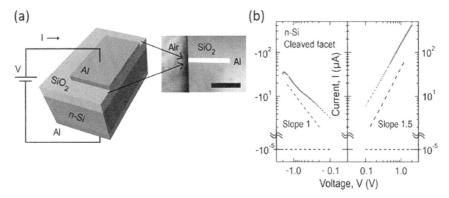

FIGURE 1.4 Emission and transport of 2DEG at a MOS capacitor with a cleaved facet. (a) A schematic drawing (left) and an optical micrograph (right: top view of Al electrode; scale bar, 200 μm) of device structure (Al/SiO_2 (23 nm)/n-Si). Coulombic repulsion among electrons in 2DEG at SiO_2/Si reduces the energy barrier at 2DEG/air. (b) Measured I–V characteristic. The cleaved sample clearly demonstrates a rectifying characteristic (forward slope of 1.5 and reverse slope of 0.5–1.0). The oxide leakage current (measured before cleaving) remains at 10 pA level (bottom, dashed). (Reprinted from Kim, M. & Kim, H. K., *J. Appl. Phys.* 118, 104504, 2015.)

Similarly the energy barrier in the metal side (metal/air interface) is lowered by the 2D positive charges induced at metal/SiO$_2$ interface, and the applied capacitor voltage appears mostly across the air gap (i.e., the vertical void channel, whose length is defined by the SiO$_2$ layer thickness).

The voltage dependence of 2DEG emission is governed by the capacitor relationship, $Q_e \sim V$. Here Q_e represents the 2DEG charge at SiO$_2$/Si and approximately equals the capacitor charge when accumulation biased. A scattering-free transport of electrons in a void channel converts the potential energy (eV) to kinetic energy ($m^* v^2/2$). The terminal velocity is expressed as $\sqrt{\frac{2eV}{m^*}}$, and the average transit time across the channel has the following voltage dependence: $\tau_{av} \sim V^{-0.5}$. The overall voltage dependence of channel current can then be expressed as $I = Q_e/\tau_{av} \sim V^{1.5}$, well matching the measurement result (Figure 1.1e): the slope of Figure 1.5 at forward accumulation bias). Here it should be mentioned that the Child–Langmuir's three-halves-power law assumes space-charge-limited emission at cathode (Child 1911, Langmui 1913, Lau et al. 1994). In our case, the electrons injected into air (i.e., a nanovoid channel) form a space charge around the anode (Al electrode) edge, and this space charge field limits emission of 2DEG at cathode edge. In reverse bias, part of the bias voltage goes to depletion region formation in Si; therefore the void-channel section receives less voltage than the accumulation case. This explains the reduced slope (0.5–1.0) in reverse bias (Figure 1.4b).

1.3.2 Photocurrent Response of GOS with a Void Channel

Now consider a graphene/SiO$_2$/p-Si (GOS) capacitor structure with a void channel (Figure 1.5). Under illumination at reverse bias, electron–hole pairs are generated in Si and separated by depletion field. The photogenerated minority carriers (electrons) drift to the Si/SiO$_2$ interface and become confined at the potential well there, forming a 2DEG inversion layer. The photoinduced 2DEG at SiO$_2$/Si is expected to be emitted into the void channel, similar to the case of accumulation 2DEG in a dark, forward-biased MOS capacitor as discussed above.

A GOS capacitor structure with a void channel was fabricated by employing electron-beam lithography (EBL), plasma reactive-ion etching (RIE), and graphene transfer processes (Figure 1.5a). In brief, a SiO$_2$ layer (23 nm thickness) was grown by thermal oxidation on p-type Si wafer (B-doped, 10 Ω-cm resistivity; (100)-oriented; 525 μm thickness). A bottom-side electrode was prepared by depositing a 150-nm-thick Al layer (5N purity) on Si by thermal evaporation, followed by Ohmic contact annealing at 350°C. Vertically etched trench or hole structures were then formed by EBL and RIE: a single or 3 × 3 holes (100–150 nm diameter, 100 nm depth, and 500 nm spacing) or a single vertical trench structure (100–440 nm width, 100–200 nm depth, and 1-mm trench length). A monolayer graphene (grown by chemical vapor deposition method, 3 mm × 4 mm size) was then transferred to the hole- or trench-etched SiO$_2$/Si substrate (Suk et al. 2011).

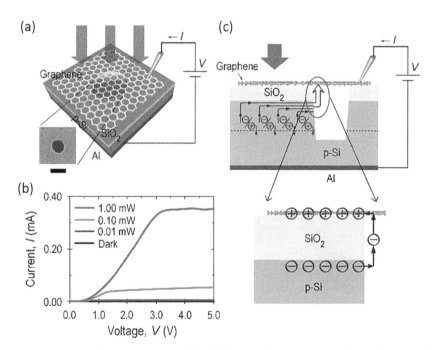

FIGURE 1.5 Photocurrent response of a graphene/oxide/Si (GOS) capacitor structure with a void channel. (a) Schematic of GOS structure (graphene/SiO$_2$ (23 nm)/p-Si) with a hole channel (150 nm diameter, 100 nm depth) covered by a suspended graphene electrode (3 mm × 4 mm). SEM image of hole channel. Scale bar, 200 nm. (b) Photo I–V characteristic measured at reverse bias under illumination with a 633 nm laser beam (diameter, 1 mm). (c) Schematic of emission and transport of photogenerated 2DEG. (Reprinted from Kim, M. & Kim, H. K., *J. Appl. Phys.* 118, 104504, 2015.)

Figure 1.5b shows the photo I–V characteristic of a GOS with a single hole measured under illumination with a 633-nm laser light (1-mm beam diameter). In reverse bias, the photocurrent saturates at 1.0–3.0 V for 0.1–1.0 mW input power. Under 1.0-mW illumination, the saturation photocurrent is read to be 0.35 mA at 3.0 V. This corresponds to a responsivity of 0.35 A/W and internal quantum efficiency (IQE) of 103%.

The absorption depth of Si is 3.0 μm at 633 nm wavelength. Since a monolayer graphene absorbs only 2.3% of incident light, most photons are absorbed in/near the depletion region in Si (910–940 nm width at 2–5 V reverse bias: see Figure 1.3c). Photogenerated carriers are separated by depletion field, and photoelectrons drift to the Si/SiO$_2$ interface forming a 2DEG inversion layer. Similar to the case of dark forward-biased MOS discussed above, Coulombic repulsion among electrons around the channel edge is expected to reduce the energy barrier enabling low-voltage emission of 2DEG into air. Emitted electrons travel ballistically in the nanovoid channel and impinge upon the suspended graphene.

Figure 1.6 depicts a band diagram that schematically illustrates electron emission, transport, and capture processes occurring at 2DEG/air, void channel, and graphene, respectively. Note that the energy band of suspended graphene is drawn flanked by the image-charge potentials on both sides, implying the presence of near-surface states. Image-potential states are a class of surface states that are widely observed on metallic surfaces, but they differ from conventional surface states in that they exist in ambient conditions outside the surface (Cole & Cohen 1969, Silkin et al. 2009, Niesner & Fauster 2014). Their origin can be traced back to the attractive potential acting on an electron at distance in front of a conductive surface. Replacing the metal surface by a conducting graphene sheet, image-potential states are expected to form in air on both sides of suspended graphene.

Some of the impinging electrons are captured/collected at the edge of 2D hole system (2DHS) in graphene, while majority initially pass through the suspended graphene (Figure 1.6b). The transmitted electrons, however, reflect back at the image-potential barrier. Greater the kinetic energy of impinging electrons, the farther they will be transmitted through the graphene. For the bias voltage (potential energy, eV) given smaller than the work function barrier, however, the transmitted electrons reflect back and are eventually captured/collected by graphene anode through back-and-forth reflections at the front and back-side image potential barriers. Assuming an image-charge potential profile, $\varphi = \frac{-1}{16\pi\varepsilon_0}e^2(\varepsilon - 1)/(\varepsilon + 1)x$ (Silkin et al. 2009, Stratton 1955), an electron impinging with 3–4 eV kinetic energy is expected to travel ~0.2 nm away from the graphene. Here ε is the dielectric constant of graphene and is assumed to be 3 (Santos & Kaxiras 2013). Also the

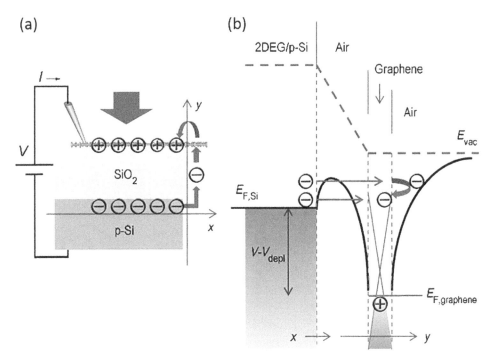

FIGURE 1.6   Emission and transport of photogenerated 2DEG in GOS. (a) Under reverse bias, photogenerated minority carriers (electrons) form a 2DEG inversion layer at SiO$_2$/Si. Electrons are emitted into air and travel toward suspended graphene electrode. (b) Schematic energy band diagram illustrating the transport of 2DEG: electron emission (along the x-direction), transport (along the y-direction), and capture/collection by graphene anode. Note that the suspended graphene is flanked by image-charge potentials at both sides. Impinging electrons can be transmitted through the graphene but are reflected back at the image potential barrier and are collected by graphene anode. (Reprinted from Kim, M. & Kim, H. K., *J. Appl. Phys.* 118, 104504, 2015.)

suspended graphene anode is assumed to be charged (by holes) for impinging electrons. It is interesting to note that the potential profile around the graphene resembles that of an ion placed near a cathode surface, which has been reported to enhance the cathode emission (Tirumala & Go 2010, Binh et al. 1992). The electrons in transit during transmission and reflection form space charge around the suspended graphene, which will affect cathode emission by its space charge field (Srisonphan et al. 2014).

Graphene is known to be transmissive to impinging electrons while being impermeable to atoms and molecules (Bunch et al. 2008). The electron transparency of graphene has been the subject of interest and debate in recent literature (Müllerová et al. 2010, Yan et al. 2011, Li et al. 2014), and the underlying mechanism is yet to be established especially in terms of understanding the interactions between impinging electrons and in-plane electrons/holes in graphene. Most studies were performed in transmission electron microscopes at relatively high energies ($\gg$100 eV) or in electron holography mode at low energies ($\sim$100 eV) demonstrating transparences >70% (Kreuzer et al. 1992, Morin et al. 1996, Mutus et al. 2011). For the case of very-low-energy electrons (<5 eV), however, reports are rare.

In this study, a suspended graphene is utilized as a collector electrode for photogenerated electrons, and the electron capture performance is characterized. Note also that, in the present study, a diode (two-terminal) configuration is employed, as opposed to a conventional triode (three-terminal) configuration (Spindt 1968, Spindt et al. 1991, Brodie 1989). In the latter case, the anode potential is usually set sufficiently high in order to collect incoming electrons transmitted through a graphene grid. In our case, no external collector is employed other than the graphene anode. Considering the confining nature of graphene's image-charge potential as discussed above, all the impinging electrons would be captured/collected by the positively biased graphene anode, and this will satisfy the charge neutrality condition of this two-terminal configuration. The closed-circuit nature (i.e., charge conservation) of this diode-mode operation was confirmed by performing measurements of anode current with the system ground connected to the bottom (cathode) or top (anode) electrode, which demonstrate the same amount of channel current for given bias voltage (Srisonphan et al. 2014). This configurational difference basically explains the apparently contradicting performances (and conflicting expectations) reported in the literature of suspended graphene, that is, widely scattered data of graphene's transparency or opacity to very-low-energy (<5 eV) impinging electrons.

1.3.3 Emission and Transport of Photocarrier 2DEG

Figure 1.7 shows the measured photo I–V of GOS with a single hole (Figure 1.7b) or a 3 × 3 array of holes (Figure 1.7c) (150 nm diameter, 100 nm depth, and 500 nm

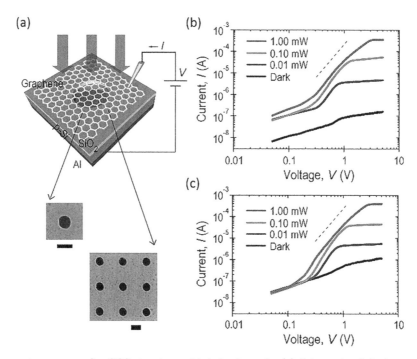

FIGURE 1.7 Photocurrent response of a GOS structure with hole channels. (a) Schematic of device structure and photocurrent measurement at 633 nm wavelength under reverse bias. SEM images of GOS samples: graphene/SiO$_2$ (23 nm)/p-Si with a single hole (bottom left) or a 3 × 3 array of holes (bottom right). (b) Measured photo I–V of a single hole sample. (c) Measured photo I–V of a 3 × 3 array of hole sample. The dashed lines indicate the V_3 dependence of photocurrent. (Reprinted from Kim, M. & Kim, H. K., *J. Appl. Phys.* 118, 104504, 2015.)

spacing, covered by a 3 mm × 4 mm monolayer graphene) at reverse bias. The measured dark $I-V$ is also shown for reference. The responsivity is measured to be 0.35 and 0.40 A/W for single and 3 × 3 holes, respectively. The saturation photocurrent is determined by the total illumination on GOS surface and is measured to remain at the same level (0.35–0.40 mA) for both samples. Note that the illumination area (laser beam size, 1-mm diameter) is much larger than the nanohole array dimension (1 μm × 1 μm). Dark current increased slightly as the number of holes increased. This indicates the dark leakage is through the void channels.

The saturation photocurrent is determined by the photocarrier generation rate in Si and is expressed as $e\eta P_{in}/h\nu$, where e is the electron charge, η is the external quantum efficiency, P_{in} is the power incident to graphene electrode, and $h\nu$ is the photon energy. For 1-mW illumination with 633-nm light (1-mm beam diameter), the photogeneration rate is calculated to be 2×10^{17} s^{-1}cm^{-2}. This corresponds to a photocarrier density of 6×10^{12} cm^{-2} when assuming a minority carrier (electron) lifetime of 30 μs in Si (Law et al. 1991). The photogenerated electrons drift to the Si/SiO$_2$ interface and flow along the 2DEG layer toward the channel edge.

It is interesting to note that the measured photocurrent rises sharply with the V^3 dependence before it reaches a saturation level: see Figure 1.7b,c, dashed. This voltage dependence differs from the $V^{1.5}$ of a cleaved MOS described above and is explained as follows. Note that, in the GOS capacitor structure, the photoinduced inversion charges (2DEG) at SiO$_2$/Si are balanced by the hole charges induced in graphene across the oxide layer. Therefore, emission of 2DEG at the channel edge is governed by the availability of hole charges in graphene (Q_h). The carrier density in graphene has a quadratic dependence on Fermi energy: $Q_h \sim E_F^2$, where E_F refers to the Dirac point. In a reverse-biased GOS capacitor, the applied voltage (V) goes to primarily three places: to support the voltage drop across the oxide layer (V_{ox}), to form and maintain a depletion region in Si (V_{depl}), and to shift the graphene's Fermi level (E_F). As the bias voltage (V) increases, the depletion voltage (V_{depl}) becomes saturated, and the oxide voltage (V_{ox}) follows closely ($V_{ox} \sim \alpha V$: $\alpha < 1$). The Fermi-level shift in graphene can then be expressed as $E_F \sim (V - \alpha V - V_{depl}) \propto (V - V_{offset})$. The hole concentration in graphene will then show a quadratic dependence on voltage with an offset, that is, $Q_h \sim (V - V_{offset})^2$. The amount of electrons being injected into a void channel is expected to show the same voltage dependence: $Q_e \sim (V - V_{offset})^2$. With enhanced injection of electrons, the carrier transport in/around graphene is expected to be scattering limited. An average velocity is then proportional to electric field; therefore, voltage $v_{av} \sim \mu\varepsilon \sim \mu(V - V_{offset})/L$. Here μ is the carrier mobility, and L is the length of scattering-limited transport section. The average transit time is then determined as $\tau_{av} = L/v_{av}$, and the overall channel current can be expressed as $I = Q_e/\tau_{av} = Q_e v_{av}/L \sim (V - V_{offset})^3$.

1.3.4 Spectral Dependence of Photocurrent Response

Photo $I-V$ of a GOS with a trench structure (340 nm width, 200 nm depth, and 1 mm length, covered by a 3 mm × 4 mm monolayer graphene) was measured at several different wavelengths in the UV-to-NIR range (325–1,064 nm). The spectral dependence of photo $I-V$ response was characterized at input power of $\sim$0.25 mW (Figure 1.8a). The photocurrent-saturation voltage (V_{sat}) shows a characteristic dependence on wavelength: a monotonic increase from 2.0 to 3.8 V at 325–850 nm, followed by a decrease to 2.0 V at 980–1,064 nm. This can be explained as follows. In the presaturation regime ($V < V_{sat}$), the channel current is limited by the electron emission/transport process in the void channel, which is dependent on bias voltage. At 325–850 nm, the photocarrier generation rate increases with increasing wavelength (i.e., with decreasing photon energy) for a given power, as can be seen from the responsivity curve (Figure 1.8b, gray). The increase of photocurrent incurs correspondingly larger saturation voltage. Near band edge (>850 nm), the absorption of light sharply drops, and therefore the photocurrent decreases, resulting in a decrease of saturation bias.

Photocurrent responsivity and quantum efficiency were calculated from the $I-V$ characteristics measured in the UV-to-NIR range (Figure 1.8b). Here the saturation photocurrent (I_{ph}) was read at the knee point (V_{sat}) of photo $I-V$ for 0.25 mW input power (P_{in}). Photocurrent responsivity (I_{ph}/P_{in}) was calculated from the measured photocurrent and input power. External quantum efficiency (EQE) was calculated from $\text{EQE} = \frac{I_{ph}/q}{P_{in}/h\nu}$. IQE was determined from $\text{IQE} = \frac{1}{1-R}\frac{I_{ph}/q}{P_{in}/h\nu} = \frac{\text{EQE}}{1-R}$. Here R is the reflectance at the GOS sample surface: graphene/SiO$_2$ (23 nm)/Si. The IQE without a carrier multiplication effect (η_{in}) was also calculated and is shown for comparison (Figure 1.8b, black, dashed): $\eta_{in} = 1 - \frac{\exp(-\alpha W_D)}{1+\alpha L_n}$. Here α is the absorption coefficient of silicon (Lide 2006, Green & Keevers 1995). L_n is the minority carrier diffusion length (Law et al. 1991). W_D is the dark depletion region width.

The measured photocurrent responsivity shows a peak value of $\sim$1 A/W at 630–858 nm. At UV (325–400 nm), a responsivity of 0.4–0.6 A/W is obtained. This corresponds to a significant improvement over those of conventional UV-enhanced photodetectors ($\sim$0.2 A/W). The high responsivity observed over a broad spectral range indicates well-passivated Si surface (by a thermally grown oxide layer), which prevents recombination loss of photocarriers at surface/defect states that would be abundant in otherwise prepared surfaces.

The internal quantum efficiency shows a three-step cascade profile: for decreasing wavelength, an initial rise to 230% level at $\sim$850 nm, followed by an increase to 300% level at $\sim$650 nm, and a ramp-up to 380% at <400 nm. The steady rise of quantum efficiency for <400 nm corresponds to the direct bandgap transition of Si and has been reported in the literature (Kolodinski et al. 1993, Geist et al. 1990).

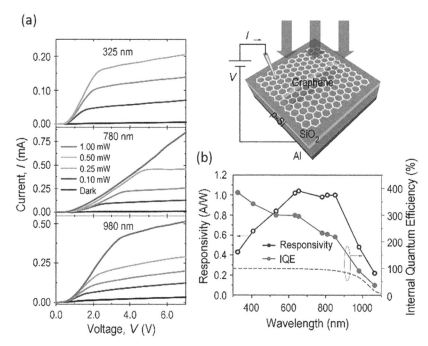

FIGURE 1.8 Spectral dependence of photocurrent responsivity and quantum efficiency of a GOS with a trench. (a) Photo I–V measured at three different wavelengths (325, 780, and 980 nm) with different input power levels. At 325 nm: (top to bottom) 0.50 mW, 0.25 mW, 0.10 mW, and dark. At 780 nm and 980 nm: (top to bottom) 1.00 mW, 0.50 mW, 0.25 mW, 0.10 mW, and dark. (b) Photo I–V was measured at 325–1,064 nm with input power of ∼0.25 mW. The responsivity (empty circles) and internal quantum efficiency (filled circles) values were calculated from saturation photocurrents. The internal quantum efficiency calculated without a carrier multiplication (or photoconductive) effect is shown for comparison (dashed). (Reprinted from Kim, M. & Kim, H. K., *J. Appl. Phys.* 118, 104504, 2015.)

The enhanced efficiency in UV is ascribed to a carrier multiplication process involving direct-bandgap transitions (Γ-point with energy gap of 3.4 eV).

At steady state, the saturation photocurrent is balanced by the photocarrier generation rate in Si. The IQE >100% indicates involvement of photocarrier multiplication or a photoconductor gain effect. The three-step-cascade profile suggests that different mechanisms are involved depending on spectral range. The NIR regime (>∼800 nm) corresponds to near-band-edge absorption. Since the amount of above-bandgap excess energy is negligible in this regime, the underlying mechanism is believed to involve a photoconductor gain effect that might be caused by charge trapping at the SiO_2/Si interface. It should be mentioned that the Si surface passivated by thermally grown oxide is known to contain a residual amount of interface states at ∼10^{11} cm^{-2} level (Deal et al. 1967, Miki et al. 1988).

In the visible-to-NIR range, photocarriers are generated both in the depletion and neutral regions, and carrier transport is governed by a drift or diffusion process, respectively. Under steady-state conditions the total photocurrent through a reverse-biased depletion layer is composed of contributions by drift and diffusion transports:

$$J_{\text{total}} = J_{\text{drift}} + J_{\text{diff}} \quad (1.6)$$

where the drift current J_{drift} is given by

$$J_{\text{drift}} = q\phi_0 \left[1 - \exp(-\alpha W_D)\right] \quad (1.7)$$

and the diffusion current J_{diff} is by

$$J_{\text{diff}} = q\phi_0 \frac{\alpha L_n}{1 + \alpha L_n} \exp\left(-\alpha W_D\right) + \frac{q n_{po} D_n}{L_n} \quad (1.8)$$

Figure 1.9 shows the contributions by drift current and diffusion current calculated in the UV-to-NIR range. At ∼530 nm or shorter wavelength, the drift transport (in the depletion region of Si) of photocurrent becomes dominant over the diffusion transport (in the quasi-neutral region of Si).

The current gain in the photoconductor mode is determined by the ratio of carrier lifetime to transit time (Sze 1981). Considering the fact that the carrier transit time in drift transport is much smaller than that of diffusion transport, the second step (rise of IQE) observed at ∼530 nm or shorter wavelength is ascribed to a photoconductive effect in conjunction with the change in carrier transit time (Figure 1.8b). An alternative explanation of the step rise at ∼530 nm might be a photoelectric effect on 2DEG at the channel edge (Fowler 1931, DuBridge 1932). This corresponds to an apparent threshold energy of ∼2.3 eV for photoemission and might have contributed to lowering the energy barrier at the 2DEG/air interface. Most incident photons, however, are absorbed deep inside the depletion region, and only a small portion falling on the 2DEG edge can contribute to this photoelectric effect. Understanding the specific contributions by different possible mechanisms requires a further study. The steep rise at <400 nm is mainly

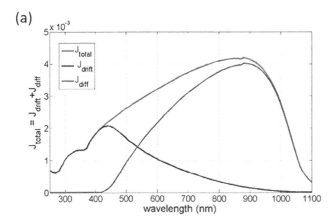

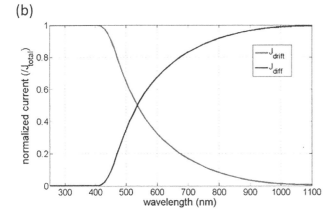

FIGURE 1.9 Photocurrent in a GOS structure. (a) Total current density composed of the drift current and the diffusion current components. (b) Contributions by the drift and diffusion transports as a function of incident wavelength. Note that a crossover occurs at ~530 nm such that the drift component becomes dominant at <530 nm.

attributed to the photocarrier multiplication effect involving direct-bandgap transitions as discussed above.

In a GOS capacitor structure with a void channel, the device characteristics can be affected by channel geometry and dimensions. First the dark current will increase as the total channel perimeter is increased (e.g., see Figure 1.7b,c). The saturation photocurrent, however, remains constant since the maximum photocurrent is determined by the incident flux. This implies that the on/off current ratio will decrease as the channel perimeter increases. See, for example, Figure 1.7b,c: the on/off ratio of ~2,000 or ~300 is observed for a single hole or a trench sample, respectively. The response time is expected to be governed by the carrier transit time and the resistance-capacitance (RC) time constant of the overall circuit. In a high-speed operation that is fully depletion biased, the transit time is expected to be of ~10 ps order or less in the depletion region and in a void channel. The device capacitance is estimated to be ~10 nF/cm^2 for 1-μm depletion width. Assuming 100 μm × 100 μm electrode (graphene) area, the device capacitance is estimated to be ~1 pF. Further assuming a device resistance (graphene electrode) of 100 Ω, the RC time constant is estimated to be ~100 ps. Overall a switching

speed of ~100 ps is then expected to be attainable with this GOS-based photodetector.

In summary, we characterized the photodetection properties of a graphene/SiO$_2$/Si structure with a nanoscale vacuum (void) channel. In this graphene/oxide/semiconductor (GOS) structure, the photogenerated 2DEG is extracted out to air, transported along the void channel, and collected by a suspended graphene anode. The photocurrent response of this GOS structure demonstrates high responsivity over a broad spectral range (UV-to-NIR), especially showing significant enhancement in UV (e.g., 0.43 A/W at 325 nm). This GOS capacitor structure with a void channel offers a potential well, well passivated by thermally grown SiO$_2$, which provides vertical confinement and lateral transport of photogenerated 2DEG in the way less prone to surface recombination losses.

1.4 Nanoscale Quantum-Dot Organic Light-Emitting Diode (QD-OLED)

With the development of quantum information technology, there has been an increasing demand for single-photon sources that can emit regulated and entangled photons on demand (Knill et al. 2001, Lounis & Orrit 2005, He et al. 2013). Incorporating inorganic quantum dots (QDs) into a semiconductor p-i-n diode structure has been intensively studied as a means of developing electrically triggered single-photon sources. However, it still remains a challenge to develop a device technology that is scalable to a nanometer range at an individual emitter level and is integrable to a chip-scale array on the same platform as driver electronics and yet providing high quantum efficiency of electrical to optical conversion.

Utilization of semiconductor QDs in OLEDs has proven to be an effective way to produce a highly efficient and cost-effective LED structure. In particular, the recent development of core–shell QDs, with shell material with larger bandgap energy, enhances carrier confinement and allows low current-spreading, high transport efficiency, and inexpensive fabrication methods when used in conjunction with OLED structure (Coe et al. 2002, Shen et al. 1997, Gustafsson et al. 1992, Veinot et al. 2002, Tang & VanSlyke 1987, Friend et al. 1999). Taking advantages of both materials systems, an improved performance has been demonstrated in terms of lifetime, brightness and energy efficiency (Sun et al. 2007, Caruge et al. 2008, Kim et al. 2011, Qian et al. 2011, Mashford et al. 2013, Dai et al. 2014). Despite their promising potential, especially in large-area emissive displays, the conventional OLED structures have drawbacks from the perspectives of scaling down device dimensions and ambient stability (Liu et al. 2002, Fiore et al. 2002). Bottom-emitting OLEDs are conventionally built on glass substrate coated with indium-tin-oxide (ITO), which is difficult to scale down to a submicron or nanometer range. Also low-work-function metal is commonly employed as cathode, which is prone to environmental degradation when exposed to moisture or oxidizing ambient. It is desirable to build OLED on a platform that is chemically inert and physically scalable (for smaller pixel/device dimensions) and yet efficient in carrier injection.

As an alternative to conventional transparent conductive oxide (TCO) substrates, we employed Si substrate for building an OLED structure that can be scaled down to a deep nanometer range. In this section, we describe a recent progress made in our lab in developing QD-OLED structures on silicon with widely scalable junction dimensions: junction diameter is scaled down from 100 μm to ~20 nm range. We observed a phenomenon that injection of accumulation 2DEG at junction periphery is the primary mechanism of carrier injection into an emissive layer, supplying a high current density at relatively low bias voltage. The resulting edge emission of light is expected to become even more dominant over area emission when the device dimension further shrinks to a nanometer scale.

Figure 1.10 shows a schematic of a nanometer-scale emitter (LED) array that employs core–shell QDs as an emissive material, confined in a nanohole array structure formed on SiO_2/Si substrate. In this structure, the Si substrate is designed to serve as an electron supplier, while the upper part of the LED structure is comprised of hole-supply/transport layers. Note that electrons are designed to get injected primarily from the edges (periphery of nanohole bottom): under forward bias, a 2DEG accumulates at the SiO_2/Si interface, and owing to the Coulombic repulsion among electrons, the 2DEG gets injected into the QD confined in a nanohole. Note also that the insulating layer (SiO_2) is designed to be thick enough to prevent tunneling of carriers through the oxide layer. The holes injected down from the upper part will get confined/trapped by the hole-barrier layer introduced on the bottom of nanohole.

This 2DEG injection is found to be more efficient than the area injection as demonstrated below. Note that the device structure also allows placing QDs on a predetermined location with nanometer precision. QDs placed outside (i.e., on SiO_2 layer) will not receive electron injection, and therefore the junction area is self-aligned to/defined by a nanohole

itself. The precise placement and tight confinement of emissive material (QDs) into a nanohole will allow high efficiency of carrier injection, suppressing a leakage current that would bypass an emissive layer.

As the junction diameter scales down, the perimeter-to-area ratio goes higher. This means almost the entire injection will be supplied with 2DEG at the edges. If the nanohole diameter is scaled down to 10 nm range, the junction will contain a single QD (Figure 1.10). This single-QD OLED structure can be envisioned as a silicon-based single-photon source on demand, which is considered to be an important optical component in future quantum information technology.

Below we describe an experimental work demonstrating that the injection of accumulation 2D electron gas (2DEG) at junction periphery is the dominant mechanism of carrier injection into an emissive layer confined in a micrometer-to-nanometer scale hole-patterned QD-OLED structure.

1.4.1 Device Fabrication: QD-OLED with Edge Injection

An n-type Si wafer was used as a substrate of QD-OLED structure (Figure 1.11). Thermally grown SiO_2 (23 nm thick) is utilized to define the OLED device area and create an isolated OLED array. An array of windows with various dimensions (20–500 μm) and geometry (circular, ring, and square) were opened in the oxide layer by photolithography and etching. A colloidal solution of QDs (CdSe/ZnS core–shell QDs, 625 nm emission wavelength, core diameter of ~7.5 nm) was spin coated on the window-etched SiO_2/Si substrate as an emissive layer. P-type organic polymers such as poly(9-vinylcarbazole) (PVK) and poly(3,4-ethylenedioxythiophene) polystyrene sulfonate (PEDOT:PSS) were used as hole-transport and electron-blocking layers. ITO was sputter deposited on top of the polymer layer as a transparent electrode.

Figure 1.11 shows an energy band diagram of the proposed device. The carriers injected into QD will remain confined due to the large-bandgap shell layer of ZnS. The 2-nm ZnS shell surrounding the CdSe core also passivates surface defects. High quantum yield of a core–shell (CdSe/ZnS) QD facilitates most of the injected electrons and holes to recombine radiatively. PVK is used as a hole-transport layer as well as an electron-blocking layer due to its low level of highest occupied molecular orbital (HOMO) and its high level of lowest unoccupied molecular orbital (LUMO). PEDOT:PSS is used as a buffer layer to increase anode work function from 4.7 eV (ITO) to 5.2 eV and also to reduce surface damage by a sputtered ITO film.

Besides the outstanding potential of nanometer-range scalability of junction dimension, the SiO_2/Si substrate offers an interesting mechanism of carrier injection into junction area. The OLED structure, when formed on SiO_2/Si substrate, can be viewed as a MOS structure. The accumulation 2DEG formed with forward bias can then serve as electron reservoir/supply that is readily available for

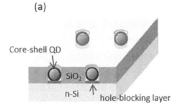
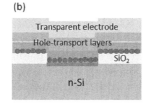

FIGURE 1.10 Schematic of nanometer-scale QD-OLED structure that contains core–shell QDs as an emissive material in a nanohole formed on SiO_2/Si substrate. By scaling the hole diameter, the number of QDs can be controlled. (a) A nanohole containing a single QD. (b) A nanohole containing multiple QDs. Note that 2DEG at SiO_2/Si is designed to get injected into QDs primarily from the edges (periphery of nanohole bottom).

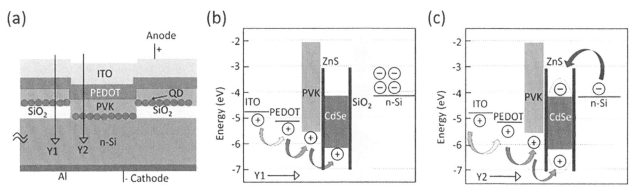

FIGURE 1.11 Proposed QD-OLED device structure on Si substrate (a), band diagram in the field oxide area (b), and band diagram in the active device area (c). (Reprinted from Emon, D. H. et al., *J. Vac. Sci. Technol. B* 34, 06KJ01, 2016.)

injection into an emissive layer through the junction edges. Figure 1.11 shows carrier flow in two different areas of OLED device, the insulating oxide area (outside the window) and the emissive area on Si (within the window), respectively.

1.4.2 Edge Emission from QD-OLED

The SiO_2/Si interface can accumulate a high concentration of 2D electron gas at proper bias, e.g., 10^{12} cm^{-2} at ~1 V forward bias. At this density level, the average spacing between electrons in the 2DEG is estimated to be ~10 nm, smaller than the oxide thickness (23 nm). This close spacing will ensure that the in-plane interaction of 2DEG electrons becomes stronger than the dipole charge interaction across the oxide layer. Due to this Coulombic repulsion at the edge, the energy barrier for electron emission into the outside is lowered, enabling electron emission at relatively low bias voltage. In the case of a MOS with a void-well channel, it has been reported that the channel current is proportional to the perimeter of the well, indicative of edge injection/emission of 2DEG. A similar behavior is expected with our OLED structure: the SiO_2/Si interface in the field oxide area under ITO electrode will accumulate 2DEG at forward bias, which will be emitted at the edge into the adjacent layer (i.e., the QD emissive layer) (Figure 1.12).

In order to test/demonstrate this edge-injection mechanism of 2DEG, we also fabricated a ring pattern (50-μm ring width and 500-μm inner diameter) of windows etched

into an oxide layer (Figure 1.13). This ring trench pattern provides two edges (inner and outer sides) and was chosen in order to test/avoid any ambiguity that may arise related to an oxide/Si step effect on lateral conduction or spin-coating process induced nonuniformity. When we start injection, the device turns on and emits visually detectable luminescence at <5 V. To capture a digital image at relatively short integration time (0.5 s), we went up to 12 V bias. At 12 V pulsed injection, short expose time captures most bright emission region, which appears in this device as the circle along the edges (Figure 1.13). But increasing the exposure time to 8 s reveals that the other bulk device area also emits light, although less bright than the edge lines (Figure 1.4). The bright circles along both the inner and outer edges of the ring trench strongly suggest quasi-1D emission induced by 2DEG injection.

Regarding the lifetime, this device was tested at elevated bias (20 V) for several minutes, and it did not show

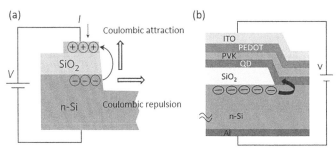

FIGURE 1.12 2DEG formation in a MOS capacitor structure. (a) Coloumbic repulsion of electrons at the edge. (b) 2DEG injection into the emissive layer (QD) at forward bias.

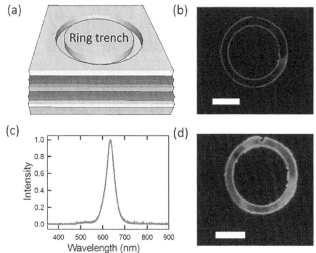

FIGURE 1.13 Ring-trench-patterned OLED device structure (a), micrograph of edge emission at short exposure (b), measured spectrum of electroluminescence of the device at 12 V (c), edge emission and area emission contrast at longer exposure time (d). Scale bar is 200 μm. (Reprinted from Emon, D. H. et al., *J. Vac. Sci. Technol. B* 34, 06KJ01, 2016.)

any degradation or thermal damage. Since the edge was very brightly emitting, indicative of high density of carrier injection, we expected thermal degradation as a form of bubble might appear at trench edges. But surprisingly and interestingly, the edge region remains very stable and degradation free, even after a long-time operation at large bias voltage (~20 V). This observation suggests that direct injection of 2DEG into the active layer has greater efficiency (than bulk area injection of electrons) in generating photons, possibly by minimizing/avoiding the nonradiative recombination processes of injected electrons which would otherwise produce heat causing thermal degradation. Figure 1.13c is a measured electroluminescence spectrum showing an emission peak at 627 nm from CdSe/ZnS QDs.

Since we observed strong and robust emission of light along the edges of trenched OLEDs, as a next step, we wanted to determine/compare contributions from area injection and perimeter injection of carriers. We patterned circular holes of 20-, 200-, and 500-μm diameter on SiO$_2$/Si substrate by photolithography and built the proposed OLED structure. The emission photo of the devices shown in Figure 1.14a reveals a strong tendency toward edge emission. A comparison of I–V characteristics of the devices shows that the injection current is proportional to the perimeter

of the hole, not the area of the hole (Figure 1.14b). Here we note that the diameter (and perimeter) is scaled up one order of magnitude, whereas the area is scaled up two orders of magnitude. The current at higher voltage follows perimeter scaling. (The larger current of 20-μm circle at lower bias is not clearly understood yet and is ascribed to nonideal leakage current.) The measured I–V characteristics also reveal relatively low turn-on voltages (1–2 V) of OLED, and this can be attributed to low-voltage injection of 2DEG at the window edges.

1.4.3 Nanoparticle-Patterned Etch Mask for Nanohole Fabrication

As an alternative to conventional nanolithography techniques (such as electron-beam or focused ion beam), we have developed a nonlithographic method that can form nanoscale hole patterns on SiO$_2$/Si substrate. In this study, we have used colloidal gold nanoparticles to create circular etching windows in a dielectric etch mask film. The etched holes have a diameter similar to that of the nanoparticle. The following sections describe the fabrication process of nanoholes and corresponding nano-OLED samples.

The overall process flow is shown in Figure 1.15. To begin with, we chose n-Si <100> wafer (phosphorous doped with a resistivity of 5 Ω cm). After solvent and acid cleaning (RCA cleaning), a thermal oxide of 25 nm thickness was grown. An Ohmic contact was prepared on the backside of bare silicon by depositing 150-nm-thick Al and subsequent annealing at 350°C.

The first step in creating nanoholes is to deposit gold nanoparticles (Au-NPs) on the substrate (Figure 1.15a). Three different sizes of citrate-stabilized Au-NPs in colloidal solution (purchased from Ted Pella) were used: 20, 50, and 200 nm diameter. A valid concern with spin coating of

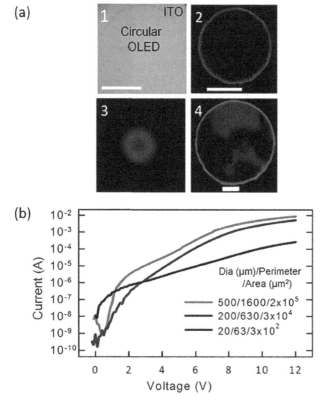

FIGURE 1.14 Circular-window-patterned OLEDs: top view (a, 1), emission photo at 8 V of 200-μm-diameter window (a, 2), 20-μm-diameter window (a, 3), and 500-μm-diameter window (a, 4) and the measured I–V characteristics of these three devices (b). Scale bar is 100 μm. (Reprinted from Emon, D. H. et al., *J. Vac. Sci. Technol.* B 34, 06KJ01, 2016.)

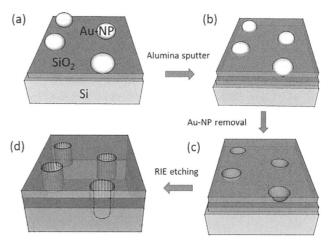

FIGURE 1.15 Process steps in fabricating nanoholes in SiO$_2$/Si substrate: (a) Deposition of Au NPs on Si. (b) Sputter deposition of alumina thin film on NPs. (c) Wet etching of NPs to create etch window. (d) Selective RIE etching of SiO$_2$/Si and followed removal of alumina by wet etching.

NPs on SiO_2 surface is that the hydroxyl group on the oxide surface lacks any affinity to citrate-passivated NPs, and therefore NPs are easily washed away without adhering to the substrate. To mitigate this surface adhesion issue, we modified the oxide substrate with an N-(2-Aminoethyl)-3-aminopropyl-trimethoxysilane (AEAPTMS) coating that binds free amine ($-NH_2$) group to the surface (Wei & Wang 2015). The silane group interacts with the hydroxyl-terminated oxide surface to form siloxane that consists of an amine ($-NH_2$) group at the tail. When this sample is immersed in the gold colloidal solution, the amine group becomes protonated forming NH_3^+ which can now bind the negatively charged citrate ions surrounding the gold nanoparticles. Process-wise, the 0.1% AEAPTMS solution in water was spin coated on the SiO_2 (25 nm)/n-Si sample at 2,000 rpm followed by a 30-min baking at 120°C. The NP colloidal solution was then spin coated on this sample. The residual chemicals/ions on the surface were removed by submerging the sample in deionized (DI) water for 5 min followed by nitrogen blow dry.

In the next step, we created a complementary hole pattern from the NP-coated sample by depositing a thin alumina film on it and then etching away the NPs (Figure 1.15b,c). Alumina (Al_2O_3) was chosen as the mask for its high selectivity of etching against gold and SiO_2 etchants. The alumina film was deposited by radio frequency (RF) sputtering using a 99.9% pure alumina target in Ar ambient with ~0.8 nm/min deposition rate. To prevent complete covering of NPs, the alumina film thickness needs to be smaller than the NP radius. In our case, the total film thickness was kept at 30, 15, and 7 nm for 200-, 50-, and 20-nm diameter NPs, respectively. Thus, the Au-NPs are expected to

create discontinuities in alumina film through which gold etchant can diffuse. Due to the conformal coating nature of sputter deposition, the etch window size formed in alumina is expected to be slightly smaller than the NP diameter. We used aqua regia as the gold etchant since it has a high etch selectivity (over 1,000 for gold over alumina). We found that, at room temperature, aqua regia etches alumina at 0.3 nm/min, while gold is etched at a rate of 10 nm/min. The sample was then ultrasonicated in aqua regia for 5 min to break loosely connected alumina film covering the upper half of the NPs. After rinsing and drying the sample, we get nanoporous alumina film where the hole diameter follows NP size.

In the fourth step (Figure 1.15d), plasma RIE was performed in CHF_3/O_2 gas to etch SiO_2 through the openings of alumina layer. The CHF_3/O_2 gas mixture was found to have an etch selectivity of ~10 for Si and SiO_2 over alumina: the etch rates for Si, SiO_2, and alumina were measured to be 0.9, 1.1, and 0.1 nm/s, respectively. We etched the nanoholes for 50 s, rendering a hole depth of 45 nm. In the final step, alumina was removed using 1:1:3 volumetric ratios of NH_4OH: H_2O_2: H_2O at 80°C temperature. At room temperature, this solution shows a high (>100) etch selectivity for alumina over Si or SiO_2. It also creates a monolayer of SiO_2 on bare Si surface, which helps to passivate the RIE etched hole surface. The SEM image of the final nanochannel samples processed from 200, 50, and 20 nm nanoparticles are shown in Figure 1.16. As stated earlier, the resultant hole in the alumina film is always smaller than the original NP diameter. In our case, the 200-nm NP results in ~160-nm hole diameter, 50-nm NP gives 40 nm, and 20-nm NP results in about the same-sized holes.

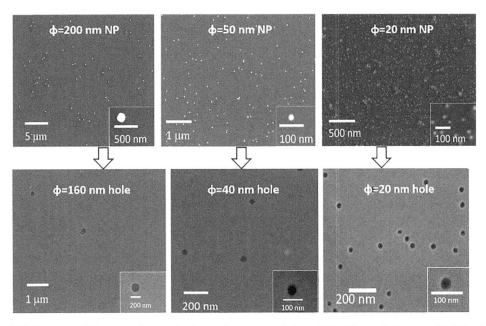

FIGURE 1.16 SEM images of final nanohole-etched samples processed from 200-, 50-, and 20-nm gold NPs: (top row) Au-NPs embedded in alumina and (bottom row) nanoholes etched into SiO_2/Si substrate. Note that the nanohole diameters are smaller than corresponding Au-NP diameters.

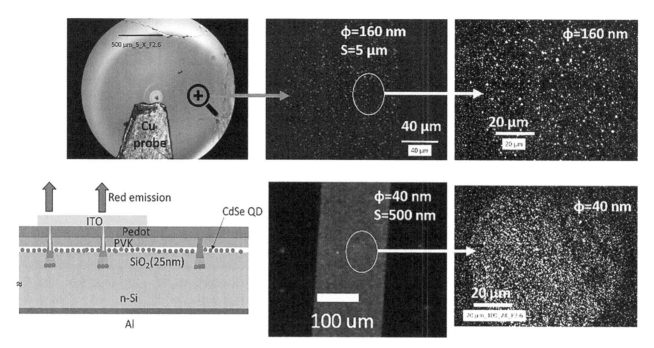

FIGURE 1.17 Optical micrograph of electroluminescence from QD-OLED samples formed on 160- and 40-nm diameter nanoholes. The bright 200-μm stripe patterns contain nanoholes, while the dark areas are planar oxide without holes.

1.4.4 Nano-QD-OLED Fabrication and Characterization

We utilized this nanohole sample to fabricate a nanoscale QD-OLED. The fabrication process of OLED structure involved the following steps: spin coating and annealing of CdSe/ZnS core–shell QDs, PVK, and PEDOT consecutively and then sputter deposition of ITO. A valid concern regarding nano-OLED is whether the polymers follow the steep steps of nanoholes or just become suspended on the holes due to their long polymer chains and high viscosity. SEM of fabricated samples confirms that the polymer layers indeed coat the nanochannels mostly following their steep steps.

Figure 1.17 shows optical micrograph of electroluminescence from 160- and 40-nm QD-OLED samples. In the micrograph, the bright 200-μm stripe patterns contain nanoholes, while the dark areas are planar oxide. We patterned the 200-μm stripes on alumina film by photolithography so that the nonstripe oxide areas do not get exposed in the RIE step. While patterning is a practical approach to defining LED pixel size, it also allows us to confirm that, even at high bias, emission originates from nanochannels, not due to oxide leakage or breakdown-induced channel emission. From the insets of Figure 1.16a, we can distinguish the luminating nano-OLEDs. The average distance between 160 nm nanoholes can be measured from the SEM image, and it matches the average spacing between nano-OLED emissions (~2–7 μm) surprisingly well, proving that most of the nano-OLEDs in this sample are active. For the case of the 40-nm hole sample, the hole spacing is 200–300 nm, and it is expected that, at this resolution, the nano-OLEDS won't be distinguishable under optical microscope.

In summary, we have developed a QD-OLED structure on Si substrate, where a 2DEG accumulated at SiO_2/Si interface is injected into a QD emissive layer at junction edges. This edge-injected OLED structure offers excellent scalability of lateral dimension (i.e., junction diameter down to the nanometer range, $<\sim$10 nm), high-efficiency carrier injection and confinement into nanoscale emissive material, and precise control (nanometer accuracy) of placing nanoscale emissive materials on predetermined locations. This 2DEG-based nano-LED structure may find applications as nanoscale emissive pixels for high-efficiency microdisplays and as single-photon sources for quantum information processing that would require precisely controlled emission on demand.

Acknowledgments

This work was supported by the Office of Naval Research (grant no. N00014-1310465) and the National Science Foundation (grant no. ECCS-0925532).

References

An, X., Liu, F., Jung, Y. J. et al., Tunable graphene–silicon heterojunctions for ultrasensitive photodetection, *Nano Lett.* 13, 909 (2013).

Ando, T., Fowler, A. B. & Stern, F., Electronic properties of two-dimensional systems, *Rev. Mod. Phys.* 54, 437 (1982).

Anikeeva, P. O., Halpert, J. E., Bawendi, M. G. et al., Electroluminescence from a mixed red–green–blue colloidal quantum dot monolayer, *Nano Lett.* 7, 2196 (2007).

Binh, V. T., Purcell, S. T., Garcia, N. et al., Field-emission electron spectroscopy of single-atom tips, *Phys. Rev. Lett.* 69, 2527 (1992).

Brinkman, A., Huijben, M., van Zalk, M. et al., Magnetic effects at the interface between non-magnetic oxides, *Nat. Mater.* 6, 493 (2007).

Brodie, I., Physical considerations in vacuum microelectronics devices, *IEEE Trans. Electron. Devices* 36, 2641 (1989).

Bunch, J. S., Verbridge, S. S., Alden, J. S. et al., Impermeable atomic membranes from graphene sheets, *Nano Lett.* 8, 2458 (2008).

Caruge, J. M., Halpert, J. E. & Wood, V. et al., Colloidal quantum-dot light-emitting diodes with metal-oxide charge transport layers, *Nat. Photonics* 2, 247 (2008).

Chen, C.-C., Aykol, M., Chang, C.-C. et al., Graphene-silicon Schottky diodes, *Nano Lett.* 11, 1863 (2011).

Child, C. D., Discharge from hot CaO, *Phys. Rev.* 32, 492 (1911).

Coe, S., Woo, W. K., Bawendi, M. et al., Electroluminescence from single monolayers of nanocrystals in molecular organic devices, *Nature* 420, 800 (2002).

Cole, M. W. & Cohen, M. H., Image-potential-induced surface bands in insulators, *Phys. Rev. Lett.* 21, 1238 (1969).

Dai, X., Zhang, Z., Jin, Y. et al., Solution-processed, high-performance light-emitting diodes based on quantum dots, *Nature* 515, 96 (2014).

Deal, B. E., Sklar, M., Grove, A. S. et al., Characteristics of the surface-state charge (Qss) of thermally oxidized silicon, *J. Electrochem. Soc.* 114, 266 (1967).

Dean, C. R., Wang, L., Maher, P. et al., Hofstadter's butterfly and the fractal quantum Hall effect in moiré superlattices, *Nature* 497, 598 (2013).

DuBridge, L. A., A further experimental test of Fowler's theory of photoelectric emission, *Phys. Rev.* 39, 108 (1932).

Emon, D. H., Kim, M., Sharbati, M. T. et al., Injection of 2D electron gas into a quantum-dot organic light-emitting diode structure on silicon substrate, *J. Vac. Sci. Technol. B* 34, 06KJ01 (2016).

Fiore, A., Chen, J. X. & Ilegems, M., Scaling quantum-dot light-emitting diodes to submicrometer sizes, *Appl. Phys. Lett.* 81, 1756 (2002).

Fowler, R. H., The analysis of photoelectric sensitivity curves for clean metals at various temperatures, *Phys. Rev.* 38, 45 (1931).

Fowler, R. H. & Nordheim, L., Electron emission in intense electric fields, *Proc. R. Soc. London* 119, 173 (1928).

Friend, R. H., Gymer, R. W., Holmes, A. B. et al., Electroluminescence in conjugated polymers, *Nature* 397, 121 (1999).

Geist, J., Gardner, J. L. & Wilkinson, F. J., Surface-field-induced feature in the quantum yield of silicon near 3.5 eV, *Phys. Rev. B* 42, 1262 (1990).

Green, M. A. & Keevers, M., Optical properties of intrinsic silicon at 300 K, *Prog. Photovoltaics* 3, 189 (1995).

Gustafsson, G., Cao, Y., Treacy, G. M. et al., Flexible light-emitting diodes made from soluble conducting polymers, *Nature* 357, 477 (1992).

Han, J.-W., Oh, J. S. & Meyyappan, M., Vacuum nano-electronics: Back to the future? Gate insulated nanoscale vacuum channel transistor, *Appl. Phys. Lett.* 100, 213505 (2012).

Han, S. & Ihm, J., Role of the localized states in field emission of carbon nanotubes, *Phys. Rev. B* 61, 9986 (2000).

Hansen, T. E., Silicon UV-photodiodes using natural inversion layers, *Phys. Scr.* 18, 471 (1978).

He, Y.-M., He, Y., Wei, Y.-J. et al. On-demand semiconductor single-photon source with near-unity indistinguishability, *Nat. Nanotechnol.* 8, 213 (2013).

Kim, M. & Kim, H. K., Ultraviolet-enhanced photodetection in a graphene/SiO_2/Si capacitor structure with a vacuum channel, *J. Appl. Phys.* 118, 104504 (2015).

Kim, T. H., Cho, K. S., Lee, E. K. et al., Full-colour quantum dot displays fabricated by transfer printing, *Nat. Photonics* 5, 176 (2011).

Knill, E., Laamme, R. & Milburn, G. J., A scheme for efficient quantum computation with linear optics, *Nature* 46, 409 (2001).

Kolodinski, S., Werner, J. H., Wittchen, T. et al., Quantum efficiencies exceeding unity due to impact ionization in silicon solar cells, *Appl. Phys. Lett.* 63, 2405 (1993).

Koppens, F. H. L., Mueller, T., Avouris, P. et al., Photodetectors based on graphene, other two-dimensional materials and hybrid systems, *Nat. Nanotechnol.* 9, 780 (2014).

Korde, R. & Geist, J., Quantum efficiency stability of silicon photodiodes, *J. Appl. Opt.* 26, 5284 (1987).

Kozuka, Y., Tsukazaki, A., Maryenko, D. et al., Insulating phase of a two-dimensional electron gas in $Mg_xZn_{1-x}O$/ZnO heterostructures below $\nu = 1/3$, *Phys. Rev. B* 84, 033304 (2011).

Kreuzer, H. J., Nakamura, K., Wierzbicki, A. et al., Theory of the point source electron microscope, *Ultramicroscopy* 45, 381 (1992).

Langmuir, I., The effect of space charge and residual gases on thermionic currents in high vacuum, *Phys. Rev.* 2, 450 (1913).

Langmuir, I. & Kingdon, K. H., Thermionic effects caused by alkali vapors in vacuum tubes, *Science* 57, 58 (1923).

Lau, Y. Y., Liu, Y. & Parker, R. K., Electron emission: From the Fowler–Nordheim relation to the Child–Langmuir law, *Phys. Plasmas* 1, 2082 (1994).

Law, M. E., Solley, E., Liyang, M. et al., Self-consistent model of minority-carrier lifetime, diffusion length, and mobility, *IEEE Electron. Device Lett.* 12, 401 (1991).

Li, C., Cole, M. T., Lei, W. et al., Highly electron transparent graphene for field emission triode gates, *Adv. Funct. Mater.* 24, 1218 (2014).

Li, X., Zhu, H., Wang, K. et al., Graphene-on-silicon Schottky junction solar cells, *Adv. Mater.* 22, 2743 (2010).

Lide, D. R. ed., *CRC Handbook of Chemistry and Physics*, 87th ed. CRC Press, Boca Raton, FL (2006).

Liu, C.-Y. & Bard, A. J., Individually addressable submicron scale light-emitting devices based on electroluminescence of solid $Ru(bpy)_3(ClO4)_2$ films, *J. Am. Chem. Soc.* 124, 4190 (2002).

Lounis, B. & Orrit, M., Single-photon sources, *Rep. Prog. Phys.* 68, 1129 (2005).

Mashford, B. S., Stevenson, M., Popovic, Z. et al., High-efficiency quantum-dot light-emitting devices with enhanced charge injection, *Nat. Photonics* 7, 407 (2013).

Mayer, A., Polarization of metallic carbon nanotubes from a model that includes both net charges and dipoles, *Phys. Rev. B* 71, 235333 (2005).

Mead, C. A., Anomalous capacitance of thin dielectric structures, *Phys. Rev. Lett.* 6, 545 (1961).

Miao, X., Tongay, S., Petterson, M. K., High efficiency graphene solar cells by chemical doping, *Nano Lett.* 12, 2745 (2012).

Miki, H., Noguchi, M., Yokogawa, K. et al., Electron and hole traps in SiO_2 films thermally grown on Si substrates in ultra-dry oxygen, *IEEE Trans. Electron. Devices* 35, 2245 (1988).

Morin, P., Pitaval, M. & Vicario, E., Low energy off-axis holography in electron microscopy, *Phys. Rev. Lett.* 76, 3979 (1996).

Müllerová, I., Hovorka, M., Hanzlíková, R. et al., Very low energy scanning electron microscopy of free-standing ultrathin films, *Mater. Transact.* 51, 265 (2010).

Mutus, J. Y., Livadaru, L., Robinson, J. T. et al., Low-energy electron point projection microscopy of suspended graphene, the ultimate 'microscope slide', *New J. Phys.* 13, 063011 (2011).

Niesner, D. & Fauster, T., Image-potential states and work function of graphene, *J. Phys. Condens. Matter.* 26, 393001 (2014).

Novoselov, K. S., Falko, V., Colombo, L., et al., A roadmap for graphene, *Nature* 490, 192 (2012).

Novoselov, K. S., Geim, A. K., Morozov, S. V. et al., Two-dimensional gas of massless Dirac fermions in graphene, *Nature* 438, 197 (2005).

Ohtomo, A. & Hwang, H. Y., A high-mobility electron gas at the $LaAlO_3/SrTiO_3$ heterointerface, *Nature* 427, 423 (2004).

Qian, L., Zheng, Y., Xue, J. et al., Stable and efficient quantum-dot light-emitting diodes based on solution-processed multilayer structures, *Nat. Photonics* 5, 543 (2011).

Santos, E. J. G. & Kaxiras, E., Electric-field dependence of the effective dielectric constant in graphene, *Nano Lett.* 13, 898 (2013).

Shen, Z., Burrows, P. E., Bulović, V. et al., Electroluminescence from single monolayers of nanocrystals in molecular organic devices, *Science* 276, 2009 (1997).

Shin, D. H., Kim, S., Kim, J. M. et al., Graphene/Si-quantum-dot heterojunction diodes showing high photosensitivity compatible with quantum confinement effect, *Adv. Mater.* 27, 2614 (2015).

Shockley, W. & Pearson, G. L., Modulation of conductance of thin films of semi-conductors by surface charges, *Phys. Rev.* 74, 232 (1948).

Silkin, V. M., Zhao, Guinea, F. et al., Image potential states in graphene, *Phys. Rev. B* 80, 121408 (2009).

Sommer, W. T., Liquid helium as a barrier to electrons, *Phys. Rev. Lett.* 12, 271 (1964).

Spindt, C. A., A thin-film field-emission cathode, *J. Appl. Phys.* 39, 3504 (1968).

Spindt, C. A., Holland, C. E., Rosengreen, A. et al., Field-emitter arrays for vacuum microelectronics, *IEEE Trans. Electron. Devices* 38, 2355 (1991).

Srisonphan, S., Jung, Y. S. & Kim, H. K., Metal–oxide–semiconductor field-effect transistor with a vacuum channel, *Nat. Nanotechnol.* 7, 504 (2012).

Srisonphan, S., Kim, M. & Kim, H. K., Space charge neutralization by electron-transparent suspended graphene, *Sci. Rep.* 4, 3764 (2014).

Stratton, R., Field emission from semiconductors, *Proc. Phys. Soc. Sect. B* 68, 746 (1955).

Suk, J. W., Kitt, A., Magnuson, C. W. et al., Transfer of CVD-grown monolayer graphene onto arbitrary substrates, *ACS Nano* 5, 6916 (2011).

Sun, Q., Wang, Y. A., Li, L. S. et al., Bright, multi-coloured light-emitting diodes based on quantum dots, *Nat. Photonics* 1, 712 (2007).

Sze, S. M. *High-Speed Semiconductor Devices.* Wiley, Hoboken, NJ (1990).

Sze, S. M., *Physics of Semiconductor Devices*, 2nd ed. Wiley, Hoboken, NJ (1981).

Tang, C. W. & VanSlyke, S. A., Organic electroluminescent diodes, *Appl. Phys. Lett.* 51, 913 (1987).

Tirumala, R. & Go, D. B., An analytical formulation for the modified Paschen's curve, *Appl. Phys. Lett.* 97, 151502 (2010).

Veinot, J. G., Yan, H., Smith, S. M. et al., Fabrication and properties of organic light-emitting "nanodiode" arrays, *Nano Lett.* 2, 333 (2002).

von Klitzing, K., Dorda, G., & Pepper, M., New method for high-accuracy determination of the fine-structure constant based on quantized Hall resistance. *Phys. Rev. Lett.* 45, 494 (1980).

Wang, Q. H. et al., Electronics and optoelectronics of two-dimensional transition metal dichalcogenides, *Nat. Nanotechnol.* 7, 699 (2012).

Wei, M. & Wang, Y., Surface attachment of gold nanoparticles guided by block copolymer micellar films and its application in silicon etching, *Materials* 8, 3793 (2015).

Yan, J.-A., Driscoll, J. A., Wyatt, B. K. et al., Time-domain simulation of electron diffraction in crystals, *Phys. Rev. B* 84, 224117 (2011).

Yan, R., Zhang, Q., Li, W. et al., Determination of graphene work function and graphene-insulator-semiconductor band alignment by internal photoemission spectroscopy, *Appl. Phys. Lett.* 101, 022105 (2012).

Yu, Y.-J., Zhao, Y., Ryu, S. et al., Tuning the graphene work function by electric field effect, *Nano Lett.* 9, 3430 (2009).

Zhang, Y., Tan, Y.-W., Stormer, H. L. et al., Experimental observation of the quantum Hall effect and Berry's phase in graphene, *Nature* 438, 201 (2005).

Zhao, J., Bardecker, J. A., Munro, A. M. et al., Efficient CdSe/CdS quantum dot light-emitting diodes using a thermally polymerized hole transport layer, *Nano Lett.* 6, 463 (2006).

Zheng, X., Chen, G., Li, Z. et al., Quantum-mechanical investigation of field-emission mechanism of a micrometer-long single-walled carbon nanotube, *Phys. Rev. Lett.* 92, 106803 (2004).

2

Polarized Nano-Optics

2.1 Introduction... 2-1
2.2 Molecular Order 2-1
2.3 Polarized Fluorescence Microscopy 2-3
    Excitation and Detection Polarization in Fluorescence: Principles •
    Application in Cell and Tissue Biology • Towards Super-Resolution
    Fluorescence Polarized Imaging
2.4 Polarized Nonlinear Microscopy.. 2-8
    Two-Photon Fluorescence • Second Harmonic Generation
2.5 Polarization Nano-Optics Provides Subdiffraction Size Information ... 2-13
2.6 Conclusion 2-14
Acknowledgments 2-15
References 2-15

Sophie Brasselet
Aix Marseille University

2.1 Introduction

Optical microscopy is an extraordinary tool for the investigation of phenomena that occur at the submicrometric scale in molecular, biological, synthetized, or nanofabricated samples. Optical imaging, and in particular fluorescence, allows to image complex processes with a spatial resolution of a few hundreds of nanometers, in real time, over large field of views (typically hundreds of micrometers). Light microscopy is a unique technique to investigate biomolecule functions (Sahl, Hell, & Jakobs, 2017) or to observe light propagating within complex nanostructures (Jiang, Zhuo, & Wang, 2018), thanks to its capacity to generate signals from local subnanometric sources, potentially with coherent properties that make them controllable in time and space.

In this chapter, we describe how the use of light polarization is able to bring one more advantage of optical microscopy. Polarization is the property of light that gives a direction to the oscillation of its associated electric field at the frequency of light radiation. Therefore, it also defines a preferential direction for its interaction with electrons, which are the important components of matter. This preferential interaction offers to optical microscopy the possibility to bring extra information on the samples investigated. Indeed, once an excitation polarization can be controlled, or a detection polarization can be measured, one can use this degree of freedom to probe locally, in a sample, directions in which bound/free electrons move preferentially. In molecules or dielectric crystals, these directions correspond to building blocks of the molecular/atomic structure, while in metallic particles, they will correspond to preferential modes of excitation of free electrons that relate to their plasmonic modes.

We will focus more particularly how light polarization can be used in optical microscopy to inform about molecular orientational organization or about structural properties of nanoobjects. The chapter addresses different optical processes, each of them having their specific properties with respect to light polarization. Fluorescence will be first described for its wide use in biological imaging. We will show how polarized fluorescence has been able to bring invaluable information on the way molecules and biomolecules are organized, which can be related to important biological functions, in particular, mechanical properties of cells and tissues. Nonlinear optical processes will then be addressed, for their larger interest in tissue imaging: we will show how second harmonic generation (SHG) can be resolved in polarization to reveal the way biomolecules are organized, even deep in tissues. At last, we will describe how these principles can be applied to nanophotonics, revealing structural properties that cannot be probed with nonpolarized imaging. A conclusion will address the current challenges of polarized fluorescence and nonlinear imaging.

2.2 Molecular Order

In what follows, we will define as molecular order the angular extent undergone by molecular/atomic directions within the focal volume of a microscope (typically 200–300 nm lateral, 600 nm axial). If molecules undergo angular fluctuations at a time faster than the integration time of the detector, molecular order also encompasses those angular fluctuations. As a result, molecular order contains two contributions: a "structural" contribution corresponding to how molecules are aligned on average over the integration timescale and a "time fluctuation" contribution coming from temporal disorder. Two examples are shown in Figure 2.1 to illustrate this specificity. Figure 2.1a shows the example of molecular probes (for instance, fluorescent) that are either inserted in a lipid membrane or attached to a membrane protein. From the intrinsic nature of lipid membranes which are essentially fluid, those probes undergo angular fluctuation, however within a certain extent imposed by the highly organized nature of lipids in membranes (Cebecauer, Owen, Markiewicz, & Magee, 2009) and their heterogeneity (Lingwood & Simons, 2010). Lipid membranes can, in addition, exhibit important morphological modifications, as for instance, in the lipid packed layer that surrounds axons to form a protective myelin sheath (Figure 2.1b) (Aggarwal, Yurlova, & Simons, 2011). At the nanometric scale, loss of adhesion of lipids layers or bending of membranes are visible on transmission electron microscopy images (Figure 2.1b), which can provoke a loss of order at the scale of optical imaging. Figure 2.1c shows an example of actin stress fibers, an assembly of actin filaments that form bundles that contribute to the mechanics of the cell cytoskeleton (Blanchoin, Boujemaa-Paterski, Sykes, & Plastino, 2014; Tojkander, Gateva, & Lappalainen, 2012). In stress fibers, actin filaments, which can be labeled by fluorophores form organized structures visible through electron microscopy (Figure 2.1d). In what follows, molecular angular distributions are modeled by an angular function of given shape $f(\theta, \varphi)$, where (θ, φ) are the angles defining a single molecule 3D orientation in the sample frame (X, Y, Z). The angular

aperture of this distribution is called molecular order. We will assume that optical interactions take place with light polarizations lying in the sample plane (X, Y), which is the most. Therefore they will involve primarily the φ variable for which we define an effective molecular distribution as a φ-dependent function $p(\varphi)$, which is related to $f(\theta, \varphi)$ in a way specific to each optical interaction process (Duboisset, Aït-Belkacem, Roche, Rigneault, & Brasselet, 2012), as described in the following sections. It is convenient to decompose a molecular angular distribution in circular functions of given symmetry orders (Figure 2.1e) (Brasselet, 2011):

$$p(\varphi) = S_0 \left[1 + \sum_{n=1}^{\infty} S_n \cos\left(n\left(\varphi - \rho_n\right)\right) \right] \qquad (2.1)$$

where the parameter S_n denotes symmetry order contributions of order n in the molecular distributions and the angle ρ_n refers to the orientations of these symmetry order in the sample plane. The S_n coefficients allow to identify specific contributions: S_0 determines the isotropic contribution of (φ), S_2 its anisotropy, and higher orders S_n define more precisely the shape of this function. Each of the symmetry order S_n has a specific orientation ρ_n, allowing to define functions of very general shape, including those that do not possess a cylindrical symmetry axis (Ferrand et al., 2014).

In what follows, we will use extensively this symmetry order decomposition as a read out of an angular distribution function, as measured by polarized optical microscopy.

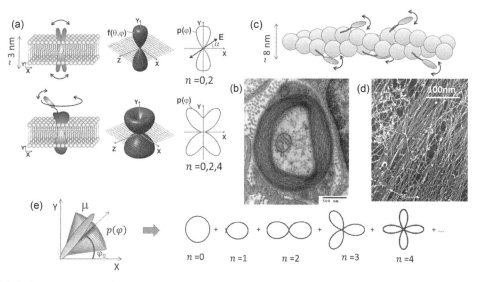

FIGURE 2.1 (a) Definition of molecular order parameters and example of different distributions corresponding to different possible situations. (a) Lipid membrane containing fluorophores (upper part, lipid probe inserted, lower part, label attached to a membrane protein), leading to different shape of molecular distribution function. (Adapted from Ferrand et al., 2014.) The gray ellipse represents the direction of the molecular dipole which is the source of emission (absorption/emission transition dipole for fluorescence, nonlinear induced dipole for nonlinear optical processes). (b) Example of lipid membrane organization in a very tight and packed way in the myelin sheath around axons. (Electron microscopy image—source: Mi et al., 2011.) (c) A similar representation for a single actin filament labeled with fluorophores (the dark gray line represents the linker to the actin protein). (d) Electron microscopy image showing both packed and disordered organization of actin in a cell. (Source: Powell, 2005.) (e) Symmetry order decomposition of a molecular angular distribution function in the sample plane, where circular functions are represented up to the order $n = 4$.

We will show in particular that an optical interaction (fluorescence, nonlinear optics) is capable of measuring only a limited quantity of symmetry orders in a molecular distribution function.

2.3 Polarized Fluorescence Microscopy

Today, fluorescence microscopy is the most widely used tool for *in vivo* imaging in cells and tissues to address current questions in biology and biomedical optics. Fluorescence is based on the emission of light using molecules used as labels (fluorophores), in which the interesting property is the ability to specifically report the behavior of chosen proteins. Using the significant potential of genetics today, and benefiting from its ability to image *in vivo* situations, in real time and with a submicrometric resolution, fluorescence imaging is still able to address open questions in cell biology, developmental biology, neurosciences, or in pathologies. Since fluorescence provides a sensitivity down the single molecule level, it is also able to tackle problems related to protein–protein interactions at specific locations in cells, bringing important elements to the understanding of fundamental biological questions. Yet, while the imaging capabilities of fluorescence give invaluable information, researchers turn to other methods to learn on the way the molecules are organized, e.g., in a more or less ordered way with respect to their orientations. These methods, typically electron microscopy or X-ray diffraction, are however incompatible with *in vivo* measurements or in-depth in tissues. In this chapter, we describe how the control of light polarization, in the excitation or detection path of a fluorescence microscope, is able to provide significant information on the way molecules are aligned.

2.3.1 Excitation and Detection Polarization in Fluorescence: Principles

One-Photon Fluorescence

Fluorescence is generated from successive steps of an absorption process and an emission process (Figure 2.2a). The first one is characterized by an absorption rate, whose efficiency depends on the absorption cross section of the molecule and the light intensity level. This absorption rate from a single molecule can be more specifically written (all vectors are written in bold in what follows) (Brasselet, 2011; Brasselet et al., 2012; Forkey, Quinlan, & Goldman, 2000; Shroder, Lippert, & Goldman, 2016):

$$P_{\mathrm{abs}} \propto |\boldsymbol{\mu}_{\mathrm{abs}} \cdot \boldsymbol{E}|^2 \qquad (2.2)$$

where $\boldsymbol{E}$ is the electric field of the electromagnetic incident wave and $\boldsymbol{\mu}_{\mathrm{abs}}$ is the absorption dipole of the molecule, which corresponds to its transition moment from the ground to the excited state (Figure 2.2b). The emission step that generates

fluorescence photons is the result of radiation of an emission dipole (which physically corresponds to the transition dipole moment from the lowest state of the excited state of the molecule down to its ground state). The emission rate of a molecule along the detection polarization direction $\boldsymbol{U}$ is proportional to the square of the emitted field:

$$P_{\mathrm{em}} \propto |\boldsymbol{E}_{\mathrm{em}} \cdot \boldsymbol{U}|^2 \qquad (2.3)$$

where $\boldsymbol{E}_{\mathrm{em}} \propto \boldsymbol{k} \times \boldsymbol{k} \times \boldsymbol{\mu}_{\mathrm{em}}$ is the radiated field by the emission dipole $\boldsymbol{\mu}_{\mathrm{em}}$ in the propagation direction by the wave vector $\boldsymbol{k}$. In the paraxial approximation, which is reasonable for low-collection numerical apertures (NA) only, the $\boldsymbol{k}$ vector is considered close to the optical axis, and the field can be simplified into a longitudinal-propagating field: $\boldsymbol{E}_{\mathrm{em}} \propto \boldsymbol{\mu}_{\mathrm{em}}$. At high NA collection, propagation directions are added up, and therefore, the polarization directions are mixed. This leads to a mixture of measured polarization states, and the emission rate is a combination of all polarizations measured in all three directions of space, with weights that depend on NA (Axelrod, 1989).

As a result of the two-step process involving absorption and emission occurring at times t and t', within the timescale of the fluorescence lifetime τ_f, the fluorescence intensity of a single molecule scales like (Florine-Casteel, 1990)

$$I_{\mathrm{1PF}}(t, t') \propto P_{\mathrm{abs}}(t) \cdot P_{\mathrm{em}}(t') \cdot e^{-(t-t')/\tau_f} \qquad (2.4)$$

The times (t, t') are the signatures of all possible time-dependent mechanisms that occur between absorption and emission processes. For an ensemble of molecules, the intensities add up since fluorescence is an incoherent process. Assume that molecules exhibit an orientation $\Omega(t) = (\theta(t), \varphi(t))$ at time t and an orientation $\Omega'(t')$ at time t', then the resulting intensity is given as (Florine-Casteel, 1990)

$$
\begin{aligned}
I_{\mathrm{1PF}} \propto N \int \int P_{\mathrm{abs}}(\Omega(t)) \cdot P_{\mathrm{em}}(\Omega'(t')) \cdot e^{-\frac{t-t'}{\tau_f}} \\
\times \, g(\Omega, \Omega', t, t') d\Omega d\Omega' d(t - t')
\end{aligned} \qquad (2.5)
$$

where N is the molecular density and $g(\Omega, \Omega', t, t')$ is the probability for each molecule to lie along the direction Ω at time t and along the direction Ω' at time t'. $d\Omega = \sin\theta d\theta d\varphi$ is the angular integral parameter for molecular orientations. The integral is performed on both angle limits of the molecular directions and, over time, within the integration time of the measurement. Assuming now that the measurement is performed at a much longer timescale than the fluorescence lifetime and the characteristics time for angular fluctuations, the fluorescence intensity can be written as

$$I_{\mathrm{1PF}} \propto N \int P_{\mathrm{abs}}(\Omega) \cdot P_{\mathrm{em}}(\Omega') \cdot f(\Omega, \Omega') d\Omega d\Omega' \qquad (2.6)$$

where Ω (respectively, Ω') defines the orientation of the absorption (respectively, emission) dipoles. The difference between both orientations is purely structural and can be negligible for very elongated molecules. Both angles are supposed to span an angular distribution (Ω, Ω') among the

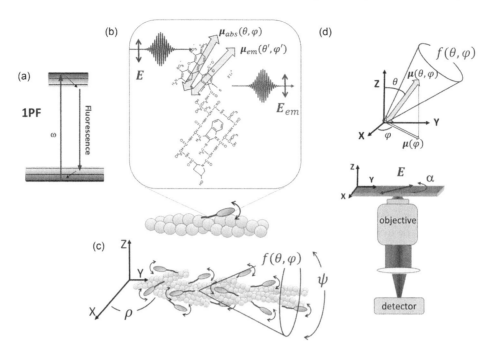

FIGURE 2.2 Principle of polarized fluorescence. (a) Ground and excited state involved in the one-photon fluorescence (1PF). (b) Schematic representation of a single Alexa-fluor 488 molecule attached to an actin filament (here via phalloidin), depicting its absorption and emission transition dipoles. The incident and emitted fields are also represented. (c) Many actin filaments together form a structural disorder of the fluorophores that contained an angular distribution (θ, φ) of aperture ψ and orientation ρ with respect to X. (d) Schematic view of a microscopy scheme, where the incident polarization is rotated in the sample plane, showing the polarization angle (α) and the dipole orientation (θ, φ) in the (X, Y) macroscopic sample plane.

assembly of molecules measured and over the timescale of measurement. Here, this average encompasses both time and spatial characteristics of the population of molecules excited within the focal volume of an objective (see for instance, Figure 2.2c for a collection of fluorophores linked to actin filaments).

From this equation, the retrieval of the angular distribution function (Ω, Ω') can be limited to either the distribution of absorption or emission dipoles. Probing absorption dipoles is feasible by the control of the excitation polarization, while the emission dipoles are probed by the control of the detected polarization. Both situations will be explored in what follows.

Consider a situation where the emission signal is not polarized, while the incident polarization is linear and rotated in the sample plane. In this case, the (X, Y, Z) coordinates of $\boldsymbol{E}$ are (Figure 2.2d): $E_0(\cos \alpha, \sin \alpha, 0)$ if we ignore longitudinal contributions of the excitation polarization. This is valid under relatively low NA excitation or for samples that lie principally in the sample plane, which is the case here (Schön, Behrndt, Aït-Belkacem, Rigneault, & Brasselet, 2010). The fluorescence intensity can therefore be written as

$$I_{1PF} \propto N \int P_{\mathrm{abs}}(\Omega) \cdot f(\Omega) d\Omega$$

$$= N \int |\boldsymbol{\mu}_{\mathrm{abs}}(\Omega) \cdot \boldsymbol{E}(\propto)|^2 \cdot f(\Omega) d(\Omega) \qquad (2.7)$$

with $d\Omega = \sin \theta d\theta d\varphi$. Renaming the in-plane projection of the dipole $\boldsymbol{\mu}_{\mathrm{abs}}(\varphi) = (\cos \varphi, \sin \varphi, 0)$ such as $\boldsymbol{\mu}_{\mathrm{abs}}(\theta, \varphi) = \boldsymbol{\mu}_{\mathrm{abs}}(\varphi) \sin \theta$, this form can be simplified into

$$I_{1PF}(\alpha) \propto N \int |\boldsymbol{\mu}_{\mathrm{abs}}(\varphi) \cdot \boldsymbol{E}(\alpha)|^2 \cdot p(\varphi) d\varphi \qquad (2.8)$$

with the in-plane integral distribution $p(\varphi) = \int f(\theta, \varphi) \sin^3 \theta d\theta$.

The goal of polarization-resolved fluorescence is to extract information on (φ) from the measurements of several polarization angles (α). Ultimately, since the α dependence of this expression does not exceed a power of two in (cos, sin) functions, the fluorescence intensity can be written in a more convenient measurable form (Kress et al., 2013):

$$I_{1PF}(\alpha) \propto a_0 + a_2 \cos 2\alpha + b_2 \sin 2\alpha \qquad (2.9)$$

This circular decomposition contains harmonic coefficients of the polarization response (a_0, a_2, b_2) that can be directly measured: $a_0 = \int I_1(\alpha) d\alpha$ is the total intensity, and the other coefficients are deduced from $a_2 = 2 \int I_{1PF}(\alpha) \cos 2\alpha \, d\alpha/a_0$ and $b_2 = 2 \int I_{1PF}(\alpha) \sin 2\alpha \, d\alpha/a_0$. These orders can be assembled into an amplitude (I_2) and orientation (φ_2):

$$I_2 = \sqrt{a_2^2 + b_2^2}, \quad \varphi_2 = \frac{1}{2} \arctan \frac{b_2}{a_2} \qquad (2.10)$$

Ultimately, (I_2, φ_2) can be related to the (φ) function. Since the intensity polarization dependence does not surpass the order two in α, the readable part of the (φ) function will be limited to the order two as well; therefore, the only measurable part of (φ) is

$$p(\varphi) = S_0 \left[1 + S_2 \cos\left(2\left(\varphi - \rho_2\right)\right)\right] \qquad (2.11)$$

A direct calculation of the intensity expression in Eq. (2.7) leads to

$$I_2 = \frac{2}{3}S_2, \; \varphi_2 = \rho_2 \qquad (2.12)$$

which permits to measure the (S_2, ρ_2) parameters that characterize the molecular angular distribution. The number of parameters that can be retrieved is limited by the fact that the optical interaction takes place at the power two of the electric field of the incident light. We will see in the next section that higher symmetry orders of this function can be retrieved by the use of higher orders of interaction, e.g., using nonlinear optics.

Note that the retrieval of the (S_2, ρ_2) parameters can be also done via a tuning of the detection polarization. In this case, the probability of absorption can be made polarization independent by the use of circular polarization illumination in the sample plane. This scheme will be described in Section 2.3.3.

Interpretation of the Retrieved Molecular Order

Interpreting the (S_2, ρ_2) parameter in terms of angular characteristics of a molecular distribution function gives a more physical picture of the organization behavior of a sample. The fact that only one symmetry parameter can be measured

using 1PF prevents, however, from measuring detailed information on the shape of this angular distribution, a part from its angular width. Assume, for instance, a cone distribution of molecules $f(\theta, \varphi)$ of aperture ψ, lying in the sample plane with an orientation ρ with respect to the direction X (Figure 2.2c), then $p(\varphi) = \int f(\theta, \varphi) \sin^3\theta \, d\theta$ and $S_2 = 2 \int p(\varphi) \cos 2\varphi \, d\varphi$ lead to a direct relation between the coefficients (S_2, ρ_2) and the molecular distribution angles (ψ, ρ):

$$S_2 = \frac{\sin\Psi}{\Psi}, \; \rho_2 = \rho \qquad (2.13)$$

Instrumentation

The purpose of polarized fluorescence microscopy is to retrieve polarization-dependent modulations of the recorded fluorescent signal, upon the variation of an incoming rotating polarization. This can be performed in both confocal scanning microscopy and parallel imaging modalities. In confocal microscopy, the excitation focal spot is scanned over the sample by galvanometric mirrors, allowing a typical image rate of 1 image/s (for about 100 µm × 100 µm field of views). The polarization is rotated by a rotation motor in which a half wave plate is mounted (Figure 2.3a). The number of polarization angles to record is minimally four but generally higher to allow for a larger number of photons to be recorded, which governs the precision of measurements (Kress et al., 2013).

To circumvent the time limitations of confocal microscopy, more recent developments have integrated faster scanning microscopy, based on a spinning disk technology that rely

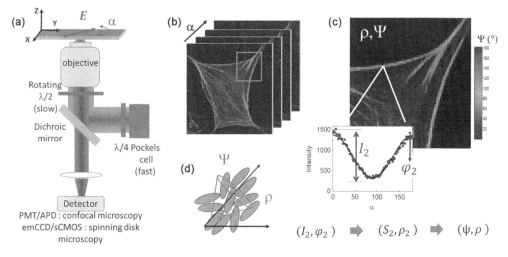

FIGURE 2.3 (a) Principle of 1PF microscopy, showing both slow (motor rotation) and fast (electro-optic modulation using a Pockels cell) modes of excitation polarization tuning. In the slow mode, the image is performed using confocal imaging, e.g., point scanning the excitation volume using a galvanometric mirror and descanning the detection, which is focused on a photomultiplier (PMT) or avalanche photodiode (APD). In the fast mode, the sample is scanned using a collection of imaged pinholes (spinning disk) that provoke a sectioned, parallel illumination that is directly imaged on an emCCD or sCMOS camera with a typical 30–100 ms integration time. (b) The recorded data are a stack of fluorescent images, taken for a large number (>4) of incident polarization angles. (c) Each pixel modulation is analyzed by projection on circular functions that directly lead to an amplitude (I_2) and a phase (φ_2), which are themselves transformed into molecular distribution parameters (S_2, ρ_2). Those parameters can be interpreted in an aperture and mean direction (ψ, ρ), assuming a cone model of disorder, for instance. (d) In the final image representation, the image is encoded for each analyzed pixel using a stick whose orientation is ρ and whose color is ψ.

Nanophotonics, Nanoelectronics, and Nanoplasmonics

on multiple pinholes illuminated at the same time (Wang, Kress, Brasselet, & Ferrand, 2013). The detection is no more performed on a point detector but on a parallel charge coupled device (CCD) camera, which allows integration times for a whole image of a few tens of milliseconds. Fast polarization rotation is performed by electro-optic modulation using, for instance, a Pockels cell, whose response time is microseconds, synchronized to the camera (Figure 2.3a). This fast polarization control has achieved a rate of about one polarimetry image (e.g., containing organization information) per second (Wang et al., 2013).

Polarization-resolved experiments lead to the measurement of a stack of images out of which each pixel is a polarization-modulated signal (Figure 2.3b). The modulation is interpreted in terms of order parameters, for instance, a cone aperture and mean direction (ψ, ρ), which are encoded in an image by sticks whose color is ψ and orientation is ρ (Figure 2.3c,d).

2.3.2 Application in Cell and Tissue Biology

Imaging Cell Membrane Molecular Order

The cell plasma membrane is made of a complex association of different lipid types, proteins, and other molecules such as cholesterol. This membrane has important functions, such as cell signaling and adhesion, which request strong modifications of the membrane at different times and spatial scales. It is, for instance, known that domains of submicrometric size are able to form or disassemble dynamically, to facilitate the functions of specific proteins at certain times (Lingwood & Simons, 2010). Polarized fluorescence imaging (Figure 2.4a) offers a way to investigate lipid order in heterogeneous systems that are still poorly known. It has been shown that fluorescent lipid probes used for polarized fluorescence imaging could reveal local membrane order properties (Florine-Casteel, 1990; Axelrod, 1979;

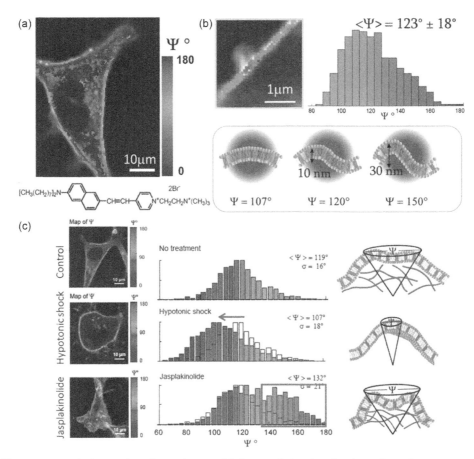

FIGURE 2.4 Fluorescence polarization in cell membranes. (a) Image of the ψ molecular order values measured on a cell (COS7) plasma membrane labeled with the lipid probe di-8-ANEPPQ. (b) Detail of the membrane molecular order, showing a region of high disorder, most probably due to the local membrane morphology. The histogram shows the large variety of ψ values measured in this region, which can be compared to models of membrane local bending. The schematic drawing shows local deformations within the focal spot size represented by a gray circle, together with expected increase in ψ, supposing a value of 107° in a flat membrane region. (c) Measured ψ values (cell maps and histograms) on cells perturbed by a hypotonic shock (middle panels) that provokes a tension of the cell membrane and by Jasplakinolide (an actin filament stabilizer, lower panels) that provokes a local ruffling of the membrane. Both effects are visibly affecting the measured molecular order. (Adapted from Kress et al., 2013.)

Benninger, Önfelt, Neil, Davis, & French, 2005; Gasecka, Han, Favard, Cho, & Brasselet, 2009), in complementarity to other methods based on spectral signatures, which probe local polarity (Sanchez, Tricerri, & Gratton, 2012). Molecular order in lipid membranes includes lipid fluctuations, local lipid conformational changes and phases, as well as nanometric scale deformations of the membrane (Figure 2.4b). Polarized fluorescence microscopy is not only able to report local changes of membrane environment, such as the formation of lipid phases (Axelrod, 1979; Florine-Casteel, 1990; Gasecka et al., 2009; Haluska et al., 2008), but also the local morphological changes due to the perturbation of the cytoskeleton (Figure 2.4c) (Kress et al., 2013) or to cell interactions in signaling processes (Benninger et al., 2009).

Fluorescence labeling of proteins in membranes has also been made possible by genetic expression of a GFP in the sequence of a protein. By engineering a linker of GFP at proper places in the protein, where the angular fluctuations are minimized, studies have allowed to investigate molecular organization of membrane proteins in cell membranes (Lazar, Bondar, Timr, & Firestein, 2011), such as nuclear pore complexes (Kampmann, Atkinson, Mattheyses, & Simon, 2011), Major Histocompatibility Complexes (Kress et al., 2011), and reporter proteins involved in immunology responses (Chouaki-Benmansour et al., 2018).

Imaging Actin Filaments Molecular Order

Fluorescence polarized microscopy has more recently been applied to the study of the organization of filaments in cells and tissues (Vrabioiu & Mitchison, 2006; Mavrakis et al., 2014). Actin, one of the most important protein of the cell cytoskeleton, is able to polymerize cells in filaments of a few nanometer diameters and micrometer lengths. Together with crosslinkers and molecular motors such as myosin, actin can form ultrastructures such as contractile stress fibers, which are at the origin of forces that are exerted in cells in adhesion, migration, or for tissue deformation in morphogenesis (Tojkander et al., 2012). While actin organization can be imaged by electron microscopy with a resolution of a few tens of nanometers, this information is invisible in an optical image due to its limited resolution of a few hundreds of nanometers. Using polarized fluorescence can provide such information by relying on the orientation of fluorescent labels to report that of actin filaments. In fixed cells and tissues, the labeling of actin by phalloidin has shown a relatively high order for attached probes such as Alexa-fluor488 (Mavrakis et al., 2014) (Figure 2.5a,b). Even though the measured molecular order aperture angle is quite large ($\psi \sim 130°$) even in stress fibers, measuring molecular order in actin via such label is an interesting readout for actin organization. This has been, for instance, used in drosophila embryos, where it was shown that the actin organization can be highly perturbed in the absence of septin, which was demonstrated to be an important actin partner in the cell cytoskeleton at the early stages of embryo formation (Figure 2.5c,d) (Mavrakis et al., 2014).

2.3.3 Towards Super-Resolution Fluorescence Polarized Imaging

The molecular order measured in actin stress fibers as revealed by phalloidin-conjugated probes appears to be low when compared with the expected high order present in

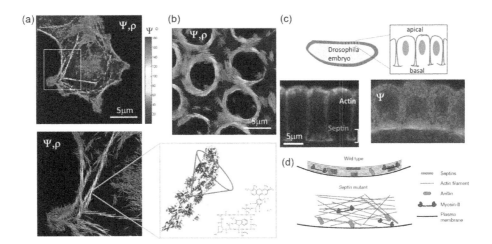

FIGURE 2.5 Polarization-resolved fluorescence microscopy in actin. (a) Molecular order in actin stress fibers in a fixed COS7 cell, labeled with Phalloidin-Alexa Fluor 488 (structure shown in the inset). The (ψ, ρ) maps show a higher molecular order along the formed stress fibers. The schematic drawing in the inset shows an ensemble of actin filaments together with the expected fluorophores molecular order. (b) Actin molecular order in a fixed drosophila embryo at the cellularization stage, labeled with Phalloidin-Alexa Fluor 488. The (ψ, ρ) map is shown at the basal surface where septin filaments are seen to be localized. (c) A schematic drawing of the drosophila embryo seen from the side, zooming on formed cells (gray ellipse: nucleus). A side view of both actin/septin fluorescence and polarized fluorescence on actin shows the colocalization of septin on a higher actin order. (d) Model derived from the molecular order measurement in wild-type embryos and a septin mutant deprived of septin. (Adapted from Mavrakis et al., 2014.)

actin stress fibers. This disorder, however, encompasses two contributions: not only the structural disorder of actin filaments themselves but also the angular fluctuations and tilt angles related to the fluorophore itself and its linker to the actin protein. In order to retrieve unambiguous information on the structural behavior of actin at the nanoscale, evaluating such fluctuations is required. A pertinent method to measure the way molecules wobble during a polarized fluorescence measurement is to look at single molecules one by one (Shroder, Lippert, & Goldman, 2016; Von Diezmann, Shechtman, & Moerner, 2017; Valades Cruz et al., 2016; Shaban, Valades-Cruz, Savatier, & Brasselet, 2017). Molecular order is then defined by the angular extent explored by single molecules during the integration time of measurement. This way, only time fluctuations are measured, and it is thus possible to access the important parameter that is missing in ensemble measurements: fluorophore wobbling (e.g., angular fluctuations due to the molecular structure flexibility). Measuring polarized information on fluorescence from single molecules can ultimately give access to both their angular fluctuations and mean orientation, which can be ultimately related to the organization of the proteins they are attached to. Achieving such measurement by the rotation of the incident polarization can be performed; however, it requires the ability to measure a single molecule for a long time, which is challenging. In addition, single molecules offer interesting capabilities for super-resolution imaging when they are made capable of switching between on (fluorescent) and off (nonfluorescent) states. This mode, which is at the origin of super-resolution imaging based on stochastic single molecule localization (Betzig et al., 2006; Rust, Bates, & Zhuang, 2006; Deschout et al., 2014; Hess, Girirajan, & Mason, 2006), imposes molecules to rapidly switch between on and off states, which discard polarization tuning methods. Measuring several states of polarization in a single shot requires a polarization control in the emission path, as mentioned earlier. The principle relies on the projection of the measured signal over several states of linear polarization in different directions, which performs a similar situation as what was previously explored when rotating the incident polarization. In this scheme, we supposed no preferential direction for the incident polarization, which imposes a circular or depolarized excitation. Under such conditions, the measured intensity in a given detection polarization state U can be written as

$$I_{\text{1PF},U} \propto \int |\boldsymbol{\mu}_{\text{em}}(\varphi) \cdot \boldsymbol{U}|^2 \cdot p(\varphi) d\varphi \qquad (2.14)$$

where $p(\varphi)$ is the distribution of a single molecule within a cone of aperture δ (wobbling extent) and mean orientation ρ relative to X in the sample plane. Note that under high NA collection conditions, the measured intensity in a polarization state U is rather a mixture of contributions from several polarizations. This can be accounted for by a rigorous writing of the integration of radiations from all emission dipole components (Axelrod, 1979) and can serve as a probe for 3D orientational properties (Patra, Gregor, & Enderlein,

2004; Backer & Moerner, 2015; Backer, Backlund, Lew, & Moerner, 2013). We will consider in what follows situations where the NA of detection is small enough (typically below 1.2) and the incident polarization mainly in the sample plane (e.g., under close to normal incidence), to ignore this polarization mixture in a first approximation (Valades Cruz et al., 2016).

Similarly, as in the previous section describing incident polarization tuning, a nonambiguous determination of both (δ, ρ) parameters requires at least four polarization states to be measured. This has been exploited in several works that reported essential ensemble information or ignored localization information (Shroder et al., 2016; Forkey, Quinlan, & Goldman, 2005; Swaminathan et al., 2017). In order to maintain a high detection efficiency that locates single molecules, less projection angles can be used. It can be shown in particular that the use of only two states of detection polarizations can be sufficient for an estimation supposing that all molecules undergo the same wobbling (Valades Cruz et al., 2016). Supposing two states of detected polarization 0° (named ∥) and 90° (named ⊥) (Figure 2.6a), we define the polarization factor as (Valades Cruz et al., 2016)

$$P(\delta, \rho) = \frac{I_{\text{1PF},\parallel} - I_{\text{1PF},\perp}}{I_{\text{1PF},\parallel} + I_{\text{1PF},\perp}} = \cos 2\rho \, \frac{\sin \delta}{\delta} \qquad (2.15)$$

which is a function of both δ and ρ. Such polarization factor can be measured for isolated molecules at each frame of a direct Stochatic Optical Reconstruction Microscopy (dSTORM) stack of images. A localization of each single molecule in both images allows to deduce both intensities $I_{\text{1PF},\parallel}$ and $I_{\text{1PF},\perp}$ (Figure 2.6b).

From the expression (δ, ρ), we see that δ principally governs the limit values reached by P over all possible orientations ρ of single molecules. Measuring a large number of single molecules is therefore a reliable first-order estimation of the upper limit of δ in a sample. This method has been applied to fixed cells in which actin stress fibers were labeled with Phalloidin-Alexa Fluor 488 (Figure 2.6c,d) (Valades Cruz et al., 2016). An average wobbling angle $\delta \sim 90° - 120°$ has been measured, which explains the large extent of molecular order measured in ensemble measurements (Figure 2.5). Once δ is estimated, the measurement of P values for each molecule can be used to infer ρ by the use of the aforementioned polarization factor expression (Figure 2.6e). This angle is an important readout of the actin molecular order, which can be interpreted as a real nanometric scale structural information, deconvolved from fluorophore wobbling of the actin label. Figure 2.6e shows, in particular, that actin filaments in stress fibers are very parallel to each other, which is expected from their structure (Valades Cruz et al., 2016).

2.4 Polarized Nonlinear Microscopy

Nonlinear optics has been introduced in microscopy imaging since the late 90s due to its unique capacities for in-depth

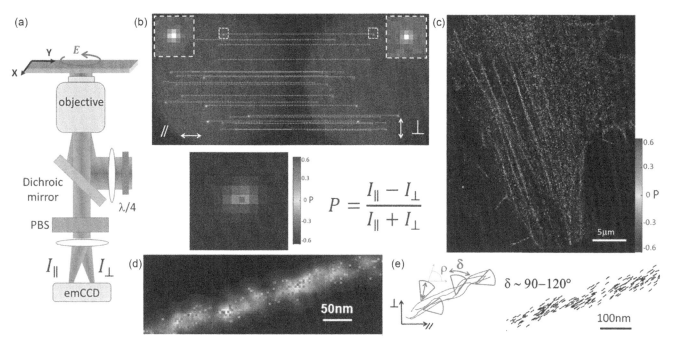

FIGURE 2.6 Polarized dSTORM. (a) Principle of polarized dSTORM: the sample is excited with a circularly polarized field, and the detection path uses a polarization beamsplitter (PBS) to project two polarization states of the same image. (b) In each STORM image, each single molecule is detected on both ∥ and ⊥ images (the light gray lines represent a pairing vector used for image analysis), resulting in a single color of the polarization factor P, per molecule. (c) A super-resolution image (here actin stress fibers labeled with Phalloidin-Alexa Fluor 488 in a fixed COS7 cell) is reconstructed from single molecule localizations, encoded in color by their P value. (d) Zoom on a single stress fiber. (e) Using statistical knowledge of P values permits to deduce an average of the upper limit of wobbling of single molecules (δ value), and thus, to deduce ρ values that are represented as stick orientations for each molecule (hereon, a zoomed part of a stress fiber). (Adapted from Valades Cruz et al., 2016.)

imaging in complex biological media where visible light is absorbed or scattered (Denk, Strickler, & Webb, 1990; Zipfel, Williams, & Webb, 2003; Campagnola et al., 2002). As multiple photon excitation is required, near infra-red excitation wavelengths are used, which penetrate deeper in biological tissues. The involved cross sections are, however, naturally lower than in one photon processes; therefore, pulsed excitation regimes are used to provide high efficiencies (Denk et al., 1990), most often using Ti:Sapphire laser oscillators that provide ~100 fs high peak power pulses with 80 MHz repetition rate. A typical nonlinear microscope uses a point scanning excitation modality with galvanometric mirrors, with the strong advantage of natural depth sectioning from the nonlinear dependence on the incident intensity (Denk et al., 1990). In what follows, polarization-resolved nonlinear imaging is described where second-order excitation processes are analyzed: two-photon fluorescence (2PF) and SHG. Higher-order processes, such as third-order harmonic generation (THG), four-wave mixing (FWM) or its vibrational resonant counterpart coherent Raman scattering (CRS) have also been recently exploited for their polarization sensitivity. The reader can refer to recent specific literature on these more complex processes (Brasselet, 2011; Bioud et al., 2014; Zimmerley, Mahou, Débarre, Schanne-Klein, & Beaurepaire, 2013; Hofer, Balla, & Brasselet, 2017).

2.4.1 Two-Photon Fluorescence

While 1PF gives access to a limited amount of information on the symmetry decomposition of a molecular distribution, accessing higher order can be made possible by an increase of the power at which light–matter interaction occurs. 2PF originates from the absorption of light by sequential two photons in a pulsed excitation regime, generating fluorescence similarly as for a 1PF emission (Figure 2.7a). In this case, the dependence of the absorption rate is written as (Brasselet, 2011; Denk et al., 1990; Gasecka et al., 2009)

$$P_{\text{abs,2PF}} \propto |\boldsymbol{\mu}_{\text{abs}} \cdot \boldsymbol{E}|^4 \qquad (2.16)$$

As a result, the 2PF intensity is given as

$$I_{\text{2PF}}(\alpha) \propto \int |\boldsymbol{\mu}_{\text{abs}}(\varphi) \cdot \boldsymbol{E}(\alpha)|^4 \cdot p(\varphi) d\varphi \qquad (2.17)$$

with $(\varphi) = \int (\theta, \varphi) \sin^5 \theta d\theta$ for a $\boldsymbol{E}$ field polarized in the sample plane. The fluorescence intensity can be decomposed in circular functions, similarly as for 1PF:

$$\begin{aligned} I_{\text{2PF}}(\alpha) &\propto a_0 + a_2 \cos 2\alpha + b_2 \sin 2\alpha + a_4 \cos 4\alpha + b_4 \sin 4\alpha \\ I_{\text{2PF}}(\alpha) &\propto a_0 + I_2 \cos 2\left(\alpha - \varphi_2\right) + I_4 \cos 4\left(\alpha - \varphi_4\right) \end{aligned} \qquad (2.18)$$

where $a_0 = \int I_{\text{2PF}}(\alpha) d\alpha$ the total intensity. The other coefficients can be measured by the relations

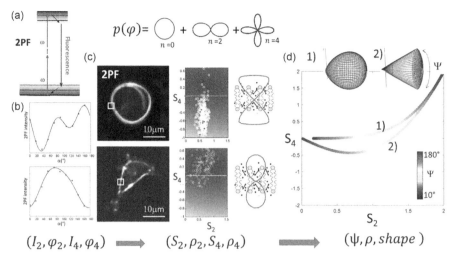

FIGURE 2.7   Polarized 2PF microscopy. (a) Principle of 2PF emission. (b) Examples of polarization-resolved 2PF signals in cell membranes, revealing a high fourth-order response of the intensity (upper graph) and a high second-order response (lower graph). (c) Those responses come from two types of samples (square region), a multilamellar vesicle (upper image) and a COS 7 cell plasma membrane (lower image), both labeled with the lipid probe di-8-ANEPPQ. Reporting the measured pixels in a (S_2, S_4) graph shows very different behaviors. The schematic drawing of the membrane shows the retrieved (φ) function, which is smoother in the lower graph case because of the negligible fourth order. (d) These responses can be compared with models for which (S_2, S_4) are retrieved for various aperture angles ψ. The measurement of S_4 permits to discriminate between 1) smooth (Gaussian) and 2) sharp (Cone) distribution functions. (Adapted from Ferrand et al., 2014.)

$a_2 = 2 \int I_{2\mathrm{PF}}(\alpha) \cos 2\alpha \ d\alpha / a_0$, $a_4 = 2 \int I_{2\mathrm{PF}}(\alpha) \cos 4\alpha \ d\alpha / a_0$, and similarly, for the (b_2, b_4) coefficient that involve projections on the $\sin 2\alpha$ and $\sin 4\alpha$ functions. These orders can be assembled into amplitudes (I_2, I_4) and orientations (φ_2, φ_4) (Ferrand et al., 2014; Réfrégier, Roche, & Brasselet, 2011):

$$I_2 = \sqrt{a_2^2 + b_2^2}, \quad I_4 = \varepsilon_4 \sqrt{a_4^2 + b_4^2}$$
$$\varphi_2 = \frac{1}{2} \arctan \frac{b_2}{a_2}, \quad \varphi_4 = \frac{1}{4} \arctan \frac{b_4}{a_4} \quad (2.19)$$

With φ_4 given modulo $\pi/2$. $\varepsilon_4 = \cos(4(\varphi_4 - \varphi_2))$ sets the sign of I_4 such that the fourth-order response lies either along the second-order $(I_4 > 0)$ response or $\pi/4$ phase shifted $(I_4 < 0)$ (Ferrand et al., 2014).

As for 1PF, the circular decomposition can be directly related to the symmetry orders of the (φ) function, and this time decomposed over higher orders:

$$p(\varphi) = S_0 \left[1 + S_2 \cos\left(2\left(\varphi - \rho_2\right)\right) + S_4 \cos\left(2\left(\varphi - \rho_4\right)\right)\right] \quad (2.20)$$

A direct calculation of the integral function leads to:

$$S_2 = \frac{3}{2} I_2, \ S_4 = 6 I_4$$
$$\text{and} \quad \rho_2 = \rho_4 = \rho = \varphi_2 = \varphi_4[\pi/2] \quad (2.21)$$

for a cylindrical symmetry distribution of in-plane orientation ρ.

The different symmetry coefficients of (φ) can thus be directly deduced from a 2PF polarization modulation measurement, with a richer amount of information since both second- and fourth-order of symmetry are now accessible.

Polarization-resolved 2PF has shown various responses in cell membranes depending on the order present in the lipids (Figure 2.7b). The extraction of readable (φ) coefficients resembles more or less sharp distributions depending on the membrane nature (Figure 2.7c). This information can be furthermore interpreted into physical interpretation of a molecular angular distribution shape. When modeling different types of distributions, for instance, Gaussian or cone-like of various angular apertures, it is visible that they strongly differ in (S_2, S_4) values, thus allowing to discriminate them (Figure 2.7d). Note that the sole second-order contribution S_2 measured by 1PF cannot allow differentiating these two distributions.

At last, the phase information of the measured polarized intensity (φ_2, φ_4) determines the orientation of both second- and fourth-order contributions. It is interesting to notice that when (φ_2, φ_4) are not equal modulo $\pi/4$, the distribution is no more of cylindrical symmetry, which can be informative on the nature of molecular interactions in complex systems (Ferrand et al., 2014).

2.4.2   Second Harmonic Generation

Fluorescence requires an absorption process (at one or multiphoton) followed by an incoherent emission, which is not related in time (within the fluorescence life timescale) with this absorption event. As a result, emitted fluorescent radiations do not possess any specific phase relation between each other, and different molecules will thus emit fluorescent light by adding up their intensities incoherently. In contrast, nonlinear coherent optical processes originate from induced dipoles that are created by the nonlinear interaction of light with matter, which makes radiating dipoles related in phase with the incident excitation field. One of the first known processes in nonlinear optics is SHG, which results from the excitation of a system at the fundamental optical frequency ω, into a scattering at the double frequency 2ω (Boyd, 2008) (Figure 2.8a). This scattering process is the result of radiation from the induced individual dipole $\propto \beta(\Omega): (\boldsymbol{E}\alpha)(\alpha)$, where β is the nonlinear susceptibility tensor of a single molecule and $\boldsymbol{E}$ is the incident field at the frequency ω, linearly polarized along an angle α with respect to X. At the ensemble level, these dipoles radiate coherently within the focal volume to lead to a macroscopic nonlinear induced dipole (Brasselet, 2011):

$$\boldsymbol{P}_{\text{SHG}} \propto N \int [\beta(\Omega): \boldsymbol{E}(\alpha)\boldsymbol{E}(\alpha)]\, f(\Omega)d\Omega$$
$$= \chi^{(2)}: \boldsymbol{E}(\alpha)\boldsymbol{E}(\alpha) \qquad (2.22)$$

where N is the molecular density and $\chi^{(2)} = N \int \beta(\Omega) f(\Omega)d\Omega$ the macroscopic nonlinear tensor of the molecular ensemble present in the focal volume of the objective. The measured SHG intensity can be deduced from the resulting radiated field: $\boldsymbol{E}_{\text{SHG}} \propto \boldsymbol{k} \times \boldsymbol{k} \times \boldsymbol{P}_{\text{SHG}}$, which leads to, in a paraxial approximation, $\boldsymbol{I}_{\text{SHG}} \propto |\boldsymbol{P}_{\text{SHG}}|^2$.

From its fourth-power dependence in (α), this intensity can thus be decomposed, similarly as for fluorescence, in contributions of different harmonics in α:

$$I_{\text{SHG}}(\alpha) \propto a_0 + a_2 \cos 2\alpha + b_2 \sin 2\alpha + a_4 \cos 4\alpha + b_4 \sin 4\alpha$$
$$I_{\text{SHG}}(\alpha) \propto a_0 + I_2 \cos 2\left(\alpha - \varphi_2\right) + I_4 \cos 4\left(\alpha - \varphi_4\right) \quad (2.23)$$

The $(a_0, a_2, \dots)$ coefficients can now be related to the way molecules are organized in a sample. These coefficients are directly related to the macroscopic tensor $\chi^{(2)}$. This tensor can be built up from the coherent addition of individual 1D dipoles, whose tensor components are nonvanishing only in one direction $\boldsymbol{e}$ (typically, the gray arrows in Figure 2.8b). This assumption is valid under nonresonant conditions (Duboisset et al., 2012; Davis, Moad, Goeken, Wampler, & Simpson, 2008). Similarly as earlier for fluorescence, the goal is thus to determine the angular distribution of the elementary dipoles $\boldsymbol{e}\,(\theta, \varphi)$, for which the orientation is (θ, φ) in the macroscopic frame. The macroscopic tensor components are thus be written as

$$\chi^{(2)}_{IJK} = N\,|\beta| \int (\boldsymbol{I}.\boldsymbol{e})(\boldsymbol{J}.\boldsymbol{e})(\boldsymbol{K}.\boldsymbol{e})(\theta, \varphi)f(\theta, \varphi)\sin\theta\, d\theta d\varphi$$
$$(2.24)$$

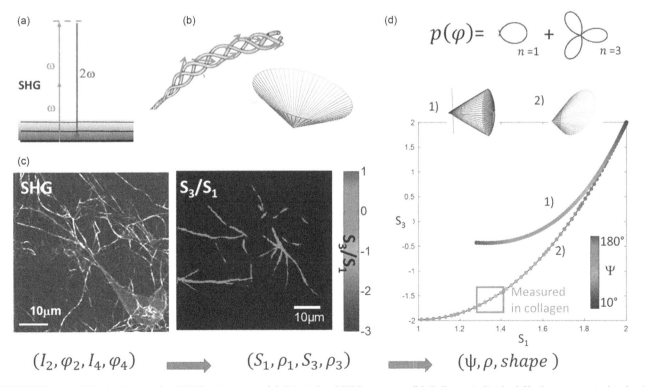

$$(I_2, \varphi_2, I_4, \varphi_4) \quad \Longrightarrow \quad (S_1, \rho_1, S_3, \rho_3) \quad \Longrightarrow \quad (\psi, \rho, shape)$$

FIGURE 2.8 Polarization-resolved SHG microscopy. (a) Principle of SHG emission. (b) Collagen individual fibrils are expected to lead to distribution of nonlinear induced dipole along a cone surface. (c) SHG image of collagen fibers (together with 2PF autofluorescence from a cell) deposited on a glass coverslip. The polarization SHG data are treated to retrieve (S_1, S_3) values, whose ratio is depicted here for a region of fibers. (d) Polarization SHG filters out only odd orders of the molecular distribution, represented here schematically. Models of two different dipole distribution 1) filled cone and 2) cone surface, showing that the measurement of (S_1, S_3) can discriminate for these two distributions in some cases. The curves represent (S_1, S_3) values for these distributions at various angular apertures ψ. Measured values in individual collagen fibers are in clear agreement with a cone surface, with an aperture of $\psi \simeq 50°$.

where $(I, J, K) = (X, Y)$ the macroscopic in-plane directions of the sample plane and $(\boldsymbol{I}.\boldsymbol{e})$ is the scalar projection of $\boldsymbol{I}$ on $\boldsymbol{e}$. $|\beta|$ is the norm of the β tensor.

Due to the θ dependence of the scalar projection functions, the macroscopic tensor can be directly related to a 2D distribution function of φ:

$$\chi^{(2)}_{IJK} = N |\beta| \int (\boldsymbol{I}.\boldsymbol{e})(\boldsymbol{J}.\boldsymbol{e})(\boldsymbol{K}.\boldsymbol{e})_{2D}(\varphi)p(\varphi)d\varphi \quad (2.25)$$

which is directly related to the distribution function (φ) mentioned in Section 2.3.1. Here, however, the power to which (cos, sin) functions appear is $n = 3$, so the only retrievable coefficients of the function (φ) will be

$$p(\varphi) = S_0 \left[1 + S_1 \cos\left((\varphi - \rho_1)\right) + S_3 \cos\left(3\left(\varphi - \rho_3\right)\right) \right] \quad (2.26)$$

Not surprisingly, the only accessible order parameters are of odd symmetry, which is expected in SHG that is only active in noncentrosymmetric structures (Boyd, 2008; Brasselet, 2011).

The relation between the $(a_0, a_2, \dots)$ intensity Fourier components and the $(S_1, S_3, \rho_1, \rho_3)$ distribution function coefficients can be found by a direct calculation (Duboisset et al., 2012), which leads to (for a cylindrical symmetry distribution):

$$I_2 = \frac{4 \left(S_1^2 + S_1 S_3\right)}{\left(5 S_1^2 + S_3^2\right)}, \quad I_4 = \frac{2 \left(S_1 S_3\right)}{\left(5 S_1^2 + S_3^2\right)}$$
$$\rho_1 = \rho_3 = \rho = \varphi_2 = \varphi_4[\pi/2] \quad (2.27)$$

for a cylindrical symmetry distribution of in-plane orientation ρ.

Visibly, the relation between the intensity coefficients and distribution coefficients is more complex than in fluorescence, due to a nonlinear dependence that originates from the nonlinear coherent nature of the optical process. Nevertheless, it is possible to resolve the inverse problem and retrieve molecular order information from polarization-resolved SHG (Duboisset et al., 2012), together with possible estimation strategies and precision that has been intensively studied (Réfrégier et al., 2011; Réfrégier, Roche, Duboisset, & Brasselet, 2012). The determination of (S_1, S_3) coefficients, complementary to 2PF, permits to deduce rich information on odd orders for a molecular distribution.

Application in Tissue Imaging: Second Harmonic Polarized Imaging in Collagen

In biological tissues, coherent SHG occurs in organized structures, such as tissues rich in collagen I (Freund, Deutsch, & Sprecher, 1986), skeletal muscles (Campagnola et al., 2002), and microtubules (Kwan, Dombeck, & Webb, 2008). SHG imaging is today exploited as a functional contrast for the diagnosis of pathological effects (Strupler et al., 2008), aging (Aït-Belkacem et al., 2012), or the understanding of the mechanics of conjunctive tissues, whose stiffness strongly depends on collagen content and structure (Gusachenko,

Tran, Houssen, Allain, & Schanne-Klein, 2012; Mansfield, Winlove, Moger, & Matcher, 2008).

The polarized nature of SHG has been in particular broadly investigated in collagen, where it has been found that peptide bounds, forming a conserved helicoidal structure (Figure 2.8b), could be at the origin of individual induced dipoles (Deniset-Besseau et al., 2009; Bancelin et al., 2014). This non centrosymmetric arrangement leads to strong SHG signals. When analyzed in polarization, strong negative third-order S_3 values are found in individual collagen fibers (Figure 2.8c) (Duboisset et al., 2012; Aït-Belkacem et al., 2012). A comparison of retrieved (S_1, S_3) with possible models shows that those values are in strong agreement with a distribution of nonlinear dipoles at the surface of a cone, rather than inside a filled distribution (Figure 2.8d). In deep tissues, this distribution is essentially broadened and filled in a more complex distribution, which also requires accounting for polarization distortions occurring from scattering and birefringence (Ait-Belkacem, Gasecka, Munhoz, Brustlein, & Brasselet, 2010; Gusachenko, Latour, & Schanne-Klein, 2010).

Subdiffraction Vectorial Imaging in Metal Nanostructures

SHG imaging has not only been employed for biological imaging. It is also an interesting contrast for the investigation of nanomaterials and crystals down to the nanometric size (Brasselet et al., 2004; Zielinski, Oron, Chauvat, & Zyss, 2009; Kim et al., 2013).

Metal nanostructures are important elements for the design of optical functions in nano-optics, which is the core field in nanoplasmonics. They are in particular able to confine optical fields at nanoscales, which is a key factor for nanophotonics applied to integrated optics, sensing, and optical devices. The optical properties of metal nanostructures are highly dependent on the excitation polarization, which has a consequence not only on their spectral scattering properties (Dopf et al., 2015) but also on the spatial and vectorial extent of the local fields they support (Shen et al., 2013; Kumar Balla et al., 2017). Controlling these properties has opened new routes for optimized biosensors, nanoantennas, and devices. Accessing vectorial properties of light–matter interactions in metal nanostructures has been so far approached using near-field scanning optical microscopy (Yurtsever & Zewail, 2012), cathodoluminescence microscopy, and photoemission electron microscopy, which can image local fields with high spatial resolution (Vesseur, de Waele, Kuttge, & Polman, 2007). An advantage of the use of nonlinear optical interactions and, in particular, SHG for the probing of vectorial local fields' properties in nanoplasmonics is the *in situ* nature of the probe, but also the background-free capabilities of the interaction, which is also intrinsically high-order dependent with respect to the incident field (Kruk et al., 2015; Bachelier, Russier-Antoine, Benichou, Jonin, & Brevet, 2008; Kujala, Canfield, Kauranen, Svirko, & Turunen, 2007; Bautista & Kauranen, 2016).

Recent works have shown that performing polarization-sensitive SHG in a scanned image can reveal important information on nonlinear vectorial information occurring at the nanoscale in metal nanostructures (Kumar Balla et al., 2017). Similarly, as in previous sections, the pixel information in a polarization-resolved scanned SHG image carries information on the local symmetry of how nonlinear-induced dipoles are organized. In Figure 2.9a, isolated gold nanoparticles of 150 nm size of various shapes are scanned in SHG microscopy, leading to images that resemble the point spread function (PSF) of the nonlinear microscope, e.g., with a resolution of about 250 nm. Resolving the SHG polarization response of each pixel for pixel sizes well below the diffraction limit (here 40 nm size per pixel) reveals very different images depending on the nanostructure shape. In Figure 2.9b, (I_2, φ_2) images of the recorded polarization SHG signal indeed reveal the directions of sticks (φ_2) that mainly follow the symmetry of the particle itself, while intense polarization modulation amplitude (I_2) appear only in certain parts of the images, in particular, for the four-arm particles. These images result from a nonlinear coupling at the overlap between the nanoparticle and the excitation diffraction-limited spot at this particular pixel position, which explains their larger spatial expansion when compared with the intensity image. The (I_2, φ_2) maps reveal information of the orientational organization of nonlinear-induced dipoles within each location of the incident focal spot. In particles larger than 100 nm, the SHG signal is dominated by their surface contribution (Bachelier et al., 2008), which is normal to the surface contour, and by the presence of localized confined field at the resonant excitation (Figure 2.9c), which explains the resemblance of the polarized SHG patterns with the particle shape themselves, even though their size is below the diffraction limit size.

The deduced (I_2, φ_2) and (I_4, φ_4) order maps are direct signatures of the symmetry of the vectorial coupling that locally takes place in the particle, occurring from local dipolar origin, or from higher orders of symmetry, including retardation effects. The presence of distinct, highly dipolar responses at the tip of the arms shows that the structure (Figure 2.9d) is expanded enough to confine optical fields away from the particle center, suggesting a negligible coupling between arms.

2.5 Polarization Nano-Optics Provides Subdiffraction Size Information

From the previous sections, it is visible that polarization microscopy offers a unique way to gain information that is not visible in traditional microscopy. Figure 2.9, in particular, shows that (I_2, φ_2) and (I_4, φ_4) maps appear as subdiffraction resolved when compared with intensity images, with patterns appearing at pixel sizes below 100 nm.

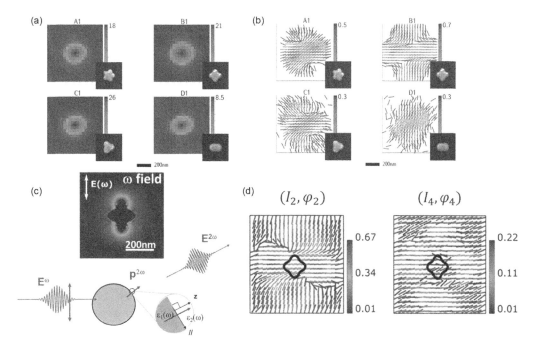

FIGURE 2.9 Polarization of SHG responses in metal nanoparticles. (a) Pure intensity of SHG images of gold nanoparticles of 150 nm size (TEM and optical SHG images are at the same scale), excited at 800 nm. (b) (I_2, φ_2) map (encoding I_2 in stick colors and φ_2 in the stick orientation) of the particles shown in (a). The five-arm structure is not sufficiently resolved to show its shape by polarized SHG. The nanorod structure evidences a transverse resonance. All images are produced using a 40 nm pixel size. (c) Simulated incident field scattered by the four-arm structure under vertically polarized excitation. The schematic drawing shows a nonlinear dipole generated at the interface metal–air, which exhibits different linear permittivities (ω) at the incident ω frequency. (d) Average polarized SHG images over many four-arm particles showing both (I_2, φ_2) and (I_4, φ_4) maps (40 nm pixel size). (Adapted from Kumar Balla et al., 2017.)

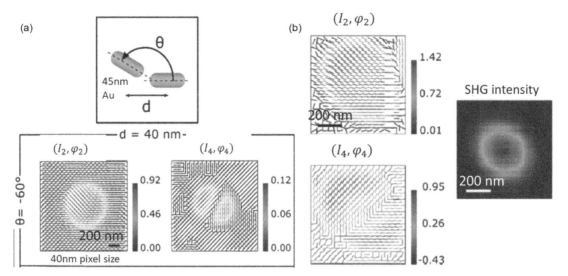

FIGURE 2.10 Polarization-sensitive SHG in nanometric objects. (a) Simulation of (I_2, I_4) and (I_2, I_4) maps from two 1D SHG-active dipoles with 40 nm size, 40 nm interdistance, and $60°$ angle between them. The presence and orientation of the two dipoles are clearly visible even though the whole object size is well below the diffraction limit (here 250 nm). (b) Experimental validation in gold nanorods of 45 nm deposited on a glass coverslip (excitation wavelength 800 nm). (Adapted from Kumar Balla et al., 2017.)

This is due to the extra degree of freedom offered by polarization modulation: even though intensity appears as a smooth decay in an image, polarization modulation can indeed change the mean direction at the pixel scale. Such a property could be exploited to render super-resolution images, if some preliminary information is known about the sample. This effect has been exploited in the so-called "polarization modulation microscopy," which has shown superior resolution gain in anisotropic media, thanks to this additional degree of freedom (Hafi et al., 2014). Nevertheless, to make it a super-resolution imaging method, hypotheses are required on the images such as sparsity of the object and its anisotropy properties.

This subdiffraction resolution capability has been probed in simple systems containing pure dipolar responses. Consider, for instance, two 1D dipoles active for SHG: when such dipoles are close to each other with an interdistance that is of a few tens of nanometers, the intensity image of the radiation through the microscope is similar to a single PSF, since the optical system cannot resolve the presence of those two objects. If those dipoles are, however, of different orientations, an excitation spot scanning the sample will see a dominance of the orientation of the first dipole in some regions, while other regions will probe the orientation of the second dipole more dominantly. Such a situation has been simulated (Figure 2.10a) and shows that for a nanometric scale dipole (for instance, 40 nm size metal nanorods) distance of 40 nm, an angle difference between the two of more than $10°$ is sufficient to demonstrate the presence of two dipoles within a PSF size (Kumar Balla et al., 2017). An experimental illustration has been realized using 45 nm length gold nanorods (aspect ratio 4.5), with a longitudinal plasmon resonance at 850 nm (Figure 2.10b). At the incident wavelength of 800 nm, the nanorod SHG emission is expected to behave as highly anisotropic with well-defined (I_2, I_4) values close to 1D dipole SHG emission (fluorescence could be similarly detected, as occurring from a two-photon resonant excitation). Among the wide collection of nanorods explored, some cases could be found where the PSF was diffraction limited, while the (I_2, I_4) maps could clearly show heterogeneities (Figure 2.10b). In those heterogeneous maps, different directions could be clearly identified and, in comparison to theoretical maps, the presence of two or more nanorods with well-defined orientations could be identified. This capability could be extended to the probing of heterogeneous nanoobjects, such as dielectric or ferroelectric nanoparticles (Kim et al., 2013).

2.6   Conclusion

In this chapter, we described the important aspects of polarized nonlinear imaging, which is the essence of polarization nano-optics and its capacity to provide subdiffraction information. We essentially concentrated on one- and two-photon excitation processes; however, numerous other optical interactions are polarization dependent and have been already used for structural imaging in biological and nanomaterial studies. An important optical contrast among those interactions is CRS, which, in addition to the advantages of coherent nonlinear imaging abilities, can target specific chemical bounds in a medium (Cheng & Xie, 2015). This process offers very rich possibilities, including cancer diagnosis. Adding polarization information to coherent Raman process, for instance, has allowed to monitor lipid loss of organization in demyelination processes (Fu et al., 2011; Imitola et al., 2011; Gasecka et al., 2017), which progressively affect the multilayer lipid sheath that protects axons in the brain or spinal cord. Some challenges however remain in the field to accommodate the complexity of biological media. First, the speed of polarization-resolved measurements needs to reach the real-time capability, which is delicate to reach in-point scanning modalities. A recent method has been proposed to circumvent this issue, based on lock-in detection of a modulated optical signal generated by a fast rotation of the incident polarization induced by an electro-optic modulator (Hofer et al., 2017). This method is promising since it is applicable to any linear or nonlinear optical contrast. Another challenge is the depolarization of light that occurs when propagating deep into tissues. Recent solutions have been proposed based on the natural polarization recovery that occurs when performing wave-front shaping in scattering media (De Aguiar, Gigan, & Brasselet, 2017). From fundamental nanoscale optics to biomedical optics, polarization microscopy offers therefore promises for a very broad range of possibilities, to see what is not visible in traditional optical microscopy.

Acknowledgments

The author would like to acknowledge all the students and colleagues who contributed to the results mentioned in this chapter: J. Duboisset (SHG); A. Kress, P. Ferrand, X. Wang, A. Gasecka, P. Gasecka (1PF, 2PF in membranes); C. Valades Cruz, H. Shaban Ahmed, C. Rimoli, M. Mavrakis (1PF in actin); C. Rendon-Barraza, N. Kumar Balla (SHG in metal nanoparticles); H. B. De Aguiar, S. Gigan (polarization in scattering media); M. Hofer (fast polarization microscopy).

References

Aggarwal, S., Yurlova, L., & Simons, M. (2011). Central nervous system myelin: Structure, synthesis and assembly. *Trends in Cell Biology*, *21*(10), 585–593. doi: 10.1016/j.tcb.2011.06.004.

Ait-Belkacem, D., Gasecka, A., Munhoz, F., Brustlein, S., & Brasselet, S. (2010). Influence of birefringence on polarization resolved nonlinear microscopy and collagen SHG structural imaging. *Optics Express*, *18*(14). doi: 10.1364/OE.18.14859.

Aït-Belkacem, D., Guilbert, M., Roche, M., Duboisset, J., Ferrand, P., Sockalingum, G., ... Brasselet, S. (2012). Microscopic structural study of collagen aging in isolated fibrils using polarized second harmonic generation. *Journal of Biomedical Optics*, *17*(8). doi: 10.1117/1.JBO.17.8.080506.

Axelrod, D. (1979). Carbocyanine dye orientation in red cell membrane studied by microscopic fluorescence polarization. *Biophysical Journal*, *26*(3), 557–573. doi: 10.1016/S0006-3495(79)85271-6.

Axelrod, D. (1989). Fluorescence polarization microscopy. *Methods in Cell Biology*, *30*, 333–352. doi: 10.1016/S0091-679X(08)60985-1.

Bachelier, G., Russier-Antoine, I., Benichou, E., Jonin, C., & Brevet, P.-F. (2008). Multipolar second-harmonic generation in noble metal nanoparticles. *Journal of the Optical Society of America B*, *25*(6), 955. doi: 10.1364/JOSAB.25.000955.

Backer, A. S., Backlund, M. P., Lew, M. D., & Moerner, W. E. (2013). Single-molecule orientation measurements with a quadrated pupil. *Optics Letters*, *38*(9), 1521–1523. doi: 10.1364/OL.38.001521.

Backer, A. S., & Moerner, W. E. (2015). Determining the rotational mobility of a single molecule from a single image: A numerical study. *Optics Express*. doi: 10.1364/OE.23.004255.

Bancelin, S., Aimé, C., Gusachenko, I., Kowalczuk, L., Latour, G., Coradin, T., & Schanne-Klein, M.-C. (2014). Determination of collagen fibril size via absolute measurements of second-harmonic generation signals. *Nature Communications*, *5*(1), 4920. doi: 10.1038/ncomms5920.

Bautista, G., & Kauranen, M. (2016). Vector-field nonlinear microscopy of nanostructures. *ACS Photonics*, *3*(8), 1351–1370. doi: 10.1021/acsphotonics.6b00052.

Benninger, R. K. P., Önfelt, B., Neil, M. A. A., Davis, D. M., & French, P. M. W. (2005). Fluorescence imaging of two-photon linear dichroism: Cholesterol depletion disrupts molecular orientation in cell membranes. *Biophysical Journal*, *88*(1), 609–622. doi: 10.1529/biophysj.104.050096.

Benninger, R. K. P., Vanherberghen, B., Young, S., Taner, S. B., Culley, F. J., Schnyder, T., ... Önfelt, B. (2009). Live Cell Linear Dichroism Imaging Reveals Extensive Membrane Ruffling within the Docking Structure of Natural Killer Cell Immune Synapses. *Biophysical Journal*, *96*(2), L13–L15. doi: 10.1016/j.bpj.2008.10.005.

Betzig, E., Patterson, G. H., Sougrat, R., Lindwasser, O. W., Olenych, S., Bonifacino, J. S., ... Hess, H. F. (2006). Imaging Intracellular Fluorescent Proteins at Nanometer Resolution. *Science*, *313*(5793), 1642–1645. doi: 10.1126/science.1127344.

Bioud, F.-Z., Gasecka, P., Ferrand, P., Rigneault, H., Duboisset, J., & Brasselet, S. (2014). Structure of molecular packing probed by polarization-resolved nonlinear four-wave mixing and coherent anti-Stokes Raman-scattering microscopy. *Physical Review A - Atomic, Molecular, and Optical Physics*, *89*(1). doi: 10.1103/PhysRevA.89.013836.

Blanchoin, L., Boujemaa-Paterski, R., Sykes, C., & Plastino, J. (2014). Actin Dynamics, Architecture, and Mechanics in Cell Motility. *Physiological Reviews*, *94*(1). Retrieved from http://physrev.physiology.org/content/94/1/235.

Boyd, R. W. (2008). *Nonlinear optics*. Academic Press.

Brasselet, S. (2011). Polarization-resolved nonlinear microscopy: Application to structural molecular and biological imaging. *Advances in Optics and Photonics*, *3*(3), 205. doi: 10.1364/AOP.3.000205.

Brasselet, S., Ferrand, P., Kress, A., Wang, X., Ranchon, H., & Gasecka, A. (2012). Imaging Molecular Order in Cell Membranes by Polarization-Resolved Fluorescence Microscopy. In D. G. Mély Y. (Ed.), *Fluorescent Methods* to Study Biological Membranes. *Springer Series on Fluorescence (Methods and Applications)*. Springer, Berlin. doi: 10.1007/4243_2012_51.

Brasselet, S., Le Floc'h, V., Treussart, F., Roch, J.-F., Zyss, J., Botzung-Appert, E., & Ibanez, A. (2004). In situ diagnostics of the crystalline nature of single organic nanocrystals by nonlinear microscopy. *Physical Review Letters*, *92*(20).

Campagnola, P. J., Millard, A. C., Terasaki, M., Hoppe, P. E., Malone, C. J., & Mohler, W. A. (2002). Three-dimensional high-resolution second-harmonic generation imaging of endogenous structural proteins in biological tissues. *Biophysical Journal*, *82*(1 Pt 1), 493–508. doi: 10.1016/S0006-3495(02)75414-3.

Cebecauer, M., Owen, D. M., Markiewicz, A., & Magee, A. I. (2009). Lipid order and molecular assemblies in the plasma membrane of eukaryotic cells. *Biochemical Society Transactions*, *37*(Pt 5), 1056–60. doi: 10.1042/BST0371056.

Cheng, J.-X., & Xie, X. S. (2015). Vibrational spectroscopic imaging of living systems: An emerging platform for biology and medicine. *Science (New York, N.Y.)*, *350*(6264), aaa8870. doi: 10.1126/science.aaa8870.

Chouaki-Benmansour, N., Ruminski, K., Sartre, A.-M., Phelipot, M.-C., Salles, A., Bergot, E., ... He, H.-T. (2018). Phosphoinositides regulate the TCR/CD3 complex membrane dynamics and activation. *Scientific Reports*, *8*(1), 4966. doi: 10.1038/s41598-018-23109-8.

Davis, R. P., Moad, A. J., Goeken, G. S., Wampler, R. D., & Simpson, G. J. (2008). Selection rules and symmetry relations for four-wave mixing measurements of uniaxial assemblies. *Journal of Physical Chemistry B*, *112*(18), 5834–5848. doi: 10.1021/JP709961K.

De Aguiar, H. B., Gigan, S., & Brasselet, S. (2017). Polarization recovery through scattering media. *Science Advances*, *3*(9). doi: 10.1126/sciadv.1600743.

Deniset-Besseau, A., Duboisset, J., Benichou, E., Hache, F., Brevet, P.-F., & Schanne-Klein, M.-C. (2009). Measurement of the second-order hyperpolarizability of the collagen triple helix and determination of its physical origin. *The Journal of Physical Chemistry B*, *113*(40), 13437–13445. doi: 10.1021/jp9046837.

Denk, W., Strickler, J. H., & Webb, W. W. (1990). Two-photon laser scanning fluorescence microscopy. *Science (New York, N.Y.)*, *248*(4951), 73–76. Retrieved from www.ncbi.nlm.nih.gov/pubmed/2321027.

Deschout, H., Zanacchi, F. C., Mlodzianoski, M., Diaspro, A., Bewersdorf, J., Hess, S. T., & Braeckmans, J. (2014). Precisely and accurately localizing single emitters in fluorescence microscopy. *Nature Methods*. doi: 10.1038/nmeth.2843.

Dopf, K., Moosmann, C., Kettlitz, S. W., Schwab, P. M., Ilin, K., Siegel, M., ... Eisler, H.-J. (2015). Coupled T-shaped optical antennas with two resonances localized in a common nanogap. *ACS Photonics*, *2*(11), 1644–1651. doi: 10.1021/acsphotonics.5b00446.

Duboisset, J., Aït-Belkacem, D., Roche, M., Rigneault, H., & Brasselet, S. (2012). Generic model of the molecular orientational distribution probed by polarization-resolved second-harmonic generation. *Physical Review A: Atomic, Molecular, and Optical Physics*, *85*(4). doi: 10.1103/PhysRevA.85.043829.

Ferrand, P., Gasecka, P., Kress, A., Wang, X., Bioud, F.-Z., Duboisset, J., & Brasselet, S. (2014). Ultimate use of two-photon fluorescence microscopy to map orientational behavior of fluorophores. *Biophysical Journal*, *106*(11). doi: 10.1016/j.bpj.2014.04.011.

Florine-Casteel, K. (1990). Phospholipid order in gel- and fluid-phase cell-size liposomes measured by digitized video fluorescence polarization microscopy. *Biophysical Journal*, *57*(6), 1199–1215. doi: 10.1016/S0006-3495(90)82639-4.

Forkey, J. N., Quinlan, M. E., & Goldman, Y. E. (2000). Protein structural dynamics by single-molecule fluorescence polarization. *Progress in Biophysics and Molecular Biology*, *74*, 1–35.

Forkey, J. N., Quinlan, M. E., & Goldman, Y. E. (2005). Measurement of single macromolecule orientation by total internal reflection fluorescence polarization microscopy. *Biophysical Journal*. doi: 10.1529/biophysj.104.053470.

Freund, I., Deutsch, M., & Sprecher, A. (1986). Connective tissue polarity. Optical second-harmonic microscopy, crossed-beam summation, and small-angle scattering in rat-tail tendon. *Biophysical Journal*, *50*(4), 693–712. doi: 10.1016/S0006-3495(86)83510-X.

Fu, Y., Frederick, T. J., Huff, T. B., Goings, G. E., Miller, S. D., & Cheng, J.-X. (2011). Paranodal myelin retraction in relapsing experimental autoimmune encephalomyelitis visualized by coherent anti- Stokes Raman scattering microscopy. *Journal of Biomedical Optics*, *16*(10), 106006. doi: 10.1117/1.3638180.

Gasecka, A., Han, T.-J., Favard, C., Cho, B. R., & Brasselet, S. (2009). Quantitative imaging of molecular order in lipid membranes using two-photon fluorescence polarimetry. *Biophysical Journal*, *97*(10). doi: 10.1016/j.bpj.2009.08.052.

Gasecka, P., Jaouen, A., Bioud, F.-Z., B de Aguiar, H., Duboisset, J., Ferrand, P., ... Brasselet, S. (2017). Lipid order degradation in autoimmune demyelination probed by polarized coherent Raman microscopy. *Biophysical Journal*, *113*(7), 1520–1530. doi: 10.1016/j.bpj.2017.07.033.

Gusachenko, I., Latour, G., & Schanne-Klein, M.-C. (2010). Polarization-resolved second harmonic microscopy in anisotropic thick tissues. *Optics Express*, *18*(18), 19339. doi: 10.1364/OE.18.019339.

Gusachenko, I., Tran, V., Houssen, Y. G., Allain, J.-M., & Schanne-Klein, M.-C. (2012). Polarization-resolved second-harmonic generation in tendon upon mechanical stretching. *Biophysical Journal*, *102*(9), 2220–2229. doi: 10.1016/j.bpj.2012.03.068.

Hafi, N., Grunwald, M., van den Heuvel, L. S., Aspelmeier, T., Chen, J.-H., Zagrebelsky, M., ... Walla, P. J. (2014). Fluorescence nanoscopy by polarization modulation and polarization angle narrowing. *Nature Methods.* doi: 10.1038/nmeth.2919.

Haluska, C. K., Schröder, A. P., Didier, P., Heissler, D., Duportail, G., Mély, Y., & Marques, C. M. (2008). Combining fluorescence lifetime and polarization microscopy to discriminate phase separated domains in giant unilamellar vesicles. *Biophysical Journal, 95*(12), 5737–5747. doi: 10.1529/biophysj.108.131490.

Hess, S. T., Girirajan, T. P. K., & Mason, M. D. (2006). Ultra-high resolution imaging by fluorescence photoactivation localization microscopy. *Biophysical Journal.* doi: org/10.1529/biophysj.106.091116.

Hofer, M., Balla, N. K., & Brasselet, S. (2017). High-speed polarization-resolved coherent Raman scattering imaging. *Optica, 4*(7). doi: 10.1364/OPTICA.4.000795.

Imitola, J., Côté, D., Rasmussen, S., Xie, X. S., Liu, Y., Chitnis, T., ... Khoury, S. J. (2011). Multimodal coherent anti-Stokes Raman scattering microscopy reveals microglia-associated myelin and axonal dysfunction in multiple sclerosis-like lesions in mice. *Journal of Biomedical Optics, 16*(2), 21109. doi: 10.1117/1.3533312.

Jiang, N., Zhuo, X., & Wang, J. (2018). Active plasmonics: Principles, structures, and applications. *Chemical Reviews, 118*(6), 3054–3099. doi: org/10.1021/acs.chemrev.7b00252.

Kampmann, M., Atkinson, C. E., Mattheyses, A. L., & Simon, S. M. (2011). Mapping the orientation of nuclear pore proteins in living cells with polarized fluorescence microscopy. *Nature Structural and Molecular Biology, 18*(6), 643–649. doi: 10.1038/nsmb.2056.

Kim, E., Steinbrück, A., Buscaglia, M. T., Buscaglia, V., Pertsch, T., & Grange, R. (2013). Second- harmonic generation of single BaTiO$_3$ nanoparticles down to 22 nm diameter. *ACS Nano.* doi: 10.1021/nn401198g.

Kress, A., Ferrand, P., Rigneault, H., Trombik, T., He, H.-T., Marguet, D., & Brasselet, S. (2011). Probing orientational behavior of MHC class i protein and lipid probes in cell membranes by fluorescence polarization-resolved imaging. *Biophysical Journal, 101*(2). doi: 10.1016/j.bpj.2011.05.021.

Kress, A., Wang, X., Ranchon, H., Savatier, J., Rigneault, H., Ferrand, P., & Brasselet, S. (2013). Mapping the local organization of cell membranes using excitation- polarization-resolved confocal fluorescence microscopy. *Biophysical Journal, 105*(1). doi: 10.1016/j.bpj.2013.05.043.

Kruk, S., Weismann, M., Bykov, A. Y., Mamonov, E. A., Kolmychek, I. A., Murzina, T., ... Kivshar, Y. S. (2015). Enhanced magnetic second-harmonic generation from resonant metasurfaces. *ACS Photonics, 2*(8), 1007–1012. doi: 10.1021/acsphotonics.5b00215.

Kujala, S., Canfield, B. K., Kauranen, M., Svirko, Y., & Turunen, J. (2007). Multipole interference in the second-harmonic optical radiation from gold nanoparticles. *Physical Review Letters, 98*(16), 167403. doi: 10.1103/PhysRevLett.98.167403.

Kumar Balla, N., Rendón-Barraza, C., Hoang, L. M., Karpinski, P., Bermúdez-Ureña, E., & Brasselet, S. (2017). Polarized nonlinear nanoscopy of metal nanostructures. *ACS Photonics, 4*(2). doi: 10.1021/acsphotonics.6b00635.

Kwan, A. C., Dombeck, D. A., & Webb, W. W. (2008). Polarized microtubule arrays in apical dendrites and axons. *Proceedings of the National Academy of Sciences of the United States of America, 105*(32), 11370–11375.

Lazar, J., Bondar, A., Timr, S., & Firestein, S. J. (2011). Two-photon polarization microscopy reveals protein structure and function. *Nature Methods, 8*(8), 684–690. doi: 10.1038/nmeth.1643.

Lingwood, D., & Simons, K. (2010). Lipid rafts as a membrane-organizing principle. *Science, 327*(5961), 46–50. doi: 10.1126/science.1174621.

Mansfield, J. C., Winlove, C. P., Moger, J., & Matcher, S. J. (2008). Collagen fiber arrangement in normal and diseased cartilage studied by polarization sensitive nonlinear microscopy. *Journal of Biomedical Optics, 13*(4), 44020. doi: 10.1117/1.2950318.

Mavrakis, M., Azou-Gros, Y., Tsai, F.-C., Alvarado, J., Bertin, A., Iv, F., ... Lecuit, T. (2014). Septins promote F-actin ring formation by crosslinking actin filaments into curved bundles.*Nature Cell Biology, 16*(4), 322–334. doi: 10.1038/ncb2921.

Mi, S., Lee, X., Hu, Y., Ji, B., Shao, Z., Yang, W., ... Pepinsky, R. B. (2011). Death receptor 6 negatively regulates oligodendrocyte survival, maturation and myelination. *Nature Medicine, 17*(7), 816–821. doi: 10.1038/nm.2373.

Patra, D., Gregor, I., & Enderlein, J. (2004). Image analysis of defocused single-molecule images for three-dimensional molecule orientation studies. *Journal of Physical Chemistry A.* doi: 10.1021/jp048188m.

Powell, K. (2005). What the cytoskeleton really looks like. *The Journal of Cell Biology, 170*(7), 1019. doi: 10.1083/jcb1707fta2.

Réfrégier, P., Roche, M., & Brasselet, S. (2011). Precision analysis in polarization-resolved second harmonic generation microscopy. *Optics Letters.* doi: 10.1364/OL.36.002149.

Réfrégier, P., Roche, M., Duboisset, J., & Brasselet, S. (2012). Precision increase with two orthogonal analyzers in polarization-resolved second-harmonic generation microscopy. *Optics Letters, 37*(20), 4173–4175.

Rust, M. J., Bates, M., & Zhuang, X. (2006). Sub-diffraction-limit imaging by stochastic optical reconstruction microscopy (STORM). *Nature Methods.* doi: 10.1038/nmeth929.

Sahl, S. J., Hell, S. W., & Jakobs, S. (2017). Fluorescence nanoscopy in cell biology. *Nature Reviews Molecular Cell Biology, 18*(11), 685–701. doi: 10.1038/nrm.2017.71.

Sanchez, S. A., Tricerri, M. A., & Gratton, E. (2012). Laurdan generalized polarization fluctuations measures membrane packing micro-heterogeneity in vivo. *Proceedings of the National Academy of Sciences of the United States of America, 109*(19), 7314–7319. doi: 10.1073/pnas.1118288109.

Schön, P., Behrndt, M., Aït-Belkacem, D., Rigneault, H., & Brasselet, S. (2010). Polarization and phase pulse shaping applied to structural contrast in nonlinear microscopy imaging. *Physical Review A, 81*(1), 13809. doi: 10.1103/PhysRevA.81.013809.

Shaban, H. A., Valades-Cruz, C. A., Savatier, J., & Brasselet, S. (2017). Polarized super-resolution structural imaging inside amyloid fibrils using Thioflavine T. *Scientific Reports, 7*(1). doi: 10.1038/s41598-017-12864-9.

Shen, H., Nguyen, N., Gachet, D., Maillard, V., Toury, T., & Brasselet, S. (2013). Nanoscale optical properties of metal nanoparticles probed by second harmonic generation microscopy. *Optics Express, 21*(10). doi: 10.1364/OE.21.012318.

Shroder, D. Y., Lippert, L. G., & Goldman, Y. E. (2016). Single molecule optical measurements of orientation and rotations of biological macromolecules. *Methods and Applications in Fluorescence, 4*(4), 42004. doi: 10.1088/2050-6120/4/4/042004.

Strupler, M., Hernest, M., Fligny, C., Martin, J.-L., Tharaux, P.-L., & Schanne-Klein, M.-C. (2008). Second harmonic microscopy to quantify renal interstitial fibrosis and arterial remodeling. *Journal of Biomedical Optics, 13*(5), 54041. doi: 10.1117/1.2981830.

Swaminathan, V., Kalappurakkal, J. M., Mehta, S. B., Nordenfelt, P., Moore, T. I., Koga, N., ... Waterman, C. M. (2017). Actin retrograde flow actively aligns and orients ligand-engaged integrins in focal adhesions. *Proceedings of the National Academy of Sciences of the United States of America, 114*(40), 10648–10653. doi: 10.1073/pnas.1701136114.

Tojkander, S., Gateva, G., & Lappalainen, P. (2012). Actin stress fibers – assembly, dynamics and biological roles. *Journal of Cell Science, 125*, 1855–1864. doi: 10.1242/jcs.098087.

Valades Cruz, C. A., Shaban, H. A., Kress, A., Bertaux, N., Monneret, S., Mavrakis, M., ... Brasselet, S. (2016). Quantitative nanoscale imaging of orientational order in biological filaments by polarized superresolution microscopy. *Proceedings of the National Academy of Sciences.* doi: 10.1073/pnas.1516811113.

Vesseur, E. J. R., de Waele, R., Kuttge, M., & Polman, A. (2007). Direct observation of plasmonic modes in Au nanowires using high-resolution cathodoluminescence spectroscopy. *Nano Letters, 7*(9), 2843–2846. doi: 10.1021/NL071480W.

Von Diezmann, A., Shechtman, Y., & Moerner, W. E. (2017). Three-dimensional localization of single molecules for super-resolution imaging and single-particle tracking. *Chemical Reviews.* doi: 10.1021/acs.chemrev.6b00629.

Vrabioiu, A. M., & Mitchison, T. J. (2006). Structural insights into yeast septin organization from polarized fluorescence microscopy. *Nature, 443*(7110), 466–469. doi: 10.1038/nature05109.

Wang, X., Kress, A., Brasselet, S., & Ferrand, P. (2013). High frame-rate fluorescence confocal angle- resolved linear dichroism microscopy. *Review of Scientific Instruments, 84*(5). doi: 10.1063/1.4807318.

Yurtsever, A., & Zewail, A. H. (2012). Direct visualization of near-fields in nanoplasmonics and nanophotonics. *Nano Letters, 12*(6), 3334–3338. doi: 10.1021/nl301643k.

Zielinski, M., Oron, D., Chauvat, D., & Zyss, J. (2009). Second-harmonic generation from a single core/shell quantum dot. *Small, 5*(24), 2835–2840. doi: 10.1002/smll.200900399.

Zimmerley, M., Mahou, P., Débarre, D., Schanne-Klein, M.-C., & Beaurepaire, E. (2013). Probing ordered lipid assemblies with polarized third-harmonic-generation microscopy. *Physical Review X, 3*(1), 11002. doi: 10.1103/PhysRevX.3.011002.

Zipfel, W. R., Williams, R. M., & Webb, W. W. (2003). Nonlinear magic: Multiphoton microscopy in the biosciences. *Nature Biotechnology, 21*(11), 1369–1377. doi: 10.1038/nbt899.

Optical Properties of Semiconductor Nanostructures

3.1 Electronic Band Structure of Semiconductor Nanostructures........... **3-1**
Bloch Functions • The Two-Band Model • More Complex Bands •
Low-Dimensional Nanostructures • Electronic Bands in SLs

3.2 Absorption Processes in Semiconductor Nanostructures............... **3-8**
Fundamental Absorption • Excitonic Absorption

3.3 Ellipsometry **3-11**

3.4 Modulation Techniques in Nanostructures............................ **3-11**

3.5 Emission Processes in Nanostructures................................. **3-12**
Band-to-Band Transitions • Bound Exciton Emission • Photoluminescence
Excitation

3.6 Raman Scattering in Semiconductor Nanostructures **3-13**
Raman Scattering Mechanisms • Raman Selection Rules • Phonons in
Low-Dimensional Structures

References **3-17**

Andres Cantarero
University of Valencia

The optical properties of an intrinsic semiconductor give us direct information on the electronic band structure. The analysis of the absorption spectrum supplies not only information on the gap, doping, or electron–phonon interaction, but also on the dimensionality of the material. The excitonic absorption is also a fingerprint of the dimensionality of the system. Some of these properties are better observed in emission (photoluminescence, PL), like exciton to bound impurity in semiconductors. There are other optical techniques used to detect small variations of the system under the application of strain or other external fields, the modulation techniques, or simply when the structures are too broad to be observable by normal absorption or emission experiments. Raman scattering is also a powerful nondestructive technique used to characterize semiconductor and semiconductor nanostructures.

3.1 Electronic Band Structure of Semiconductor Nanostructures

3.1.1 Bloch Functions

An electron in a semiconductor satisfies the Schrödinger equation

$$\mathcal{H}\Psi = E\Psi \qquad (3.1)$$

where the Hamiltonian contains the coordinates of all atoms and electrons in the solid. The adiabatic approximation takes into account that the atomic movement is much slower than electron movement. In this way, to study the electronic

properties in a solid, we can assume that the atoms are Pikus at rest, allowing the existence of translational symmetry. The movement of atoms or ions can be considered afterwards in a semiclassical way (lattice dynamics), and finally, the influence of the lattice on the electrons is introduced through the electron–phonon Hamiltonian as a perturbation.

The Bloch theorem takes into account the translational invariance of our system and shows that the wave function of an electron in a periodic potential $V(\boldsymbol{r} + \boldsymbol{R}) = V(\boldsymbol{r})$ ($\boldsymbol{R}$ being the vector of the atomic lattice while $\boldsymbol{r}$ is the electron coordinate) can be written as the product

$$\psi_{n\boldsymbol{k}}(\boldsymbol{r}) = u_n(\boldsymbol{r})e^{i\boldsymbol{k}\cdot\boldsymbol{r}}, \qquad (3.2)$$

called Bloch function, where the function $u_k(\boldsymbol{r})$ has the property $u_k(\boldsymbol{r} + \boldsymbol{R}) = u(\boldsymbol{r})$. On the other hand, the periodic function is multiplied by an exponential function (envelope function). Due to the translational symmetry, the electron is characterized by two indices, the wave number $\boldsymbol{k}$ and the index of the band n. From Bloch theorem, the electrons are defined by the wave function $\psi_{n\boldsymbol{k}}(\boldsymbol{r})$, not only by $\boldsymbol{k}$ as in a free electron system but also by n, the band index. Due to the existence of lattice periodicity (translational symmetry), the electronic band structure can be folded into the so-called first Brillouin zone (BZ), which in a cubic lattice is given by $-\pi/a < k_x, k_y, k_z < \pi/a$ in the three directions of space. The actual BZ depends on the symmetry of the crystal, where the first BZ is usually given in terms of the Wigner-size cell in the reciprocal space, which conserves all the symmetry operations of the point group.

(a)

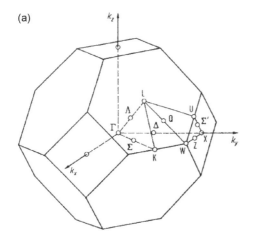

(b)

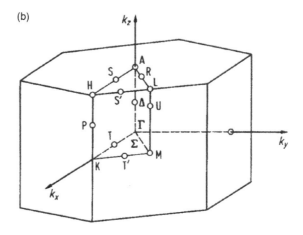

FIGURE 3.1 BZ of ZB (a) and WZ (b) structures showing high symmetry points.

In this chapter, we are going to talk mainly on technological important materials either with the zinc blende (ZB) or wurtzite (WZ) structures. In Figure 3.1, we have represented the two BZs and the main direction and symmetry points within the BZ.

3.1.2 The Two-Band Model

In the simplest model to analyze the optical properties of a semiconductor, we consider two parabolic bands: the valence band (VB) and the conduction band (CB). Actually, the most important properties in the absorption or emission are close to the gap, *i.e.*, the bottom of the CB and the top of the VB. In Figure 3.2, we show the absorption and emission properties in a simple semiconductor with two bands. We will explain the process in detail in Sec. 3.5. Now, let us extract some information using the two-band model from the Schrödinger equation, introducing the Bloch functions in the equation.

The $k \cdot p$ Method and Effective Mass Approximation

Taking the solution given by the Bloch theorem, the one-electron Hamiltonian can be written as

$$\left[\frac{p^2}{2m} + V(r) - \frac{\hbar k \cdot p}{m} \right] u_{nk}(r) = \left(E - \frac{\hbar^2 k^2}{2m} \right) u_{nk}(r)$$
(3.3)

for the periodic part of the wave function, where $V(r)$ is the lattice periodic potential and $p = -i\hbar\nabla$. The earlier equation gives us a complete set of solutions for a given k ($k = 0$ or, more generally, $k = k_0$) and the real-wave function can be written in terms of a combination of all solutions in the form

$$\phi_{nk}(r) = \sum_{n'} C_{n,n'}(k,0) u_{k,0}$$
(3.4)

By introducing $\phi_{nk}(r)$ into the Schrödinger equation, we can write the energy as

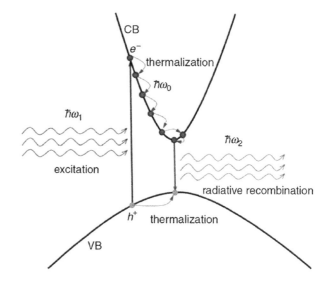

FIGURE 3.2 Absorption and emission properties in a semiconductor with two single bands.

$$E_{nk} = E_{n0} + \frac{\hbar^2 k^2}{2m} + \frac{\hbar^2}{m^2} \sum_{n' \neq n} \frac{|\langle u_{n0}|k \cdot p|u_{n'0}\rangle|^2}{E_{n0} - E_{n'0}}$$
(3.5)

Considering only two bands and calling $P_{cv}^2 \equiv |\langle c|P_x|v\rangle|^2$, we arrive to the simple expression

$$\frac{m}{m^*} = 1 + \frac{2P_{cv}^2}{mE_0}$$
(3.6)

with E_0 being the gap of the material and m^* is called effective mass, since we can write

$$E = E_0 + \frac{\hbar^2 k^2}{2m^*}$$
(3.7)

and the electron in the semiconductor behaves as a particle with a mass m^*.

To study the transport or optical properties of a semiconductor, we can consider that they move like electrons, but with a mass m^*. In GaAs, for instance, $m_e^* = 0.065m$. The rule that the effective mass is proportional to the gap fits well with many semiconductors [1].

The Tight-Binding Method

The tight-binding method or linear combination of atomic orbitals arise to explain the electronic bands in insulators, where the wave functions are supposed to be atomic-like functions [2]. We start from the solution of the free atoms forming the crystal,

$$\mathcal{H}_A(\boldsymbol{r} - \boldsymbol{r}_n)\phi_i(\boldsymbol{r} - \boldsymbol{r}_n) = E_i\phi_i(\boldsymbol{r} - \boldsymbol{r}_n) \qquad (3.8)$$

The subindex A corresponds to "free atoms" (like if there is no overlapping) and $\boldsymbol{r}_n = n_1\boldsymbol{a}_1 + n_2\boldsymbol{a}_2 + n_3\boldsymbol{a}_3$, a point of the lattice. Let us assume that the real Hamiltonian is the atomic one plus a perturbation due to the existence of the remaining atoms (overlapping plus any other interaction). This perturbation can be written as

$$V(\boldsymbol{r} - \boldsymbol{r}_n) = \sum_{m \neq n} V_A(\boldsymbol{r} - \boldsymbol{r}_m) \qquad (3.9)$$

The wave function must also be a linear combination of atomic wave functions, with the form of Bloch functions:

$$\Phi(\boldsymbol{r}) = \sum_n e^{i\boldsymbol{k} \cdot \boldsymbol{r}_n}\phi_i(\boldsymbol{r} - \boldsymbol{r}_n) \qquad (3.10)$$

to fulfill the Bloch theorem. The energy will be

$$E(\boldsymbol{k}) = \frac{1}{N}\sum_{n,m} e^{i(\boldsymbol{r}_n - \boldsymbol{r}_m)} \int \phi_i^*(\boldsymbol{r} - \boldsymbol{r}_n)[E_i + V(\boldsymbol{r} - \boldsymbol{r}_n)]$$
$$\times \phi_i(\boldsymbol{r} - \boldsymbol{r}_m)d^3\boldsymbol{r} \qquad (3.11)$$

In the simplest case of only one wave function per atom (spherical symmetry) and limiting the interaction to first neighbors,

$$E(\boldsymbol{k}) = E_i - A - BF(\boldsymbol{r}) \qquad (3.12)$$

where

$$F = \sum_m e^{i\boldsymbol{k} \cdot (\boldsymbol{r}_m - \boldsymbol{r}_n)} \qquad (3.13)$$

a sum over the first neighbor to atom n, called structure factor, and

$$A = -\int \phi^*(\boldsymbol{r} - \boldsymbol{r}_n)V(\boldsymbol{r} - \boldsymbol{r}_n)\phi(\boldsymbol{r} - \boldsymbol{r}_n)d^3\boldsymbol{r} \qquad (3.14)$$

and

$$B = -\int \phi^*(\boldsymbol{r} - \boldsymbol{r}_n)V(\boldsymbol{r} - \boldsymbol{r}_n)\phi(\boldsymbol{r} - \boldsymbol{r}_m)d^3\boldsymbol{r} \qquad (3.15)$$

3.1.3 More Complex Bands

Degenerate Bands: $\boldsymbol{k} \cdot \boldsymbol{p}$ Method

Most of the technologically important materials do not have a simple lattice. For instance, GaAs has a face centered cubic (FCC) lattice with two atoms in the unit cell or GaN and a WZ hexagonal lattice with four atoms in the unit cell. The BZs of these two lattices have been shown in Figure 3.1.

The high symmetry points are also shown in the figure. The Γ–point corresponds to $k = 0$.

The VB of diamond (group IV) or GaAs (III–V) are triply degenerate at Γ ($\boldsymbol{k} = 0$), *i.e.*, they are usually written as $|X\rangle$, $|Y\rangle$ and $|Z\rangle$ ($l = 1$ and $\langle p_x|X\rangle = \langle p_y|Y\rangle = \langle p_z|Z\rangle$), while the CB has s character ($l = 0$,) and it does not present degeneracy (actually all the bands present spin degeneracy, *i.e.* there are two CBs and six VBs). The triple degeneracy can be removed by introducing the spin–orbit interaction, which is important in most of the III–V compounds. As it is well known, more heavy elements contribute to the increase of the spin–orbit coupling. The spin can be introduced in the Hamiltonian as a perturbation, being the perturbation Hamiltonian

$$\mathcal{H}_{so} = \frac{\hbar}{4c^2m^2}(\boldsymbol{\nabla}V(\boldsymbol{r}) \times \boldsymbol{p}) \cdot \boldsymbol{\sigma} \qquad (3.16)$$

where $V(\boldsymbol{r})$ is the crystal potential and $\boldsymbol{\sigma}$ the Pauli matrices. In ZB materials (also diamond materials), the spin removes partially the degeneracy partially. Two of the three VBs remain degenerate at the Γ– point, although the degeneracy is removed at $k \neq 0$. The third VB is split, and this band is called split-off band or spin–orbit split-off band, and the value of the splitting is given by Δ or Δ_{so}. As mentioned earlier, in InP, for instance, has $\Delta = 0.108$ eV, while InSb has a much larger value, $\Delta = 0.75$ eV due to the larger mass (actually the amount of electrons) of Sb.

In Figure 3.3, we show the band structure of CdTe calculated with the *ab initio* self-consistent fully relativistic linear-muffin-tin orbital method [3]. The most interesting

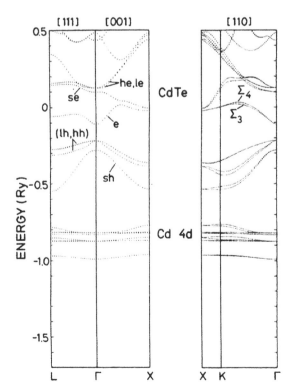

FIGURE 3.3 Electronic band structure of CdTe calculated by the *ab initio* muffin-tin orbital method [3].

structure is obviously around the gap. There is a single CB called "e" (actually, all the bands are doubly degenerate due to the spin as mentioned earlier, and the degeneracy can be removed by the application of a magnetic field), the light-hole ("lh") and heavy-hole ("hh") bands, which are degenerate at $\boldsymbol{k} = 0$, below is the split-off ("sh") band, and there is a similar structure in the CBs, "se", and the light-electron and heavy-electron ("le", "he") bands, also degenerate at $\boldsymbol{k} = 0$. Although in the extended work of Cardona et al. [3] they work with 16 bands, in most ZB semiconductors, it is adequate to consider the VB (lh, hh, and spin-orbit, so) and CBs. Actually, the number of bands considered, for instance, to calculate the effective masses depend on the spin–orbit splitting (coupling) constant. In Table 3.1, we can see how small is the spin–orbit constant in InP when compared with its gap, while in the case of InSb, it is just the other way round, the gap is very small while the spin–orbit constant is very large. Depending on these values, the electronic band (and effective masses) calculations can be simplified. For instance, in the case of InSb, to calculate the effective masses, it is necessary to consider a 6×6 Hamiltonian, including the lh, hh, and e bands (doubly degenerate due to the spin). However, in the case of InP, we also need a 6×6 Hamiltonian, but with the lh, hh, and so bands.

The band structure of GaAs is basically the same, since it has the same symmetry (ZB). The CB is a single band with a minimum at the Γ−point of BZ, while the VBs are split in the Γ−point. The first two are the hh and lh bands due to the fact that their effective masses are still degenerate at Γ, while the split-off band is 0.34 eV below. The CB has a minimum at the $X-$ and $L-$points of the BZ. In some semiconductors like GaP, the $X-$valley is below the Γ−point, and it becomes an indirect semiconductor. In GaAs, it is enough to consider a 4×4 Hamiltonian, since both so and e bands are far away. $\boldsymbol{k} \cdot \boldsymbol{p}$ theory predicts an effective mass proportional to the gap in a first approximation. But due to the different spin–orbit couplings (more important in heavy elements), there are some variations as can be observed in Table 3.2. Although all the bands are in principle degenerate, there is actually a small lack of degeneracy coming from the fact that ZB materials do not have a center of inversion. This is translated into the $\boldsymbol{k} \cdot \boldsymbol{p}$ theory in the appearance of linear terms in k [3,4].

The earlier results and the matrix or matrices shown in Ref. [3] are valid for bulk semiconductors. In a

superlattice (SL) along z, k_z is quantized, and actually, we have to substitute in the Hamiltonian by the term $-i\partial/\partial z$.

In the case of GaAs, as commented earlier, it is enough to consider a 4×4 matrix. Considering spin–orbit coupling, the wave function can be written as [5]

$$
\begin{aligned}
u_1 &= |3/2, +3/2\rangle = \frac{1}{\sqrt{2}} (X + iY) \uparrow \\
u_2 &= |3/2, +1/2\rangle = \frac{i}{\sqrt{6}} [(X + iY) \downarrow -2Z \uparrow] \\
u_3 &= |3/2, -1/2\rangle = \frac{1}{\sqrt{6}} [(X - iY) \uparrow +2Z \downarrow] \\
u_4 &= |3/2, -3/2\rangle = \frac{i}{\sqrt{2}} (X - iY) \downarrow
\end{aligned}
\tag{3.17}
$$

The first and last wave functions correspond to the hh band and the $|3/2, \pm 1/2\rangle$ to the lh band. For completeness, the split-off band wave functions can be written as

$$
\begin{aligned}
u_5 &= |1/2, +1/2\rangle = \frac{1}{\sqrt{3}} [(X - iY) \downarrow +Z \uparrow] \\
u_6 &= |1/2, -1/2\rangle = \frac{i}{\sqrt{3}} [-(X - iY) \uparrow +Z \downarrow].
\end{aligned}
\tag{3.18}
$$

As commented previously, the $\boldsymbol{k} \cdot \boldsymbol{p}$ method can be used, including upper CBs [3], giving more precise values of gaps and effective masses. Luttinger reduced the dimension of the matrix to 4×4 (only hh and lh bands) [6], including the effect of the other bands by means of Löwdin perturbation theory [7]. He introduced three parameters, γ_1, γ_2, and γ_3, related to the effective masses, which allows us to have a very good approximation close to the Γ−point. In the axial approximation $\gamma_2 \approx \gamma_3 = \bar{\gamma}$ and the Hamiltonian

$$
\mathcal{H}_{LK} = \begin{pmatrix} H_{11} & H_{12} & H_{13} & H_{14} \\ H_{21} & H_{22} & H_{23} & H_{24} \\ H_{31} & H_{32} & H_{33} & H_{34} \\ H_{41} & H_{42} & H_{43} & H_{44} \end{pmatrix}
\tag{3.19}
$$

can be written in terms of γ as

$$
\begin{aligned}
H_{11} &= \frac{\hbar^2}{2m_0} (\gamma_1 - 2\bar{\gamma}) k_z^2 - \frac{\hbar^2}{2m_0} (\gamma_1 + \bar{\gamma}) k_\perp^2 \\
H_{22} &= \frac{\hbar^2}{2m_0} (\gamma_1 + 2\bar{\gamma}) k_z^2 - \frac{\hbar^2}{2m_0} (\gamma_1 - \bar{\gamma}) k_\perp \\
H_{12} &= -\frac{\hbar^2}{2m_0} \frac{\sqrt{3}}{2} \bar{\gamma} (k_x - ik_y)^2 \\
H_{13} &= \frac{\hbar^2}{m_0} \sqrt{3} \bar{\gamma} (k_x - ik_y) k_z
\end{aligned}
\tag{3.20}
$$

where we have the following relationships:

$$
\begin{aligned}
H_{3,3} &= H_{2,2}, \quad H_{44} = H_{11} \\
H_{24} &= -H_{13}, \quad H_{14} = H_{23} = 0
\end{aligned}
\tag{3.21}
$$

In quantum wells, we used the same Hamiltonian, with the simple substitution of k_z by $-i\partial/\partial z$. In nanowires (NWs), the lateral side is never thin enough to be quantized and, in quantum dots, depending on the shape/size sometimes, we can consider quantization in only one direction, they are usually flat on top of a wetting layer (strained

TABLE 3.1 Electron Energy Gap and Spin–Orbit Splitting in Some III-V ZB Semiconductors

	GaAs	InP	GaSb	InSb
E_0 (eV)	1.519	1.424	0.813	0.235
Δ_0 (eV)	0.340	0.108	0.75	0.803

TABLE 3.2 Effective Masses in Some III-V ZB Semiconductors

	GaAs	InP	GaSb	InSb
m_{lh}^* (eV)	0.085	0.13	0.044	0.017
m_{hh}^* (eV)	0.73	0.66	0.45	0.49
m_{so}^* (eV)	0.18	0.20	0.13	0.098

quantum dots or self-assembled quantum dots). In the case of nanoparticles, unless they are small enough, we can neglect electron confinement. Electron confinement becomes important for sizes smaller than 10 nm.

In the case of WZ crystals, the lh VB is not degenerate any more due to the so-called crystal field splitting, and we miss the cubic symmetry and now have hexagonal symmetry. Thus, there are three VBs, in general, very close to each other. From up to down in energy, bands A, B, and C, giving rise to three excitonic states. If we call Δ_1 the separation between bands A and B, and Δ_2 the separation between bands B and C, the crystal field and spin–orbit splitting can be written in terms of energy separation [8] as

$$\Delta_{cr} = \frac{1}{2}\left[\Delta_1 + \Delta_2 + \sqrt{(\Delta_1 + \Delta_2)^2 - 16\Delta_1\Delta_2}\right] \quad (3.22)$$

and

$$\Delta_{so} = \frac{1}{2}\left[\Delta_1 + \Delta_2 - \sqrt{(\Delta_1 + \Delta_2)^2 - 16\Delta_1\Delta_2}\right] \quad (3.23)$$

In the case of InP, from Density Functional Theory (DFT) calculations [8] we obtain, using the experimental data from Ref. [9] using the earlier equations $\Delta_{cr} \approx 0.147$ eV and $\Delta_{so} \approx 0.084$ eV. This last value is very close to that calculated by DFT [8]: $\Delta_{so}(\text{WZ}) \approx 0.084 \approx \Delta_{so}(\text{ZB}) \approx 0.083$ eV (the experimental value in the case of ZB InP is 0.108 eV).

The equivalent to the Luttinger Hamiltonian for WZ is given in Ref. [10]. They follow the same procedure as Luttinger, considering

$$H_{ij}(\boldsymbol{k}) = H_{ij}(0) + D_{ij} \quad (3.24)$$

but now H_{ij} is a 6×6 Hamiltonian and D_{ij} is calculated using Löwdin perturbation theory [7].

Degenerate Bands: Tight-Binding Method

In the simplest case of a hydrogen molecule, as soon as we approach the two hydrogen atoms to form the bonding, the two wave functions overlap, and the wave function is a combination of the two wave functions of the free atom

$$\Psi = a\psi_1 + b\psi_2, \quad (3.25)$$

where the functions ψ_1 and ψ_2 are orthogonal. The energy of the molecule is obtained by solving the equation

$$\mathcal{H}|\Psi\rangle = a\mathcal{H}|\psi_1\rangle + b\mathcal{H}|\psi_2\rangle = E|\Psi\rangle = E(a|\psi_1\rangle + b|\psi_2\rangle)). \quad (3.26)$$

By multiplying the equation by $\langle\psi_1|$ and $\langle\psi|$, we obtain a system of equations whose energy is obtained by solving the determinant

$$\begin{pmatrix} H_{11} - E & H_{12} \\ H_{12}^* & H_{22} - E \end{pmatrix} = 0 \quad (3.27)$$

giving the bonding and antibonding states of the hydrogen molecule. In the case of the VB of a ZB semiconductor, we have four electrons from the cation and four electrons from the anion forming a covalent tetrahedral bond. In this case, the Hamiltonian matrix has the form [11]:

$$\begin{bmatrix} E_s^c & E_{ss}g_0 & 0 & 0 & 0 & E_{sp}g_1 & E_{sp}g_2 & E_{sp}g_3 \\ E_{ss}g_0^* & E_s^a & -E_{sp}g_1^* & -E_{sp}g_2^* & -E_{sp}g_3^* & 0 & 0 & 0 \\ 0 & -E_{sp}g_1 & E_p^c & 0 & 0 & E_{xx}g_0 & E_{xy}g_3 & E_{xy}g_2 \\ 0 & -E_{sp}g_2 & 0 & E_p^c & 0 & E_{xy}g_3 & E_{xx}g_0 & E_{xy}g_1 \\ 0 & -E_{sp}g_3 & 0 & 0 & E_p^c & E_{xy}g_2 & E_{xy}g_1 & E_{xx}g_0 \\ E_{sp}g_1^* & 0 & E_{xx}g_0^* & E_{xy}g_3^* & E_{xy}g_2^* & E_p^a & 0 & 0 \\ E_{sp}g_2^* & 0 & E_{xy}g_3^* & E_{xx}g_0^* & E_{xy}g_1^* & 0 & E_p^a & 0 \\ E_{sp}g_3^* & 0 & E_{xy}g_2^* & E_{xy}g_1^* & E_{xx}g_0^* & 0 & 0 & E_p^a \end{bmatrix} \quad (3.28)$$

For an SL, we do not have two atoms in the unit cell, but a large number of atoms, thus the matrix could have a dimension of $8n \times 8n$, n being the number of atomic layers in the unit cell. Although it is a large matrix, since it is a sparse matrix, there are several algorithms that efficiently solve this kind of matrices. In a work of A. R. Goni et al. [12], they have calculated the limiting case of one monolayer of InAs into a matrix of GaAs. Since there is only one monolayer (it is not an SL), a very large chain of Ga and As atoms must be introduced in the calculations (in the other two directions, the periodicity is like the ZB unit cell). In Figure 3.4, it is interesting to observe how the lh wave function extends more than the electron wave function. This kind of calculations, with only one monolayer, cannot be done using the effective mass approximation. In one monolayer the concept of effective mass lacks its sense.

In the case of WZ-type semiconductors, the solutions are similar [13], but now we have four atoms in the unit cell. The starting matrix (for bulk) is 16×16. The spin–orbit coupling has not been introduced in any of the earlier equations. A simple way to introduce it is to change from $|X\rangle$, $|Y\rangle$, $|Z\rangle$, to the wave functions mentioned when discussing the $\boldsymbol{k}\cdot\boldsymbol{p}$ method. In this way, the spin is diagonal, and we have to add the spin–orbit constant only in the diagonal terms [14].

3.1.4 Low-Dimensional Nanostructures

The only strictly two-dimensional (2D) material is graphene. There is a group of layered materials, among them transition metal dichalcogenides, consisting of a few atomic layers bound by van der Waals forces. It is possible to isolate one layer, but since there is more than one atomic layer, sometimes they are named as quintuple layer (Bi_2Te_3), quadruple layer (GaSe), and so on. However, when the confinement

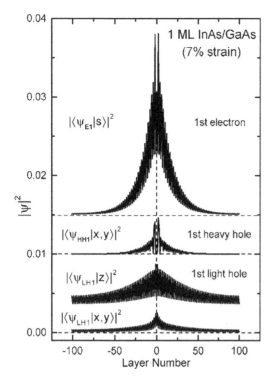

FIGURE 3.4 Square modulus of the main atomic-orbital components of the electron and hole wave function at the Γ-point.

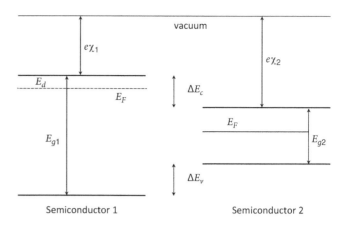

FIGURE 3.5 Band alignment between semiconductors 1 and 2. 1 is doped while 2 is intrinsic.

in one dimension is much larger than in the other two, materials are called 2D even if they are not strictly 2D. Typical cases are δ-doped semiconductors, quantum wells, multiple quantum wells, and SLs. In the junction of a doped semiconductor with an intrinsic one, we can also have a nearly 2D behavior. We will go through some basic concepts to understand the junctions.

Metal–Semiconductor Junctions

In a metal, the energy necessary to extract an electron to vacuum is called work function $e\phi_m$. This is the energy difference between the Fermi level E_F of the metal, defined as the last populated electronic level, and the vacuum. In a semiconductor, the work function is not a well-defined quantity because the Fermi energy can change with doping and temperature. With the same purpose, we introduce a different amount called electron affinity, $e\chi$ (see Figure 3.5). The electron affinity is the energy difference between the bottom of the CB and the vacuum. We can also define the work function $e\phi_s$ of a semiconductor, but as mentioned earlier, this is a not well-defined amount since it depends on several factors. The difference between the metal work function, and the electron affinity defines the behavior of the metal–semiconductor junction. Basically, depending on the work function of the semiconductor, when we produce a metal–semiconductor junction, the bands of the semiconductor at the junction will bend up or down to level the Fermi energies. In the book of D. A. Neamen [15], there is an extended description of metal–semiconductor and semiconductor–semiconductor junctions.

Heterojunctions and Two-Dimensional Electron Gas 2DEG

Epitaxy allows to grow a material B on top of a material A. Let us assume that material A is the substrate, where we plan to grow material A epitaxially, to increase the crystal quality. After a large number of monolayers, we can stop growing material A and grow a material B. We have formed a heterostructure if materials A and B are different. If they are the same, for instance, n-Si$^+$/p-Si, we have a homojunction. It can be used to build a *pn* diode [15].

Heterojunctions or homojunctions are also called "metallurgical" junctions. In the case of two different semiconductors, they have different energy gaps and electron affinities. In Figure 3.5, we compare the band scheme in real space (energy as a function of distance) of two semiconductors. The one at the left has a larger gap and that at the right has a larger electron affinity ($e\chi_1 < e\chi_2$). The properties of the junction will be defined by the electron affinities and the gaps if they are intrinsic semiconductors, but in the case of doped materials, the Fermi level will be closer to the impurity band, and in practice, there will be an electron (or hole) transfer from the more doped region to the less doped one at the junction, producing a curvature of bands. As an example, let us consider undoped AlGaAs growth on a GaAs buffer layer and, after a thick enough AlGaAs region, an intrinsic GaAs semiconductor. In that case, the CB of GaAs will be below the CB of GaAs and the VB of GaAs will be on top of the VB of AlGaAs, with a larger gap. But, let us consider the same structure but with an n-doped AlGaAs region. Since in equilibrium, the Fermi energy must be the same along the whole semiconductor structure, electrons from AlGaAs will move into the undoped GaAs creating an electric field in the junction and a depletion layer (a region without carriers). Since we are dealing with an *n*-doped material, the interesting physics happens in the CB. Figure 3.6 shows the CB region of the heterostructure before and after the junction is performed. Since the Fermi energy is the same as mentioned earlier, we have at the junction, but in the GaAs side, a triangular quantum well in a first approximation and discrete electronic levels, a 2D electron gas (in the plane

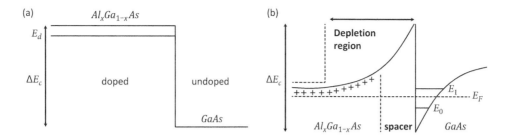

FIGURE 3.6   (a) Energy bands of AlGaAs and GaAs and (b) energy bands forming a 2D electron gas.

parallel to the junction). Since the free electrons are in the GaAs region, which is undoped, the mobility of this kind of heterostructures is several orders of magnitude larger than that of GaAs. This is the basis of the high electron mobility transistors (HEMT), nowadays mainly based on GaN due to the high power capability [16].

Type I and II Quantum Wells

Depending on the electron affinity (considering intrinsic semicoductors), when we have several heterojunctions as shown in Figure 3.7, we can have electron and hole confinement in the same semiconductor (type I quantum well) or confinement of the electrons in semiconductor 1 and confinement of the hole in semiconductor 2. In that case, we have a type-II quantum well (QW). The physics is completely different, because in type-II QWs, the electrons and holes are spatially separated. The classical example of a type-I QW is that of AlGaAs/GaAs. An example of type-II QW is that formed by the junction of InP QW with GaAs [17]. The quality of a QW depends on the mismatch between the lattice parameters. In practice, the limit in the mismatch is ~7%. InAs on GaAs is very close to this limit, and we can grow only one or two monolayers of InAs in GaAs [12], but in the case of AlAs/GaAs, they are practically lattice match, and we can build multiple QWs (MQWs) of hundreds of periods. One of the most interesting materials for the fabrication of blue and violet lasers is GaN, but it is difficult to find a suitable substrate to grow high-quality GaN (bulk GaN is usually grown

by high-pressure/high-temperature techniques). The typical substrates for molecular beam epitaxy (MBE) or molecular organic chemical vapor deposition (MOCVD) growth are sapphire, silicon carbide, and silicon [111], and this last is the most interesting for obvious reasons, since it allows the integration of microelectronics. SiC is the best substrate (smaller lattice mismatch), although it is the most expensive. The mismatch between the substrates however has allowed the growth of strained quantum dots (also called self-assembled quantum dots), where they can be aligned on top of the other to optimize the strain (the energy).

3.1.5   Electronic Bands in SLs

An SL is an engineered material grown epitaxially on a substrate (cubic for simplicity) by depositing n_1 monolayers of material 1, n_2 monolayers of material 2, and so on. In this way, the lattice parameter is not a, the bulk lattice parameter in a cubic material, but $d = (n_1 + n_2)a$. To simplify the explanation, we can imagine that the two materials are GaAs and AlAs (they have very similar lattice constants) and the growth direction is the (001). In the plane, the lattice constant is a, but in the (001)−direction, it is $d \gg a$. In the reciprocal space, the size of the BZ is $2\pi/a$ in the plane, but $2\pi/d$ in the (001)−direction, i.e., $n_1 + n_2$ times smaller than in the plane. The new BZ is usually called mini BZ. Since the region is $n_1 + n_2$ times smaller, the number of electronic bands is $n_1 + n_2$ times the bulk bands. SLs can be seen as a large number of quantum wells, where there is an overlap between them.

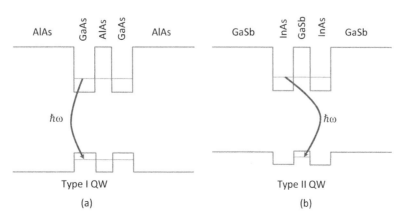

FIGURE 3.7   Type I (a) and type II (b) quantum wells.

Due to the overlapping and Pauli's principle, all states cannot have the same energy, and a miniband arises. As soon as the energy of the electronic level increases, the width of the miniband increases. A book of Gerard Bastard [18] on the subject became very famous in the 90s. In this book, he applied normal quantum mechanics to semiconductors. A complete description of the electronic band structure of QWs and SLs can be found in this book.

3.2 Absorption Processes in Semiconductor Nanostructures

3.2.1 Fundamental Absorption

Many of the properties of a solid are given by the shape of the top of the VB and the bottom of the CB. Since close to these maxima and minima, the functions are nearly parabolic, in the more general case, the energy can be expanded into a Taylor series, being the first term parabolic (the first derivative in an extremal point is zero):

$$E_c(\boldsymbol{k}_c) = E_c(0) + \frac{\hbar^2 k_c^2}{2m_c^*}; \quad E_v(\boldsymbol{k}_v) = E_v(0) - \frac{\hbar^2 k_v^2}{2m_v^*} \quad (3.29)$$

for the CB and VB, respectively, where m_c^* and m_v^* are the electron and hole effective masses. The dynamics of electrons or holes in a solid is defined by their effective masses. The simplest band model is given by two parabolic bands, the VB and the CBs (CB), which can be assumed to be isotropic. Thus, in a two-band model, we can write the energy as a function of $\boldsymbol{k}$ as

$$E_c(\boldsymbol{k}) - E_v(\boldsymbol{k}) = E_g(\boldsymbol{k}_0) + \frac{\hbar^2 k^2}{2\mu^*} \quad (3.30)$$

where μ^* is the reduced effective mass and $E_g = E_c - E_v$ is the gap of the material, and we have assumed $\boldsymbol{k}_c = \boldsymbol{k}_v$ (we will come back to this point later). We will assume a direct bandgap material, but the gap is not always at the center of the BZ ($\Gamma-$point), for instance, in InSe is at Z [19].

In a simple two-band model, an absorption process is produced when a photon impinges on a crystal and is able to promote an electron from the VB to the CB, leaving a hole in the VB. The transition probability for photon absorption (dipole allowed) can be written in terms of the Fermi Golden rule:

$$R = \frac{2\pi}{\hbar} \sum_{\boldsymbol{k}_c} \sum_{\boldsymbol{k}_v} |P_{cv}|^2 \delta(E_c(\boldsymbol{k}_c) - E_v(\boldsymbol{k}_v) - \hbar\omega(\boldsymbol{\kappa})) \quad (3.31)$$

where P_{cv} is the matrix element of the transition and $\delta(E)$ stands for the energy conservation between initial and final states. In the initial state, there is a photon with energy $\hbar\omega$ and wave number $\boldsymbol{\kappa}$, and in the final state, there is an electron in the CB and a hole in the VB. This is called a direct bandgap transition. The matrix element is

$$P_{cv} = \langle v|\mathcal{H}_{eR}|c\rangle = -\frac{e}{m}\langle v|\boldsymbol{A} \cdot \boldsymbol{p}|c\rangle \quad (3.32)$$

and we will assume that the wave number of the light is $\boldsymbol{\kappa} = 0$ (dipole allowed transition). This is the common case of III–V and II–VI semiconductors. $\langle v|\boldsymbol{A} \cdot \boldsymbol{p}|c\rangle$ is

$$\int u_v^*(\boldsymbol{r})e^{-i\boldsymbol{k}_v \cdot \boldsymbol{r}} \boldsymbol{A}_0 e^{i\boldsymbol{\kappa} \cdot \boldsymbol{r}} \boldsymbol{p} u_c(\boldsymbol{r}) e^{i\boldsymbol{k}_c \cdot \boldsymbol{r}} d^3\boldsymbol{r} \quad (3.33)$$

As commented, we will consider $\boldsymbol{\kappa} = 0$ (dipole-allowed transition). Since $\boldsymbol{p}$ is odd, one of the wave functions must be odd to have an even integral (nonzero). In III–V semiconductors the VBs have basically $p-$character, while the CB $s-$character and the integral will be nonzero. Since $\boldsymbol{p} = -i\hbar\boldsymbol{\nabla}$,

$$-i\hbar\boldsymbol{\nabla} u_c(\boldsymbol{r})e^{i\boldsymbol{k}_c \cdot \boldsymbol{r}} = e^{i\boldsymbol{k}_c \cdot \boldsymbol{r}}(\boldsymbol{p} + \boldsymbol{k}_c)u_c(\boldsymbol{r}) \quad (3.34)$$

but the second term does not contribute because u_c and u_v are orthogonal. Since the periodic part of the Bloch functions vary only within a unit cell, the integral can be divided into an integral in the unit cell times an integral over the remaining space. The result is more clear if we write it in terms of a sum. The integral becomes

$$\frac{1}{NV_c} \sum_{\boldsymbol{R}} e^{i(\boldsymbol{k}_v - \boldsymbol{k}_c) \cdot \boldsymbol{R}} \int_{V_c} d^3\boldsymbol{r} \, u_v(\boldsymbol{r}) \boldsymbol{p} u_c(\boldsymbol{r})$$

$$= \frac{\delta_{\boldsymbol{k}_v, \boldsymbol{k}_c}}{V_c} \int_{V_c} u_v(\boldsymbol{r}) \boldsymbol{p} u_c(\boldsymbol{r}) d^3\boldsymbol{r} \quad (3.35)$$

since the exponential varies very little inside the unit cells, it can be extracted from the integral. On the other hand, the matrix element does not depend on $\boldsymbol{k}$. Introducing this result into Eq. (3.31),

$$R = \frac{2\pi}{\hbar}|P_{cv}|^2 \sum_{\boldsymbol{k}} \delta(E_g + \hbar^2 k^2/2\mu^* - \hbar\omega) = \frac{2\pi}{\hbar}|M_{cv}|^2\rho(E) \quad (3.36)$$

The density of state (DOS) can be obtained easily from the DOS in the $\boldsymbol{k}$ space. In 3D,

$$\rho(\boldsymbol{E})dE = \rho(\boldsymbol{k})d^3\boldsymbol{k} = \frac{2}{(2\pi)^3}4\pi k^2 dk \quad (3.37)$$

$$\rho(\boldsymbol{E}) = \frac{1}{\pi^2}\frac{k^2}{dE/dk} = \frac{1}{\pi^2}\frac{\mu^* k}{\hbar^2} \quad (3.38)$$

Substituting

$$k = \sqrt{\frac{2\mu^*}{\hbar^2}}\sqrt{\hbar\omega - E_g} \quad (3.39)$$

The absorption coefficient can be derived by taking into account that the power loss due to the absorption is $R\hbar\omega$. The power loss can be related to α, ε_2, or the extinction coefficient n_2, $\alpha = 2\omega n_2/c$ ($\varepsilon_2 = 2n_1 n_2$) [1].

$$\alpha = \frac{1}{c\omega n}\left(\frac{2\pi e}{m}\right)^2 |P_{cv}|^2 \sum_{\boldsymbol{k}} \delta(E_c(\boldsymbol{k}) - E_v(\boldsymbol{k}) - \hbar\omega)$$

$$\propto f_{cv}\sqrt{\hbar\omega - E_g} \quad (3.40)$$

where we have used the analogy with the dielectric constant of a group of oscillators, with the oscillator strength being defined as

$$f_{cv} = \frac{2|P_{cv}|^2}{m\hbar\omega} \quad (3.41)$$

Since the absorption is proportional to the DOS, as shown in Eq. (3.31), we can extend our results to lower dimensional semiconductors, always within the two-band model. In the case of strictly 2D semiconductors, the energy DOS can be obtained from

$$\rho(E)dE = \frac{2}{(2\pi)^2}d^2k = \frac{2}{(2\pi)^2}2\pi k dk \qquad (3.42)$$

giving the well-known result

$$\alpha_{2D} \propto f_{cv}\frac{\mu^*}{\pi\hbar^2} \qquad (3.43)$$

a constant value, energy independent. In one dimension,

$$\rho(E)dE = \frac{2}{2\pi}dk \qquad (3.44)$$

and

$$\rho(E) = \frac{1}{2\pi}\left(\frac{2m*}{\hbar^2}\right)^{3/2}\frac{1}{\sqrt{E}} \qquad (3.45)$$

and the absorption coefficient is

$$\alpha_{1D} \propto f_{cv}\frac{1}{\sqrt{\hbar\omega - E_g}} \qquad (3.46)$$

In zero dimensions, the levels are completely quantized, and the absorption coefficient is a set of delta functions. The DOS is shown in Figure 3.8.

There are several techniques to grow semiconductor nanostructures and high-quality thick layers (3D materials). However, the common technique used to obtain the absorption in bulk can only be used in 3D and some 2D materials: transmission experiments. In the case of quasi-one-dimensional nanostructures (NWs) and quasi-zero-dimensional nanostructures (quantum dots), α can be indirectly obtained through photoluminescence (PL) excitation, which will be explained later.

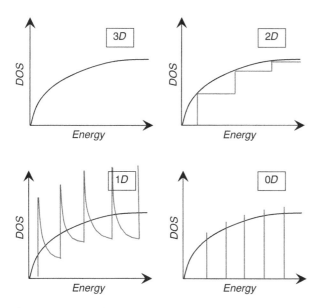

FIGURE 3.8 DOSs of 3D, 2D, 1D, and 0D semiconductors.

3.2.2 Excitonic Absorption

In intrinsic semiconductors (or slightly doped), when an electron is promoted to the CB by a photon, the hole remaining in the VB interacts electrostatically with the electron. They form an exciton. Actually, a Wannier exciton. In a bulk semiconductor, an exciton behaves as a hydrogen atom, but with energy

$$E_n = -\frac{\mu R}{n^2\varepsilon^2} = -\frac{E_0}{n^2}, \qquad (3.47)$$

where R is the Rydberg of the hydrogen atom, $\mu^{-1} = m_e^{-1} + m_h^{-1}$ the exciton effective reduced mass (we remove the star to indicate effective mass for convenience) and ε the relative permittivity of the medium. With the same philosophy, we can define

$$a_B = \frac{\varepsilon a_0}{\mu}, \qquad (3.48)$$

as the exciton Bohr radius in terms of the hydrogen Bohr radius. In a typical semiconductor, where $\mu \approx m_e \approx 0.1$ in terms of the electron mass and $\varepsilon \approx 10$, instead of 13.6 eV, we are talking about a binding energy 1,000 times smaller and a Bohr radius about 100 times larger than the hydrogen Bohr radius. In GaAs, for instance, $E_1 = 4.2$ meV and $a \approx 6$ nm. Optically, only the s states (1s, 2s, ..., which correspond to $n = 1, 2, ...$) are accessible.

But the hydrogen atom (in that case, the exciton) does not have only bound states. From quantum mechanics, we know that the hydrogen atom not only has bound states, with negative energy, but also a continuum of states, with positive energy. In the states with positive energy, the electron and holes still fill the electrostatic interaction between them (they are correlated or not independent). Actually, the band scheme does not consist of a VB and CB with wave numbers $\boldsymbol{k}_e$ and $\boldsymbol{k}_h$, and they have a total wave number $\boldsymbol{K} = (m_e\boldsymbol{k}_e + m_h\boldsymbol{k}_h)/M$, being $M = m_e + m_h$ the total mass of the exciton, which must be conserved in an optical transition, *i.e.* $\boldsymbol{K} = \boldsymbol{\kappa} \approx 0$ (exactly 0 in the dipole approximation). The positive energy states are a continuum with $\boldsymbol{k} = \boldsymbol{k}_e - \boldsymbol{k}_h$. The band scheme for excitons is given later in Figure 3.9.

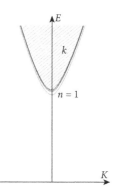

FIGURE 3.9 Excitonic band states in a bulk material. The energy is represented as a function of the center of mass wave number of the exciton, $\boldsymbol{K}$. Only states with $\boldsymbol{K} = 0$ can be accessed optically.

Actually, these expressions can be derived formally adding the interband polarization into the electron Hamiltonian [20]. The result is the Wannier equation

$$\left[-\frac{\hbar^2 \boldsymbol{\nabla_r}}{2\mu^*} + V(\boldsymbol{r}) \right] \psi(\boldsymbol{r}) = E\psi(\boldsymbol{r}) \qquad (3.49)$$

where $\boldsymbol{r}$ is the electron–hole relative vector. This equation has solutions for the 3D and 2D case [20].

The optical constants at the band edge (n and α) are usually measured through transmission and reflectance experiments using a thin layer. The contrast between the transmission maxima T_M and minima T_m can be written as

$$C_T(\alpha) = 2\frac{T_M - T_m}{T_M + T_m} = \frac{4r}{e^{\alpha d} + r^2 e^{-\alpha d}} \qquad (3.50)$$

where d is the thickness of the sample, α the absorption coefficient, and

$$r = \left(\frac{n-1}{n+1} \right)^2 \qquad (3.51)$$

is the reflectivity (reflectance over a surface of an infinitely thick sample), neglecting the extinction coefficient or imaginary part of the refractive index. When $\alpha = 0$, below the band edge, the transmission coefficient is maximum and

$$C_T(0) = \frac{4r}{1 + r^2} \qquad (3.52)$$

In this way, we can assign a numerical value to the $C_T(\alpha(\lambda))$ and obtain $\alpha(\lambda)$, neglecting the variations of the refractive index. By following the maxima of the transmission, we can also measure the variations of the refractive index. In Ref. [21], they show in Figure 3 the small variation of the refractive index at the band edge (from 2.9 to 3). Figure 4 of Ref. [21] shows the 1s, 2s, 3s, and 4s excitons.

The absorption coefficient can be shown to be [21,22]:

$$\alpha(\hbar\omega) = \frac{C_0 \mathcal{R}^{1/2}}{\hbar\omega_0} \left[\sum_{m=1}^{\infty} \frac{2\mathcal{R}}{m^3} \frac{\Gamma_m}{(\hbar\omega - E_m)^2 + \Gamma_m^2} \right.$$
$$\left. + \int_{E_g}^{\infty} dE \frac{1}{1 - e^{-2\pi z}} \frac{\Gamma_c^2}{(\hbar\omega - E)^2 + \Gamma_c^2} \right] \qquad (3.53)$$

where

$$z^2 = \frac{\mathcal{R}}{E - E_g}; \quad E_m = E_g - \frac{\mathcal{R}}{m^2}, \quad m = 1, 2, 3, \dots \qquad (3.54)$$

and the oscillator strength f_{cv} can be written in terms of the matrix element

$$f_{cv} = \frac{2|M_R|^2}{m_0 \hbar\omega} \qquad (3.55)$$

and the constant

$$C_0 = \frac{4\pi\mathcal{R}^{1/2}(2\mu)^{3/2} e^2 |M_R|^2}{n(\omega)c\hbar^2 m_0} \qquad (3.56)$$

and

$$\Gamma_m = \Gamma_c - \frac{\Gamma_c - \Gamma_1}{m^2} \qquad (3.57)$$

We were able to solve the integral corresponding to the excitons continuum many years ago [22], converting it in a sum over a few components and the calculation of some basic functions. The excitonic continuum can be written as

$$\alpha(\hbar\omega) = \frac{C_0 \mathcal{R}^{1/2}}{\hbar\omega} \left\{ \frac{1}{2} \left[\frac{\pi}{2} - \arctan \frac{\hbar\omega - E_g}{\Gamma_c} \right] \right.$$
$$- \sum_{m=1}^{\infty} \frac{\mathcal{R}}{m^3} \frac{\Gamma_c}{(\hbar\omega - E_m)^2 + \Gamma_c^2}$$
$$\left. + \frac{\pi}{2} \frac{\sinh 2u^+}{\cosh 2u^+ - \cosh 2u^-} \right\} \qquad (3.58)$$

where

$$u^{\pm} = \pi\sqrt{\frac{\mathcal{R}}{2}} \sqrt{\frac{\sqrt{(\hbar\omega - E_g)^2 + \Gamma_c^2} \pm (\hbar\omega - E_0)}{(\hbar\omega - E_g)^2 + \Gamma_c^2}} \qquad (3.59)$$

The shape of the refractive index spectra can be envisaged from Kramers–Krönig relations. We show in Figure 3.10 the excitonic spectrum of GaAs, a more interesting material from its technological point of view [23].

In strictly 2Ds, the Schrödinger equation for the hydrogen atom also has a solution and the energy states can be written as

$$E_m = E_g - \frac{\mathcal{R}}{(m + 1/2)^2}, \qquad (3.60)$$

but with $m = 0, 1, 2, \dots$, which means the Rydberg of the exciton in a strictly 2D semiconductor is four times its bulk value. In strictly one dimension [24], the singularity is too strong and there is not a complete set of solutions. To reduce the singularity, the potential can be considered to be

$$V(x) = -\frac{e^2}{a + |x|} \qquad (3.61)$$

Through a change of variable,

$$w = \begin{cases} \dfrac{2(x+a)}{a_0\alpha} & x > 0 \\ \dfrac{2(x-a)}{a_0\alpha} & x < 0 \end{cases} \qquad (3.62)$$

where a_0 is the Bohr radius and α is a parameter calculated through the boundary conditions. The energy has the form, for negative values,

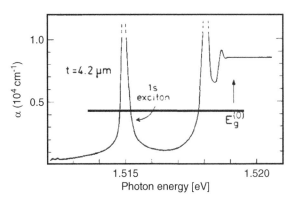

FIGURE 3.10 Excitonic absorption in a high-mobility GaAs sample, from Ref. [23]. The frame and labels have been polished to improve the quality of the figure.

$$E = E_g - \frac{\mathcal{R}}{\alpha^2} \qquad (3.63)$$

similar to the 3D case. The parameter α must be deduced from the boundary conditions [20]. In practice, since a NW is usually thick enough to have confinement, the 3D expression is used. But it is often observed in PL experiments, since it is very difficult to measure absorption in a single NW.

3.3 Ellipsometry

Spectroscopic ellipsometry is a very useful technique to measure the real and imaginary parts of the dielectric function, although it is often used to measure thin film thicknesses [25]. It analyses the polarization of the light once reflected in a surface. An ellipsometer consists of a light source (usually a Xe lamp), a polarizer, and an analyzer. Depending on the system, sometimes, the polarizer is rotating, but the polarizer can be fixed and the analyzer can be rotated. The reflected electric field components (parallel and perpendicular to the incidence plane) are

$$\begin{pmatrix} \tilde{E}_p^r \\ \tilde{E}_s^r \end{pmatrix} = \begin{pmatrix} \tilde{R}_{pp} & \tilde{R}_{sp} \\ \tilde{R}_{ps} & \tilde{R}_{ss} \end{pmatrix} \begin{pmatrix} \tilde{E}_p^i \\ \tilde{E}_s^i \end{pmatrix} \qquad (3.64)$$

$\tilde{R}_{pp}$ and $\tilde{R}_{ss}$ are the complex Fresnel reflection coefficients for parallel and perpendicular ("s" from senkrecht, from German) polarized light. The crossed terms correspond to the coupling of perpendicular and parallel polarizations. They are zero in the case of isotropic materials. Actually, from the analysis of ellipsometric spectra, we cannot directly obtain the real and imaginary parts of the dielectric function. Experimentally, we measure two parameters, Ψ and Δ, related to the Fresnel coefficients as

$$\tilde{p} = \frac{\tilde{R}_{pp}}{\tilde{R}_{ss}} = \tan \Psi e^{i\Delta} \qquad (3.65)$$

The modulus of $\tilde{p}$ is written as $\tan \Psi$ for convenience while Δ is the phase difference between the reflection coefficients for perpendicular and parallel polarizations. One of the handicaps of ellipsometry when compared with the absorption coefficient (or transmission measurements) is the fact that the extinction coefficient n_2 appears in the reflectivity, remember that

$$R = \frac{(n_1 - 1)^2 + n_2^2}{(n_1 + 1)^2 + n_2^2} \qquad (3.66)$$

and cannot be negligible when compared with n_1. That means, ellipsometry is a technique that can be used only above the band edge of semiconductors. On the other hand, since we do not measure directly $\varepsilon_1(\omega)$ and $\varepsilon_2(\omega)$, but Ψ and Δ, we need a mathematical model to extract the dielectric functions. These models were developed in the 80s in the group of Cardona for high T_c superconductors and SLs [1]. More recently (2009) the models have been extended to systems at the nanometer scale [26]. Unfortunately, although it is a very powerful technique, a model must be developed for a specific case. In Figure 3.11,

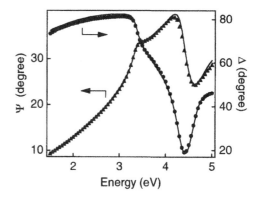

FIGURE 3.11 Spectroscopy ellipsometry scan for the multilayer film of PEI/Ti$_{1-\delta}$O$_2^{\delta-}$(PDDA/Ti$_{1-\delta}$O$_2^{\delta-}$)$_9$ on a Si wafer, where PEI is the short name of polyethylenimine and PDDA is poly(diallyldimethylammonium chloride). Observed and fit data are shown as symbols and lines, respectively.

we show the ellipsometric raw data of a sandwich of PEI/Ti$_{1-\delta}$O$_2^{4\delta-}$/(PDDA/Ti$_{1-\delta}$O$_2^{4\delta-}$)$_9$ on a Si wafer. Once the signal of Si was removed, an effective medium model was used for the polycation while a dispersion relation based on quantum mechanics was necessary to model the Ti$_{1-\delta}$O$_2^{4\delta-}$ [27].

There are some recent measurements to study by ellipsometry the plasmon coupling of silver nanoparticle arrays on Si [28], but the technique is basically limited to thin films, SLs [29], or other periodic structures. In Ref. [29], they perform spectroscopic ellipsometric measurements during growth. In the last years, spectroscopic imaging ellipsometry has been used for the search of flakes of 2D materials [30].

3.4 Modulation Techniques in Nanostructures

Once the expitaxial techniques were optimized for the growth of heterostructures or SLs, new kind of structures like SL on (n11) surfaces or corrugated SL were available to produce anisotropies or, at the end, one-dimensional structures. There are well-known modulation techniques like electroreflectance or piezoreflectance, which, although very powerful, need a sample manipulation. A simple and useful technique to analyze this kind of structures is the modulated ellipsometry or reflectance difference spectroscopy. In a [311]-SL, for instance, modulated ellipsometry measures the difference of dielectric functions $\Delta \varepsilon = \varepsilon_2([1\bar{1}1]) - \varepsilon_2(\bar{2}33)$ [31] while reflectance difference measures the relative difference in reflectivity

$$\frac{\delta r}{r} = 2 \frac{r_{[0\bar{1}1]} - r_{[\bar{2}33]}}{r_{[0\bar{1}1]} + r_{[\bar{2}33]}} \qquad (3.67)$$

Both measurements were performed in 4.2/4.7 nm GaAs/AlAs (311)-oriented SLs [31]. The sharp peaks found in the spectra are mainly due to confinement effects, as supported by the tight-binding calculations performed in the same work.

Another very simple modulation technique is the photoreflectance spectroscopy (PR). The fundamental is to pulse a laser on a sample surface (laser energy over the gap) and, using a chopper, measure the difference of reflectance with and without the laser spot and perform a spectra, and in this way, we obtain ΔR and R at the same time. The laser creates electron–hole pairs that modify the carrier population on the surface. Actually, the laser power must be very low to get a good PR signal. The PR is measured as

$$\frac{\Delta R}{R} = \frac{R - R'}{R} \qquad (3.68)$$

where R is the reflectance without illumination and R' the reflectance while the sample is illuminated. Using PR it is possible to take the second or third derivative of the spectra to get a better contrast.

The sample whose data are shown in Figure 3.12 has a E_{hh} confinement of 338 and 417 meV for the lh.

Recently, reflectance difference (RD) has been used to measure thicknesses in the nanometer scale. In Ref. [32], a setup has been developed with this purpose. The anisotropy produced at the boundary of the nanostructures makes this technique appropriate to measure thicknesses with nanometer accuracy.

3.5 Emission Processes in Nanostructures

3.5.1 Band-to-Band Transitions

In the simplest case of a VB and CB (two-band model), the process of emission is that shown in Figure 3.2. After the absorption of a photon in the material, as we know, an electron is promoted to the CB leaving a hole in the VB. Electrons and holes are scattered by phonons (optical phonons

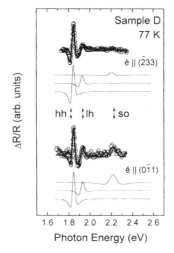

FIGURE 3.12 Third-derivative line-shape fitting for both polarizations in a GaAs/AlAs sample with 2.4/2.2 nm period. Open circles represent experimental data while the lines are a result of fitting.

dominate the scattering process whenever is possible, otherwise acoustic phonons), and they relax until they are at the bottom of the band. The only option is then the radiative recombination, giving rise to the phenomenon called photoluminescence. The lifetime corresponding to the radiative recombination is much higher than that corresponding to phonon scattering.

In the case of a direct transition and intrinsic semiconductor, the emission process is related to the absorption through the Roosbroeck–Shockley relations. The emission rate R_{cv} is

$$R_{cv} = \frac{\alpha(\omega) 2\omega^2 n(\omega)^2}{\pi c^2} N_{\text{photon}} \qquad (3.69)$$

where $\hbar\omega$ is the photon energy, α the absorption coefficient, and n the refractive index. c is the speed of light in vacuum. In doped semiconductors, there are additional factors influencing emission.

There are many kinds of emission peaks, like donor–acceptor pair emission, free to bound transitions, etc. [1], but we are going to discuss only those important in nanostructures. In the emission spectra (also in the absorption), we can observe resonant Raman peaks corresponding to longitudinal optical (LO) phonons. Many times, it is difficult to distinguish between Raman and PL. In Ref. [33], the multipeak emission in ZnTe produced by the resonance was distinguished from PL by measuring the spectra with several laser lines. Those fixed in a wave number scale (referred to the laser) are Raman lines, while those fixed in an energy scale are PL peaks.

3.5.2 Bound Exciton Emission

As we have discussed in previous sections, when light interacts with a semiconductor, an electron–hole pair or exciton is created. In that case, in the PL emission, we observed the free exciton emission. But if we have a small amount of impurities, which at low temperature are neutral, the exciton is attracted by the impurity via van der Waals forces and we have a bound exciton. The emission peak is labeled $D^0 X$, where D^0 means a neutral donor impurity. If we know the impurity giving rise to the peak, for instance silicon, we can label the transition $Si^0 X$. In a recent work of Callsen et al. [34], they study the excited states of neutral donors bound to excitons in GaN. As we know, in GaN there are three excitons, A, B, and C, but usually only A and B are observed. Actually, they use both PL and PLE (see next section) to study the emission. While in PL they observe basically the $Si^0 X_A$ peak, in PLE, they can see the $Si^0 X_A^{n=2}$, allowing to estimate the silicon donor binding energy as 28.8 meV. They have also assigned the X_C exciton at 3.4995 eV, a transition in between the $X_A^{n=2}$ and $X_B^{n=2}$ (Figure 2 of Ref. [34]).

3.5.3 Photoluminescence Excitation

The usual way to measure the absorption coefficient is through transmission measurements. These measurements are not possible in the case of thin films, since usually there is a substrate or any other semiconductor nanostructure, because they do not fill the whole space. In that case, the PL is excited with a continuous wavelength (cw) laser or a halogen lamp, fixing the collection wavelength close but below the bandgap. When we tune the excitation wavelength, if the absorption increases, the number of photons emitted will also increase, and the excitation of PL is given as a qualitative measurement (not in absolute value) of absorption. By using a halogen lamp, we have shown that the gap of InN is at 0.7 eV [35], in agreement with a previous measurement, since the gap was previously establish to be at 1.4 eV. In the reference discussed in the previous section, [34], in their Figure 2, we can see the amount of information provided by the PLE spectrum when compared with the PL. The PLE spectra are proportional to the absorption only in intrinsic samples or weakly doped.

3.6 Raman Scattering in Semiconductor Nanostructures

Raman scattering or inelastic light scattering consists of the emission of light after being scattered by optical phonons in ordinary semiconductors. In magnetic materials or metals, there are other kinds of excitations that scatter the light inelastically, but in semiconductors or semiconductor nanostructures, one usually refer to the light scattering by phonons. In infrared absorption, there is a direct interaction of light with the lattice (the electric field of the light couples with the electric field of phonons), while in Raman scattering, the electrons are virtual intermediate states of the process [36]. The total process consists of the absorption of a photon, *i.e.*, the creation of an electron–hole pair or exciton, the scattering of the phonon either by the electron or the hole of the exciton, and finally the electron–hole recombination. The electronic system returns to the fundamental state. There are six possible combinations in a first-order Raman process [37], but only one of them is important near the gap, *i.e.*, when the photon energy is close to the gap. The Feynman diagram is shown in Figure 3.13.

In the diagram, ω_l is the laser frequency, ω_s the scattered frequency, ω_0 the phonon frequency, and $|p\rangle$ and $|q\rangle$ are excitonic wave functions. The discontinuous line represents the photon, the continuous line the exciton, and the curly line the phonon. The solid circle represents the exciton–photon interaction while the solid square the exciton–phonon interaction.

The convenient amount to be measured in a Raman experiment is the Raman efficiency per unit solid angle Ω (Raman cross section per unit volume and solid angle), which can be defined as [38]:

$$\frac{dS}{d\Omega} = \frac{\omega_s^3 n_s^3 n_l}{h^2 c^4 \omega_l} \left| \sum_{p,q} \frac{\langle v|\mathcal{H}_{eR}|p\rangle\langle p|\mathcal{H}_{eL}|q\rangle\langle q|\mathcal{H}_{eR}|c\rangle}{(E_p - \hbar\omega_l + i\gamma_p)(E_q - \hbar\omega_s + i\gamma_q)} \right|^2$$
$$\times (N_0 + 1), \tag{3.70}$$

where γ_p and γ_q are the broadening of excitonic states and N_0 the phonon population. We have written the Stokes component, the anti-Stokes has a N_0 instead of $N_0 + 1$, and it is very weak at low or room temperature. The initial (v) and final (c) states are the VB and CB wave functions, and the internal sum is over the intermediate states. From Eq. (3.70), we realize that the denominator can be very large if the intermediate states are close either to the laser or scattered photon. In that case, we name the phenomenon resonant Raman scattering (RRS). If they are very close to the gap, the broadening avoids the divergence, but in any case, the efficiency becomes very large.

As previously commented, excitons are the intermediate states in the Raman process close to the gap of a semiconductor or one of the transitions of the nanostructure. The photon-radiation Hamiltonian is responsible for the absorption (or the emission) and it has been analyzed in the previous section, thus we need to know the exciton–phonon Hamiltonian to completely understand the Raman process. We will write both in the second quantization formalism in the next section.

3.6.1 Raman Scattering Mechanisms

In the case of long wavelength optical modes, the perturbation on the crystal potential is separable into two parts [39], a short-range component and a long-range component. The short-range component, called deformation potential, is directly connected with the theory of strain-induced effects in semiconductors [39] and acts on the short-range component of Bloch functions, $u_{nk}(r)$. Different matrices correspond to the different symmetry properties of the crystal [40]. The long-range component corresponds to the long-range electrostatic interaction, and it is related to the envelope part of the Bloch function. In the case of excitonic states, the envelope part is the exciton wave function.

The exciton-radiation Hamiltonian is [41]

$$\mathcal{H}_{eR} = \sum_{p,\boldsymbol{K}} \sum_{e,\boldsymbol{\kappa}} T_{cv}^p(\boldsymbol{K}) D_{p,\boldsymbol{K}}^\dagger \left(a_{\boldsymbol{\kappa},e} + a_{-\boldsymbol{\kappa},e}^\dagger \right) + c.c. \tag{3.71}$$

where e is the photon polarization, $D_{p,\boldsymbol{K}}^\dagger$ $(D_{p,\boldsymbol{K}})$ the creation (annihilation) exciton operator, $a_{\boldsymbol{\kappa},e}^\dagger$ $(a_{\boldsymbol{\kappa},e})$ the creation (annihilation) operator for photons, and the coupling constant

$$T_{cv}^p = -\frac{e}{m}\sqrt{\frac{h}{\omega n^2}}\, e \cdot \langle c|\boldsymbol{p}|v\rangle \psi_p(0)\delta_{\boldsymbol{K},\boldsymbol{\kappa}} \tag{3.72}$$

being $\psi_p(0)$ the exciton wave function calculated in zero [42] (indicating that the electron and hole are produced at the

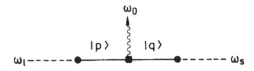

FIGURE 3.13 Feynman diagram of the main contribution to a Raman process in semiconductors.

relative radius vector $\boldsymbol{r} = 0$). In III-V and II-VI semiconductors, $|\langle c|\boldsymbol{p}|v\rangle| = P_{cv}$ is the same for all transitions. The reason is that the VB is triply degenerate, the degeneracy being removed by spin–orbit coupling or crystal splitting (in WZ-type crystals). In the absorption or emission process, the wave number is conserved ($\delta_{\boldsymbol{K},\boldsymbol{\kappa}}$), while the exciton–phonon Hamiltonian can be written as

$$\mathcal{H}_{eL} = \sum_{\boldsymbol{q},\nu} \sum_{\substack{p,q \\ \boldsymbol{K},\boldsymbol{K}'}} S_{p,q}^{\boldsymbol{K},\boldsymbol{K}'}(\boldsymbol{q}) D_{q,\boldsymbol{K}'}^{\dagger} D_{p,\boldsymbol{K}} \left(b_{\boldsymbol{q},\nu}^{\dagger} + b_{-\boldsymbol{q},\nu} \right),$$

(3.73)

where $b^{\dagger}_{\boldsymbol{q},\nu}$ ($b_{\boldsymbol{q},\nu}$) is the creation (annihilation) operator of a phonon of wave number $\boldsymbol{q}$ belonging to the branch ν. The coupling constant is

$$S_{qp}^{\boldsymbol{K}',\boldsymbol{K}}(\boldsymbol{q}) = \frac{u_0 \sqrt{3}}{2a_0} [D_e^{\nu}(\boldsymbol{r}) I_{qp}(-\boldsymbol{q}_h)$$
$$- D_h^{\nu}(\boldsymbol{r}) I_{qp}(\boldsymbol{q}_e)] \, \delta_{\boldsymbol{K}',\boldsymbol{K}+\boldsymbol{q}}.$$

(3.74)

Here, the relative displacement $u_0 = \sqrt{\hbar V_c / 2V M^* \omega_0}$, a_0 is the lattice constant, and $D_e^{\nu}(\boldsymbol{r})$ and $D_h^{\nu}(\boldsymbol{r})$ are the deformation potential matrices defined by Bir and Pikus [39]. These matrices are the different components of the Raman tensor. In ZB and WZ semiconductors, only the matrix $D_h \neq 0$. In the case of Fröhlich interaction,

$$S_{qp}^{\boldsymbol{K}',\boldsymbol{K}}(\boldsymbol{q}) = \frac{C_F^*}{\sqrt{V} q} [I_{qp}(-\boldsymbol{q}_h) - I_{qp}(\boldsymbol{q}_e)] \, \delta_{\boldsymbol{K}',\boldsymbol{K}+\boldsymbol{q}},$$

(3.75)

where $C_F = -ie\sqrt{2\pi\hbar\omega_0(1/\varepsilon_\infty - 1/\varepsilon_0)}$ and

$$I_{pq}(\boldsymbol{q}_e) = \int d^3\boldsymbol{r} \, \psi_q^*(\boldsymbol{r}) e^{i\boldsymbol{q}\cdot\boldsymbol{r}} \psi_q(\boldsymbol{r}),$$

(3.76)

being $q_e = (m_e^*/M^*)q$, the same for the hole by changing m_e^* by m_h^*.

Coming back to the Raman efficiency corresponding to the optical phonon ν (at $\boldsymbol{q} = 0$), we can write it in the form

$$\left(\frac{dS}{d\Omega} \right)_{\nu} = \frac{\omega_s^3 \omega_l}{c^4} \frac{\hbar}{2V_c M^* \omega_\nu} \frac{n_s}{n_l} (N_0 + 1) \sum_{\alpha} |e_l \mathcal{R}_{\Gamma} e_s|^2,$$

(3.77)

where n_l and n_s are the refractive indices corresponding to the laser and scattered frequency, respectively, and M^* the total mass of the exciton. $\mathcal{R}_{\Gamma}$ are the Raman tensors of the corresponding modes. In the case of Fröhlich interaction, the Raman tensor is diagonal, since the Coulomb interaction is a scalar. As we will comment later, Fröhlich interaction is important in resonance.

3.6.2 Raman Selection Rules

In ZB crystals, there are three optical phonon modes, all 3D,

$$\mathcal{R}(T_2(x)) = \begin{pmatrix} 0 & 0 & 0 \\ 0 & 0 & d \\ 0 & d & 0 \end{pmatrix} \mathcal{R}(T_2(y)) = \begin{pmatrix} 0 & 0 & d \\ 0 & 0 & 0 \\ d & 0 & 0 \end{pmatrix}$$

$$\mathcal{R}(T_2(z)) = \begin{pmatrix} 0 & d & 0 \\ d & 0 & 0 \\ 0 & 0 & 0 \end{pmatrix},$$

(3.78)

where d is called Raman polarizability. The selection rules are introduced using the Porto notation [43], which indicate the direction of the photon and polarization, $\boldsymbol{\kappa}_l(\boldsymbol{e}_l, \boldsymbol{e}_s)\boldsymbol{\kappa}_s$. ZB crystals have three optical phonons, one LO and two transversal optical (TO) modes, at the $\Gamma-$ point. These are 3D modes and are called F_2 or T_2. In backscattering geometry along the [100] direction, the LO phonon can be observed in crossed polarization. Since it is LO, the $\boldsymbol{q}$ must be along $\boldsymbol{\kappa}$ following the wave number conservation, *i.e.*, along x. Thus, only the Raman component $\mathcal{R}(T_2(x))$ gives a nonzero result in the configuration $x(y,z)\bar{x}$. The TO modes cannot be observed in the [100] direction in backscattering configuration. Thus, the selection rules for backscattering along the [001] tell us that the LO mode is allowed while the TO modes are forbidden. More examples are given in the book by Yu and Cardona [1].

In WZ-type crystals, there are nine optical phonons, but only seven are Raman active. There are two silent modes ($2B_1$). The Raman active modes are $\Gamma = A_1 + E_1 + 2E_2$. The A_1 is a one-dimensional mode and the E modes are 2D. Following group theory, the first E_2 mode consists of the vibration of the anion with the cation at rest, while the second mode corresponds to the vibration of the cation with the anion at rest. But these two modes have the same symmetry and they mix. For instance in GaN, there is a lower frequency mode, E_{2l} consisting of the movement of Ga, but with a small displacement of N, while in the high frequency mode E_{2h}, the main vibration is due to the N (lower mass, higher frequency), but there is a small component of Ga in the phonon eigenvector. The Raman tensor corresponding to the one-dimensional mode is

$$\mathcal{R}(A_1(z)) = \begin{pmatrix} a & 0 & 0 \\ 0 & a & 0 \\ 0 & 0 & b \end{pmatrix}$$

(3.79)

The asymmetry is due to the fact that the crystal is uniaxial. If the phonon is propagating along z, it is an $A_1(LO)$ mode, and it is observable with parallel polarization $z(x,x)\bar{z}$ or $z(y,y)\bar{z}$, while in the case of $A_1(TO)$, the mode can be observed in $x(z,z)\bar{x}$ polarization, for instance. The Raman tensor for the E_1 and E_2 modes are

$$\mathcal{R}(E_1(x)) = \begin{pmatrix} 0 & 0 & c \\ 0 & 0 & 0 \\ c & 0 & 0 \end{pmatrix} \mathcal{R}(E_1(y)) = \begin{pmatrix} 0 & 0 & 0 \\ 0 & 0 & c \\ 0 & c & 0 \end{pmatrix}$$

$$\mathcal{R}(E_2) = \begin{pmatrix} d & d & 0 \\ d & -d & 0 \\ 0 & 0 & 0 \end{pmatrix}.$$

(3.80)

Some times, silent modes can be observed for several reasons. In particular, we have recently observed [48] the B_{1l} mode due to the isotopical disorder of Ga (Ga^{69} and Ga^{71}, while the B_{1h} cannot be observed since N has basically one isotope. The isotopical disorder produces a beating and the appearance of a small dipole moment and the mode appears only under resonant conditions (Fröhlich interaction).

The Raman selection rules in semiconductor nanostructures are basically the same as that in bulk, although due to confinement effects, the number of phonons at the $\Gamma-$point could depend on the dimensionality of the system, and new modes can appear due to the interface (IF) or surfaces. These modes will be described in the next section.

3.6.3 Phonons in Low-Dimensional Structures

In an SL, we have, as in the case of electrons, a mini BZ in the growth direction. But, depending on the materials forming the SL, the optical modes can be propagated through the SL or confined in one of the materials. A typical example is an AlAs/GaAs SL. The optical dispersion of GaAs phonons is in the 280–320 cm^{-1} range, while that for AlAs is in the 330–400 cm^{-1} range. Thus, an optical mode excited in AlAs cannot propagate through the GaAs part. These modes are thus confined to the AlAs sublattice. The same occurs at the GaAs mode. However, the acoustic phonons are folded into the mini BZ, and since we can always find another mode at the same energy in both sublattices, the acoustic modes can propagate through SL. However, since they are folded in the mini BZ, new modes appear in the zone center, which can be observed with Raman spectroscopy. A third kind of mode can appear in an SL, the so-called IF modes, due to the dissimilitude of the dielectric permittivities of the two media [44,45]. In Figure 3.14, we can see the confined optical phonons in the SL and the appearance of an IF mode. In quantum wires or NWs, usually we observe the bulk phonons, since the material is relaxed. There are also surface optical (SO) phonons, usually due to surface imperfection [46,47], although in general they are less intense due to the smaller contribution of the surface. Sometimes, they can be observed in resonance. In WZ-type crystals, we can not only observe the SO corresponding to the $A_1(LO)$ mode but also that corresponding to the $E_1(LO)$ [46]. However, in very high quality NWs of 12 nm diameter and 100 nm length,

we have only observed bulk phonons of WZ GaN. There was no shift since the sample was completely unstrained [48] or even SO modes, in spite of being a very thin NW.

There is a nice work by Gupta et al. [49], where they observe the SO mode and confirm that the observed mode is an SO mode. The frequency of the mode can be written as [38,50]:

$$\omega_{SO}^2 = \frac{\varepsilon_0 + \varepsilon_M}{\varepsilon_\infty + \varepsilon_M}\omega_{TO}^2 \qquad (3.81)$$

In Ref. [49], they use two different media and check that the SO mode moves actually according to the earlier expression. Unfortunately, they try to apply a wrong theoretical model and did not succeed in the explanation. The right model was developed more than 40 years before [50].

But let us pay attention to the Raman spectrum of a single NW. The difference with bulk is that the electric field only fulfills the boundary conditions when the polarization is along the NW, in case the diameter is small enough, and we can find differences in the relative amplitude between the modes. In a work of Xong et al. [51], they claim that there is a shift in the LO-phonon and an antenna effect in very thin NWs of GaP. There is clear evidence, shown in the framework of a macroscopic model based on finite differences, that the boundary conditions depend on the polarization direction (along or perpendicular to the NW axis) and that the Raman emission depends on a $\cos^2\theta$ factor, what they call antenna effect because it is the behavior of a dipole emission. However, the splitting can be due to the appearance of the WZ phase along the growth direction. It is well known that III-V materials like GaAs or InP can grow in the WZ phase in the form or NWs and, usually, there is a twin of the WZ and ZB phases. This can be an explanation for the splitting [52–54]. On the other hand, as we can observe in Figure 3.15, it is clear that, in large diameter NWs, the electrostatic boundary conditions can be satisfied for an electric field along the NW direction or perpendicular to it, but in the case, we have a very thin NW, when the electric field is along the NW direction, the tangential component can be the same outside and inside the NW, which is not the case if the light is polarized perpendicularly. In that case, the normal components of the displacement vector are the same, unless there is a surface charge in the NW.

We have seen that the phonons in NW are basically that of the bulk. The reason is that, while the electron confinement is produced below 10 nm, to have phonon confinement (not phonon shift due to stress), we need a NW of 1-2 nm radius. Comas et al. [55] calculated the confinement in GaAs NWs as a function of radius. There is an additional structure related either to SO or IF phonons (in core–shell NWs). From the work of Comas et al. [47,55] we can see how, if we take into account the phonon broadening, in 3-4 nm diameter NWs, we can see as much an asymmetry in the phonon, but not confinement. In conclusion, there is no evidence of phonon confinement in NWs, since for the NWs available (with diameter higher than 10 nm), the size is not small enough to produce confinement.

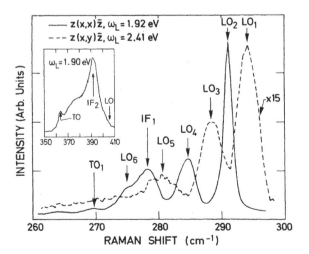

FIGURE 3.14 Confined optical modes and IF modes in GaAs-AlAs SLs. From Ref. [44].

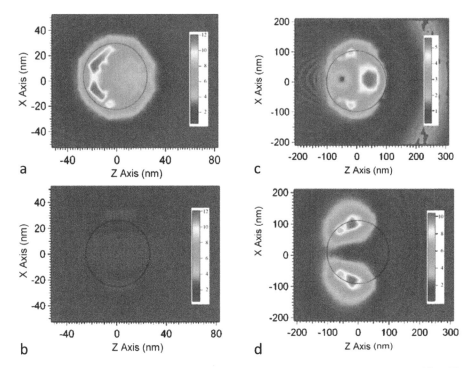

FIGURE 3.15 Calculated electric field intensity maps in the cross section of the middle of a GaP NW ($d = 50$ nm). (a) Transversal magnetic (TM) and (b) Transversal electric (TE) polarizations. In the right panels, (c) and (d), the same mapping is seen for a $d = 200$ nm NW.

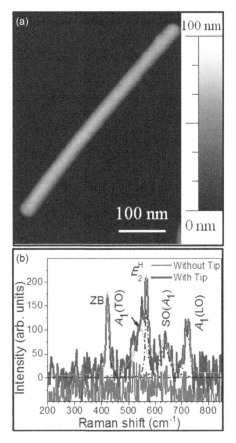

FIGURE 3.16 TERS spectrum of oxygen-rich GaN NWs. Comparison of a Raman measurement and the measurement with the tip. After Ref. [45]. At the top, atomic force microscopy image of the NW.

Another recent technique used to perform Raman spectroscopy measurements in bad quality NWs (with impurities) or NWs with very small signals is the tip-enhanced RS (TERS). This technique consists of touching the NW with the metallic tip of an atomic force microscope (AFM) and at the same time to take the Raman spectrum. It is a technique similar to surface-enhanced Raman scattering (SERS), but we do not have to find an appropriate substrate for the experiment. In Figure 3.16, we show the TERS spectrum of an oxygen-rich GaN NW. The light gray data correspond with the Raman spectrum and the dark gray one to the same spectrum once the metallic tip is touching the sample.

Quantum dots can actually be divided into two kinds. The first one is a nanoparticle (not necessarily spherical) that can be embedded into a different material or can be diluted in a liquid. They have sizes from a few nanometers to 40–50 nm. Although there is a theoretical work predicting confinement in quantum dots (as an example GaAs embedded in AlAs) [56], when compared with the confinement in quantum wires, it is clearly more difficult to have confinement in quantum dots. If the radius is in the range between 1 and 2 nm, we have confinement; otherwise, we have a continuum of levels and the possible confinement cannot be observed. Connecting with a previous comment on confinement, in the work of Cheng et al. [57], we can see how the electrons are confined, there is a shift of electronic levels of ZnO quantum dots (QDs) from 3.3 to 3.4 eV (12–3.5 nm diameter), while there is no shift in the Raman spectra, *i.e.*, there is no phonon confinement.

The second kind of quantum dots, probably more important from a technological point of view, is the so-called self-assembled QDs. They are grown using the Stranski–Krastanov method [58], which is usually based on the strain between a buffer layer and the material constituting the dot. Since it is strongly stressed, it is difficult to separate the strain from the confinement [59].

References

1. P. Yu and M. Cardona. *Fundamentals of Semiconductors: Physics and Materials Properties*. 4th edition, Springer, Berlin Heidelberg, 2010.

2. H. Ibach and H. Lüth. *Solid-State Physics: An Introduction to Principles of Materials Science*, 4th edition. Springer: Berlin Heidelberg, 2009.

3. M. Cardona, N. Christensen, and G. Fasol. Relativistic band structure and spin-orbit splitting of zinc-blende-type semiconductors. *Phys. Rev. B*, 38:1806–1827, 1988.

4. P. V. Santos, M. Willatzen, M. Cardona, and A. Cantarero. Tight-binding calculation of spin splittings in semiconductor superlattices. *Phys. Rev. B*, 51:5121–5130, 1995.

5. J. M. Luttinger and W. Kohn. Motion of electrons and holes in perturbed periodic fields. *Phys. Rev.*, 97:869–883, 1955.

6. J. M. Luttinger. Quantum theory of cyclotron resonance in semiconductors: General theory. *Phys. Rev.*, 102:1030–1041, 1956.

7. P. O. Löwdin. A note on the quantum-mechanical perturbation theory. *J. Chem. Phys.*, 19:1396–1401, 1951.

8. L. C. O. Dacal and A. Cantarero. Ab initio electronic band structure calculation of InP in the wurtzite phase. *Sol. State Commun.*, 151:781–784, 2011.

9. E. G. Gadret, G. O. Dias, L. C. O. Dacal, M. M. de Lima Jr., C. V. R. S. Ruffo, F. Iikawa, M. J. S. P. Brasil, T. Chiaramonte, M. A. Cotta, L. H. G. Tizei, D. Ugarte, and A. Cantarero. Valence-band splitting energies in wurtzite InP nanowires: Photoluminescence spectroscopy and ab initio calculations. *Phys. Rev. B*, 82:125327, 2010.

10. S. L. Chuang and C. S. Chang. $k \cdot k$ method for strained wurtzite semiconductors. *Phys. Rev. B*, 54:2491–2504, 1996.

11. W. A. Harrison. *Electronic Structure and the Properties of Solids, The Physics of the Chemical Bond*. Dover Publications Inc., New York., 1989.

12. A. R. Goni, A. Cantarero, H. Scheel, S. Reich, C. Thomsen, P. V. Santos, F. Heinrichsdorff, and D. Bimberg. Different temperature renormalization s for heavy and light hole states of monolayer-thick heterostructures. *Sol. State Commun.*, 116:121–124, 2000.

13. A. Kobayashi, O. F. Sankey, S. M. Volz, and J. D. Dow. Semi-empirical tight-binding band structures of wurtzite semiconductors: AlN, CdS, CdSe, ZnS, and ZnO. *Phys. Rev. B*, 28:935–945, 1983.

14. M. D. Jaffe and J. Singh. Inclusion of spin-orbit coupling into tight-binding band structure calculations for bulk and superlattice semiconductors. *Sol. State Commun.*, 62:399–402, 1987.

15. D. A. Neamen. *Semiconductor Physics and Devices: Basic Principles*, 4th edition. Richard D. Irwin, Inc., Homewood, IL, 2012.

16. A.S.A. Fletcher and D. Nirmal. A survey of gallium nitride HEMT for RF and high power applications. *Superlatt. Microstr.*, 109:519–537, 2017.

17. F. Iikawa, V. Donchev, T. Ivanov, G. O. Dias, L. H. G. Tizei, R. Lang, E. Heredia, P. F. Gomes, M. J. S. P. Brasil, M. A. Cotta, D. Ugarte, J. P. M. Pastor, M.M. De Lima Jr., and A. Cantarero. Spatial carrier distribution in InP/GaAs type II quantum dots and quantum posts. *Nanotechnology*, 22:065703, 2011.

18. G. Bastard. *Wave Mechanics Applied to Semiconductor Heterostructures*. Wiley-Interscience, 1991.

19. D. Olguin, A. Rubio-Ponce, and A. Cantarero. Ab initio electronic band structure of III-VI layered compounds. *Eur. Phys. J. B*, 86:350, 2013.

20. H. Haug and S. W. Koch. *Quantum Theory of the Optical and Electronic Properties of Semiconductors*. World Scientific Publishing Co. Pte. Ltd., Singapore, 2004.

21. R. Le Toullec, N. Piccioli, and J. C. Chervin. Optical properties of the band-edge exciton in GaSe at 10 K. *Phys. Rev. B*, 22:6162–6170, 1980.

22. A. R. Goni, A. Cantarero, K. Syassen, and M. Cardona. Effect on pressure on the low-temperature exciton absorption in GaAs. *Phys. Rev. B*, 41:10111–10119, 1990.

23. G. W. Fehrenbach, W. Schäfer, and R. G. Ulbrich. Excitonic versus plasma screening in highly excited gallium arsenide. *J. Lumin.*, 30:154–161, 1985.

24. R. Loudon. One-dimensional hydrogen atom. *Am. J. Phys.*, 27:649–655, 1959.

25. M. S. Diware, K. Park, J. Mn, H. G. Park, W. Chegal, Y. J. Cho, H. M. Cho, J. Park, H Kim, S. W. Kang, and Y. D. Kim. Characterization of wafer-scale mos$_2$ and wse$_2$ 2d films by spectroscopic ellipsometry. *Curr. Appl. Phys.*, 17:1329–1334, 2017.

26. M. Losurdo, M. Bergmair, G. Bruno, D. Cattelan, C. Cobet, A. de Martino, K. Fleischer, Z. Dohcevic-Mitrovic, N. Esser, M. Galliet, R. Gajic, D. Hemzal, K. Hingerl, J. Humlicek, R. Ossikovski, Z. V. Popovic, and O. Saxl. Spectroscopic ellipsometry and polarimetry for materials and systems analysis at the nanometer scale: State-of-the-art, potential, and perspectives. *J. Nanopart. Res.*, 11:1521–1554, 2009.

27. T. Sasaki, Y. Ebina, T. Tanaka, M. Harada, and M. Watanabe. Layer-by-layer assembly of titania nanosheet/polycation composite films. *Chem. Mater.*, 13:4661–4667, 2001.

28. D. Gkogkou, T. Shaykhutdinov, T. W. H. Oates, U. Gernert, B. Schreiber, S. Facsko, P. Hildebrandt,

I. M. Weidinger, N. Esser, and K. Hinrichs. Characterization of anisotropically shaped silver nanoparticle arrays via spectroscopic ellipsometry supported by numerical optical modeling. *Appl. Surf. Sci.*, 421:460–464, 2017.

29. C. Simbrunner, T. Li, A. Bonanni, A. Kharchenko, J. Bethke, K. Lischka, and H. Sitter. In situ growth observation of gan/algan superlattice structures by simultaneous x-ray diffraction and ellipsometry. *J. Chryst. Growth*, 308:258–262, 2007.

30. S. Funke, U. Wurstbauer, B. Miller, A. Matković, A. Green, A. Diebold, C. Röling, and P. H. Thiesen. Spectroscopic imaging ellipsometry for automated search of flakes of mono- and n-layers of 2d-materials. *Appl. Surf. Sci.*, 421:435–439, 2017.

31. P. V. Santos, A. Cantarero, M. Cardona, R. Nötzel, and K. Ploog. Optical properties of (311)-oriented GaAS/AlAs superlattices. *Phys. Rev. B*, 52:1970–1977, 1995.

32. C. Hu, S. Huo, W. Shen, Y. Li, and X. Hu. Reflectance difference microscopy for nanometre thickness microstructure measurements. *J. Microsc.*, 270:318–325, 2017.

33. J. Camacho, A. Cantarero, I. Hernández-Calderón, and L. González. Raman spectroscopy of znte thin films grown on gaas. *J. Appl. Phys.*, 92:6014–6018, 2002.

34. G. Callsen, T. Kure, M. R. Wagner, R. Butté, and N. Grandjean. Excited states of neutral donor bound excitons in gan. *J. Appl. Phys.*, 123:215702, 2018.

35. J. Segura-Ruiz, A. Molina-Sánchez, N. Garro, A. García Cristóbal, A. Cantarero, F. Iikawa, C. Denker, J. Malindretos, and A. Rizzi. Inhomogeneous free-electron distribution in inn nanowires: Photoluminescence excitation experiments. *Phys. Rev. B*, 82:125319, 2010.

36. R. Loudon. The Raman effect in crystals. *Adv. Phys.*, 13:423–474, 1964.

37. Achintya K. Ganguly and Joseph L. Birman. Theory of lattice Raman scattering in insulators. *Phys. Rev.*, 162:806–816, 1967.

38. A. Cantarero. Optical techniques for nanostructure characterization. In B. Bhushan, editor, *Encyclopedia of Nanotechnology*. Springer, Dordrecht, 2015.

39. G. L. Bir and G. E. Pikus. *Symmetry and Strain Induced Effects in Semiconductors*. Translated from Russian by P. Shelnitz. Translation edited by D. Louvish. Wiley, New York, 1974.

40. Bilbao Crystallographic Server. www.cryst.ehu.es.

41. A. Cantarero, C. Trallero-Giner, and M. Cardona. Excitons in one-phonon resonant Raman scattering: Deformation potential interaction. *Phys. Rev. B*, 39:8388–8397, 1989.

42. R. J. Elliot. Intensity of optical absorption by excitons. *Phys. Rev.*, 108:1384–1389, 1957.

43. D. L. Rousseau, R. P. Bauman, and S. P. S. Porto. Normal modes determination in crystals. *J. Raman Spectr.*, 10:253–290, 1981.

44. A. K. Sood, J. Menéndez, M. Cardona, and K. Ploog. Second-order Raman scattering by confined optical modes and interface vibrational modes in GaAs-AlAs superlattices. *Phys. Rev. B*, 32:1412–1414, 1985.

45. A. Patsha, S. Dhara, and A. K. Tyagi. Localized tip enhanced Raman spectroscopic study of impurity incorporated single GaN nanowire in the sub-diffraction limit. *Phys. Chem. Chem. Phys.*, 18:29864, 2016.

46. S. Bhattacharya, A. Datta, S. Dhara, and D. Chakravorty. Surface optical Raman modes in GaN nanoribbons. *J. Raman Spectrosc.*, 42:429–433, 2009.

47. A. Cantarero. Review on Raman scattering in semiconductor nanowires: I. Theory. *J. Nanophoton.*, 7:071598, 2013.

48. C. Rodriguez-Fernandez, M. Almokhtar, W. Ibarra-Hernandez, M. M. de Lima Jr., A. H. Romero, and A. Cantarero. Isotopic heft on the B_{1l} silent mode in ultra-narrow gallium nitride nanowires. *Nano Lett.*, 18:5091–5097, 2018.

49. Rajeev Gupta, Q. Xiong, G. D. Mahan, and P. C. Eklund. Surface optical phonons in gallium phosphide nanowires. *Nano Lett.*, 3:1745–1750, 2003.

50. R. Ruppin and R. Englman. Optical phonons of small crystals. *Rep. Prog. Phys.*, 33:149–196, 1970.

51. Q. Xiong, G. Chen, H. R. Gutierrez, and P. C. Eklund. Raman scattering studies of individual polar semiconductor nanowires: Phonon splitting and antenna effect. *Appl. Phys. A*, 85:299–305, 2006.

52. E. G. Gadret, M. M. de Lima Jr., J. R. Madureira, T. Chiaramonte, M. A. Cotta, F. Iikawa, and A. Cantarero. Optical phonon modes of wurtzite InP. *Appl. Phys. Lett.*, 102:122101, 2013.

53. A. J. Lohn, T. Onishi, and N. P. Kobayashi. Optical properties of indium phosphide nanowire ensembles at various temperatures. *Nanotechnology*, 21:355702, 2010.

54. L. C. O. Dacal and A. Cantarero. An ab initio study of polytypism in InP. *Sci. Rep.*, 6:33914, 2016.

55. F. Comas, C. Trallero-Giner, and A. Cantarero. Optical phonons and electron-phonon interaction in quantum wires. *Phys. Rev. B*, 47:7602–7605, 1993.

56. E. Roca, C. Trallero-Giner, and M. Cardona. Polar optical vibrational modes in quantum dots. *Phys. Rev. B*, 49:13704–13711, 1994.

57. H. M. Cheng, K. F. Lin, H. C. Hsu, and W. F. Hsiehb. Size dependence of photoluminescence and resonant

Raman scattering from ZnO quantum dots. *Appl. Phys. Lett.*, 88:261909, 2006.

58. C. Teichert. Self-organization of nanostructures in semiconductor heteroepitaxy. *Phys. Rep. Rev. Sect. Phys. Lett.*, 365(5-6):335–432, 2002.

59. W. Lei, H. H. Tan, C. Jagadisha, Q. J. Ren, J. Lu, and Z. H. Chen. Strain relaxation and phonon confinement in self-assembled InAsSb/InP (001) quantum dashes: Effect of deposition thickness and composition. *Appl. Phys. Lett.*, 97:223108, 2010.

Hybrid Phase-Change Nanophotonic Circuits

4.1 Introduction 4-1
     Integrated Photonics and Phase-Change Materials • Historical Background
4.2 Nanophotonic Circuits 4-2
     Dielectric Waveguides and Refractive Index Contrast • Material Platforms
     and Waveguiding Properties • Loss Performance and Optical Resonators
4.3 Phase-Change Materials 4-4
     Overview of Different Materials • Origin of the Property Contrast and
     Switching Mechanism
4.4 Phase-Change Photonics 4-5
     Waveguide-Coupled PCMs • Far-Field Coupled PCMs
4.5 Applications and Outlook 4-9
     All-Optical Multilevel Memory • All-Optical Switching and Routing •
     Photonic Arithmetic Processing • Neuromorphic Computing
4.6 Conclusions 4-11
References 4-12

Johannes Feldmann and
Wolfram Pernice
WWU Münster

4.1   Introduction

4.1.1   Integrated Photonics and Phase-Change Materials

Phase-change nanophotonic devices are promising candidates for devising reconfigurable integrated optical components. Such systems combine the outstanding properties of phase-change materials (PCMs), which offer extraordinary contrast in the refractive index when switched between their amorphous and crystalline states, with the advantages of integrated photonic circuits. Both topics on their own already have huge technological impacts. PCMs have been a commercial success since the 1990s, as they form the basis for well-established optical data-storage media, such as rewritable compact disk (CD RW), digital versatile disks (DVD), and Blu-Ray RE (recordable erasable) (Yamada & Matsunaga 2000). More recently, phase-change random access memories (PCRAM) have been extensively studied to compete with conventional random access memories (Wong et al. 2010; Burr 2010). Similarly, integrated photonic circuits are widely used for implementing active and passive elements for telecommunication applications. Merging these two technologies in a hybrid platform leads to many novel applications in all-optical signal processing and offers reliable, fast, and low-energy operation. Hereby, the integrated photonics platform provides scalability and reliability of the fabricated devices, whereas the outstanding properties of PCMs guarantee fast operation, energy efficiency, and reversibility.

4.1.2   Historical Background

Phase-Change Materials

PCMs were first discovered by Stanford R. Ovshinsky, who described "reversible electrical switching phenomena in disordered structures" (Ovshinsky 1968) back in the 1960s. He observed switching between a conductive and a highly resistive state by applying an electric field to a PCM consisting of tellurium, arsenic, silicon, and germanium and suggested first ideas of microscopic mechanisms to explain this phenomenon. Ovshinsky, therefore, paved the way for many further innovations based on PCMs, exploiting not only the extraordinary electrical but also the optical properties. He also contributed to the work of Feinleib et al., who first suggested optical switching of PCMs in 1971 (Feinleib et al. 1971). It was Gordon Moore, cofounder of Intel and eponym of the famous law about increasing transistor density in semiconductor industry (Moore 1965), who first recognized the great potential of these newly discovered materials and developed an electrical memory based on PCMs already in 1970 (Welnic & Wuttig 2009). However, since then, power consumption was too high, the size of the memory too large, and the approach was not commercialized.

In the 1990s, the optical properties again leveraged PCMs from scientific research to an application in everyday life. It was discovered that not only the electrical resistance of a PCM can be altered by more than an order of magnitude but also the optical properties, especially the refractive index. By applying optical pulses, a phase transition between the crystalline and amorphous phase of a PCM can be induced, which greatly changes the reflection properties. This property finds multiple applications in optical data storage media such as CD RWs and DVDs.

In the past years, the electrical properties of PCMs again come into focus in the so-called PCRAMs that serve as nonvolatile memories and compete with the well-established FLASH memory technology with the advantage of better endurance and random access operation (Lankhorst et al. 2005; Wuttig 2005).

Nowadays, PCMs have developed into a thriving field of research, including the search for better materials in terms of switching times and energy efficiency, and into a wide field of technical applications. Due to their outstanding properties, PCMs find diverse implementations in display technology or arithmetic and neuromorphic computing (Hosseini et al. 2014; Schmitt et al. 2017; Merolla et al. 2014).

Integrated Photonics

In parallel, the field of integrated nanophotonics was started in the 1980s when the first optical waveguides on silicon in the telecommunication wavelength regime were built (Soref & Lorenzo 1985, 1986). Based on these early works, the main fundamental questions about waveguide geometries, different material platforms, and basic optical components like filters, multiplexers, and switches have been answered in the 1990s (Soref et al. 2006). Nowadays, integrated implementations of optical tools commonly found in free-space setups such as mirrors, beam splitters, or wavelength multiplexers are readily available on chip and can be arbitrarily combined in optical circuits to perform various operations. These devices form the basis for phase-change nanophotonic circuits and will be discussed in detail in the following.

4.2 Nanophotonic Circuits

4.2.1 Dielectric Waveguides and Refractive Index Contrast

Dielectric waveguides are the main ingredients for nanophotonic circuits and are used to direct the flow of light in analogy to conductive paths for the current on electronic circuit boards. Guiding of light to arbitrary positions on a chip is achieved by locally varying the dielectric constant. The easiest and most intuitive approach to understand the underlying principles of nanophotonic waveguides is given in the picture of ray optics and total internal reflection. When traveling through an interface from medium 1 with refractive index n_1 to a second medium n_2 with $n_1 > n_2$ under a certain angle θ_{in}, an optical beam is reflected towards the interface with an angle of θ_{out} according to Snell's law: $n_1 \sin(\theta_{\text{in}}) = n_2 \sin(\theta_{\text{out}})$. If the angle of incidence exceeds a critical angle $\theta_{\text{c}} = \arcsin(n_1/n_2)$, total internal reflection occurs, and the beam is fully reflected into the medium with higher refractive index. This principle is, for example, exploited in optical fibers that are used to confine and guide light over large distances with low losses and builds the basis of the data transfer in our modern information age.

The on-chip analogy of optical fibers is a ridge-type waveguide shown in Figure 4.1a. Waveguides with rectangular cross section are built out of a high refractive index medium, surrounded by a low refractive index medium by lithographic means. For understanding the exact relations between geometry and waveguiding properties, the ray optics picture is no longer suitable, and Maxwell's equations have to be solved directly. Figure 4.1b,c show solutions for the electrical field of an optical waveguide composed of silicon nitride on a silicon dioxide substrate. The field is confined in the middle of the high refractive index medium and propagates along the waveguide in an optical mode. Two general types of modes can be distinguished, transverse electric (TE)-like (Figure 4.1a) and transverse magnetic (TM)-like (Figure 4.1b) modes, depending on the direction of electric field inside the waveguide. TE-like modes have almost no electric field component in the direction of propagation, whereas for TM-like modes, the magnetic field component in the propagation direction almost vanishes. The properties of the waveguide (single mode or multimode, operation wavelength, effective refractive index, etc.) are entirely determined by its width and height and the refractive indices of the employed materials. As can be already seen from the mode simulation, the mode leaks out of the waveguide, thus offering the ability of near-field coupling, which is essential for realizing hybrid phase-change photonics, as will be explained in Section 4.4.1.

4.2.2 Material Platforms and Waveguiding Properties

Before designing photonic integrated circuits, a suitable material platform must be chosen. The main criterion for this is the wavelength regime that the circuit will be

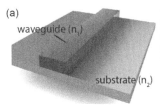

FIGURE 4.1 Integrated waveguides. (a) Ridge-type waveguide: the light is confined in the higher refractive index medium ($n_1 > n_2$). (b) First-order TE-like mode. (c) First-order TM-like mode.

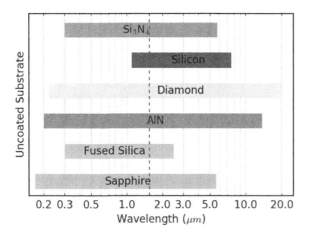

FIGURE 4.2 Transparency windows for different materials.

operated in. As shown in Figure 4.2, several materials offer a finite transparency window that can be exploited for optical waveguiding. The aforementioned silicon, for example, provides a high refractive index (all refractive indices in the following are given for a wavelength of 1,550 nm) of around $n_{Si} = 3.5$ (Li 1980), which leads to a very good confinement when used on a silicon dioxide substrate with a refractive index of $n_{SiO_2} = 1.44$ (Malitson 1965). This enables sharp bends of the waveguide and, therefore, small structures, but prohibits applications in the visible regime, as silicon is only transparent down to wavelengths of around 1.1 μm. The transparency window is defined by the optical bandgap of a material, determining the energies where no light is absorbed. Instead, Si_3N_4 can be used in the visible regime with the drawback of a lower index contrast ($n_{Si_3N_4} = 1.98$) (Philipp 1973), leading to larger footprints of the photonic circuits. Diamond offers the ability to work with very large wavelength in the far infrared up to wavelengths of 20 μm that are especially interesting for sensing applications, as these are in the fingerprint region of many biological molecules (Petersen et al. 2014). Yet, large-area substrates are still a limiting factor.

Besides the operating wavelength, the strength of higher order nonlinearities is an important criterion when choosing the right material platform, as they, for example, offer the ability of second harmonic generation or electro-optic modulation. While silicon nitride shows weak nonlinearity of the refractive index, aluminium nitride offers the ability to exploit the electro-optical and piezoelectric effect, opening a huge range of applications (Xiong et al. 2012; Taniyasu et al. 2006; Piazza et al. 2006). Thus, when choosing a material for circuit manufacture, the different advantages depending on the application have to be weighed appropriately.

4.2.3 Loss Performance and Optical Resonators

One of the most important characteristics of a material platform besides the transparency window is the propagation loss obtainable in waveguides after nanofabrication.

The main contributions to the loss in a waveguide are absorption and scattering losses. The latter are mainly introduced by surface roughness due to fabrication imperfections. Table 4.1 gives an overview of the propagation loss per centimeter for several material platforms.

An important tool to characterize the loss of a waveguide and therefore the fabrication quality are optical ring resonators that consist of a straight-feeding waveguide and a ring situated next to it separated by a small gap. Light traveling down the feeding waveguide can couple into and interfere constructively inside the ring if the wavelength λ matches the resonance condition $2\pi r = n\lambda$, with r being the ring radius and n an integer number. A theoretical study of the optical coupling between the feeding waveguide and the ring and its properties are detailed in Yariv and Yeh (2006) and Rabus (2007).

An important property of resonators is the optical quality factor (Q-factor) that is a direct measure of the time the light stays inside of the ring and can be used to determine the propagation loss of a waveguide. The optical loss per round trip is determined by two factors: the propagation loss caused by absorption in the material and scattering out of the waveguide (true waveguide losses) and the light that is coupled back from the ring to the feeding waveguide. The absorption is an intrinsic material parameter that cannot be changed. As mentioned earlier, the scattering losses are mainly caused by surface roughness and can be reduced using advanced fabrication techniques. The coupling losses depend only on the geometry and can be tuned by choosing the size of the gap between the waveguide and the ring. If the ring is in close vicinity to the waveguide, the coupling between the waveguide and the ring dominates the loss factor (overcoupled), as a large-mode overlap between feeding waveguide and ring resonator provides efficient coupling between the two. When reaching a certain gap size (critical coupling), both loss factors (intrinsic and coupling) are equal, and no light that fulfills the resonance condition is transmitted to the output due to destructive interference. To estimate the internal propagation loss, the weakly coupled regime is of interest, where the coupling loss can be neglected. By placing the ring further away from the waveguide, the coupling efficiency decreases, and the intrinsic loss factor starts to dominate. Now the time (and therefore, the Q-factor of the resonator) the light stays inside the ring is only determined by the propagation loss while traveling inside the resonator. Because the equation $Q = \lambda_r/\Delta\lambda_r$ holds for the resonant wavelength λ_r and the full-width at half-maximum (FWHM) $\Delta\lambda_r$ of the resonance, the quality factor can be easily measured

TABLE 4.1 Propagation Loss for Waveguides of Different Material Fabricated on Silicon Oxide Substrate

Material	Waveguide Propagation Loss (dB/cm)
Silicon nitride	0.1 (Xuan et al. 2016)
Silicon	1.3 (Wu et al. 2018)
Diamond	0.34 (Hausmann et al. 2014)
Aluminum nitride	0.6 (Pernice et al. 2012)

in a transmission measurement by fitting a Lorentzian to a ring resonance. The propagation loss can now be calculated from the quality factor and the group index n_g using the equation $\alpha[\text{dB}/\mu\text{m}] = 10/\ln(10) \times 2\pi n_g/(\lambda\,[\mu\text{m}]\,Q)$ (Note that additional bending losses can be neglected, if the ring radius is chosen big enough.). Advanced fabrication techniques lead to a very good standard in integrated photonics with very low propagation losses, as summarized in Table 4.1.

4.3 Phase-Change Materials

4.3.1 Overview of Different Materials

To tune the optical properties of nanophotonic device materials, a strong change in the refractive index is desired. PCMs are materials that show an extraordinary change in their properties between two different phases of matter as, for example, a change in the refractive index or the electrical resistance (Wuttig & Yamada 2007; Raoux et al. 2014; Burr et al. 2016). The most common PCMs are chalcogenides and can be found in the ternary phase diagram of Ge:Sb:Te as, for example, $Ge_2Sb_2Te_5$ (GST) (see Figure 4.3a). These are also the compounds used for optical data storage media

(Wuttig & Yamada 2007). Other common materials are doped Sb_2Te as AIST (AgInSbTe) (Matsunaga et al. 2011) that exhibit similar properties but differ in the crystallization mechanism. More recently, interfacial PCMs that are artificial superlattices of the ordinary PCMs have been developed to improve the switching and energy characteristics of PCMs and specifically design the properties desired for an application (Simpson et al. 2011).

A different class of PCMs is based on vanadium dioxide (Ryckman et al. 2013). These materials show a similar property contrast between their phases. Differing from the aforementioned materials, their response is volatile (if not fabricated in a special process (Zhang et al. 2017)), and therefore, they do not store information over a long period of time if the excitation is turned off. Therefore, we focus solely on nonvolatile chalcogenide-based PCMs.

4.3.2 Origin of the Property Contrast and Switching Mechanism

An interesting question of research is what mechanism leads to the huge property contrast between the amorphous and crystalline phases of matter of a PCM. In the amorphous

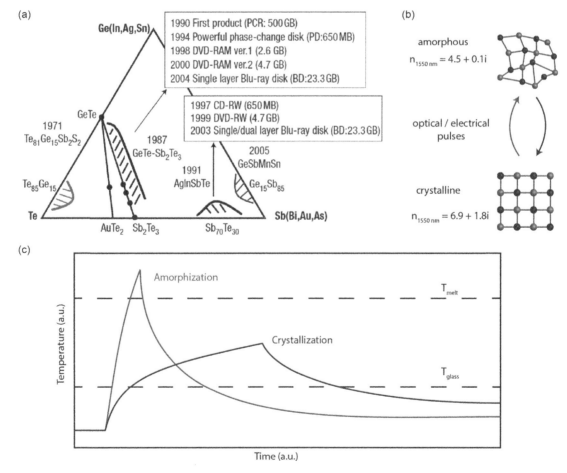

FIGURE 4.3 Overview of common PCMs and switching mechanism. (a) Ternary phase diagram showing different PCMs and their commercial applications. (From Wuttig & Yamada 2007) (b) Sketch of the phase transformation. (c) Temperature and time dependence of the switching mechanism.

state, no long-range order is observed in the material and covalent (Zhou et al. 2016; Lucovsky & White 1973) bonds dominate the structure. As the electrons are highly localized in these bonds, the electrical resistance is low, and the material in its amorphous state can be considered an isolator.

In contrast, during the transition to the crystalline material, the local atomic order changes and the so-called "resonant bonds" are formed, characterized by a strong delocalization of electrons as they are now shared between different atoms despite being localized in a covalent bond. The electronic polarizability increases, and therefore, the optical bandgap of the material decreases leading to an extraordinary change in the refractive index between the amorphous and crystalline states.

An important fact making PCMs so valuable for real applications like data storage and PCRAM is the fast switching time between the two states, which will be discussed in detail in Section 4.4.1.2. Although the exact mechanisms are not yet fully understood, fast switching times provide strong evidence, that only small local changes in the arrangement of atoms are necessary for undergoing the phase transformation.

The switching between the two states can be induced by short heat stimuli. These can be provided either electrically by joule heating or optically by absorption of light. The underlying process is illustrated in Figure 4.3c. To amorphize the crystalline PCM, it is first heated above its melting temperature (approximately 600°C), thereby destroying the crystalline order. If the material is now cooled down fast enough below the glass transition temperature (approximately 200°C), the unordered amorphous state is frozen and conserved, because the atoms are not mobile enough anymore to form the energetically preferred crystalline structure. Cooling rates in the range of 10^9 K/s are necessary to prevent crystallization (Wuttig et al. 2017). The metastable amorphous state is at room temperature preserved for many years (Wuttig & Yamada 2007; Lee et al. 2007). If in turn, the PCM is heated above the glass transition temperature but below the melting temperature, the diffusion process is enhanced. Hence, the atoms become more mobile and rearrange into their crystalline order.

In general, two crystallization mechanisms can be distinguished: growth-dominated and nucleation-dominated crystallization. In a growth-dominated material such as AIST, the crystal mainly forms from one initial nucleus, whereas in a nucleation-dominated material such as GST, many crystalline nuclei form and make up the final crystal (Salinga et al. 2013; Coombs et al. 1995). Although both materials can be switched on the same timescale employing similar energies, the underlying crystallization kinetic can have important implications for a certain application, as for example nucleation-dominated materials tend to experience filamentation when switched electrically (Liang et al. 2011). Nevertheless, both types of PCMs are highly promising for complementing traditional nanophotonic circuits via heterogeneous integration as discussed in the following.

4.4 Phase-Change Photonics

Phase-change photonics is the field of integrating PCMs with a nanophotonic platform. Basically, two ways of implementation can be distinguished. The first is PCMs coupled directly to the near field on top of a waveguide suitable for all-optical routing, signal processing, and computation. The second implementation is far-field coupled PCMs that can, for example, be used for reconfigurable metasurfaces, reflective displays, or smart windows. Both approaches will be discussed in the following sections.

4.4.1 Waveguide-Coupled PCMs

Experimental Techniques and Measurement Platforms

The basic PCM cell used in near-field coupled devices is shown in Figure 4.4a and consists of a photonic waveguide with a patch of a PCM deposited on top. When light is traveling down the waveguide, it evanescently couples to the PCM and therefore the transmission depends on the phase state. The phase state of the material can be detected in a simple transmission measurement. If the PCM is in the crystalline state, the imaginary part of the refractive index corresponding to the absorption coefficient is high, and most of the light is absorbed, depicted by the first readout pulse in Figure 4.4a. If now a higher-power write pulse (white) is sent to amorphize the PCM, the absorption changes by more than an order of magnitude and most of the light is transmitted, indicated by a second (higher) readout pulse in the figure.

Considering only the fully amorphous and fully crystalline state, the transmission through the basic cell can be switched between on and off. However, it is additionally possible to tune the transmission to any level in between by only partly crystallizing/amorphizing the PCM. In this case, crystalline and amorphous domains coexist, and an average absorption is observed in the transmission measurement.

Switching Events—Timescale, Energy, and Endurance Considerations

When aiming for real-world applications beyond scientific research, a material platform for optical signal processing and routing has to satisfy three major requirements. These comprise fast operation times, energy efficiency, and endurance. Optically operated PCMs offer very good properties in all three categories and are able to compete in several applications with their traditional counterparts.

To understand what influences and limits the PCMs behavior, it is important to understand the underlying physics when a PCM is switched by an optical pulse. When a pulse travels down a waveguide with a PCM on top, it is absorbed and locally heats up the PCM. From the PCM, the heat now diffuses away to the surrounding material, which is, in this case, the waveguide and the substrate. Hence, shorter pulses lead to a better energy efficiency, because they only heat up the PCM. When applying longer pulses, the total

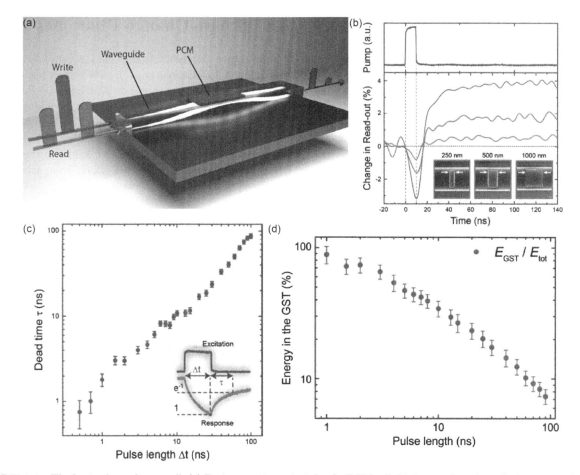

FIGURE 4.4 The basic phase-change cell. (a) Basic operation principle of a PCM cell. Light traveling through the waveguide couples evanescently to the PCM. The transmittance is a measure for the phase state. (b) Time dependence of a switching event induced by a 10 ns write pulse. After an initial drop caused by the thermo-optic effect, the transmission settles quickly to a new level. (c) Dead time of the PCM cell as a function of the pulse length. Subnanosecond pulses allow operation in the GHz regime. (d) Energy in the PCM in dependence of the pulse length. Shorter pulses yield higher efficiency, as only the PCM and not the surrounding waveguide are heated up (Ríos et al. 2015).

energy that is involved in the switching process becomes less as the system is given more time to relax to the equilibrium state and heat diffuses to the waveguide and the substrate. Using even longer pulses and heating up the surrounding material too much even prevents fast melt quenching to the amorphous phase. This can also be seen in Figure 4.4d, pointing out that for shorter pulses the energy stored in the PCM is maximized. In terms of energy consumption, it is important to note that after switching the PCM cell, it remains stable for years, and no further energy is needed to preserve its state.

A second requirement for PCM is the fast switching time. The time it takes for a PCM cell to settle to a new level after switching is determined by the crystallization or amorphization process and the thermal relaxation time. Amorphization and crystallization occur on picosecond and sub-nanosecond timescales (Loke et al. 2012; Bruns et al. 2009) and are determined by the PCM itself. Figure 4.4c shows how the thermal relaxation (dead time of the PCM cell) is related to the pulse width pointing towards operation in the gigahertz regime when using pulses in the sub-nanosecond range. As

illustrated in the inset, the dead time is defined as the 1/e decay time after the excitation pulse.

Besides amorphizing or crystallizing a cell, the readout of a PCM cell is even faster and less energy consuming. Because there is no phase transition involved and the material is not heated to elevated temperatures, the readout only depends on the pulse width and can occur on a picosecond timescale or below, only limited by the detection speed and efficiency.

A further advantage of PCMs employed as memory elements lies in the cyclability and reversibility of the switching process. Compared with FLASH memories commonly used in USB drives that can only be switched up to 10^5–10^6 times, endurance of 10^{12} cycles has been already shown (Raoux et al. 2014) for PCMs and a cyclability up to 10^{15} is predicted (Kim et al. 2010).

Networks of PCM Devices

Based on the single photonic PCM cell described earlier, the next step is to combine these elementary units into larger networks. This can, for example, be achieved in waveguide crossing arrays (Figure 4.5), where the PCM is deposited on top of each waveguide intersection. This structure, reminiscent of electronic crossbar memories, enables random access of the individual elements by selecting a certain crossing point with two perpendicular optical pulses. Only the PCM cell where both pulses overlap in time receives sufficient energy to switch the material between its phase states. This network of PCM cells can not only be used for implementing all-optical random access memory but also provides the possibility of in-memory computing and circumventing the von Neumann bottleneck, as described in Section 4.5.3.

4.4.2 Far-Field Coupled PCMs

Besides coupling PCMs evanescently to a waveguide, far-field coupling in an out-of-plane fashion can also be exploited. In this case, the PCM is deposited onto a substrate and by writing a certain pattern of amorphous and crystalline areas into the layer, the optical properties of the surface can be engineered towards desired applications. Differing from waveguide-coupled structures, these devices are illuminated under vertical incidence and used either as transmissive or reflective optical elements.

Reconfigurable Metasurfaces

Reconfigurable metasurfaces based on PCMs offer the ability of configuring the surface properties of an object in a reversible and nonvolatile manner. A gray-scale pattern corresponding to the amount of crystallinity is written into the PCM by applying optical pulses, and therefore, the refractive index can be varied along the surface direction. This leads to diverse applications for photonic devices, such as reconfigurable lenses, wavelength-selective filters, or engineering the surface reflectance and absorption (Karvounis et al. 2016; Wang et al. 2015).

An example of a tunable metamaterial (Wang et al. 2015) is shown in Figure 4.6a. Each inclusion is composed of two overlapping crystalline marks, fabricated via laser writing, and Figures 4.6b,c present the transmission and reflection spectra obtained for light polarized along the long axis (b) and short axis (c) of inclusion. It can be seen that only in one polarization direction the reflection shows a resonance dip at around 2 μm, while the other direction has no resonance features. As the structure does not diffract light with wavelengths longer than 1.78 μm, it can be concluded that this is a true metamaterial effect where the light experiences an effective medium. In the same way, lenses for wavelength multiplexing and different focal length have also been fabricated. Due to the reversible nature of the switching process, this technique offers reconfigurable photonic structures that do not need a constant bias to preserve their state. Using a combined electro-optical platform, optical beam steering induced by electrical pulses can also be achieved offering possibilities, for example, in light detection and ranging (LIDAR) systems for autonomous vehicles (de Galarreta et al. 2018).

Reflective Displays

Another interesting application of PCMs can be found in new display technologies. The basic principle is depicted in Figure 4.6d. A PCM (GST in this case) is deposited between two transparent and conductive indium tin oxide films on a reflective surface. By switching the PCM between the amorphous and crystalline phases, the color of the reflected incoming white light can be modulated, as the refractive index of the PCM film is varied (Hosseini et al. 2014; Ríos et al. 2016). Depending on the exact film thickness of the ITO and PCM films, the color of the whole stack can be tuned. If the layer stack is now cut to individual pixels as shown in Figure 4.6e that can individually be controlled via electric contacts, a high-resolution display with pixel sizes below 1 μm x 1 μm can be fabricated. Employing conductive tip atomic force microscopy (CAFM) pictures, as shown in Figure 4.6f, the optical micrograph can be transferred to the PCM on a very small scale.

Because of the extremely thin layer structure, these phase-change displays can also be used as flexible displays, especially interesting for wearable electronics. A further application in this context are PCM-based smart windows,

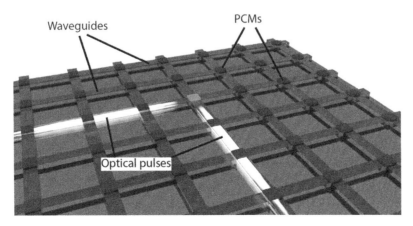

FIGURE 4.5 Illustration of a hybrid phase-change photonic network. Each PCM cell on top of a waveguide intersection can be selected individually using two perpendicular optical pulses.

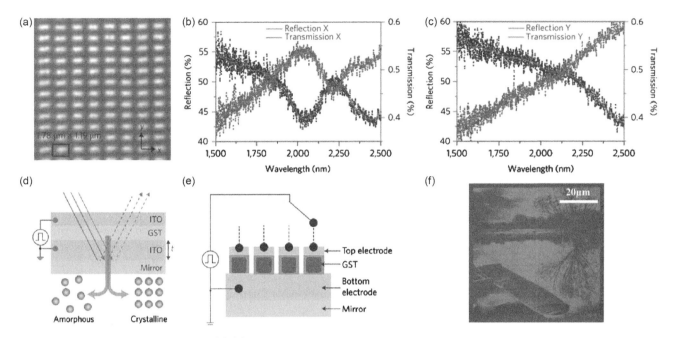

FIGURE 4.6   Far-field coupled PCMs. (a)–(c) By writing a pattern onto a PCM-coated substrate via optical excitation, metamaterials can be fabricated. Here, polarization-dependent reflection and transmission characteristics are shown. Light polarized along the long axis of the marks in (a) exhibits a resonance dip in the reflectance at around 2.0 μm, (b) whereas the orthogonal polarization is unaffected (c). (d) Using a layer stack as, for example, indium tin oxide (ITO)/GST/ITO on a reflecting substrate, the surface color can be continuously modulated by switching the PCM between its states. Separating these stacks and adding individual control to each PCM cell, as shown in (e), offers the ability to fabricate displays from phase-change pixels. (f) Image recorded in a PCM-coated substrate. (Wang et al. 2015.)

that can, for example, adjust the optical transmission of wavelengths in the near infrared of the solar spectrum (where the property contrast of PCMs is particularly large) and, therefore, modulate the heat transport to the inside. Also, windows with integrated displays or applications in virtual and augmented reality can be envisaged.

PCM Plasmonics

A drawback of conventional integrated optics is that downscaling the geometry is limited by the diffraction limit imposed through the wavelength of the light used. A way to overcome this limitation and thus to build smaller optical devices is given by the thriving field of plasmonics. Here, quantized oscillations of the charge carrier density—so-called plasmons—that can be excited by light are employed for signal processing (Ozbay 2006; Gramotnev & Bozhevolnyi 2010; Schuller et al. 2010). Especially, surface plasmon polaritons (SPP) that are bound to a dielectric–metal interface and exponentially decay into the surrounding media can propagate in waveguides along the interface. Because the light is strongly confined to the boundary between the metal and the dielectric, tightly arranged devices with subwavelength dimensions can be build. In the same way as for conventional optics, PCMs provide a way to control and modulate the SPPs.

An example of a surface plasmon waveguide is shown in Figure 4.7 (Rudé et al. 2015). The cross section of the device is shown in Figure 4.7a illustrating the layer stack.

The plasmonic waveguide consists of a 60 nm gold film that is covered by a buffer layer of silicon oxide used to thermally isolate the PCM (here GST) from the gold layer. The oxide also prevents the PCM film from interacting strongly with the surface plasmon and thereby prohibits its propagation. Silicon nitride is used to cover the GST and prevent oxidation. To tune the waveguide properties and the number of propagating modes, a poly methyl methacrylate layer is employed. Figure 4.7b shows an optical microscope image from the top of a fabricated plasmonic waveguide. Light incident normal to the surface is coupled to the plasmonic mode via the grating in the middle of a large gold pad (Bragg reflection) and is subsequently coupled out on the other side. Between both couplers, a GST strip lies on top of the waveguide that is switched using optical pump pulses at 975 nm. The probe light is sent with a wavelength of 1,550 nm. Figure 4.7c,d show the intensity distribution of the light scattered by the couplers and transmitted through the plasmonic waveguide with the PCM strip in the amorphous and crystalline phases, respectively. The light is coupled in on the left and transmitted to the right, where the light intensity is collected and background corrected in the indicated rectangular areas. Figure 4.7e illustrates the difference in transmission between the two states of the PCM, which exhibits a modulation contrast of about 30%. While plasmonic waveguides are a promising technology for miniaturizing optical circuits, up to now, the propagation loss is still too high compared with conventional optical integrated circuits.

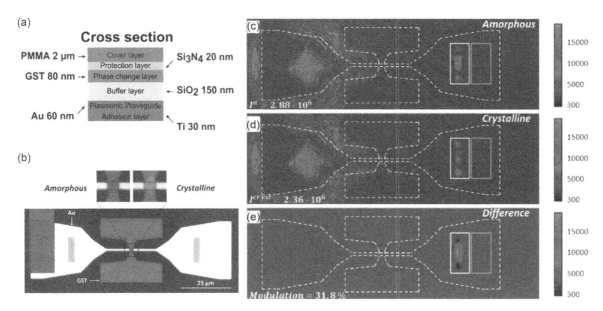

FIGURE 4.7 Plasmonic waveguides. (a) Layer stack of the waveguide in the interaction area. The plasmonic waveguide made out of 60 nm gold film is thermally isolated from the PCM by a 150 nm buffer layer of silicon oxide. (b) Optical microscope image of the top view of a fabricated device. At the large gold pads with the vertical grooves, incident light at 1,550 nm is coupled to the plasmonic waveguide. A small strip of PCM is deposited across the waveguide. (c) and (d) show the measured intensity distribution of the transmitted and scattered light with the PCM in the amorphous and crystalline states. (e) The difference in transmission between the amorphous and crystalline phase could be modulated up to 3.1%. (Rudé et al. 2015)

4.5 Applications and Outlook

4.5.1 All-Optical Multilevel Memory

After the first commercial success of PCMs in the optical domain, recently, PCMs got into the focus in electronic PCRAMs to compete with established technology like FLASH memories. Since working in the optical domain is advantageous in many respects (speed, energy efficiency, bandwidth), bringing the concept back to an all-optical operation is desirable. An example for this approach is the multilevel memory cell presented by Ríos et al. that can store up to eight transmission levels in a single PCM cell and thus allows for storing three bits of information. While a classical binary memory stores values up to 256 in eight bits, this type of memory needs only three elements (multilevel bits) to store values up to 512, paving the way for a significant increase in memory density. Each level is defined by a specific pulse energy and can be set in a reversible manner not depending on the previous state of the material. The number of levels distinguishable in a phase-change cell depends on the length of the PCM strip, the exact pulse parameters, and the readout noise, enabling the use of even higher memory densities. Combined with the waveguide crossing array introduced in Section 4.4.1.3, this memory holds promise for realizing an all-optical multi-level random access memory. All-optical nonvolatile memories have the potential to remove the need for electro-optical conversions featuring faster and more efficient computation, thus circumventing the von Neumann bottleneck (Ríos et al. 2015).

4.5.2 All-Optical Switching and Routing

An important component for fully optical signal processing on an integrated platform is a device for all-optical switching and routing. Differing from most optical switches that need a constant bias power to preserve their state (Reed et al. 2010), phase-change photonics passively offers the possibility to create switches by exploiting nonvolatility. An example of such an all-optical switch (Stegmaier et al. 2016) is illustrated in Figure 4.8a. The device consists of an optical ring resonator in an add-drop configuration with a patch of PCM deposited on top of the ring. Depending on the state of the material, light is guided from the input to the through or to the drop port, effectively resembling an 1×2 switch.

In Figure 4.8b, the transmission to the through and to the drop port is plotted as a function of wavelength in the amorphous as well as crystalline states, yielding an on–off switching contrast for this device of about 5 dB in both ports. Switching was induced by optical picosecond pulses, and the transition times were shown to be shorter than 200 ps.

4.5.3 Photonic Arithmetic Processing

Signal processing using PCMs can further lead to a new class of computing architectures. In contrast to traditional computers based on the von Neumann architecture, where processor and memory are physically separated and the data has to be continuously transferred between the two limiting the computational speed, PCM processors are based on in-memory computing. Here, the calculations are directly

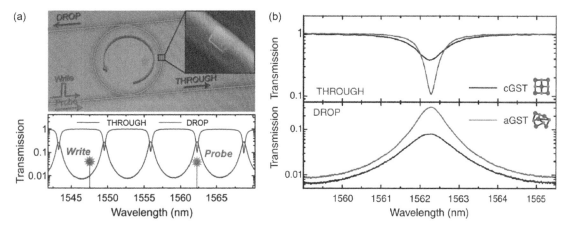

FIGURE 4.8 Ring resonator in add–drop configuration used as a 1 × 2 switch. (a) Scanning electron micrograph of the switch illustrating the basic operation principle. The inset shows a close-up of a PCM strip deposited on top of the ring resonator. (b) Optical transmission in the through and drop port with the PCM in the crystalline and amorphous states, respectively (Stegmaier et al. 2016).

carried out in the memory, thus circumventing the need of writing the results back to a physically separate place. Wright et al. have shown how arithmetic computation using PCMs can be carried out with free-space optics (Wright et al. 2011), and the same concepts can also be transferred to the integrated platform using the basic phase-change cells described in Section 4.4.1. Figure 4.9a shows a sketch of how such a cell is operated in analogy to an abacus, and Figure 4.9b,c show experimental examples (Feldmann et al. 2017).

Starting from the amorphous (high transmittance) state, successive optical picosecond pulses induce quanta of crystallization in the material, leading to distinguishable steps in the transmission. In the present example, ten pulses are used for full crystallization, exhibiting the ability of calculating directly in base ten. Each phase-change cell represents a different place value (i.e., rod of an abacus). By choosing appropriate pulse energies, working in other bases (for example hexadecimal) is possible when more suitable for the specific problem.

Figure 4.9a shows how the addition "7 + 3 = 10" is performed on this platform. The "dark and bright lines" indicate the state of crystallinity of two different phase-change cells representing the tens and ones, respectively. In a first step, seven identical optical pulses are sent to the red cell, leaving it on the seventh crystallization level. This corresponds to a shifting of seven beads of the lowest rod of an abacus to the right, as indicated in the sketch. The second step consists of adding the remaining three pulses, leaving the first cell in the fully crystalline bottom state. Equivalent to using an abacus where all beads of one rod are shifted to the right, a carryover has to be performed, and one single bead of the next higher place value is shifted to the right, whereas all ten beads of the first rod are moved to their initial position. This is achieved by amorphizing the first cell to its initial state and inducing a single crystallization step in the second cell representing the tens. Now the result can be read out. Cell one is in state "0" and cell two in

state "1," yielding the correct result "10." It is important to note here that the result is now directly stored in the same physical location that it was calculated in, as the PCM is nonvolatile and will keep the information, circumventing the von Neumann bottleneck. Because subtraction, multiplication, and division can be reduced to addition tasks, the phase-change cell is capable of all four basic arithmetic tasks. Figure 4.9b,c show experimental examples of multiplication and subtraction where the change in optical transmission during the operation is plotted as a function of the operation time. Multiplication thereby is simply carried out as a sequential addition, where "4 × 3" is stripped down to "3 + 3 + 3 + 3." Subtraction relies on the concept of the nine's complement. The nine's complement of a given number can be calculated by subtracting $10^N - 1$ from the initial value, N being the number of digits. This way, adding the nine's complement of the minuend to the subtrahend yields the nine's complement of the subtraction. Figure 4.9c shows how "79 − 14 = 65" is solved using the PCM cell. As a first step, the complement of the minuend "'79," which is "'20," is added to the cell. Inserting also the subtrahend now results in three carryovers, and the PCM cell shown (representing ones) on level four, resembling the result "34," is the nine's complement of "65."

The pulse energies used in these experiments are below 20 pJ for each step. It has to be noted that the timescale is only limited by the crystallization and thermal relaxation time, as described in Section 4.4.1.2, giving rise to computation in the gigahertz regime, and the time between individual events on the order of seconds, as shown here, is chosen for demonstration purposes only.

4.5.4 Neuromorphic Computing

A completely different approach to traditional computers are brain-inspired computing techniques called neuromorphic computers implementing the idea of mimicking the

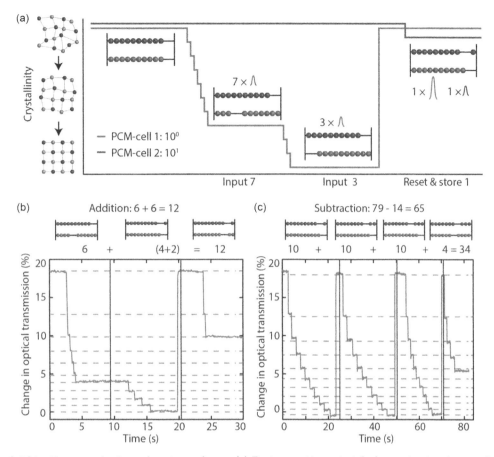

FIGURE 4.9 Arithimetic processing in analogy to an abacus. (a) Basic operation principle demonstrating the use of ten crystallization levels to perform calculations in base ten. (b) and (c) show experimental demonstrations of addition and subtraction performed in a phase-change cell (Feldmann et al. 2017).

highly efficient and parallel processing of huge amounts of data accomplished by biological brains. Individual neurons connected by synapses process the information and are especially suitable for classification tasks, pattern, and speech recognition (Lane et al. 2017). Due to the serial approach used by traditional computers based on the von Neumann architecture that process one command at a time, these tasks cannot be efficiently solved (Preissl et al. 2012). Neuromorphic computers implementing the massively parallel processing of brains can instead efficiently deal with such problems and outperform conventional computers in terms of speed and energy efficiency by orders of magnitude.

The basic building block of the so-called artificial neural networks (ANNs) are individual neurons that are grouped on different layers. Neurons take several inputs, weighted by synapses, sum them up, and generate an output signal if a certain threshold power is succeeded. The output is passed to the next layer of neurons until an output layer is reached. The information in neural networks is stored in the connections between the neurons, i.e., the synapses that weight the input signals of a neuron.

This makes PCMs a promising candidate for synapse mimics, as they intrinsically offer a weighting mechanism through their large change in the imaginary part of the refractive index and therefore the absorption. Several designs for hardware implementations of optical phase-change synapses have already been demonstrated, showing a huge potential in neuromorphic computing (Cheng et al. 2017; Kim et al. 2015; Kuzum et al. 2012).

4.6 Conclusions

Hybrid phase-change nanophotonic circuits are a promising concept for all-optical signal processing. Even though commercially in use since the 1990s, only in the last years, more light was shed onto the underlying physics behind the switching mechanism of PCMs, leading to deeper insight in the material properties. Ongoing research paves the way for even faster and less energy-consuming materials. Combined with integrated photonics, PCMs offer a robust platform for all-optical signal processing and routing. The nonvolatile and reversible nature of the phase transition enables applications in data storage as well as in conventional and unconventional computations. Being able to access multiple levels adds another degree of flexibility. An integrated photonic platform offers scalability and robustness of the fabricated circuits. Using plasmonic waveguides, the footprint of such devices could in the future be significantly reduced. Besides integrated nanophotonic solutions, engineering surfaces covered with PCM gives rise not only to reconfigurable photonic components but also to new display and smart window technologies. Hybrid phase-change photonic circuit is therefore a versatile tool for various applications.

References

Bruns, G. et al., 2009. Nanosecond switching in GeTe phase change memory cells. *Applied Physics Letters*, 95(4), pp. 14–17.

Burr, G.W., 2010. Phase change memory technology. *Journal of Vacuum Science*, 2004(2), p. 903.

Burr, G.W. et al., 2016. Recent progress in phase-change memory technology. *IEEE Journal on Emerging and Selected Topics in Circuits and Systems*, 6(2), pp. 146–162. Available at: http://ieeexplore.ieee.org/document/7453199/.

Cheng, Z. et al., 2017. On-chip photonic synapse. *Science Advances*, 2(September), pp. 1–7. Available at: http://advances.sciencemag.org/content/advances/3/9/e1700160.full.pdf.

Coombs, J.H. et al., 1995. Laser-induced crystallization phenomena in GeTe-based alloys. I. Characterization of nucleation and growth. *Journal of Applied Physics*, 78, pp. 4906–4917.

Feinleib, J. et al., 1971. Rapid reversible light-induced crystallization of amorphous semiconductors. *Applied Physics Letters*, 18(6), pp. 254–257.

Feldmann, J. et al., 2017. Calculating with light using a chip-scale all-optical abacus. *Nature Communications*, 8. Available at: www.nature.com/articles/s41467-017-01506-3.pdf.

de Galarreta, C. et al., 2018. Non-volatile reconfigurable phase-change metadevices for beam steering in the near-infrared. *Advanced Functional Materials*, 1704993, pp. 1–9.

Gramotnev, D.K. & Bozhevolnyi, S.I., 2010. Plasmonics beyond the diffraction limit. *Nature Photonics*, 4(2), pp. 83–91. doi: 10.1038/nphoton.2009.282.

Hausmann, B.J.M. et al., 2014. Diamond nonlinear photonics. *Nature Photonics*, 8(5), pp. 369–374.

Hosseini, P., Wright, C.D. & Bhaskaran, H., 2014. An optoelectronic framework enabled by low-dimensional phase-change films. *Nature*, 511(7508), pp. 206–211. Available at: www.ncbi.nlm.nih.gov/pubmed/25008527.

Karvounis, A. et al., 2016. All-dielectric phase-change reconfigurable metasurface. *Applied Physics Letters*, 109(5), p. 051103.

Kim, I.S. et al., 2010. High performance PRAM cell scalable to sub-20 nm technology with below 4F2 cell size, extendable to DRAM applications. *Digest of Technical Papers: Symposium on VLSI Technology*, Honolulu, Hawaii, pp. 203–204.

Kim, S. et al., 2015. NVM neuromorphic core with 64k-cell (256-by-256) phase change memory synaptic array with on-chip neuron circuits for continuous in-situ learning. *2015 IEEE International Electron Devices Meeting (IEDM)*, Washington, DC, pp. 17.1.1–17.1.4.

Kuzum, D. et al., 2012. Nanoelectronic programmable synapses based on phase change materials for brain-inspired computing. *Nano Letters*, 12, pp. 2179–2186.

Lane, N.D. et al., 2017. Squeezing deep learning into mobile and emdedded devices. *IEEE Pervasive Computing*, 16(3), pp. 82–88.

Lankhorst, M.H.R., Ketelaars, B.W.S.M.M. & Wolters, R.A.M., 2005. Low-cost and nanoscale non-volatile memory concept for future silicon chips. *Nature Materials*, 4(4), pp. 347–352. Available at: www.nature.com/doifinder/10.1038/nmat1350.

Lee, S.-H., Jung, Y. & Agarwal, R., 2007. Highly scalable non-volatile and ultra-low-power phase-change nanowire memory. *Nature Nanotechnology*, 2(10), pp. 626–630.

Li, H.H., 1980. Refractive index of silicon and germanium and its wavelength and temperature derivatives. *The Journal of Physical and Chemical Reference Data*, 9(3), pp. 561–658.

Liang, G. et al., 2011. Comparison of optical and electrical transient response during nanosecond laser pulse-induced phase transition of $Ge_2Sb_2Te_5$ thin films. *Chemical Physics Letters*, 507(1–3), pp. 203–207. doi: 10.1016/j.cplett.2011.03.076.

Loke, D. et al., 2012. Breaking the speed limits of phase-change memory. *Science*, 336(6088), pp. 1566–1569.

Lucovsky, G. & White, R.M., 1973. Effects of resonance bonding on the properties of crystalline and amorphous semiconductors. *Physics Review B*, 8(2), pp. 660–667.

Malitson, I.H., 1965. Interspecimen comparison of the refractive index of fused silica. *Journal of the Optical Society of America*, 55(10), p. 1205. Available at: www.osapublishing.org/abstract.cfm?URI=josa-55-10-1205.

Matsunaga, T. et al., 2011. From local structure to nanosecond recrystallization dynamics in AgInSbTe phase-change materials. *Nature Materials*, 10(2), pp. 129–134. doi: 10.1038/nmat2931.

Merolla, P.A. et al., 2014. A million spiking-neuron integrated circuit with a scalable communication network and interface. *Science*, 345(6197), pp. 668–673.

Moore, G.E., 1965. Cramming more components onto integrated circuits. *Electronics*, 38, pp. 114–117.

Ovshinsky, S.R., 1968. Reversible electrical switching phenomena in disordered structures. *Physical Review Letters*, 21(20), pp. 1450–1453.

Ozbay, E., 2006. Plasmonics: Merging photonics and electronics at nanoscale dimensions. *Science*, 311(5758), pp. 189–193.

Pernice, W.H.P. et al., 2012. Second harmonic generation in phase matched aluminum nitride waveguides and micro-ring resonators. *Applied Physics Letters*, 100(22), P. 223501.

Petersen, C.R. et al., 2014. Mid-infrared supercontinuum covering the 1.4-13.3 μm molecular fingerprint region using ultra-high NA chalcogenide step-index fibre. *Nature Photonics*, 8(11), pp. 830–834. doi: 10.1038/nphoton.2014.213.

Philipp, H.R., 1973. Optical properties of silicon nitride. *Journal of the Electrochemical Society: Solid-State Science and Technology*, 120(2), pp. 295–300.

Piazza, G., Stephanou, P.J. & Pisano, A.P., 2006. Piezoelectric aluminum nitride vibrating contour mode MEMS resonators. *Journal of Microelectromechanical Systems*, 15(6), pp. 1406–1418.

Preissl, R. et al., 2012. Compass: A scalable simulator for an architecture for cognitive computing. *International Conference for High Performance Computing, Networking, Storage and Analysis, SC*, Salt Lake City, UT.

Rabus, D.G., 2007. *Integrated Ring Resonators*, Springer-Verlag: Berlin Heidelberg.

Raoux, S. et al., 2014. Phase change materials and phase change memory. *MRS Bulletin*, 39(08), pp. 703–710.

Reed, G.T. et al., 2010. Silicon optical modulators. *Nature Photonics*, 4(8), pp. 518–526. Available at: http://epubs.surrey.ac.uk/7096/4/licence.txt.

Ríos, C. et al., 2015. Integrated all-photonic non-volatile multi-level memory. *Nature Photonics*, 9(11), pp. 725–732. Available at: www.nature.com/doifinder/10.1038/nphoton.2015.182.

Ríos, C. et al., 2016. Color depth modulation and resolution in phase-change material nanodisplays. *Advanced Materials*, 28, pp. 4720–4726.

Rudé, M. et al., 2015. Active control of surface plasmon waveguides with a phase change material. *ACS Photonics*, 2(6), pp. 669–674. Available at: http://pubs.acs.org/doi/abs/10.1021/acsphotonics.5b00050.

Ryckman, J.D. et al., 2013. Ultra-compact silicon photonic devices reconfigured by an optically induced semiconductor-to-metal transition. *Optics Express*, 21(9), pp. 10753–10763. Available at:www. osapublishing.org/irectPDFAccess/F3EFA96F-B225-678D-C53DF95C5C808622_253055/oe-21-9-10753.pdf?da=1&id=253055&seq=0&mobile=no.

Salinga, M. et al., 2013. Measurement of crystal growth velocity in a melt-quenched phase-change material. *Nature Communications*, 4, p. 2371.

Schmitt, S. et al., 2017. Neuromorphic hardware in the loop: Training a deep spiking network on the BrainScaleS wafer-scale system. *In Proceedings of the International Joint Conference on Neural Networks*, Anchorage, Alaska. pp. 2227–2234.

Schuller, J.A. et al., 2010. Plasmonics for extreme light concentration and manipulation. *Nature Materials*, 9(3), pp. 193–204. doi: 10.1038/nmat2736.

Simpson, R.E. et al., 2011. Interfacial phase-change memory. *Nature Nanotechnology*, 6, pp. 501–505.

Available at: www.nature.com/doifinder/10.1038/nnano.2011.96.

Soref, R., Fellow, L. & Paper, I., 2006. The past, present, and future of silicon photonics. *IEEE Journal of Selected Topics in Quantum Electronics,* 12(6), pp. 1678–1687. doi: 10.1109/jstqe.2006.883151.

Soref, R.A. & Lorenzo, J.P., 1985. Single-crystal silicon: A new material for 1.3 and 1.6 μm integrated-optical components. *Electronics Letters*, 21(21), pp. 953–954.

Soref, R.A. & Lorenzo, J.P., 1986. All-silicon active and passive guided-wave components for λ = 1.3 and 1.6 μm. *IEEE Journal of Quantum Electronics*, 22(6), pp. 873–879.

Stegmaier, M. et al., 2016. Nonvolatile all-optical 1 × 2 switch for chipscale photonic networks. *Advanced Optical Materials*, 5, p. 1600346.

Taniyasu, Y., Kasu, M. & Makimoto, T., 2006. An aluminium nitride light-emitting diode with a wavelength of 210 nanometres. *Nature*, 441(7091), pp. 325–328.

Wang, Q. et al., 2015. Optically reconfigurable metasurfaces and photonic devices based on phase change materials. *Nature Photonics*, 10(1), pp. 60–65. doi: 10.1038/nphoton.2015.247.

Welnic, W. & Wuttig, M., 2009. Phasenwechsel-Materialien als universale Speichermedien. *Physik in unserer Zeit*, 40(4), pp. 189–195. doi: 10.1002/piuz.200801205.

Wong, H.-S.P. et al., 2010. Phase change memory. *Proceedings of the IEEE*, 98(12), pp. 2201–2227. Available at: http://ieeexplore.ieee.org/ielx5/5/5628287/05609179.pdf?tp=&arnumber=5609179&isnumber=5628287%5Cn http://ieeexplore.ieee.org/xpls/abs˙all.jsp?arnumber=5609179.

Wright, C.D. et al., 2011. Arithmetic and biologically-inspired computing using phase-change materials. *Advanced Materials*, 23(30), pp. 3408–3413.

Wu, B. et al., 2018. Silicon integrated interferometric optical gyroscope. *Scientific Report*, 8, pp. 1–7.

Wuttig, M., 2005. Phase-change materials: Towards a universal memory? *Nature Materials*, 4(4), pp. 265–266. Available at: www.nature.com/doifinder/10.1038/nmat1359.

Wuttig, M., Bhaskaran, H. & Taubner, T., 2017. Phase-change materials for non-volatile photonic applications. *Nature Photonics*, 11. Available at: www.nature.com/nphoton/journal/v11/n8/pdf/nphoton.2017.126.pdf [Accessed August 4, 2017].

Wuttig, M. & Yamada, N., 2007. Phase-change materials for rewriteable data storage. *Nature Materials*, 6(11), pp. 824–832. Available at: www.nature.com/nmat/journal/v6/n11/abs/nmat2009.html%5Cn.

Xiong, C., Pernice, W.H.P. & Tang, H.X., 2012. Low-loss, silicon integrated, aluminum nitride photonic circuits and their use for electro-optic signal processing. *Nano Letters*, 12(7), pp. 3562–3568.

Xuan, Y. et al., 2016. High-Q silicon nitride microresonators exhibiting low-power frequency comb initiation. *Optica*,

3(11), p. 1171. Available at: www.osapublishing.org/abstract.cfm?URI=optica-3-11-1171.

Yamada, N. & Matsunaga, T., 2000. Structure of laser-crystallized $Ge_2Sb_{2+x}Te_5$ sputtered thin films for use in optical memory. *Journal of Applied Physics*, 88(12), p. 7020. Available at: http://scitation.aip.org/content/aip/journal/jap/88/12/10.1063/1.1314323.

Yariv, A. & Yeh, P., 2006. Photonics: *Optical Electronics in Modern Communications*. Oxford University Press, New York, p. 848.

Zhang, Z. et al., 2017. Evolution of metallicity in vanadium dioxide by creation of oxygen vacancies. *Physical Review Applied*, 7(3), pp. 1–13.

Zhou, X. et al., 2016. Phase-change memory materials by design: A strain engineering approach. *Advanced Materials*, 28, pp. 3007–3016. doi: 10.1002/adma.2015 05865.

5

X-Ray Nanophotonics Based on Planar X-Ray Waveguide-Resonator

5.1 Introduction.. 5-1
5.2 Procedure of Waveguide Unit Preparation 5-2
5.3 Experimental Arrangement for the Study of Radiation Intensity
Distribution.. 5-4
5.4 Angular Radiation Intensity Distribution in X-Ray Beams 5-5
5.5 X-Ray Beam Reflection on a Material Interface 5-14
5.6 Features of X-Ray Flux Total Reflection Phenomenon 5-16
5.7 Model of X-Ray Flux Waveguide-Resonance Propagation............... 5-19
5.8 Problem of PXWR Integral Intensity Increasing 5-24
5.9 Problem of Angular Divergence Reduction for a PXWR Emergent
Beam... 5-25
5.10 Common Approach to PXWR Emergent Beam Characteristics
Improving 5-30
5.11 Waveguide-Resonance Propagation of Particles Characterized
by Nonzero Rest Mass... 5-33
5.12 Conclusion 5-34
Acknowledgments 5-34
References 5-34

V. K. Egorov
IMT RAS

E. V. Egorov
IMT RAS
Financial University under Government of RF

5.1 Introduction

The fundamental task of any spectrometric technique is to get at its disposal radiation fluxes with variable parameters, and X-ray spectrometric and diffractometric material methods are not an exception. Testing X-ray beams can be formed by the Bragg reflection [1], the total external reflection [2], Borrmann effect [3], material-focusing lens application [4], and the use of restricting devices. All these methods are characterized by a direct influence of X-ray flux. On the basis of these methods in the past few decades, much effort has been directed towards generation of narrow low-divergent X-ray photon fluxes. The first step in this field was taken by P. Hirsch and J. Keller, who suggested to form micron-size X-ray beams by employing a glass capillary [5]. In a short time, planar thin film waveguides have been offered for X-ray beam transportation [6]. In this work, the resonant beam coupling of CuK_α radiation onto a $Al_2O_3/BN/Al_2O_3$ planar film structure was used. The thickness of the core layer was nearly 50 nm. As a result of the study, it was shown that the waveguide could transport X-ray radiation. According to the information received in the work, the mode structure of X-ray flux in the core layer had a discrete character. Subsequent development of this direction has manifested the flux intensity enhancement

for the waveguide emergent beam. Experimental investigations [7] have revealed the 20-fold flux intensity, increasing for the synchrotron monochromatic radiation $\lambda = 0.1048$ nm on the output of a $SiO_2/polyimide/Si$ waveguide with a core layer width of $d = 123$ nm. The important result of the work was an experimental observation of mode mixing explained by an existence of inhomogeneities and interface roughness. Similar measurements were carried out in a number of works later [8–15]. The authors of these investigations have the aim to obtain X-ray beams with a width near 100 nm, a height of some millimeters, and a total intensity near 5×10^7 photon/s in using the resonant synchrotron radiation coupling. A significant progress of these researches was connected with the switching over of a study from coupling emergent beams to the ones transported by the core layer up to the waveguide outlet. On analysis of the flux mode structure, the phenomenon of an X-ray standing wave (XSW) arising was mentioned [9]. At the same time, reference on the initial work discovered and studied the effect of a standing wave arising in the vacuum space under interface accompanied the X-ray total external reflection of beams [16] was absent. Earlier experimental works described the specific technologies of the influence on X-ray beams [17,18].

Properties and peculiarities of X-ray beams formed by polycapillary optics systems have been intensively studied for the past 25 years [19–26]. X-ray optics of the system is based on the phenomenon of X-ray beam multiple total external reflection on the inner surface of a quartz capillary. Diameter of capillaries is greater on five orders than the wavelength of the transporting radiation. Therefore, authors of initial works on the capillary optics assumed that the geometrical optics approach is enough for the description of polycapillary properties [19–22]. On the basis of this approach, a detailed X-ray tracing code took into account the effects of inner surface roughness/waviness and the effects of photon penetration through the capillary walls [22]. A similar algorithm is very effective for the description of X-ray guide structures with core sizes more than 10 μm [23]. However, the experimental explorations have showed that the description in framework of the geometrical optics is not universal owing to the disregard of interference and diffraction effects. Because of this, some elements of wave theory have been adopted in previous years [24]. In a specific review, the polycapillary optics application in tandem with the liquid metal-jet X-ray tube is described in detail [27].

While traditional approaches to the micron and submicron X-ray beam formation are described earlier, it has been discovered that the specific technique to prepare superfine beams by using the so-called "slitless" collimator is depicted in Figure 5.1 [28–30]. The slitless collimator of X-ray radiation is formed by two quartz plane-polished plates mated together. The visible light is not propagated through the interface area between these plates. It was found that the slitless device allows to pass an X-ray flux, and the emergent beam intensity was compared with the incident one [28]. However, the injection of any liquid into the space between plates led to a total block of X-ray flux propagation. A systematic study of this effect and attempts of its practical application were undertaken later [31,32]. In fact, the slitless X-ray collimator represents the planar waveguide with a practicable minimum size of an air slit. The width of the slit is defined by roughness and waviness level of the collimator plane surfaces. At the same time, the air core between guide claddings is the ideal waveguide channel for the decrease in radiation flux attenuation [6]. Similar waveguides with fixed and tunable air gaps have begun to find practical applications in the end of the twentieth century [33,34]. These waveguides with Cr claddings and air core (width 500 nm) can produce an emergent beam with width $d = 500$ nm, height $h = 0.1$ mm, and total intensity $J = 2.4 \times 10^7$ photon/s [33]. The waveguide testing study has been carried

out at the undulator beam line ID-10A ($\lambda = 0.0751$ nm) of an European Synchrotron Radiation Facility (ESRF) in Grenoble. Very important result of this work is the experimental observation of the neighboring mode mixing at the off-axes incidence of an initial beam. At the same time, these authors did not include into a consideration the simplest interpretation of its results connected with the division possibility for emergent beam on two components with the summary width being equivalent to the initial beam and variation intensities at the waveguide rotation. In Ref. [35], the waveguide with a narrow extended air slit with width $s = 88$ nm was used for the beam formation in tandem with a conventional X-ray tube (Cu) in diffractometry measurements of an Au (115 nm)/SiO$_2$ (100 nm)/Si multilayer structure. At the tube regime $U = 20$ keV and $I = 10$ mA, the waveguide has produced a beam with width 88 nm, height 10 mm, and total characteristic deposit intensity $I = 2 \times 10^6$ photon/s. The work presented the necessary prerequisites for the formulation of a pseudospecular model describing the X-ray beam propagation through a narrow extended slit on the basis of the standing wave arising conception. The next step has manifested the preliminary account about the pseudospecular model and presented first the experimental data that described X-ray radiation flux passing through narrow extended slits with fixed width [36]. The suggestion of the waveguide-resonance mechanism for the X-ray flux propagation through narrow space areas was also expressed. Unfortunately, the materials devoted to a waveguide-resonator description and discussion of its practical application are not mentioned in the book about X-ray spectrometry [37] and other reference books [38–41].

This paper is devoted to a systematic investigation of the waveguide slit width influence on the particularities of X-ray propagation, study of specific conditions that allowed to manipulate the flux parameters, presentation of a model, the main features of the waveguide-resonance mechanism, and description of the beam use formed by such devices.

5.2 Procedure of Waveguide Unit Preparation

Reflectors for the studied planar X-ray waveguides (PXW) were prepared from optical-polished amorphous quartz glass plates of $100 \times 33 \times 2.5$ mm^3 size. Reflector surfaces were polished using a standard optical technology up to a roughness near 5 nm. Two reflectors were placed at a fixed distance between them, and the distance was defined by the thickness of metallic strips deposited on the edges of one reflector (Figure 5.2). Strips were deposited by using the electron beam evaporation method on the L-560 Leybold AG apparatus at the 0.1 nm/s deposition rate (Figure 5.3). The distance between Ti target and quartz reflector positions was selected experimentally, on the basis of the condition that provided uniformity in strip thickness along its length. The control of this parameter was realized by testing of coatings on SiO$_2$/Si diagnostical plates with $10 \times 10 \times$

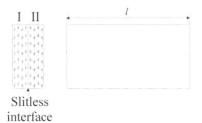

FIGURE 5.1 A schematic picture of a slitless collimator. I and II are plane-polished dielectric plates.

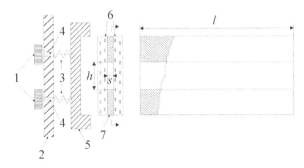

FIGURE 5.2 A schematic design of a waveguide with air core was used to study X-ray beam propagation through a narrow extended slit. 1, Handles for waveguide aligning; 2, installation plate; 3, springs; 4, fine adjustment screws; 5, waveguide holder; 6, quartz reflectors; and 7, Ti strips.

0.1 mm^3 sizes that have been disposed along Al sheet that covered the waveguide channel on a reflector surface. During the deposition procedure, ten diagnostical plates were used in the process of metal film deposition together with the reflector. The thickness control was done by the method of Rutherford backscattering (RBS) of He^+ ion ($E_0 = 1$ MeV) by using the Sokol-3 ion beam investigation complex built on the basis of the ESA-2 Van de Graaff generator [42]. The wide peak is responsible for ion scattering on the film material and the step is connected with the scattering on a substrate. A sample of such measurements is presented in Figure 5.4. The results of the plate set study have shown that the thickness variation along the reflector length ($l = 100$ mm) did not exceed 1% when the distance between Ti target and the quartz reflector was 1,000 mm. The strip thickness defined the waveguide air slit width. The height of the slit was selected to be equal to 10 mm, similar to the height of an X-ray tube focal area. A set of waveguides with slit width that varied in the size interval from 30 nm up to 120 micrometers was studied in the process of the first investigation step.

Reflector pairs were housed into the duralumin holder and were placed on a horizontal three-circle diffractometer HZG-4 (Karl Zeiss Jena Production) instead of a cut-slit system for the X-ray beam formation similar to the setup used in previous works [35,36]. Reflectors were uniformly pressed on one another by means of a holder cover. Four microscrews located near waveguide reflector corners enabled the operator to align waveguides in the diffractometer arrangement. They provide one translational

and two angular degrees of freedom for the holder. The microscrews were supplied by the springs to fix the waveguide position chosen. Figure 5.2 shows a schematic design of waveguides used to study X-ray propagation through a narrow extended slit. Experimental exploitation of such construction devices showed that its possibilities are enough for an accurate aligning of waveguides in the arrangement of a digital diffractometer.

Great attention has been given to the determination of an effective slit width in a slitless collimator, because the data presented in early works [28–30] was not submitted for precision analysis. The magnitude of evaluation of the slit width was obtained using a slitless collimator in comparison with the RBS testing waveguides on the basis of the method of attenuated total internal reflection (ATIR) [43,44]. The small-power laser diode has been used as a light beam source ($\lambda = 680$ nm). The principle measurement scheme is shown in Figure 5.5, and the real measurement procedure was shown next. We squeezed the waveguide with a black light absorber using a special press cartridge that provided a press effort similar to the one effected in the waveguide holder in view of the cover compress. Thereupon, the quartz prism was glued on the outside waveguide surface using a specific oil ($n = 1.45$), and the light source and detector were hardly connected with the prism. The prism device can be moved along the waveguide lightly. The conventional planar photodiode equipped using a $\varnothing = 0.5$ circular aperture was used as a detector. In measurements, the light beam underwent ATIR on the lower surface of an upper reflector, and the reflection intensity was recorded using a luxmeter. The light yield normalization was carried out at a slit width $s = 0.12$ mm. The measurement principles of the procedure were described in detail in [45] and was applied, in first, in [46]. The measurement data obtained for several waveguides with different slit widths are presented in Figure 5.6.

The approximation of these data by the method of least squares allows to build a relationship between the waveguide slit width and the factor of light beam reflectivity. On the basis of the relationship, we have evaluated the air gap width of a slitless collimator. As it has turned out, the reflectivity magnitude varied in area and corresponded to a gap width change in the $0 \div 60$ nm size interval at the prism transference along the slitless unit. By this means, as a first approximation, the slitless collimator effective gap width may be thought as half of the interval extent ($s = 30 \pm 30$ nm). Great variation of the distance between

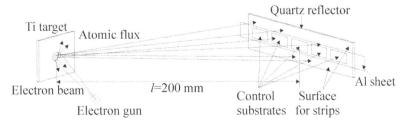

FIGURE 5.3 A principle scheme for the deposition of Ti strips on waveguide reflector edges and small targets for thickness control of the strips.

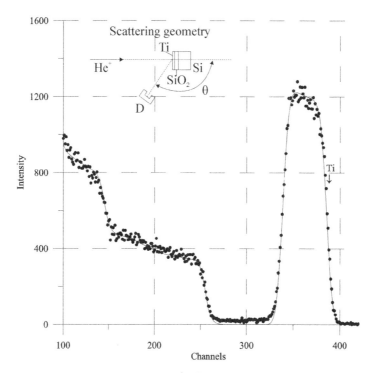

FIGURE 5.4 Experimental and theoretical RBS spectra of He$^+$ ($E_0 = 0.9$ MeV) for one of the control target that has been placed on Al sheet at the deposition of Ti strips with a thickness of 105 nm. Channel price 1.9 keV/channel.

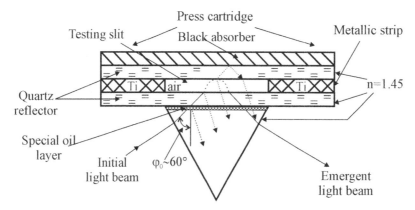

FIGURE 5.5 A principle scheme for direct measurement of a waveguide slit width by ATIR. I and II are quartz reflectors of a waveguide. The scheme was first published in Ref. [36].

reflectors surfaces in the slit collimator for short local positions in the gap we connect with an appearing in during of the reflector polishing process the local stressed states on it's surface layers. It is interesting that the prism translation along quartz waveguide units is accompanied by the reflective intensity variation corresponding to the statistical derivation.

5.3 Experimental Arrangement for the Study of Radiation Intensity Distribution

The main device to study the spatial intensity distribution of X-ray radiation was the HZG-4 diffractometer modified by the increase of detector circle radius up to 500 mm. This modification allowed to improve the spatial resolution of our measurements. Spectroscopic circuit was completed using the Nuclear Instruments Modules (NIM) standard modules of an Ortec firm [47]. The amplifier shaping time was selected to be equal to 0.5 μs. As a result, the registration system could collect information at the count rate up to 100 kHz. Figure 5.7a shows the design of our apparatus setup. The diffractometer permits to carry out a step-by-step scanning with a minimum step $\delta(2\theta) = 0.001°$. The width and height of a detector slit were 0.1 and 10 mm, respectively, for all measurements. A slit-cut forming system and a detector unit of the diffractometer were equipped with Soller slits, providing an angular divergence in a vertical direction less than 2°. A tube BSW-24 with immobile Cu anode was used

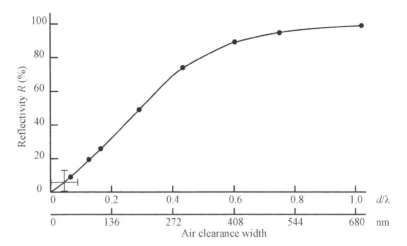

FIGURE 5.6 Experimental data presented the relationship between the waveguide slit width and the reflectivity magnitude obtained by ATIR method. The point characterized that the slitless collimator has a specific design.

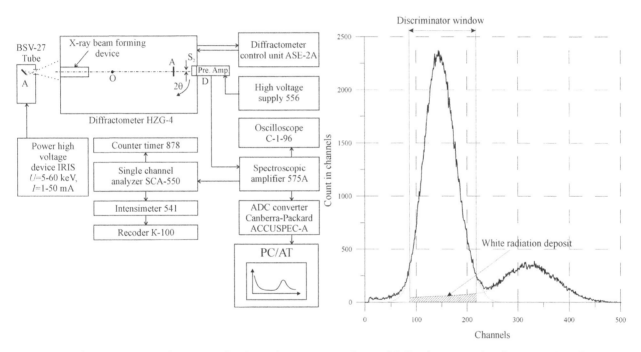

FIGURE 5.7 An arrangement of apparatus for X-ray characteristic radiation ($CuK_{\alpha\beta}$) intensity distribution study in beams formed by PXW and conventional slit-cut systems (a) and pattern of the energy distribution in an X-ray beam generated by BSW-24 (Cu) radiation source that functioned at $U = 20$ keV, $I = 10$ mA (b). The distribution was collected using a scintillator detector equipped with an Al attenuator.

as an X-ray source. (In some experiments, tubes with Fe, Mo, and Ag anodes were exploited.) The anode spot was 1×10 mm^2. The takeoff angle was near $6°$. The first experiments were carried out with Cu anode at a constant tube regime $U = 20$ keV, $I = 10$ mA. The data collection for radiation intensity distribution was performed with a Cu film attenuator (A) of $K_a = 200$ attenuation factor for $CuK_{\alpha\beta}$ X-rays. Energy spectrum for the waveguide emergent beam was also measured using a scintillation detector that is presented in Figure 5.7b. A single-channel analyzer allowed to cut out a $CuK_{\alpha\beta}$ characteristic line with an X-ray white component deposit on levels smaller than 5%. This fact surely allowed that the registered intensity of radiation beams reflects

the properties of its characteristic parts. The measurement setup was equipped with a multilayer analyzer and an Analogue Digital Converter (ADC)-board ACCUSPEC Canberra-Packard used for spectral measurements.

5.4 Angular Radiation Intensity Distribution in X-Ray Beams

Figure 5.8 shows (on top position) a scheme of an experimental arrangement to study the spatial intensity distribution of X-ray radiation in the beams formed by PXWs. The waveguide position in the scheme was fixed

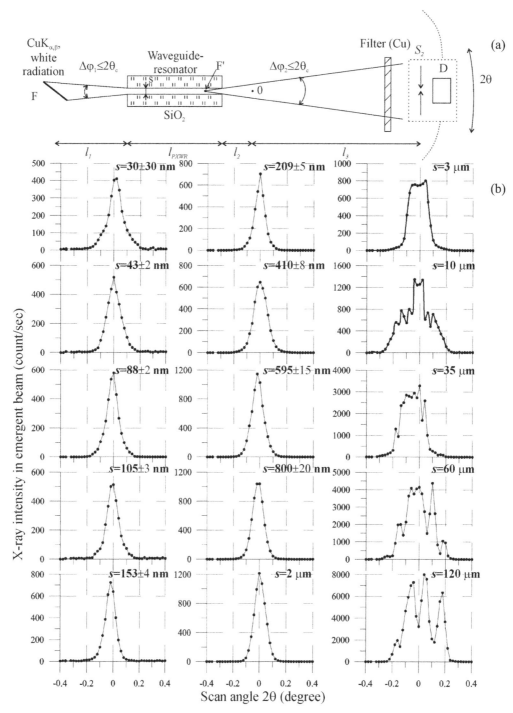

FIGURE 5.8 A scheme of an experimental layout for the measurement of an X-ray characteristic radiation ($CuK_{\alpha\beta}$) intensity distribution in the beams behind waveguides. (a) $l_1 = 75$ mm, $l_{PXW} = 100$ mm, $l_2 = 60$ mm, $l_3 = 400$ mm, $S_2 = 0.1$ mm, A-attenuator (factor for $CuK_{\alpha\beta}$ $K = 200$). X-ray characteristic intensity distributions collected at the tube (Cu) regime $U = 20$ keV, $I = 10$ mA (b) in beams behind PXWs with different slit widths.

during all measurement procedures. The distance between the detector slit and the waveguide outlet was equal to 460 mm, and the distance between the tube focal position and the waveguide inlet is 75 mm. Because of the width of an X-ray tube, the focus projection was near 0.1 mm, and the waveguide radiation capture angle was approximately equal to 0.08°. The angular acceptance of a scintillation detector slit was equal to 0.01°, and the measurement angular step was equal to 0.02°. Because of this, the total intensities of peaks shown in the bottom position of Figure 5.8 are the total sum of the peak channel's content multiplied by the normalization factor ($K = 2$). During experiments, the single-channel analyzer transmitted only pulses connected with a characteristic Cu radiation. Experimental data were collected for 10 s and then accompanied by normalization to improve the spectrum's statistics. The waveguides, i.e., the experimental distributions of which is shown in Figure 5.8, have slit sizes of $s = 43, 88, 105, 153, 209, 410, 600, 800,$ 2,000, 3,000, 10,000, 35,000, 60,000, and 120,000 nm. Moreover, the distribution behind the slitless collimator is also presented.

An intensity distribution outline for X-ray beams formed by waveguides with slit sizes $0 \leq s \leq 2$ mkm does not undergo serious transformations. These distributions are characterized by an almost Gauss profile with some Full Width at Half Maximum (FWHM) variations. The variations are likely connected with a technological inaccuracy appearing at surface polishing. Its major sources are the surface waviness and local surface microstresses. FWHM of these distributions for the slit size interval is approximately

equal to the waveguide capture angle. At slit size, increase in the peak outline is taking a new configuration, and the profile contains a set of lines. The calculation of the peak total intensity allows to observe additional peculiarities. The experimental dependence of the X-ray beam total intensity on the slit width shown in Figure 5.9 can be presented by three typical intervals: $s \leq 200$ nm, $200 \leq s \leq 3,000$ nm, and $s \geq 3$ mkm. For comparison, the figure presents the functions of distribution of integral intensities collected for beams formed by a single slit-cut unit and a double slit-cut system.

It is possible to offer that the emergent beam involves some independent deposits. In use, the conception of geometric optics can assume that one of them—the direct beam generated by X-ray tube—can propagate through a waveguide slit without interactions with a reflector's surface

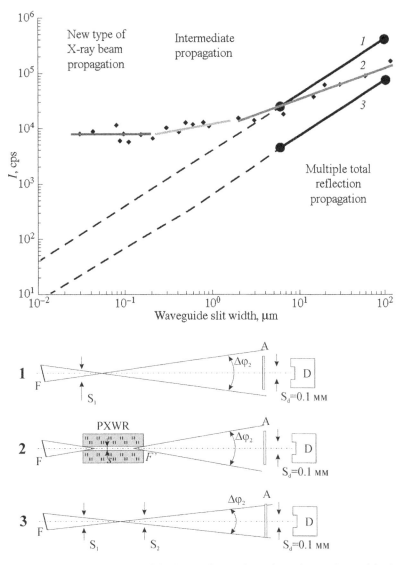

FIGURE 5.9 The scheme for comparison formation of $CuK_{\alpha\beta}$ radiation beam by a slit-cut device (1), planar extended slit clearance (2), and double slit-cut former system (3), presented on the upper part (a), and experimental dependence of CuK_{α} radiation beam intensity on slit width of the set formers. Functions of (1) and (3) types are characterized by a linear dependence and are explained by free propagation of the direct beam. Function of (2) type demonstrates three ranges of inequivalent mechanisms of radiation propagation.

at any slit width. At the ideal arrangement of an X-ray optical system, the direct beam deposit must form one's own partial peak at the zero detector position and shows a linear dependence of the distribution of integral intensity on the slit width. Moreover, the intensity magnitude at zero slit size must aspire to zero. However, the experimental data presented in Figure 5.9 show that the total intensity of an X-ray beam formed by a planar waveguide is approximately constant within the nanosize slit width region. Evaluations show that the direct beam intensity for this slit size interval can constitute a fraction smaller than 1% from the experimental value of the total intensity of beams. The experimental magnitude of PXW emergent beam in the intensity intermediate region begins to increase monotonously with the increase in slit size. Owing to this, the direct radiation beam can claim on an appreciable deposit in the magnitude of summary X-ray beam intensity for this slit width interval. When a slit size exceeds the value $s \approx 3$ mkm, the emergent beam intensity begins to grow sharply with an increase in slit width. Specific contributions appears in the output profile in addition to the direct beam. Additional independent deposits are connected with the appearance of total X-ray beam reflections on both reflectors of waveguides characterized by a wide slit. The macroslit size interval is characterized by independent increases of direct and total reflection beams at the slit widening, until its magnitude achieves the size of a tube anode focus projection.

The results can be interpreted using the following method. PXW with a small slit size is distinguished by the resonance (or quasi-resonance) manner of an X-ray radiation transportation, and such waveguides are called as PXW resonators (PXWR) [36]. The radiation beam formed by PXWR constitutes an indivisible ensemble irrespective of its slit width. The mode structure of such ensemble was actively studied [48,49], and according to our ultimate opinion, the resonance propagation of an X-ray flux is characterized by a continuous (or quasi-continuous) mode structure [50]. The waveguide-resonance mechanism of an X-ray radiation transportation by narrow extended slit effects a maximum of X-ray radiation density and minimizes irreversible losses. The input angular aperture of PXWR is the same as the output one, and they cannot exceed double the magnitude of the critical angle of total reflection [49].

The intermediate interval of the slit width is described by two independent deposits (direct and quasi-resonance propagating beams). Input and output angular apertures continue to remain alike. The interval shows a single-component form of the radiation distribution in an emergent beam and the absence of individual deposits related to the total X-ray reflection phenomenon (Figure 5.8b). These deposits appear in area $s \geq 3$ mkm, and the output angular aperture achieves a maximum angular magnitude $\Delta\theta = 2\theta_c$ independent of the input aperture value. The intensity distribution in the emergent beam created by PXW that belonged to the multiple total reflection region shows a multipeak composition. Since the anode of our X-ray tube had two separated active zones with sizes near 0.05 mm divided by a nonemitting area, the

peak component quantity was equal to six. These components are connected with two direct beam's deposits and four peaks associated with the total external reflection on PXW reflectors. This slit size area is characterized by free passing and the multiple-consequence total reflection mechanisms of X-ray propagation through a slit.

The efficiency evaluation of PXW was based on the knowledge of characteristics of the radiation source used in the experiment. Furthermore, we must have the possibility to compare the properties of beams formed by conventional formation systems and planar waveguides. X-ray beam forming systems based on the single and double slit-cut application are best suited to the competitive experiments. Moreover, the beam forming system equipped by two slit-cut units situated one after another on the distance equivalent to the waveguide length allows one to analyze by direct way the intensity of X-ray flux propagated through the waveguide by the free flow (direct beam).

To display a high efficiency for waveguide-resonators, we carried out a direct experimental comparison of characteristic X-ray radiation density magnitudes in beams formed by PXW and a conventional single-slit forming unit. Summary of the characteristic intensities for the distributions behind slit-cut unit are significantly higher in comparison with the ones for the distribution corresponding to waveguides forming beams at the same X-ray source regime. But the waveguide beam formation wins in a radiation density parameter. Figure 5.10 shows that the beam radiation density in beams formed by PXW is considerably greater than the radiation density measured and expected in beams featured for slit-cut systems emergent beams. It is connected with the capture of X-ray radiation by PXW in the angle interval $\Delta\varphi \leq 2\theta_c$ into the narrow waveguide slit. The direct comparison shows that the radiation density for beams corresponding to first area (Figure 5.10) is approximately 1,000 times higher than the expected one in beams formed by slit-cut conventional systems. The high radiation density in the beam formed by PXWR is connected with a great size of X-ray source projection in comparison with the width of a waveguide-resonator slit. It is significant that the maximum of X-ray radiation density must correspond to the use of a slitless collimator, but it is characterized by additional absorption on surface imperfections and great instability connected with temperature, humidity, and vibration.

It is very important to experimentally investigate some features of PXWRs. Specifically, on the basis of symmetry ideas, one can assume that the divergence of the waveguide-resonator emergent beam must be equal to the radiation capture angle for any PXWR. This assumption can be confirmed or refused, resulting in simple experimental execution. The magnitude of a waveguide capture angle cannot be precisely defined in direct manner, since the width of X-ray tube focal projection is the variable parameter that depends on exploitation conditions. But we can build a parametric relation that connects the distance between the tube focal projection and the waveguide inlet position (l_1 on

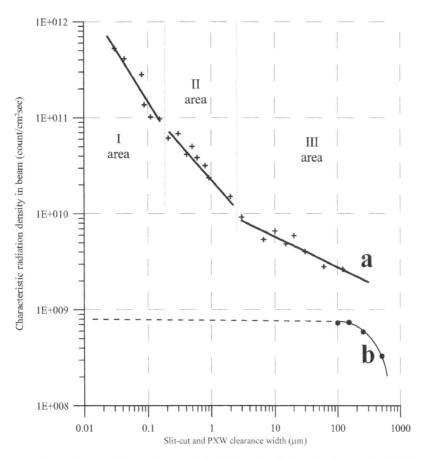

FIGURE 5.10 Experimental dependences of X-ray characteristic ($CuK_{\alpha\beta}$) radiation density on slit width for the waveguide emergent beam (a) and for the beam formed by a single slit-cut system (b). Radiation density values are reduced to the space position corresponding to the waveguide outlet place (for the distance $l = 175$ mm from the tube's focus).

Figure 5.8) and the magnitude of the radiation capture angle and then compare it with the experimental data collected for the X-ray radiation distribution in the waveguide emergent beam obtained for a set of different distances l_1. Similar experimental distribution for FeK_α characteristics radiation and comparison of experimental data about the divergence magnitudes obtained from these distributions with parametrical calculated curves are shown in Figure 5.11. The comparison shows that the experimental points take up positions near the parametric curve corresponding to the tube focal projection width $F = 0.13$ mm. The positioning of experimental values near the definite parametric curve is the validation of suggestion about the divergence angle equivalence for X-ray flux on the inlet and outlet of PXWR.

It is very important for practice to receive a qualitative evaluation about the energy spectrum modification in X-ray flux after it passes through a narrow slit of a waveguide-resonator. This task can be reduced to compare the X-ray spectra of the initial flux and the beam formed by PXWR and solved using the perfect monocrystal as a crystal analyzer. We executed these measurements using the diffractometer and SiO_2 (101) monocrystal. Diffraction patterns presented the energy distribution in spectrum created by an X-ray source, and one of the waveguide-resonator emergent beam is shown in Figure 5.12. The initial spectrum

(Figure 5.12a) is characterized by the presence of a set of characteristics lines (FeK_α, FeK_β, CoK_α, CoK_β, CuK_α, CuK_β) and an appreciable deposit of white radiation. The X-ray flux passing through a waveguide-resonator does not lead to a significant modification of the energy spectrum. The normalization of these spectra was carried out by equalizing the main characteristic line intensities (FeK_α and FeK_β). This equalizing was achieved as a result of source power variation. The source power was three times higher for the collection of energy spectrum characterized for the beam after PXWR.

Spectra presented in Figure 5.12 demonstrate clearly that all components of the initial beam spectrum pass through the waveguide-resonator slit. Intensities of characteristic lines presented in the spectrum are almost unchanged after its transportation by PXWR. However, the intensity of the white component falls approximately two times in all its spectral range. So, a PXWR cannot be considered as a restrictive filter for the hard white radiation that was suggested earlier [36]. It is connected with the fact that the coherence length of white tube radiation in accordance with Kramer–Wentzel approach [51] is not determinated. Our experiments showed that the decrease in slit clearance width leads to restriction of the white component intensity in all energy ranges of the spectrum.

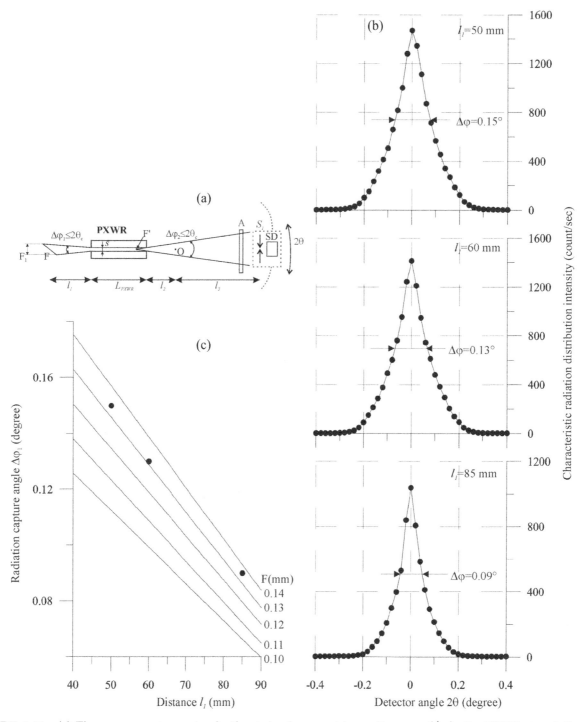

FIGURE 5.11 (a) The measurement geometry for the study of emergent beam divergence ($\Delta\varphi$) after PXWR at variation of the distance between X-ray tube focal spot position and one of PXWR inlet (l_1). $L_{\mathrm{PXWR}} = 100$ mm, $s = 93$ nm, $l_3 = 235$ mm, $S_1 = 0.1$ mm. (b) The X-ray characteristic (FeK$_\alpha$) radiation distribution in beams formed by PXWR at different distances l_1. (c) A set of parametric curves evaluated the size of the X-ray tube focal spot. Curves are plotted for different magnitudes of the focal spot size. Evaluation value is near 0.13 mm.

In addition to our investigations of the CuK$_{\alpha\beta}$ radiation flux transportation by planar extended slit clearance in conditions of its width variation, we carried out the systematic investigation of the MoK$_{\alpha\beta}$ radiation beam intensity dependence on the width of the extended slit clearance formed by quartz reflectors in the wide range of its variation.

These investigations were executed by using the same HZG-4 goniometer fabricated by Carl Zeiss Jena firm. The construction of X-ray extended slit clearance used for these investigations was the same as we applied in the work with CuK$_{\alpha\beta}$ radiation flux. All measurements were executed by using a BSW-24 (Mo) X-ray source in regime $U = 25$ keV,

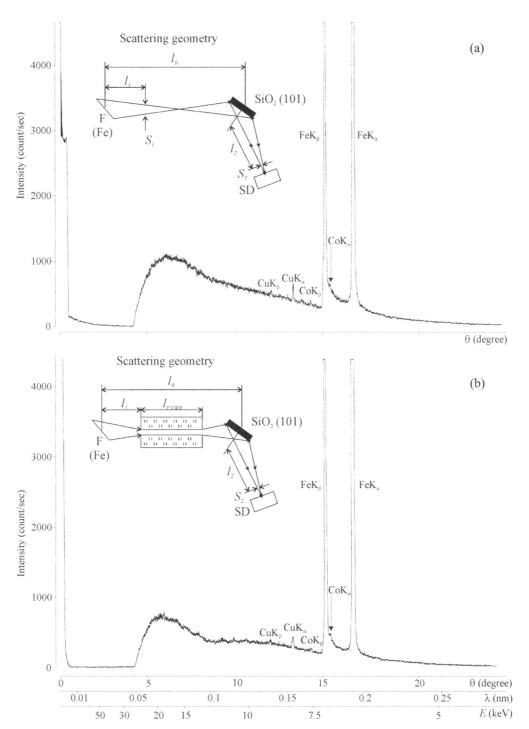

FIGURE 5.12 Diffractometry patterns of SiO_2 (101) monocrystal obtained by conditions of a conventional Bragg–Brentano geometry (a) and of a waveguide-resonator used for the incident beam formation (b). The normalization conditions for these patterns were obtained on the basis of the equivalence of characteristical line intensities. Pattern (a) was registered at a source regime $U = 25$ keV, $I = 3$ mA, pattern (b) $U = 25$ keV and $I = 9$ mA. $l_0 = 235$ mm, $l_2 = 235$ mm, $l_1 = 50$ mm, $S_1 = S_2 = 0.1$ mm; (b) $l_0 = 235$ mm, $l_2 = 155$ mm, $l_1 = 50$ mm, $l_3 = 85$ mm, $S_2 = 0.1$ mm.

$I = 10$ mA. $MoK_{\alpha\beta}$ line in the emergent energy spectrum was selected by SCA-550-Ortec single channel analyzer. The X-ray white deposit in the collected intensity did not exceed 7%.

Figure 5.13 shows the experimental scheme for the study of $MoK_{\alpha\beta}$ radiation flux parameters by application of the X-ray guide device presented in Figure 5.2 and spatial intensity distributions collected in conditions of clearance width variation. Unlike analogical data obtained for the $CuK_{\alpha\beta}$ radiation beam propagation through planar extended slit clearance, the spatial intensity distributions for $MoK_{\alpha\beta}$ beams formed by the slit clearance device

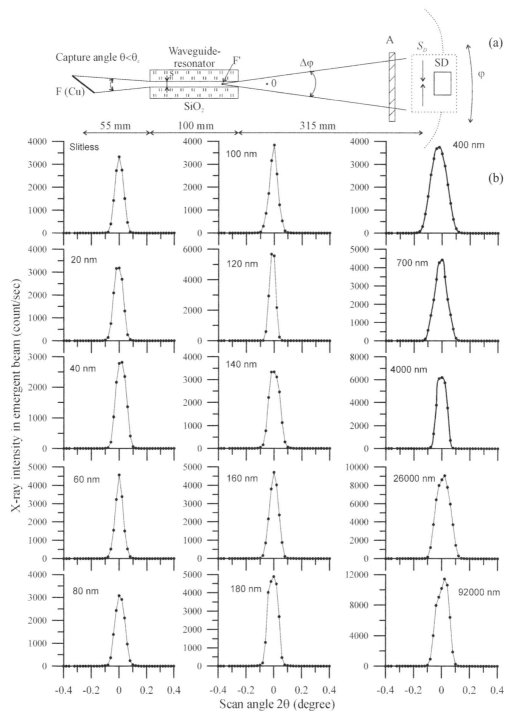

FIGURE 5.13 An experimental scheme for the spatial intensity distribution of MoK$_{\alpha\beta}$ radiation beam formed by the device showed in Figure 5.10 the conditions of the slit clearance width variation (a) and diagrams of MoK$_{\alpha\beta}$ intensity distributions registered by the experimental setup (b). Data received at work of BSW-24 (Mo) X-ray source $U = 25$ keV, $I = 10$ mA. Attenuation factor of A filter $K = 700$.

do not demonstrate noticeable changes at the slit width variation. Experimental data show that the MoK$_{\alpha\beta}$ beam spatial intensity distribution envelopes are close to a Gauss outline in all slit clearance size ranges. It is likely that it is connected with using a new X-ray source, which has a standard solitary focal spot with thickness near 0.1 mm. The integral intensity of MoK$_{\alpha\beta}$ radiation beams formed

by the slit clearance device in the width magnitude range of 0–110 nm remained constant. This result is beautifully illustrated in Figure 5.14. It presents the total experimental results about the integral intensity dependence of MoK$_{\alpha\beta}$ radiation beam on the width slit clearance magnitude. This experimental dependence can be presented in a similar manner as for Cu radiation by three typical

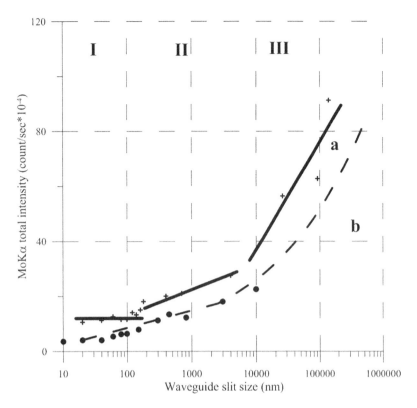

FIGURE 5.14 Experimental dependence of the MoK$_{\alpha\beta}$ radiation flux total intensity on the slit clearance width formed by a device built on a scheme showed in Figure 5.2. First dimension interval can be interpreted as answering to the waveguide-resonance mechanism of MoK$_{\alpha\beta}$ flux propagation. For comparison, a similar dependence obtained for XWR with polycrystalline reflectors is presented (b).

intervals: $S \leq 110$ nm (a), $110 \leq S \leq 3,000$ nm (b), and $S \geq 3$ μm (c). Similar to the experimental data interpretation obtained for the CuK$_{\alpha\beta}$ radiation, it may be concluded that the first size interval permits the waveguide-resonance propagation of the MoK$_{\alpha\beta}$ radiation beam. The third size interval is characterized by multiple total reflection mechanisms of the molybdenum-characteristic radiation beam propagation. The second size interval is connected with the change of X-ray beam propagation mechanism. The slit clearance structure corresponding to the first size interval was designated as the PXWR. For comparison, the Figure presents the dependence of X-ray integral intensity formed by an X-ray Waveguide-Resonator (XWR) built on the basis of the polycrystalline reflectors on the slit clearance width. It is clear that the polycrystalline reflectors do not allow the waveguide-resonance propagation of X-ray quasi-monochromatic fluxes.

Figure 5.15 shows the experimental dependence of the radiation density parameter of MoK$_{\alpha\beta}$ radiation beams formed by the slit clearance device (a) and the double slit-cut system (b) on the slit width magnitude. Comparison of these dependences as similar as Cu radiation shows that the beam radiation density formed by the waveguide device is very high in comparison with the one after the slit-cut system. The experimental values presented in the Figure correspond to the position near the outlet of these beam formers. The radiation density parameter magnitude of MoK$_{\alpha\beta}$ radiation beam formed by the waveguide-resonance unit is

approximately 10,000 times higher than the expected one in the X-ray beam formed by the conventional slit-cut system.

Our systematic experimental investigations showed that the waveguide-resonance effect can be realized in case of reflector application with amorphous and monocrystalline structures only. In this plane, it is interesting to present the results published by our colleagues from Japan [52].

This work devoted to transport properties study of the angular structure shown Figure 5.16. This facility was designed for the investigation of the characteristic MoK$_\alpha$ flux intensity dependence on the magnitude of the taper angle between two Si monocrystalline reflectors. Experimental results demonstrated by the Figure is characterized by the diffuse extremum near the value $\theta \approx 0.1°$ and the second extremum near $\theta \approx 0.007°$. First extremum corresponds to reaching of the total external reflection critical angle for MoK$_\alpha$ radiation on the Si surface. Second extremum cannot be explained without the waveguide-resonance idea attraction. The growth of X-ray radiation transport efficiency connected with this maximum reflects the transformation effect from the multiple total reflection propagation featured for great angular magnitudes to the mechanism of waveguide-resonance flux stream. The emergent beam intensity increasing characteristic for this angular range with the reduction of X-ray flux attenuation is featured for the waveguide-resonance propagation mechanism. The intensity of the second extremum is half of the first one. At the same time, it is clear that the width of the X-ray

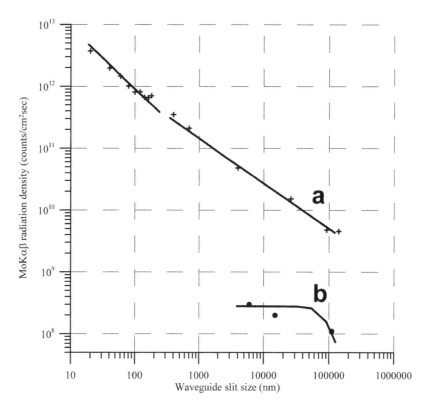

FIGURE 5.15 Experimental dependence of the $MoK_{\alpha\beta}$ flux radiation density on the slit clearance width for emergent beam formed by the device showed in Figure 5.2 (a) and a similar dependence for the X-ray beam formed by double slit-cut system (b).

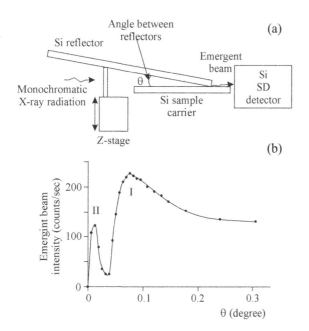

FIGURE 5.16 The experimental scheme for the MoK_{α} radiation transportation efficiency study by the specific angular structure use (a) and the diagram of the structure emergent beam intensity dependence on the taper angle magnitude (b) [52].

beam corresponding to the second maximum is smaller and more than one order. As a result, the X-ray beam formed in conditions of small angles between reflectors will be characterized by an enhanced value of the radiation density.

5.5   X-Ray Beam Reflection on a Material Interface

The problem of X-ray interaction with a real material occupies a specific area in the electrodynamics of continuous media owing to the wavelength of this radiation, which is similar to the interatomic distance [53]. This is the reason why a conventional macroscopic formalism of the material structure as the continuous medium with average values of parameters describing the properties is no more valid in principle. The interaction of X-rays with a vacuum and material interface must be based on the photon scattering effects on individual charged particles, i.e., electrons distributed in the material.

Let the electric field of X-ray flux that fall on the interface has a harmonic form:

$$\vec{E} = \vec{E}_0 e^{i\omega t} \tag{5.1}$$

where $\vec{E}_0$ is the vector amplitude and ω the radiation frequency. Each of the electrons in the material feels this electric field and is compelled to induce oscillations with respect to the electromagnetic field frequency near its own equilibrium position. One can assume that the electrons are fastened elastically to the positions and obey the following equation of motion:

$$m\frac{d^2x}{dt^2} + m\gamma\frac{dx}{dt} + m\omega_0^2 x = -e\vec{E} - \frac{2}{3}\frac{e^2}{c^3}\frac{d^3x}{dt^3}, \tag{5.2}$$

where x is the electron displacement from the equilibrium position, m the electron mass, ω_0 the resonant frequency of the electron, and γ the damping factor. The last term means the influence of electron for the field induced by its own radiation [54]. In a first approximation, this term is negligible owing to its trifle magnitude. It is well known that Eq. (5.2) has the following steady-state solution:

$$\vec{x} = \frac{-e\vec{E}}{m\left(\omega_0^2 - \omega^2 + i\omega\gamma\right)}, \qquad (5.3)$$

if the field of a falling radiation flux obeys the harmonic law. The frequency of a resonant electron oscillation in material is *lower than the frequency* of the X-ray radiation. Thus, the solution of the equation of electron motion in the high frequency field can be expressed in a simplified form:

$$\vec{x} = \frac{e\vec{E}_0}{\omega^2 m}\left(1 + i\gamma\right)e^{i\omega t} \qquad (5.4)$$

taken into account a small magnitude of γ. Under the influence of the field, every electron in the material structure acquires a velocity of motion:

$$\vec{v} = \frac{d\vec{x}}{dt} = \frac{e\vec{E}_0}{m\omega}\left(1 - \gamma\right)e^{i\omega t} \qquad (5.5)$$

The local electron density in the material is $n(x,y,z)$. The electron motion under the high-frequency radiation field leads to a local electron current described by the expression:

$$\vec{j} = en\vec{v} = \frac{e^2 n\vec{E}_0}{m\omega}\left(1 - \gamma\right)e^{i\omega t} \qquad (5.6)$$

This local current creates an internal local field, which is the response to the high-frequency field perturbation of material, and expresses the effect of an X-ray flux scattered by the electron in the crystal subsystem. The real picture of the local *field-in* material excited by *high-frequency radiation* can be described by Maxwell's microscopic equations:

$$\operatorname{rot}\vec{E} = i\frac{\omega}{c}\vec{H}, \qquad (5.7)$$

$$\operatorname{rot}\vec{H} = \frac{i\omega\vec{E}}{c} + \frac{4\pi}{c}\vec{j} = -\frac{i\omega}{c}\left[1 - \frac{4\pi e^2 n\left(x,y,z\right)}{m\omega^2}\left(1 - i\gamma\right)\right]\vec{E}. \qquad (5.8)$$

Because the electrons in material stimulated the forced oscillations in the frequency area between remote and the resonant state, one can expect that the scattering effect will not be greater and the second term in square brackets in Eq. (5.8) will be characterized by a small magnitude.

The expression in brackets is similar to the conventional one for dielectric permittivity ε. But this parameter is a local characteristic of the material media for the specific point with (x,y,z) coordinates and does not express average structural properties. By analogy to the conventional approach, one can introduce the local material refractive index n_l for X-ray radiation:

$$n_l = \sqrt{\varepsilon_l} = 1 - \delta_l - i\beta_l = 1 - \frac{e^2\lambda^2 n\left(x,y,z\right)}{2\pi mc^2}\left(1 - i\gamma\right). \quad (5.9)$$

Experimental measurements showed that the local magnitude calculated for materials does not greatly differ from the values obtained from the experimental data interpreted on the basis of the average media [51]. In all likelihood, it is due to the small magnitudes of the local polarization factor ($\delta \approx 10^{-5}$) and damping parameter ($\gamma \approx 10^{-2}$) [55]. This fact allows us to use the conventional approach developed for the interaction of long-wavelength radiation with material for the case when the radiation wavelength is equal to the interatomic distance. Using the deductions of this approach, however, one must note that, in several conditions, they can lead to an erroneous conclusion.

The polarization factor is approximately equal to the square root from the value at the critical total reflection θ_c [51]:

$$\delta = \frac{\theta_{NA}^2}{2} \qquad (5.10)$$

The interaction of X-rays with material media is, in fact, the total effect of the individual X-ray photon scattered by the electron subsystem in the media. The motion of a photon with energy $E = \hbar\omega$ ($\omega \sim 10^{19}\,\mathrm{s}^{-1}$) in the field with a potential $V(x,y,z)$ can be described by the Schrödinger equation for the vector potential $\vec{A} = \vec{A}_0 \exp\left(i\left(kz - \omega t\right)\right)$ [54]:

$$\nabla^2\vec{A} + \frac{\omega^2}{c^2}\vec{A} = V\vec{A} \qquad (5.11)$$

Let the individual X-ray photon that falls on the vacuum/material interface with the geometrical boundary at $z = 0$ (Figure 5.17) and the potential V is mostly defined by z coordinate. The potential dependence from two other coordinates is connected with the electron density irregularity in the material, which leads to a modulation in the potential. Equation (5.11) can be reduced to an ordinary differential equation except, but a modulation factor, assuming the vacuum potential V_1 is zero as well as the plane scattering

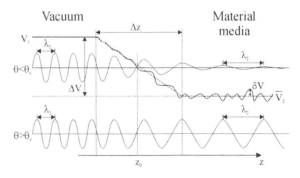

FIGURE 5.17 The behavior of electromagnetic *s*-wave at two angles for incidence on the vacuum–material interface. Vacuum potential is equal 0, the potential of material–$V(x,y,z)$. Dashed line presents the average potential position. z_0 is the interface geometrical boundary, Δz is the area of potential changing. λ_1 and λ_2 are the wavelengths in vacuum and material medium, accordingly. δV is the value of potential modulation by the electron density irregularity.

picture, is applied [56]. The Figure shows that the sharp geometrical interface will be characterized by some near-surface transition layer for the electromagnetic radiation propagation through the interface area. The existence of the transition layer on the material surface was first mentioned in a book by Drude [57] for the explanation of Brewster effect with some peculiarities.

Further, field investigation models and detailed description for the surface transition layer in cases of optical radiation interactions with material [58–60] and X-ray beams are described [61,62]. These investigations show that one can expect on the reflector surface a transition layer with thickness from 0.5 to 2.0 nm. The transition layer existence reckoning is an important factor for radiation total reflection phenomena, but it is unclear how the layer existence influences the results of similar investigations.

5.6 Features of X-Ray Flux Total Reflection Phenomenon

The phenomenon of light flux total internal reflection is common knowledge for the optic science [43,44]. When the light flux is incident on an interface from the medium with high refractive index to the one with smaller index magnitude, one can observe the effect of light flux propagation absence through the interface in some angular range of its incidence. It is the effect of light internal total reflection. Similar effect of X-ray flux reflection on the vacuum/material interface was observed by A.H. Compton [2] and was studied by other scientists [63–65]. Angular range of X-ray external total reflection is not great. For example, the critical total reflection angle θ for MoK_α radiation on the quartz surface is equal to $0.1°$.

Experimental functions of MoK_α flux total external reflections on reflectors building from other materials are shown in Figure 5.18. θ_c magnitude corresponds to the angle of maximum total external reflection half. It is well known that

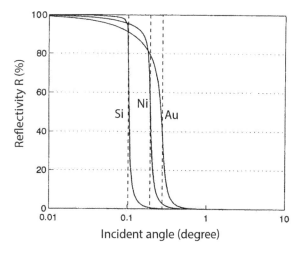

FIGURE 5.18 Reflectivity functions of MoK_α flux for Si, Ni, and Au surface materials.

the limited radiation flux undergoes a specific displacement, which is named as Goos–Hanchen shift [66]:

$$d = \frac{\lambda_0}{\pi} \frac{1}{\sqrt{n_0^2 \cos^2 \theta - n_1}} \qquad (5.12)$$

where n_0 is the vacuum refraction index, n_1 is the refraction index of the reflector material, and θ is the radiation flux incident angle. It is interesting to evaluate the magnitude of this parameter. It is clear that the parameter will attain its maximum at the sliding angle $\theta \sim 0$. The sliding incidence of CuK_α radiation flux on a quartz surface leads to a shift of the reflecting flux of 28 nm. This value is smaller as the CuK_α coherence length L, which defined the longitudinal size of this radiation photons [67]:

$$L = \frac{\lambda_0^2}{\Delta\lambda} \qquad (5.13)$$

where $\Delta\lambda$ is the radiation monochromatism degree. For example, this parameter for CuK_α has a magnitude of 430 nm [38]. In comparison of these parameters, one can expect that incident and reflected fluxes will be able to interfere. And the interference effect in conditions of X-ray flux total external reflection was indeed exhibited experimentally. Bedzyk's group had registered the interference field of an XSW at the total external reflection of quasi-monochromatic radiation flux on the material interface [16]. They suggested a simplest scheme of the interference field arising above a flat material interface. The scheme is presented in Figure 5.19a.

At the same time, the simplest model does not take into account the fundamental principle—the interference field of an XSW cannot abruptly terminate the material—vacuum interface, and, especially, the interface area is the transient zone with the width comparable to the wavelength of the reflected radiation flux. The conventional total external reflection model suggests that the field of the electromagnetic radiation is attenuated in the material reflector volume according to the exponential law [51,68]. The field continuity principle demands a modification of this approach by multiplication of the exponential function with the interference term. The modified model is presented in Figure 5.18b. The standing wave period in the vacuum is described using a conventional expression:

$$D = \frac{\lambda_0}{2 \sin \theta} \cong \frac{\lambda_0}{2\theta} \qquad (5.14)$$

This period in the material differs insignificantly. The expansion of the standing wave interference field in vacuum above the interface is finite. It is known that the interference disappears when the phase shift between the interactive waves achieves the value π [67]. Because of this, the expansion area size for the interference field in the longitudinal direction is restricted by the criterion:

$$\delta \leq 2\pi \frac{\lambda_0^2}{2\Delta\lambda} \qquad (5.15)$$

On the basis of this criterion, one can conclude that this expression presents the upper limit for the longitudinal

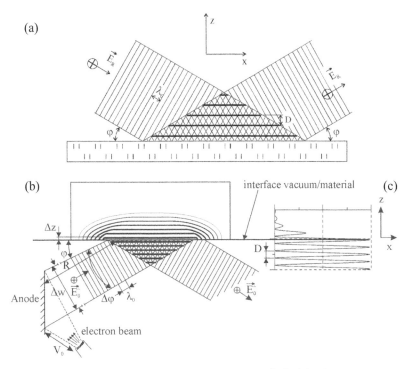

FIGURE 5.19 (a) A simplest scheme of an XSW interference field arising [16]. (b) The improving model of the standing wave interference field arising at external total reflection of X-ray quasi-monochromatic flux generated by the extended radiation source. (c) The space distribution of the interference field over and under interface vacuum/material. Δz, penetration depth [68].

dimension of the interference field. For this purpose, instead of the parameter δ, one can use the concept of path length l [67], which is the distance through which a wave front reduces when the phase achieves a δ value:

$$l = \frac{\lambda_0}{2\pi}\delta = \frac{\lambda_0^2}{2\Delta\lambda} = \frac{L}{2} \qquad (5.16)$$

This distance corresponds to half the coherence length of incidence radiation. The transverse size of the field is defined by the experimental conditions [69]. The quantum–mechanical consideration of the total external reflection phenomenon allows to conclude that the longitudinal dimension and the transverse one are the same [70].

The magnitude of transverse size area can be evaluated on the basis of the geometrical model. For this evaluation, we must calculate the tangent of the incident angle under the condition that the angle is smaller than θ_c, i.e., the critical angle of total reflection of the X-ray flux. We obtained that the calculated value of the cross section of the interference area for quasi-monochromatic radiation of the Fe group of atoms at total external reflection on a quartz surface is approximately equal to 2 nm. This magnitude does not correlate with the experimental values of the diameter of the interference area obtained in [16] and, in further investigations [71], with the magnitude of the coherence length of the reflecting radiation. These experimental results led to attempts for model development based on the concept of radiation spatial coherence [72].

The spatial coherence of radiation includes both the interference of photon itself and the interference of different quasi-monochromatic photons radiated by the same place of the X-ray source [72]. The spatial interference of the quasi-monochromatic radiation emitted by an extended X-ray source can arise if the product of the flux divergence angle ($\Delta\varphi$) and the size projection of source focal spot (Δw) will be smaller than the wavelength of radiation [69]:

$$\Delta\varphi\Delta w < \lambda_0 \qquad (5.17)$$

This condition means that the spatial coherence for the characteristic X-ray flux at the total external reflection phenomenon will be achieved when the wavelength of the radiation exceeds the product of transverse size of the interference area (s) and the width of the source focus projection divided on the distance between the source focus and the flux radiation position (R):

$$\lambda_0 > s\Delta w/R \qquad (5.18)$$

Every local element of the extended X-ray source can be presented as a point emitter characterized by the absence of azimuthal anisotropy. So it is possible to pick out some surface on the irradiating object that correlates with the condition (5.18), which is called "the coherence square" [72] in optics. This parameter, as applied to the interference area, represents a square with sides whose sizes are equal to the transverse size:

$$\sigma_{\text{coh}} = s^2 = \lambda_0 R^2/\Delta w^2 \qquad (5.19)$$

The product of the radiation coherence square and its coherence length characterizes the space coherence volume [69]:

$$V_{\text{coh}} = (R/\Delta w)^2 \left(\lambda_0^4/\Delta\lambda\right) \qquad (5.20)$$

The next step in the idea development can be realized between the coherence volume with a quantum mechanically defines cell of the photon phase space. Let a quasi-monochromatic X-ray flux propagate in the direction (x) normal to the plane of a source spot projection with width Δw. In this case, the photons of different momentum and polarization will be independent in the flux. A similar flux can be described in terms of the "one-photon" phase space [69], and the volume of an elementary cell in the phase space will be presented by the following expression:

$$\Delta p_x \Delta p_y \Delta p_z \Delta q_x \Delta q_y \Delta q_z = h^3 \qquad (5.21)$$

where Δp_i and Δq_i are uncertainties in the components of the momentum and position coordinates of every photon in the flux. On the basis of geometrical considerations, one can calculate

$$\Delta p_y = \Delta p_z = (h/\lambda_0)\,(\Delta w/R) \qquad (5.22)$$

The uncertainty of the photon momentum Δp_x is denoted by its energy uncertainty:

$$\Delta E = \Delta p_x c = h \Delta v_0 = h \Delta \lambda / \lambda_0^2 \qquad (5.23)$$

On substituting Eqs (5.22) and (5.23) into Eq. (5.21), one obtains

$$\Delta q_x \Delta q_y \Delta q_z = \left(\frac{R}{\Delta w}\right)^2 \left(\frac{\lambda_0}{\Delta \lambda}\right) \lambda_0^3 \qquad (5.24)$$

which is seen to be equivalent to Eq. (5.20) for the spatial coherence volume. This accordance allows the introduction of a specific interpretation for the model of an X-ray flux total external reflection as an achievement of conditions of occupation of local interference space by the photon elementary cells of the phase space. Also, it is very important to notice that the space uncertainty $(\Delta q_x \Delta q_y \Delta q_z)$ m in contrast to the momentum uncertainty is not characterized by a preferential direction. The existence of this space indifference allowed authors of Refs [16,71] to register the appearance of interference field in the experiment of the XSW.

Yet, another sample of X-ray flux space coherence observation was described in the work devoted to laser interferometry of soft X-ray radiation [73,74]. The work demonstrated the beautiful interference pattern under conditions of extended X-ray source, using a total reflection mirror. Though the aim of the work was plasma diagnostics by X-ray beam application, its results are an additional convincing proof of our approach to the transverse size evaluation of XSW local interference field.

Executed experimental and model investigations allow to expect that the transverse size and the longitudinal one for the interference area of XSW arising at the total external reflection of X-ray quasi-monochromatic flux on material/vacuum interface in the vacuum space are approximately equivalent. At the same time, the X-ray flux reflection excites the whole volume of the reflector (Figure 5.18). XSW penetrates into the reflector material and undergoes attenuation according to the exponential law [68]. The penetration effect of XSW interference field into the reflector volume is described by the penetration depth parameter, which signifies the interference field intensity decreasing on "e" times. This parameter Δz is defined by the expression [68]:

$$(\Delta z)^2 = \frac{\lambda_0^2}{8\pi^2} \frac{1}{\sqrt{\left(\theta_c^2 - \theta^2\right)^2 + 4\beta^2} + (\theta_c - \theta^2)} \qquad (5.25)$$

Evaluation of this parameter magnitude for X-ray beam interaction with different materials shows that its size is varied in the range of 3–5 nm at the θ scattering angle near zero. This parameter is more with the expected value of the surface transition layer. So, in the first approximation, the total external reflection phenomenon is same as the secular reflection [16,62,64,65].

When an X-ray flux is incident on a material interface, some of the flux is reflected, and the remaining part travels into the material volume, regardless of its angle of incidence. According to Figure 5.19, which models the total external reflection (TER) phenomenon, the incident and reflected fluxes, as well as the flux penetrating into the depth of the material, may be associated with wave vectors $\mathbf{e_0}$, $\mathbf{e_R}$, and $\mathbf{e_T}$, respectively. If it is supposed that these vectors lie in the x–z plane and the z axis corresponds to the normal to the surface, the electric fields of the running incident wave, the reflected wave, and the wave that passed through the interface can be described [64] by the expressions

$$\vec{e}_0\left(\vec{r},t\right) = \vec{E}_0^\perp \exp i\left[\omega t - 2\pi\left(k_{0x}x - k_{0z}z\right)\right]$$

$$\vec{e}_R\left(\vec{r},t\right) = \vec{E}_R^\perp \exp i\left[\omega t - 2\pi\left(k_{1x}x - k_{1z}z\right)\right]$$

$$\vec{e}_T\left(\vec{r},t\right) = \vec{E}_T^\perp \exp i\left[\omega t - 2\pi\left(k_{Tx}x - k_{Tz}z\right)\right] \qquad (5.26)$$

For convenience of consideration, the interface coordinate along the z axis corresponds to zero, and only the σ-polarization is considered, i.e., vectors $\vec{E}_0^\perp$, $\vec{E}_R^\perp$, and $\vec{E}_T^\perp$ perpendicular to the x–z plane. According to the Fresnel approach [64], the complex coefficients of the reflection and refraction amplitudes have the form

$$\frac{E_R^\perp}{E_0^\perp} = \left|\frac{\vec{E}_R^\perp}{\vec{E}_0^\perp}\right| e^{ir} = \frac{\theta - \sqrt{\theta^2 - 2\delta - 2i\beta}}{\theta - \sqrt{\theta^2 + 2\delta - 2i\beta}} e^{ir}$$

$$\frac{E_T^\perp}{E_0^\perp} = \left|\frac{\vec{E}_T^\perp}{\vec{E}_0^\perp}\right| e^{ir} = \frac{2\sqrt{\theta^2 - 2\delta - 2i\beta}}{\theta - \sqrt{\theta^2 + 2\delta - 2i\beta}} e^{ir} \qquad (5.27)$$

Following the authors of [51,68], who introduced the simplifying notation

$$\sqrt{\theta - 2\delta - 2i\beta} = a + ib \qquad (5.28)$$

where parameters a and b have the form

$$a^2 = \frac{1}{2}\left(\sqrt{\left(\theta^2 - 2\delta\right)^2 + 4\beta^2} + \left(\theta^2 - 2\delta\right)\right)$$

$$b^2 = \frac{1}{2}\left(\sqrt{\left(\theta^2 - 2\delta\right)^2 + 4\beta^2} - \left(\theta^2 - 2\delta\right)\right) \qquad (5.29)$$

One can transform expression (5.27) as follows:

$$\frac{\vec{E}_R^{\perp}}{\vec{E}_0^{\perp}} = \frac{(\theta - a) - bi}{(\theta + a) + bi}$$

$$\frac{\vec{E}_T^{\perp}}{\vec{E}_0^{\perp}} = \frac{2(a + bi)}{(\theta + a) + bi} \tag{5.30}$$

Using the factor of smallness of TER angles and transformation procedure performed in [64], one can represent the expression for an electric field passing through the flux interface in the following form:

$$\vec{e}(x, z, t) = \left| \vec{E}_0^{\perp} \right| \frac{2(a + ib)}{(\theta + a) + ib} \exp\left(-\frac{2\pi i Z}{\lambda} (a + bi) \right)$$
$$\times \exp\left(i\left(\omega t - k_x x \right) \right) \tag{5.31}$$

This expression shows that the field decreases along the z axis experimentally, but, at the same time, oscillates along the x axis. Therefore, one should expect that an X-ray photon incoming on the vacuum/material interface at the TER angle will be emitted by the material.

5.7 Model of X-Ray Flux Waveguide-Resonance Propagation

The conventional approach assumes four mechanisms of X-ray flux propagation. There are free fluxes spreading, multiple total reflections [20], the multiple Bragg reflection, and the anomalous X-ray flux transmission or the Bormann effect [3]. For realization of the multiple external total reflection propagation of an X-ray quasi-monochromatic flux, it is a need the extended slit clearance formed by parallel disposition of two flat reflectors on large distance between its (Figure 5.20a). The phenomenon of consecutive multiple total reflections is characterized by the appearance of a set of local interference areas. Since every reflection act of X-ray flux excites reflector volumes, the next second reflection will lead to a recurrent volume excitation. We can find some incident angles for X-ray flux, which will be characterized by the phasing of consecutive reflections on every reflector (magic angles). In conditions of the flux multiple total reflection, the existence or absence of the phasing factor will lead to an essential difference in the magnitude of X-ray flux intensity for the waveguide emergent beam. The phasing effect allows the transport of X-ray flux by the planar double reflector unit with small attenuation owing to minimization of energy losses at the recurrent excitation of reflector volumes. As the result of this, the mechanism of X-ray flux multiple total reflection is characterized by a discrete mode structure.

If the flux radiation incident angle is not correlated with one of the magic angles, an influence of consecutive reflections will not be in phase, and every reflection will be led to overexcitation of the reflector volume. The overexcitation effect is connected with energy losses, and magic flux will not be presented in the waveguide emergent beam in small deposits. In the result, the multiple total reflection mechanism of X-ray fluxes is characterized by a visible attenuation of the resulting flux working with wide angular fluxes. This mechanism was the choice for the explanation of high radiation transport properties of the X-ray

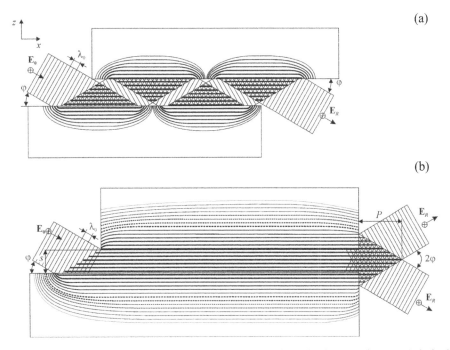

FIGURE 5.20 Schemes of quasi-monochromatic X-ray flux propagation through planar wide extended slit by the multiple total reflection mechanism (a) and its stream through planar narrow extended slit by the waveguide-resonance mechanism (b). P, interference field protrusion.

beam slitless collimator [28–30]. But the similar approach is not able to describe integral intensity dependences on width of the planar extended slit clearance collected for the CuK$_{\alpha\beta}$ flux propagation (Figure 5.9) and similar data obtained for MoK$_{\alpha\beta}$ flux propagation (Figure 5.14). Moreover, the multiple total reflection mechanism cannot present a distinct explanation for the appearance of a supersmall angular peak demonstrated by Figure 5.16. Observing the discrepancies between experimental data and model representations following the multiple total reflection theory allowed us to assume the existence of a specific mechanism for X-ray characteristic radiation fluxes propagation featured for planar extended slit clearances with nanosize width [75]. The specific type of X-ray flux propagation was called as the waveguide-resonance mechanism. The simplest construction device that functioned in the frame of the mechanism was called the PXWR [76].

Model of the multiple total external reflection mechanism is correct when the distance between planar reflectors forming the waveguide device is higher as the transverse size of local interference fields appears in the process of X-ray flux multiple total reflections (Figure 5.20a). But when the distance between reflectors is of smaller magnitude, the process of X-ray flux propagation will be changed. In this case, local interference fields will mix and form the uniform interference field of XSW. It is apparent that a similar transformation of the interference field is possible for quasi-monochromatic (characteristic) X-ray fluxes only. The transverse size of local interference fields defined as the half coherence length of radiation flux transported by wide extended slit clearance is the upper boundary of the slit clearance width, admitting the appearance of a uniform interference field of XSW. Comparison of the magnitudes of copper and molybdenum radiation fluxes with value of nanosize range upper boundaries for experimental dependencies of emergent beam integral intensity on slit clearances width presented in Figures 5.9 and 5.14 showed that these sizes are approximately the same. Similar concurrence gives grounds to maintain that nature allows to propagate radiation fluxes in specific conditions by the mechanism of waveguide-resonance propagation. Standing wave period of the uniform interference field is defined by the expression (5.14). This expression allows to establish the lower size boundary for possibility of the waveguide-resonance mechanism realization. The boundary can be defined from the condition when the period of XSW will be equal to the slit clearance width magnitude. The period achieves its minimum possible value when the incident beam angle reaches a critical total external reflection angle θ_c. In the result, we can conclude that the waveguide-resonance mechanism can be realized when the slit clearance size s is correlated with proportion:

$$\frac{\lambda_0^2}{2\Delta\lambda} > S > \frac{\lambda_0}{2\theta_c} = \frac{1}{8}\sqrt{\frac{\pi mc^2 A}{e^2 N \rho Z}} \qquad (5.32)$$

where c is the light velocity in vacuum, m and e are the mass and charge of electrons, A, Z, and ρ are the atomic mass of reflector material, its atomic number, and its density. The minimum size of waveguide-resonance slit clearance does not depend on transported radiation wavelength and is defined by reflector material properties. In frame of the Fresnel approach, this parameter value for all materials is near 20 nm. Upper parameter value for the copper characteristic radiation for any material is equal to 204 nm (for Mo, 110 nm). At the same time, the waveguide-resonator radiation capture angle is very sensitive to the reflector material density. It is limited by double magnitude of the critical angle of X-ray flux total external reflection for the reflector material. For example, the maximum capture angle for CuK$_{\alpha\beta}$ radiation is equal to 0.42° and for MoK$_{\alpha\beta}$ – 0.21° in case of quartz used as the reflector material. The capture radiation angle can be increased by application of the material with enhanced density. The minimum capture angle $\theta_{\min}$ can be defined from left correlation of the expression:

$$\theta_{\min} = \frac{2\Delta\lambda}{\lambda_0} \qquad (5.33)$$

This parameter does not depend on the reflector material density. For CuK$_{\alpha\beta}$ radiation, it is equal to 0.043° and for MoK$_{\alpha\beta}$ – 0.046°. Existence of the PXWR minimum capture angle means that X-ray flux propagating along the waveguide-resonance axis will not be captured by its extended slit clearance. For similar fluxes capture, there is a need to create some disorientation. As a whole, the waveguide-resonance device function can be presented as the flux radiation capture in some angular range $\Delta\theta$, its transportation by the nanosize extended slit clearance and formation of X-ray emergent beam with $\Delta\theta$ the same angular divergence. It is important to evaluate X-ray flux attenuation as a result of transportation of PXWR nanosize extended slit clearance.

The excitation of a standing wave in a waveguide-resonator slit gives rise to a stationary distribution of the interference field intensity, both along the slit channel (along axis x) and crosswise (along axis z). This distribution pictured in Figure 5.21 is fitted to the input of a PXWR ($x = 0$). The distribution is plotted for a plane X-ray flux (CuK$_\alpha$) impinging into slit clearance of the quartz waveguide-resonator under the angle $\theta = 0.92\cdot\theta_c$ on the reflector surface. The reflection conditions correspond to a value of the phase variation $\psi \cong 45°$ and the standing wavelength period $D \approx 1.1 D_{\min}$. The X-ray incident wave is characterized by the penetration depth $\Delta z = 8.6$ nm. Hence, the distribution shown in Figure 5.21 is consistent with the expression $\Delta z \approx 0.4 D_{\min}$.

The standing wave within PXWR is characterized by the intensity distribution described by expression (5.31):

$$W(x, z, t) = \left| \vec{e}(x, z, t) \right|^2 \qquad (5.34)$$

The total integral intensity of an X-ray beam in the inlet slit area ($x = 0$) can be represented in the form of two terms:

$$W(\theta, x = 0) = L(\theta, x = 0) + M(\theta, x = 0) \qquad (5.35)$$

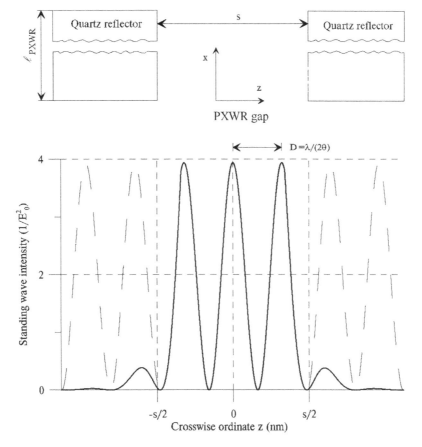

FIGURE 5.21 Intensity distribution function for an XSW in a waveguide-resonator slit clearance and top layers of quartz reflectors for an X-ray beam impinging on the slit under a certain angle of the total external reflection θ for quartz reflectors. The function without absorption upon the total reflection is shown by a dashed line. The function reflects the picture for $\lambda = 0.1541$ nm (CuK$_\alpha$); $\theta = 0.92\ \theta_c$; $s = 97$ nm.

where $L(\theta, x = 0)$ will characterize the total integral intensity portion concentrated in the slit space, and $M(\theta, x = 0)$ is connected with the portion percolating into waveguide reflector's volumes. The functions $L(\theta, x = 0)$ and $M(\theta, x = 0)$ can be represented by the expressions

$$L(\theta, x = 0) = \int_{-\frac{s}{2}}^{+\frac{s}{2}} I(\theta, z)\, dz;$$

$$M(\theta, x = 0) = 2 \int_{\frac{s}{2}}^{\infty} I(\theta, z)\, e^{-\frac{z}{\Delta z}}\, dz \qquad (5.36)$$

where s is the slit width. As the X-ray moves in the slit along the x axis, its total integral intensity $W(\theta, x)$ together with its components $L(\theta, x)$ and $M(\theta, x)$ will demagnify. But the ratios between the components and between every component and the total integral intensity $W(\theta, x)$ must be saved at any x-magnitude because the standing wave interference field produces the equality of boundary conditions between the slit space and the reflector volumes along the whole slit. Consequently, we can write

$$\alpha = \frac{L(\theta, x = 0)}{W(\theta, x = 0)} = \frac{L(\theta, x)}{W(\theta, x)} = \text{const} \qquad (5.37)$$

If the total integral intensity of an X-ray beam for the slit inlet position ($x = 0$) is represented by expression (5.35),

the intensity for the distance $x = \delta x$ from the inlet will be described as

$$W(\theta, x = \Delta x) = L(\theta, x = \Delta x) + M(\theta, x = \Delta x) \qquad (5.38)$$

X-ray beam does not suffer attenuation in the slit space. The X-ray radiation undergoes absorption and scattering inside the reflector's volume only. Hence, the common beam attenuation can be represented in the form:

$$\Delta W(\theta, x = \delta x) = W(\theta, x = 0) - W(\theta, x = \delta x) = \qquad (5.39)$$
$$= M(\theta, x = 0) - M(\theta, x = \delta x)$$
$$= M(\theta, x = 0)\left(1 - e^{-\mu \delta x}\right)$$

Dividing the left and right terms in expression (5.39) on $W(\theta, x = 0)$ and directing $\delta x \to 0$, we can get a simple differential equation:

$$\frac{dW}{W} = -\frac{M(\theta, x = 0)}{W(\theta, x = 0)}\mu dx = -\alpha \mu dx \qquad (5.40)$$

The integration of the equation with the initial condition $W_{\text{total}} = W(\theta, x = 0)$ leads to the inference of the main waveguide-resonator equation:

$$W(x) = W_0 e^{-\alpha \mu x} \qquad (5.41)$$

The magnitude of α depends on the wavelength of incident radiation, the reflector material properties, the angle of radiation impinging, and the width of the waveguide slit. The dependence of α on the incident angle can be evaluated by calculating its values for some incident angles of CuK$_\alpha$ radiation reflected on one quartz waveguide reflectors. This gives the values $\alpha(\theta_c) = 0.8$ and $\alpha(\theta_c/2) = 0.05$. Using formula (5.41) with the values of α for the incident angle $\theta = 0.92\theta_c$ and the condition $s_{\text{slit}} \approx 4D$, we can calculate the total intensity attenuation for an X-ray beam of CuK$_\alpha$ radiation in the planar quartz extensible waveguide-resonator. The total intensities corresponding to an X-ray beam after passing the way $\ell = 100$ mm in the waveguide are $W(\theta_c)=0.3\,W_0$ and $W(\theta_c/2) = 0.6\,W_0$. It is interesting to compare these values with the magnitudes characterized for the model of a multiple total X-ray reflection. Such calculations require the knowledge of the total X-ray beam reflections quantity N in the long slit [20], which can be evaluated on the basis of the expression:

$$N = \frac{\ell}{\Delta x + sc\tan\theta} \qquad (5.42)$$

where ℓ is the waveguide length, Δx is Goos–Hanchen shift Eq. (5.12), and s is the size of the clearance. For the quartz reflectors, length $\ell = 100$ mm, the clearance size $s = 97$ nm, and CuK$_\alpha$ radiation, we can get $N(\theta_c) \approx 200$ and $N(\theta_c/2) = 100$. Using this magnitude, we can calculate the real intensity for the model of the multiple X-ray total reflection: $W(\theta_c) \cong 0.008\,W_0$; $W(\theta_c/2) \approx 0.1\,W_0$. At the sizes characterized for a slitless collimator, the difference between the values calculated for those models is higher.

The PXWR uses the total X-ray reflection phenomenon for the standing wave generation in the slit space of the device. But it is well known that the excitation of an XSW is possible for the Bragg geometry too [77,78]. The XSW can be excited in the narrow slit between two parallel polished reflectors if the reflectors are perfect monocrystals being oriented mutually. In this case, we shall get a new device: PXWR monochromator (PXWRM). The mechanism of the standing wave formation in Bragg reflection conditions (Figure 5.22) somewhat differs from the one in the case of a X-ray total external reflection,

and the standing wave period will be close to the lattice plane distance d_{cryst} satisfied to Bragg conditions. A formal consideration of a standing wave formation in Bragg PXWR requests the replacement of the X-ray depth penetration parameter Δz by the parameter of primary extinction length z_{ext} [79]:

$$z_{\text{ext}} = \frac{1}{\sigma} = \frac{\sin\theta_{\text{b}}}{2\lambda\,|s|}\frac{mc^2}{e^2}\frac{v}{|F_h|} \qquad (5.43)$$

where σ is the extinction factor, s is the polarization factor being equal to unity for a σ-polarization, c is the light velocity, m and e are the electron characteristics, v is the unit cell volume for the reflector material, and F_h is the relative structure factor of the chosen reflection. The z_{ext} magnitude defines the crystal thickness attenuated by the X-ray beam intensity falling on the crystal under Bragg angle with "e" factor. It is significant that the primary extinction length does not depend on the wavelength of the X-ray radiation. The magnitudes of the primary extinction length are usually two orders higher as values of depth penetration are at an X-ray total external reflection.

For example, the magnitude of z_{ext} for (200) NaCl reflection is equal to 660 nm [79]. Because of this, the practical size of a PXWR slit may visibly differ from the one calculated by the procedure adopted for a PXWR, and its efficiency must be lower when compared with a PXWR built on the total reflection phenomenon. However, the X-ray radiation density in its emergent beam will be significantly higher as one for the conventional Bonse–Hart monochromator [80,81]. The PXWR monochromator function is conceptually identical with the Borrmann effect manifested to perfect crystals [3]. The minimum clearance size for the Bragg reflection case can be evaluated on the basis of the expression:

$$D^b_{\text{min}} = \frac{\lambda}{2\sin\theta_b} = d_{\text{cryst}} \qquad (5.44)$$

where θ_b is the Bragg reflection angle. Intensities of incident and emergent X-ray beams will be approximately equal, but the energy width of the line will be different. The line narrowing will be accompanied with the corresponding enhancement of line amplitude (Figure 5.22):

$$\Delta A_i\Delta\lambda_i = A_e\delta\lambda_e \qquad (5.45)$$

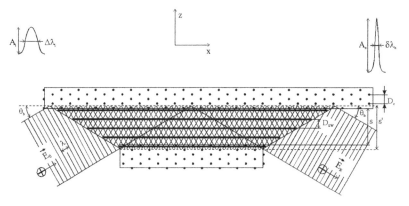

FIGURE 5.22 Principle scheme of a standing wave picture formation in the planar slit of a resonance waveguide-monochromator. D_{SW} (standing wave length) is equal to d_{cryst}. Intensity of an emergent beam is equal to one for an incident beam ($I_i = A_i\Delta\lambda_i = I_e = A_e\delta\lambda_e$).

where A_i and A_e are the amplitudes of the incident and emergent X-ray beams and $\Delta\lambda_i$ and $\delta\lambda_e$ are FWHMs of radiation inn the inlet and outlet of the PXWRM.

Unlike waveguide-resonance devices functioned in the frame of the X-ray flux total external reflection phenomenon, the waveguide-resonator monochromator is not a unit of X-ray nanophotonics.

Nanodimension and small intensity attenuation of X-ray flux in the process of its transportation are not exhaustive advantages of PXWR. These devices are characterized by the ability to increase the flux radiation density in comparison with X-ray former based on slit-cut systems (Figures 5.10 and 5.15). The radiation concentration effect is connected with difference of X-ray sources focuses projection width and width of PXWR slit clearances. Waveguide-resonance devices capture photons flux emitted by X-ray source into the nanosize slit clearance. As a result, the flux radiation density in the slit increases three to four orders.

The paramount factor of the waveguide-resonance mechanism realization for X-ray flux propagation is arising of the XSW uniform interference field. It appeared at the X-ray quasi-monochromatic flux transportation by nanosize extended slit clearance with a constant width in all its length. It is important to evaluate the possible PXWR reflector's nonparallelism admitting the arising uniform interference field.

The scheme showed the formation of XSW in local interference field areas in conditions of some reflectors nonparallelism that are presented in Figure 5.23. The scheme reflects X-ray flux transportation by the wide-slit guide. Angle ψ is the factor of the reflector's nonparallelism. We postulate that $\psi \ll \theta$, where θ is the flux incident angle. Suppose that a planar front of a quasi-monochromatic radiation with the wavelength λ_0 and the monochromatism degree $\Delta\lambda$ falls on the X-ray guide inlet at an angle θ to the surface of its first reflector ($\theta < \theta_c$, where θ_c is the critical angle of total external reflection for the reflector material). The first local area of the interference field will be characterized by the XSW

period D_1 defined by the expression (5.14). XSW period is not a constant for the quasi-monochromatic radiation, and its variation can be described by the following formula that can be received by differentiation of the expression (5.14):

$$\Delta D_1 = \frac{\Delta\lambda}{2\theta} \tag{5.46}$$

Similar expressions can be written for the XSW period (D_2) in the second local area of interference field, which arises near the surface of a second reflector:

$$D_2 = \frac{\lambda_0}{2\sin(\theta + \psi)} \approx \frac{\lambda_0}{2(\theta + \psi)} \tag{5.47}$$

And for the period variation:

$$\Delta D_2 = \frac{\Delta\lambda}{2(\theta + \psi)} \approx \frac{\Delta\lambda}{2\theta} = \Delta D_1 \tag{5.48}$$

A decreasing of slit clearance width can lead to a change of X-ray flux propagation mechanism. But in conditions of existence of some nonparallelism ($\psi \neq 0$), the fact that the local interference areas overlap will not be enough for the mechanism change. In this case, the additional demand appears. There is a need that the standing wave periods in consecutive local areas D_i and D_{i+1} must correlate between it. Consequently, the magnitude of the standing wave period discrepancy ($D_1 - D_2$) can be presented by the expression:

$$(D_1 - D_2) = \frac{\lambda_0}{2}\left(\frac{1}{\theta} - \frac{1}{\theta + \psi}\right) \approx \frac{\lambda_0\psi}{2\theta^2} \tag{5.49}$$

And it must not exceed the dispersion of the period ΔD_i. In such manner, the specific condition for realization of the waveguide-resonance mechanism for X-ray flux propagation through a narrow extended slit at the existence of some reflector's nonparallelism can be written by the equation:

$$\frac{\Delta\lambda}{2\theta} = \frac{\lambda_0\psi}{2\theta^2} \tag{5.50}$$

or

$$\psi = \theta\frac{\Delta\lambda}{\lambda_0} \tag{5.51}$$

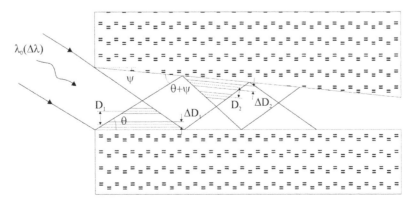

FIGURE 5.23 The scheme showed the formation of XSW local interference field areas in conditions of some nonparallelism of PXWR reflectors. D_1 and D_2 are periods of XSW for two consecutive local areas of the interference field. ΔD_1 and ΔD_2 are expected values for the standing wave periods difference controlled by the radiation monochromatization degree $\Delta\lambda$. ψ, the nonparallelism ($\psi \ll \theta$).

X-ray flux will be transported by PXWR, according to Eq. (5.32), when the incident angle of the flux will be smaller as θ_c will be higher as the minimum capture angle. The quantitative evaluation demands an analysis, including both limiting boundaries defined by Eq. (5.32). If the flux incident angle is near θ_c, Eq. (5.51) takes the form:

$$\psi_1 = \Delta\lambda \sqrt{\frac{e^2 N Z \rho}{2\pi m c^2 A}} \qquad (5.52)$$

For the quasi-monochromatic radiation $FeK_{\alpha 1}$ ($\lambda_0 = 0.1936$ nm, $\Delta\lambda = 0.1 \times 10^{-3}$ nm [38]) and for quartz reflector use, we can receive that the waveguide-resonance mechanism can be realized when $\psi_1 < 1.5 \times 10^{-6}$ rad ($\sim 9 \times 10^{-5}$ degree). If the flux incident angle is near the minimum capture one, the Eq. (5.51) will be transformed into the expression

$$\psi_2 = \frac{1}{2}\left(\frac{\Delta\lambda}{\lambda_0}\right)^2 \qquad (5.53)$$

In this case, we can receive that the quartz PXWR at $FeK_{\alpha 1}$ flux transportation must not be characterized by a higher nonparallelism $\psi_2 = 1.3 \times 10^{-7}$ rad (8×10^{-6} degree). In the result, we can establish that the nonparallelism critical angle depends on the incident angle and the critical angular value is very high. On the basis of this value, we can evaluate the admissible error for reflector coating thickness during its preparation. The maximum thickness discrepancy is expected between the edge and middle points of the coating. Taking into account PXWR length ($l = 100$ mm) and tangent for ψ boundary values, we receive that the calculated value for the critical thickness discrepancy changes from 2,000 to 400 nm. These values are higher than the critical width of the extended slit clearance, allowing realization of the flux waveguide-resonance propagation. So, in first approximation, it is expected that the possible technological errors must not influence the radiation-transportation efficiency of PXWR for the quasi-monochromatic radiation generated by conventional X-ray tubes. Moreover, this exception was supported by direct experimental observation of X-ray flux waveguide-resonance propagation through the small angular structure (Figure 5.16).

5.8 Problem of PXWR Integral Intensity Increasing

On the basis of the results presented in Figure 5.9, one can first confirm that the absolute magnitude of the PXWR emergent beam intensity is smaller than the intensity of beams formed by the other type of devices because the size of its slit clearance is limited by the demands of the waveguide-resonance mechanism realization. So, the first vital task of PXWR modernization is the increase in beam integral intensity in the conservation of main waveguide-resonator advantages. A set of simplest schematic constructions, which can be background for the task solution, is presented in

Figure 5.24. Constructions based on the additional planar total reflection mirrors the application (Figure 5.24a) and use of mirrors with a specific surface form (Figure 5.24b) are most suitable for an increase in intensity of the emergent beam, because similar constructions can be theoretically calculated with a high accuracy. But its experimental realization demands for the specific expensive equipment utilization. Attempts for building similar systems by application of the improvised means demonstrated a small efficiency (near 10%).

Figure 5.24c presents a simplest construction for the increase in emergent beam intensity, which can be created on the basis of improvised means. This construction uses the input-skewed concentrator, which can be prepared by the creation of a planar chamfer on one end of the reflector completing PXWR. Angle and length of the chamfer are chosen on the basis of total reflection critical angle magnitude for the reflector material. Calculation showed that a similar construction of PXWR is able to increase the unit radiation gathering power three to five times. For effect amplification from the radiation-skewed concentrator utilization, we deposited HfO_2 coating on the reflector chamfer planes. The coating application allowed to increase the efficiency of the skewed concentrator.

The efficiency of the radiation-skewed concentrator application was evaluated on the basis of the emergent beams comparison measurements formed by the simplest PXWR and waveguide-resonator equipped by the input-skewed concentrator. Moreover, these investigations were supplemented by a similar study of a beam formed by the double slit-cut system with slit width $s_2 = s_3 = 6$ μm, with the distance between them being equal to 100 mm. Measurement schemes and experimental spatial intensity distributions of beams formed by simple, modified PXWRs and the double slit-cut system are displayed in Figure 5.25. The Figure contains the integral intensity data of the beams obtained with and without the attenuator (A)

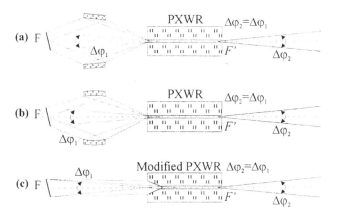

FIGURE 5.24 Schemes of possible constructing solutions for PXWR radiation gathering power increasing built on the basis of the planar total reflection mirrors set (a), the parabolic mirrors set (b), and the input-skewed concentrator (c).

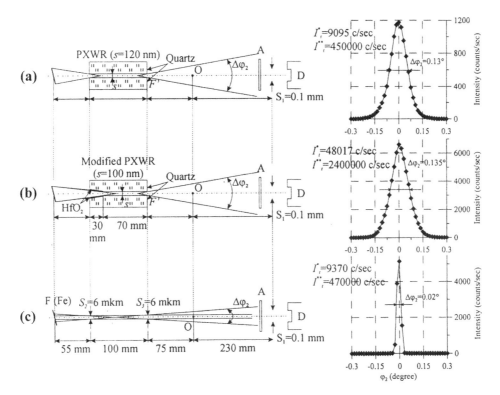

FIGURE 5.25 Measurement arrangements for the study of spatial intensity distribution and the experimental distribution envelopes for FeK$_{\alpha\beta}$ beams formed by a simplest PXWR (a), the waveguide-resonator equipped by the skewed concentrator (b), and the double slit-cut system (c). F' is the imaginary focus formed by PXWRs, I^* and I^{**} are the beams integral intensities registered with and without taking into account the attenuating plate (A).

application. The attenuating factor was approximately equal to 50. All measurements were carried out at the same X-ray source function regime. Slit and clearance heights were 4 mm.

The Figure shows that the envelope form and FWHM of the spatial intensity distributions of beams formed by the simple PXWR and the modified waveguide-resonator are alike. At the same time, the integral intensity of a beam formed by PXWR equipped by the skewed concentrator is higher in comparison with the intensity of a beam on the output of the simple waveguide-resonator. So, we can establish that the input-skewed concentrator that functioned on the basis of the total reflection phenomenon increases the radiation gathering power without worsening the other parameters of PXWR, indeed. In the bottom part of the figure, the spatial intensity distribution of the beam formed by the double slit-cut system is presented. The angular divergence of the beam is smaller as the ones featured for PXWRs are approximately older. But the integral intensity of the beam is approximately equal to the magnitude of this parameter characterized for the beam formed by a simple quartz PXWR. In consequence of the source regime invariability, it is possible to conclude that the simple PXWR with a slit width $s = 0.12$ µm is equivalent to the double slit-cut system with slit width $s_2 = s_3 = 6$ µm on the absolute integral intensity of the beam. By analogy, we can say that the modified waveguide-resonator with slit width $s = 0.1$ µm is approximately equivalent to the double slit-cut system with slit width $s = 30$ µm.

In our investigation, we used the radiation-skewed concentrator with planar chamfer only, owing to our technological scarcity. At the same time, it is well known that best results can be achieved by using a specific form of chamfer [82]. The application of PXWR-skewed concentrator specific form is able to increase its emergent beam integral intensity to 1–1.5 orders. However, Figure 5.25 shows that the emergent beam angular divergence is higher in comparison with the one featured for beams formed by slit-cut systems. So, the second vital task of PXWR modernization is the angular divergence decrease of its emergent beam.

5.9 Problem of Angular Divergence Reduction for a PXWR Emergent Beam

The search for this problem solution allowed to elaborate some constructions for PXWR modification. Figure 5.26 presents examples of waveguide-resonance devices that allowed to decrease the angular divergence of the formed beam with the preservation of its integral intensity. The most obvious way to decrease the angular divergence of the output flux by half is a PXWR construction based on using reflectors with different lengths (Figure 5.26a). In this case, one can

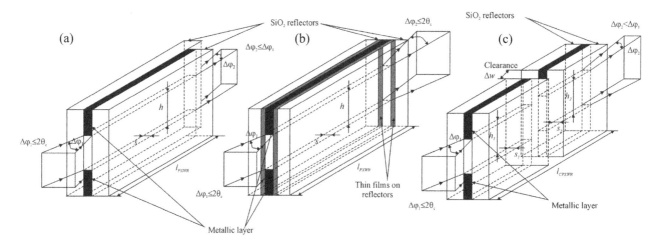

FIGURE 5.26 Schemes of waveguide-resonators of quasi-monochromatic X-radiation fluxes with different design: (a) a PXWR formed by quartz ref lectors with different lengths, (b) a waveguide-resonance structure with a thin film gradient coating on quartz plane-polished films, and (c) a composite waveguide-resonator.

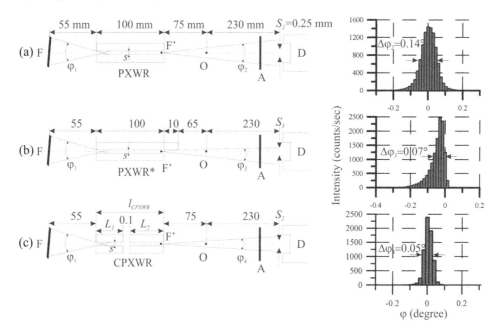

FIGURE 5.27 Scheme of formation of quasi-monochromatic X-radiation fluxes by (a) a simplest quartz PXWR, (b) a waveguide-resonator constructed based on different-sized reflectors, and (c) CPXWRs as well as the spatial distribution of the CuK$_\alpha$ radiation intensity in fluxes formed by these devices. The distributions were recorded under the following operation conditions of the BSV-24 (Cu) radiation source: $U = 20$ keV and $J = 10$ mA. Absorption coefficient of filter A $K = 50$. Slit width of the waveguide-resonator device $s = 40$ nm.

expect a significant change in the spatial distribution of the intensity in the X-ray flux, from a shape close to Gaussian to a semiGaussian distribution. The experimental data shown in Figure 5.27 corroborated these expectations.

One more construction giving grounds to expect the solution of the posed problem is a waveguide-resonance structure with gradient film coatings on polished quartz reflectors (Figure 5.26b). It was intended for obtaining film coatings on the basis of an amorphous HfO$_2$–Al$_2$O$_3$ solid solution with a thickness of several tens of nanometers; the composition of these coatings continuously changed from almost stoichiometric hafnium oxide with a density of 9.7 g/cm^3 to

Al$_2$O$_3$ with a density of 3.9 g/cm^3. A PXWR constructed based on such reflectors would make it possible to have the angular divergence of the formed flux without making changes in the spatial distribution of the intensity. Unfortunately, due to technical difficulties, attempts to construct it are yet unsuccessful.

The construction of a specific PXWR shown in Figure 5.26c is a very interesting finding. This construction consists of two sequentially mounted and mutually adjusted waveguide-resonators of the simplest design with gap Δw between them. The gap size corresponds to the following relationship [83]:

$$\frac{\lambda_0}{2\theta_c^2} \leq \Delta w \leq \frac{\lambda_0^3}{8\Delta\lambda^2} \qquad (5.54)$$

The boundaries of this interval for the waveguide-resonance structure called a "composite PXWR" (CPXWR) formed by a set of quartz reflectors during the transportation of the MoK_α flux take values from 9 to 490 μm (for CuK_α, from 5 to 600 μm). Studying the operation of this device shows that it captures the radiation in the same angular interval as the simplest PXWR; at its output (i.e., at the output of the second PXWR forming the CPXWR), a flux with a lesser angular divergence is formed when compared with the angle of radiation capture by the first PXWR with the preservation of the integral intensity and almost Gaussian shape of the spatial distribution of the X-radiation intensity. This is informatively illustrated in Figure 5.27c. Moreover, according to investigations, varying the gap size Δw between subsequently mounted simplest waveguide-resonators has an effect on the degree of the angular divergence of the formed X-ray flux. Unfortunately, mutual adjustment of PXWRs with nanosize-extended slit clearances is difficult and the formation of the output flux, as it was revealed in investigations, is influenced by the shape of the ends of reflectors constituting the PXWR and by the quality with which they are treated. For these reasons, an undistorted dependence of the degree of the decrease in the angular divergence of the formed flux on gap size Δw is still not constructed. However, the obtained data made it possible to state that, if relationship Eq. (5.54) is satisfied, the angular divergence of the formed X-ray flux will decrease when compared with the capture angle of this flux with the preservation of integral intensity. As a working hypothesis explaining the observed experimental results, it was proposed that the effect of partial angular tunneling of the X-ray flux in the space of gap Δw between the waveguide-resonators takes place [83].

If the distance between subsequently mounted PXWRs exceeds $\lambda_0^3/8\Delta\lambda$ (Figure 5.28a), this structure forms a low-divergence X-ray beam. However, its integral intensity turns out to be significantly less than the intensity of the flux formed at the output of the first waveguide-resonator. In this case, uniform interference fields of the X-ray standing wave that are excited in the first and second PXWRs turn out to be independent, and the intensity of the X-ray flux formed at the output of the second waveguide-resonator is determined only by the geometric factor. At the same time, if the distance between sequentially mounted PXWRs becomes less than the aforementioned value, the situation at the output of the second waveguide-resonator will change cardinally (Figure 5.28b). The occurring change is related to the fact that the uniform interference field of the X-ray standing wave appearing in the slit clearance of the first PXWR is characterized by a certain protrusion, the size of which is determined by the angle of partial X-ray–flux capture (quantity $\lambda_0^3/8\Delta\lambda$ corresponds to the radiation capture with a minimum possible angular value). In this case, the second waveguide-resonator will capture into its slit clearance during the initial time period the partial X-ray flux, the intensity of which will be purely determined by geometric parameters. This flux will form the interference field of the XSW in correspondence to the intensity of this partial flux. However, due to the fact that the field of the first PXWR turns out to penetrate into the slit clearance of the second waveguide-resonator, the interference fields begin to interact and reach a certain new state. As a consequence of their interaction, a portion of the X-ray flux formed at the output of the first PXWR is not able to fall into the slit clearance of the second waveguide-resonator and, according to the geometric factor, changes its propagation direction and tunnels into the slit clearance of the second PXWR. The effect of tunneling leads to an increase in the intensity of the flux in the slit clearance of the second PXWR and, correspondingly, to an increase in the intensity of the interference field of the X-ray standing wave in its slit clearance. This, in turn, causes a change in the parameter of the interaction between interference fields of the first and second PXWRs and leads to a gradual establishment of the stationary regime of the X-ray flux transportation. At the same time, according to experimental investigations (e.g., data of Figure 5.27c), the achieved stationary regime is characterized by preservation of the integral intensity of the quasi-monochromatic X-ray flux and decrease in the angular

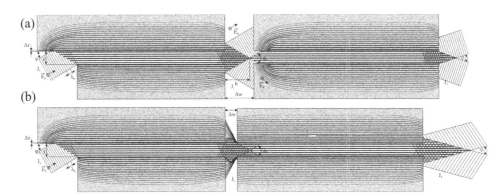

FIGURE 5.28 Scheme of formation of a quasi-monochromatic X-radiation flux by a pair of subsequently mounted and mutually adjusted PXWRs with slit clearance width $s = 40$ nm with a distance Δw between the PXWRs: (a) $\Delta w > \lambda_0^3/8\Delta\lambda^2$, which does not admit the partial angular tunneling of the quasi-monochromatic X-radiation flux and (b) $\Delta w > \lambda_0^3/8\Delta\lambda^2$, which implements the radiation angular tunneling.

divergence of this flux when compared with the angle of radiation capture by the first waveguide-resonator. Therefore, the proposed speculative model yields a logical explanation of the experimentally observed effect, but, at the same time, puts aside the most important postulate of Liouville's theorem of the constancy of the volume of the phase space describing any statistical continuum [84]. Experimental data demonstrate that a flux of X-ray quasi-monochromatic radiation preserves its integral intensity after passing through the composite waveguide-resonator, but, in addition, is characterized by lower angular divergence. At the same time, it is clear that the characteristic of the X-ray flux formed at the CPXWR output is not restricted by the two aforementioned parameters and, to refine the proposed model of CPXWR operation, it is necessary to reveal the flux parameter that, according to Liouville's theorem, must undergo the supposed changes. Analysis of possible parametric changes demonstrates that the deterioration of monochromaticity, i.e., an increase in $\Delta\lambda$ and, correspondingly, a decrease in the coherence length of the quasi-monochromatic X-ray flux formed at the CPXWR output, is the most probable factor accompanying the decrease in the angular divergence of this flux. This hypothesis was discussed in detail in our

report at the 25th International Symposium "Nanostructures: Physics and Technology" [85] and was verified experimentally.

Figure 5.29 presents fragments of X-ray-diffraction patterns of single-crystalline silicon cut perpendicularly to the (111) plane in the region of the (111) reflex. They were obtained using a source of BSV-24 (Cu) radiation in the regime with $U = 20$ keV and $J = 10$ mA under conditions of the formation of the primary radiation flux with (a) a simplest PXWR and (b) a composite waveguide-resonator. The X-ray diffraction patterns were recorded in the regime of high angular resolution on an HZG-4 goniometer (with a minimum scanning step of $\delta\varphi = 0.0005°$) in the Bragg–Brentano geometry with measurement radius $R = 235$ mm and detector slit width $S_{\rm d} = 0.1$ mm. The measurements were carried out using waveguide-resonance devices with planar quartz slit clearance $s = 40$ nm and length $l = 100$ mm. The distance between the sequentially mounted waveguide-resonators in the CPXWR was $\Delta w = 0.1$ mm. The integral and angular characteristics of fluxes at the PXWR and CPXWR outputs correspond to the data presented in Figure 5.27. The comparative diffractometric data presented in Figure 5.29 indicate a small increase in the

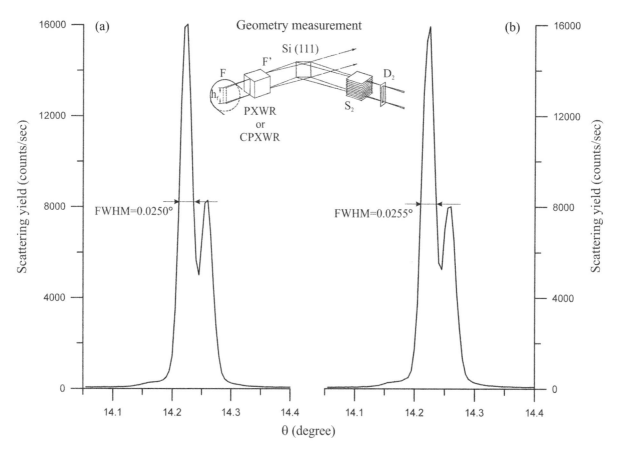

FIGURE 5.29 Fragments of an X-ray diffraction pattern of the (111) reflex of a silicon single crystal. The cross section is parallel to the (111) plane. The initial X-ray beam was formed using (a) a very simple waveguide-resonator with slit clearance width $s = 40$ nm and (b) a CPXWR with a similar slit clearance width. The fragments of the X-ray-diffraction pattern were obtained using a BSV-24 (Cu) radiation source in the regime of $U = 20$ keV and $J = 10$ mA on an HZG-4 digital goniometer, radius $R = 235$ mm, and measurement step $\Delta\theta = 0.005°$.

reflex width when the initial X-ray flux is formed by using a composite waveguide-resonator. The observed broadening is a direct confirmation of the preliminary hypothesis about the decrease in the monochromaticity degree of the radiation flux after its transportation by the CPXWR, although the presented measurements are insufficient for a quantitative estimation of this fact. However, the obtained qualitative data allow one to state that, according to requirements of Liouville's theorem, the decrease in the angular divergence of a quasi-monochromatic X-radiation flux is compensated by deterioration of its monochromaticity. In spite of the fact that these two parameters do not commute, one can try to find an analytical relation between their mutual changes.

Energy and pulse of a single photon in an electromagnetic radiation flux E_ρ and P_ρ are determined [54] by the expressions:

$$E_\rho = \hbar\omega = \frac{hc}{\lambda_0} = P_\rho c \qquad (5.55)$$

$$P_\rho = \frac{E_\rho}{c} = \frac{h}{\lambda_0} \qquad (5.56)$$

where $h = 2\pi\hbar$ is the Planck constant, $\omega = 2\pi f$ and λ_0 are the angular frequency and average radiation wavelength, and c is the speed of light in vacuum. Differentiating formulas (5.55) and (5.56), one can obtain the relation between changes in the energy and pulse of an individual photon with variation in the radiation wavelength:

$$\Delta E_\rho = -\frac{hc}{\lambda_0^2}\Delta\lambda = \frac{P_\rho^2 c}{h}\Delta\lambda \qquad (5.57)$$

$$\Delta P_\rho = -\frac{h}{\lambda_0^2}\Delta\lambda = -\frac{P_\rho}{\lambda_0}\Delta\lambda = \frac{P_\rho^2}{h}\Delta\lambda \qquad (5.58)$$

As was described earlier, a portion of the quasi-monochromatic X-radiation flux formed by the first PXWR constituting the composite waveguide-resonator falls into the second PXWR and initiates a uniform interference field of the X-ray standing wave in it. As a result of the interaction between the interference fields of the first and second PXWRs, a stationary regime of radiation transport is established; a portion of the flux formed by the first PXWR is geometrically unable to get into the second PXWR tunnels in the gap Δw to the second PXWR. Let us suppose that photons of this portion of the X-ray flux formed by the first PXWR undergo angular tunneling and are characterized by the pulse vector $\mathbf{P}_1$, equal to $\mathbf{P}_p$, and directed at angle φ_1 to the CPXWR axis. After passing the second PXWR, this portion of the flux will by characterized by vector $\mathbf{P}_2$ equal to $\mathbf{P}_\rho + \delta\mathbf{P}_\rho$, and directed at angle φ_2 with respect to the axis of the composite waveguide-resonator (Figure 5.30). Experimental data show that $\varphi_1 > \varphi_2$, and the change in the angular aperture for this portion of the flux is represented as

$$\Delta\varphi = \varphi_1 - \varphi_2 \qquad (5.59)$$

Since X-ray waveguide-resonance structures operate in the scope of the total external reflection by the phenomenon of

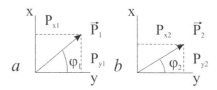

FIGURE 5.30 A schematic representation of the photon pulse (a) before and (b) after tunneling. P_1 corresponds to the initial value of the pulse P_p, and P_2 corresponds to the new values $P_p + \delta P_p$ and $P_p - \delta P_p$.

X-radiation fluxes on a material surface, values of angles φ_1 and φ_2, as well as their differences, are small. Therefore, for tangents of these angles, expansion into series [86] can be used without higher-order terms:

$$\Delta\varphi = (\varphi_1 - \varphi_2) \approx \frac{P_{y1}}{P_{x1}} - \frac{P_{y2}}{P_{x2}} \qquad (5.60)$$

Variations in values of the small angles slightly change coordinate x of the pulse vector of photons undergoing the tunneling effect, because the coordinate is described by the cosine of these angles. Therefore, expression (5.60) can be reduced to the form

$$\Delta\varphi \approx \frac{P_{y1} - P_{y2}}{P_{x1}} \approx \frac{\Delta P_y}{P_p} \qquad (5.61)$$

At the same time, the scalar value of the pulse of these photons can be represented in the commonly accepted format:

$$P_p^2 = P_x^2 + P_y^2 \qquad (5.62)$$

As a result of differentiation, we obtain the relation between the total change in the pulse and changes in its projections:

$$P_p\frac{\Delta P}{\Delta\varphi} = P_x\frac{\Delta P_x}{\Delta\varphi} + P_y\frac{\Delta P_y}{\Delta\varphi} \qquad (5.63)$$

Since the change in the pulse component with respect to the x coordinate due to tunneling of photons is considerably less than the change in its component with respect to the second coordinate, expression (5.61) in the first approximation can be represented in the form

$$\Delta P_p = P_y\Delta\varphi = \Delta\varphi P_p \sin\varphi_1 \approx P\varphi_1\Delta\varphi \qquad (5.64)$$

Comparing expressions (5.58) and (5.64), one can obtain a real relation between the decrease in the angular divergence of a portion of quasi-monochromatic X-radiation flux $\Delta\varphi$ and deterioration of this portion of the flux by quantity $\delta\lambda$:

$$\delta\lambda = \varphi_1\lambda_0\Delta\varphi \qquad (5.65)$$

It is clear that this expression is valid only for small angles typical for observations of the TER of X-ray fluxes on a material interface. In addition, as applied to the flux formed by the composite waveguide-resonator, it is necessary to remember that deterioration in radiation monochromaticity is typical only for the flow portion that underwent tunneling in interwaveguide-resonance gap Δw.

In any event, it seems useful to estimate the change in the degree of monochromaticity of an X-ray quasi-monochromatic radiation flux after its tunneling in the gap Δw. According to our experimental investigations (Figure 5.27c), the CPXWR we have tested decreased the angular divergence in the formed flux almost by three times when compared with the angle of capture. For the estimates, one can imagine an ideal case in which a collinear X-ray flux is formed at the CPXWR output. This corresponds to the maximum possible value of $\Delta\varphi$. Therefore, the maximum decrease in the degree of monochromaticity due to tunneling of the flux will be characterized by the quantity

$$\delta\lambda_{\max} = 2\varphi_1\lambda_0 \qquad (5.66)$$

For the CuK$_\alpha$ radiation ($\lambda_0 = 0.1541$ nm), our estimate of quantity $\delta\lambda_{\max}$ amounts to 4.0×10^{-4} nm. For comparison, one mentions that the value of $\Delta\lambda$ for the characteristic CuK$_\alpha$ X-radiation generated by a standard laboratory source amounts to 0.44×10^{-4} nm [38]; i.e., radiation tunneling in gap Δw of the composite waveguide-resonator is able to reduce the degree of monochromaticity of radiation generated by laboratory X-ray sources by almost an order of magnitude. In the experiment presented in Figure 5.27c, the tunneled portion of the X-ray flux is characterized by triple deterioration of the degree of monochromaticity, which indeed led to some broadening of the diffraction pattern. Therefore, for all practical purposes, use of composite waveguide-resonance structures for the formation of excitation fluxes for X-ray fluorescence analysis seems quite justifiable, while such fluxes application in diffractometry can lead to the appearance of certain difficulties.

As mentioned earlier, our experimental investigations demonstrate that varying the width of gap Δw between the waveguide-resonators in the CPXWR has an effect on the divergence of formed flux $\Delta\varphi$. However, it became clear from the same measurements that the real divergence of the flux is also influenced by a series of technological parameters. For this reason, obtaining a quantitative relation between quantities Δw, $\Delta\varphi$, and $\Delta\lambda$ at present seems to be problematic. At the same time, the proposed model of angular tunneling of a portion of the X-radiation flux relates to the decrease in the angular divergence of the flux formed by the CPXWR with a decrease in the degree of monochromaticity of a portion of the flux undergoing angular tunneling and seems to be quite plausible and justifiable. Here, it is worth noting an important corollary of the proposed model. A portion of the formed flux undergoes tunneling, and the average wavelength of this portion preserves its value as λ_0. At the same time, in this portion of the flux, there appeared photons that changed their wavelengths by $\pm\delta\lambda/2$. Therefore, there appeared photons with the energy and pulse increased and decreased due to tunneling by the values

$$\delta E = \pm\frac{hc}{\lambda_0}\varphi_1\Delta\varphi \qquad (5.67)$$

$$\delta P = \pm P_p\varphi_1\Delta\varphi \qquad (5.68)$$

Certainly, changes δE and δP related to tunneling of a portion of an X-ray flux will be superposed by the presence of the natural energy and pulse spread in the flux of characteristic radiations generated by laboratory sources. The spread is determined by the initial value $\Delta\lambda$ and, correspondingly, by ΔE and ΔP. The value of $\Delta\lambda$ itself and, therefore, ΔE in the initial flux can be considerably reduced by using quality monochromators. In this case, it is apparently possible to distinguish partial fluxes with energy changed due to tunneling. The mere fact of the appearance of photons with decreased and increased energies resembles to some extent the Raman scattering with the appearance of the Stokes and anti-Stokes components in the resulting spectrum. However, a more detailed investigation of the phenomenon of partial angular tunneling of the quasi-monochromatic X-radiation flux needs at present developing a more perfect technology of preparing X-ray TER reflectors used for the formation of planar extended waveguide-resonance structures used for CPXWR assemblies.

5.10 Common Approach to PXWR Emergent Beam Characteristics Improving

X-ray nanophotonics development on the basis of the PXWR design modification showed that it is possible by traditional way and through discovery of early uncertain effects. Common approach allows to unite efforts for formation of X-ray nanobeams with enhanced properties. Figure 5.31 presents the possibilities to compare the radiation density parameter for MoK$_{\alpha\beta}$ flux formed by PXWRs with different designs and double slit-cut systems built on the basis of two slit-cuts with width 6 μm. All waveguide-resonance devices presented in the Figure were characterized by slit clearances $s = 40$ nm. It is clear that the MoK$_{\alpha\beta}$ radiation density on the output of all waveguide-resonance devices will be much higher in comparison with one in the beam for the position near S_3 of slit-cut system. But on the great distances from X-ray formers output, the emergent beam of slit-cut system has some advantages in comparison with beams formed by z simplest PXWR and waveguide-resonator built on nonequivalent length reflectors. The modified PXWR application allowed to elicit some ways for X-ray nanophotonic development. At the same time, suggested approaches are partial solutions of the problem. We suppose that the fundamental development of X-ray nanophotonics will be connected with the elaboration of multiclearance systems and units for the X-ray beam parameters, steering by influence of the interference field of XSW initiated by the beam.

Figure 5.32 presents a sample of the multiclearance waveguide-resonance hypothetical construction with two slit clearances. Central reflector of the construction with thickness C is the chief element of the waveguide-resonator. Interference fields of XSW excited in S_1 and S_2 slit clearances are interacted in the volume of reflector C. In the result, the phase relation between these interference fields appears, and

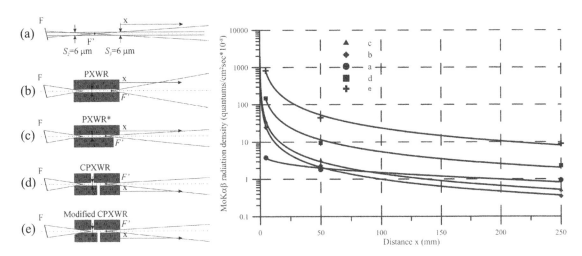

FIGURE 5.31 Original composite PXWR is the best exciting beam former for Total X-ray Reflection Fluorescence (TXRF) spectrometry. Data corresponding to BSW-27 (Mo) X-ray source regime $U = 25$ keV, $I = 10$ mA.

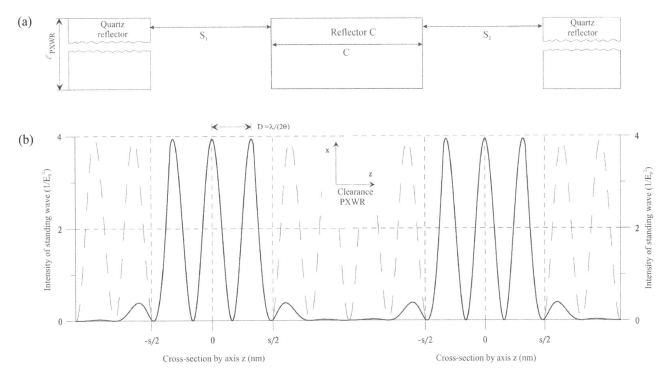

FIGURE 5.32 Scheme of double slit clearance waveguide-resonator built on the basis of quartz reflectors (a) and XSW intensity distribution at the radiation capture emanating from the same X-ray source (b).

its magnitude is defined by the reflector thickness. In its turn, it evokes the phase shift between emergent beams formed by every slit clearance η, which influence the parameters of the common emergent beam. Its intensity will be described by conventional expression [51,68]:

$$J_{\sum} = J_1 + J_2 + 2\sqrt{J_1 J_2}\cos\eta \qquad (5.69)$$

Our pioneer experimental investigations of similar structures showed that the integral intensity of common emergent beam is very sensitive to the width magnitude of central reflector and quality treatment of surfaces and butt-ends of reflectors. The construction parameter variation led to

changing of the common emergent beam intensity from zero to values increasing simplest sub of J_1 and J_2. Our technological possibilities are not capable to ensure the required quality of reflector preparation for stable multiclearance waveguide-resonance devices built today. At the same time, it is possible to propose some hypothetical schemes for the emergent beam formation with enhanced parameters. Schemes of nanophotonic devices presented in Figure 5.33, built on the basis of similar constructions, will allow to form X-ray quasi-monochromatic beams with nanosize cross section and enhanced radiation intensity.

Specific direction of X-ray nanophotonic development is connected with building of active X-ray units. This direction

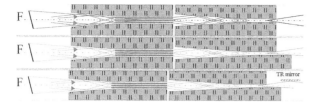

FIGURE 5.33 Hypothetical schemes of X-ray beam formation with enhanced parameters based on the multiclearance conception appear with the use of skewed concentrator, tunneling effect, and nonequivalent length reflector application ideas. Total reflection mirror application can allow to focus on X-ray emergent beam.

appeared owing to disclosure of the composite waveguide-resonator mechanism function. Investigations of its operation showed that nature admits possibility of interaction between interference fields of XSWs. Moreover, similar interaction allows to change X-ray emergent beam parameters formed by structure admitting the interference field interaction. In frame of this paradigm, we attempted to elaborate two types of units for realization of an external influence on the interference field of XSW. Principle schemes of these ideas implementation are presented in Figure 5.34.

The first type of expected actuation on the interference field of XSW was connected with possible modulation of the interference field (Figure 5.34a). There were built X-ray waveguide-resonance cell on the basis of ferrite monocrystal

reflectors for the high frequency electromagnetic field influence on XSW interference field and $LiNbO_3$ monocrystal reflectors for realization of surface acoustic wave influence on the interference field. Unfortunately, our electromagnetic and acoustic investigations were not let to avail results in spite of our repeated attempts. It is possible that our experimental facility was inadequate to raised tasks, and investigations in this field will be continued.

Our investigations in frame of the second direction showed better results. The second type of the expected actuation on the XSW interference field was planned as a result of the standing wave interference field interaction excited by independent radiation sources. It was interesting to find conditions promoting interaction between independent fluxes of X-ray and optical radiation. For experimental investigation, CuK_α radiation with wavelength 0.154 nm and green light with wavelength 532 nm were chosen. The experimental setup for the study of possible interactions of X-ray and light-independent fluxes is presented in Figure 5.35. Based on the unit for investigation, the simplest construction of PXWR was formed by quartz reflectors with slit clearance width 190 nm. X-ray flux emitted by BSW-24 (Cu) radiation source and formed by Si monochromator fell on PXWR input under angle ξ, which can be varied in wide limits. The X-ray beam provoked in the PXWR slit clearance space the interference field of XSW with the period defined by expression (5.14). XSW interference field appearance is possible in the angular interval of X-ray incident beam $0.02° \leq \theta \leq 0.21°$.

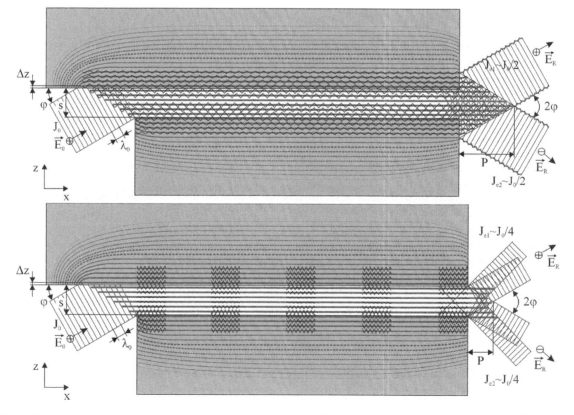

FIGURE 5.34 Principle schemes of the waveguide-resonance units building for possibility of implementation of beam parameters steered by external actuation.

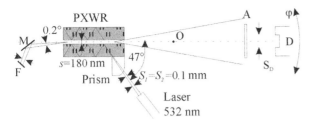

FIGURE 5.35 Scheme of an experimental setup for the study of possible interactions between independent fluxes of X-ray and light beams through mutual influence of standing wave interference fields excited by these fluxes. A schematic unit is PXWR formed by quartz reflector with slit clearance size $s = 190$ nm. Light flux hits into the PXWR reflector through a semicircular prism P. θ and χ angles can be varied.

It is able to excite the interference field of XSW with the period from 180 to 21 nm. Simultaneously, the light flux is able to excite the interference field of an optical standing wave in the PXWR reflector with period 360–1,027 nm at the total internal reflection realization in the angular range $47.5° \leq \chi \leq 75°$. The upper value of correlation is connected with technical limitations. The interference field interaction was registered near ends of both angular ranges of the total reflection effect realization. Its arising was fixed by varying the spatial intensity distribution of the X-ray emergent beam presented in Figure 5.36. The influence of the radiation spatial intensity distribution was registered at divisible values of standing wave periods: $D_{XSW} = 180$ nm, $D_{ISW} = 360$ nm only. Moreover, the resonance was very sharp and disappeared at small variations of experiment conditions. Owing to experimental difficulties it is not clear whether a degree influences the interaction effect of the relation between

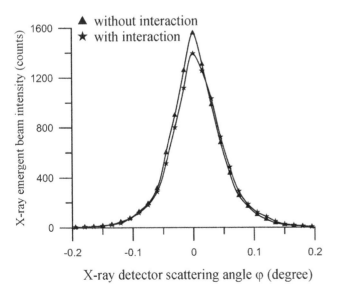

FIGURE 5.36 Experimental patterns demonstrating the mutual interaction between standing wave interference fields provoked by CuK$_\alpha$ radiation flux and green light flux in the unit built on the basis of the simplest design of PXWR. The measurements showed that the interaction effect has a resonance character.

interference field intensities and polarization factor of X-ray and light fluxes. Moreover, we could not establish spatial areas that are responsible for the interaction. It is likely that the interaction takes place in the transition layer of surface reflector that underwent the light interference field arising. But the effect of registration shows that the interaction between independent wave radiation fluxes is possible and is characterized likely by its resonant nature.

5.11 Waveguide-Resonance Propagation of Particles Characterized by Nonzero Rest Mass

Our investigations were connected with the study of the quasi-monochromatic (characteristic) X-ray flux propagation through planar extended slit clearances. It was ascertained that the reflection phenomenon can be accompanied in some conditions by appearance of a uniform interference field of standing wave. The nature of wave process can be of any origin. It is well known that the neutron standing wave arising phenomenon is used in practical investigations [87]. In specific conditions, the polarized quasi-monochromatic neutron flux can excite the uniform interference field of neutron standing wave, and the flux will propagate according to the waveguide-resonator mechanism. It is known experimental investigation about the high-energy proton beam scattering on the dielectric surface [88]. And any material particle with nonzero rest mass can display wave properties at moving with some velocity v. Wavelength of the particle moving is defined by the expression [89]:

$$\lambda_0 = \frac{h}{p} = \frac{h}{mv} \qquad (5.70)$$

So, one can expect that any nuclei can be characterized in its moving by some wavelength and, in specific conditions, can excite the interference field of the standing wave. In development of this hypothesis, one can assume that fluxes of different nuclei can interact through mutual influence of standing wave interference fields provoked by its fluxes. As a result of the influence, we can expect to get flux of the hybrid nuclei, which means that the nuclear reaction occurred. Degree of the reaction efficiency will be defined by conditions of the interference field's mutual influence. For example, we can expect the flux of ^{3}He^{++} ions if we shall find technology for H$^+$ and D$^+$ ion beam interaction through mutual influence of interference standing waves excited by these fluxes. A similar approach can be applied for the solution of the philosopher's stone problem, consisting of the search of methods for the gold preparation by using conventional ingredients [90]. We are believed that the gold atom's flux can be received by interaction of standing waves of interference fields excited by fluxes of ^{11}B^5 and ^{186}W^{74} ion beams. But it is clear that conditions for the reaction realization are not evident. At the same time, the experimental investigations about low-energy nuclear reactions are known [91].

TABLE 5.1 Table of Difference Radiation Fluxes Parameters and Factors of Its Interaction of Quartz Reflectors Formed by PXWR

Radiation	E_0 (keV)	λ_0 (nm)	$\Delta\lambda$ (nm)	D_{min} (nm)	D_{max} (nm)	θ_c	$x_e(\theta_c)$ (nm)
$CaK_{\alpha 1}$	3.691	0.3358	1.6×10^{-4}	21	176	$0.46°$	20
$CrK_{\alpha 1}$	5.414	0.2290	1.03×10^{-4}	21	127	$0.32°$	28
$FeK_{\alpha 1}$	6.403	0.1936	1.01×10^{-4}	21	93	$0.27°$	33
$CoK_{\alpha 1}$	6.929	0.1789	0.81×10^{-4}	21	99	$0.25°$	35
$CuK_{\alpha 1}$	8.046	0.1541	0.58×10^{-4}	21	102	$0.21°$	41
$GeK_{\alpha 1}$	9.885	0.1254	0.43×10^{-4}	21	93	$0.17°$	48
$\gamma_m Fe^{57}$	14.39	0.0862	0.28×10^{-13}	21	6.6×10^{10}	$0.12°$	66
$MoK_{\alpha 1}$	17.476	0.0709	0.29×10^{-4}	21	43	$0.10°$	83
$AgK_{\alpha 1}$	22.159	0.0559	0.21×10^{-4}	21	21	$0.077°$	103
$\gamma_m Sn^{119}$	23.80	0.0521	0.55×10^{-13}	21	1.2×10^{10}	$0.072°$	110
$\gamma_m Sb^{121}$	37.15	0.0353	1.17×10^{-13}	21	2.7×10^{9}	$0.047°$	161
$\gamma_m I^{127}$	59.0	0.0210	0.89×10^{-13}	21	1.2×10^{9}	$0.030°$	197
$\gamma_m Hf^{177}$	113.0	0.0110	0.51×10^{-16}	21	5.9×10^{11}	$0.016°$	170
$\gamma_m Os^{188}$	155.0	0.0080	0.43×10^{-13}	21	3.7×10^{9}	$0.011°$	156
$\gamma_1 Sr^{85}$	514.0	0.0024	0.25×10^{-16}	21	5.8×10^{11}	$0.003°$	109
$\gamma_1 Cs^{137}$	662.0	0.0019	0.25×10^{-12}	21	3.6×10^{6}	$0.0026°$	108
$\gamma_1 Zn^{65}$	1112.0	0.0011	0.2×10^{-14}	21	1.5×10^{8}	$0.0016°$	89
$\gamma_1 Cu^{64}$	7910.0	0.00016	0.4×10^{-11}	21	1.6×10^{3}	$0.0002°$	50

It is quite possible that experimental procedures of the cold fusion create conditions for the interference field standing waves formation, and nuclear reactions occur owing to the interference field interaction.

5.12 Conclusion

Our investigation is devoted to description of mechanism features of X-ray fluxes propagation through nanosize-extended slit clearances that can be background for X-ray nanophotonics. Owing to the reflection phenomenon typical for any wave fluxes, we expect that the mechanism has common significance. This mechanism forms the basis of the interaction effect of interference field standing waves excited by independent wave fluxes. Its application can be used for some problem solutions of nuclear physics. In conclusion of our work, we present the table with X-ray and gamma radiation parameters and slit clearance dimensions admitting the waveguide-resonance regime of flux radiation propagation for quartz PXWR, critical angles values of radiation fluxes on quartz surface, and radiation depth penetration on the surface (Table 5.1).

Acknowledgments

We convey extreme gratitude to J. Kawai and A.G. Revenko for the interest to PXWR idea development. This work was produced by the Russian government task #075-00920-20-00 by the Ministry of Education and Science of the Russian federation (the agreement 02.a03.008).

References

1. Bragg WL. Diffraction of short electromagnetic waves by a crystal. *Proc. Camb. Phil. Soc.* 1912; **17**: 43–57.

2. Compton AH. The total reflection of X-ray. *Phil. Mag.* 1923; **45**, #270: 1121–1131.

3. Borrmann G. User extinktion der rontgenstrahlen von quartz. *Physic Z.* 1941; **42**: 157–162.

4. Underwood JH, Attwood DT. The renaissance of X-ray optics. *Phys. Today* 1984; **37**, #4: 44–52.

5. Hirsch PB, Kellar J. An X-ray micro-beam technique: I-collimation. *Proc. Phys. Soc. London Scr.* 1951; **B64**: 369–374.

6. Spiller E, Segmuller A. Propagation of X-rays in waveguide. *Appl. Phys. Lett.* 1974; **24**, #2: 60–61.

7. Feng YP, Sinha SK, Deckman HW, Hasting JB, Siddons DP. X-ray flux enhancement in thin film waveguides using resonant beam couplers. *Phys. Rev. Lett.* 1993; **71**, #4: 537–540.

8. Feng YP, Sinha SK, Fullerton EE, Grubel G, Abernathy D, Siddons DP, Hastings JB. X-ray fraunhofer diffraction patterns from a thin film waveguides. *Appl. Phys. Lett.* 1995; **67**, #24: 3647–3649.

9. Lagomarsino S, Jark W, Di Fonzo S, Cedola A, Mueller B, Engstrom P, Riekel C. Submicrometer X-ray beam production by a thin film waveguide. *Appl. Phys.* 1996; **79**, #8: 4471–4473.

10. Jark W, Di Fonzo S, Lagomarsino S, Cedola A, Di Fabrizio E, Bram A, Riekel C. Properties of a submicrometer X-ray beam at the exit of a waveguide. *Appl. Phys.* 1996; **80**, #9: 4831–4836.

11. Di Fonzo S, Jark W, Soullie G, Cedola A, Lagomarsino S, Cloetens P, Riekel C. Submicrometre resolution phase-contrast radiography with a beam from an X-ray waveguide. *Synchrotron Radiat.* 1998; **5**: 376–378.

12. Jark W, Cedola A, Di Fonzo S, Fiordelisi M, Lagomarsino S, Kovalenko NV, Chernov VA. High gain beam compression in new generation thin film X-ray waveguides. *App. Phys. Lett.* 2001; **78**, #9: 1192–1194.

13. Pfeiffer F, Saldit T, Hoghoj P, Anderson I, Schell N. X-ray waveguide with multiple guide layers. *Phys. Rev.* 2000; **B62**: 16939–16942.

14. Prudnikov IR. X-ray waveguides based on Bragg scattering of multilayers. *Phys. Rev.* 2003; **B67**: 233303–233304.

15. Boungaerts JHH, Zwannenburg MJ, Zontone F, van der Veen JF. Propagation of coherent X-ray in a

multistep-index X-ray waveguide. *Appl. Phys.* 2001; **V90**, #1: 94–100.

16. Bedzyk M, Bommarito G, Schildkraut J. X-ray standing waves at a reflecting mirror surface. *Phys. Rev. Let.* 1989; **69**: 1376–1379.

17. Pellisia D. X-ray photonics: Bending X-ray with nanochannels. *Nat. Photonics* 2016; **v10**, #2: 75–77.

18. Hoffman-Urlaub S, Salditt T. Miniaturization beam-splittered realized by X-ray waveguides. *Acta Crystallogr.* 2016; **72**, #5: 515–522.

19. Marton J. The glass tube as X-ray guide. *App. Phys. Lett.* 1966; **9**, #2: 194–197.

20. Kumakhov M, Komarov F. Multiple reflection from surface X-ray optics. *Phys. Rep.* 1990; **191**: 289–352.

21. Engstrom P, Larsson S, Rinby A. A submicron synchrotron X-ray beam generated by capillary optics. *NIM.* 1991; **A302**: 547–552.

22. Vincze L, Janssens K, Adams F, Ridby A. Detail ray-tracing code for capillary optics. *X-Ray Spectrom.* 1995; **24**, #1: 27–37.

23. Alexandrov YuM, Dabagov SB, Kumakhov MA, Murashova VA, Fedui DA. Peculiarities of photon transmission through capillary systems. *NIM.* 1998; **B134**, #2: 174–180.

24. Dabagov SB. Wave theory of X-ray scattering in capillary structures. *X-Ray Spectrom.* 2003; **32**: 179–185.

25. Yonehara T, Yamaguchi M, Tsuji K. X-ray fluorescence imaging with polycapillary optics. *Spectrochim. Acta, Part B* 2010; **65**, #6: 441–444.

26. Romanov AYu. X-ray microbeam based on Kumahov polycapillary optics and its application. *Pramana J. Phys.* 2011; **76**, #2: 339–344.

27. Lindquist M. Polycapillary X-ray optics for liquid-metal-jet X-ray tube. Examinsarbete 30hp, UPTEC F 1719, Master Thesis, Uppsala University, Sweden. 2017: 55 p.

28. Mingazin T, Zelenov V, Lejkin V. Slitless collimator for X-ray beams. *Instrum. Exp. Tech.* 1981; **24**, #1, part 2: 244–247.

29. Leykin V, Mingazin T, Zelenov V. Collimating device for X-ray radiation. *Pribori i Tekhnika Experimenta.* 1981; **24**, #3: 208–211. (In Russian).

30. Lejkin V, Mingazin T, Zelenov V. X-ray beam forming by using of a slitless collimator. *Instrum. Exp. Tech.* 1984; **27**, #6, part 1: 1333–1336.

31. Egorov V, Zuev A, Maljukov B. Surface contamination diagnostics of silicon wafers by total reflection X-ray fluorescence spectrometry. *Journal Izvestija Vuzov, Tsvetnaja Metallurgija.* 1997; #5: 54–69. (In Russian).

32. Egorov V, Kondratiev O, Zuev A, Egorov E. The modification of TXRF method by use of X-ray slitless collimator. *Adv. X-Ray Anal.* 2000; **43**: 406–417.

33. Zwanenburg MJ, Peters JF, Bongaerts JHH, de Vries A, Abernathy DL, van de Veen JF. Coherent propagation of X-ray in a planar waveguide with a turnable air gap. *Phys. Rev. Lett.* 1999; **82**, #8: 1696–1699.

34. Zwanenburg MJ, van de Veen JF, Ficke HG, Neerings H. A planar X-ray waveguide with a turnable air gap for the structural investigation of confined fluids. *Rev. Sci. Inst.* 2000; **71**, #4: 1723–1732.

35. Egorov V, Egorov E. Application of an planar X-ray waveguide for structure study of thin film coating. *Thin Solid Films* 2001; **398–399**: 405–412.

36. Egorov V, Egorov E, Bil'chik T. Planar X-ray waveguide-resonator and some aspects of it's practical application. *Invest. Russ.* 2002; **3**: 423–441. http://zhurnal.ape.relarn.ru/articles/2002/040e.pdf.

37. Tsuji K, Injuk J, van Grieken R (eds). *X-Ray Spectrometry; Recent Technological Advances.* Wiley: Europa, 2004; 616.

38. Zschornack G. *Handbook of X-Ray Data.* Springer: Berlin, 2007; 967.

39. Erko A, Idir M, Krist T, Michet AG, *Modern Developments in X-Ray and Neutron Optics.* Springer: Berlin, 2008; 533.

40. Benedictovitch A, Feranchuk I, Ulyanenkov A. *Theoretical Concepts of X-Ray Nanosize Analysis.* Springer: Berlin, 2014; 318.

41. Guo T. *X-Ray Nanochemistry, Concepts and Development.* Springer: Cham, 2018; 513.

42. Egorov V, Egorov E. Ion beams for material analysis: Conventional and advanced approaches. In: *Ion Beam Application,* Ahmad I (ed.) InTechOpen: London, 2018; 37–71.

43. Harrick N. *Internal Reflection Spectrometry.* Wiley: New York, 1967; 282.

44. Milosevic M. *Internal Reflection and Art Spectroscopy.* Wiley: New York, 2012; 256.

45. Turner A. Modern state of multilayer optical films field. *J. Phys. Rad.* 1950; **11**: 444–462. (In French).

46. Egorov V, Egorov E. X-ray waveguide-resonator as a future competitor for synchrotron facilities. *Proceeding of the Conference X-Ray Optics,* 2003 11–14 March. IPM Press: Nizniy Novgorod, 2003; 34–43. (In Russian).

47. *EG&G Ortec Catalog.* Applied Nuclear Spectroscopy. Ortec Inc. Press: Oak Ridge, USA, 1997/1998: 2.360.

48. Egorov V, Egorov E. Experimental study of mode structure peculiarities in X-ray beams formed by planar waveguide-resonator. *Report on the 52nd Annual Denver X-ray Conference,* 4–8 August, Demver USA. In *Conference Abstract Book.* 2003: 137.

49. Egorov V, Egorov E. Planar X-ray waveguide-resonator features. *Trends Appl. Spectrosc.* 2010; **8**: 67–83.

50. Egorov V, Egorov E. *Planar X-Ray Waveguide-Resonators, Realization and Perspectives.* Lambert Academic Publishing: Saarbrucken, 2017; 388. (In Russian).

51. Compton AH, Alisson SK. *X-Ray in Theory and Experiment.* Princeton: New York, 1935; 828.

52. Tsuji K, Delalieux F. Characterization of X-ray emerging from between reflectors and sample carrier in reflector-assisted TXRF analysis. *X-Ray Spectrom.* 2004; **34**: 281–284.

53. Landau LD, Lifshitz EM. *Electrodinamics of Continuous Medium.* Addison-Wesley: Reading, MA, 1965; 586.

54. Bohm D. *Quantum Theory.* Prentice-Hall Inc.: New York, 1952; 755.

55. Henke BL, Lee P, Tanaka TY, et al. The atomic scattering factor, $f_1 + if_2$, for 94 elements for the 100 to 2000 eV photon energy region. In: *Low Energy X-Ray Diagnostic*, Attwood DT, Henke BL (eds) AIP Conference Proceeding. American Institute of Physics: New York, 1981; **75**: 340–388.

56. Lekner J. *Theory of Reflection of Electromagnetic and Particle Waves.* Martinus Nijhoff Publisher: Dordrecht, 1987; 279.

57. Drude P. *Lenburch der Optics.* Verlag von S. Hirzel: Leipzig, 1912; 488.

58. Sivukhin DV, To the eliptical polarization theory at light flux reflection on isotropic medium. *Zh. Exp. Theor.* 1956; **30**, #2: 374–381. (In Russian).

59. Gadomskii OM, Sukhov SV. Microscopic theory of a transition layer on the ideal surface of semiinfinite dielectric media and the near field effect. *Opt. Spectrosc.* 2000; **89**, #2: 261–267.

60. Odiego W, Tempelman K, Nipolitano S, Benes NE. Evidence of a transition layer between the free surface and the bulk. *J. Phys. Chem. Let.* 2018; **9**, #6: 1195–1199.

61. Artioukov IA, Asadchikov VE, Kozhevnikov IV. Effects of a near-surface transition layer on X-ray reflection and scattering. *J. X-Ray Sci. Tech.* 1996; **6**: 223–243.

62. Mazuritskiy MI, Novakovich AA. Surface transition-layer model used for study the fine structure of X-ray reflection spectra. *J. Surf. Invest.* 2014; **8**, #6: 1291–1296.

63. Alikhnov AI. *X-Ray Radiation Optics.* GTTI Press: Moscow, 1933; 104. (In Russian).

64. Parrat LG. Surface studies of solids by total reflection of X-rays. *Phys. Rev.* 1954; **95**, #2: 359–369.

65. Vinogradov AV, Zorev NN, Kozhevnikov IV, Yakushkin IG. Phenomenon of total external reflection of X-ray. *Zh. Exp. Teor. Fiz.* 1985; **89**: 2124–2132. (In Russian).

66. Jakiel J, Kantor W. Quantum (not frustrated) theory of the total internal reflection as the source of the Goos-Hanchen shift. *Eur. Phys. J. D Ser.* 2014; **68**, #10: 305, 11 pp.

67. Born M, Wolf E. *Principles of Optics, Electromagnetic Theory of Propagation Interference and Diffraction of Light.* Pergamon Press: Oxford, 1993; 808.

68. Blochin M. *Physik der Runtgenstrahlen.* Verlag der Technik: Berlin, 1957; 535.

69. Mendel L, Wolf E. *Optical Coherence and Quantum Optics.* Cambridge University Press: Cambridge, 1995; 837.

70. Egorov VK, Egorov EV. Background of X-ray nanophotonics based on the planar air waveguide-resonator. *X-Ray Spectrosc.* 2007; **36**: 381–397.

71. Zheludeva S, Kovalchuk M, Novikova N. Total reflection X-ray fluorescence study of organic nanostructures. *Spectrochim. Acta.* 2001; **56B**, #11: 2019–2026.

72. Lanterborn W, Kurz T, Wiesenfeldt M. *Coherent Optics.* Springer: Berlin, 1995; 294.

73. Rossa JJ, Moreno CH, Markoni MC, Kanizay. Soft X-ray laser interometry of a plasma with a table for laser and Lloyd's mirror. *Opt. Lett.* 1999; **24**, #6: 420–422.

74. Bonse U, Graf W. X-ray and neutron interferometry. In: *X-Ray Optics Application to Solids*, Queser HJ (ed.) Springer: Berlin, 1977: 93–143.

75. Egorov V, Egorov E. Waveguide-resonance mechanism for X-ray beam propagation: Physics and experimental background. *Adv. X-Ray Anal.* 2003; **46**: 307–313.

76. Egorov V, Egorov E. Planar waveguide-resonator: A new device for X-ray optics. *X-Ray Spectrom.* 2004; **33**: 360–371.

77. Batterman BW. Dynamic diffraction of X-ray by perfect crystals. *Rev. Mod. Phys.* 1964; **36**, #3: 681–717.

78. Authier A. *Dynamic Theory of X-Ray Diffraction.* Oxford University Press: Oxford, 2001; 661.

79. James R. The dynamic theory of X-ray diffraction. *Solid State Phys.* 1963; **15**: 53–147.

80. Bonse U, Hart M. Tailess X-ray single-crystal reflection curves obtained by multiple reflection. *Appl. Phys. Lett.* 1965; **7**, #9: 238–240.

81. Sanchez del Rio M, Ferrero C, Freudent AK. Pandellosung supersupression in the diffraction pattern of a set of thin perfect crystals in Bonse-Hart camera. *Rev. Sci. Instr.* 1995; **66**, #2: 2206–2209.

82. Underwood JH. Glancing incidence optics in X-ray astronomy: A short review. *Space Sci. Instrum.* 1975; **1**: 289–304.

83. Egorov V, Egorov E. Composite X-ray waveguide-resonator as a background for the new generation of the material testing equipment for films on Si substrate. *Mat. Res. Soc. Symp. Proc.* 2002; **716**: 189–195.

84. Frenkel YI. *Statistical Physics.* Academii Nauk SSSR Press: Moscow, 1948; 760 p. (In Russian).

85. Egorov V, Egorov E. X-ray development in light of planar waveguide-resonator modification. *Proceeding of 25th International Symposium "Nanostructures, Physics and Technology".* Academic University Press, St. Petersburg, 2017: 264–265.

86. Bronshtein IN, Semendyaev KA. *Handbook on Mathematics for Engineers and Technical Students.* Springer: New York, 2004; 610.

87. Nikitenko YuV. Neutron standing waves in layered systems: Formation, detection and application for neutron physics and for investigations of nanostructures. *Phys. Part. Nucl.* 2009; **40**, #6: 890–996.

88. Zhilyakov LA, Kostanovski AV, Iferov IG, et al. Experimental study the glassing interaction of high energy protons with surface of dielectric plate. *Abstract Book of Reports on 31 International Conference on Physics of Change Particles Beams Interaction with Crystals.* Moscow State University Press, Moscow, 2001: 22.

89. de Broglic L. The reinterpretation of wave mechanics. *Found. Phys.* 1970; **1**, #1: 5–15.

90. Jehane J. The philosopher's stone. Alchemy and chemistry. *J. Comp. Poetics* 1992; **12**: 55–77.

91. Storms E. *The Science of Low Energy Nuclear Reaction.* World Scientific: New Jersey, USA, 2007: 312.

6

Optical Tweezers

6.1 Introduction, Principle, and Application................................. **6-1**
6.2 Experimental Setup, Operation, and Measurement **6-2**
Descriptions of Optical Tweezers Setup • Optical Trap Stiffness Calibration • Optical Force Measurement • Novel Optical Tweezers
6.3 Physical Principle of Optical Tweezers................................... **6-6**
Small Particles and Rayleigh Scattering Theory • Large Particles and Geometrical Optics Method • Mesoscopic Particles and Electromagnetic Scattering Theory
6.4 Manipulation of Micro-/Nanoobjects................................... **6-8**
Dielectric Particles • Metal Particles • Dielectric-Metal Particle (Janus Particles)
6.5 Future Prospects in Biology and Physics **6-11**
Optical Tweezers in Biology • Optical Tweezers in Physics
References **6-12**

Jing Liu
South-Central University for Nationalities

Zhi-Yuan Li
South China University of Technology

6.1 Introduction, Principle, and Application

Light carries energy and momentum. Thus, light can exert force on objects and control the motions of objects remotely and wirelessly. However, this is not an intuitive and obvious concept due to the fact that the force is extremely weak in usual situations. Until the 1960s, the invention of lasers led to the discovery of an ever-increasing number of applications that utilize optical force. In 1986, A. Ashkin, the pioneer of optical tweezers, creatively used a highly focused laser beam by a high numerical aperture (NA) objective lens to implement three-dimensional (3D) trapping of dielectric particles around the focus spot [1], as shown in Figure 6.1.

To understand the principle of optical tweezers, the total optical force acting on a particle can be decomposed into two contributions: (i) the radiation pressure, known as the scattering force, which is proportional to the Poynting vector of optical field and points along the direction of the incident beam, tends to destabilize the trap; (ii) the gradient force, which is proportional to the gradient of the light intensity and points toward the trap focus, confines the particle near the focal spot. In optical tweezers, scattering and refraction of light occurs when an incident light illuminates the particle. For symmetric particles, the resulting scattering forces to cancel in all but the forward direction, and the resulting gradient forces to pull the particles to the highest intensity region of optical field, as illustrated in Figure 6.2. For 3D stable optical trapping, the axial gradient force pulling the particle towards the focal region must overwhelmingly exceed the axial scattering force pushing it

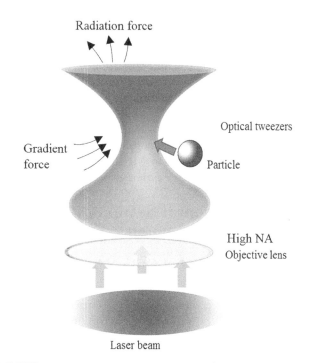

FIGURE 6.1 Basic optical tweezers. A single laser beam (usually Gaussian beam) is focused to a diffraction-limited spot by a high NA objective lens.

away from that region. This condition requests a very steep gradient in the light beam, and this can be produced by sharply focusing a laser beam (usually Gaussian beam) to a diffraction-limited spot with an objective lens of high NA.

In 1987, A. Ashkin et al. first demonstrate that optical tweezers could be utilized to manipulate bacteria and virus

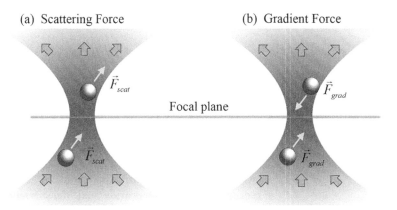

FIGURE 6.2   Optical force exerted on particles in the optical tweezers can be decomposed into (a) scattering and (b) gradient forces.

in biosciences [2]. Moreover, the optical tweezers technology has been further developed to cool and trap atoms [3] by Nobel Prize winner Steven Chu. In recent years, researches on optical tweezers have begun to spread out and expand. As is well known, optical tweezers can trap various objects as small as nanometer-size particles and can exert optical forces with controllable amount at the femtonewton resolution. When combined with microscope, optical tweezers has become a promising tool to observe and manipulate a living sample remotely with noncontact mode and a unique tool to quantify various interaction forces with high-resolution and strong analytical capability by balancing them with precisely controllable optical force. Thus, optical tweezers technology has very important applications and values in diversified research fields, such as biology, physics, optics, nanophotonics, nanoscience, and nanotechnology. For example, optical tweezers has been widely applied to investigate the mechanical properties of cell cytometry, implement force measurements of DNA- and RNA-based motors [4–6], and even make microsurgery within cells [7,8].

6.2   Experimental Setup, Operation, and Measurement

6.2.1   Descriptions of Optical Tweezers Setup

Optical tweezers is a 3D gradient force trap generated by highly focused laser beam. However, to capture and manipulate micro-/nanoscale particles and biological cells, and simultaneously monitor and record mechanical motions of these objects in real time, optical tweezers system has the capabilities of mechanical manipulation, observation, and measurement. Four main components used to construct an optical tweezers system are the laser beam, high NA microscope objective lens, image monitoring, and displacement measurements.

Laser. The selection of the laser system should take into consideration factors of wavelength, power, stability, quality, and noise to meet specific demands. To form a stable optical trap, the laser should be a continuous wave beam that can

exert a sustained optical force on objects. When trapped objects are physical samples, the strongest trap stiffness (namely, the force gradient against displacement) occurs for objects of the same size as the beam waist, which determines the optimum wavelength of laser beam. When trapped objects are biological samples, optical damage should be concerned. In general, in absorption spectroscopy, biological sample is more transparent at longer wavelengths, and water is more transparent at shorter wavelengths. Therefore, it is an important issue to avoid the absorption band of biological samples and water as far as possible when one builds an optical tweezers system. On the other hand, the laser beam will be received by the image monitoring system while exerting on samples. When the sample has a strong capability to scatter the visible laser, the diffuse scattering light of visible laser will be complete with the entire field of view (FOV) to impair the imaging quality of samples. Comparatively speaking, the near-infrared (IR) laser is a relative perfect laser resource for low absorption and invisible scattering light. In addition, the stability and quality of laser is critically relevant to the accuracy of force and position measurements, which are two crucial quantities for precise characterization and investigation of biological properties. Note that the change in microradians of the incident angle of laser beam will cause the nanometers in shifting the focus spot on the sample plane. Furthermore, the noise of laser beam also limits the position-detection resolution, such as mechanical vibrations, statistical fluctuation in light intensity, and electronic noise.

Objective lens. To form an effective optical trap, an objective lens with an NA of 1.0 or greater is necessary to highly converge the laser beam onto the diffraction limit spot. Generally, the high NA objective works at a high refractive index working medium (like oil, water, or glycerol immersion objective). The NA of oil immersion objective can be up to 1.4, which has better effect on imaging and manipulation. However, the large refractive index mismatch between the immersion oil ($n \sim 1.51$) and the aqueous medium ($n \sim 1.33$) leads to significant spherical aberration. Thus, to minimize spherical aberration, a water immersion objective of high NA (1.0, even 1.2) will be more suitable to observe a living specimen. There can be a significant concern for calibration measurement that most objectives will absorb a considerable amount of the laser light entering the back aperture, especially for IR light. Most high NA objectives contain several individual lenses that are antireflection coated for optimal visible and ultraviolet transmission, but internally reflected and absorbed for IR light. For this reason, the objective being considered for optical trapping should be characterized at the wavelength of the laser beam.

Monitoring device. Image monitoring plays a key role in observing the motion of particles or life process of biological samples. In general, an inverted microscope has been a necessary and desirable part of optical tweezers. Therefore, a video camera mounted on a microscope can serve as a monitor to image the motion of trapped particles. With regard to nanoscale objects or the specimen with a refractive index similar to the surrounding medium, the images on bright-field microscopy cannot be visualized clearly. To visualize these objects well, dark-field microscopy is utilized instead to increase image contrast and sharpness. Simultaneously, the high NA condenser should be designed to form a hollow cone of light to collect scattering light of the specimen for better optical observation.

Displacement measurement. The camera-generated images can steer the illuminated objects into digital value images. The motions of trapped objects can be extracted via a barycenter method from the digital value images. To achieve high spatial resolution and precise position measurements, the images of trapped objects can be magnified and then focused onto a photodetector at the conjugate sample plane. It is important to note that, by such a method, the diffraction-limited resolution can be exceeded, even up to 1 nm in the lateral measurement resolution.

6.2.2   Optical Trap Stiffness Calibration

As a high precision mechanical detector, optical tweezers is usually applied to measure biomolecular interaction, elastic modulus of cell wall, or the interaction force of colloidal particles. Therefore, it is critical to precisely calibrate the force exerted on samples by optical tweezers. As particles in optical tweezers are similar to a harmonic oscillator, when the distance from trapped particle center to optical trap center (namely the displacement) is close to Rayleigh distance, optical force exerted on particles is proportional to the displacement and can be described by $\vec{F}_{\text{optical}} = \kappa_x S_x \hat{x} + \kappa_y S_y \hat{y} + \kappa_z S_z \hat{z}$, where $\vec{F}_{\text{optical}}$ is the optical force, $\vec{\kappa} = [\kappa_x, \kappa_y, \kappa_z]$ is the trap stiffness, and $\vec{S} = [S_x, S_y, S_z]$ is the displacement. Moreover, the trap stiffness depends on the particle size, shape, the refractive index of particles and the surrounding medium, and the laser power. At present, there are two widely used methods to calibrate trap stiffness, namely hydrodynamics drag method and power spectrum method.

Hydrodynamic Drag Method

The hydrodynamics drag method is a common, efficient calibration method of the trap stiffness of optical tweezers. In this method, the well-established frictional viscous drag force is exploited to balance the optical force exerted on a sphere, so that $\vec{F}_{\text{drag}} = \vec{F}_{\text{optical}}$. An optically trapped sphere with radius r moves with a velocity v relative to the surrounding liquid of viscosity η, which is produced by driving the sample stage in a constant velocity. Due to the constant lateral drag force, the particle moves to a new equilibrium position where the trapping optical force and the drag force balance with each other. The optical force and trap stiffness can be expressed by

$$\vec{F}_{\text{optical}} = \vec{F}_{\text{drag}} = 6\pi\eta r\,\vec{v}, \qquad (6.1)$$

$$k_x = 6\pi\eta r v_x/S_x, k_y = 6\pi\eta r v_y/S_y. \qquad (6.2)$$

However, it should be noticed that this trap stiffness measurement neglects the axial displacement calibration of trapped objects. Differing from the lateral displacement calibration, the axial displacement depends on the object stage of a microscope driven by a piezoelectric transducer with an alternating electric voltage. The axial position corresponding to each image of the trapped particle can be determined by comparing the experimental image with the standard image.

Power Spectrum Method

The power spectrum analysis of the position of the trapped objects to calibrate optical tweezers is usually considered to be the most reliable option. By measuring and analyzing the Brownian motion of trapped objects in experiments, the trap stiffness $\vec{\kappa}$ can be obtained, independent of the detector calibration. This method is a passive measurement of optical trap stiffness and does not need to exert an external force. Therefore, this method is more suitable for real-time monitoring of the change of optical trap stiffness in biological experiments.

A particle that can move freely in a viscous fluid suffers from Brownian motion due to the continuous bombardment of solvent molecules. The diffusional motion of particle can be predicted using Einstein's expression and the free diffusion coefficient D:

$$D = \frac{k_B T}{\gamma}, \qquad (6.3)$$

where the hydrodynamic drag coefficient $\gamma = 6\pi r\eta$. Moreover, the random diffusion is proportional to the absolute temperature T. In terms of D, each coordinate $S_x(t)$ of a diffusing particle is described by

$$\text{Var}\left[S_x\right] = \overline{S_x^2} - \overline{S}_x^2 = 2Dt, \qquad (6.4)$$

And for 3D diffusion, the squared distance from the origin grows as $\vec{S}(t)^2 = 6Dt$. A particle in an optical trap experiences not only a thermal fluctuation force $\vec{F}_\xi$ from solvent molecules but also a restoring optical force

$\vec{F}_{\text{optical}} = \kappa_x S_x \hat{x} + \kappa_y S_y \hat{y} + \kappa_z S_z \hat{z}$. An approximate equation of each coordinate $S_x(t)$ for the trapped sphere is a Langevin equation, which can be expressed by

$$m\frac{d^2 S_x}{dt^2} + \gamma\frac{dS_x}{dt} + \kappa_x S_x = F_\xi(t) \quad (6.5)$$

where S_x represents the x-axis displacement of a sphere with mass m. Furthermore, in an optical tweezers system with a low Reynolds number, viscous drag is dominant over inertial forces as expressed by $m\frac{d^2 S_x}{dt^2}$, which can thus be ignored. Then, Eq. (6.5) can be written by

$$\gamma\frac{dS_x}{dt} + \kappa_x S_x = F_\xi(t). \quad (6.6)$$

If $X(f)$ is the Fourier transform of $S_x(t)$, then the Fourier transform of both sides of Eq. (6.6) is given accordingly

$$2\pi\gamma(f_c - if)X(f) = F_\xi(f), \quad (6.7)$$

where $f_c = \frac{\kappa_x}{2\pi\gamma}$ denotes the characteristic frequency of the trap. By taking their squared modulus, one obtains

$$4\pi^2\gamma^2\left(f_c^2 + f^2\right)|X(f)|^2 = |F_\xi(f)|^2. \quad (6.8)$$

The thermal fluctuation force F_ξ has an average value of zero, and its power spectrum is constant, namely $|F_\xi(t)| = 0$ and $|F_\xi(f)|^2 = 4\gamma k_B T$. According to Eq. (6.8), $S_x(f)$ can be written as follows:

$$S_x(f) = |X(f)|^2 = \frac{|F_\xi(f)|^2}{4\pi^2\gamma^2\left(f_c^2 + f^2\right)} = \frac{k_B T}{\pi^2\gamma\left(f_c^2 + f^2\right)} \quad (6.9)$$

Equation (6.9) shows that a Lorentzian power spectral density function describes how fluctuations are distributed over different frequencies f. The characteristic frequency f_c divides the Brownian motion into two regimes. For frequencies $f \ll f_c$, the power spectrum is approximately a constant, $S_x(f) \approx 4\gamma k_B T/\kappa_x$, referring to the confinement of particles. On the contrary, when $f \gg f_c$, $S_x(f)$ falls off like $1/f^2$, which indicates free diffusion and cannot confine the particles.

Once $S_x(f)$ and f_c are measured, κ_x and γ can be obtained by fitting Eq. (6.9). If γ is known in advance (for instance, for a sphere), the value of κ_x can be obtained directly from $\kappa_x = 2\pi\gamma f_c$.

6.2.3 Optical Force Measurement

Tracking the sphere position with nanometer-level resolution is important for both the calibration of trap stiffness and force measurement. The position of trapped particles has been traditionally measured using video-based position detection or photodetector-based position detection, such as a quadrant photodiode (QD) [9] and position-sensitive detector (PSD). The difference is that the video-based position detection achieves the image of trapped particles in the bright field illumination, but the photodetector-based position detection achieves the image of trapped particles in the laser. The details of the two methods are as follows.

Video-based position detection. This method requires a Charge Coupled Device (CCD), bright field illumination, and assistance with data acquisition software. In addition, it has the ability to record multiparticles even at the speed of several kilohertz by a high-speed camera, but in keeping with most CCD devices, the FOV reduces when higher frame rates are used. In the experiments, first, one uses the CCD camera to record the videos of trapped particles in optical tweezers. Second, one digitizes the signal acquired from the camera and calculates the size subtended by a single pixel. Third, one determines the position of trapped particles geometrically with subpixel accuracy using the standard weighted centroid algorithm. Finally, one analyzes a sequence of snapshots with a data acquisition software and tracks the motions of trapped particles with this algorithm in real time. However, this approach is restricted to low video acquisition rates (typically ~25–120 Hz), and the precision is ultimately limited by video pixel and variations in illumination source. Furthermore, due to limited precision, it cannot well-suit the measurement of the relative position of an object with respect to the trap center and will further complicate force measurement.

Photodetector-based position detection. This method has the ability to measure smaller forces with much higher resolution. Unlike the video-based position detection, the photodetector-based position detection requires the incorporation of a dichroic mirror and high NA condenser to couple out the laser light scattered by the sample. The photodetector and its associated optics together collect the output laser light, with the related setup, as shown in Figure 6.3. Note that the photodetector should be placed at the conjugate plane to the back focal plane of the condenser. This method takes advantage of the fact that a small displacement of a trapped particle at the focal plane results in a well-pronounced change in the interference pattern (intensity distribution of laser) on the back focal plane of the condenser. For a bare trap without particles, the intensity distribution is axis symmetric. Once a particle is trapped, the intensity distribution will shift its center of mass in proportion to the motion of the particle in the trap. As a result, the motion of the particles can be recorded by a photodetector in terms of output voltages with sampling rates of 10 kHz or higher. Moreover, this method records the displacement data of trapped particles in real time, and not the images. Thus, it is suitable for measuring displacement with nanometer resolution (Figure 6.4).

Once the optical trap stiffness and displacement of trapped particles are measured, the optical force exerted on the particles can be calculated by $\vec{F}_{\text{optical}} = \kappa_x S_x \hat{x} + \kappa_y S_y \hat{y} + \kappa_z S_z \hat{z}$.

6.2.4 Novel Optical Tweezers

As is well known, optical tweezers can limit objects bound by the optical potential well in a small space range. It can trap various objects as small as nanometer-size particles and can exert optical forces with controllable amount at

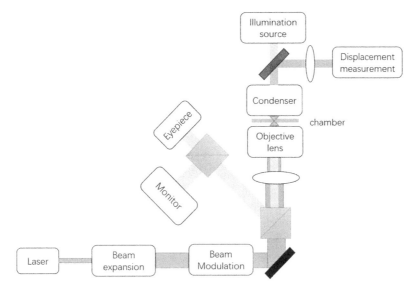

FIGURE 6.3 The basic setup of an optical tweezers system.

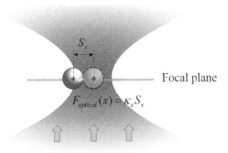

FIGURE 6.4 Illustration of the displacement and optical force of trapped objects.

the femtonewton resolution. However, optical forces on the captured objects crucially depend on the interaction between the laser field and the objects themselves. Therefore, multiple degrees and freedoms of control over optical force can be created by shaping either the laser beam or the particles. Especially, laser beam shaping, i.e., the art of controlling the amplitude, phase, and polarization profile of a laser output, is an extremely useful means to enrich optical tweezers technologies. In particular, the advent of holographic optical tweezers (HOTs) [10] greatly advances the development of optical tweezers. HOTs provides a straightforward means to create a large number of optical traps in arbitrary 3D configurations, to move them freely and independently in 3D space, and to transform them into optical vortices, optical bottles, Bessel traps, and a host of other all-optical tools [11]. Moreover, it was demonstrated that quasi-statically fast temporal control of computer-generated hologram [12], together with spatial light modulators (SLM), could achieve controlling various motions of multiple particles [13]. As a promising tool for biology and physics, HOTs will facilitate new approaches to cell sorting, 3D cellular constructs, individual particle control, assembly of structures, and integrated sensor array as well as many other possibilities.

As the conventional optical tweezers works based on the far-field technique with optical lens or microscope objectives were used to focus an incident laser beam into a tiny spot, the spatial confinement of these far-field optical tweezers is inevitably limited by the diffraction effect of light, namely the Rayleigh diffraction limit of microscope resolution. In addition, the gradient force is proportional to the third power of the particle radius (or simply speaking the volume of particle), and this intrinsically limits the ability to capture and position nanoscale objects, because in these situations, the optical trapping force is not sufficiently large to overcome the escape force imposed by the random Brownian motion of water environment. Simply using higher NA lenses or increasing the laser power does not significantly improve the optical gradient trap. In 1997, Lukas Novotny proposed a method to irradiate the metal tip with laser and achieve significant enhancement of evanescent field [14]. Unlike propagating fields used in conventional optical tweezers, the energy of evanescent field is spatially concentrated in the vicinity of the light source and extends from the interface up to several hundred nanometers away in distance. As the intensity distribution of light decays rapidly with a length far smaller than half the wavelength (the scale of conventional optical trap) this generates a very strong gradient force enabling to capture nanoscale particles [15–17]. In addition, with the development of nanofabrication technology, plasmon nanotweezers, which depends on the surface plasmon polaritons (SPPs) of a metal-dielectric interface, is particularly efficient in confining and localizing light down to the nanometer scale [18]. Utilizing the strong gradient force achieves 3D trap nanometer dielectric particles or micron metal particles [19,20]. Since the concept of using near-field effects to enhance the optical gradient force was introduced for optical trapping, various configurations have been proposed, such as nanowire waveguide, well-designed bowtie apertures, and wrapping dielectric particles with black phosphorene layers [21–23]. Meanwhile, trapping scale will be further extended to nanometers, dielectric particles, and micrometer metal particles. Rapidly developing near-field optical tweezers may help to reveal new light-matter interactions in optical tweezers and create abundant opportunities in bioscience.

6.3 Physical Principle of Optical Tweezers

Optical force is a consequence of momentum transfer from light fields to particles due to scattering, absorption, or emission of photons. Although the underlying physical principles may be simple, the implementation of precise quantitative optical force is not. In essence, it depends on many parameters, of which the most important ones are the material properties and geometric structures of particles, and the intensity distribution and polarization of the trapping laser beam. Rigorous analytical solution of light-particle interaction can be obtained from Mie's theory and its generalized formulation in the framework of classical electromagnetics and electrodynamics. Although the theory and solution are rigorous and reliable, they are quite tedious and troublesome for larger diameters and higher refractive indices of the particles, and thus, some simplification is highly desirable. Therefore, according to the size of particle in reference to the wavelength of the incident laser, the calculation of optical forces can be categorized into the following three cases. Case (1), the size of particle is far less than laser wavelength; case (2), the size of particle is much larger than laser wavelength; and case (3), the size of particle is comparable to laser wavelength. To illustrate how optical forces arise, we take a homogeneous dielectric sphere as an example to discuss. The corresponding methods are introduced and discussed in detail as follows.

6.3.1 Small Particles and Rayleigh Scattering Theory

The Rayleigh scattering theory is appropriate where the particle's radius is far smaller than the wavelength of the trapping light, $R \ll \lambda$. In the optical electromagnetic field with a wavelength far larger than the size of small particles, these particles are often regarded as electric dipoles to account for their optical and mechanical responses [24–26]. The resulting optical force $\vec{F}$ that acts on a dipole is composed of the scattering force $\vec{F}_{scat}$ and gradient force $\vec{F}_{grad}$, namely $\vec{F} = \vec{F}_{scat} + \vec{F}_{grad}$.

The induced electric dipole will oscillate synchronously with the time-harmonic electric field and emit secondary scattering waves. This process will cause the change of energy flux and amplitude of the incident laser beam, and then results in momentum transfer between light and dipole. Hence the scattering force acting on small particles can be calculated as

$$\vec{F}_{scat} = \frac{n_1}{c} C_{scat} \left\langle \vec{S} \right\rangle. \tag{6.10}$$

Here n_1 is the refractive index of the surrounding medium, $\left\langle \vec{S} \right\rangle$ is the Poynting vector, and C_{scat} is the scattering cross section for the radiation pressure of the particle. In the case of a small spherical dielectric particle with radius R that scatters light isotropically, the scattering cross section can be written by $C_{scat} = 8/3\pi k R^6 ((m^2 - 1)/(m^2 + 2))^2$, where $m = n_2/n_1$ denotes the relative refractive index of the particle and $k = 2\pi n_1/\lambda$ is the wave number of light. For particles positioned on the optical axis of the laser beam, the scattering forces will point along the propagation of the laser beam.

The second component, the gradient force, is proportional to the intensity gradient of the optical field and can be given by

$$\vec{F}_{grad} = \frac{1}{4} \varepsilon_0 \alpha \nabla \vec{E}^2 \tag{6.11}$$

where $\vec{E}$ is the electric field, α is the polarizability of the particle, and it can be written as $\alpha = 4\pi n_1^2 R^3 \left(\left(m^2 - 1 \right) / \left(m^2 + 2 \right) \right)^2$.

Notice that a common optical tweezers is strongly dependent on the NA of the objective lens. The formed optical intensity is highest at the focus spot and decays with the distance away from this spot. Obviously, the gradient force acts along the gradient in intensity, which is always pointing towards the diffraction-limited focal spot in a trap. On the other side, the repulsive scattering force balances the attractive gradient force, so that a particle will be located beyond the focus spot along the propagation of laser beam. In addition, it is worthwhile to emphasize that the scattering force is proportional to R^6 while the gradient force is proportional to R^3, $\vec{F}_{scat}$ decreases much faster than $\vec{F}_{grad}$ when the size of the nanosphere decreases. Thus, it is easier to trap a small sphere than does a large sphere. Another point is that the gradient force is proportional to the intensity at the focus spot, thus a higher power of laser beam and a smaller focus spot (e.g., realized via a higher NA objective lens) would be beneficial for reaching a larger gradient force and stronger optical trapping.

6.3.2 Large Particles and Geometrical Optics Method

When the size of the particle is much larger than the wavelength of the trapping laser, $R \gg \lambda$, the geometrical optics method is suitable and widely applied for calculating the optical forces [27,28]. In short, the incident light beam is decomposed into a bundle of individual rays, each with appropriate intensity, direction, and state of polarization. Each ray can change direction, intensity, and polarization when it reflects and refracts at dielectric interfaces according to the Fresnel formulas. Moreover, reflection and refraction can happen many times before the intensity of these rays eventually decays to zero. At each reflection and refraction, exchange of momentum between light and particle takes place, and this creates an optical force according to Newton's law of mechanics. The total force on the particle is the sum of all individual forces at each account of interface reflection and refraction for all the rays comprising the incident laser beam. The ray-optics model for calculation is shown in Figure 6.5a. The total light beam is decomposed into a lot of individual rays, and the initial propagation directions of these rays are denoted by $\vec{k}_1, \vec{k}_2, \vec{k}_3, \ldots, \vec{k}_n, \ldots$ before they interact with the sphere. Then, the overall force $\vec{F} = [F_x, F_y, F_z]$ can be given by $\vec{F} = \vec{Q} \cdot Pn_1/c$, where $\vec{Q} = [Q_x, Q_y, Q_z]$ is a dimensionless factor that describes the momentum exchange coefficient of each ray interacting with the sphere and is associated with the overall reflection of light beam.

A single ray of power P hitting the sphere at the propagation direction $\vec{k}_1$ with an incident momentum Pn_1/c per second (n_1 is the refractive index of the aqueous solution and c is the speed of light in vacuum) gives rise to multiple reflection and refraction events, as shown in Figure 6.5b in detail for clarity. Note that each time there will be a momentum transfer and exchange between light and sphere, generating optical force and torque. The factor $\vec{Q}_{k_1}$ is the sum of the reflected ray with directional strength $\vec{k}_{1r}R_1$ and the infinite number of emergent refracted rays of directional strength $\vec{k}_{2t}T_1T_2, \vec{k}_{3t}T_1R_2T_2, \ldots, \vec{k}_{nt}T_1T_n (R_2 \cdot R_3 \cdot \ldots \cdot R_{n-1}), \ldots$ The quantities T_n and R_n represent the Fresnel reflection and transmission coefficients at the nth intersection event of the transport ray with the particle surface.

Hence, the overall force contributed by this ray can be calculated via the principle of the exchange of momentum as

$$\vec{F}_{k_1} = \frac{Pn_1}{c}\left(\vec{k}_1 - \vec{k}_{1r}R_1 - \vec{k}_{2t}T_1T_2 - \right. \tag{6.12}$$

$$\left.\sum_{n=3}^{\infty} \vec{k}_{nt}T_1T_n (R_2 \cdot R_3 \cdot \ldots \cdot R_{n-1})\right) = \frac{Pn_1}{c}\vec{Q}_{k_1}.$$

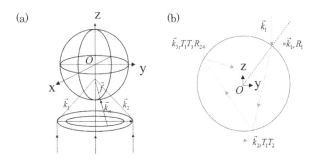

(a)   (b)

FIGURE 6.5 (a) A schematic diagram of a Gaussian laser beam tightly focused by a high-NA lens and illuminating the sphere for optical trapping and manipulation. In the ray-optics model, the laser beam is decomposed into a large amount of rays of light denoted by their directional unit vectors $\vec{k}_1, \vec{k}_2, \ldots, \vec{k}_n, \ldots$; and (b) schematic diagram of the ray tracking for a specific ray $\vec{k}_1$ within the sphere, where multiple events of ray reflection and refraction take place.

The total force imposed upon the sphere by the focused laser beam is simply the vector sum over the force of all rays of light,

$$\vec{F} = \sum_{i=1}^{N} W_{k_i}\vec{F}_{k_i} = \frac{Pn_1}{c}\sum_{i=1}^{N} W_{k_i}\vec{Q}_{k_i}. \tag{6.13}$$

Here W_{k_i} is the weight of the contribution parameter of the k_i ray of light, which is proportional to the intensity profile of the incident laser beam in the entrance pupil of the high-NA lens, and N is the total number of rays considered in the calculation.

In the geometrical optics approach, the total force can be calculated by integrating over a set of light rays representing the trapping beam, applicable for any configuration optical field. This method intuitively gives a simple while quantitative description of optical forces, originated from the exchange of energy and momentum between light and particle. Despite the neglect of wave character of light, this ray-optical model performs quite well and, especially for larger particles, seems to be in both qualitative and quantitative agreements with the forces actually measured, for example, optical forces acting on cells [29], deformation of microbubbles [30], and Kramers transitions between two optical traps [31].

6.3.3 Mesoscopic Particles and Electromagnetic Scattering Theory

For mesoscopic particles with diameter equivalent to or comparable with the laser wavelength, $R \sim \lambda$, neither ray optics nor the Rayleigh approximation is appropriate. Of course, for a spherical particle, Mie's theory can be used to yield analytical solution, albeit still very troublesome. Instead, electromagnetic scattering theory has to be used. The principle is to calculate the electromagnetic field, construct the Maxwell electromagnetic tensor, use the electromagnetic energy and momentum conservation law, and calculate the optical forces. However, this turns out to be a cumbersome procedure. At present, modeling within optical tweezers has benefited from the availability for high-performance computing. Various numerical algorithms have been developed to handle Maxwell's equations, such as finite difference time domain method (FDTD) [32], discrete dipole approximation method (DDA) [33], T-matrix method [34], and so on. By using the earlier methods to directly obtain the scattering optical field, the optical force exerted on particles can be written by

$$\vec{F} = \oint_S \overleftrightarrow{T}(r, t)dS \tag{6.14}$$

where S represents the closed surface surrounding the particles and $\overleftrightarrow{T}$ denotes the Maxwell's electromagnetic tensor. Physically, $\overleftrightarrow{T}$ accounts for the connection between optical forces and mechanical momentum exchange [35] through the following formula,

$$\vec{T} = \frac{1}{4\pi}\left[\varepsilon\vec{E}\vec{E} + \vec{H}\vec{H} - \frac{1}{2}\left(\varepsilon\vec{E}^2 + \vec{H}^2\right)\vec{I}\right] \quad (6.15)$$

The earlier paragraphs briefly discuss the theoretical approaches to understand and calculate the optical forces of objects with different sizes in optical tweezers. This knowledge becomes the physical basis and crucial point for studying the mechanical mechanism of the controlled motion of mesoscopic particles in optical tweezers. Generally speaking in physics, optical forces not only depend on the optical field but also on the particle's geometry and physical properties. Modeling the interaction of a laser beam upon a single trapped particle is just a beginning. There are many other forces that are also at work within optical tweezers, like hydrodynamic force, thermodynamic force, thermophoretic force, and various interaction forces from the walls of a sample cell. Although the concept of optical tweezers and the origin of the force produced are easy to explain by the earlier models, the precise calculation of the forces operating in any trap is complicated. At a more subtle level, the detailed and precise calculation of optical force exerted on a particle is still an area of ongoing study. On this basis, it is an urgent need to advance and develop optical tweezers theory to a high level that can achieve extremely close agreement with the experimental observations and handle the static and dynamic motions of various particles in diversified optical tweezers with high precision, light computation burden, and deep insights.

6.4 Manipulation of Micro-/Nanoobjects

6.4.1 Dielectric Particles

Optical tweezers has been rapidly developed to manipulate various objects, including biological cells and particles. Moreover, optical forces produced by commonly available lasers lie in the piconewton range, which is just right to handle experiments with individual molecules. In addition, as most biological materials absorb only weakly in the near IR region of the electromagnetic spectrum, Nd:YAG ($\lambda = 1,064$ nm) lasers are well suited to optical tweezers applications in biology. Because of the diffraction limit, the resolution of conventional optical microscopes ($\sim$200 nm) is too low for observing single protein molecules, even their small assembly. Therefore, researchers have developed a variety of techniques and methodologies that enable indirect access to these mechanical parameters, allowing for a more comprehensive understanding of the behavior of certain biological components. Especially, the magnitude of optical forces in optical tweezers system is generally insufficient to stably trap biological macromolecules themselves but more than adequate for manipulating microscopic dielectric objects, such as polystyrene spheres or silica particles, which can be utilized as carriers or handles to be biochemically linked to these molecules. To date, the greatest remarkable progress of optical tweezers is to revolutionize biology by investigating

the mechanical properties of a single molecule [36,37]. This technology has revealed important information about molecular structures and mechanisms, and their biggest impact has been in studies of molecular motors that convert chemical potential energy into mechanical work. It opens up new fields in the biological and biophysical sciences to understand how the components of life behave, not only biochemically but also mechanically.

In the majority of single-molecule optical tweezers experiments, exerting calibrated forces usually requires that the biomolecule should be attached at the other end. The common experimental setups for trapped particles linked to biosample are illustrated in Figure 6.6. In the first method, typically, one end of molecule needs to attach to the modified glass surface, and then the modified particle solution is injected into the chamber, which leads to the other end of attached molecules linked to the modified particle with a certain probability specificity. In the second method, the other attachment point is the second bead to be held on the top of a micropipette by suction, as seen in Figure 6.6. In the third method, one end of the molecule is tethered to the trapped bead and the other end to the second bead is trapped by another optical trap. In this configuration, the system can be stretched by moving the optical trap relative to the attachment point.

In the actual experiment, the first method is the simplest. However, they are susceptible to relative drift of the surface and the trap (e.g., due to thermal heating of the objective, stage settling, laser pointing fluctuations, or acoustic noise coupling). By independently measuring the relative positions of the trap and surface, noise from the surface drift can be reduced. The second method, which involves attaching one end of the molecule to a bead held in the optical tweezers and the other to a bead held by a micropipette, allows for a versatile manipulation of the sample and increased position sensitivity. This geometry, however, is still subject to the same noise sources as the first method, because of the mechanical connection through the micropipette. In the third method, using a second optical trap instead of the micropipette greatly reduces the mechanical noise, providing the most stable configuration. However, this configuration is relatively hard to achieve.

6.4.2 Metal Particles

Dielectric particles such as polystyrene or silica particles are often used as carriers or handles to measure the mechanical properties of biomolecules. However, in many experiments, especially *in vivo*, the size of the probe might be of concern. Decreasing the particle size also decreases the polarizability of particles, as already stated in Eq. (6.11). By contrast, the enhanced polarizability of metallic nanoparticles compared with dielectric nanoparticles with the same size can lead to stronger gradient forces and therefore a stronger trap stiffness. However, the trap strength increases more slowly than expected as a result of an increasing scattering force. Especially, when the size of a metal particle is close to or larger

(a)

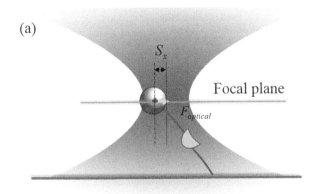

(b)

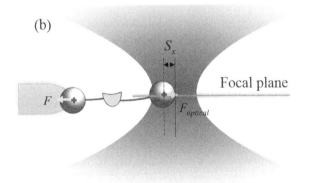

(c)

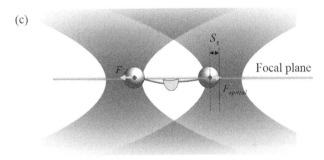

FIGURE 6.6 Three common experimental setups for the mechanical properties of biosamples in optical tweezers. One end of the biosample is directly attached to the trapped particle in an optical trap, and the other end is attached to (a) the bottom surface of the sample chamber, (b) a second particle suctioned onto the end of a micropipette, or (c) a second particle trapped in a second optical trap.

than the laser wavelength, the scattering force increases with particle size much faster than the gradient force. Therefore, these mesoscopic metal particles tend to escape from the optical trap, and the small metal particles with sizes well below the Rayleigh diffraction limit can be stably trapped.

The first trapping of gold nanoparticles was reported by Svoboda and Block in 1994 with optical tweezers [38]. Since then, a lot of experiments and theories have been developed, such as trapping of nanorods or usage of nanoparticles as a probe or handle, to monitor the motion of membrane proteins and to investigate the interaction between metal particles with optical field [33,39,40]. However, the key to the experiment study about metal particles lies in the high-resolution precision position and force detection. As

the size of trapped metal particles is below the Rayleigh diffraction limit, the spatial information about particle size and geometry are lost under an optical microscope. Therefore, the interferometric position sensing technique with high resolution has been employed, which detects the interference between the scattered laser beam by particles and the unscattered laser beam. This technique has two operation modes, the forward-scattered detection method (FSD) and back-scattered detection method (BSD), as shown in Figure 6.7. In the most common setup, the position of a trapped particle with respect to the trap center is accurately described by measuring the deflection of the forward-scattered light transmitting through the particle. The photosensitive surface of PSD should be conjugated with the rear focal plane of condenser lens. Generally, this FSD method is applied to calibrate stiffness of the optical tweezers by power spectrum method. However, in the FSD method, suspended matter will influence the interference when the transmitting laser light passes through the surrounding environment. By contrast, the BSD method effectively reduces the influence because the back-scattered light is collected by an objective lens without passing through the surrounding environment. The photosensitive surface of PSD is conjugated with the rear focal plane of an objective lens. Furthermore, the setup of BSD no longer requires a condenser, making it possible to combine the optical tweezers with other techniques, like visualizing particles by dark-field (DF) imaging. Note that it is relatively insensitive to the background mechanical noise, since lights in and out share the same objective and most optical paths. Compared with FSD, the disadvantage of BSD lies in the weaker interference signal.

As metal nanoparticles support Surface Plasmon Resonance (SPR) [41,42], this characteristic has a profound influence on their interaction with light, including strongly enhanced light absorption and light scattering. For example, plasmon excitation induces strongly amplified optical near-fields near the metal surface, so that the trapped nanoparticles can be used as antennas to effectively couple light with molecules or other objects. Moreover, the SPR properties of metal nanoparticles, including the resonance wavelength, charge displacement feature, near-field enhancement, and plasmon damping characteristics, can be precisely tuned via particle properties design, in particular, by selecting the particle shape, size, and composition. Moreover, due to the strong optical heating effect and excellent compatibility to biological molecules, metal particles also give unique possibilities for extremely localized delivery of thermal energy in life sciences. Finally yet importantly, metal particles also offer immense opportunities for light-driven research and other optomechanical applications. For example, gold nanoparticles can rotate at high frequencies of several kilohertz [43,44] in water by absorbing spin angular momentum from circularly polarized light, much faster than previously reported results of optical spinning. This ultrafast rotation of particles is highly dependent on the surrounding environment, so it could be useful for probing localized viscosity and temperature [45].

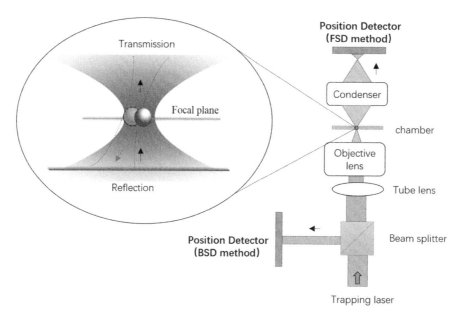

FIGURE 6.7   The sketches of FSD and BSD methods in an optical tweezers system.

6.4.3   Dielectric-Metal Particle (Janus Particles)

Due to the symmetric structure of particles, both dielectric microspheres and metal nanospheres can be firmly captured in the conventional optical tweezers and rest eventually right at the focus spot or more precisely at the minimum position of optical potential despite their initial position and velocity. In 1991, de Gennes first demonstrated the term two-faced Roman god "Janus" in his noble lecture to describe the characteristics and performance of such particles with two different components districted on an individual microobject. Here, in optical tweezers system, we introduce this special asymmetry particle, namely Janus particle involving dielectric and metal materials simultaneously, for investigating the optical and mechanical interactions between matter and light.

Until now, the optical studies on Janus particles have experienced a fast development. It has been demonstrated that the controlled stable rotation of a Janus particle by introducing patterned metal coating in a linearly polarized optical trap could be achieved [46]. The fabrication of Au-PS Janus particles by half coating polystyrene spheres (several micrometers in diameter) with a gold thin film (several nanometers in thickness) via magnetron sputtering technique is illustrated in Figure 6.8a. In experiments, two kinds of Janus particles have been obtained by controlling the concentrations of polystyrene spheres [46]. These are Janus particles with patterned and flat dividing lines, as shown in Figure 8b–d. The fact that the scattering force of the metal film layer is much higher than the dielectric layer in a strongly focused laser beam indicates the thicker the metal film, the weaker the trap efficiency. Therefore, rational design of the Janus particle with morphology and composites is a core concern for their performance.

In the experiments, when illuminated by a focused laser, the patterned Janus particles in water can stably rotate around the optical axis. A series of snapshots of clockwise and counterclockwise rotations of a Janus particle are displayed in Figure 6.9. The brighter part of the sphere corresponds to the uncoated hemisphere of polystyrene. The Au-coated hemisphere appears darker in the images, as it transmits less light. Both the rate and the direction of Janus particles can be flexibly controlled by adjusting the position or intensity of the focused laser beam. In a dramatic contrast, a Janus particle with flat dividing lines captured by the optical tweezers does not show any sustained directional rotation, but randomly vibrates near the trap center. Numerical calculations reveal that the spontaneous symmetry breaking plays a critical role to the rotation of patterned Janus particles.

Different to pure metal or pure dielectric particles, Janus particles naturally provide a route to break down the structure symmetry and have the full potential of strong coupling in rotational and translational motion. The self-propelled cyclic round-trip motion of a metallodielectric Janus particle in a static line optical tweezers (LOT) has been reported [47]. In short, this complicated mechanical motion can be described more accurately by the dynamical motion state that the Janus particle moved back and forth around two special positions. Thus, Janus particles are very good candidates to demonstrate fruitful mechanical motion modes for micro-/nanoparticles in various types of optical traps, due to the generation of complicated but controllable optical forces and torques. These unique mechanical motion properties also make Janus particles as a promising component to build self-adaptive-optic microdevices.

Although the number of reported experiments and theories about Janus particles in optical tweezers has increased steadily, the field is still relatively unexplored. However, accurate calculations about optical and thermophoresis

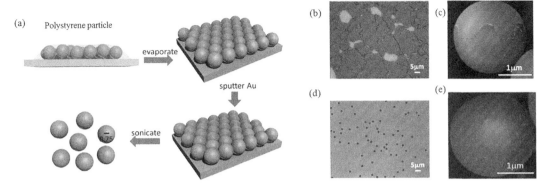

FIGURE 6.8 (a) Fabrication of Janus particle via magnetron sputtering methods. (b,d) show the micrographs of dried polystyrene particles at high and low concentrations on a glass substrate. (c,e) Scanning Electron Microscope (SEM) images of Janus particles with patterned and flat dividing lines.(Reprinted with permission from [47]. Copyright 2018 American Chemical Society.)

forces exerted on Janus particles still face intractable problems, due to the drastic contrast in space scale for optically distinct different metal and dielectric materials. Continuous developments in computational electromagnetics simulations will improve this situation and may give possibilities for the detailed dynamic process of Janus particles in various optical traps. Last but not least, with the advantage of amphipathic property, the compound Janus particles can promote magnificent prospects in diverse areas of high current interest, including targeted drug delivery, light-driven micromachines, self-assembly of particles [47–50], and more.

6.5 Future Prospects in Biology and Physics

In the past few decades, the science and technology of optical tweezers have made great progress, yet, with the input of knowledge from other research fields, such as physics, optics, nanophotonics, nanoscience, and nanotechnology, optical

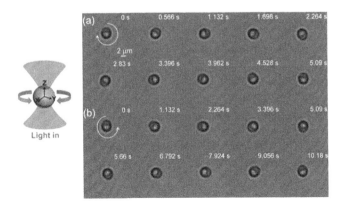

FIGURE 6.9 The bistable rotation of patterned Janus particles in an optical tweezers [47]. (a) The clockwise rotation, laser power $I = 57$ mW, objective NA = 0.7. (b) The counterclockwise rotation, laser power $I = 28.5$ mW, objective NA = 0.7. (Reprinted with permission from [47]. Copyright 2018 American Chemical Society.)

tweezers is still energetic in research and has continuously shown the power to break the physical limitations to this technology one by one. These breakthroughs include manipulating nanometer-sized dielectric particles and micrometer-sized metal particles, capturing particles in both static and dynamical ways, and achieving complicated coupling motions. As a rapidly ongoing technique, the application prospects of optical tweezers mainly include two fronts: biology and physics.

6.5.1 Optical Tweezers in Biology

With the advantageous features of nanometer accuracy, high-precision force transducer (ranging from 0.1 to 100 pN), and compatibility with liquid medium environments, optical tweezers has been one of the most prosperous tools in the biological field. One of the most remarkable achievements is that Wang et al. utilized a trapped particle in optical tweezers as a handle to achieve the elastic modulus measurement of DNA with high accuracy [51]. After that, optical tweezers has stimulated many related scientific researches of single molecules like Ribonucleic Acid Polymerase (RNAP), Small interfering Ribonucleic Acid (siRNA), DeoxyriboNucleic Acid (DNA) polymerase [52–55] and has boosted technical demand with high stability and high resolution, e.g., with a displacement control even up to 0.1 nm scale in precision. Unfortunately, the biological application in optical tweezers is basically limited to single molecule and single cell research *in vitro*. Until nowadays, measurements of biomolecular mechanical properties under physiological conditions have been recognized as magnificent goals in academic research.

Although the measurement and operation of these living organisms *in vivo* have faced great challenge and difficulty, optical tweezers can no doubt provide the opportunity for characterizing the dynamics of these life systems. In 2013, Yinmei Li's research group first demonstrated that they achieved an optical capture of red blood cells in living animals [56] with optical tweezers technology. This achievement provides a completely new technical method for living research and clinical diagnosis and has shown a great potential of optical tweezers in studying life science *in vivo*.

6.5.2 Optical Tweezers in Physics

In the early years, the technique of optical tweezers was successfully applied to capture and manipulate micrometer particles [57] and even atoms [58]. However, optical forces on the captured objects depend crucially on the interaction between the laser field and the objects themselves. Therefore, multiple degrees and freedoms of control over optical force can arise by shaping either the laser beam or the particles. To achieve various motions of trapped micro-objects, two typical schemes have been successfully adopted in an optical tweezers system.

One scheme is the laser beam shaping realized by controlling the amplitude, phase, and polarization profile of the laser to enrich optical tweezers technologies. Various novel laser beams, like optical vortex beams [59], Bessel beams [60], Laguerre-Gaussian beams (LG beam) [61], have been applied to manipulate microparticles. For instance, "nondiffracting" Bessel beams have been exploited to trap atoms and microscopic particles in multiple planes and to construct conveyor belts for them [62]. When embedded within a quasi-perfect optical vortex beam, low-refractive-index hollow dielectric particles will rotate around the propagation of laser via the transfer of optical orbital angular momentum of an optical vortex [63].

Other schemes make use of the special properties of particles, for instance, introducing precisely fabricated asymmetric micro-objects [64], such as microturbines or gammadion-shaped microrotors, birefringent particles [65,66], and Janus particles [47] into an optical tweezers system. Besides capturing particles as a handle for molecule force measurement, fabrication of needle-like particles has been illustrated as a scanning probe to detect surface morphology wirelessly with the lateral resolution of 200 nm and the depth resolution of ~10 nm [67]. Inspired by the complex organization of biological machines, researchers have developed similar artificial micromotors, driven by light field, to achieve self-driven motions. Because of the small size and strong loading capacity, micromotors have emerged and advanced quickly in the fields of drug delivery [68], biomedicine [69], and chemical analysis [70].

With considerable advances, optical tweezers has opened up exciting possibilities for assembly [71], force transducer [72], morphology probe [67], and the embryo of micromachines [47,73]. Despite the evolving scientific research, there is still plenty of room for further development. The improvement of optical tweezers setup with higher spatial resolution can greatly promote our capability to monitor various physical parameters, giving the precise space-time configuration and evolution of particles in focused optical field. Moreover, taking the influence of external factors into account, such as Brownian motion of particles, thermal effect, and interference of other particles, the light–particle interaction in optical tweezers can be analyzed comprehensively for exploring novel ways to control the dynamics of particles. The last but not the least, the general, accurate, and efficient theoretical and numerical calculations [26,74,75] of optical forces and torques remain a challenge. Therefore, to calculate all the forces and torques exerted on particles, modeling rigorous numerical approach in an accurate way at the nanometer scale needs to be made, and they can play a vital role in pushing flourishing applications for controllable motions of micromachines.

References

1. A. Ashkin, J. M. Dziedzic, J. E. Bjorkholm, and S. Chu, Observation of a single-beam gradient force optical trap for dielectric particles, *Optics Letters* **11**, 288 (1986).
2. A. Ashkin, and J. M. Dziedzic, Optical trapping and manipulation of viruses and bacteria, *Science* **235**, 1517–1520 (1987).
3. S. Chu, J. E. Bjorkholm, A. Ashkin, and A. Cable, Experimental observation of optically trapped atoms, *Physical Review Letters* **57**, 314–317 (1986).
4. Z. Bryant, D. Altman, and J. A. Spudich, The power stroke of myosin VI and the basis of reverse directionality, *Proceedings of the National Academy of Sciences of the United States of America* **104**, 772–777 (2007).
5. C. G. Baumann, V. A. Bloomfield, S. B. Smith, C. Bustamante, M. D. Wang, and S. M. Block, Stretching of single collapsed DNA molecules, *Biophysical Journal* **78**, 1965–1978 (2000).
6. G. L. Liu, Y. Yin, S. Kunchakarra, B. Mukherjee, D. Gerion, S. D. Jett, D. G. Bear, J. W. Gray, A. P. Alivisatos, L. P. Lee, and F. F. Chen, A nanoplasmonic molecular ruler for measuring nuclease activity and DNA footprinting, *Nature Nanotechnology* **1**, 47–52 (2006).
7. M. M. Wang, E. Tu, D. E. Raymond, J. M. Yang, H. Zhang, N. Hagen, B. Dees, E. M. Mercer, A. H. Forster, I. Kariv, P. J. Marchand, and W. F. Butler, Microfluidic sorting of mammalian cells by optical force switching, *Nature Biotechnology* **23**, 83 (2004).
8. A. Clement-Sengewald, K. Schütze, A. Ashkin, G. A. Palma, G. Kerlen, and G. Brem, Fertilization of bovine oocytes induced solely with combined laser microbeam and optical tweezers, *Journal of Assisted Reproduction and Genetics* **13**, 259–265 (1996).
9. R. M. Simmons, J. T. Finer, S. Chu, and J. A. Spudich, Quantitative measurements of force and displacement using an optical trap, *Biophysical Journal* **70**, 1813–1822 (1996).
10. A. Jesacher, S. Fürhapter, C. Maurer, S. Bernet, and M. Ritsch-Marte, Holographic optical tweezers for

object manipulations at an air-liquid surface, *Optics Express* **14**, 6342 (2006).

11. D. G. Grier, A revolution in optical manipulation, *Nature* **424**, 810–816 (2003).

12. G.-Z. Yang, B.-Z. Dong, B.-Y. Gu, J.-Y. Zhuang, and O. K. Ersoy, Gerchberg–Saxton and Yang–Gu algorithms for phase retrieval in a nonunitary transform system: A comparison, *Applied Optics* **33**, 209–218 (1994).

13. J. A. Rodrigo, and T. Alieva, Freestyle 3D laser traps: Tools for studying light-driven particle dynamics and beyond, *Optica* **2**, 812–815 (2015).

14. L. Novotny, R. X. Bian, and X. S. Xie, Theory of nanometric optical tweezers, *Physical Review Letters* **79**, 645–648 (1997).

15. A. Maimaiti, V. G. Truong, M. Sergides, I. Gusachenko, and S. Nic Chormaic, Higher order microfibre modes for dielectric particle trapping and propulsion, *Scientific Reports* **5**, 9077 (2015).

16. G. Rui, X. Wang, and Y. Cui, Manipulation of metallic nanoparticle with evanescent vortex Bessel beam, *Optics Express* **23**, 25707–25716 (2015).

17. H. Ito, K. Sakaki, W. Jhe, and M. Ohtsu, Atomic funnel with evanescent light, *Physical Review A* **56**, 712–718 (1997).

18. M. L. Juan, M. Righini, and R. Quidant, Plasmon nano-optical tweezers, *Nature Photonics* **5**, 349–356 (2011).

19. Y. Zhang, X. Dou, Y. Dai, X. Wang, C. Min, and X. Yuan, All-optical manipulation of micrometer-sized metallic particles, *Photonics Research* **6**, 66–71 (2018).

20. G. Volpe, R. Quidant, G. Badenes, and D. Petrov, Surface plasmon radiation forces, *Physical Review Letters* **96**, 238101 (2006).

21. A. H. Yang, S. D. Moore, B. S. Schmidt, M. Klug, M. Lipson, and D. Erickson, Optical manipulation of nanoparticles and biomolecules in sub-wavelength slot waveguides, *Nature* **457**, 71–75 (2009).

22. R. A. Jensen, I. C. Huang, O. Chen, J. T. Choy, T. S. Bischof, M. Lonar, and M. G. Bawendi, Optical trapping and two-photon excitation of colloidal quantum dots using bowtie apertures, *ACS Photonics* **3**, 423–427 (2016).

23. Y. Yang, X. Jiang, B. Ruan, X. Dai, and Y. Xiang, Tunable optical forces exerted on a black phosphorus coated dielectric particle by a Gaussian beam, *Optical Materials Express* **8**, 211 (2018).

24. J. P. Barton, D. R. Alexander, and S. A. Schaub, Internal and near-surface electromagnetic fields for a spherical particle irradiated by a focused laser beam, *Journal of Applied Physics* **64**, 1632 (1988).

25. J. P. Barton, D. R. Alexander, and S. A. Schaub, Theoretical determination of net radiation force and torque for a spherical particle illuminated by a focused laser beam, *Journal of Applied Physics* **66**, 4594–4602 (1989).

26. Y. Harada, and T. Asakura, Radiation forces on a dielectric sphere in the Rayleigh scattering regime, *Optics Communications* **124**, 529–541 (1996).

27. A. Ashkin, Forces of a single-beam gradient laser trap on a dielectric sphere in the ray optics regime, *Biophysical Society* **61**, 14 (1986).

28. J. Liu, C. Zhang, Y. Zong, H. Guo, and Z.-Y. Li, Ray-optics model for optical force and torque on a spherical metal-coated Janus microparticle, *Photonics Research* **3**, 265–274 (2015).

29. Y.-R. Chang, L. Hsu, and S. Chi, Optical trapping of a spherically symmetric sphere in the ray-optics regime: A model for optical tweezers upon cells, *Applied Optics* **45**, 3885–3892 (2006).

30. S. E. Skelton, M. Sergides, G. Memoli, O. M. Maragó, and P. H. Jones, Trapping and deformation of microbubbles in a dual-beam fibre-optic trap, *Journal of Optics* **14**, 075706 (2012).

31. A. Callegari, M. Mijalkov, A. B. Gököz, and G. Volpe, Computational toolbox for optical tweezers in geometrical optics, *Journal of the Optical Society of America B* **32**, B11 (2015).

32. Y. Zhang, X. Dou, Y. Dai, X. Wang, C. Min, and X. Yuan, All-optical manipulation of micrometer-sized metallic particles, Photonics Research **6**, 66 (2018).

33. L. Ling, H. L. Guo, X. L. Zhong, L. Huang, J. F. Li, L. Gan, and Z. Y. Li, Manipulation of gold nanorods with dual-optical tweezers for surface plasmon resonance control, *Nanotechnology* **23**, 215302 (2012).

34. F. Borghese, P. Denti, R. Saija, M. A. Iatì, and O. M. Maragò, Radiation torque and force on optically trapped linear nanostructures, *Physical Review Letters* **100**, 163903 (2008).

35. O. M. Marago, P. H. Jones, P. G. Gucciardi, G. Volpe, and A. C. Ferrari, Optical trapping and manipulation of nanostructures, *Nature Nanotechnology* **8**, 807–819 (2013).

36. H. Guo, and Z. Li, Optical tweezers technique and its applications, *Science China Physics, Mechanics and Astronomy* **56**, 2351–2360 (2013).

37. M. C. Zhong, Z. Q. Wang, and Y. M. Li, Aberration compensation for optical trapping of cells within living mice, *Applied Optics* **56**, 1972–1976 (2017).

38. K. Svoboda, and S. M. Block, Optical trapping of metallic Rayleigh particles, *Optics Letters* **19**, 930–932 (1994).

39. M. Pelton, M. Z. Liu, H. Y. Kim, G. Smith, P. Guyot-Sionnest, and N. E. Scherer, Optical trapping and alignment of single gold nanorods by using plasmon resonances, *Optics Letters* **31**, 2075 (2006).

40. Y. Sako, and A. Kusumi, Barriers for lateral diffusion of transferrin receptor in the plasma membrane as characterized by receptor dragging with laser tweezers: Fence versus tether, *Journal of Cell Biology* **129**, 1559–1574 (1995).

41. L. Huang, H. Guo, K. Li, Y. Chen, B. Feng, and Z.-Y. Li, Three dimensional force detection of gold

nanoparticles using backscattered light detection, *Journal of Applied Physics* **113**, 113103 (2013).

42. L. Ling, L. Huang, J. Fu, H. Guo, J. Li, H. D. Ou-Yang, and Z. Y. Li, The properties of gold nanospheres studied with dark field optical trapping, *Optics Express* **21**, 6618–6624 (2013).

43. A. Lehmuskero, R. Ogier, T. Gschneidtner, P. Johansson, and M. Kall, Ultrafast spinning of gold nanoparticles in water using circularly polarized light, *Nano Letters* **13**, 3129–3134 (2013).

44. L. Shao, Z. J. Yang, D. Andren, P. Johansson, and M. Kall, Gold nanorod rotary motors driven by resonant light scattering, *ACS Nano* **9**, 12542–12551 (2015).

45. L. Shao, and M. Käll, Light-driven rotation of plasmonic nanomotors, *Advanced Functional Materials*, 1706272 (2018).

46. Y. Zong, J. Liu, R. Liu, H. Guo, M. Yang, Z. Li, and K. Chen, An optically driven bistable Janus rotor with patterned metal coatings, *ACS Nano* **9**, 10844–10851 (2015).

47. J. Liu, H.-L. Guo, and Z.-Y. Li, Self-propelled round-trip motion of Janus particles in static line optical tweezers, *Nanoscale* **8**, 19894–19900 (2016).

48. Z. Zhao, F. Zhu, X. Qu, Q. Wu, Q. Wang, G. Zhang, and F. Liang, pH-Responsive polymeric Janus containers for controlled drug delivery, *Polymer Chemistry* **6**, 4144–4153 (2015).

49. R. S. Larysa Baraban, D. Makarov, L. Han, D. Karnaushenko, O. G. Schmidt, and G. Cuniberti, Fuel-free locomotion of Janus motors: Magnetically induced thermophoresis, *ACS Nano* **7**, 1360–1367 (2013).

50. A. Walther, M. Drechsler, S. Rosenfeldt, L. Harnau, M. Ballauff, V. Abetz, and A. H. E. Mueller, Self-assembly of Janus cylinders into hierarchical superstructures, *Journal of the American Chemical Society* **131**, 4720–4728 (2009).

51. M. D. Wang, H. Yin, R. Landick, J. Gelles, and S. M. Block, Stretching DNA with optical tweezers, *Biophysical Journal* **72**, 1335–1346 (1997).

52. M. D. Wang, M. J. Schnitzer, H. Yin, R. Landick, J. Gelles, and S. M Block, Force and velocity measured for single molecules of RNA polymerase, *Science* **282**, 902–907 (1998).

53. T. Hong, H. Guo, and Y.-Q. Xu, Carbon nanotube-mediated siRNA delivery for gene silencing in cancer cells, *Biosensing and Nanomedicine IV* **8099**, 80990H (2011).

54. J. Qiao, T. Hong, H. Guo, Y. Q. Xu, and D. H. Chung, Single-walled carbon nanotube-mediated small interfering RNA delivery for gastrin-releasing peptide receptor silencing in human neuroblastoma, *Methods in Molecular Biology* **1026**, 137–147 (2013).

55. G. J. L. Wuite, S. B. Smith, M. Young, D. Keller, and C. Bustamante, Single-molecule studies of the effect of template tension on T7 DNA polymerase activity, *Nature* **404**, 103 (2000).

56. M. C. Zhong, X. B. Wei, J. H. Zhou, Z. Q. Wang, and Y. M. Li, Trapping red blood cells in living animals using optical tweezers, *Nature Communications* **4**, 1768 (2013).

57. A. Ashkin, Acceleration and trapping of particles by radiation pressure, *Physical Review Letters* **24**, 156–159 (1970).

58. A. Ashkin, Atomic-beam deflection by resonance-radiation pressure, *Physical Review Letters* **25**, 1321–1324 (1970).

59. J. Ng, Z. Lin, and C. T. Chan, Theory of optical trapping by an optical vortex beam, *Physical Review Letters* **104**, 103601 (2010).

60. J. Arlt, V. Garces-Chavez, W. Sibbett, and K. Dholakia, Optical micromanipulation using a Bessel light beam, *Optics Communications* **197**, 239–245 (2001).

61. L. Allen, M. W. Beijersbergen, R. J. C. Spreeuw, and J. P. Woerdman, Orbital angular momentum of light and the transformation of Laguerre-Gaussian laser modes, *Physical Review A* **45**, 8185–8189 (1992).

62. V. Garcés-Chávez, D. McGloin, H. Melville, W. Sibbett, and K. Dholakia, Simultaneous micromanipulation in multiple planes using a self-reconstructing light beam, *Nature* **419**, 145 (2002).

63. Y. Liang, M. Lei, S. Yan, M. Li, Y. Cai, Z. Wang, X. Yu, and B. Yao, Rotating of low-refractive-index microparticles with a quasi-perfect optical vortex, *Applied Optics* **57**, 79–84 (2018).

64. T. Asavei, V. L. Y. Loke, M. Barbieri, T. A. Nieminen, N. R. Heckenberg, and H. Rubinsztein-Dunlop, Optical angular momentum transfer to microrotors fabricated by two-photon photopolymerization, *New Journal of Physics* **11**, 093021 (2009).

65. M. E. J. Friese, T. A. Nieminen, N. R. Heckenberg, and H. Rubinsztein-Dunlop, Optical alignment and spinning of laser-trapped microscopic particles, *Nature* **394**, 348–350 (1998).

66. S. H. Simpson, and S. Hanna, Application of the discrete dipole approximation to optical trapping calculations of inhomogeneous and anisotropic particles, *Optics Express* **19**, 16526–16541 (2011).

67. D. B. Phillips, M. J. Padgett, S. Hanna, Y. L. D. Ho, D. M. Carberry, M. J. Miles, and S. H. Simpson, Shape-induced force fields in optical trapping, *Nature Photonics* **8**, 400–405 (2014).

68. W. Gao, and J. Wang, Synthetic micro/nanomotors in drug delivery, *Nanoscale* **6**, 10486–10494 (2014).

69. L. K. E. A. Abdelmohsen, F. Peng, Y. Tu, and D. A. Wilson, Micro- and nano-motors for biomedical applications, *Journal of Materials Chemistry B* **2**, 2395–2408 (2014).

70. F. Wong, K. K. Dey, and A. Sen, Synthetic micro/nanomotors and pumps: Fabrication and applications, *Annual Review of Materials Research* **46**, 407–432 (2016).

71. J. Liu, and Z.-Y. Li, Light-driven crystallization of polystyrene micro-spheres, *Photonics Research* **5**, 201 (2017).

72. M. A. Taylor, J. Janousek, V. Daria, J. Knittel, B. Hage, H.-A. Bachor, and W. P. Bowen, Biological measurement beyond the quantum limit, *Nature Photonics* **7**, 229–233 (2013).

73. S. Kuhn, B. A. Stickler, A. Kosloff, F. Patolsky, K. Hornberger, M. Arndt, and J. Millen, Optically driven ultra-stable nanomechanical rotor, *Nature Communications* **8**, 1670 (2017).

74. S. Chang, and S. S. Lee, Optical torque exerted on a homogeneous sphere levitated in the circularly polarized fundamental-mode laser beam, *Journal of the Optical Society of America B* **2**, 1853 (1985).

75. A. Rahimzadegan, M. Fruhnert, R. Alaee, I. Fernandez-Corbaton, and C. Rockstuhl, Optical force and torque on dipolar dual chiral particles, *Physical Review B* **94**, 125123 (2016).

7

Metal Nanostructures with Plasmonically Enhanced Raman and Photoluminescence Signals

7.1 Introduction... 7-1
7.2 Fundamentals of Plasmonic Metal Nanostructures 7-2
   Maxwell's Equation in a Medium • EM Wave Equations • Localized Surface
   Plasmons • Mie Theory • Plasmon Hybridization Model • Dielectric Function
7.3 Plasmonic Effects on Enhanced Spectroscopies 7-5
   Molecular Plasmonics • Direct PL of Metal Nanostructures
7.4 Molecular Raman Enhancement 7-10
   Single Particles • Inter-Nanogap-Based Plasmonic Particles •
   Nanocrevice-Based Plasmonic Particles • Intra-Nanogap-Based Plasmonic
   Particles
7.5 Direct PL Enhancement of Metals...................................... 7-22
   Single Nanoparticles and Suggested Mechanisms • Plasmonic Coupling Effect
7.6 Future Directions... 7-33
References 7-33

Jeong-Eun Park, Minho Kim,
Jiwoong Son, Chungyeon Lee,
Sung Min Ko, and
Jwa-Min Nam
Seoul National University

7.1 Introduction

Shrinking light at the nanoscale with plasmonic nanomaterials has yielded numerous intriguing outcomes. In particular, the considerable local field enhancement that arises from the strong light confinement result in significantly improved optical responses such as surface-enhanced Raman scattering (SERS) and metal-enhanced fluorescence (MEF) (Lakowicz 2005, Le Ru and Etchegoin 2012, Dong et al. 2015, Ding et al. 2016, Nam et al. 2016, Park et al., 2017a). SERS is one of the most distinguished and quickly evolving fields. Raman spectroscopy was developed as early as 1928, and this technique involves the inelastic scattering of light and provides information about the molecular structure of chemicals (Raman and Krishnan 1928). It has, however, not been heavily utilized for practical applications because of the small molecular Raman cross section. Since the measurement of the Raman spectra of submonolayer molecules on roughened metal surfaces (Albrecht and Creighton 1977, Jeanmaire and Van Duyne 1977) and the possibility of single-molecule SERS (SMSERS) (Kneipp et al. 1997, Nie and Emory 1997), recent advances have enabled SERS as an effective tool for chemical sensing and bioimaging (Ding et al. 2016, Kumar et al. 2016)· When the plasmonic nanostructure couples with fluorophores, the florescence signal increases by plasmonic effect (Drexhage 1974). Recent advances in structural design for proper distance control

and flexible modulation of plasmonic resonance to maximize the fluorescence signal allow enhancement factors (EFs) over several thousand times for fluorophores emitting visible light (Punj et al. 2013, Puchkova et al. 2015). Also, infrared (IR)-emitting fluorophores, whose quantum yield is usually very low showed more than one order of magnitude enhancement (Ayala-Orozco et al. 2014). While SERS and MEF are caused by the interaction between molecules and plasmonic nanostructures, the direct emission of light by metals is called photoluminescence (PL). This phenomenon can be divided into two classes: one-photon-induced PL and two-photon-induced PL (Boyd et al. 1986). One-photon PL is a process involving a single photon, while in the other, two photons are sequentially absorbed in the two-photon PL process and photons of higher energy are then emitted (Mooradian 1969, Imura et al. 2005). Recently, plasmonic effects on one-photon PL have attracted increasing attention. Although light emission from metals is much weaker than that from molecular systems because of fast and strong electron–electron interactions and electron–phonon interactions (Hartland 2011), light emission by PL does not undergo photobleaching or photoblinking, and can be further enhanced via molecular and nanostructural engineering, and can therefore be distinguished from the PL of organic dyes or quantum dots.

This chapter attempts to introduce and discuss metal nanostructures that have shown enhanced Raman scattering and single-photon PL with plasmonic effects in a comprehensive manner. We first introduce the fundamentals of plasmonic metal nanostructures. Then we introduce the concepts of enhanced spectroscopic techniques of SERS, tip-enhanced Raman scattering (TERS), MEF, and surface-enhanced IR absorption (SEIRA) and PL of metallic nanostructures. Then, we highlight investigations and advances in plasmonically enhanced Raman and PL with metal nanostructures in terms of design principles and structural features. Finally, future directions in these fields are discussed, including both challenges and opportunities.

7.2 Fundamentals of Plasmonic Metal Nanostructures

Surface plasmons, which are coherent oscillations of free electrons in a metal, have an intrinsic wave nature. This naturally requires describing light as an electromagnetic (EM) wave and surface plasmons as its response, based on Maxwell's equation. Classical electrodynamics can provide plausible explanations of macroscopic phenomena, including surface plasmon resonance, the large absorption/scattering cross sections of metal nanostructures, or the strong light-molecule interactions induced by surface plasmon-enhanced electric fields. However, Maxwell's equation does not divulge the microscopic origin of the interactions between materials and an external EM field. These properties of metals are phenomenologically described by complex dielectric functions. In this section, we start by revisiting Maxwell's equation and discuss how Maxwell's equation reflects the characteristics of surface plasmons in metal nanostructures. Lastly, a brief introduction of the microscopic origins of phenomenological dielectric functions will be presented.

7.2.1 Maxwell's Equation in a Medium

To discuss the EM wave equations and surface plasmons, we start with Maxwell's equations in matter

$$\nabla \cdot \mathbf{D} = \rho \tag{7.1}$$

$$\nabla \cdot \mathbf{B} = 0 \tag{7.2}$$

$$\nabla \times \mathbf{E} = -\frac{\partial \mathbf{B}}{\partial t} \tag{7.3}$$

$$\nabla \times \mathbf{H} = \frac{\partial \mathbf{D}}{\partial t} + \boldsymbol{J} \tag{7.4}$$

where $\mathbf{E}$ is the electric field, $\mathbf{B}$ is the magnetic flux density, $\mathbf{D}$ is the electric displacement field, $\mathbf{H}$ is the magnetic field, ρ is the charge density, and $\boldsymbol{J}$ is the current density. Here, we consider a linear, dispersive medium. Therefore, $\mathbf{D}, \mathbf{E}, \mathbf{B}$, and $\mathbf{H}$ have the following constitutive relations

$$\mathbf{D} = \epsilon_0 \mathbf{E} + \mathbf{P} = \epsilon \mathbf{E} \tag{7.5}$$

$$\mathbf{H} = \frac{\mathbf{B}}{\mu_0} + \mathbf{M} = \frac{\mathbf{B}}{\mu} \tag{7.6}$$

where $\mathbf{P}$ is the polarization, $\mathbf{M}$ is the magnetization, ϵ is the permittivity, and μ is the permeability of the

medium. The permittivity and permeability subscripted with a "naught" symbol denote values in vacuum. The typical plasmonic resonance of metal nanostructures lies within the optical frequencies, where the medium shows a nonmagnetic behavior ($\mu = \mu_0$), and thus the permittivity solely defines the medium properties. In literature on nanoparticle plasmonics, the permittivity is more often referred to as the dielectric function, implying its dependence on the angular frequency ($\epsilon = \epsilon(\omega)$) of oscillating EM fields.

7.2.2 EM Wave Equations

To derive the propagation of EM waves through a medium, we first consider Maxwell's equation with a source-free condition ($\rho = 0$, $\boldsymbol{J} = 0$).

By substituting $\mathbf{D}, \mathbf{B}$ in Eqs. (7.3) and (7.4) and by taking the curl of Eq. (7.3), we obtain

$$\nabla \times \nabla \times \mathbf{E} = -\mu \left(\nabla \times \frac{\partial \mathbf{H}}{\partial t} \right) = -\mu \frac{\partial}{\partial t} (\nabla \times \mathbf{H})$$

The left-hand side can be reduced with the identity ($\nabla \times \nabla \times \mathbf{E} = \nabla (\nabla \cdot \mathbf{E}) - \nabla^2 \mathbf{E} = -\nabla^2 \mathbf{E}$). The divergence of the electric field vanishes because of the source-free condition. By substituting the right-hand side with Eqs. (7.3) and (7.6), we obtain

$$-\frac{\partial}{\partial t} (\nabla \times \mathbf{B}) = -\frac{1}{\mu} \frac{\partial}{\partial t} (\nabla \times \mathbf{H}) = -\frac{1}{\mu} \frac{\partial^2 \mathbf{D}}{\partial t^2} = -\frac{1}{\mu \epsilon} \frac{\partial^2 \mathbf{E}}{\partial t^2}$$

Removing the negative sign on each term yields the final EM wave equation

$$\nabla^2 \mathbf{E} = \frac{1}{c^2} \frac{\partial^2 \mathbf{E}}{\partial t^2}$$

where $c = \frac{1}{\sqrt{\mu \epsilon}}$ is the velocity of light in a medium. It is possible to obtain the wave equation for the magnetic field $\mathbf{H}$ with the same formulation by taking the curl of Eq. (7.4), and the straightforward conclusion is

$$\nabla^2 \mathbf{H} = \frac{1}{c^2} \frac{\partial^2 \mathbf{H}}{\partial t^2}$$

For the time-harmonic fields ($\mathbf{E}, \mathbf{H} \propto \mathrm{e}^{\mathrm{i}\omega t}$), $\frac{\partial}{\partial t} \rightarrow i\omega$, the two wave equations become

$$\nabla^2 \mathbf{E} + k^2 \mathbf{E} = 0$$

$$\nabla^2 \mathbf{H} + k^2 \mathbf{H} = 0$$

where $k^2 = \omega^2 \epsilon \mu$. This equation is called the *Helmholtz equation*.

7.2.3 Localized Surface Plasmons

Typically, metal nanoparticles synthesized from a bottom-up approach have dimensions of about 5–100 nm. This size is comparable to the skin depth of metals, and all the electrons in the metal nanoparticles can interact with the penetrating EM field. Under an oscillating EM field, the electron cloud of metal nanostructures oscillates along the electric field in a coherent fashion. This coherent oscillation of the electron cloud is called a "localized surface plasmon" (LSP) (Willets and Van Duyne 2007, Mayer and Hafner 2011). The LSP is responsible for the exceptionally strong absorption and scattering of small metal nanoparticles. The typical resonance wavelength of surface plasmons lies in the visible region, and it is possible to distinguish resonances with the naked eye by identifying the color of nanoparticle solutions.

For extremely small particles compared with EM wavelengths (e.g., 30 nm), the electric field acting on the particles can be considered to be uniform. We can thus regard the EM field as having only harmonic time dependence and can ignore the spatial dependence. This regime is called *quasi-static*.

Consider a sphere of radius R with a dielectric function ϵ embedded in a medium with dielectric function ϵ_m under a uniform electric field $\mathbf{E} = E_0 \hat{z}$ (Figure 7.1). From the static condition, the problem is reduced by solving the Laplace equation

$$\nabla^2 V = 0$$

where V is the electric potential. We can evaluate $\mathbf{E}$ from V with the relation $\mathbf{E} = -\nabla V$. Using spherical coordinates, the electric potential inside and outside the sphere is

$$V_{\text{in}}(r,\theta) = \sum_{l=0}^{\infty} A_l r^l P_l(\cos\theta) \, (r \le R)$$

$$V_{\text{out}}(r,\theta) = \sum_{l=0}^{\infty} \frac{B_l}{r^{l+1}} P_l(\cos\theta) \, (r \ge R)$$

The r^l term in V_{out} and the $\frac{1}{r^{l+1}}$ term in V_{in} are not taken into account, because they diverge as $r \to \infty$ and $r \to 0$, respectively. The boundary conditions require that (i) the tangential component of $\mathbf{E}$ is continuous at $r = R$,

$$-\frac{1}{R}\frac{\partial V_{\text{in}}}{\partial \theta} = -\frac{1}{R}\frac{\partial V_{\text{out}}}{\partial \theta}(\text{at } r = R)$$

and (ii) the normal component of $\mathbf{D}$ is continuous.

$$-\epsilon \frac{\partial V_{\text{in}}}{\partial r} = -\epsilon_m \frac{\partial V_{\text{out}}}{\partial r}(\text{at } r = R)$$

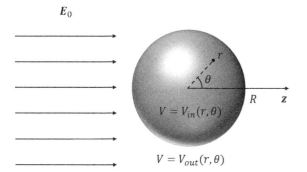

FIGURE 7.1 Small spherical metal nanoparticle of radius R under uniform electric field E_0 along the z direction. The electrostatic potential inside and outside the sphere can be evaluated by solving the Laplace equation with appropriate boundary condition at $r = R$.

These boundaries lead to the following equations.

$$V_{\text{in}} = -\frac{3\epsilon_m}{\epsilon + 2\epsilon_m} E_0 r \cos\theta$$

$$V_{\text{out}} = -E_0 r \cos\theta + \frac{\epsilon - \epsilon_m}{\epsilon + 2\epsilon_m}\frac{R^3}{r^2}\cos\theta$$

We can rearrange V_{out} into

$$V_{\text{out}} = -E_0 r \cos\theta + \frac{\boldsymbol{p}\cdot\hat{\boldsymbol{r}}}{4\pi\epsilon r^2}$$

with $\boldsymbol{p} = 4\pi\epsilon R^3 \frac{\epsilon-\epsilon_0}{\epsilon+2\epsilon_m} \boldsymbol{E}_0$. The second term of V_{out} is interpreted as the electric potential from the dipole $\boldsymbol{p}$. Thus, the excited surface plasmon is called the electric dipole mode. In a linear medium, the polarization is proportional to the incident electric field, and the proportionality constant is called the *polarizability*. Here, the polarizability of the sphere is $4\pi\epsilon R^3 \frac{\epsilon-\epsilon_0}{\epsilon+2\epsilon_m}$. At resonance, the polarizability has the maximum value. We can easily determine the resonance by calculating the frequency for which $|\epsilon + 2\epsilon_m|$ has a minimum value.

We now focus on the EM field created by the induced dipole $\boldsymbol{p}$ near the particle surface and far from the interaction region. The EM field radiating from the dipole $\boldsymbol{p}$ is given by

$$\mathbf{H} = \frac{ck^2}{4\pi}(\mathbf{n}\times\mathbf{p})\frac{e^{ikr}}{r}\left(1 - \frac{1}{ikr}\right)$$

$$\mathbf{E} = \frac{1}{4\pi\epsilon}\left\{k^2(\mathbf{n}\times\mathbf{p})\times\mathbf{n}\frac{e^{ikr}}{r}\right.$$
$$\left. +[3\mathbf{n}(\mathbf{n}\cdot\mathbf{p})-\mathbf{p}]\left(\frac{1}{r^3} - \frac{ik}{r^2}\right)e^{ikr}\right\}$$

where $k = \frac{2\pi}{\lambda}$ and $\mathbf{n}$ is the unit vector along the direction of $\boldsymbol{r}$ (Jackson 1962). In the near-field zone ($kr \ll 1$), the electrostatic field is recovered; in the far-field zone ($kr \gg 1$), the field forms a spherical wave. We can calculate the optical cross sections by evaluating the Poynting vector $\mathbf{S} := \frac{1}{2}\text{Re}(\mathbf{E}^* \times \mathbf{H})$. The absorption and scattering cross sections are

$$C_{\text{sca}} = \frac{k^4}{6\pi}|\alpha|^2 = \frac{8\pi}{3}k^4 R^6 \left|\frac{\epsilon-\epsilon_m}{\epsilon+2\epsilon_m}\right|^2$$

$$C_{\text{abs}} = k\text{Im}[\alpha] = 4\pi k R^3 \text{Im}[\frac{\epsilon-\epsilon_m}{\epsilon+2\epsilon_m}]$$

where α denotes the polarizability (Bohren and Huffman 1998). For small particles for which the quasi-static approximation holds, the absorption cross section scales as R^3 and the scattering cross section scales as R^6. For small particles, absorption is dominant. As the particle size increases, the scattering increasingly dominates.

7.2.4 Mie Theory

Although the quasi-static regime is simple to treat and easy to understand, it is no longer valid when the particle size increases and a surface plasmon resonance other than the electric dipole mode is excited. In this case, we should solve the full Maxwell's equation. For spherical particles, an exact solution of Maxwell's equation exists and was proposed by Gustav Mie in 1908. The derivation is quite lengthy and beyond the scope of this chapter. Readers interested in details on the Mie theory are led to another textbook (Bohren and Huffman 1998). We only briefly discuss the logical flow of the Mie theory here. First, the vector functions $\mathbf{M}, \mathbf{N}$ are defined as

$$\mathbf{M} = \nabla \times (\mathbf{c}\psi)$$

$$\mathbf{N} = \frac{\nabla \times \mathbf{M}}{k}$$

where $\mathbf{c}$ is a constant vector and ψ is a scalar function.

If ψ satisfies the scalar Helmholtz equation

$$\nabla^2 \psi + k^2 \psi = 0$$

then $\mathbf{M}, \mathbf{N}$ satisfy the Helmholtz equation

$$\nabla^2 \mathbf{M} + k^2 \mathbf{M} = 0$$

$$\nabla^2 \mathbf{N} + k^2 \mathbf{N} = 0$$

and

$$\nabla \times \mathbf{N} = k\mathbf{M}$$

$\mathbf{M}, \mathbf{N}$ satisfy all the requirements of EM waves. ψ is a generating function and is given as

$$\psi = \frac{\cos(m\phi)}{\sin(m\phi)} P_n^m (\cos\theta) z_n(kr)$$

where z_n is any type of four spherical Bessel functions. The $\mathbf{M}, \mathbf{N}$ fields generated from ψ are called vector spherical harmonics and form a complete set. The normal modes are interpreted as electric/magnetic dipole, quadrupole, etc.

7.2.5 Plasmon Hybridization Model

When two metal nanoparticles approach each other, the scattered waves of the two particles interact strongly. This interaction alters the amplitude and shape of the surface plasmon of each particle and leads to a new collective resonance condition. This phenomenon is called *plasmon hybridization* and is analogous to molecular orbital hybridization. Two plasmon resonances should be spectrally close to each other for effective hybridization. This concept was first proposed (Prodan et al. 2003) with a spherical nanoshell as an example (Figure 7.2). Spherical nanoshells have two metal/dielectric interfaces and can be decomposed into a metal sphere and a cavity embedded in the metal. The hybridization of the sphere plasmon and cavity plasmon yields two resonances; one is called a bonding plasmon, whose two hybridizing plasmons are in phase, and the other is an antibonding plasmon, whose two hybridizing plasmons are out of phase (π difference). Analogous to the molecular hybridization theory, the bonding plasmon lies at a lower frequency than the antibonding plasmon.

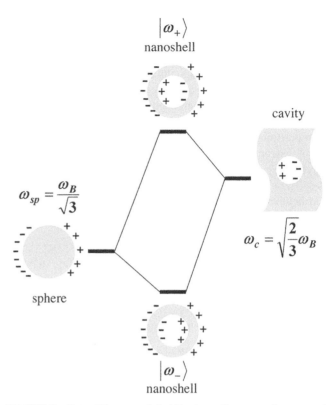

FIGURE 7.2 Plasmon hybridization diagram of a metal nanoshell. The sphere plasmon and the cavity plasmon hybridize and result in bonding and antibonding plasmon modes. (Reprinted with permission from Prodan et al. 2003, Copyright 2003, AAAS.)

7.2.6 Dielectric Function

We now move on to the microscopic viewpoint of the interaction between a metal and an EM wave and consider how these microscopic interactions are represented as the dielectric function. Most unique properties of the metal, including surface plasmons, arise from the free electrons. For example, when a surface plasmon is excited, the almost-free electrons oscillate in phase and gradually lose their coherence through collisions. We can use the simple *free electron gas model* to describe the optical response of electrons. In the free electron gas model, the electron cloud is represented as a uniform gas of density n spread over the nuclei with the same density. The internal region of the metal maintains a neutral charge, and the surface has a net nonzero charge due to polarization. The nuclear potential and interactions between electrons are not explicitly calculated. Instead, electrons are considered to move freely with an *effective mass* m and experience some friction γ. The equation of motion of the electrons under a time-harmonic electric field $\mathbf{E} = \mathbf{E}_0 e^{-i\omega t}$ is given by

$$m\frac{d^2\mathbf{r}}{dt^2} + m\gamma\frac{d\mathbf{r}}{dt} = -e\mathbf{E}_0 e^{-i\omega t}$$

A particular solution to this equation is $\mathbf{r} = \mathbf{r}_0 e^{-i\omega t}$. By inserting this particular solution in the original equation, we obtain

$$\mathbf{r} = \frac{e}{m(\omega^2 + i\gamma\omega)}\mathbf{E}_0 e^{-i\omega t}$$

Using the fact that $\mathbf{P} = -ne\mathbf{r}$ and $\epsilon_0\mathbf{E} + \mathbf{P} = \epsilon\mathbf{E}$, the dielectric function ϵ becomes

$$\epsilon(\omega) = 1 - \frac{\omega_p^2}{\omega^2 + i\gamma\omega}$$

$$\omega_p = \frac{ne^2}{\epsilon_0 m}$$

where ω_p is called the plasma frequency, which corresponds to the natural frequency of the plasma. This is called the Drude dielectric function.

The Drude function is in excellent agreement with experiments in the low-frequency region (e.g., the near-IR (NIR)), where EM waves mostly interact with electrons in conduction bands. However, when the frequency increases enough to interact with lower band electrons, which are strongly bound to the atomic nucleus, the Drude function fails. This phenomenon thus requires an additional term to describe such electrons. We can phenomenologically describe the equation of motion of the bound electrons as a damped harmonic oscillator. For a more rigorous calculation of the interband term, we may use a quantum mechanical treatment. By treating the potential from the electric field as a perturbation, the rate of interband transition can be calculated by Fermi's golden rule. Only direct transitions are considered because the momentum of photons is much lower than that of electrons. The response returns the imaginary part of the dielectric function, and the real part of the dielectric function is evaluated from the imaginary part by the Kramers–Kronig relation. The interband transition plays an important role in metal PL, and this phenomenon will be discussed later in this chapter.

7.3 Plasmonic Effects on Enhanced Spectroscopies

There are several enhanced spectroscopies that utilize plasmonic effects. Some of them are based on the interaction between molecules and photons. If the molecule is placed in an enhanced field, the spectroscopic signal can be modulated and further enhanced according to the characteristics of and coupling with the enhanced field. Thus, as subtopics of molecular plasmonics, we briefly introduce several enhanced spectroscopies, such as SERS, TERS, MEF, and SEIRA. Then we briefly introduce the concept of single-photon PL of metal nanostructures.

7.3.1 Molecular Plasmonics

Enhanced Raman Scattering: SERS, TERS, NERS

When molecules are irradiated by light (especially a monochromatic laser), most of the light will be elastically scattered with a wavelength corresponding to that of the incident light (Rayleigh scattering). However, a small portion of light will be inelastically scattered with different wavelengths that correspond to the vibrational and rotational modes of the molecules. This inelastic scattering of

photons by molecules that are excited to higher vibrational or rotational energy levels is called Raman scattering (Stöckle et al. 2000, Huh et al. 2009, Schlücker 2014). Owing to structural differences, different molecules have different vibrational and rotational modes; therefore, the spectra of the inelastically scattered light are uniquely different for each molecule. Due to this high specificity, Raman scattering can be considered a molecular fingerprint (Huh et al. 2009, Cialla et al. 2012). In addition, the Raman scattering signal is relatively sharp compared with the optical signals of conventional fluorescent molecules and quantum dots, whose full width at half maximum is 50–100 and 25–40 nm, respectively (Isola et al. 1998, Zhang and Johnson 2006). Therefore, although the different types of Raman scattering signals overlap over a wide spectral range, individual spectra can be easily and specifically distinguished, which facilitates multiplexing even at a single excitation wavelength. Despite this high specificity and multiplexing capability, the use of Raman scattering has been limited because of the relatively weak signal intensity originating from the very low optical cross section and the low efficiency of the inelastic scattering process (Campion and Kambhampati 1998, Huh et al. 2009).

When the LSP resonates with the incident light (LSP resonance, LSPR), the EM field is extremely amplified near the surface of the metallic nanostructures, which significantly enhances the intensity of the light inelastically scattered by molecules near the surface of the metal. This phenomenon, in which the intensity of the Raman scattering signal is highly enhanced at the surface of metallic nanostructures, is called SERS (Kneipp et al. 2008, Qian and Nie 2008, Le Ru and Etchegoin 2012). In the 1970s, the SERS effect was first observed by Fleischmann and coworkers when detecting pyridine adsorbed on roughened Ag electrodes (Fleischmann et al. 1974). In 1997, the pioneering works independently conducted by Kneipp and Nie confirmed that the inherent weak Raman scattering signals could be detected even at the single-molecule and single-particle level via the SERS mechanism (Kneipp et al. 1997, Nie and Emory 1997). Since then, the interest and use of SERS has increased exponentially, and the in-depth scientific advances (e.g., the design of metallic nanostructures to improve SERS efficiency, the optimization of various parameters affecting signal amplification, etc.) have allowed SERS to tremendously amplify the weak Raman scattering signals to 10^{14}–10^{15} times (Michaels et al. 1999, 2000, Xu et al. 1999, 2000, Haslett et al. 2000, Moyer et al. 2000, Bosnick et al. 2002).

In SERS, two different mechanisms contribute simultaneously to the amplification of Raman scattering signals: EM enhancement and chemical enhancement (Metiu and Das 1984, Moskovits 1985, Otto et al. 1992, Schatz and Duyne 2002, Park and Kim 2010, Xia et al. 2014, Cialla-May et al. 2017). The EM enhancement mechanism arises from the optical excitation of LSPR in metallic nanostructures, leading to the localization and amplification of the EM field at the surface of nanostructures. The amplified EM field, in turn, significantly enhances the Raman scattering signals of molecules located within the localized EM field via the so-called antenna effect or lightning rod effect (Schatz 1984, Crozier et al. 2003, Hering et al. 2008, Zhang et al. 2010, Schumacher et al. 2011). If the molecules are adsorbed at certain defect sites (e.g., terraces, steps, etc.) on the metal surface, they form metal–molecule complexes, which causes changes in the electronic structure of the molecules by charge transfer between the metal and the molecule (Qian and Nie 2008, Park and Kim 2010, Sharma et al. 2012). The chemical enhancement of the Raman scattering signals occurs when the frequency of the excitation laser is resonant with that of the electronic states in the metal–molecule complexes (Otto et al. 2003, Jensen et al. 2008). In general, the EM enhancement mechanism is considered to play a major role in the overall SERS enhancement due to its significantly high EF of approximately 10^8-10^{11}, whereas the chemical enhancement mechanism possessing an EF of only 10^1-10^3 only contributes to a small extent to the overall SERS enhancement (Metiu and Das 1984, Moskovits 1985, Otto et al. 1992, Xu et al. 1999, Morton and Jensen 2009, Park and Kim 2010, Xia et al. 2014).

Other outstanding techniques to significantly enhance the intensity of Raman scattering signals, such as TERS and nanogap-enhanced Raman spectroscopy (NERS), have been intensively developed and widely utilized in various fields, including material science, biochemistry, catalysis, and electrochemistry (van Schrojenstein Lantman et al. 2012, Kurouski 2017, Zhong et al. 2017). Fundamentally, the signal obtained from SERS originates from the sum of signals from a relatively large number of molecules owing to the focal size of the laser; thus, the spatial resolution is relatively low. However, TERS, which combines Raman spectroscopy and apertureless near-field scanning optical microscopy, possesses high spatial resolution below the diffraction limit owing to the use of a sharp metal tip to acquire signals in a highly localized area (Stöckle et al. 2000, Bailo and Deckert 2008, Deckert 2009, Zhang et al. 2016b, Richard-Lacroix et al. 2017). In addition, TERS is a highly sensitive technique that allows for single-molecule-level detection while maintaining the signal EF due to the field-enhancing effect between the metallic tip resonating with the local mode of the surface plasmon and the metallic substrate (Bailo and Deckert 2008, Deckert 2009, Zhang et al. 2016b, Richard-Lacroix et al. 2017). In NERS, the EM field is greatly amplified by strong plasmonic coupling in the nanogap between the plasmonic nanoparticles (Lee et al. 2013b, Nam et al. 2016). The nanogap-based plasmonic properties are highly altered by changes in the size and structure of the nanogap, resulting in highly enhanced and controllable Raman scattering signals. Generally, as the size of the nanogap decreases, the EM field generated in the nanogap becomes stronger (Shanthil et al. 2012). However, if the size of the nanogap becomes as small as or smaller than 1 nm, the EM field in the nanogap can be greatly weakened because electrons can flow across the nanogap via quantum tunneling (Savage et al. 2012). Therefore, the uniform and precise control of the nanogap is extremely important to amplify and control the EM field within the

nanogap to obtain a strong, controllable, and reproducible NERS signal.

Metal-Enhanced Fluorescence

Fluorophores near metallic nanostructures take advantage of the larger effective extinction coefficient of the metal nanostructures (Evanoff and Chumanov 2004, Ni et al. 2008, Anderson et al. 2010). Upon exposition to incident light, the metallic nanostructures act as an optical antenna that concentrates the electric field around them. These enhanced local fields can lead to increases in the spontaneous emission rate and quantum yield of the fluorophore and can ensure its photostability (Fort and Grésillon 2008). Such metal-induced improvement in emission is called MEF (Geddes and Lakowicz 2002). In the typical emission process of a fluorophore without plasmonic nanocrystals, the observed fluorescence intensity can be described by $I^0 = \Gamma_{ex}^0 \cdot QY^0$ (the superscript "0" indicates that the molecule is in free space). The fluorescence intensity (I^0) of a fluorophore is determined by its excitation rate (Γ_{ex}^0) and intrinsic quantum yield (QY^0), which is defined as $QY^0 = \Gamma_{rad}^0/\Gamma_{tot}^0$. The total decay rate Γ_{tot}^0 of the fluorophore is the sum of the intrinsic radiative rate Γ_{rad}^0 and the nonradiative relaxation rate $\Gamma_{non-rad}^0$ in free space. When a fluorophore molecule is present near a metallic nanostructure, its excitation rate and emission quantum yield can be modulated by modifying its local EM field (Figure 7.3). Such a modification can be described by

$$I = \Gamma_{ex} \cdot \left(\frac{\Gamma_{rad}}{\Gamma_{tot}} + \frac{\Gamma_{pl}}{\Gamma_{tot}} \cdot \frac{\Gamma_{rad}^{pl}}{\Gamma_{rad}^{pl} + \Gamma_{non-rad}^{pl}} \right)$$

In the presence of a metallic nanostructure, the excitation rate (Γ_{ex}), radiative decay rate (Γ_{rad}), and nonradiative decay rate ($\Gamma_{non-rad}$) of the fluorophore can either increase or decrease. Specifically, the modified nonradiative decay of the fluorophore consists of quenching by the metal at the rate of Γ_q and decay by exciting the plasmons in the metal at the rate of Γ_{pl}. Plasmons excited by Γ_{pl} undergo reradiative and nonradiative decay at both rates of Γ_{rad}^{pl} and $\Gamma_{non-rad}^{pl}$, respectively. The modification can be explained by two mechanisms. First, the fluorescence excitation rate (Γ_{ex}) can be increased. As the enhancement of the excitation rate is proportional to the square of the local electric field strength (Guerrero et al. 2012), the excitation rate of a fluorophore molecule situated in the enhanced local field near the metallic nanostructure increases. Second, the emission rate can be accelerated via the Purcell effect, by which spontaneous emission rates can be increased by the environment (Zhang et al. 2012). The magnitude of enhancement is expressed as the ratio of the modified ($\Gamma_{rad} + \Gamma_{pl}$) and free-space ($\Gamma_{rad}^0$) emission rates and has come to be known as the Purcell factor $F_P = (\Gamma_{rad} + \Gamma_{pl})/\Gamma_{rad}^0$ (Pelton 2015). In the presence of a metallic nanostructure, the decay route of the fluorophore can be modified: the plasmon-enhanced radiative decay rate (Γ_{rad}), the nonradiative decay rate ($\Gamma_{non-rad}$), rate of quenching by the metal (Γ_q), and decay rate by

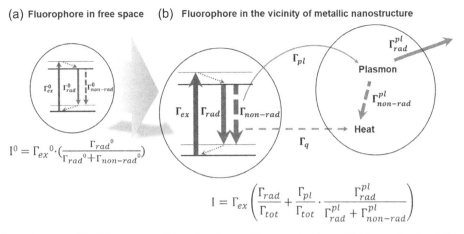

(a) Fluorophore in free space **(b)** Fluorophore in the vicinity of metallic nanostructure

$$I^0 = \Gamma_{ex}^0 \cdot \left(\frac{\Gamma_{rad}^0}{\Gamma_{rad}^0 + \Gamma_{non-rad}^0} \right)$$

$$I = \Gamma_{ex} \left(\frac{\Gamma_{rad}}{\Gamma_{tot}} + \frac{\Gamma_{pl}}{\Gamma_{tot}} \cdot \frac{\Gamma_{rad}^{pl}}{\Gamma_{rad}^{pl} + \Gamma_{non-rad}^{pl}} \right)$$

FIGURE 7.3 Schematics in modified fluorescence intensity via metallic nanostructure. (a) Energy level and fluorescence intensity of fluorophore in free space. (b) Modified fluorescence excitation and emission process when fluorophore couples to a metallic nanostructure.

exciting plasmons in the metal (Γ_{pl}). The total decay rate ($\Gamma_{tot} = \Gamma_{rad} + \Gamma_{non-rad} + \Gamma_{pl} + \Gamma_q$) of the fluorophore is always higher than the decay rate in free space. As a result, the lifetime is reduced and fluorophore photostability is improved (Cang et al. 2013). Then, the energy transferred to the metallic nanostructure can be reradiated to the far field at a rate of Γ_{rad}^{pl} that is determined by the absorption (C_{abs}) and scattering (C_{scat}) cross sections of the metallic nanostructure (Tam et al. 2007). Finally, in fluorophore–metal coupled systems, the enhanced total radiative decay rate to the far field (Γ_{far}) and QY can be expressed as $\Gamma_{far} = \Gamma_{rad} + \Gamma_{pl} \cdot \Gamma_{rad}^{pl} / (\Gamma_{rad}^{pl} + \Gamma_{non-rad}^{pl})$ and QY $= \Gamma_{far}/\Gamma_{tot}$, respectively. A relationship between the plasmon resonance and fluorescence wavelengths is a major precondition for MEF. Plasmon resonance spectral tuning toward the absorption or emission of the fluorophore can be engineered via the shape and size of the particle: the more the overlap, the higher the fluorescence intensity. This phenomenon can be explained by the wavelength dependence of the local electric field enhancement and the emission modification for a given localized plasmon resonance (Lu et al. 2011). The best way to enhance the fluorescence intensity consists of matching both the absorption and emission wavelengths of the fluorophore to the plasmon resonance. However, because of the difference between the absorption and emission wavelengths caused by Stokes shifts, the plasmon resonance cannot overlap both wavelengths simultaneously. Therefore, the position of the plasmon resonance peak between the absorption and emission wavelengths of the fluorophore should be optimized to maximize the fluorescence intensity (Chen et al. 2007, Munechika et al. 2010). The ability to concentrate light in metallic nanostructures does not always imply the ability to increase the emission rate. Anger et al. studied the effect of the distance between metal and fluorophore with well-controlled fluorophore–metal spacing. They controlled the location of a gold nanosphere attached to the scanning probe tip from a single molecule embedded in the polymer layer, and obtained maximum fluorescence intensity at 5 nm separation (Anger et al. 2006). In contrast to SERS, fluorophore

molecules in close proximity to metal surfaces (<2–3 nm) sacrifice the energy from the fluorophore to the metal, and the increased Γ_q leads to a decrease in the overall quantum yield. To avoid fluorescence quenching, additional layers that optimize molecule–metal separation have been developed. This was accomplished with SiO_2, Al_2O_3, or DNA strands that have controllable layer thicknesses (Fu and Lakowicz 2006, Reineck et al. 2013, Abadeer et al. 2014, Hao et al. 2014). At the optimal overall quantum yield and excitation rate, the fluorescence intensity will reach a maximum. The optimal distance between the fluorophore and the surface will differ with the metallic nanostructure, material, and environment but generally ranges between 5 and 20 nm, with most reported results agreeing on a displacement of ~10 nm (Chhabra et al. 2009, Acuna et al. 2012). Based on the understanding of the mechanism and the rational design of systems integrating the aforementioned factors, MEF performances have been further improved. Achieving fluorescence enhancement of several thousand times enables a single molecule level of sensitivity at physiologically relevant concentrations (Punj et al. 2013, Puchkova et al. 2015). Fluorescence enhancement of NIR-emitting dye using core–gap–shell type of nanomatryoshka nanoparticles suggests the potential of utilizing MEF-active plasmonic nanoparticles in biomedical applications (Ayala-Orozco et al. 2014). Besides fluorescence enhancement from typical organic fluorophores, increase in the emission from quantum dots or phosphors have been reported (Mishra and Geddes 2014, Hoang et al. 2015, 2016, Planas et al. 2016). Alongside these developments, MEF will allow advances in various fields, such as photochemistry, solar cells, organic photovoltaic devices, biomedicines, and biosensors (Cade et al. 2009, Webb and Bardhan 2014, Li et al. 2015a, Ahn et al. 2016, Mandal and Sharma 2016).

Surface-Enhanced Infrared Absorption

When free molecules are attached to metallic nanostructures, their optical properties are dramatically changed.

SERS is the most well-known example, in which the Raman scattering of molecules is enhanced by orders of magnitude in the visible/NIR spectral range (Schlücker 2014). A similar effect can be observed in the mid-IR region. Molecules located in the near field of resonantly excited metallic nanostructures exhibit amplified IR absorption compared with those measured in metal-free environments. This phenomenon is known as SEIRA (Hartstein et al. 1980). The fingerprint (800–1,800 cm^{-1}) of the IR spectra provides characteristic information about the molecular constituents, their chemical bonds, as well as their configuration. Although IR spectroscopy is extensively used in many fields to precisely identify individual molecular species, a large amount of material is required because of the low absorption cross-section σ_{abs} of mid-IR vibrations ($\sim$10–20 cm^2). Surface-enhanced IR spectroscopy using resonant metal nanoantennas overcomes this limitation, which extends its application to sensing, imaging, and protein detection (Adato and Altug 2013, Chen et al. 2014). The collective electron oscillations resonantly excited by the metal nanostructures provide huge EM fields on a nanometer scale. The IR vibrations of molecules located in

these fields are enhanced by 10 or more orders of magnitude, because the absorption scales with the intensity of the local field, which enables spectroscopic characterization with exceptional sensitivity (Weber et al. 2017). Until now, to understand and optimize SEIRA, micrometer-long and nanometer-wide (cross-section) nanorods fabricated by lithography techniques have been studied as basic experimental models (Crozier et al. 2003, Neubrech et al. 2008b). The resonance frequency or the linear nanorod antennas can be easily adjusted to the IR spectral region by altering their length, because the plasmon resonance frequency is proportional to the antenna length (Novotny 2007). In 2008, Neubrech et al. demonstrated the interaction of molecular vibrations with a plasmon resonance frequency in the IR region of an Au nanorod antenna (Figure 7.4a) (Neubrech, et al. 2008a). The IR signals of the chemisorb octadecanethiol on the Au surface were notably strengthened upon exposure to the locally enhanced EM near field in the excited plasmonic antennas. The local field enhancement effect on the SEIRA was evidenced using parallel polarized light (E$_\|$) on the nanorod antenna, which was excited by plasmons, and an enhanced vibration signal was obtained.

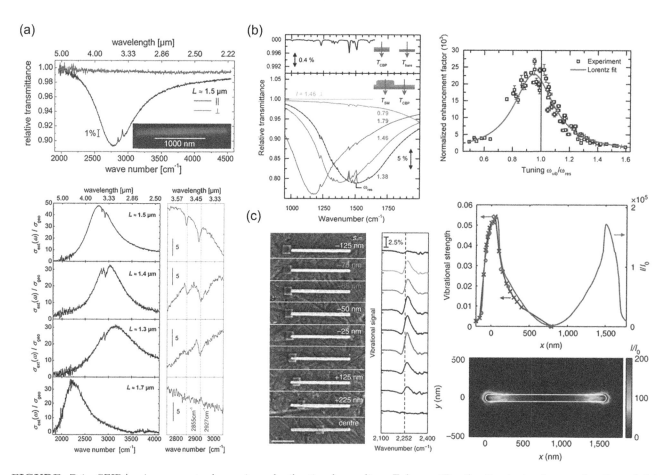

FIGURE 7.4 SEIRA via resonant plasmonic and vibrational coupling. Enhance IR vibrational signal as a function of (a) parallel/perpendicular polarization of the IR radiation, (b) spectral tuning ratio between the molecular vibrational modes (ω_{vib}) and the far-field plasmonic resonance of nanoantennas (ω_{res}), and (c) near-field distribution and intensity of plasmonic antennas. (Reproduced with permission from (a) Neubrech et al. 2008a, Copyright 2008 American Physical Society; (b) Vogt et al. 2015, Copyright 2015 The PCCP Owener Societies; (c) Dregely et al. 2013, Copyright 2013 Macmillan Publishers Ltd. Nature Communications.)

On the other hand, with vertically polarized light ($E_\perp$), no vibrational bands were found. In addition, the strength of the enhanced vibrational signal remarkably relies on the plasmon resonance frequency. The strongest signal enhancement can be obtained at the point where the plasmon and vibrational frequency match best. To detail this behavior, Vogt et al. (2015) used nanorod antennas of different lengths covered with a layer of 4,4′-bis(N-carbazolyl)-1,1′-biphenyl (Figure 7.4b). The signal strength of the molecular vibration frequency was strongly dependent on the plasmon resonance frequency. The vibration EF was estimated at the measured vibration signal, and the strength was plotted against the ratio of the plasmon resonance frequency and the respective vibrational frequency ($\omega_{vib}/\omega_{res}$). A tuning ratio of 1 indicates a perfect match of the far-field extinction maximum to the vibrational resonance; any other value smaller or larger than 1 suggests detuning. Interestingly, the maximum enhancement is observed at a ratio slightly smaller than 1 and not exactly at 1. It is known that a spectral shift occurs between the near- and far-field responses (Zuloaga and Nordlander 2011), which explains the scale of the near-field vibrational signal enhancement and the maximized enhancement where the molecular and plasmon resonance frequencies are well matched. In 2013, Dregely et al. confirmed the influence of the plasmonic near-field intensity on the vibrational signal enhancement with an analyte (hydrogen silsesquiosance) selectively positioned at nanometer-scale intervals on a plasmonic nanoantenna (Figure 7.4c) (Dregely et al. 2013). By observing the enhanced vibrational signal strength in terms of the molecular positions, the most enhanced SEIRA spectra were revealed to be those of the molecules in the strongest near-field region (at the antenna tips). Additionally, the results prove that the vibrational signal of the molecule increases with the intensity of the square root of the electrical near field. Interestingly, in SEIRA studies, the enhanced vibrational spectra show significantly different line shapes, such as asymmetries, dips, and peaks, which are the typical signature of resonant coupling. The resonance coupling can be understood as a coupled harmonic two-oscillator model (Adato et al. 2013), which is applied to IR vibrational modes to understand the SEIRA mechanism. As the molecular absorption cross section is small, the resonance does not efficiently interact with IR light. When the plasmonic surface and molecule are close, the molecular vibrations can be excited by coupling with the plasmonic resonance, which has a relatively strong interaction with the incident light. In general, when the plasmonic mode is coupled with the incident light, it loses energy via both external radiation and intrinsic material absorption. In resonance conditions, the amplitude of the absorption peak can be adjusted by the ratio of the external to intrinsic loss rate. When a weak absorber is introduced to the system, additional intrinsic damping of the plasmon occurs, which leads to a change in the initial damping rate. Therefore, by introducing a molecule, the plasmon resonance and molecule can be undercoupled or overcoupled from the maximum absorption peak,

where the ratio of the external to intrinsic loss rate becomes 1. Therefore, the electromagnetically induced transparency-like and electromagnetically induced absorption-like line shapes found in the SEIRA spectra are caused by the modified ratio of the external to internal losses based on resonance coupling. Therefore, a consideration of the initial plasmon damping rates and the effect of the additional absorber can lead to optimum SEIRA enhancement. In addition, as the asymmetric profile is similar to Fano resonances, which are based on the constructive and destructive interference of two excitation pathways, they could be applied to understand the SEIRA mechanism (Osley et al. 2013). Additional deep understanding of the SEIRA mechanism is attainable by integrating the temporal coupled mode theory (Adato and Altug 2013), the analogy to Fano-resonances (Giannini et al. 2011), the point-dipole model (Rezus and Selig 2016), and other approaches. This precise understanding of the mechanism provides insight for the design of nanostructures for optimum SEIRA enhancement.

7.3.2 Direct PL of Metal Nanostructures

The first observation of one-photon PL was reported by Mooradian (1969). When Au, Cu, and Au–Cu alloy films were irradiated with 488 or 514 nm lasers, light was emitted (Figure 7.5a), which was attributed to the direct recombination of d-band holes with sp-band electrons below the Fermi level (Figure 7.5b). About 20 years later, the discovery of Raman enhancement on roughened metal surfaces inspired the investigation of PL from roughened surfaces (Boyd et al. 1986). According to the experimental results comparing the PL intensity of smooth and roughened noble metal surfaces of Au, Ag, and Cu, the surface roughness caused an increase in PL intensity. Furthermore, this increase was explained by the enhancement of optical fields by surface plasmon resonance. Nevertheless, the quantum yield from the noble metal film was only $\sim 10^{-10}$. This low PL efficiency may be caused by electrons relaxing through a faster nonradiative process such as Coulomb carrier scattering, which quenches the efficient PL, before radiative recombination (Hartland 2011). This PL is modulated and further enhanced by nanomaterials with the aid of surface plasmons. In 2000, Mohamed et al. reported PL from Au nanorods of the same width and varying aspect ratios of 2.0–5.4 (Mohamed et al. 2000). As the aspect ratio increases, the scattering peaks were redshifted, and the PL peaks showed a similar aspect-ratio-dependent redshift. Most importantly, the quantum yield of the Au nanorods was much higher than that of Au films. Thus, the authors discovered the origin of the enhancement of the local fields at the tip of Au nanorods, which enhance the incoming excitation and the outgoing radiation. A few years later, Dulkeith et al. reported plasmon-modulated light emission from spherical Au nanoparticles of 1–30 nm size (Dulkeith et al. 2004). The peak of the PL band directly follows the peak of the LSPR in the scattering spectra, which varied from 2.3 to 2.4 eV (Figure 7.5c). Because of the weak spectral overlap between the excitation

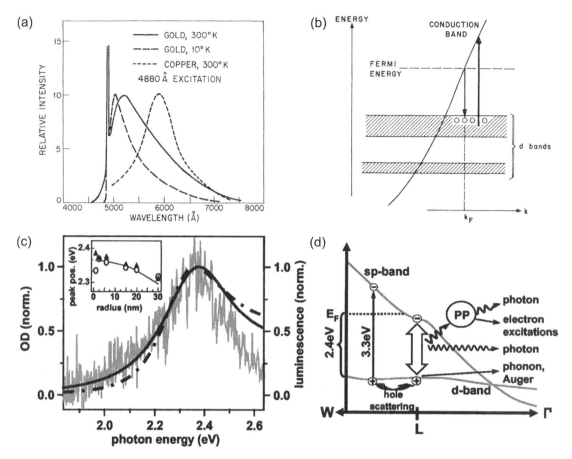

FIGURE 7.5 One-photon PL from metal. (a) PL spectra of gold and copper. (b) Schematic illustration of the band structure of a noble metal and excitation and recombination transitions for PL. (c) PL (gray) and extinction (black) spectra of 6 nm gold nanoparticles. The dashed-dotted black line indicates a calculated extinction spectrum using Mie theory. The inset shows the peak positions of PL (open circles) and extinction (triangles) spectra from gold nanoparticle solutions of different radii. The solid line represents calculated maximum peak position using Mie theory. (d) Schematic representation of interband excitation and subsequent hole relaxation close to the L symmetry point of the band structure of gold, followed by the excitation of a particle plasmon that subsequently decays radiatively. (Images were reproduced with permission from (a,b) Mooradian 1969, Copyright 1969 American Chemical Society; (c,d) Dulkeith et al. 2004, Copyright 2004 American Chemical Society.)

source and the peak of the scattering spectra, the authors suggest that the enhancement of excitation and recombination is not sufficient to account for such a large emission enhancement. There is no enhancement in the excitation process of spherical nanoparticles, as only interband transitions can be excited at the energy of 3.3 eV. In addition, the relatively small increase in the near field could not account for the several orders of enhancement of the quantum yield compared with Au films. Moreover, the possibility that the PL spectrum arises from the simple recombination of the d-band holes and sp-band electrons could not account for the observed phenomena and was excluded. Therefore, the authors have developed a new three-step model for nanostructured PL, in which the optically excited *d*-band holes are relaxed in the d-band and then scattered into conduction bands that stimulate surface plasmons that collapse through radiation emission (Figure 7.5d). Since its discovery, the tremendous enhancement of the PL of nanostructures compared with that of bulk metal has drawn much attention,

and its underlying mechanism has been a major subject in the field. However, the principle is still incomplete, and vigorous debates are ongoing. We will discuss the suggested mechanisms and experimental results in Section 7.5.

7.4 Molecular Raman Enhancement

The metal surface-based enhancement of Raman scattering signal is very important for Raman spectroscopy, because the magnitude of Raman scattering signal is very weak due to its small cross sections compared with fluorescence. Despite the first discovery that nanometer-scale metal surface can strongly enhance the magnitude of Raman scattering signal (i.e., SERS), several studies on the design of plasmonic nanostructure for strong amplification and control of Raman scattering signal have been steadily investigated. Fundamentally, the SERS intensity relies on the strongly amplified and focused EM field near the metal surface, which varies greatly depending on the size, shape, material, and composition of the plasmonic metal nanostructures. In addition, SERS properties can also be further enhanced by plasmonic coupling between spaced apart at nanometer-scale intervals such as inter-nanogap, nanocrevice, and intra-nanogap. In this section, we introduce representative studies of synthesizing and designing various plasmonic nanostructures and discuss how their plasmonic properties (especially Raman enhancement) change according to the nanostructure's morphology/composition affecting LSPR and the nanogap/nanocrevice, leading to plasmonic coupling.

7.4.1 Single Particles

The spectral position and shape of plasmon modes are determined by LSPR conditions (Willets and Van Duyne 2007, Cialla et al. 2012, Cialla-May et al. 2017). Because the intensity and wavelength of the LSPR bands are highly affected by the electron density on the surface of plasmonic nanostructures, the LSPR band is highly tunable depending on the size, shape, material, and composition of the plasmonic metal nanostructures as well as the dielectric constant of the surrounding medium (Haes et al. 2005, Xia and Halas 2005, Willets and Van Duyne 2007, Cialla et al. 2012, Cialla-May et al. 2017).

When the particle size is smaller than 10 nm, the LSPR band of Au nanoparticles is greatly damped owing to the phase changes induced by the higher rate of electron-surface collisions in comparison with the larger nanoparticles (Huang and El-Sayed 2010). On the other hand, as the size of the nanoparticles increases, the LSPR wavelength is redshifted, the peak intensity increases, and the LSPR band is broadened due to the dominant contributions from the higher-order electron oscillations (Huang and El-Sayed 2010). From the theoretically calculated results, Jain and coworkers reported that LSP's absorption and scattering are largely affected by the nanoparticle size (Jain et al. 2006). For Au nanoparticles of 20 nm size, the total extinction was dominantly caused by absorption (Figure 7.6a), whereas the contribution of scattering to the total extinction appeared as the particle size was increased to 40 nm (Figure 7.6b). As the particle size was further increased to 80 nm, the scattering increased, resulting in a similar contribution of absorption and scattering to the total extinction (Figure 7.6c). The size-dependent changes in the plasmonic properties suggest that an appropriate choice of nanoparticles is required based on the desired application: larger nanoparticles for imaging or SERS, which require higher scattering efficiency, and smaller nanoparticles for photothermal therapy or hot electron-based engineering, which require higher absorption efficiency (Huang and El-Sayed 2010). In a similar context, Zhang and coworkers reported size-dependent SERS properties using Au nanoparticles with sizes from 31 to 577 nm (Figure 7.6d) (Zhang et al. 2016a). When the Au nanoparticles were excited by a laser with a 633 nm wavelength, the nanoparticles with a size of ~97 nm showed the largest SERS enhancement (Figure 7.6e), whereas the nanoparticles with a size

of ~408 nm showed the largest SERS enhancement under exposure to a 785 nm laser (Figure 7.6e). Both experimental results directly demonstrate the size dependence of SERS; they also imply that the choice of an appropriate laser to excite the plasmon modes based on the particle size is also important.

To date, shape-dependent changes in SERS properties have been intensively studied with plasmonic nanostructures of various shapes, such as nanospheres, nanorods, nanowires, nanocubes, nanocages, nanoplates, nanoprisms, and nanoshells, and branched structures, including nanostars, nanoflowers, nanourchins, bipyramids, etc (Tao et al. 2003, Orendorff et al. 2006, Xie et al. 2008, Rycenga et al. 2009, 2012, Wang et al. 2010, Wu et al. 2011, Liu et al. 2013, 2014, Tan et al. 2013, Indrasekara et al. 2014, Yang et al. 2014a, Li et al. 2015b). Li and coworkers synthesized Au nanoparticles of different shapes, such as nanospheres, nanorods, and nanostars, and demonstrated the effect of the shape of the plasmonic nanoparticle on SERS properties (Figure 7.7a) (Li et al. 2012). To conduct the SERS measurement and to endow long-term stability and good solubility in water, malachite green isothiocyanate (MGITC)-labeled nanoparticles were encapsulated in silica (Figure 7.7a). To exclude the effect of resonance Raman enhancement, 532 and 785 nm lasers, which do not match the optical absorption of MGITC, were used. Under both excitation lasers, the nanostars generated stronger SERS signals than the nanospheres and nanorods. This experimental result is mainly caused by the sharp geometrical features of the nanostars, which can highly concentrate and amplify the EM field (Figure 7.7b) (Khoury and Vo-Dinh 2008, Dondapati et al. 2010). In addition, three-dimensional finite-difference time domain (FDTD) simulation results obtained by normalization with the surface area also showed that nanostars exhibited the strongest SERS enhancement (Figure 7.7c). Therefore, it is inferred that the SERS enhancement is strongly influenced by the geometry of plasmonic nanostructures rather than its size. As the structure of each nanoparticle can be varied depending on the synthesis conditions (e.g., the aspect ratio of nanorods, corner curvature of nanocubes), even the same type of structure can display different SERS characteristics. In particular, anisotropic nanostructures (such as nanorods and bipyramids), which exhibit multiple extinction spectra owing to the different shape-dependent plasmon modes generated in the nanoparticles (Orendorff et al. 2006, Li et al. 2015b), can exhibit different SERS properties depending on the position on the nanoparticle as well as the wavelength and orientation of the excitation laser.

As the density (number) of surface electrons differs depending on the type of metal, the LSPR characteristics of materials are very different (Sharma et al. 2012). Among the various metals, the coin metals such as Au, Ag, and Cu are mainly used as SERS substrates because their strong LSPR covers most of the visible and NIR spectral range that most Raman measurements use (Johnson and Christy 1972, Huang and El-Sayed 2010, Cialla et al. 2012). However, other metals such as Al, Pt, or Pd exhibit weak and broad

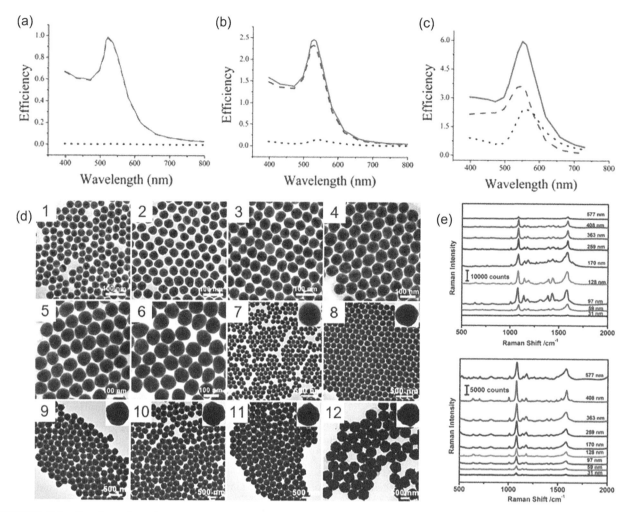

FIGURE 7.6 Size-dependent changes in optical properties of plasmonic nanoparticles. (a–c) Calculated spectra of the efficiency of absorption (dashed line), scattering (dotted line), and extinction (solid line) for Au nanospheres of different diameters (D): (a) $D = 20$ nm, (b) $D = 40$ nm, and (c) $D = 80$ nm. (d) Transmission electron microscope (TEM) images of Au nanoparticles of different sizes: 1) 31 ± 2 nm, 2) 40 ± 2 nm, 3) 52 ± 4 nm, 4) 59 ± 4 nm, 5) 85 ± 5 nm, 6) 97 ± 10 nm, 7) 128 ± 11 nm, 8) 170 ± 15 nm, 9) 259 ± 20 nm, 10) 363 ± 20 nm, 11) 408 ± 25 nm, and 12) 577 ± 30 nm. Insets show high-magnification TEM images of the corresponding Au nanoparticles. (e) SERS spectra of 4-aminothiophenol molecules on the films of Au nanoparticles of different sizes at an excitation wavelength of 633 nm (upper panel) and 785 nm (lower panel). The size of the nanoparticles is 31, 59, 97, 128, 170, 259, 363, 408, and 577 nm, respectively, from bottom to top. (Reproduced with permission from (a–c) Jain et al. 2006, Copyright 2006 American Chemical Society; (d,e) Zhang et al. 2016a, Copyright 2016 Wiley-VCH.)

LSPR bands in the ultraviolet region; they are thus used as SERS substrates in the ultraviolet spectral range (Dörfer et al. 2007, Cui et al. 2008, 2009). In addition, because the LSPR characteristics can also be changed according to the metal composition, plasmonic nanoparticles composed of various metals (e.g., core–shell or alloy nanoparticles) are also utilized as tunable SERS-active substrates (Pande et al. 2007, Wu et al. 2012, Fan et al. 2013, Paramanik and Patra 2014, Yang et al. 2014b, Han et al. 2016). Notably, unlike noble metals (e.g., Au, Ag, Cu) that are chemically stable and whose surface can be easily modified, some metals (e.g., Al, Li, Na) are highly reactive in air or water and may be detrimental to stable, reliable, and reproducible SERS measurements (Sharma et al. 2012).

7.4.2 Inter-Nanogap-Based Plasmonic Particles

When plasmonic nanoparticles are brought close to each other on a scale smaller than the nanoparticle size, a strong plasmonic coupling phenomenon occurs in the nanometer-scaled gap between the plasmonic nanoparticles (inter-nanogap) (Alvarez-Puebla et al. 2010). This plasmonic coupling phenomenon strongly localizes and amplifies the EM field in the inter-nanogap region, and the Raman scattering signal is greatly enhanced by the SERS mechanism (Shanthil et al. 2012). From a classical EM field point of view, the intensity of the EM field becomes stronger as the size of the inter-nanogap decreases, especially around 1 nm (Lim et al. 2010, Lee et al. 2012a, 2013a,

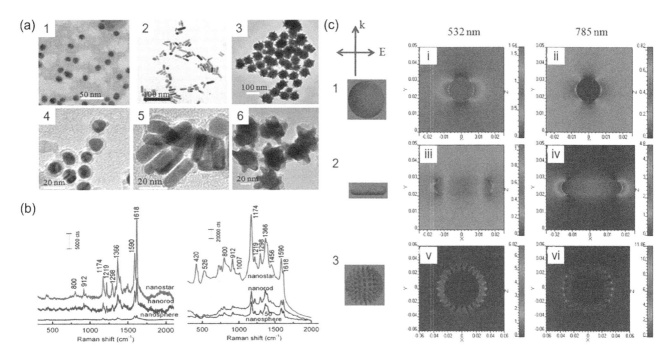

FIGURE 7.7 Shape-dependent changes in optical properties of plasmonic nanoparticles. (a) TEM images of Au nanoparticles of different shapes: 1) nanospherers, 2) nanorods, 3) nanostars, 4) Au-nanosphere-MGITC-SiO$_2$, 5) Au-nanorod-MGITC-SiO$_2$, and 6) Au-nanostar-MGITC-SiO$_2$. (b) Raman spectra of MGITC on Au nanosphere, nanorod, and nanostar solutions under excitation by the 532 nm laser (left) and the 785 nm laser (right). (c) Three-dimensional FDTD simulated EM field distributions of Au nanoparticles of different shapes: 1) nanosphere, 2) nanorod, and 3) nanostar under excitations by the 532 nm laser (i, iii, v) and the 785 nm laser (ii, iv, vi). TEM, transmission electron microscope; MGITC, malachite green isothiocyanate. (Reproduced with permission from Li et al. 2012, Copyright 2012 IOP Publishing.)

Nam et al. 2016). However, in ultranarrow inter-nanogaps whose sizes are much smaller than 1 nm, electrons can flow across the inter-nanogap (i.e., by quantum tunneling) due to the surface charge density oscillations in the inter-nanogap region, which significantly weakens the local electric field (Savage et al. 2012). Therefore, the uniform and precise control of the inter-nanogap is an essential challenge to achieve strong and reproducible SERS signals based on the plasmonic coupling phenomenon.

To realize well-defined plasmonic inter-nanogap structures with strongly amplified EM fields, Lim and coworkers reported a unique method to synthesize SERS-active gold–silver core–shell nanodumbbells (GSNDs) (Figure 7.8a) (Lim et al. 2010). In this method, the DNA-modified Au nanoparticles were systemically linked via linker DNA, resulting in dimeric nanostructures. By stoichiometrically controlling the ratio of the two different DNA sequences (linker-capturing DNA and particle surface-protecting DNA), the surface of each Au nanoparticle was functionalized with a maximum of one linker-capturing DNA strand. Eventually, this ratiometric control could maximize the synthetic yield of the dimeric nanostructures while preventing the formation of trimeric or multimeric nanostructures. Finally, the GSNDs were synthesized by forming Ag nanoshells on the Au dimers (Figure 7.8b). The Ag shell thickness of the GSNDs was controlled with nanometer-scaled precision by adjusting the amount of Ag precursor, which allowed

engineering the inter-nanogap size (i.e., the interparticle distance) with high structural controllability and reproducibility (Figure 7.8c). The SMSERS signals obtained with an atomic force microscopy (AFM)-correlated nano-Raman instrument experimentally demonstrated that the GSNDs with a 5 nm Ag shell produced strong and reliable SMSERS signals due to the ultranarrow inter-nanogap structures (<1 nm), thus generating a highly amplified EM field within the nanogap (Figure 7.8d,e). In addition, those GSNDs exhibited narrowly distributed EF values (1.9×10^{12} to 5.9×10^{13}, average value of 1.8×10^{13}) due to the strong plasmonic coupling by the inter-nanogap structure and the high-yielding and controlled synthesis of GSNDs (Lim et al. 2010, Lee et al. 2012a). However, when the Ag shell thickness was smaller than 3 nm, the SERS signals became undetectable or irreproducible, because the thin Ag layer (i.e., a large inter-nanogap) is inadequate to form proper junctions for sufficient EM field amplification (Figure 7.8d,e). These experimental results obviously suggest that ultranarrow inter-nanogap structures are crucial to obtain strong, reproducible, and reliable SERS signals. In a similar manner, Lee and coworkers experimentally and theoretically studied the relationships between the inter-nanogap size of GSNDs, the SMSERS intensities, EF distributions, nanoparticle size/shape/composition, and the excitation laser wavelength to maximize and tune the SMSERS property of GSNDs (Lee et al. 2012a). For GSND-I with varying inter-nanogap

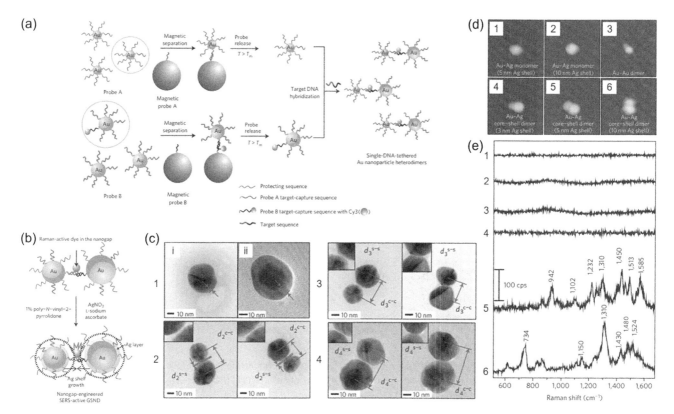

FIGURE 7.8 Synthesis and SERS properties of GSNDs. (a) Schematic illustration of a high-yield synthesis for the Au nanoparticle heterodimers using magnetic purification and stoichiometric DNA modification. (b) Formation of SERS-active GSND by Ag-shell growth. (c) Representative HRTEM images of the nanoparticles: 1-i) Au–Ag core–shell monomer with 5 nm Ag shell, 1-ii) Au–Ag core–shell monomer with 10 nm Ag shell, GSND particles with 2) ∼3 nm, 3) ∼5 nm, and 4) ∼10 nm Ag shell thicknesses, respectively. The d^{s-s} and d^{c-c} indicate the distance between two particle surfaces and cores, respectively. (d) AFM images of the nanoparticles: 1) Au–Ag core–shell monomer with 5 nm Ag shell, 2) Au–Ag core–shell monomer with 10 nm Ag shell, 3) Au–Au dimer, 4) Au–Ag core–shell dimer with 3 nm Ag shell, 5) Au–Ag core–shell dimer with 5 nm Ag shell, and 6) Au–Ag core–shell dimer with 10 nm Ag shell. (e) The corresponding SERS spectra taken from the nanostructures in (d). All spectra were obtained using a 514.5 nm excitation laser. HRTEM, high-resolution transmission electron microscope; AFM, atomic force microscope. (Reproduced with permission from Lim et al. 2010, Copyright 2009 Springer Nature.)

sizes (from ∼4.8 to <1 nm) and similar Au core size (20–30 nm pair) (Figure 7.9a), they found that <2 nm inter-nanogap is essential to generate detectable SMSERS signals, and <1 nm inter-nanogap significantly boosts the SMSERS signal intensity due to the strong plasmonic coupling (Figure 7.9b). Importantly, the GSND-I with <1 nm inter-nanogap provided the strongest SMSERS signals as well as a most narrow distribution of high EF values, while the GSND-I with a 2 nm inter-nanogap and coalesced GSND-I with no inter-nanogap generated wide EF value distributions (Figure 7.9c). For GSND-II with tuned Au core sizes (from 13–20 nm pair to 40–50 nm pair) and a fixed inter-nanogap size (∼0.9 nm) and Ag shell thickness (∼5 nm) (Figure 7.9d), they found that a combination of >50 nm Au cores and 514.5 nm laser wavelength, which matches well with the Ag shell, produced the strongest SMSERS signals with narrowly distributed EF values compared with <50 nm Au cores with a 514.5 nm laser or the GSND-II with 632.8 nm laser (Figure 7.9e). As the relative volume of Au and Ag within the GSNDs affects their plasmon resonance energy

(Chuntonov et al. 2012), it is critical to choose a proper laser wavelength to produce strong SMSERS signals. These results offer large flexibility for the design of inter-nanogap-based plasmonic couplings and insight on the importance of precisely engineering <1 nm inter-nanogaps with a narrow distribution of high EF values for sensitive and quantitative SERS detection.

Unlike dimeric nanostructures with geometrically linear inter-nanogaps, trimeric nanostructures with inter-nanogaps exhibit various plasmonic properties according to their geometries (Lee et al. 2015). Lee and coworkers experimentally and theoretically analyzed the quantitative plasmon modes and SERS properties of strongly coupled plasmonic nanotrimers with diverse geometries, such as acute triangular, obtuse triangular, nearly linear, and linear geometric configurations (Figure 7.10a) (Lee et al. 2015). The DNA-tethered trimeric nanostructure was synthesized using a strategy similar to that of GSNDs (i.e., the formation of an Ag shell on the trimeric Au cores), and a single Raman dye molecule was placed in both 1 nm inter-nanogaps.

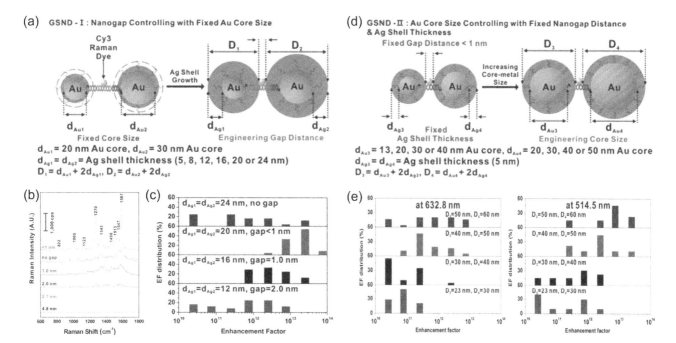

FIGURE 7.9 Inter-nanogap size and gold core size-dependent SERS properties of GSNDs. (a) Schematic illustration of GSND-I structures, controlling the size of inter-nanogap with the fixed Au core size. (b) SERS spectra of GSND-I structures with different sizes of inter-nanogap. All spectra were acquired from Cy3 Raman molecules located within inter-nanogap of GSND-I structures and measured using a 514.5 nm excitation laser. (c) Distributions of EF of the GSND-I structures according to the size of inter-nanogap. (d) A schematic illustration of GSND-II structures having fixed Ag shell thickness and inter-nanogap size, with control in the Au core size. (e) Distributions of EF of GSND-II structures at two different excitation laser wavelengths of 632.8 nm (left) and 514.5 nm (right), respectively. (Reproduced with permission from Lee et al. 2012a, Copyright 2012 American Chemical Society.)

Using Rayleigh scattering and SERS measurements with laser-scanning-assisted dark-field (DF) microscopy and finite element method (FEM) calculations, they found that the longitudinally polarized and axially polarized plasmon modes were increased and decreased, respectively, as the geometric configuration of the trimers continuously changed from an acute triangle to a linear shape (Figure 7.10b–e). Importantly, the systematic single-particle-level correlation between the near field, far field, and SERS revealed that the SERS signals of the trimers were predominately determined by the largely excited coupled plasmon between the two-competing longitudinal and axial modes. Furthermore, from the FEM calculations, the authors found that a discrepancy smaller than 0.5 nm in the size of the two inter-nanogaps of the linear trimer could cause a larger tenfold difference in the SERS signals. Similarly, the correlation between LSP and SERS properties of dimeric nanostructures with an ~1 nm inter-nanogap was also studied (Lee et al. 2013a). These in-depth studies on the structure-dependent variations in the plasmon properties of inter-nanogaps provide a fundamental understanding of the various types of coupled plasmon modes as well as suggest an appropriate design to enable the systematic control of plasmonic responses and SERS signals.

Conventional electrolyte (or salt)-induced nonspecific nanoparticle aggregation cannot produce inter-nanogaps in a controllable manner (Camden et al. 2008, Zhang et al. 2013b). Until now, to control the structure of inter-nanogaps, various structures such as DNA-linked nanostructures, DNA origami-based assemblies of nanoparticle dimers, one-dimensional ultralong nanochains, two-dimensional homo- and heterodimers, three-dimensional nanoclusters, self-assembled polyhedral nanoparticles in linear or circular templates, and superlattice nanostructures have been reported (Macfarlane et al. 2011, Henzie et al. 2013, Urban et al. 2013, Kühler et al. 2014, Thacker et al. 2014, Kim et al. 2015, Zhang et al. 2015a); however, the precise control of these nanostructures and their inter-nanogaps at the nanometer level as well as their synthesis in high yields still remain challenging.

7.4.3 Nanocrevice-Based Plasmonic Particles

As the size of the inter-nanogap in plasmonic nanodimers decreases—until the two nanoparticles eventually contact or fuse—strong plasmonic coupling does not occur, causing the EM field to no longer be amplified at the inter-nanogap (Savage et al. 2012). However, the EM field can be strongly amplified in small "crevices" around these conductive junctions (Nam et al. 2016). Unlike the inter-nanogap structures where the degree of amplification of the EM field depends on the size of the gap, the nanocrevice-based EM field amplification varies greatly with the shape of the conductive junction

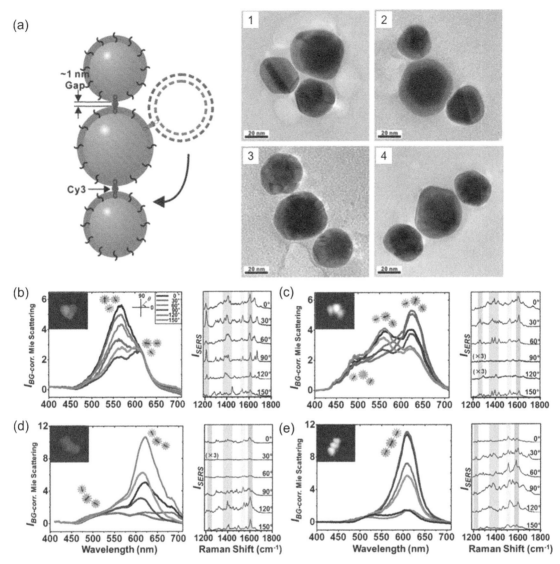

FIGURE 7.10 Synthesis and optical properties of strongly coupled gold–silver core–shell plasmonic nanotrimers with diverse geometries. (a) (left) Schematic illustration of the rotatable DNA-tethered trimeric nanostructure with a single Raman dye molecule placed in each of two inter-nanogap whose size is 1 nm. (right) Representative TEM images of the trimers with different geometric configurations: 1) acute triangle, 2) obtuse triangle, 3) nearly linear, and 4) linear. (b,c) Polarization-resolved Rayleigh scattering (left) and SERS spectra (right) obtained from the trimers with different geometric configurations: (b) acute triangle, (c) obtuse triangle with the central vertex angle in the range of 90–120°, (d) nearly linear, and (e) linear. Insets show AFM images of corresponding trimeric nanostructures. The LSPR peaks in the Rayleigh scattering spectra were obtained by FEM calculation. TEM, transmission electron microscope; AFM, atomic force microscope; LSPR, localized surface plasmon resonance; FEM, finite element method. (Reproduced with permission from Lee et al. 2015, Copyright 2015 American Chemical Society.)

structures (e.g., the curvature or sharpness) (Lee et al. 2014). To synthesize the plasmonic conductive junction structures and to control the morphology of the nanocrevices, Lee and coworkers reported a salt-tuned synthetic strategy for the formation of Au−Ag head−body-type bimetallic nanostructures called nanosnowmen (Lee et al. 2012b). When a small amount of salt was present near the DNA-modified Au nanoparticles, the Au−Ag head−body nanosnowmen were synthesized in high yield (>95%) through asymmetric Ag deposition, while Au−Ag core−shell nanospheres were formed in the presence of larger amounts of salt around the

DNA-modified Au nanoparticles (Figure 7.11a,b). It is well known that poly(vinylpyrrolidone) (PVP) can form coordination complexes with Ag ions (i.e., Ag−PVP) in aqueous solution (Zhang et al. 1996). In the presence of PVP and at high salt concentrations, the Ag−PVP complexes can difficultly penetrate through the salt layer, which leads to slow Ag growth kinetics. Given the uniformity of DNA at high salt concentrations, multiple nucleation sites form simultaneously and slowly, resulting in the formation of Au−Ag core−shell nanospheres (Figure 7.11c). On the other hand, less uniform DNA structures and imperfect salt layers are

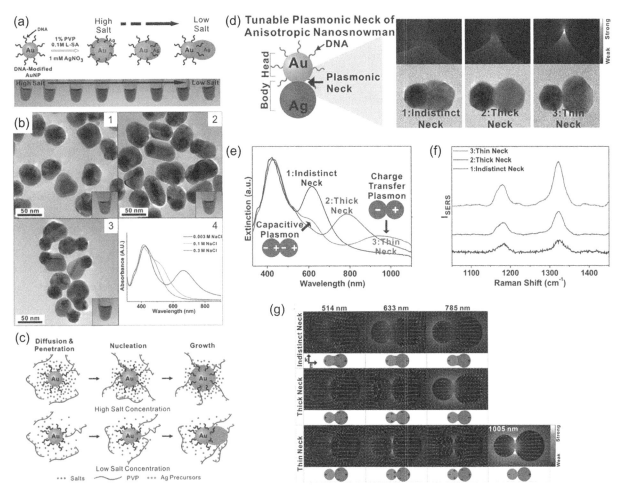

FIGURE 7.11 Synthesis of Au–Ag head–body nanosnowmen with plasmonic nanocrevice, and changes in their optical properties according to the morphologies of plasmonic nanocrevice. (a) A schematic illustration and solution color image of Au–Ag nanostructures with different salt concentrations. (b) TEM images of Au–Ag nanostructures fabricated with different salt concentrations of 1) 0.3 M, 2) 0.1 M, 3) 0.003 M, and 4) corresponding ultraviolet–vis spectra. Insets show solution color corresponding Au–Ag nanostrcutures. (c) Proposed reaction mechanism for the synthesis of Au–Ag nanostructures at different salt concentrations. (d) Changes in the morphology of the nanocrevices around the plasmonic neck of nanosnowmen. (e–g) Extinction spectra (e), SERS spectra (f), and three-dimensional FEM calculation-based EM field enhancements and distributions (g) of the Au–Ag head–body nanosnowmen with different plasmonic neck morphologies. In (f), SERS spectra were obtained from FAM dyes on each nanosnowmen and measured using a 633 nm excitation laser. TEM, transmission electron microscope; FAM, fluorescein amidite; FEM, finite element method. (Reproduced with permission from (a–c) Lee et al. 2012b, Copyright 2012 American Chemical Society; (d–g) Lee et al. 2014, Copyright 2014 American Chemical Society.)

formed on the Au nanoparticle surface at low salt concentrations. Therefore, the Ag–PVP complexes approach the Au nanoparticle surface more easily. Once Ag nucleation sites form on the Au nanoparticle surface, Ag ions are preferentially deposited on these sites, which eventually leads to fast Ag growth kinetics and the anisotropic growth of Ag nanostructures on the Au nanoparticle surface, which forms Au–Ag head–body nanosnowmen (Figure 7.11c). Furthermore, the morphology of the nanocrevices around the plasmonic neck between the Au head and Ag body could be systematically tuned by the controllable salt-dependent Ag growth kinetics (Figure 7.11d) (Lee et al. 2014). These structural changes highly altered the charge transfer and capacitive coupling plasmon modes of the

plasmonic nanosnowmen (Figure 7.11e) and dramatically affected the magnitude of the EM fields and SERS signals at the nanocrevices owing to the different plasmon couplings and broken symmetry (Figure 7.11f,g). The nanocrevice around the thin conductive junction (i.e., the sharp neck junction) of the nanosnowmen generated the strongest EM fields and SERS signals and exhibited charge-transfer mode-based NIR signals on the optical spectra.

In a similar manner, Shen and coworkers synthesized highly SERS-active Au–Ag nanomushrooms containing a plasmonic nanocrevice with 1–2 nm gaps between the Au head and Ag cap (Figure 7.12a) (Shen et al. 2015). In the absence of NaCl around the DNA-modified Au nanoparticles, single-strand DNA protects the surface of

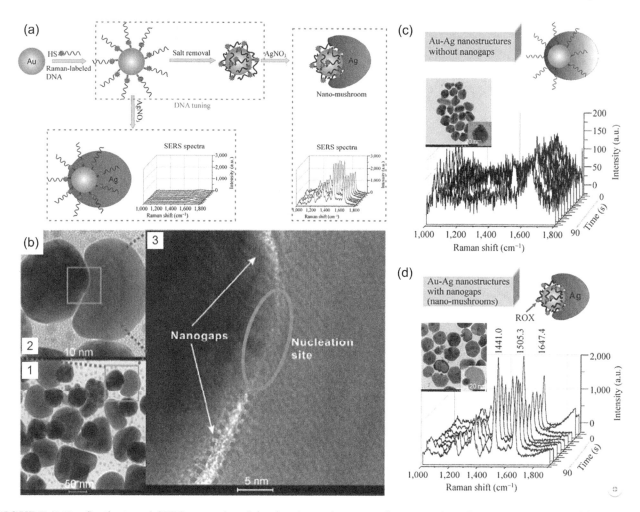

FIGURE 7.12 Synthesis and SERS properties of Au–Ag plasmonic nanomushrooms with a plasmonic nanocrevice. (a) Schematic illustration of the synthesis of nanomushrooms. Au–Ag nanostructures without nanogap were formed in the presence of NaCl, while Au–Ag nanomushrooms containing a plasmonic nanocrevice with 1–2 nm gaps between the Au head and Ag cap were formed in the absence of NaCl. (b) TEM images of Au–Ag nanomushrooms: 1) low magnification image, 2) enlarged high magnification image, and 3) HRTEM image. Gray circle area shows the Ag nucleation site and light gray dashed area indicates the plasmonic nanocrevice with 1–2 nm gaps. (c,d) SERS spectra of the as-synthesized Au–Ag nanostructures: (c) without nanocrevice, (d) with nanocrevice. All spectra were taken using a 632.8 nm excitation laser. Insets show TEM images of the corresponding Au–Ag nanostructures. NaCl, sodium chloride; TEM, transmission electron microscope; HRTEM, high resolution transmission electron microscope. (Reproduced with permission from Shen et al. 2015, Copyright 2015 Springer Nature.)

the Au nanoparticle due to the binding affinity between the DNA bases and Au or the electrostatic attraction between the sugar-phosphate backbone of DNA and the surface of Au nanoparticles (Demers et al. 2000, 2002, Hurst et al. 2006). This phenomenon eventually prevents the direct formation of Ag onto the Au surface, and Ag is thus initially only partly grown on the nucleation site. Then, Ag is further grown over the DNA layer on the Au nanoparticles, resulting in the formation of a nanocrevice between the Au head and the Ag cap connected by nucleation sites (Figure 7.12b). The mushroom-like plasmonic nanostructures generated highly enhanced SERS signals due to amplification of the EM field in the nanocrevice formed by the junction between the Au head and Ag cap (Figure 7.12c,d). In addition, the same group synthesized

multicolor Au–Ag nanomushrooms containing different Raman molecules within the nanocrevices and utilized them as ready-to-use SERS probes for ultrasensitive and multiplex DNA/miRNA detection (Su et al. 2017). Recently, unlike the conductive junction structure consisting of two metals (e.g., nanosnowmen and nanomushrooms), the plasmonic coupling in the heterojunctions of trimetallic (Au, Ag, and Cu) nanoparticles has also been studied (Lin et al. 2017b).

The nanocrevice structures can be implemented in anisotropic nanostructures as well as isotropic nanostructures (Kumar et al. 2014, Liu et al. 2016). Nanoporous structures provide nanometer-sized pores between ligaments, which act as plasmonic nanocrevices to strongly amplify the EM fields. Liu and coworkers synthesized porous Au–Ag nanospheres with a high density of nanocrevices throughout the plasmonic nanostructure (Figure 7.13a) (Liu et al. 2016). The highly porous nanostructures were formed by a dealloying process that selectively removes the less stable metal component in alloys, and the porosity (i.e., pore size and number) can be controlled by adjusting both the relative ratio of Au to Ag in the alloys and the dealloying temperature. Importantly, the abundant nanopores served as hotspots that extremely amplified the EM field and enhanced the Raman scattering signal (Figure 7.13b,c), thereby exhibiting excellent SERS activity with an EF of $\sim 1.3 \times 10^7$. In addition to the nanocrevices formed between the caved structures (e.g., nanoporous structures), nanocrevices can be also formed between the protruded structures (e.g., nanobranch structures). Kumar and coworkers reported a highly controllable, oxidative nanopeeling chemistry-based strategy for the synthesis of plasmonic core–petal nanoparticles (CPNs) with massively branched and plasmonically coupled nanostructures (Kumar et al. 2014). By both Au chloride-induced oxidative disruption and the peeling of the polydopamine layer on the Au core, the petal structures were anisotropically grown through the disrupted polydopamine layer, resulting in the formation of CPNs in high yield (Figure 7.13d). In addition, the density of nanopetals on the Au core was highly controllable by simply tuning the amount of Au precursor and modifying the reduction kinetics (Figure 7.13d). The plasmonic CPN structure provided a high number of nanocrevices between the branching nanopetals, thus acting as an excellent SERS substrate, which allowed the SERS signal-based monitoring of reactive oxygen species-mediated primary and secondary structural changes in isolated DNA (Kumar et al. 2014).

7.4.4 Intra-Nanogap-Based Plasmonic Particles

The EM field can also be strongly enhanced in nanometer-scaled interior nanogaps via plasmonic coupling (Nam et al. 2016). Unlike inter-nanogap and nanocrevice structures, the nanogaps in intra-nanogap structures generate highly amplified and relatively uniform and controllable EM fields, resulting in strongly enhanced and tunable SERS signals. Because of the structural features of the plasmonic nanogaps formed inside the particles, the Raman molecules are uniformly embedded inside the nanoparticles and are not exposed to the external environment. Thus, stable and uniform SERS signals are obtained, with no noticeable loss and changes over extended time periods. To generate reproducible and reliable SERS signals from intra-nanogap structures, the high-precision control of the nanogap at the nanometer-scale-level is a key factor and challenge.

So far, various strategies have been reported for the synthesis of plasmonic intra-nanogap structures (Bardhan et al. 2010, Lim et al. 2011, Oh et al. 2014, Song et al. 2014, Lin et al. 2015, Zhang et al. 2015b, Lee et al. 2016, Li et al. 2016, Kim et al. 2018a). In 2011, Lim and coworkers reported a facile DNA-based synthetic strategy to obtain highly SERS-active plasmonic particles with a

well-defined nanobridged hollow intra-nanogap in high yields ($\sim 95\%$), which are called Au-nanobridged nanogap particles (Au-NNPs) (Figure 7.14a) (Lim et al. 2011). At the early stage of the formation of Au-NNPs, small budding spheres appeared and grew laterally on the Au core surface that was modified with a Raman dye (Cy3)-labeled DNA. Further lateral growth eventually formed a shell-like structure around the Au core, and the shell and core surfaces were connected via nanobridges, resulting in a uniform and hollow intra-nanogap ~ 1 nm in size (Figure 7.14a). Other Au cores whose surfaces were modified with citrate, bis(p-sulphonatophenyl)phenylphosphane dehydrate, or methoxy poly(ethylene glycol) could not form intra-nanogaps between the core and shell, indicating that DNA is a key surface material to form the intra-nanogaps. Specifically, the binding affinities of different four DNA bases (adenine (A), cytosine (C), guanine (G), and thymine (T)) to the Au core as well as the grafted DNA density on the Au core are critical to the formation and control of the Au shell and plasmonic intra-nanogap inside the Au-NNPs (Oh et al. 2014). When poly(A) and poly(C) sequences were attached on the Au core, a wider intra-nanogap and a smoother Au shell were formed, while poly(G) and poly(T) sequences created narrower and irregular nanohole-like intra-nanogaps with rougher Au shells (Figure 7.14b) (Oh et al. 2014). Adenine and cytosine have relatively stronger binding affinities for the Au surface, while thymine is weakly bound to the Au surface (Storhoff et al. 2002, Kimura-Suda et al. 2003, Brown et al. 2008, Erdmann et al. 2010, Wang et al. 2013). Therefore, poly(A) and poly(C) sequences can cover larger areas of the Au core surface; however, the poly(T) sequence is vertically stretched out and densely packed on the Au surface. Therefore, the strong bonds between the DNA bases (i.e., poly(A) and poly(C)) and the Au surface can effectively protect the Au core surface and prevent the direct formation of an Au shell on the Au core surface, resulting instead in the formation of intra-nanogaps. Interestingly, the thickness of single-stranded DNA and the intra-nanogap size of the Au-NNPs are similar (~ 1 nm), which is another supporting evidence that the intra-nanogap is formed due to the binding between DNA bases and the Au core surface (Lim et al. 2011, Oh et al. 2014).

Because the uniformly formed intra-nanogaps provide a highly amplified EM field by plasmonic coupling between the core and shell, the Au-NNPs generate a very strong, stable, and reproducible SERS signal (Figure 7.14c) (Lim et al. 2011, Oh et al. 2014). The SERS signals of Au-NNPs were linearly proportional to the particle concentration and were sensitively detected down to the 10 fM level (Lim et al. 2011). In addition, the single-particle Raman analysis of individual Au-NNPs showed that most particles ($>90\%$) generated high EF values ($>1.0 \times 10^8$), and the EF values were uniformly and narrowly distributed, ranging from 1.0×10^8 to 5.0×10^9, which reveals that Au-NNPs can produce strong, reliable, and reproducible SERS signals (Lim et al. 2011). Interestingly, the roughened surface of the Au-NNPs could further enhance the EM field inside the

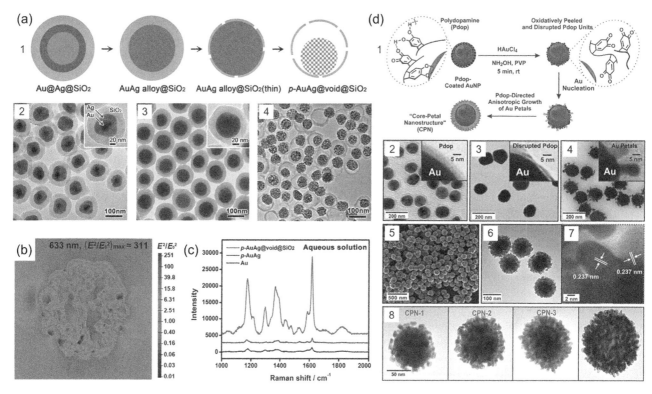

FIGURE 7.13 Synthesis and optical properties of porous Au–Ag nanospheres and oxidative nanopeeling chemistry-based synthesis of plasmonic core–petal nanostructures. (a) Synthesis and optical properties of the porous Au–Ag nanospheres: 1) A schematic illustration of the synthesis of porous Au–Ag nanospheres by a dealloying method. TEM images of nanoparticles: 2) Au@Ag@SiO₂, 3) AuAg alloy@SiO₂, and 4) p-AuAg@void@SiO₂. (b) Simulated near-field EM field distribution of the p-AuAg@void@SiO₂ using an incident plane wave with a wavelength of 633 nm. (c) SERS spectra were recorded for aqueous solutions of CV with different suspensions of the as-synthesized nanoparticles. All spectra were obtained using a 633 nm excitation laser. (d) Synthesis and structural characterization of the plasmonic CPNs: 1) A schematic illustration of the synthesis of plasmonic CPNs, 2) TEM image of pdop-AuNPs with 80 nm Au core and ~5 nm pdop layer, 3) TEM image of pdop-AuNPs immediately after adding HAuCl₄, 4) TEM image after 1 min from adding HAuCl₄ to pdop-AuNPs, 5) scanning electron microscope (SEM) image of CPNs, 6) TEM image of CPNs, 7) Magnified HRTEM image of petal structure and 8) TEM images of CPNs synthesized by increasing amounts of HAuCl₄ from left to right. TEM, transmission electron microscope; CV, crystal violet; pdop, polydopamine; AuNPs, gold nanoparticles; HAuCl₄, tetrachloroauric acid; SEM, scanning electron microscope. (Reproduced with permission from (a–c) Liu et al. 2016, Copyright 2016 American Chemical Society; (d) Kumar et al. 2014, Copyright 2014 American Chemical Society.)

intra-nanogap (Lee et al. 2016). As the ratio of HA to Au precursor increased during the formation of the Au shell, the number of nucleation sites on the DNA-modified Au core surface increased, and the shell surface roughness of the Au-NNPs could be controlled (Figure 7.14d). Importantly, the particle surface roughness could be associated with the EM field inside the intra-nanogap, and stronger NERS signals could be produced from particles with the roughest surface (Figure 7.14e). Both the shell surface roughness and the number and size of nanobridges in the Au-NNPs could affect the EM field between the Au core and the rough shell; thus, these factors should be adjusted in a highly controllable manner to generate reproducible NERS signals and should be considered simultaneously to maximize the NERS signals (Oh et al. 2014, Lee et al. 2016).

In addition to DNA, other materials such as silica, polymers, and small organic molecules were utilized to form intra-nanogap particles (Bardhan et al. 2010, Song

et al. 2014, Lin et al. 2015, Zhang et al. 2015b, Li et al. 2016). Bardhan and coworkers reported spherically concentric plasmonic nanoparticles ("nanosphere-in-a-nanoshell"), in which the intra-nanogap was formed by introducing a silica layer between the Au nanosphere core and the Au shell layer (Au/SiO₂/Au nanoshells) (Figure 7.15a,b) (Bardhan et al. 2010). By controlling the interlayer (SiO₂) and the shell thickness, the relative absorption magnitude and scattering cross section were strongly modified, and the resonance frequencies were shifted according to the coupling between the plasmon modes of the inner Au core and the outer Au shell (Figure 7.15c,d). Another approach to form plasmonic intra-nanogaps was conducted by Song et al. (2014). They synthesized Raman tag-encoded plasmonic intra-nanogaps based on the nanoparticle-templated self-assembly of amphiphilic block copolymers and the localized reduction of metal precursors by redox-active polymer brushes (Figure 7.15e). The intra-nanogap size could be

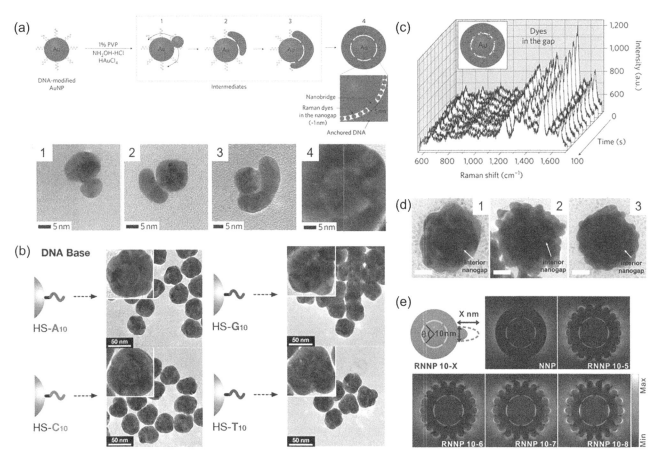

FIGURE 7.14 DNA-based controllable synthesis of gold NNPs and control of their plasmonic properties due to structural changes. (a) A schematic illustration of the synthesis of gold NNPss (Au-NNPs). TEM images indicate intermediate (panels 1–3) and Au-NNPs (panel 4). (b) TEM images of Au nanostructures after Au shell formation on an Au core functionalized with different thiolated DNA strands. (c) Time-dependent Raman profiles of Au-NNP. All spectra were taken using a 633 nm excitation laser. (d) TEM images of Au surface roughness-controlled NNPs (Au-RNNPs) synthesized by different amounts of hydroxylamine (HA) as a reducing agent: 1) HA/Au^{3+} = 2, 2) HA/Au^{3+} = 5, and 3) HA/Au^{3+} = 30. All the scale bars are 20 nm. (e) Three-dimensional FEM-based EM filed distributions of Au-RNNPs with different surface morphologies containing six nanobridges at a 633 nm excitation laser wavelength. TEM, transmission electron microscope; FEM, finite element method. (Reproduced with permission from (a,c) Lim et al. 2011, Copyright 2011 Springer Nature; (b) Oh et al. 2014, Copyright 2014 American Chemical Society; (d,e) Lee et al. 2016, Copyright 2016 Wiley-VCH.)

tailored by controlling the thickness of the polymer coating, and SERS-active plasmonic nanoparticles with 1.5 nm intra-nanogaps exhibited the highest SERS enhancement (Figure 7.15f). In addition, plasmonic Au core–shell nanoparticles with subnanometer intra-nanogaps (∼0.7 nm) could be directly synthesized by forming an Au shell on an Au core, whose surface was modified with a monolayer of small organic molecules (Figure 7.15g) (Lin et al. 2015). The self-assembled monolayer of 1,4-benzenedithiol molecules played the role of a spacer between the Au core and the shell to uniformly and reproducibly form an intra-nanogap. Galvanic replacement reactions, which are redox processes occurring between metals and metal ions due to differences in their electrical reduction potentials, have also been widely utilized to synthesize plasmonic intra-nanogaps (Figure 7.15h) (Zhang et al. 2015b, Li et al. 2016).

Kim and coworkers developed a facile dealloying-based strategy for the synthesis of highly SERS-active plasmonic nanoparticles containing a uniformly confined intra-nanogap in high yield (∼95%), called dealloyed intra-nanogap particles (DIPs) (Kim et al. 2018a). In the Au−Ag alloy shell formed on the surface of the Au core to which the Raman molecules are introduced, the Ag atoms mainly located near the surface of the Au core were selectively dissolved by the dealloying reaction, which resulted in uniform, controllable, and reproducible plasmonic intra-nanogaps of ∼2 nm (Figure 7.16a,b). In the early stage of the dealloying reaction, the Ag atoms on the Au−Ag alloy shell surface are selectively dissolved by the Ag etchant (Fe(NO$_3$)$_3$), and the Ag etchant penetrates the Au−Ag alloy shell through the etched Ag sites (Figure 7.16a). As the dealloying reaction progresses, the vacancies created at the etched Ag sites diffuse inward, while the metal atoms (mainly Ag atoms) inside the alloy shell diffuse outward due to the nanometer-scale Kirkendall effect (González et al. 2011) and are continuously removed by the Ag etchant (Figure 7.16a).

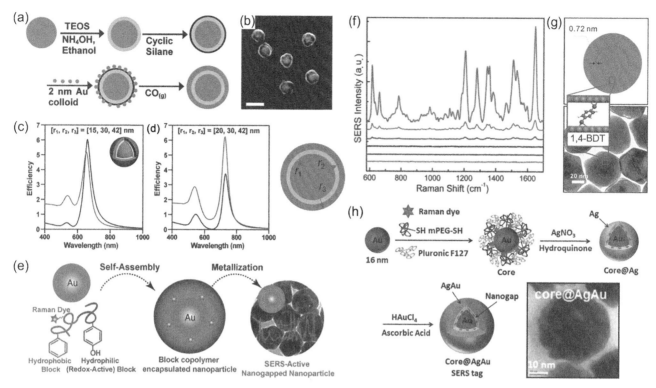

FIGURE 7.15 Synthesis of plasmonic intra-nanogap particles using interlayer (silica, polymer, and small organic molecules) or galvanic replacement reaction. (a) A schematic illustration of the synthesis of $Au/SiO_2/Au$ nanoshells. (b) SEM image of $Au/SiO_2/Au$ nanoshells. (c,d) Calculated efficiencies of absorption (gray line) and scattering (black line) of $Au/SiO_2/Au$ nanoshells with different sizes: (c) $[r_1, r_2, r_3] = [15, 30, 42]$ nm, and (d) $[r_1, r_2, r_3] = [20, 30, 42]$ nm. (e) A schematic illustration of the synthesis of SERS-encoded nanogapped plasmonic Au nanoparticles, based on the nanoparticle-templated self-assembly of amphiphilic block copolymers and the localized reduction of Au precursors by redox-active polymer brushes. (f) SERS spectra of as-synthesized nanoparticles based on the growth of metallic nanoshell by templating redox-active polymer brushes. Each spectrum (shown from top to bottom) was obtained from Au-nanogapped nanoparticles with 1.5 nm gap, 5 nm gap, and 11 nm gap; 40 nm Au nanoparticles; hollow Au nanoshells, and double-layered Au nanoparticles of ~43 nm with RhB tags positioned between the two layers. All spectra were acquired using RhB-modified nanoparticles and measured using a 632.8 nm excitation laser. (g) 1,4-BDT spacer layer-mediated synthesis of intra-nanogap structure (nanomatryoshkas) with a gap size of 0.7 nm. (h) A schematic illustration and TEM image of Au@AgAu nanoparticles with intra-nanogap, produced by galvanic replacement reaction. SEM, scanning electron microscope; RhB, rhodamine B; 1,4-BDT, 1,4-benzenedithiol. (Reproduced with permission from (a–d) Bardhan et al. 2010, Copyright 2010 American Chemical Society; (e,f) Song et al. 2014, Copyright 2014 American Chemical Society; (g) Lin et al. 2015, Copyright 2015 American Chemical Society; (h) Li et al. 2016, Copyright 2016 American Chemical Society.)

At the same time, the Au atoms bind to each other to minimize their surface energy. As this reaction continues, the interconnected Au atoms form a shell, and the accumulated vacancies between the core and shell eventually form an intra-nanogap (Figure 7.16a). This interlayer-free synthetic strategy for fabricating intra-nanogaps is simple, cheap, rapid, controllable, and scalable compared with the previous methods requiring interlayers (e.g., DNA, silica, and polymer) (Bardhan et al. 2010, Lim et al. 2011, Oh et al. 2014, Song et al. 2014, Lee et al. 2016). Due to the strong EM field generated in the intra-nanogap as well as the uniform, reproducible, and high-yield synthesis of the targeted structure, DIPs generated highly strong, long-term stable, and quantitatively reproducible SERS signals (Figure 7.16c,d) and exhibited highly enhanced and narrowly distributed EF values, ranging from 1.1×10^8 to 2.5×10^9, for 90.0%

of the particle population. Importantly, biofunctionalized DIP probes provided practical opportunities as SERS-based versatile probes for ultrasensitive target-DNA detection at the attomolar level (Figure 7.16e) and for long-term stable target-cell imaging with low laser power and short exposure time (Figure 7.16f,g).

7.5 Direct PL Enhancement of Metals

Since the first discovery of the tremendous enhancement in PL from nanostructures, the one-photon PL of metal nanoparticles has gained increasing attention. For a few decades, the development of single-nanoparticle measurement techniques such as TEM and AFM-correlated

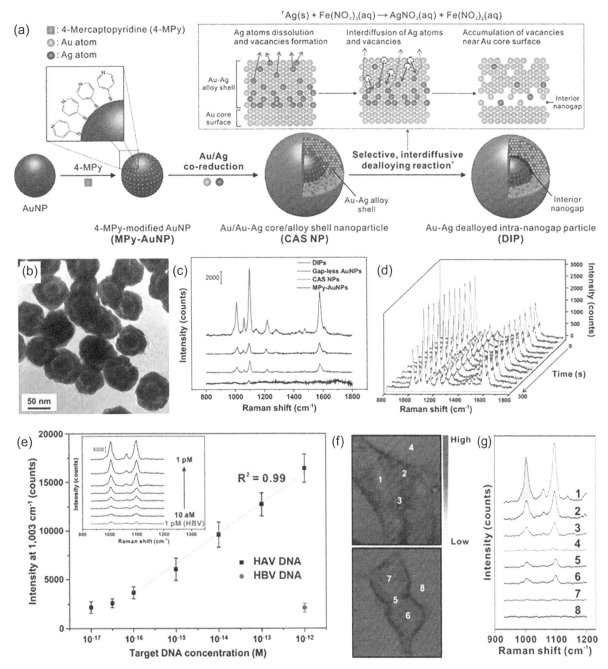

FIGURE 7.16 Synthesis and SERS properties of plasmonic intra-nanogap particles using dealloying reaction, and their applications for SERS-based DNA detection and cell imaging. (a) Schematic illustration of the dealloying-based strategy for the synthesis of Au–Ag DIPs. (b) TEM image of DIPs. (c) SERS spectra of as-synthesized nanoparticles in solution. Each spectrum (shown from bottom to top) was obtained from MPy-modified Au nanoparticles (MPy-AuNPs), Au/Au-Ag core/alloy shell nanoparticles (CAS NPs), Au/Au core/shell nanoparticles without intra-nanogap (gapless AuNPs), and dealloyed intra-nanogap particles (DIPs). All spectra were acquired from nanoparticles synthesized using 4-MPy-modified Au core and measured using a 633 nm excitation laser. (d) Time-dependent Raman profiles of DIPs. (e) SERS-based DNA detection assay results with DNA-modified DIPs. Inset shows the changes in SERS spectra obtained using different concentrations of target DNA. All spectra were acquired using 785 nm excitation laser. (f,g) SERS-based integrin $\alpha_V \beta_3$-specific cell imaging using peptide-functionalized DIPs. (f) DIP-based SERS maps of U87MG cells (high integrin $\alpha_V \beta_3$ expression, upper panel) and MCF-7 cells (integrin $\alpha_V \beta_3$ negative, lower panel). (g) SERS spectra acquired from the numerically marked positions in (f). All spectra were obtained using a 785 nm excitation laser. TEM, transmission electron microscope; 4-Mpy, 4-mercaptopyridine. (Reproduced with permission from (a–g) Kim et al. 2018a, Copyright 2018 American Chemical Society.)

spectroscopy has allowed for the observation and analysis of the structure and properties of single nanoparticles. Notably, the effect of structural heterogeneity can be removed by single-particle experiments. Therefore, one could study the PL properties of simple nanostructures without the need for advanced synthesis techniques. Researchers have thus focused on revealing the underlying mechanism of PL and further enhancing the PL performance by applying plasmonic effects such as strong local fields or plasmonic couplings. In this section, we introduce representative studies of single nanoparticle PL and discuss the suggested mechanism followed by investigations of the plasmonic coupling effect on PL.

7.5.1 Single Nanoparticles and Suggested Mechanisms

Most experiments on single nanostructures have been carried out on basic nanoparticles, such as Au nanorods and Au nanospheres. The differences in PL for various nanoparticles irradiated with the same excitation wavelength (Wu et al. 2010, Hu et al. 2012, Yorulmaz et al. 2012, Walsh and Negro 2013, Huang et al. 2015, Rao et al. 2015, Lin et al. 2016, Loumaigne et al. 2016, Park et al. 2016, Sivun et al. 2016, Dvoynenko et al. 2017, Mertens et al. 2017), the evolution of the PL when nanoparticles with the same structure are excited to different wavelengths (Tcherniak et al. 2011, Fang et al. 2012, Wackenhut et al. 2013, Lumdee et al. 2014, Zhang et al. 2014, Cheng et al. 2016, Lin et al. 2016, Cai et al. 2018), and the PL dependence on the excitation of emission polarization (Tcherniak et al. 2011, Hu et al. 2013, Zhang, et al. 2013a, 2014, Andersen et al. 2015, Wan et al. 2015, Yin et al. 2016, Vidal et al. 2018) have been studied.

The most distinctive feature of PL is that the PL spectrum can be modulated by plasmon resonance. When single Au nanoparticles with three different geometries were compared, their PL spectra were distinctly different but resembled their respective scattering spectra (Figure 7.17a–c) (Hu et al. 2012). In particular, the close correlation of the peak positions and spectral line width between the PL and scattering spectra implies the critical role of plasmons in the generation of PL. The PL spectra of coupled 80 nm Au nanodisks with gap sizes of 0–90 nm also show similar shifts (Figure 7.17d–f) (Hu et al. 2012). It is therefore unlikely that PL arises from the direct recombination of d-holes and sp-electrons, because emission from direct recombination would result in a shape-independent spectrum. This set of experiments was the first demonstration that PL in coupled structures follows the resonance shifts of plasmons. The authors also reported an incomplete depolarization of PL and suggested that PL from nanostructures originates from hot electrons that are inhomogeneously distributed in the nanostructures.

In the same year, two groups reported a similar study on the PL of Au nanorods (Fang et al. 2012, Yorulmaz et al. 2012). Nanorods have two distinct plasmon resonances; a transverse mode and a longitudinal mode. The resonance energy of the longitudinal mode can be readily engineered

from the visible to the NIR region by changing the aspect ratio of the nanoparticles. Thus, flexible tunability makes Au nanorods the most heavily investigated nanoparticles in terms of the role of plasmons in PL generation. Their studies share similar points. First, the PL spectra of Au nanorods resemble their scattering spectra, whether or not the excitation wavelength can only excite interband transitions. Thus, whether irradiated at 532 nm, which excites the transverse surface plasmon resonances, or at 785 nm, which excites the longitudinal surface plasmon resonances, the photoluminescent spectra of the Au nanorods follow the DF scattering spectra of the same nanorods, especially at a wavelength close to the longitudinal surface plasmon resonance (Figure 7.18a) (Fang et al. 2012, Yorulmaz et al. 2012). This phenomenon is observed in almost all nanostructures, including Ag nanorods (Lin et al. 2016). To compare the PL intensity of a nanoparticle for different plasmon energies, slow thermal reshaping was conducted, and the structural transformation of the nanorod to a sphere-like structure occurred. When the plasmon energy was shifted to 540 nm, the QY dropped sixfold, implying that Au nanorods definitely show higher PL intensity than Au nanospheres and that the PL is affected by the localized plasmon resonance through emission enhancements (Figure 7.18b) (Yorulmaz et al. 2012). Given the absorption coefficient of Au nanorods, the two studies reported similar quantum yields in the order of $\sim 10^{-5}$ to 10^{-6} (Fang et al. 2012, Yorulmaz et al. 2012). Moreover, the PL emission is polarized along the axis of anisotropic nanoparticles. When the Au nanorods are excited, the PL emission has a perfect dipole response along the long axis of the Au nanorod (Figure 7.18c) (Yorulmaz et al. 2012). This phenomenon was also observed with another anisotropic nanoparticle, an Au nanobipyramid. By correlating the PL spectra, DF scattering spectra, emission pattern, and AFM data, the spatial orientation and radiation characteristics of a single Au nanobipyramid could be unambiguously determined (Zhang et al. 2013a). The emission pattern revealed that the Au bipyramids excited by incident light generate efficient PL emission, which can be simplified as a single dipole radiation with the oscillation direction along the longitudinal axis.

Single nanoparticles with different shapes can provide insight on the local field effect. When the PL properties of individual Au nanobipyramid structure were investigated and compared with that of Au nanorods, the Au nanobipyramids showed a quantum yield of PL double that of the Au nanorods in a similar surface plasmon resonance range, which is presumed to be due to the local field enhancement at the sharp end, inducing a more pronounced lightning rod effect (Figure 7.19a,b) (Rao et al. 2015). In a similar manner, the stronger local field enhancement of Au nanocubes with edges and corners results in higher PL quantum yields than that of Au nanorods under irradiation at the same excitation wavelength (Wu et al. 2010).

Recently, a quantitative analysis of SERS spectra with PL has been reported. Such a strategy arose from the understanding of the origin of the physical background of SERS

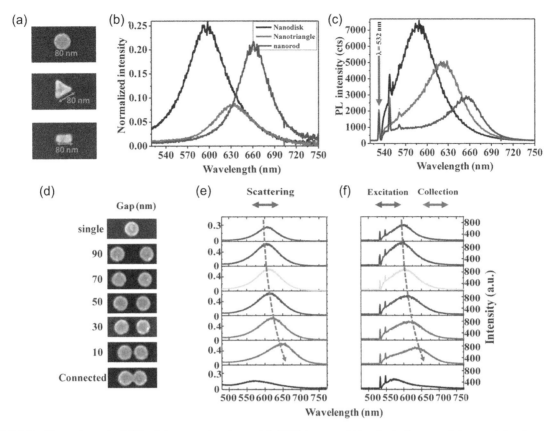

FIGURE 7.17 Shape-dependent PL of Au nanostructures. (a) SEM images of the single Au nanodisk, nanotriangle, and nanorod. (b,c) DF scattering spectra and PL spectra of these particles, respectively. The arrow in (c) indicates the Rayleigh scattering of the excitation laser. (d) SEM images of 80 nm Au nanodisks with varied gap size (left) and scattering and PL spectra of coupled structures. (e) Scattering spectra of the structures excited by horizontally polarized white light. (f) PL spectra of the structures obtained with both horizontally polarized excitation and collection. (Reproduced with permission from Hu et al. 2012, Copyright 2012 American Chemical Society.)

spectra. The authors found that the SERS background originates from the LSPR-modulated PL, which also shares the local field information like SERS. Thus, by multiplying the background-subtracted SERS spectrum of each Au nanorod with the PL_{bulk} (i.e., the PL of an Au(111) single crystal) and then dividing by the corresponding SERS background, the authors could interpret the relative strength of the Raman peaks and retrieve the intrinsic Raman signal of the molecules (Lin et al. 2017a). This result suggests that understanding the physical origin of the spectral information enables more reliable and reproducible spectral analysis, providing quantitative information on analytes.

Suggested Mechanisms

As we already mentioned in Section 7.2, the optical properties of Au are characterized by the interband absorption due to the penetration of EM fields onto the surface of the metal. The 5d → 6sp interband transition occurs near the X- and L-symmetry points of the Brillouin zone. The transition energy values near each symmetry point are ~1.84 and ~2.45 eV, respectively (Figure 7.20a) (Ngoc et al. 2015). The selection rules for common dipole transitions

between d-band and sp-band states are forbidden, but the large d-band electron density near the X-symmetry point compensates for the small transition matrix element from the Fermi's golden rule. It is also known that for plasmon-assisted interband transitions, the conventional selection rules for dipole processes may not strictly apply due to the large field gradients (Beversluis et al. 2003). As the absorption cross section near X-symmetry points is much weaker than the near L-symmetry points, the interband transition near the X-symmetry point has been more frequently studied. The interband PL process, which was initially proposed for the PL mechanism of noble bulk metals, occurs in three steps (Mooradian 1969). In the first step, the d-band electrons above the Fermi level are excited to the sp-band by radiation. Then, thermalization, consisting of electron–electron and electron–phonon scattering, results in fast energy relaxation. Finally, the radiative recombination of the conduction-band electrons below the Fermi level with holes in the d-band near the X- and L-symmetry points, where the bands are relatively flat in the momentum space, causes light emission. Therein, the authors pointed out that the recombination is direct because changes in momentum are negligible with photons. The observed broad line width

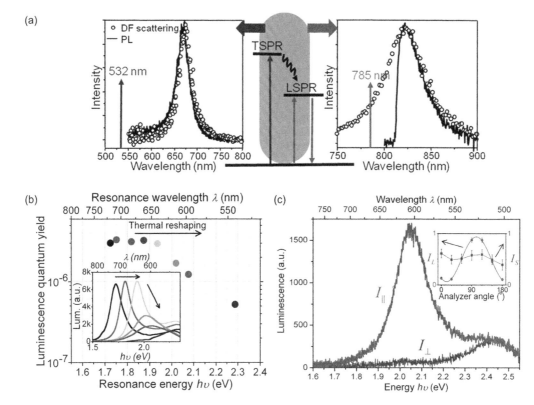

FIGURE 7.18 PL spectra of gold nanorods. (a) A schematic illustration of the PL process of a gold nanorod and the corresponding PL spectra excited by 532 nm (left) and 785 nm (right), respectively. Open circles indicate the scattering spectra of each nanorod. (b) Reshaping of a single gold nanorod. Each dot represents the quantum yield of the same nanorod obtained after each thermal reshaping step as a function of its resonance energy. The inset shows the emission spectra of the nanorod after each reshaping step. Arrows in the graph indicate the direction of spectral and intensity change during the melting process of the nanorod. (c) PL spectra of a gold nanorod excited by a circularly polarized 476 nm laser and recorded with a differently polarized detection: parallel ($I_\parallel$) and perpendicular ($I_\perp$) to the longitudinal axis of the nanorod. The inset shows the detection polarization dependence of the low-energy peak (I_L) and the high-energy peak (I_S). (Reproduced with permission from (a) Fang et al. 2012, Copyright 2012 American Chemical Society; (b,c) Yorulmaz et al. 2012, Copyright 2012 American Chemical Society.)

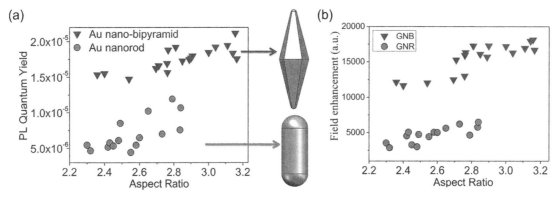

FIGURE 7.19 PL of gold nanobipyramids. (a) PL quantum yield of individual gold nanorods (circles) and gold nanobipyramids (triangles) as a function of aspect ratio. (b) Normalized field intensity integrated over the volume surrounding each nanoparticle. (Reproduced with permission from Rao et al. 2015, Copyright 2015 American Chemical Society.)

in the PL spectrum can be accounted for by the scattering of excited holes in relatively flat, broad regions of the 5d-bands of noble metals.

After the discovery of the relationship between PL and LSPR, researchers started to include the role of LSP (Dulkeith et al. 2004, Tcherniak et al. 2011, Fang et al. 2012, Hu et al. 2012, Wackenhut et al. 2013, Jiang et al. 2015, Wang et al. 2015, Yin et al. 2016). When the excitation energy simultaneously corresponds to the interband transition energy and the plasmon energy, both excitations are assumed to be possible. For example, a 488 nm laser only excites the interband transition in typical Au

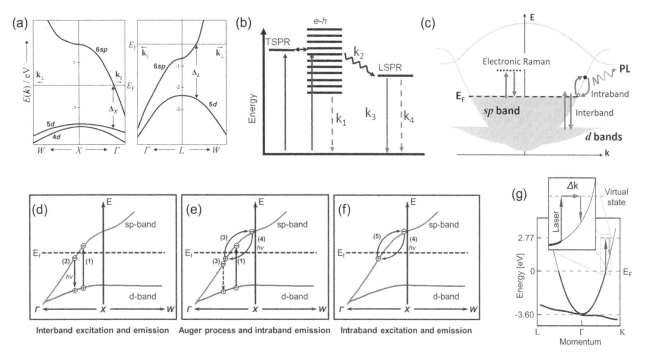

FIGURE 7.20 Suggested mechanisms for PL of metal nanostructures. Light emission through (a) interband transition near the X- and L-symmetry points, (b) radiative decay of plasmons, (c) intraband transition, (d–f) a Purcell effect enhanced radiative recombination of hot carriers, and (g) electronic Raman scattering. (Reproduced with permission from (a) Ngoc et al. 2015, Copyright 2015 American Chemical Society; (b) Fang et al. 2012, Copyright 2012 American Chemical Society; (c) Lin et al. 2016, Copyright 2016 American Chemical Society; (d–f) Cai et al. 2018, Copyright 2018 American Chemical Society; (g) Hugall and Baumberg 2015, Mertens et al. 2017, Copyright 2017 American Chemical Society.)

nanostructures, but a 532 nm energy excites both d → sp interband transition of bound electrons and the collective oscillation of sp-band free electrons of Au. Moreover, it is hypothesized that the relaxation of excited e–h pairs excites plasmons in nanostructures. Sometimes, the interconversion between LSPR and e–h pairs is also considered (Fang et al. 2012, Wackenhut et al. 2013). Therefore, the energy of the transverse SPR excitation can be transferred to the excitation of longitudinal SPR via e–h pairs. Therefore, in this mechanism, instead of the direct recombination of excited electron–hole pairs, the plasmons are excited and their radiative decay emits light (Figure 7.20b) (Fang et al. 2012). Of course, the dominant decay channel of e–h pairs or excited plasmons is nonradiative relaxation. Thus, although emission occurs mainly through LSPR and through interband transition, their efficiency is limited, resulting in low quantum yields.

On the other hand, sp-band electrons can directly undergo intraband transition, which is a transition between electronic states within the same sp-band. At first, intraband transition has been suggested to explain the observation of the IR emission of a rough Au film excited at 780 nm (1.59 eV), which is much lower than the bandgap for interband transition (~1.8 eV near the X-symmetry point and ~2.4 eV near the L-symmetry point) (Beversluis et al. 2003). Thus, it is possible to explain light emission in a situation where interband transition is involved. Although the intraband transition of smooth metal surfaces is momentum-forbidden in

principle, the selection rules are expected to be relaxed in metal nanoparticles by the additional momentum, which enables efficient intraband transition and radiative recombination (Figure 7.20c) (Lin et al. 2016). Nevertheless, PL by intraband transitions was assumed to be weaker than that by interband transitions because of the breaking of momentum. One of the main reasons for suggesting intraband transitions as a mechanism for PL arose from the observation of wavelength-dependent PL. To subtract the plasmon resonance effect from the PL, the PL spectra of nanorods were normalized to their own scattering spectra. The ratio of PL to scattering spectra is consistent for each nanorod at 532 or 633 nm excitation. However, the ratios of the spectra at 532 and 633 nm excitation were quite different, indicating the excitation wavelength dependence of PL generation of the Ag nanorods (Lin et al. 2016). The electronic Raman scattering should be independent of the excitation wavelength, and the intraband transition PL should be dependent on the excitation wavelength. Therefore, the authors suggested that the PL of the nanorods does not seem to follow the first model.

Most recently, Cai et al. suggested that PL could be viewed as a Purcell effect enhanced radiative recombination of hot carriers (Figure 7.20d–f) (Cai et al. 2018). The authors first prepared two Au nanorods, and the DF scattering measurement showed resonance wavelengths of 704 and 871 nm. Then, the PL spectra of a single Au nanorod were obtained using 405, 488, 532, 633, and 785 nm excitation wavelengths. Of the five excitation wavelengths, the 785 nm light was the only one below the threshold for exciting interband transitions in Au (1.7–1.8 eV). As in the previous report, the PL spectra of all Au nanorods showed major peaks with resonance energies and line shapes similar to the longitudinal surface plasmon resonance observed in the correlated DF scattering spectra. While the PL spectra for other four excitation wavelengths exhibit two peaks, the PL spectrum for the 785 nm excitation showed a single peak, with the absence of a peak at higher energies. As the hot carriers excited by the 785 nm laser cannot be released through interband transition, the lower energy peak only involves conduction band electrons, resulting in emission by intraband transition (Figure 7.20f). The PL emitted at 785 nm excitation is consistent with the previous reports of Au and Ag nanorods emitting light through the radiative recombination of charge carriers in the conduction band without the need to include d-band holes. In addition, the high-energy peak that was absent in the 785 nm excitation spectra was assigned to the direct recombination of interband transitions in Au (Figure 7.20d). In particular, the energy of the higher energy peak was similar for each excitation. Upon excitation, most hot carriers tend to have energies similar to the energy of the incident photon. Hence, the energy of the emission through such recombination also matches the excitation energy. It is worth noting that, when interband transition is possible, intraband transition following the nonradiative Auger scattering of d-band holes and sp-band electrons can simultaneously occur. Therefore, a peak at lower energies, which was attributable to intraband transition, can result from pure intraband recombination or interband transition followed by Auger relaxation and intraband recombination (Figure 7.20e). When we neglect the difference between direct intraband excitation and Auger-assisted intraband excitation, the sum of the interband and intraband emissions results in a larger QY of interband excitations. The authors also estimated that up to half the total interband excitation emission was relaxed by intraband PL with a QY ratio of 488 and 785 nm excitation. They also simulated the PL spectra by taking into consideration the hot carrier distribution generated by the optical excitation and the radiative photonic density of states. The simulated PL spectra were very close to the experimental results. The density of states indicates the average number of states available for occupation in an electronic system, and the radiative photonic density of state describes the number of EM states available to the nanoparticle for the emission of a photon. The shape of the PL peak is ultimately determined by the photonic density of states, which originates from hot carriers with different energies rather than being directly emitted from a certain electronic excited state. Given the much smaller photonic density of states of the transverse plasmon mode of the Au nanorods, the lack of influence on the PL spectra near the transverse plasmon mode was understandable. The results reported therein also demonstrated the excitation power-dependent and excitation wavelength-dependent blueshift of the PL from the scattering spectra.

The authors explained these phenomena by the fact that hot carriers with high energies can increase the magnitude of the shift, thus supporting the proposed mechanism of the radiative recombination of hot carriers.

In contrast to the absorption–recombination process, the Baumberg group has proposed an electronic Raman scattering mechanism (Figure 7.20g) (Hugall and Baumberg 2015, Mertens et al. 2017). Therein, the PL is the light emission from the excitation of s-band electrons to virtual states in the metal followed by prompt de-excitation. The main difference between electronic Raman scattering and electron recombination lies in the way emission is perceived. In the former, emission is considered rapid, while the latter considers much longer times. Electronic Raman scattering is very similar to the SERS from molecules in hotspots but differs in that it involves an intraband electron continuum instead of a molecular vibration state. In film-coupled nanoparticle systems in which single nanoparticles are placed atop a metal film covered with self-assembled molecular layers, the system can be easily diversified with different Au and Ag compositions and nanoparticle structures, such as spheres and cubes. Continuous-wave laser radiation at 447 nm causes the adatom to drift inside the gap, thus shifting the DF scattering and light emission (Mertens et al. 2017). These two spectra are mostly similar, however, and interestingly, the emissions are almost lost in the later stages of the conductive connection inside the gap despite strong DF scattering. In this stage, the plasmon field is localized around the pillar structure instead of in the gap between the substrate and the nanoparticle. The authors attributed the origin of the dramatic drop in the light emission to the changes in the near-field decay length inside the metal pillar, which is increased tenfold compared with that in the original nanogap. From these results, they developed an analytical model to describe the findings. The electrons inside the metal are excited into a virtual state by the incoming plasmon-coupled photon followed by de-excitation to an empty state, thereby emitting a lower energy photon. Due to the quadratic dispersion of s-band electrons in Au, a difference in the momentum Δk of the electron arises. The localized spatial field distribution of the plasmon supplies this momentum difference and enhances the PL by coupling the incoming and outgoing photons. Based on the difference in the complex wavelength-dependent dielectric permittivity of Au and Ag, the model explains why Ag nanostructures generate light more efficiently than the equivalent Au nanostructures. Moreover, based on the acceleration of the hot virtual electrons in the strongly confined plasmonic near-field gradient at the metal surface, the gap-size dependent PL can be described. The PL from ERS is similar to the intraband PL process, except that ERS involves virtual intermediate states, whereas the intraband PL process involves real intermediate states. Hence, the latter process is constrained by the dispersion relation of the energy and momentum, which is one of the main reasons for ERS being more probable than intraband PL, as claimed by the authors.

7.5.2 Plasmonic Coupling Effect

As the unprecedented enhancement of Raman scattering by plasmonically coupled nanostructures was established, researchers started to investigate the coupling effect on PL. In the previous single particle study cited earlier, if the same nanoparticles are tested under various conditions, the PL changes depending on the coupling conditions can be confirmed by using plasmonically coupled nanostructures. Various physical properties of coupled nanostructured systems, such as absorption, scattering, near field, and hot electron distribution, can be modeled and predicted relatively accurately. Therefore, many groups have investigated the correlation between these physical properties and the observed PL and obtained clues about the underlying mechanisms.

Top-down approaches such as lithography are simple and powerful methods to fabricate simple nanostructures in a controllable manner. A dolmen-like plasmonic nanostructure, such as a dolomite, consists of a single rod (monomer) that supports a vertically arranged dipole mode with two parallel bars (dimers) that in turn support the quadrupolar mode. Under excitation at 532 nm, the polarization is parallel, and a single Lorentz-like peak appears in the scattering and PL spectra (Figure 7.21a) (Yin et al. 2016). For polarizations perpendicular to the long axis of the dimer, Fano-like PL and DF scattering spectra with similar dips at ~660 nm are obtained (Figure 7.21b). Interestingly, both PL spectra showed a blueshift of similar magnitude of ~20 nm relative to the DF scattering peak. To understand these phenomena, a physical model of the excitation of collective free electrons followed by relaxation into the radiative density of plasmon states, which are represented in the scattering spectrum, was presented. Thus, the product of the redistributed excited electron population during the relaxation process and the density of the plasmonic states produced similar blueshifted PL peaks in both emissions (sp excitations, here dipole and quadrupole). Another facile way to achieve interplasmonic coupling is through nanoparticle film-coupled nanostructures systems such as Ag nanowires, spherical Au nanoparticles, Au nanodisks, and Au tips (Hu et al. 2013, Lumdee et al. 2014, Andersen et al. 2015, Wang et al. 2015). These systems have the advantage of tunable gaps and thus tunable couplings (at a more sophisticated level than that achieved with lithography) by controlling the interlayer thickness on the metal film. When 80 nm Au nanoparticles were placed on 3 nm Al_2O_3-coated Au films, a stronger PL was detected near the gap plasmon resonance compared with systems without these nanoparticles (Lumdee et al. 2014). In particular, irradiation at 633 nm caused a significantly increased PL compared with 532 nm excitation (Figure 7.21c). Such excitation efficiently excited the gap plasmon mode and induced a strong electric field near the particle–substrate junction with interband absorption near the X-symmetry point, which increased the efficiency of electron–hole pair generation (Figure 7.21d). Furthermore, the coupling between the nanoparticle and

shell forms a strong gap plasmon at 652 nm, through which the PL emission at 660 nm can be greatly increased with higher emission efficiency. However, in this study, they only compared the PL of the system with that of a film; thus, additional experimentation is needed to assess the extent of increase compared with the PL of nanoparticles. Similarly, the coupling between Au nanodisks and Au films with a dielectric spacer strengthens the PL signal (Andersen et al. 2015). As the disk size increases, the PL intensity abruptly decreases, although the radiative decay of plasmons dominates over the nonradiative decay channel in larger particles. The authors suggested that this drop is a consequence of decreasing the spatial overlap between the absorption distribution at the excitation wavelength and the mode intensity at the plasmon wavelength, resulting in diminished LSP excitation in the metal. Another interesting paper reports the amplification of PL at the junction between the tip and the substrate, where a nanogap is formed, in a scanning tunneling microscope by inelastic electron tunneling, and the degree of induced electron tunneling can be controlled by a bias voltage between the tip and the sample (Figure 7.21e) (Wang et al. 2015). Under laser irradiation at 634 nm, the PL is derived from inelastic electron tunneling-based PL with typical interband transitions. When the laser is removed, the peaks from the interband transition disappear, and only the inelastic electron tunneling-based PL remains. As the bias voltage increases, the intensity of these peaks increases. Consequently, the photon emission between a sharp Au tip and an Au substrate can be enhanced by the radiated decay of hot electrons by inelastic tunneling in the gap plasmon mode. Therefore, this result reveals that PL from such nanostructures can be divided into two processes; first, the typical photon-induced electron–hole recombination; and second, the radiative decay of the local plasmon resonances produced by inelastic electron tunneling (Figure 7.21f).

Plasmonic coupling can generate a super-radiant plasmon mode, which yields the highest PL intensity and quantum yield reported for metallic nanostructures. Controlled galvanic void formation produced a core–gap–shell-type of cube-in-cube (CiC) nanoparticles with controllable nanogaps from Au nanocubes in high yield (Figure 7.22a–c) (Park et al. 2016). Plasmonic coupling between the core and the shell generates two plasmon modes that were interpreted with the plasmon hybridization model, which is analogous to the molecular orbital theory. The parallel coupling of two dipoles increases the total dipole moments, resulting in a super-radiant plasmon mode, whereas antiparallel coupling exhibits a subradiant character (Figure 7.22d). The super-radiance concept was first developed for the cooperative emission of fluorescence emitters when they interfere with each other in a coherent manner. If the collective oscillation of free electrons in a metal can be regarded as analogous to an oscillating dipole with the plasmon mode becoming super-radiative, the degree of radiant damping of the plasmon is greatly increased, as in the cooperative emission of fluorescent molecules. The measured PL

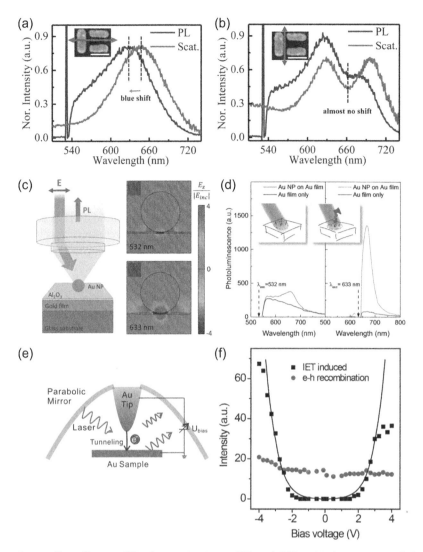

FIGURE 7.21 Plasmonic coupling effect on PL of nanostructures. PL and DF scattering spectra of the individual dolmen-like nanostructures with excitation–collection polarization of 0–0° (a) and 90–90° (b), respectively, as specified by gray arrows in the inset SEM images with a scale bar of 100 nm. (c) Experimental configuration for single-particle PL spectroscopy (left). Snapshots of the simulated electric field distributions at 532 and 633 nm excitation. (d) PL spectra collected at 532 nm (left) and 633 nm excitation (right) from areas containing a gold nanoparticle and no gold nanoparticle (shown from top to bottom, respectively). The insets schematically show that a gold nanoparticle adds PL signal at the particle resonance wavelength (dark gray arrows) to the weak background PL. (e) Sketch of the Au-tip Au-sample junction experimental configuration. A bias voltage is applied between the tip and the sample to induce electron tunneling with a radially polarized laser beam at 634 nm is focused by a parabolic mirror on the junction. (f) PL intensity contributions from Inelastic Electron Tunneling (IET) (black solid squares) and electron–hole combination (solid circles) as a function of bias voltages under laser illumination. (Reproduced with permission from (a,b) Yin et al. 2016, Copyright 2016 American Chemical Society; (c,d) Lumdee et al. 2014, Copyright 2014 American Chemical Society; (e,f) Wang et al. 2015, Copyright 2015 American Chemical Society.)

intensity of the 9 nm gap CiC nanoparticles differed greatly from that of the two Au nanocubes or that of the CiC nanoparticles with different gap sizes. In particular, the larger Au nanocubes (NCs) exhibited markedly lower intensities, suggesting the role of the inner gap for intense PL. The 16-fold increase in the quantum yield and a twofold increase in absorption cross section caused a 31 times higher PL intensity for CiC nanoparticles (Figure 7.22e). The authors attribute this behavior to the super-radiant plasmon mode, which significantly increases the radiative damping of the

corresponding plasmon, and these observations agreed well with calculation results. Moreover, an optical signal stable over 1 h was observed by total internal reflection microscopy, which indicates that nanoparticles can be used as stable optical labels in plasmonic nanomaterial-based bioapplications (Figure 7.22f).

In other studies, it has been reported that plasmonic coupling has little effect on PL enhancement, and sometimes, a strong coupling reduces PL. For instance, spherical Au nanodisk arrays with varying sizes and separations were

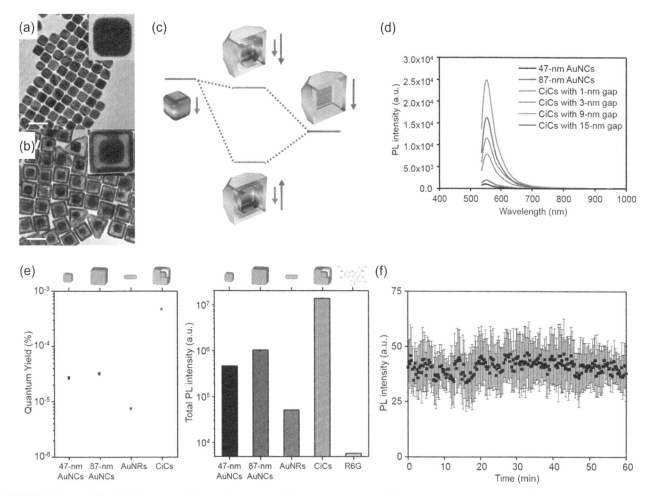

FIGURE 7.22 PL of CiC nanoparticles with engineerable interior nanogap. TEM images of gold nanocube (a) and CiC (b) nanoparticles; the insets show representative single-particle images. The scale bar is 100 nm. (c) A diagram of plasmon hybridization indicating super-radiant plasmon modes at a higher energy and subradiant plasmon modes at a lower energy. Inset images are the surface charge distributions for the corresponding plasmon modes. (d) PL spectra of suspensions of 47 and 87 nm Au nanocubes and the CiCs with varying interior gap sizes measured using a 532 nm excitation laser. Comparison of (e) PL QY (left) and total PL intensity (right) of from CiCs with 47 and 87 nm Au nanocubes, an Au nanorod and Rhodamine 6G (R6G). Error bars are standard deviations for three independent measurements. (f) PL intensity for a single CiC measured using total internal reflection fluorescence (TIRF) microscopy under continuous illumination from a 532 nm laser for 1 h. Images were taken every 30 s. The signal and error bars are averages for 20 nanoparticles. (Reproduced with permission from Park et al. 2016, Copyright 2016 American Chemical Society.)

prepared to investigate the scaling of PL efficiency (Figure 7.23a–c) (Walsh and Negro 2013). As the scattering spectrum is redshifted with the increase in disk size at fixed edge-to-edge interparticle separations, the PL spectra excited by 405 nm light is simultaneously redshifted (Figure 7.23d,e). When the edge-to-edge separation decreases from 300 to 150 nm, inhomogeneous broadening of the PL spectrum occurs as the particle size increases. Therefore, the broadening of the scattering spectrum bandwidth with decreasing distance is due to the increased nonradiative LSP damping, which results in a larger near-field interparticle coupling between the nanoparticles. When a set of nanoparticle arrays in which the edge-to-edge separation varied from 100 to 300 nm was investigated, the PL spectra normalized per unit volume of the arrays of the less-coupled nanoparticles showed the highest PL intensity. According to the authors, the increase

in the nonradiative damping of the LSP resonance due to near-field coupling is the key to the lower PL intensity, which limits the PL efficiency as the particles approach each other. Bottom-up-based assembly strategy makes more precisely coupled nanoparticles. The PL spectra of 50 nm spherical Au nanoparticle dimers suggested the minor role of the strong local electric field in PL enhancement (Huang et al. 2015). When excited by a 488 nm laser, dimer nanostructures showed about thrice stronger PL intensities. However, given the magnitude of the increase in absorption, the final quantum yield of the dimer was found to be similar to that of the constituent monomers, even with a strong local electric field (Figure 7.23f). Another study with two Au bipyramids in close proximity also showed that strong plasmonic coupling caused the PL signal to decrease (Sivun et al. 2016). In this system, the tip-to-tip distance of the

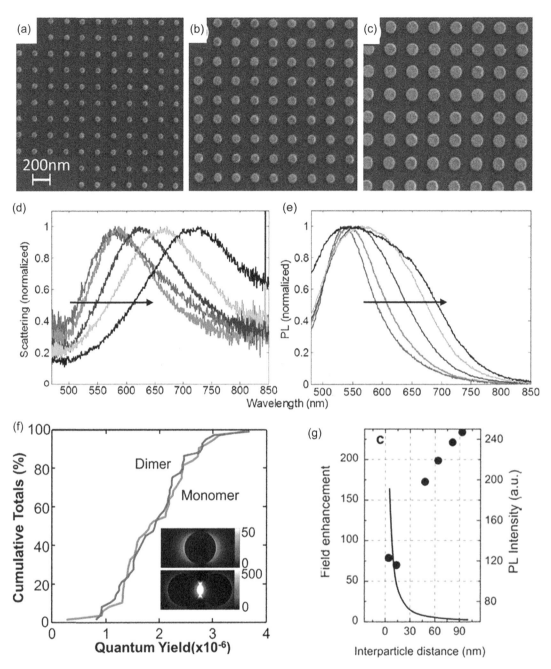

FIGURE 7.23 Negligible effect of plasmonic coupling on PL enhancement. SEMs of arrays of Au nanodisks of 60 nm (a), 100 nm (b), and 140 nm (c) in diameter with 150 nm edge-to-edge separations. (d) DF scattering spectra and (e) PL spectra excited by 405 nm laser of gold nanodisks arrays with center-to-center interparticle separation of 400 nm and particle diameters of 60 nm, 80 nm, 100 nm, 120 nm, and 140 nm (from left to right). (f) Cumulative distributions of the measured quantum yield for gold nanosphere monomers and dimers. Inset: EM field enhancements for a monomer and a dimer calculated at 550 and 640 nm, respectively. (g) Anticorrelation between the calculated field enhancement at the center between the two adjacent tips (solid lines) and the measured PL intensity (dots) of gold bipyramids with up–up (left), up–down (middle), and down–down (right) configurations. (Reproduced with permission from (a–e) Walsh and Negro 2013, Copyright 2013 American Chemical Society; (f) Huang et al. 2015, Copyright 2015 American Chemical Society; (g) Sivun et al. 2016, Copyright 2015 American Chemical Society.)

two Au bipyramids was manipulated by AFM and systematically varied on the nanometer scale. The PL and scattering spectra were recorded for each separation distance. Interestingly, the PL spectrum was anticorrelated with the scattering intensity (Figure 7.23g). Such anticorrelation was explained by a theoretical model that considered the probability of d-band hole recombination and a decrease in the field strength inside the nanostructure. Further investigations of plasmon-modulated or plasmon-enhanced PL phenomena are required in a systematic manner.

7.6 Future Directions

So far, we have presented fundamentals, representative advances, and numerous applications in plasmonically enhanced Raman and PL signals, but there are still some hurdles that need to be overcome to transform the aforementioned intriguing outcomes into practical applications.

Reliable signal generation is a primary goal to fully harness the advantageous plasmonic properties. It is well known that SERS-based plasmonic nanostructures have an ultrasensitive biosensing potential even at the single-molecule level (Le Ru and Etchegoin 2012). To realize a practical platform for detecting and imaging molecules in a reliable and quantifiable manner at very low concentrations, it is essential to develop plasmonic nanostructures that generate amplified optical signals with minimal changes to the particles (Lim et al. 2011, Lee et al. 2012a, Kim et al. 2018a). Therefore, the synthesis of plasmonic nanostructures with high structural reproducibility and precision should be achieved first. As most plasmonically coupled nanostructures involve ultrasmall nanogap regions (Ding et al. 2016, Nam et al. 2016) and as the degree of near-field enhancement can greatly vary in the order of a nanometer (Le Ru and Etchegoin 2012), it is imperative to control the plasmon nanostructure at the nanometer or subnanometer scale. A recent study has used a hinged DNA origami to place molecules with the smallest displacement step of 0.04 nm (Funke and Dietz 2016). The use of tools such as functionalized oligonucleotides, DNA origami, or the host–guest chemistry of organic molecules can lead to precisely controlled plasmonic nanostructures (Lim et al. 2010, Chikkaraddy et al. 2016). It is known that highly accelerated electron beams or light irradiation can induce atomic displacement on metal surfaces. Therefore, theoretical models and results need to be rigorously correlated with various experimental results to convincingly analyze these type of nanostructures with nm and subnanometer features. Further quantum mechanical effects such as the tunneling effect and nonlocality have to be adequately considered, as they might cause discrepancies between the ideal model system and the measured plasmonic properties (Ciracì et al. 2012, Tan et al. 2014). We should also note that the plasmonic performance is affected not only by the nanogap morphology and laser wavelength but also by the molecular environment (Lim et al. 2010, Park and Kim 2010, Le Ru et al. 2011, Lee et al. 2012a, Chikkaraddy et al. 2016). Thus, a precise control of the molecular position or orientation is another important challenge to fully exploit the plasmonic effects, which can be addressed by using ultrasmall nanogaps to position dye molecules at specific locations or macrocyclic cucurbit[n]uril molecules to align molecules of interest (Lim et al. 2010, Kasera et al. 2012, Chikkaraddy et al. 2016).

The PL phenomenon can also be significantly improved as SERS has been over the past decade. The main difficulty in this area is the low quantum yield, which mitigates the advantages of nonphotobleaching and nonphotoblinking (Park et al. 2017a). Therefore, an in-depth understanding of the PL enhancement mechanism through systematic investigations and the synthetic realization of rational designs for strong PL measurements are the primary challenges in this field. Fortunately, the low quantum yield can be partially compensated by the large absorption cross section of plasmonic nanostructures, which yield a much higher total PL intensity than conventional organic fluorophores (Park et al. 2016), thereby demonstrating the potential of photoluminescent plasmonic nanostructures as optical labels for bioimaging. Thus, further studies on PL enhancement will deliver potential next-generation luminescent materials that produce a stable optical signal over extended periods of time without the need for other labeling molecules.

Recently, NIR-II windows (1,000–1,400 nm) have been reported to have several advantages over conventional NIR-I windows (750–900 nm) for increased tissue penetration and low autofluorescence background along with a high maximum permissible exposure to laser radiation (Park et al. 2017b). Thus, the development of nanostructures capable of interacting with light in the NIR-II region and amplifying the optical signal of interest would be helpful for future biomedical applications. Last, although most studies introduced in this chapter mainly focused on noble plasmonic materials, we should broaden our palette of materials with feasible alternatives such as non-noble metals, conductive oxides and nitrides, and even graphene (Grigorenko et al. 2012, Naik et al. 2013, Kim et al. 2018b). First, a material with the necessary characteristics would be selected, and then nanostructures with desired features and appropriate surface modifications would be synthesized to exploit the plasmonic effects fully and cost-effectively. In this manner, the broadening of spectral ranges can be achieved at lower costs, and a broader range of applications can be promoted.

Given the earlier results and future improvements, plasmonic materials have a myriad of possibilities in fundamentals and applications, promising a brighter future in the years to come.

References

Abadeer, N.S., Brennan, M.R., Wilson, W.L., and Murphy, C.J., 2014. Distance and plasmon wavelength dependent fluorescence of molecules bound to silica-coated gold nanorods. *ACS Nano*, 8 (8), 8392–8406.

Acuna, G.P., Bucher, M., Stein, I.H., Steinhauer, C., Kuzyk, A., Holzmeister, P., Schreiber, R., Moroz, A., Stefani, F.D., Liedl, T., Simmel, F.C., and Tinnefeld, P., 2012. Distance dependence of single-fluorophore quenching by gold nanoparticles studied on DNA origami. *ACS Nano*, 6 (4), 3189–3195.

Adato, R. and Altug, H., 2013. In-situ ultra-sensitive infrared absorption spectroscopy of biomolecule interactions in real time with plasmonic nanoantennas. *Nature Communications*, 4, 2154.

Adato, R., Artar, A., Erramilli, S., and Altug, H., 2013. Engineered absorption enhancement and induced transparency in coupled molecular and plasmonic resonator systems. *Nano Letters*, 13 (6), 2584–2591.

Ahn, S., Rourke, D., and Park, W., 2016. Plasmonic nanostructures for organic photovoltaic devices. *Journal of Optics*, 18 (3), 033001.

Albrecht, M.G. and Creighton, J.A., 1977. Anomalously intense Raman spectra of pyridine at a silver electrode. *Journal of the American Chemical Society*, 99 (15), 5215–5217.

Alvarez-Puebla, R., Liz-Marzán, L.M., and García De Abajo, F.J., 2010. Light concentration at the nanometer scale. *Journal of Physical Chemistry Letters*, 1 (16), 2428–2434.

Andersen, S.K.H., Pors, A., and Bozhevolnyi, S.I., 2015. Gold photoluminescence wavelength and polarization engineering. *ACS Photonics*, 2 (3), 432–438.

Anderson, L.J.E., Mayer, K.M., Fraleigh, R.D., Yang, Y., Lee, S., and Hafner, J.H., 2010. Quantitative measurements of individual gold nanoparticle scattering cross sections. *The Journal of Physical Chemistry C*, 114, 11127–11132.

Anger, P., Bharadwaj, P., and Novotny, L., 2006. Enhancement and quenching of single-molecule fluorescence. *Physical Review Letters*, 96 (11), 3–6.

Ayala-Orozco, C., Liu, J.G., Knight, M.W., Wang, Y., Day, J.K., Nordlander, P., and Halas, N.J., 2014. Fluorescence enhancement of molecules inside a gold nanomatryoshka. *Nano Letters*, 14(5), 2926–2933.

Bailo, E. and Deckert, V., 2008. Tip-enhanced Raman scattering. *Chemical Society Reviews*, 37 (5), 921.

Bardhan, R., Mukherjee, S., Mirin, N.A., Levit, S.D., Nordlander, P., and Halas, N.J., 2010. Nanosphere-in-a-nanoshell: A simple nanomatryushka. *Journal of Physical Chemistry C*, 114 (16), 7378–7383.

Beversluis, M., Bouhelier, A., and Novotny, L., 2003. Continuum generation from single gold nanostructures through near-field mediated intraband transitions. *Physical Review B: Condensed Matter and Materials Physics*, 68 (11), 115433.

Bohren, C.F. and Huffman, D.R., 1998. *Absorption and Scattering of Light by Small Particles*. Research supported by the University of Arizona and Institute of Occupational and Environmental Health New York WileyInterscience 1983 541 p. Weinheim: Wiley

Bosnick, K.A., Jiang, J., and Brus, L.E., 2002. Fluctuations and local symmetry in single-molecule Rhodamine 6G Raman scattering on silver nanocrystal aggregates. *Journal of Physical Chemistry B*, 106 (33), 8096–8099.

Boyd, G.T., Yu, Z.H., and Shen, Y.R., 1986. Photoinduced luminescence from the noble metals and its enhancement on roughened surfaces. *Physical Review B*, 33 (12), 7923–7936.

Brown, K.A., Park, S., and Hamad-Schifferli, K., 2008. Nucleotide-surface interactions in DNA-modified Au-nanoparticle conjugates: Sequence effects on reactivity and hybridization. *Journal of Physical Chemistry C*, 112 (20), 7517–7521.

Cade, N.I., Ritman-Meer, T., Kwakwa, K.A., and Richards, D., 2009. The plasmonic engineering of metal nanoparticles for enhanced fluorescence and Raman scattering. *Nanotechnology*, 20 (28), 285201.

Cai, Y.Y., Liu, J.G., Tauzin, L.J., Huang, D., Sung, E., Zhang, H., Joplin, A., Chang, W.S., Nordlander, P., and Link, S., 2018. Photoluminescence of gold nanorods: Purcell effect enhanced emission from hot carriers. *ACS Nano*, 12 (2), 976–985.

Camden, J.P., Dieringer, J.A., Wang, Y., Masiello, D.J., Marks, L.D., Schatz, G.C., and Van Duyne, R.P., 2008. Probing the structure of single-molecule surface-enhanced Raman scattering hot spots. *Journal of the American Chemical Society*, 130 (38), 12616–12617.

Campion, A. and Kambhampati, P., 1998. Surface-enhanced Raman scattering. *Chemical Society Reviews*, 27 (4), 241.

Cang, H., Liu, Y., Wang, Y., Yin, X., and Zhang, X., 2013. Giant suppression of photobleaching for single molecule detection via the Purcell effect. *Nano Letters*, 13 (12), 5949–5953.

Chen, C.K., Chang, M.H., Wu, H.T., Lee, Y.C., and Yen, T.J., 2014. Enhanced vibrational spectroscopy, intracellular refractive indexing for label-free biosensing and bioimaging by multiband plasmonic-antenna array. *Biosensors and Bioelectronics*, 60, 343–350.

Chen, Y., Munechika, K., and Ginger, D.S., 2007. Dependence of fluorescence intensity on the spectral overlap between fluorophores and plasmon resonant single silver nanoparticles. *Nano Letters*, 7 (3), 690–696.

Cheng, Y., Lu, G., He, Y., Shen, H., Zhao, J., Xia, K., and Gong, Q., 2016. Luminescence quantum yields of gold nanoparticles varying with excitation wavelengths. *Nanoscale*, 8 (4), 2188–2194.

Chhabra, R., Sharma, J., Wang, H., Zou, S., Lin, S., Yan, H., Lindsay, S., and Liu, Y., 2009. Distance-dependent interactions between gold nanoparticles and fluorescent molecules with DNA as tunable spacers. *Nanotechnology*, 20 (48), 5201.

Chikkaraddy, R., De Nijs, B., Benz, F., Barrow, S.J., Scherman, O.A., Rosta, E., Demetriadou, A., Fox, P., Hess, O., and Baumberg, J.J., 2016. Single-molecule strong coupling at room temperature in plasmonic nanocavities. *Nature*, 535 (7610), 127–130.

Chuntonov, L., Bar-Sadan, M., Houben, L., and Haran, G., 2012. Correlating electron tomography and plasmon spectroscopy of single noble metal core-shell nanoparticles. *Nano Letters*, 12 (1), 145–150.

Cialla, D., März, A., Böhme, R., Theil, F., Weber, K., Schmitt, M., and Popp, J., 2012. Surface-enhanced Raman spectroscopy (SERS): Progress and trends. *Analytical and Bioanalytical Chemistry*, 403 (1), 27–54.

Cialla-May, D., Zheng, X.-S., Weber, K., and Popp, J., 2017. Recent progress in surface-enhanced Raman spectroscopy for biological and biomedical applications: From cells to clinics. *Chemical Society Reviews*, 46 (13), 3945–3961.

Ciracì, C., Hill, R.T., Mock, J.J., Urzhumov, Y., Fernández-Domínguez, A.I., Maier, S.A., Pendry, J.B., Chilkoti, A., and Smith, D.R., 2012. Probing the ultimate limits of plasmonic enhancement. *Science*, 337 (6098), 1072–1074.

Crozier, K.B., Sundaramurthy, A., Kino, G.S., and Quate, C.F., 2003. Optical antennas: Resonators for local field enhancement. *Journal of Applied Physics*, 94 (7), 4632–4642.

Cui, L., Mahajan, S., Cole, R.M., Soares, B., Bartlett, P.N., Baumberg, J.J., Hayward, I.P., Ren, B., Russell, A.E., and Tian, Z.Q., 2009. UV SERS at well ordered Pd sphere segment void (SSV) nanostructures. *Physical Chemistry Chemical Physics*, 11 (7), 1023–1026.

Cui, L., Wang, A., Wu, D.Y., Ren, B., and Tian, Z.Q., 2008. Shaping and shelling Pt and Pd nanoparticles for ultraviolet laser excited surface-enhanced Raman scattering. *Journal of Physical Chemistry C*, 112 (45), 17618–17624.

Deckert, V., 2009. Tip-enhanced raman spectroscopy. *Journal of Raman Spectroscopy*, 40 (10), 1336–1337.

Demers, L.M., Mirkin, C.A., Mucic, R.C., Reynolds, R.A., Letsinger, R.L., Elghanian, R., and Viswanadham, G., 2000. A fluorescence-based method for determining the surface coverage and hybridization efficiency of thiol-capped oligonucleotides bound to gold thin films and nanoparticles. *Analytical Chemistry*, 72 (22), 5535–5541.

Demers, L.M., Östblom, M., Zhang, H., Jang, N.H., Liedberg, B., and Mirkin, C.A., 2002. Thermal desorption behavior and binding properties of DNA bases and nucleosides on gold. *Journal of the American Chemical Society*, 124 (38), 11248–11249.

Ding, S.-Y., Yi, J., Li, J.-F., Ren, B., Wu, D.-Y., Panneerselvam, R., and Tian, Z.-Q., 2016. Nanostructure-based plasmon-enhanced Raman spectroscopy for surface analysis of materials. *Nature Reviews Materials*, 1 (6), 16021.

Dondapati, S.K., Sau, T.K., Hrelescu, C., Klar, T.A., Stefani, F.D., and Feldmann, J., 2010. Label-free biosensing based on single gold nanostars as plasmonic transducers. *ACS Nano*, 4 (11), 6318–6322.

Dong, J., Zhang, Z., Zheng, H., and Sun, M., 2015. Recent progress on plasmon-enhanced fluorescence. *Nanophotonics*, 4 (1), 472–490.

Dörfer, T., Schmitt, M., and Popp, J., 2007. Deep-UV surface-enhanced Raman scattering. *Journal of Raman Spectroscopy*, 38 (11), 1379–1382.

Dregely, D., Neubrech, F., Duan, H., Vogelgesang, R., and Giessen, H., 2013. Vibrational near-field mapping of planar and buried three-dimensional plasmonic nanostructures. *Nature Communications*, 4, 1–9.

Drexhage, K.H., 1974. IV Interaction of light with monomolecular dye layers. *Progress in Optics*, 12, 163–232.

Dulkeith, E., Niedereichholz, T., Klar, T.A., Feldmann, J., Von Plessen, G., Gittins, D.I., Mayya, K.S., and Caruso, F., 2004. Plasmon emission in photoexcited gold nanoparticles. *Physical Review B: Condensed Matter and Materials Physics*, 70 (20), 205424.

Dvoynenko, M.M., Wang, H.H., Hsiao, H.H., Wang, Y.L., and Wang, J.K., 2017. Study of signal-to-background ratio of surface-enhanced Raman scattering: Dependences on excitation wavelength and hot-spot gap. *Journal of Physical Chemistry C*, 121 (47), 26438–26445.

Erdmann, M., David, R., Fornof, A.R., and Gaub, H.E., 2010. Electrically induced bonding of DNA to gold. *Nature Chemistry*, 2 (9), 745–749.

Evanoff, D.D. and Chumanov, G., 2004. Size-controlled synthesis of nanoparticles. 2. Measurement of extinction, scattering, and absorption cross sections. *Journal of Physical Chemistry B*, 108 (37), 13957–13962.

Fan, M., Lai, F.-J., Chou, H.-L., Lu, W.-T., Hwang, B.-J., and Brolo, A.G., 2013. Surface-enhanced Raman scattering (SERS) from Au:Ag bimetallic nanoparticles: The effect of the molecular probe. *Chem. Sci.*, 4 (1), 509–515.

Fang, Y., Chang, W.S., Willingham, B., Swanglap, P., Dominguez-Medina, S., and Link, S., 2012. Plasmon emission quantum yield of single gold nanorods as a function of aspect ratio. *ACS Nano*, 6 (8), 7177–7184.

Fleischmann, M., Hendra, P.J., and McQuillan, A.J., 1974. Raman spectra of pyridine adsorbed at a silver electrode. *Chemical Physics Letters*, 26 (2), 163–166.

Fort, E. and Grésillon, S., 2008. Surface enhanced fluorescence. *Journal of Physics D: Applied Physics*, 41 (1), 013001.

Fu, Y. and Lakowicz, J.R., 2006. Enhanced fluorescence of Cy5-labeled DNA tethered to silver island films: Fluorescence images and time-resolved studies using single-molecule spectroscopy. *Analytical Chemistry*, 78 (17), 6238–6245.

Funke, J.J. and Dietz, H., 2016. Placing molecules with Bohr radius resolution using DNA origami. *Nature Nanotechnology*, 11 (1), 47–52.

Geddes, C.D. and Lakowicz, J.R., 2002. Metal-enhanced fluorescence. *Journal of Fluorescence*, 12 (2), 121–129.

Giannini, V., Francescato, Y., Amrania, H., Phillips, C.C., and Maier, S.A., 2011. Fano resonances in nanoscale plasmonic systems: A parameter-free modeling approach. *Nano Letters*, 11 (7), 2835–2840.

González, E., Arbiol, J., and Puntes, V.F., 2011. Carving at the nanoscale: Sequential galvanic exchange and Kirkendall growth at room temperature. *Science (New York, N.Y.)*, 334 (6061), 1377–1380.

Grigorenko, A.N., Polini, M., and Novoselov, K.S., 2012. Graphene plasmonics. *Nature Photonics*, 6, 749–758.

Guerrero, A.R., Zhang, Y., and Aroca, R.F., 2012. Experimental confirmation of local field enhancement determining far-field measurements with shell-isolated silver nanoparticles. *Small*, 8 (19), 2964–2967.

Haes, A.J., Haynes, C.L., McFarland, A.D., Schatz, G.C., Van Duyne, R.P., and Zou, S., 2005. Plasmonic materials for surface-enhanced sensing and spectroscopy. *MRS Bulletin*, 30 (5), 368–375.

Han, Q., Zhang, C., Gao, W., Han, Z., Liu, T., Li, C., Wang, Z., He, E., and Zheng, H., 2016. Ag-Au alloy nanoparticles: Synthesis and in situ monitoring SERS of plasmonic catalysis. *Sensors and Actuators, B: Chemical*, 231, 609–614.

Hao, Q., Du, D., Wang, C., Li, W., Huang, H., Li, J., Qiu, T., and Chu, P.K., 2014. Plasmon-induced broadband fluorescence enhancement on Al-Ag bimetallic substrates. *Scientific Reports*, 4, 1–5.

Hartland, G. V., 2011. Optical studies of dynamics in noble metal nanostructures. *Chemical Reviews*, 111 (6), 3858–3887.

Hartstein, A., Kirtley, J.R., and Tsang, J.C., 1980. Enhancement of the infrared absorption from molecular monolayers with thin metal overlayers. *Physical Review Letters*, 45 (3), 201–204.

Haslett, T.L., Tay, L., and Moskovits, M., 2000. Can surface-enhanced Raman scattering serve as a channel for strong optical pumping? *Journal of Chemical Physics*, 113 (4), 1641–1646.

Henzie, J., Andrews, S.C., Ling, X.Y., Li, Z., and Yang, P., 2013. Oriented assembly of polyhedral plasmonic nanoparticle clusters. *Proceedings of the National Academy of Sciences*, 110 (17), 6640–6645.

Hering, K., Cialla, D., Ackermann, K., Dörfer, T., Möller, R., Schneidewind, H., Mattheis, R., Fritzsche, W., Rösch, P., and Popp, J., 2008. SERS: A versatile tool in chemical and biochemical diagnostics. *Analytical and Bioanalytical Chemistry*, 390 (1), 113–124.

Hoang, T.B., Akselrod, G.M., Argyropoulos, C., Huang, J., Smith, D.R., and Mikkelsen, M.H., 2015. Ultrafast spontaneous emission source using plasmonic nanoantennas. *Nature Communications*, 6 (1), 7788.

Hoang, T.B., Akselrod, G.M., and Mikkelsen, M.H., 2016. Ultrafast room-temperature single photon emission from quantum dots coupled to plasmonic nanocavities. *Nano Letters*, 16 (1), 270–275.

Hu, H., Akimov, Y.A., Duan, H., Li, X., Liao, M., Tan, R.L.S., Wu, L., Chen, H., Fan, H., Bai, P., Lee, P.S., Yang, J.K.W., and Shen, Z.X., 2013. Photoluminescence via gap plasmons between single silver nanowires and a thin gold film. *Nanoscale*, 5 (24), 12086.

Hu, H., Duan, H., Yang, J.K.W., and Shen, Z.X., 2012. Plasmon-modulated photoluminescence of individual gold nanostructures. *ACS Nano*, 6 (11), 10147–10155.

Huang, D., Byers, C.P., Wang, L.Y., Hoggard, A., Hoener, B., Dominguez-Medina, S., Chen, S., Chang, W.S., Landes, C.F., and Link, S., 2015. Photoluminescence of a plasmonic molecule. *ACS Nano*, 9 (7), 7072–7079.

Huang, X. and El-Sayed, M.A., 2010. Gold nanoparticles: Optical properties and implementations in cancer diagnosis and photothermal therapy. *Journal of Advanced Research*, 1 (1), 13–28.

Hugall, J.T. and Baumberg, J.J., 2015. Demonstrating photoluminescence from Au is electronic inelastic light scattering of a plasmonic metal: The origin of SERS backgrounds. *Nano Letters*, 15(4), 2600-2604.

Huh, Y.S., Chung, A.J., and Erickson, D., 2009. Surface enhanced Raman spectroscopy and its application to molecular and cellular analysis. *Microfluidics and Nanofluidics*, 6 (3), 285–297.

Hurst, S.J., Lytton-Jean, A.K.R., and Mirkin, C.A., 2006. Maximizing DNA loading on a range of gold nanoparticle sizes. *Analytical Chemistry*, 78 (24), 8313–8318.

Imura, K., Nagahara, T., and Okamoto, H., 2005. Near-field two-photon-induced photoluminescence from single gold nanorods and imaging of plasmon modes. *Journal of Physical Chemistry B*, 109 (27), 13214–13220.

Indrasekara, A.S.D.S., Meyers, S., Shubeita, S., Feldman, L.C., Gustafsson, T., and Fabris, L., 2014. Gold nanostar substrates for SERS-based chemical sensing in the femtomolar regime. *Nanoscale*, 6 (15), 8891–8899.

Isola, N.R., Stokes, D.L., and Vo-Dinh, T., 1998. Surface-enhanced Raman gene probe for HIV detection. *Analytical Chemistry*, 70 (7), 1352–1356.

Jackson, W.D., 1962. *Classical Electrodynamics*. Wiley, New York.

Jain, P.K., Lee, K.S., El-Sayed, I.H., and El-Sayed, M.A., 2006. Calculated absorption and scattering properties of gold nanoparticles of different size, shape, and composition: Applications in biological imaging and biomedicine. *Journal of Physical Chemistry B*, 110 (14), 7238–7248.

Jeanmaire, D.L. and Van Duyne, R.P., 1977. Surface raman spectroelectrochemistry. *Journal of Electroanalytical Chemistry and Interfacial Electrochemistry*, 84 (1), 1–20.

Jensen, L., Aikens, C.M., and Schatz, G.C., 2008. Electronic structure methods for studying surface-enhanced Raman scattering. *Chemical Society Reviews*, 37 (5), 1061.

Jiang, L., Yin, T., Dong, Z., Hu, H., Liao, M., Allioux, D., Tan, S.J., Goh, X.M., Li, X., Yang, J.K.W., and Shen, Z., 2015. Probing vertical and horizontal plasmonic resonant states in the hotoluminescence of gold nanodisks. *ACS Photonics*, 2 (8), 1217–1223.

Johnson, P.B. and Christy, R.W., 1972. Optical constants of the noble metals. *Physical Review B*, 6 (12), 4370–4379.

Kasera, S., Biedermann, F., Baumberg, J.J., Scherman, O.A., and Mahajan, S., 2012. Quantitative SERS using the sequestration of small molecules inside precise plasmonic nanoconstructs. *Nano Letters*, 12 (11), 5924–5928.

Khoury, C.G. and Vo-Dinh, T., 2008. Gold nanostars for surface-enhanced Raman scattering: Synthesis, characterization and optimization. *Journal of Physical Chemistry C*, 112 (48), 18849–18859.

Kim, M., Ko, S.M., Kim, J.M., Son, J., Lee, C., Rhim, W.K., and Nam, J.M., 2018a. Dealloyed intra-nanogap particles with highly robust, quantifiable surface-enhanced Raman scattering signals for biosensing and bioimaging applications. *ACS Central Science*, 4 (2), 277–287.

Kim, P.Y., Oh, J.W., and Nam, J.M., 2015. Controlled co-assembly of nanoparticles and polymer into ultralong and continuous one-dimensional nanochains. *Journal of the American Chemical Society*, 137 (25), 8030–8033.

Kim, S., Kim, J.-M., Park, J.-E., and Nam, J.-M., 2018b. Nonnoble-metal-based plasmonic nanomaterials: Recent advances and future perspectives. *Advanced Materials*, 9, 1704528.

Kimura-Suda, H., Petrovykh, D.Y., Tarlov, M.J., and Whitman, L.J., 2003. Base-dependent competitive adsorption of single-stranded DNA on gold. *Journal of the American Chemical Society*, 125 (30), 9014–9015.

Kneipp, J., Kneipp, H., and Kneipp, K., 2008. SERS—a single-molecule and nanoscale tool for bioanalytics. *Chemical Society Reviews*, 37 (5), 1052.

Kneipp, K., Wang, Y., Kneipp, H., Perelman, L.T., Itzkan, I., Dasari, R.R., and Feld, M.S., 1997. Single molecule detection using surface-enhanced raman scattering (SERS). *Physical Review Letters*, 78 (9), 1667–1670.

Kühler, P., Roller, E.M., Schreiber, R., Liedl, T., Lohmüller, T., and Feldmann, J., 2014. Plasmonic DNA-origami nanoantennas for surface-enhanced Raman spectroscopy. *Nano Letters*, 14 (5), 2914–2919.

Kumar, A., Kim, S., and Nam, J.M., 2016. Plasmonically engineered nanoprobes for biomedical applications. *Journal of the American Chemical Society*, 138 (44), 14509–14525.

Kumar, A., Kumar, S., Rhim, W.K., Kim, G.H., and Nam, J.M., 2014. Oxidative nanopeeling chemistry-based synthesis and photodynamic and photothermal therapeutic applications of plasmonic core-petal nanostructures. *Journal of the American Chemical Society*, 136 (46), 16317–16325.

Kurouski, D., 2017. Advances of tip-enhanced Raman spectroscopy (TERS) in electrochemistry, biochemistry, and surface science. *Vibrational Spectroscopy*, 91, 3–15.

Lakowicz, J.R., 2005. Radiative decay engineering 5: Metal-enhanced fluorescence and plasmon emission. *Analytical Biochemistry*, 337 (2), 171–194.

Lee, H., Kim, G.H., Lee, J.H., Kim, N.H., Nam, J.M., and Suh, Y.D., 2015. Quantitative plasmon mode and surface-enhanced Raman scattering analyses of strongly coupled plasmonic nanotrimers with diverse geometries. *Nano Letters*, 15 (7), 4628–4636.

Lee, H., Lee, J.H., Jin, S.M., Suh, Y.D., and Nam, J.M., 2013a. Single-molecule and single-particle-based correlation studies between localized surface plasmons of dimeric nanostructures with ∼1 nm gap and surface-enhanced Raman scattering. *Nano Letters*, 13 (12), 6113–6121.

Lee, H.M., Jin, S.M., Kim, H.M., and Suh, Y.D., 2013b. Single-molecule surface-enhanced Raman spectroscopy: A perspective on the current status. *Physical Chemistry Chemical Physics*, 15 (15), 5276.

Lee, J.-H., Nam, J.-M., Jeon, K.-S., Lim, D.-K., Kim, H., Kwon, S., Lee, H., and Suh, Y.D., 2012a. Tuning and maximizing the single-molecule surface-enhanced Raman scattering from DNA-tethered nanodumbbells. *ACS Nano*, 6 (11), 9574–9584.

Lee, J.H., Kim, G.H., and Nam, J.M., 2012b. Directional synthesis and assembly of bimetallic nanosnowmen with DNA. *Journal of the American Chemical Society*, 134 (12), 5456–5459.

Lee, J.H., Oh, J.W., Nam, S.H., Cha, Y.S., Kim, G.H., Rhim, W.K., Kim, N.H., Kim, J., Han, S.W., Suh, Y.D., and Nam, J.M., 2016. Synthesis, optical properties, and multiplexed Raman bio-imaging of surface roughness-controlled nanobridged nanogap particles. *Small*, 12 (34), 4726–4734.

Lee, J.H., You, M.H., Kim, G.H., and Nam, J.M., 2014. Plasmonic nanosnowmen with a conductive junction as highly tunable nanoantenna structures and sensitive, quantitative and multiplexable surface-enhanced Raman scattering probes. *Nano Letters*, 14 (11), 6217–6225.

Le Ru, E.C. and Etchegoin, P.G., 2012. Single-molecule surface-enhanced Raman spectroscopy. *Annual Review of Physical Chemistry*, 63 (1), 65–87.

Le Ru, E.C., Grand, J., Sow, I., Somerville, W.R.C., Etchegoin, P.G., Treguer-Delapierre, M., Charron, G., Félidj, N., Lévi, G., and Aubard, J., 2011. A scheme for detecting every single target molecule with surface-enhanced raman spectroscopy. *Nano Letters*, 11 (11), 5013–5019.

Li, C.Y., Meng, M., Huang, S.C., Li, L., Huang, S.R., Chen, S., Meng, L.Y., Panneerselvam, R., Zhang, S.J., Ren, B., Yang, Z.L., Li, J.F., and Tian, Z.Q., 2015a. 'Smart' Ag nanostructures for plasmon-enhanced spectroscopies. *Journal of the American Chemical Society*, 137 (43), 13784–13787.

Li, J., Zhu, Z., Zhu, B., Ma, Y., Lin, B., Liu, R., Song, Y., Lin, H., Tu, S., and Yang, C., 2016. Surface-enhanced Raman scattering active plasmonic nanoparticles with ultrasmall interior nanogap for multiplex quantitative detection and cancer cell imaging. *Analytical Chemistry*, 88 (15), 7828–7836.

Li, M., Cushing, S.K., Zhang, J., Lankford, J., Aguilar, Z.P., Ma, D., and Wu, N., 2012. Shape-dependent surface-enhanced Raman scattering in gold-Raman-probe- silica sandwiched nanoparticles for biocompatible applications. *Nanotechnology*, 23 (11), 115501.

Li, Q., Zhuo, X., Li, S., Ruan, Q., Xu, Q.H., and Wang, J., 2015b. Production of monodisperse gold nanobipyramids with number percentages approaching 100% and evaluation of their plasmonic properties. *Advanced Optical Materials*, 3 (6), 801–812.

Lim, D.K., Jeon, K.S., Hwang, J.H., Kim, H., Kwon, S., Suh, Y.D., and Nam, J.M., 2011. Highly uniform and reproducible surface-enhanced Raman scattering from DNA-tailorable nanoparticles with 1-nm interior gap. *Nature Nanotechnology*, 6 (7), 452–460.

Lim, D.K., Jeon, K.S., Kim, H.M., Nam, J.M., and Suh, Y.D., 2010. Nanogap-engineerable raman-active nanodumbbells for single-molecule detection. *Nature Materials*, 9 (1), 60–67.

Lin, K., Yi, J., Zhong, J., Hu, S., Liu, B., Liu, J., Zong, C., Lei, Z., Ren, B., Wang, X., and Aizpurua, J., 2017a. Chemical information from surface-enhanced Raman scattering. *Nature Communications*, 8, 14891.

Lin, K.Q., Yi, J., Hu, S., Sun, J.J., Zheng, J.T., Wang, X., and Ren, B., 2016. Intraband hot-electron photoluminescence from single silver nanorods. *ACS Photonics*, 3 (7), 1248–1255.

Lin, L., Zapata, M., Xiong, M., Liu, Z., Wang, S., Xu, H., Borisov, A.G., Gu, H., Nordlander, P., Aizpurua, J., and Ye, J., 2015. Nanooptics of plasmonic nanomatryoshkas: Shrinking the size of a core-shell junction to subnanometer. *Nano Letters*, 15 (10), 6419–6428.

Lin, M., Kim, G.H., Kim, J.H., Oh, J.W., and Nam, J.M., 2017b. Transformative heterointerface evolution and plasmonic tuning of anisotropic trimetallic nanoparticles. *Journal of the American Chemical Society*, 139 (30), 10180–10183.

Liu, K., Bai, Y., Zhang, L., Yang, Z., Fan, Q., Zheng, H., Yin, Y., and Gao, C., 2016. Porous Au-Ag nanospheres with high-density and highly accessible hotspots for SERS analysis. *Nano Letters*, 16 (6), 3675–3681.

Liu, M., Wang, Z., Zong, S., Zhang, R., Zhu, D., Xu, S., Wang, C., and Cui, Y., 2013. SERS-based DNA detection in aqueous solutions using oligonucleotide- modified Ag nanoprisms and gold nanoparticles. *Analytical and Bioanalytical Chemistry*, 405 (18), 6131–6136.

Liu, Z., Yang, Z., Peng, B., Cao, C., Zhang, C., You, H., Xiong, Q., Li, Z., and Fang, J., 2014. Highly sensitive, uniform, and reproducible surface-enhanced Raman spectroscopy from hollow Au-Ag alloy nanourchins. *Advanced Materials*, 26 (15), 2431–2439.

Loumaigne, M., Laurent, G., Werts, M.H. V., and Débarre, A., 2016. Photoluminescence spectra and quantum yields of gold nanosphere monomers and dimers in aqueous suspension. *Physical Chemistry Chemical Physics*, 18 (48), 33264–33273.

Lu, G., Zhang, T., Li, W., Hou, L., Liu, J., and Gong, Q., 2011. Single-molecule spontaneous emission in the vicinity of an individual gold nanorod. *Journal of Physical Chemistry C*, 115 (32), 15822–15828.

Lumdee, C., Yun, B., and Kik, P.G., 2014. Gap-plasmon enhanced gold nanoparticle photoluminescence. *ACS Photonics*, 1 (11), 1224–1230.

Macfarlane, R.J., Lee, B., Jones, M.R., Harris, N., Schatz, G.C., and Mirkin, C.A., 2011. Nanoparticle superlattice engineering with DNA. *Science (New York, N.Y.)*, 334 (6053), 204–208.

Mandal, P. and Sharma, S., 2016. Progress in plasmonic solar cell efficiency improvement: A status review. *Renewable and Sustainable Energy Reviews*, 65, 537–552.

Mayer, K.M. and Hafner, J.H., 2011. Localized surface plasmon resonance sensors. *Chemical Reviews*, 111 (6), 3828–3857.

Mertens, J., Kleemann, M.E., Chikkaraddy, R., Narang, P., and Baumberg, J.J., 2017. How light is emitted by plasmonic metals. *Nano Letters*, 17 (4), 2568–2574.

Metiu, H. and Das, P., 1984. The electromagnetic theory of surface enhanced spectroscopy. *Annual Review of Physical Chemistry*, 35 (1), 507–536.

Michaels, A.M., Jiang, and Brus, L., 2000. Ag nanocrystal junctions as the site for surface-enhanced Raman scattering of single rhodamine 6G molecules. *The Journal of Physical Chemistry B*, 104 (50), 11965–11971.

Michaels, A.M., Nirmal, M., and Brus, L.E., 1999. Surface enhanced Raman spectroscopy of individual rhodamine 6G molecules on large Ag nanocrystals. *Journal of the American Chemical Society*, 121 (43), 9932–9939.

Mishra, H. and Geddes, C.D., 2014. Metal-enhanced S 1 and alpha- S 1 fluorescence: Effects of far-field excitation irradiance on enhanced fluorescence. *The Journal of Physical Chemistry C*, 118 (49), 28791–28796.

Mohamed, M.B., Volkov, V., Link, S., and El-Sayed, M.A., 2000. The 'lightning' gold nanorods: Fluorescence enhancement of over a million compared to the gold metal. *Chemical Physics Letters*, 317 (6), 517–523.

Mooradian, A., 1969. Photoluminescence of metals. *Physical Review Letters*, 22 (1), 185–187.

Morton, S.M. and Jensen, L., 2009. Understanding the molecule-surface chemical coupling in SERS. *Journal of the American Chemical Society*, 131 (11), 4090–4098.

Moskovits, M., 1985. Surface-enhanced spectroscopy. *Reviews of Modern Physics*, 57 (3), 783–826.

Moyer, P.J., Schmidt, J., Eng, L.M., and Meixner, A.J., 2000. Surface-enhanced Raman scattering spectroscopy of single carbon domains on individual Ag nanoparticles on a 25 ms time scale. *J. Am. Chem. Soc.*, 122, 5409–5410.

Munechika, K., Chen, Y., Tillack, A.F., Kulkarni, A.P., Plante, I.J. La, Munro, A.M., and Ginger, D.S., 2010. Spectral control of plasmonic emission enhancement from quantum dots near single silver nanoprisms. *Nano Letters*, 10 (7), 2598–2603.

Naik, G. V., Shalaev, V.M., and Boltasseva, A., 2013. Alternative plasmonic materials: Beyond gold and silver. *Advanced Materials*, 25 (24), 3264–3294.

Nam, J.M., Oh, J.W., Lee, H., and Suh, Y.D., 2016. Plasmonic nanogap-enhanced Raman scattering with nanoparticles. *Accounts of Chemical Research*, 49 (12), 2746–2755.

Neubrech, F., Pucci, A., Cornelius, T.W., Karim, S., García-Etxarri, A., and Aizpurua, J., 2008a. Resonant plasmonic and vibrational coupling in a tailored nanoantenna for infrared detection. *Physical Review Letters*, 101 (15), 2–5.

Neubrech, F., Weber, D., Lovrincic, R., Pucci, A., Lopes, M., Toury, T., and De La Chapelle, M.L., 2008b. Resonances of individual lithographic gold nanowires in the infrared. *Applied Physics Letters*, 93 (16), 163105.

Ngoc, L.L.T., Wiedemair, J., van den Berg, A., and Carlen, E.T., 2015. Plasmon-modulated photoluminescence from gold nanostructures and its dependence on plasmon resonance, excitation energy, and band structure. *Optics Express*, 23 (5), 5547.

Ni, W., Kou, X., Yang, Z., and Wang, J., 2008. Tailoring longitudinal surface plasmon wavelengths, scattering and absorption cross sections of gold nanorods. *ACS Nano*, 2 (4), 677–686.

Nie, S. and Emory, S.R., 1997. Probing single molecules and single nanoparticles by surface-enhanced Raman scattering. *Science (New York, N.Y.)*, 275 (5303), 1102–1106.

Novotny, L., 2007. Effective wavelength scaling for optical antennas. *Physical Review Letters*, 98 (26), 1–4.

Oh, J.W., Lim, D.K., Kim, G.H., Suh, Y.D., and Nam, J.M., 2014. Thiolated DNA-based chemistry and control in the structure and optical properties of plasmonic nanoparticles with ultrasmall interior nanogap. *Journal of the American Chemical Society*, 136 (40), 14052–14059.

Orendorff, C.J., Gearheart, L., Jana, N.R., and Murphy, C.J., 2006. Aspect ratio dependence on surface enhanced Raman scattering using silver and gold nanorod substrates. *Phys. Chem. Chem. Phys.*, 8 (1), 165–170.

Osley, E.J., Biris, C.G., Thompson, P.G., Jahromi, R.R.F., Warburton, P.A., and Panoiu, N.C., 2013. Fano resonance resulting from a tunable interaction between molecular vibrational modes and a double continuum of a plasmonic metamolecule. *Physical Review Letters*, 110 (8), 1–5.

Otto, A., Bruckbauer, A., and Chen, Y.X., 2003. On the chloride activation in SERS and single molecule SERS. *Journal of Molecular Structure*, 661–662 (1–3), 501–514.

Otto, A., Mrozek, I., Grabhorn, H., and Akemann, W., 1992. Surface-enhanced Raman scattering. *Journal of Physics: Condensed Matter*, 4 (5), 1143–1212.

Pande, S., Ghosh, S.K., Praharaj, S., Panigrahi, S., Basu, S., Jana, S., Pal, A., Tsukuda, T., and Pal, T., 2007. Synthesis of normal and inverted gold-silver core-shell architectures in β-cyclodextrin and their applications in SERS. *Journal of Physical Chemistry C*, 111 (29), 10806–10813.

Paramanik, B. and Patra, A., 2014. Fluorescent AuAg alloy clusters: Synthesis and SERS applications. *Journal of Materials Chemistry C*, 2 (16), 3005–3012.

Park, J.-E., Kim, J., and Nam, J.-M., 2017a. Emerging plasmonic nanostructures for controlling and enhancing photoluminescence. *Chemical Science*, 8 (7), 4696–4704.

Park, J.-E., Kim, M., Hwang, J.-H., and Nam, J.-M., 2017b. Golden opportunities: Plasmonic gold nanostructures for biomedical applications based on the second near-infrared window. *Small Methods*, 1 (3), 1600032.

Park, J.E., Kim, S., Son, J., Lee, Y., and Nam, J.M., 2016. Highly controlled synthesis and super-radiant photoluminescence of plasmonic cube-in-cube nanoparticles. *Nano Letters*, 16 (12), 7962–7967.

Park, W.-H. and Kim, Z.H., 2010. Charge transfer enhancement in the SERS of a single molecule. *Nano Letters*, 10 (10), 4040–4048.

Pelton, M., 2015. Modified spontaneous emission in nanophotonic structures. *Nature Photonics*, 9 (7), 427–435.

Planas, O., Macia, N., Agut, M., Nonell, S., and Heyne, B., 2016. Distance-dependent plasmon-enhanced singlet oxygen production and emission for bacterial inactivation. *Journal of the American Chemical Society*, 138 (8), 2762–2768.

Prodan, E., Radloff, C., Halas, N.J., and Nordlander, P., 2003. A hybridization model for the plasmon response of complex nanostructures. *Science (New York, N.Y.)*, 302 (5644), 419–422.

Puchkova, A., Vietz, C., Pibiri, E., Wünsch, B., Sanz Paz, M., Acuna, G.P., and Tinnefeld, P., 2015. DNA origami nanoantennas with over 5000-fold fluorescence enhancement and single-molecule detection at 25 μM. *Nano Letters*, 15 (12), 8354–8359.

Punj, D., Mivelle, M., Moparthi, S.B., van Zanten, T.S., Rigneault, H., van Hulst, N.F., García-Parajó, M.F., and Wenger, J., 2013. A plasmonic 'antenna-in-box' platform for enhanced single-molecule analysis at micromolar concentrations. *Nature Nanotechnology*, 8 (7), 512–516.

Qian, X.-M. and Nie, S.M., 2008. Single-molecule and single-nanoparticle SERS: From fundamental mechanisms to biomedical applications. *Chemical Society Reviews*, 37 (5), 912.

Raman, C. V. and Krishnan, K.S., 1928. A new type of secondary radiation. *Nature*, 121 (3048), 501–502.

Rao, W., Li, Q., Wang, Y., Li, T., Wu, L., and Nanorods, G., 2015. Comparison of photoluminescence quantum yield of single gold nanobipyramids and gold nanorods. *ACS Nano*, 9 (3), 2783–2791.

Reineck, P., Gómez, D., Ng, S.H., Karg, M., Bell, T., Mulvaney, P., and Bach, U., 2013. Distance and wavelength dependent quenching of molecular fluorescence by Au@SiO$_2$ core-shell nanoparticles. *ACS Nano*, 7 (8), 6636–6648.

Rezus, Y.L.A. and Selig, O., 2016. Impact of local-field effects on the plasmonic enhancement of vibrational signals by infrared nanoantennas. *Optics Express*, 24 (11), 12202.

Richard-Lacroix, M., Zhang, Y., Dong, Z., and Deckert, V., 2017. Mastering high resolution tip-enhanced Raman spectroscopy: Towards a shift of perception. *Chemical Society Reviews*, 46 (13), 3922–3944.

Rycenga, M., Hou, K.K., Cobley, C.M., Schwartz, A.G., Camargo, P.H.C., and Xia, Y., 2009. Probing the surface-enhanced Raman scattering properties of Au–Ag nanocages at two different excitation wavelengths. *Physical Chemistry Chemical Physics*, 11 (28), 5903.

Rycenga, M., Langille, M.R., Personick, M.L., Ozel, T., and Mirkin, C.A., 2012. Chemically isolating hot spots on concave nanocubes. *Nano Letters*, 12 (12), 6218–6222.

Savage, K.J., Hawkeye, M.M., Esteban, R., Borisov, A.G., Aizpurua, J., and Baumberg, J.J., 2012. Revealing the quantum regime in tunnelling plasmonics. *Nature*, 491 (7425), 574–577.

Schatz, G.C., 1984. Theoretical studies of surface enhanced Raman scattering. *Accounts of Chemical Research*, 17 (10), 370–376.

Schatz, G.C. and Duyne, R.P.V.D., 2002. Electromagnetic mechanism of surface-enhanced spectroscopy. In: *Handbook of Vibrational Spectroscopy*, J. M. Chalmers and P. R. Griffiths (eds) Wiley, New York, 759–774.

Schlücker, S., 2014. Surface-enhanced raman spectroscopy: Concepts and chemical applications. *Angewandte Chemie: International Edition*, 53 (19), 4756–4795.

van Schrojenstein Lantman, E.M., Deckert-Gaudig, T., Mank, A.J.G., Deckert, V., and Weckhuysen, B.M., 2012. Catalytic processes monitored at the nanoscale with tip-enhanced Raman spectroscopy. *Nature Nanotechnology*, 7 (9), 583–586.

Schumacher, T., Kratzer, K., Molnar, D., Hentschel, M., Giessen, H., and Lippitz, M., 2011. Nanoantenna-enhanced ultrafast nonlinear spectroscopy of a single gold nanoparticle. *Nature Communications*, 2 (1), 333.

Shanthil, M., Thomas, R., Swathi, R.S., and George, T.K., 2012. Ag@SiO$_2$ core-shell nanostructures: Distance-dependent plasmon coupling and SERS investigation. *Journal of Physical Chemistry Letters*, 3 (11), 1459–1464.

Sharma, B., Frontiera, R.R., Henry, A.I., Ringe, E., and Van Duyne, R.P., 2012. SERS: Materials, applications, and the future. *Materials Today*, 15 (1–2), 16–25.

Shen, J., Su, J., Yan, J., Zhao, B., Wang, D., Wang, S., Li, K., Liu, M., He, Y., Mathur, S., Fan, C., and Song, S., 2015. Bimetallic nano-mushrooms with DNA-mediated interior nanogaps for high-efficiency SERS signal amplification. *Nano Research*, 8 (3), 731–742.

Sivun, D., Vidal, C., Munkhbat, B., Arnold, N., Klar, T.A., and Hrelescu, C., 2016. Anticorrelation of photoluminescence from gold nanoparticle dimers with hot-spot intensity. *Nano Letters*, 16 (11), 7203–7209.

Song, J., Duan, B., Wang, C., Zhou, J., Pu, L., Fang, Z., Wang, P., Lim, T.T., and Duan, H., 2014. SERS-encoded nanogapped plasmonic nanoparticles: Growth of metallic nanoshell by templating redox-active polymer brushes. *Journal of the American Chemical Society*, 136 (19), 6838–6841.

Stöckle, R.M., Suh, Y.D., Deckert, V., and Zenobi, R., 2000. Nanoscale chemical analysis by tip-enhanced Raman spectroscopy. *Chemical Physics Letters*, 318 (1–3), 131–136.

Storhoff, J.J., Elghanian, R., Mirkin, C.A., and Letsinger, R.L., 2002. Sequence-dependent stability of DNA-modified gold nanoparticles. *Langmuir*, 18 (17), 6666–6670.

Su, J., Wang, D., Nörbel, L., Shen, J., Zhao, Z., Dou, Y., Peng, T., Shi, J., Mathur, S., Fan, C., and Song, S., 2017. Multicolor gold-silver nano-mushrooms as ready-to-use SERS probes for ultrasensitive and multiplex DNA/miRNA detection. *Analytical Chemistry*, 89 (4), 2531–2538.

Tam, F., Goodrich, G.P., Johnson, B.R., and Halas, N.J., 2007. Plasmonic enhancement of molecular fluorescence. *Nano Letters*, 7 (2), 496–501.

Tan, S.F., Wu, L., Yang, J.K.W., Bai, P., Bosman, M., and Nijhuis, C.A., 2014. Quantum plasmon resonances controlled by molecular tunnel junctions. *Science*, 343 (6178), 1496–1499.

Tan, T., Tian, C., Ren, Z., Yang, J., Chen, Y., Sun, L., Li, Z., Wu, A., Yin, J., and Fu, H., 2013. LSPR-dependent SERS performance of silver nanoplates with highly stable and broad tunable LSPRs prepared through an improved seed-mediated strategy. *Physical Chemistry Chemical Physics*, 15 (48), 21034.

Tao, A., Kim, F., Hess, C., Goldberger, J., He, R., Sun, Y., Xia, Y., and Yang, P., 2003. Langmuir-Blodgett silver nanowire monolayers for molecular sensing using surface-enhanced Raman spectroscopy. *Nano Letters*, 3 (9), 1229–1233.

Tcherniak, A., Dominguez-Medina, S., Chang, W.-S., Swanglap, P., Slaughter, L.S., Landes, C.F., and Link, S., 2011. One-photon plasmon luminescence and its application to correlation spectroscopy as a probe for rotational and translational dynamics of gold nanorods. *The Journal of Physical Chemistry C*, 115 (32), 15938–15949.

Thacker, V. V., Herrmann, L.O., Sigle, D.O., Zhang, T., Liedl, T., Baumberg, J.J., and Keyser, U.F., 2014. DNA origami based assembly of gold nanoparticle dimers for surface-enhanced Raman scattering. *Nature Communications*, 5 (1), 3448.

Urban, A.S., Shen, X., Wang, Y., Large, N., Wang, H., Knight, M.W., Nordlander, P., Chen, H., and Halas, N.J., 2013. Three-dimensional plasmonic nanoclusters. *Nano Letters*, 13 (9), 4399–4403.

Vidal, C., Sivun, D., Ziegler, J., Wang, D., Schaaf, P., Hrelescu, C., and Klar, T.A., 2018. Plasmonic horizon in gold nanosponges. *Nano Letters*, 18 (2), 1269–1273.

Vogt, J., Huck, C., Neubrech, F., Toma, A., Gerbert, D., and Pucci, A., 2015. Impact of the plasmonic near- and far-field resonance-energy shift on the enhancement of infrared vibrational signals. *Physical Chemistry Chemical Physics*, 17 (33), 21169–21175.

Wackenhut, F., Failla, A.V., and Meixner, A.J., 2013. Multi-color microscopy and spectroscopy reveals the physics of the one-photon luminescence in gold nanorods. *Journal of Physical Chemistry C*, 117 (34), 17870–17877.

Walsh, G.F. and Negro, L.D., 2013. Engineering plasmon-enhanced Au light emission with planar arrays of nanoparticles. *Nano Letters*, 13 (2), 786–792.

Wan, A., Wang, T., Yin, T., Li, A., Hu, H., Li, S., Shen, Z.X., and Nijhuis, C.A., 2015. Plasmon-modulated photoluminescence of single gold nanobeams. *ACS Photonics*, 2 (9), 1348–1354.

Wang, F., Liu, B., Huang, P.J.J., and Liu, J., 2013. Rationally designed nucleobase and nucleotide coordinated nanoparticles for selective DNA adsorption and detection. *Analytical Chemistry*, 85 (24), 12144–12151.

Wang, X., Braun, K., Zhang, D., Peisert, H., Adler, H., Chassé, T., and Meixner, A.J., 2015. Enhancement of radiative plasmon decay by hot electron tunneling. *ACS Nano*, 9 (8), 8176–8183.

Wang, Y., Lee, K., and Irudayaraj, J., 2010. Silver nanosphere SERS probes for sensitive identification of pathogens. *Journal of Physical Chemistry C*, 114 (39), 16122–16128.

Webb, J.A. and Bardhan, R., 2014. Emerging advances in nanomedicine with engineered gold nanostructures. *Nanoscale*, 6 (5), 2502.

Weber, K., Nesterov, M.L., Weiss, T., Scherer, M., Hentschel, M., Vogt, J., Huck, C., Li, W., Dressel, M., Giessen, H., and Neubrech, F., 2017. Wavelength scaling in antenna-enhanced infrared spectroscopy: Toward the far-IR and THz region. *ACS Photonics*, 4 (1), 45–51.

Willets, K.A. and Van Duyne, R.P., 2007. Localized surface plasmon resonance spectroscopy and sensing. *Annual Review of Physical Chemistry*, 58 (1), 267–297.

Wu, H.L., Tsai, H.R., Hung, Y.T., Lao, K.U., Liao, C.W., Chung, P.J., Huang, J.S., Chen, I.C., and Huang, M.H., 2011. A comparative study of gold nanocubes, octahedra, and rhombic dodecahedra as highly sensitive SERS substrates. *Inorganic Chemistry*, 50 (17), 8106–8111.

Wu, L., Wang, Z., Zong, S., Huang, Z., Zhang, P., and Cui, Y., 2012. A SERS-based immunoassay with highly increased sensitivity using gold/silver core-shell nanorods. *Biosensors and Bioelectronics*, 38 (1), 94–99.

Wu, X., Ming, T., Wang, X., Wang, P., Wang, J., and Chen, J., 2010. High-photoluminescence-yield gold nanocubes: For cell imaging and photothermal therapy. *ACS Nano*, 4 (1), 113–120.

Xia, L., Chen, M., Zhao, X., Zhang, Z., Xia, J., Xu, H., and Sun, M., 2014. Visualized method of chemical enhancement mechanism on SERS and TERS. *Journal of Raman Spectroscopy*, 45 (7), 533–540.

Xia, Y. and Halas, N.J., 2005. Shape-controlled synthesis and surface plasmonic properties of metallic nanostructures. *MRS Bulletin*, 30 (05), 338–348.

Xie, J., Zhang, Q., Lee, J.Y., and Wang, D.I.C., 2008. The synthesis of SERS-active gold nanoflower tags for in vivo applications. *ACS Nano*, 2 (12), 2473–2480.

Xu, H., Aizpurua, J., Käll, M., and Apell, P., 2000. Electromagnetic contributions to single-molecule sensitivity in surface-enhanced Raman scattering. *Physical Review E: Statistical Physics, Plasmas, Fluids, and Related Interdisciplinary Topics*, 62 (3 B), 4318–4324.

Xu, H., Bjerneld, E.J., Käll, M., and Börjesson, L., 1999. Spectroscopy of single hemoglobin molecules by surface enhanced raman scattering. *Physical Review Letters*, 83 (21), 4357–4360.

Yang, J.K., Kang, H., Lee, H., Jo, A., Jeong, S., Jeon, S.J., Kim, H.I., Lee, H.Y., Jeong, D.H., Kim, J.H., and Lee, Y.S., 2014a. Single-step and rapid growth of silver nanoshells as SERS-active nanostructures for label-free detection of pesticides. *ACS Applied Materials and Interfaces*, 6 (15), 12541–12549.

Yang, Y., Liu, J., Fu, Z.W., and Qin, D., 2014b. Galvanic replacement-free deposition of au on ag for core-shell nanocubes with enhanced chemical stability and SERS activity. *Journal of the American Chemical Society*, 136 (23), 8153–8156.

Yin, T., Dong, Z., Jiang, L., Zhang, L., Hu, H., Qiu, C.W., Yang, J.K.W., and Shen, Z.X., 2016. Anomalous shift behaviors in the photoluminescence of dolmen-like plasmonic nanostructures. *ACS Photonics*, 3 (6), 979–984.

Yorulmaz, M., Khatua, S., Zijlstra, P., Gaiduk, A., and Orrit, M., 2012. Luminescence quantum yield of single gold nanorods. *Nano Letters*, 12 (8), 4385–4391.

Zhang, C.Y. and Johnson, L.W., 2006. Quantum-dot-based nanosensor for RRE IIB RNA-Rev peptide interaction assay. *Journal of the American Chemical Society*, 128 (16), 5324–5325.

Zhang, M., Large, N., Koh, A.L., Cao, Y., Manjavacas, A., Sinclair, R., Nordlander, P., and Wang, S.X., 2015a. High-density 2D homo- and hetero- plasmonic dimers with universal sub-10-nm gaps. *ACS Nano*, 9 (9), 9331–9339.

Zhang, P., Li, Y., Wang, D., and Xia, H., 2016a. High-yield production of uniform gold nanoparticles with sizes from 31 to 577 nm via one-pot seeded growth and size-dependent SERS property. *Particle and Particle Systems Characterization*, 33 (12), 924–932.

Zhang, T., Lu, G., Li, W., Liu, J., Hou, L., Perriat, P., Martini, M., Tillement, O., and Gong, Q., 2012. Optimally designed nanoshell and matryoshka-nanoshell as a plasmonic-enhanced fluorescence probe. *Journal of Physical Chemistry C*, 116 (15), 8804–8812.

Zhang, T., Lu, G., Shen, H., Shi, K., Jiang, Y., Xu, D., and Gong, Q., 2014. Photoluminescence of a single complex plasmonic nanoparticle. *Scientific Reports*, 4 (1), 3867.

Zhang, T., Shen, H., Lu, G., Liu, J., He, Y., Wang, Y., and Gong, Q., 2013a. Single bipyramid plasmonic antenna orientation determined by direct photoluminescence pattern imaging. *Advanced Optical Materials*, 1 (4), 335–342.

Zhang, W., Huang, L., Santschi, C., and Martin, O.J.F., 2010. Trapping and sensing 10 nm metal nanoparticles using plasmonic dipole antennas. *Nano Letters*, 10 (3), 1006–1011.

Zhang, W., Rahmani, M., Niu, W., Ravaine, S., Hong, M., and Lu, X., 2015b. Tuning interior nanogaps of double-shelled Au/Ag nanoboxes for surface-enhanced Raman scattering. *Scientific Reports*, 5 (1), 8382.

Zhang, Y., Wang, F., Yin, H., and Hong, M., 2013b. Nonuniform distribution of capping ligands promoting aggregation of silver nanoparticles for use as a substrate for SERS. *Advances in Nanoparticles*, 2013 (May), 104–111.

Zhang, Z., Sheng, S., Wang, R., and Sun, M., 2016b. Tip-enhanced Raman spectroscopy. *Analytical Chemistry*, 88 (19), 9328–9346.

Zhang, Z., Zhao, B., and Hu, L., 1996. PVP protective mechanism of ultrafine silver powder synthesized by chemical reduction processes. *Journal of Solid State Chemistry*, 121 (1), 105–110.

Zhong, J.H., Jin, X., Meng, L., Wang, X., Su, H.S., Yang, Z.L., Williams, C.T., and Ren, B., 2017. Probing the electronic and catalytic properties of a bimetallic surface with 3 nm resolution. *Nature Nanotechnology*, 12 (2), 132–136.

Zuloaga, J. and Nordlander, P., 2011. On the energy shift between near-field and far-field peak intensities in localized plasmon systems. *Nano Letters*, 11 (3), 1280–1283.

GaN Nanoflowers: Growth to Optoelectronic Device

8.1 Introduction to GaN Nanoflowers... 8-1
8.2 Growth of GaN NFs by MBE 8-2
  Growth Kinetics and Nucleation Mechanism • Characteristics of GaN NFs
8.3 Photodetection Aspect from GaN NFs................................... 8-6
  Background of a PD • Performance Evaluation Parameters and Their
  Dependence • Response upon UV Illumination • Self-Driven Operation
8.4 Potential Applications 8-13
8.5 Summary 8-13
Acknowledgments 8-13
References 8-13

Neha Aggarwal, Shibin Krishna,
and Govind Gupta
CSIR-National Physical Laboratory

8.1 Introduction to GaN Nanoflowers

A huge curiosity prevails about the use of semiconductor nanostructures (NSs) for active nanoscale electronic and photonic devices [1–5]. III-Nitride semiconductors were recognized to be potential candidates due to tunability of bandgap, strong interatomic bonds, and sustainability in harsh environmental conditions [6]. Among them, Gallium nitride (GaN) has become a remarkable semiconducting material in the industry of optoelectronic devices due to its magnificent properties, such as wide direct bandgap, good thermal stability, strong piezoelectricity, and high thermal conductivity [7]. However, its interest has been extended because of the material potential to fabricate solar cells, lasers, sensors, high frequency electronics, etc. [8–13]. A diversity of substrates (sapphire, SiC, silicon, GaN) and different growth techniques, such as plasma enhanced chemical vapor deposition (PECVD) [14], metal–organic chemical vapor deposition (MOCVD) [15], metal–organic vapor phase epitaxy (MOVPE) [16], and plasma-assisted molecular beam epitaxy (PAMBE) [17], have been used to grow extremely high crystalline quality GaN films. Among these, PAMBE offers precise control over the growth parameters, deep understanding of each growth step, in situ growth monitoring, and growth repeatability [17,18]. Moreover, the choice of the substrate for GaN epitaxy is crucial as the performance of an optoelectronic device is directly affected by the quality of grown GaN films. Si has attracted considerable attention as a substrate material by virtue of low cost, accessibility, and wide availability in large

diameter in order to integrate the GaN-based electronics with well-established Si technology [7]. Besides, growth of low-dimensional structures is predicted as a promising design used for enhancing the device properties due to high surface-to-volume ratio [19]. Thus, to fabricate high-efficiency devices based on GaN NSs, we need GaN layer and GaN nanoflower (NF)-like structure over it, which can be realized by following Stranski–Krastanov growth mode (which will be discussed in Section 8.2.1) [20]. Now, there exist two vapor phase growth mechanisms of NSs: vapor–solid (VS) and vapor–liquid–solid (VLS). Contrary to the restricted growth of NS from the metal catalysts by VLS mechanism, the VS mechanism can give more freedom in the formation of NFs. Several studies for growing catalyst-free NSs have been reported. For example, Kumar et al. proposed a qualitative mechanism for the growth of self-assembled flower-like NSs of InN and GaN using droplet epitaxy (DE) and molecular beam epitaxy (MBE) [21]; a compact and nanocolumnar growth regime for the growth of GaN nanocolumns has been elaborated by Fernández-Garrido et al., which indicates that adatom diffusion length and actual Ga/N ratio on the growing surface are the key factors to achieve desired growth [22]; Kang et al. established the growth of special morphology of InN NF pattern on a c-plane (0001) sapphire by MOCVD via self-catalysis VLS process, which modulates the interface kinetics and thermodynamics among the sapphire substrate, indium, and InN [23]; and Debnath et al. demonstrated a catalyst-free growth of GaN nanowires on Si (111) substrate by PAMBE, where the deposition time longer than the nucleation stage will decrease the nanowire length [24].

In this chapter, we review the journey of GaN NF from epitaxial growth to highly efficient optoelectronic devices. In Section 8.2, we elaborate the nucleation and growth mechanisms followed by the analysis of various structural, morphological, and optical properties of GaN NFs. Section 8.3 focused on the photodetection prospect of GaN NFs, which elucidate a brief understanding about the performance parameters and self-driven operation of fabricated ultraviolet (UV) photodetectors (PDs). Later, the potential applications of low-dimensional GaN-based photodetection devices are presented in Section 8.4, followed by the conclusion in Section 8.5.

8.2 Growth of GaN NFs by MBE

Till date, a wide variety of GaN NFs has been grown by various growth techniques. A brief review about the NF grown by various techniques is tabulated in Table 8.1. On the

TABLE 8.1 Representative List of Developed Nitride-Based NFs

S. No.	Material	Growth Technique	Substrate	Field-Emission Scanning Electron Microscopy Image	References
1.	AlN	Chemical Vapor Deposition (CVD)	Au-coated silicon		[25]
2.	GaN	CVD	Silicon		[26]
			AlN template		
3.	GaN	PAMBE	Silicon		[27]

(Continued)

TABLE 8.1 (*Continued*) Representative List of Developed Nitride-Based NFs

S. No.	Material	Growth Technique	Substrate	Field-Emission Scanning Electron Microscopy Image	References
4.	InGaN	MOVPE	Sapphire		[28]
5.	GaN	Hydride–Vapor Phase Epitaxy	Silicon		[29]
6.	AlGaN	Halide –CVD	Silicon		[15]
7.	InN	DE-MBE	Silicon		[21]
	GaN	Radio Frequency-MBE			

other hand, huge research efforts have been devoted in the formation of catalyst-free GaN NSs. In this section, we provide a brief review of the work carried out by the present authors in the area of III-nitride semiconducting materials growth. The ability to maximize detection device performance depends critically on two aspects: the enhanced material properties and the device fabrication procedure. Thus, the focus was mainly on the growth of single crystalline GaN and to develop high surface-to-volume ratio NSs that can be utilized efficiently for PD applications, based on the principle of photoelectric effect (which will be discussed in Section 8.3.1).

8.2.1 Growth Kinetics and Nucleation Mechanism

The growth kinetics is affected by a variety of factors, including substrate material, impinging flux of adatoms, the growth parameters, and surface interaction with adatoms. The growth mechanism can be explained by adatoms residing on the surface of a substrate, which involve existence of several surface phenomena. These are presented in Figure 8.1, which includes adatom adsorption on the surface, desorption from the surface, surface diffusion, and diffusion towards nucleation site having lowest surface free energy. The surface free energy quantifies the disturbance in intermolecular bonds when any surface is formed. Since the adatoms have the tendency to acquire the site with lowest surface free energy, the adatom diffusion occurs towards the nucleation site, which was created by defects/dislocations in the lattice and exhibit low surface free energy.

Depending upon the interaction between film adatoms and the substrate, the epitaxial growth processes can be divided into three modes, namely (i) Frank–van der Merwe (FM) growth mode (two-dimensional (2D) morphology, layer-by-layer growth); (ii) Volmer–Weber (VW) growth mode (island type growth), and (iii) Stranski–Kranstanov (SK) growth mode (initially 2D, after critical thickness, three-dimensional (3D) morphology, layer-plus-island growth) [30]. The SK growth mode has been adopted in this chapter, which deals with effective charge transport from 3D morphology of GaN NF leading to efficient device fabrication (Figure 8.1). Further, a

schematic diagram illustrating the stepwise structural evolution and growth mechanism of GaN NF-like NS on Si (111) substrate is represented in Figure 8.2. Initially, to reduce the high lattice and thermal mismatch of 17% and 56%, respectively, between Si (111) and GaN (0002), a high-temperature AlN buffer layer is grown on the atomically cleaned 7 × 7 reconstructed Si (111) surface (which can be achieved by flashing the Si substrate at high temperatures for a few seconds). Also, the introduction of AlN buffer layer can change the tensile stress of GaN grown on Si into compressive stress, which will result in crack-free growth [31].

A GaN epitaxial layer has been grown on AlN-buffered Si (111) substrate. Here, the hexagonal pits emerged on the surface of epitaxial GaN propagated from the AlN buffer layer due to the screw dislocations encountered by a huge lattice mismatch between AlN and Si (111) substrate that will act as nucleation centers for GaN NFs. These dislocations was noted to act as "seed" and help in promulgating the growth of GaN NFs. Hence, with optimized growth conditions, GaN nuclei are found to cultivate in a direction vertical to the substrate to form NF-like structures, and its petals were found to nurture upon the sides of the hexagonal pits. This unique NF-like morphology has been realized due to high density of dangling bonds at a region near the sidewalls of hexagonal pits with lower surface free energy to adsorb the impinging adatoms of Ga metal. Thus, the formation of vertical arrays of GaN NFs was possible due to dislocation-mediated growth mechanism.

8.2.2 Characteristics of GaN NFs

Growth of GaN NFs appears to ensue through a complex interplay between surface free energy, surface migration, and lattice strain. This relationship challenged the development of high-quality GaN for successful device fabrication. The crystallinity, surface morphology, optical, and structural properties of the grown GaN NSs need to be explored in detail using various characterization tools, as discussed later in this section.

The crystallinity of the nurtured GaN NFs can be analyzed by high-resolution X-ray diffraction (HRXRD). Figure 8.3 shows a 2θ-ω profile of the GaN NFs grown on AlN-buffered Si (111) substrate using PAMBE system. Here, the 2θ peaks positioned at 34.6° and 35.9° correspond to GaN (0002) and AlN (0002) diffractions, respectively, which represents the growth of wurtzite structures. Moreover, the peaks positioned at 28.45° and 72.9° belong to Si (111) and second harmonic diffraction of GaN (i.e., 0004 diffraction), respectively.

Here, the peak position of GaN along (0004) has been used to evaluate the interplanar spacing d using the Bragg's law [32]. Further, the d value has been used to estimate the lattice constant, c. The calculated lattice constant (c) was found to be 0.5184 nm, which is almost near to the lattice constant value of 0.5185 nm for strain-free GaN [7]. This insignificant change in the lattice constant of the grown

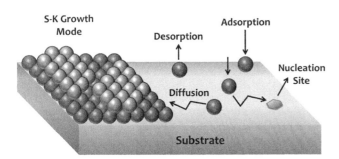

FIGURE 8.1 A schematic representation of various surface phenomena involved in the growth process.

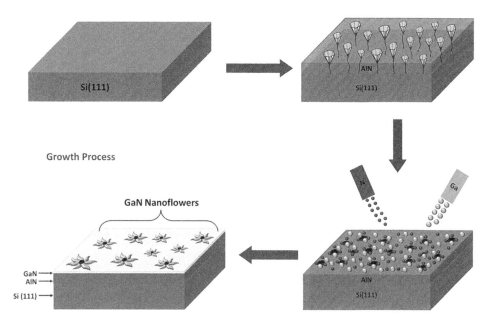

FIGURE 8.2 A schematic diagram of the growth mechanism of GaN NF-like structure on AlN-buffered Si (111) substrate.

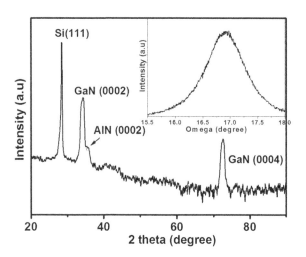

FIGURE 8.3 2θ-ω HRXRD scan of the GaN NFs grown on Si(111) substrate. Inset shows the RC of GaN (0002).

NFs indicates the growth of single crystalline NFs possessing negligibly small strain (ε), which was assessed using the following relation [33]:

$$\text{Strain}, \varepsilon = \frac{(c - c_o)}{c_o} \qquad (8.1)$$

The strain present in NS was evaluated to be 0.017%. This relaxation of strain is due to the growth of NS despite of thin film. Further, the stress quantification can be done by converting the determined strain into stress by using the following relation [32],

$$\text{Stress}, \sigma = M \times \varepsilon \qquad (8.2)$$

where σ is the biaxial stress, ε is the calculated strain, and M is the Young's modulus ($M_{\text{Si}(111)} = 165.6$ GPa) [7].

The stress perceived in the grown structure is found to be tensile in nature. Subsequently, the full-width half-maximum (FWHM) of the (0002) peak in the rocking curve (RC) was utilized to evaluate the screw dislocation density [34]. The high intensity peak of GaN in the RC along the symmetric (0002) plane of diffraction is shown in the inset of Figure 8.3, which reveals that highly crystalline NSs are grown along the c-direction. The FWHM of GaN (0002) plane of diffraction was measured to be 0.8° and the screw dislocation density was calculated to be 8.1×10^9 cm^{-2}.

Furthermore, the topographical study can be carried out by field-emission scanning electron microscopy (FESEM) and atomic force microscopy (AFM). The FESEM image (Figure 8.4a) illustrates that the deposited GaN possesses evenly distributed NF-like structures, where the average NF density was observed to be 8.8×10^7 cm^{-2}. The inset of Figure 8.4a displays the AFM image of the top surface of the unique structure grown by PAMBE. The average height of GaN NFs can be gauged using a depth profiling tool, and the elemental distribution on a single NF can be performed using energy dispersive X-ray spectroscopy (EDX). A typical EDX spectrum of the grown GaN NS is shown in Figure 8.4b. The inset of Figure 8.4b displayed a higher magnified FESEM image of a single GaN NF, which clearly manifests six petals cultivated along the sides of hexagonal pits.

For examining the state of residual stress, the GaN NF was probed by Raman spectroscopy (RS). Figure 8.5 shows the Raman spectra obtained at room temperature (RT) in backscattering geometry with an incident laser spot size of 1 μm normal to the surface of the film. The peak located at 567.69 cm^{-1} is associated with the E_2 (high) phonon mode of grown GaN, while those marked with an asterisk (*) at around 618 and 670 cm^{-1} belongs to phonon modes from Si (111) substrate. We eminently emphasize on the E_2 (high) phonon mode peak as its FWHM, and the intensity divulges

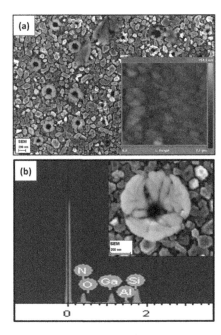

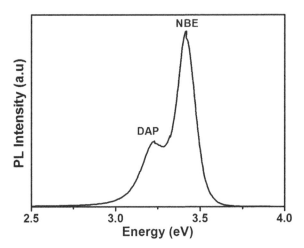

FIGURE 8.6   RT-PL spectra of the PAMBE-grown GaN NF on AlN/Si (111) substrate.

FIGURE 8.4   (a) FESEM image represents the surface morphology of the grown GaN NFs. Inset shows its AFM image and (b) the EDX spectrum representing the elemental composition. Inset shows the highly magnified image of a single GaN NF.

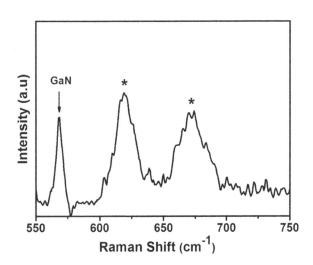

FIGURE 8.5   Raman spectra of a GaN NF-like structure grown by PAMBE.

on the quality of grown NFs and its shift in wave number are assigned to monitor the stress/strain present in it [35]. The E_2 (high) phonon mode is perceived at 568 ± 0.1 cm^{-1} for the stress-free GaN [36], which has been taken as a reference for stress calculation. Thus, a shift observed here specifies that the nature of stress applied on grown NF is tensile, which is in agreement with the tensile stress projected from HRXRD measurements. The following relation is exploited to evaluate the residual stress from the shift in E_2 (high) phonon mode [37],

$$\text{Residual Stress } \sigma = \frac{\Delta\omega}{4.3} (in\ GPa) \qquad (8.3)$$

where $\Delta\omega$ corresponds to the observed Raman frequency shift from the unstressed peak position. The tensile stress evaluated using Eq. (8.3) is found to be 0.072 GPa.

The optical properties of GaN NFs were explored using photoluminescence (PL) spectroscopy. The RT-PL spectra of GaN NFs, as shown in Figure 8.6, was observed to have peak energies centered at 3.41 and 3.23 eV, attributed to near band-edge (NBE) emission and donor–acceptor-pair (DAP) emission, respectively. The acquired sharp and narrow NBE emission emerges due to the radiative recombination of excitons bound to neutral donors. The value of NBE peak for unstressed bulk GaN film was taken to be 3.40 eV [38]. Thus, an upshift of 18 meV in band-edge emission has been observed against a strain-free GaN film. However, the NBE emission peak of GaN NFs is slightly upshifted compared with bulk GaN, which was attributed to the quantum confinement effect in GaN NFs [39,40]. While DAP emission band appears due to the transition from shallow donor to shallow acceptor levels existing within the bandgap region resulting in Ga vacancy-related complex [41].

8.3   Photodetection Aspect from GaN NFs

8.3.1   Background of a PD

A PD is an optoelectronic device that converts optical signals into electrical impulses. The fundamental operating principle of a photodetection device is based on photoelectric effect. Figure 8.7 demonstrates a band-diagram schematic illustrating the basic principle of a PD along with all the favorable transitions activated by incident photon energy. Transition I represents the basic principle where a photon with energy greater than or equal to the bandgap energy interacts with the semiconducting material and electrons present in the valence band comes to an excited state by absorbing the energy from incoming photons, thereby

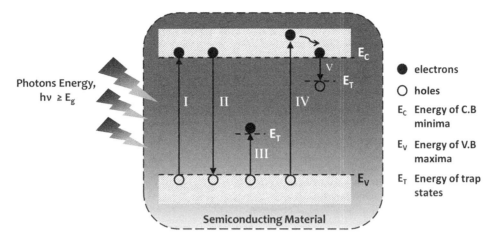

FIGURE 8.7 A band-diagram schematic illustrating the operating principle of a PD along with related feasible transitions.

creating electron–hole (e–h) pairs. These photogenerated e–h pairs are then collected by their respective electrodes to accomplish current conduction.

Eventually, if the generated e–h pairs are not collected quickly, then they will recombine, giving up the extra energy in the form of photons, as depicted in transition II of Figure 8.7. In Transition III, the excited electron may jump from the valence band to a trap state existing within the bandgap (E_T). Further, if a photon has much higher energy, the bandgap energy enters the semiconducting material, and then the electron may reach a high energy state in the conduction band, which then relaxes by thermalization process in which it releases energy in the form of phonons and then reaches the bottom of the conduction band (Transition IV). Transition V depicts trapping of electron in a trap state during the recombination process.

In recent times, the necessity to detect UV radiation (being the highly harmful and energetic radiation) has stimulated much interest in the field of research and development of UV PDs. There are a number of potential applications of UV PDs, such as, air/water purification, ozone layer monitoring, forest fire prevention, missile detection, submarine oil leakage monitoring, chemical sensing, interspace communication, etc. An ideal PD for such applications must possess high sensitivity, fast response, minimum noise, high detectivity, insensitive to temperature variations, long operating lifetime, high spectral selectivity, high radiation hardness, and low cost. However, the growing need of miniaturized and reliable UV detection systems for portable applications has driven the development of semiconductor-based UV PDs. Earlier, narrow bandgap semiconductors were first considered to perform UV detection, such as silicon and some III–V compounds. However, the well-established Si technology has some limitations in the UV region because filters are required to block low energy photons (visible and infrared light), and degradation is produced by highly energized UV photons (shorter device lifetime). Moreover, SiC-based photodiodes have a relatively narrow range of sensitivity, and GaP photodiodes have a good signal-to-noise ratio, but it is impossible to create solar-blind and visible-blind photodiodes based on

them without using filters. To avoid the use of filters and achieve better device operation, UV detectors based on wide bandgap semiconductors have been studied during the last decade. The first GaN-based UV detector fabricated from as-deposited single-crystal GaN films was demonstrated in 1992 [42]. Thereafter, several research groups have contributed to the development of UV PDs.

A block diagram of a complete experimental setup for acquiring the response from a PD device is briefly elucidated in Figure 8.8. Broadly, a PD experimental setup consists of a laser source to illuminate the sample, highly precise probes on the sample to minimize the noise power in the data, a source meter to measure current flowing through the circuit upon illumination and apply bias for better charge carrier collection, and finally, a graphical user interface (GUI) to provide user-friendly output data. After analyzing the output data, various parameters exist to gauge the performance of a PD that is mainly affected by an active area of the detection device. This leads to a recent curiosity that has emerged in the use of semiconductor NSs for active nanoscale electronic and photonic devices [2,4,43]. Boruah and Misra demonstrated that the light harvesting efficiency can be drastically incremented using NSs [44]. A comprehensive study by Sang et al. discussed about the semiconductor UV PDs for thin films as well as NSs [45]. Thus, by increasing the surface-to-volume ratio, the detection ability of a PD can be significantly enhanced. Various types of GaN-based photodetection devices have been developed so far, such as p-n junction diode, p-i-n diode, metal–semiconductor–metal (MSM) photodiodes, and Schottky barrier diodes [46–48]. Among these, MSM geometry-based devices have captivated ample consideration owing to their high operation speed, low noise, and easy to fabricate technology. Researchers have fabricated PDs using such structural designs to obtain good reliability and sensitivity, even at very low or no external bias (self-driven) [49]. Recently, Peng et al. have fabricated a flexible and self-powered GaN membrane-based UV photoswitch that possesses a high on–off ratio and excellent sensitivity [50]. However, in earlier reports, Zhao's group was able

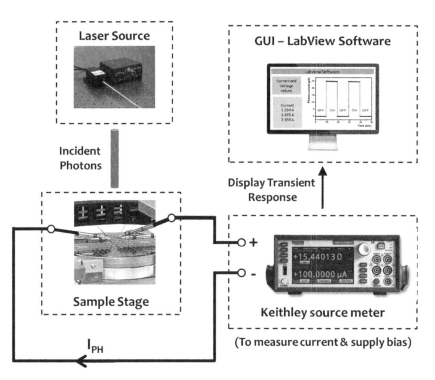

FIGURE 8.8 A block diagram representing the experimental setup for acquiring response from the photodetection device.

to fabricate a GaN MSM photoconductive detector on Si (111) substrate, which achieved a maximum responsivity of 6.9 A W^{-1} at 357 nm with a 5 V bias [47]. Further, a GaN UV PD has been developed, which delivers a responsivity of 0.192 and 0.15 A W^{-1} at 3 V applied bias using TiW and W electrodes, respectively, while detectivity and noise equivalent power (NEP) were in the order of 10^9 Jones and 10^{-10} W Hz$^{-1/2}$, respectively [51]. A high responsivity of 7.2 A W^{-1} at 5 V has been reported from a GaN MSM UV sensor developed by Su et al. [52].

8.3.2 Performance Evaluation Parameters and Their Dependence

In this section, we will discuss the performance parameters along with the physical properties that influence them. The performance of a PD can be evaluated by various parameters that are explained as follows:

Responsivity (R) is one of the important performance evaluation parameters that tell how the device is responding to a particular wavelength of light. It is defined as the ratio of photocurrent (I_p) and incident optical power intensity (P), where P is evaluated by the product of power density (P_d) and active area of the device (A). The unit of R is Amperes/Watts.

$$R = \frac{I_p}{P_d \cdot A} = \frac{I_p}{P} \tag{8.4}$$

Further, quantum efficiency (η) is defined as the ratio of fractional number of carriers generated, which contribute towards photocurrent and the incident photon flux (Φ).

$$\eta = \frac{\text{Fractional number of carriers generated that contribute to photocurrent}}{\text{Incident photon flux } (\Phi)} \tag{8.5}$$

As we know, incident photon flux is defined as the number of photons incident per unit time.

$$\Phi = \frac{P}{h\nu} \tag{8.6}$$

where h is the Planck's constant, ν is the frequency corresponding to the incident wavelength, and $h\nu$ is the energy of one electron. Thus, using Eq. (8.6) in (8.5), quantum efficiency can be written as

$$\eta = \frac{I_p/e}{P/h\nu}; (0 \leq \eta \leq 1) \tag{8.7}$$

Here, I_p/e is the fractional number of carriers generated and e is the electronic charge. Using Eqs (8.4), (8.6), and (8.7), we have

$$\eta = \frac{I_p}{P}\frac{h\nu}{e} = \frac{R \cdot h\nu}{e} \tag{8.8}$$

From Eq. (8.8), we can observe that $R \propto \eta$

Therefore, if the responsivity needs to be enhanced for any semiconductor-based photodetection device, then η should be maximized. Now, we need to know the ways to maximize the efficiency of PD.

Let us consider a PD made from a semiconducting material of thickness d, and the photon flux Φ is incident upon it (Figure 8.9). Then, due to difference in refractive index of the air and the semiconducting material, some part of the

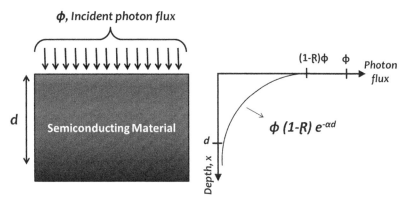

FIGURE 8.9 A schematic diagram representing illumination of a semiconductor surface and decay plot of the photon flux in the semiconductor with respect to its depth.

photon flux entering on the surface of the semiconductor will be reflected back.

If R is the reflectivity of the surface, then the $R\Phi$ amount of flux will be reflected back, and the amount of flux entering the detector material will be $(1-R)\Phi$.

As the intensity of photon flux will decay exponentially with the depth in the material according to Beer–Lambert law (Figure 8.9), and it can state that the output intensity of flux from a material of thickness d with absorption coefficient α will be

$$\Phi_{\text{output}} = \Phi \cdot e^{-\alpha d} \qquad (8.9)$$

Thus, the intensity of flux absorbed into the material is given by

$$\Phi_{\text{absorbed}} = \Phi(1 - e^{-\alpha d}) \qquad (8.10)$$

Now, the fractional amount of photons absorbed into the semiconducting material will be

$$N = \Phi(1-R)(1 - e^{-\alpha d}) \qquad (8.11)$$

We need to consider various factors that can affect photocurrent generation. So, considering that some of these absorbed photons (N) may transfer energy to the phonons, all absorbed photons will not result in e–h pair generation. Further, all the e–h pairs generated might not contribute towards the current generated in the external circuit, as they may immediately recombine due to trap states that exist in the bandgap region because of defects in the material. Let us assume ξ be a factor that depends upon the material property and deals with the aforementioned factors. Now, $N\xi$ will be the fractional number of photons absorbed that contribute to current in the external circuit. Thus, using Eq. (8.5), the quantum efficiency can be evaluated as

$$\eta = \frac{\text{Fractional number of carriers generated contributing to photocurrent}}{\text{incident photon flux } (\Phi)} = \frac{N\xi}{\Phi}$$

Therefore, from Eq. (8.11), we get

$$\eta = (1-R)\left(1 - e^{-\alpha d}\right)\xi \qquad (8.12)$$

Thus, to maximize the quantum efficiency (η), three factors need to be considered:

i. Reflectivity (R) of the surface of the material should be minimized to reduce the photon losses due to reflection. This can be done by using antireflective coatings on the device.

ii. Optimize the thickness (d) of the material according to its absorption coefficient so as to minimize the factor $e^{-\alpha d}$ to zero, which implies that "αd" has to be large.

iii. The value of ξ should be maximized, which can be achieved by applying external bias to effectively collect the generated e–h pairs before they recombine and enhance the material growth properties, such that it pertains to minimal structural/optical defects.

Another performance parameter NEP has been considered, which implies signal power that yield a signal-to-noise ratio of 1 in a 1 Hz output bandwidth and can be calculated using

$$\text{NEP} = \frac{\{2eI_d\}^{\frac{1}{2}}}{R} \qquad (8.13)$$

where R is the responsivity evaluated from the device under test and I_d is the dark current, i.e., current value obtained in the absence of UV light.

Also, the detectivity (D) for a PD is an important figure of merit which, is used to characterize its performance and is inversely proportional to NEP, normalized per square root of the device's active area.

$$D = \frac{A^{\frac{1}{2}}}{\text{NEP}} \qquad (8.14)$$

Further, for a detection device, the response time τ (rise and decay time) holds great significance, as it will determine the suitability of a device for a specific application. This parameter characterizes the speed of response, i.e., measures the time required for a PD to respond to a specific light impulse and can be evaluated by fitting the rise (τ_r) and decay (τ_d) time equations, with I_0 being the maximum current value at a particular time t:

$$I = I_0(1 - e^{-t/\tau_r}) \qquad (8.15)$$

$$I = I_0(e^{-t/\tau_d}) \qquad (8.16)$$

Thus, these performance parameters will evaluate the ability of a photodetection device to detect a specific light signal.

8.3.3 Response upon UV Illumination

The electronic behavior of GaN NF-based device under UV light conditions needs to be explored to realize a highly efficient PD. A schematic structure of the developed device is shown in Figure 8.10a, where UV illumination triggers e–h pairs in the device leading to current conduction.

To quantify the electrons contributing towards photocurrent in the external circuitry, electrical transport behavior has been investigated using current-voltage (I–V) measurements. The linear plot shown in Figure 8.10b implies the quasi-Ohmic nature from the GaN NF-based device fabricated using Au metal electrodes. It was found that, under dark conditions, the leakage current acknowledged from the device is due to the presence of thermionic emissions [53]. While upon UV exposure, the photon-assisted excited electron present in the conduction band will contribute toward light current, which was affirmed by the increment in current collected from the device as shown in Figure 8.10b. Thus, the GaN NF-based fabricated device exhibit an effective photodetection aspect. Often, an external bias needs to be applied to a device for collecting the photogenerated e–h pairs rapidly before they recombine. So, the effect of increasing applied bias on the transient photoresponse of a PD should be a crucial concern. Figure 8.11 shows a significant increase in photoresponsivity with increasing bias, which recurred from a prominent photocurrent rise upon turning-on the laser and decayed after turn-off under an external bias variation from 0 to 3 V, inferring good stability and repeatability with switching time (inset of Figure 8.11). A linear dependence of external bias with photocurrent is observed with around four times increment as bias varied from 0 to 3 V.

Since fast detection of low-intensity UV radiation is highly desired for various photodetection applications, the demand

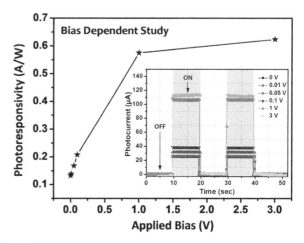

FIGURE 8.11 The bias-dependent photoresponsivity study and inset shows the time-correlated photoresponse from the fabricated GaN NF-based photodetection device.

of low-power operable devices is intensified due to the miniaturization of electronics in the emerging devices. This has prompted the necessity for devices that are highly responsive towards weak UV signals and should be integrated with the existing PD technology. So, the power-dependent transient response on fabricated GaN NF-based PD has been explored. The time-correlated response under the illumination of a laser source operating at a low optical power of 1 mW and an applied bias of 1 V has been shown in Figure 8.12. Here, a continuous four on–off cycles representing sharp rise and decay signifies the development of highly responsive and stable UV PD. The device fabricated for this study using GaN-based NFs yields the highest photoresponsivity of 10.5 A W^{-1} at an optical power of 1 mW, which is even higher among commercially available UV PDs. The significant enhancement in photoresponsivity could be due to the presence of high photoconductive gain in the fabricated PD device [54].

The photoconductive gain (G) can be directly correlated to photocurrent by the given relation, $G = h\nu I_p/e\eta P_{in}$, which governs good photoresponsivity from the device pertaining to a high gain value [55]. The device

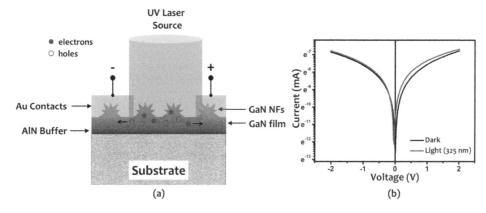

FIGURE 8.10 (a) A schematic structure illustrating generation of e–h pairs upon UV illumination and (b) I–V characteristics from the fabricated GaN NF-based UV PD device.

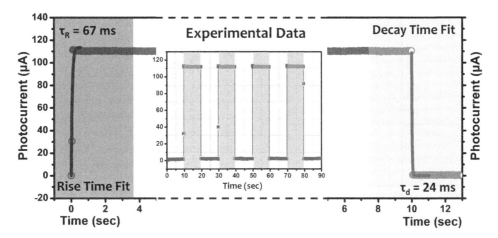

FIGURE 8.12 Rise time and decay time fitted curves for response time constants of the GaN NF-based UV PD in the experimental data represent the stable transient response at low optical power of 1 mW under a bias of 1 V.

characteristics from such GaN NF-like structure can be extremely responsive with remarkably high responsivity of 10.5 A W^{-1} at 1 V compared with the stated responsivities of 6.9 and 7.2 A W^{-1} at 5 V in earlier reports on GaN-based UV PDs [47,52]. Since the response time determines the speed of a detection device, the experimental data can be fitted for the rising as well as falling curves (Figure 8.12) and was found to be in the range of few milliseconds, which reveals very fast switching. Although the device witnessed a very high photoresponsivity, it has still been limited by the carrier collection ability and the UV exposed area of PD, which further need to be enhanced by augmenting the productivity of charge carrier generation as well as collection using upgraded designing of interdigitated metal electrodes.

To compare the effect of surface-to-volume ratio on the device properties, another PD device was fabricated using the planar GaN thin film and its device properties were briefly examined to analyze the advantage of NSs over planar structures in optoelectronic devices. It was observed that the maximum responsivity of 1.51 A W^{-1} at an optical power of 1 mW under 1 V bias photocurrent has been drawn from the planar GaN-based PD. However, the photoresponse from the device fabricated using low-dimensional structures (GaN NF) proves to be much faster and extremely responsive with around ten times increment

in the responsivity under similar illumination conditions (as shown in Figure 8.13a). This is accredited to very high surface-to-volume ratio, which increases photon absorption and, thereby, enhances charge carrier generation. When the effect of noise power was investigated (Figure 8.13b), a prominent 16-fold reduction in NEP for the PD designed using NF-like structures has been witnessed. Thus, an extremely high reduction in noise power can be realized due to the elimination of stress and defect states in GaN NF-like structures, such that the photodetection device yielded faster and highly responsive device characteristics. This section establishes that the operation of nanostructured GaN-based devices can be highly responsive at low optical power signals.

8.3.4 Self-Driven Operation

Upon applying higher bias to the device, the carrier collection speeds up by reducing e–h pair recombination and enhancing the current conduction. Consequently, thermionic field emissions also intensify, which results in high photocurrent generation. Thus, there is always a difficulty in increasing the applied bias, as it may increment the leakage current due to the augmented thermionic field emissions. So, it is preferred for an efficient device to operate at lower bias or zero bias (i.e., self-driven condition).

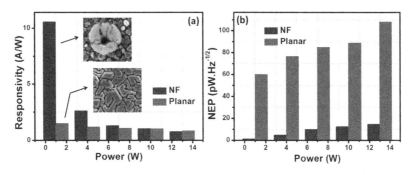

FIGURE 8.13 The power-dependent analysis elucidating a clear comparison of planar GaN film vs. GaN NF-based photodetection device as a function of (a) photoresponsivity and (b) noise power under a 1 V bias.

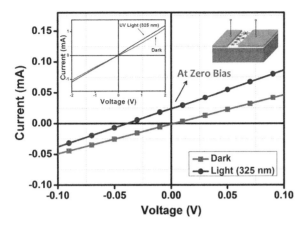

FIGURE 8.14   The magnified I–V characteristics to analyze self-driven behavior (under no bias) from the GaN NF-based photodetection device. Inset shows the complete I–V curve [27].

Captivatingly, a careful analysis of I–V characteristics (earlier discussed in Figure 8.11b) shown in the inset of Figure 8.14 discloses the presence of photocurrent even at no bias, which was magnified (shown in Figure 8.14) to validate the self-driven ability of GaN NF-based UV PD. To deeply analyze this self-driven nature, the band theory under dark and light conditions is elaborated here. A schematic band diagram of an energy band theory illustrating the fundamental mechanism associated with the self-driven behavior is presented in Figure 8.15.

The energy band represented in Figure 8.15a explains the case of dark conditions where the metal (Au) and semiconductor (GaN) are placed in contact with each other and reached a thermal equilibrium. The device is exposed to UV radiation in Figure 8.15b, where the energy from incident photons results in the generation of e–h pairs. As the band of metal is supposed to be completely filled with electrons, the holes generated in the valence band of GaN tends to diffuse toward the Au/GaN interfaces as shown in Figure 8.15b. The accumulated holes at the interfaces amend its local potential, subsequently lowering the effective barrier height [56,57]. In a metal–semiconductor junction, the effect

of metal electrode size on the distribution of built-in electric field has been explained by Freeouf et al., suggesting that the depletion region at the metal semiconductor interface with higher width of metal electrode should be larger [58]. Since the metal electrodes are of different sizes, the quantity of accumulated holes will be different at both the interfaces due to the difference in distribution of localized electric potential. Therefore, the metal electrode with more contact area will accommodate more number of holes compared with the other electrode, resulting in lowering of the potential barrier on that metal–semiconductor interface (Figure 8.15c). Now, electrons that are free to move in the conduction band can easily overcome the reduced barrier at the metal–semiconductor interface, pursuing increased photocurrent, yielding the photoresponse even without any bias.

Moreover, the switching behavior of GaN NF-based self-driven PD demonstrating a quick and stable on–off time-correlated photoresponse at a complete cycle time period of 20 s was depicted in Figure 8.16. Noticeably, a very low dark current in the range of nanoampere and a very high light-to-dark current ratio of ≈260 was obtained, which implies low noise power and better device performance. The nanostructured self-driven device can be assessed by evaluating its performance parameters, which were discussed in Section 8.3.2. The photoresponsivity value of 132 mAW^{-1} stated at no bias from the NF-based device is significantly higher than the device developed from planar GaN-based PDs. Sang et al. developed a GaN-based planar device yielding photoresponsivity of 0.037, 0.083, and 0.104 A W^{-1} using distinct electrodes and a flexible self-powered GaN UV photoswitch developed by Sun et al. possesses photoresponsivities of 0.03 and 0.0116 A W^{-1} at the power density of ≈3.5 and 35 mW cm^{-2}, respectively [50,59]. The low dark current followed high detection ability of the device, which was evaluated to be 2.4 × 10^{10} Jones, and is high enough to produce an efficiently responsive photodetection device. This section illustrates theoretical as well as experimental analysis of self-driven behavior from low-dimensional GaN-based UV photodetection device.

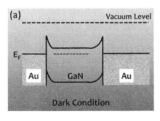
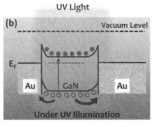
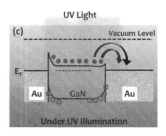

FIGURE 8.15   The schematic of energy band diagram demonstrating the charge carrier transport in the self-driven UV photodetection device [27].

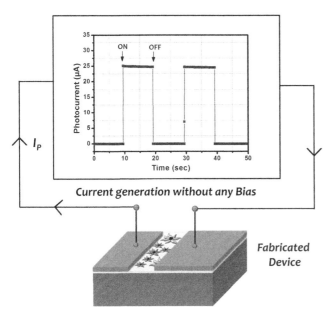

FIGURE 8.16 A transient response from the fabricated UV PD operating under self-driven condition.

8.4 Potential Applications

The photodiodes based on low-dimensional nitride NSs have acquired fascinating achievements in the last few decades due to their great potential applications in electronic and photonic devices. However, their practical implications are still endless and complex. High-performance PD requires high responsivity and low dark current to enable detection capability of low-strength optical signals. Nanostructured (nanorod, nanowire, nanowalls, nanoneedles, and NFs) detectors are attracting more attention owing to the improved sensitivity towards light due to their large surface-to-volume ratio and Debye length comparable to their small size [60]. Recently, Mishra et al. demonstrated the nonpolar surface-engineered GaN MSM PD by chemical etching, which offers a drastically reduced response time for fast switching applications [5]. Further, Son et al. [61] reported a single GaN nanorod p–n junction device that was grown using PAMBE and illustrated a rectifying behavior under dark conditions with the photoresponse on/off ratio estimated to be 14 under a reverse bias of 30 mV. The visible-blind PD based on p-i-n junction GaN nanowire ensembles were achieved by Andres de Luna et al. [62]. The detector presented a high peak responsivity of 0.47 A W^{-1}under a reverse bias of 1 V, which was higher than the thin film GaN p-i-n PDs. The spectral response of the detector was restricted to a UV range with a UV-to-visible rejection ratio of more than 10^2. So, there is still a plenty of room for the development of one-dimensional (1D) nitride NSs and their PD applications. We believe that future work in this direction should continue to focus on generating NSs in a more controlled, predictable, reliable, and simple way, and enhancing their photoconductor properties up to the level desirable in real industrial applications. The hopes are high that significant practical PD devices will soon arise due to

the integration of 1D nanostructure into conventional micro-electronics.

8.5 Summary

This chapter provided an overview on the developments in the growth of low-dimensional structures, specifically GaN NFs, their structural, morphological, and optical characteristics, and their applications in UV photodetection devices. With the enhanced surface-to-volume ratio in self-assembled GaN nanoflowered structures, it has been found that the developed device exhibits a strong response towards UV illumination of low-power optical signals. More importantly, it has been found that the evolution of NFs on the surface resulted in a 60-fold reduction in noise power and four times increment in the performance of the device compared with their planar counterparts. Further, a self-driven operation of GaN NFs was explored, which yields dark current as low as few nA and a huge increment in light-to-dark current ratio, which can provide a viable path towards realization of high-efficiency UV optoelectronic devices. With these remarkable progresses, we have concluded with a brief review about such low-dimensional structures offering a great promise for a broad range of potential applications that may open up new prospects in advanced nanoscale optoelectronic devices.

Acknowledgments

We sincerely acknowledge the director of CSIR-NPL, New Delhi, India for his perpetual encouragement and support.

References

1. J. Müßener, J. Teubert, P. Hille, M. Schäfer, J. Schörmann, M. de la Mata, J. Arbiol, M. Eickhoff, Probing the internal electric field in GaN/AlGaN nanowire heterostructures, *Nano Letters*, 14 (2014) 5118–5122.

2. P. Tchoulfian, F. Donatini, F. Levy, A. Dussaigne, P. Ferret, J. Pernot, Direct imaging of p–n junction in core–shell GaN wires, *Nano Letters*, 14 (2014) 3491–3498.

3. L. Goswami, R. Pandey, G. Gupta, Epitaxial growth of GaN nanostructure by PA-MBE for UV detection application, *Applied Surface Science*, 449 (2018) 186–192.

4. F. Qian, H. Wang, Y. Ling, G. Wang, M.P. Thelen, Y. Li, Photoenhanced electrochemical interaction between shewanella and a hematite nanowire photoanode, *Nano Letters*, 14 (2014) 3688–3693.

5. M. Mishra, A. Gundimeda, S. Krishna, N. Aggarwal, L. Goswami, B. Gahtori, B. Bhattacharyya, S. Husale, G. Gupta, Surface-engineered nanostructure-based efficient nonpolar GaN ultraviolet photodetectors, *ACS Omega*, 3 (2018) 2304–2311.

6. S. Krishna, N. Aggarwal, M. Mishra, K.K. Maurya, M. Kaur, G. Sehgal, S. Singh, N. Dilawar, B.K. Gupta, G. Gupta, Epitaxial growth of high In-content In0.41Ga0.59N/GaN heterostructure on (11–20) Al_2O_3 substrate, *Journal of Alloys and Compounds*, 658 (2016) 470–475.

7. L. Liu, J.H. Edgar, Substrates for gallium nitride epitaxy, *Materials Science and Engineering: R: Reports*, 37 (2002) 61–127.

8. E. Matioli, C. Neufeld, M. Iza, S.C. Cruz, A.A. Al-Heji, X. Chen, R.M. Farrell, S. Keller, S. DenBaars, U. Mishra, S. Nakamura, J. Speck, C. Weisbuch, High internal and external quantum efficiency InGaN/GaN solar cells, *Applied Physics Letters*, 98 (2011) 021102.

9. A.K. Pantazis, G. Konstantinidis, E. Gizeli, Characterization of a GaN lamb-wave sensor for liquid-based mass sensing applications, *IEEE Sensors Journal*, 14 (2014) 908–911.

10. J. Wu, When group-III nitrides go infrared: New properties and perspectives, *Journal of Applied Physics*, 106 (2009) 011101.

11. J. Millán, P. Godignon, X. Perpiñà, A. Pérez-Tomás, J. Rebollo, A survey of wide bandgap power semiconductor devices, *IEEE Transactions on Power Electronics*, 29 (2014) 2155–2163.

12. M. Martens, J. Schlegel, P. Vogt, F. Brunner, R. Lossy, J. Würfl, M. Weyers, M. Kneissl, High gain ultraviolet photodetectors based on AlGaN/GaN heterostructures for optical switching, *Applied Physics Letters*, 98 (2011) 211114.

13. E. Cicek, R. McClintock, C.Y. Cho, B. Rahnema, M. Razeghi, Al_xGa_{1-x}N-based back-illuminated solar-blind photodetectors with external quantum efficiency of 89%, *Applied Physics Letters*, 103 (2013) 191108.

14. J.W. Zhao, Y.F. Zhang, Y.H. Li, C.H. Su, X.M. Song, H. Yan, R.Z. Wang, A low cost, green method to synthesize GaN nanowires, *Scientific Reports*, 5 (2015) 17692.

15. F. Chen, X. Ji, Q. Zhang, Morphology-controlled synthesis and structural characterization of ternary Al_xGa_{1-x}N nanostructures by chemical vapor deposition, *CrystEngComm*, 17 (2015) 1249–1257.

16. J. Johansson, C.P.T. Svensson, T. Mårtensson, L. Samuelson, W. Seifert, Mass transport model for semiconductor nanowire growth, *The Journal of Physical Chemistry B*, 109 (2005) 13567–13571.

17. S. Krishna, N. Aggarwal, G.A. Reddy, P. Dugar, M. Mishra, L. Goswami, N. Dilawar, M. Kumar, K.K. Maurya, G. Gupta, Probing the correlation between structure, carrier dynamics and defect states of epitaxial GaN film on (112 [combining macron] 0) sapphire grown by rf-molecular beam epitaxy, *RSC Advances*, 5 (2015) 73261–73267.

18. X. Wang, A. Yoshikawa, Molecular beam epitaxy growth of GaN, AlN and InN, *Progress in Crystal Growth and Characterization of Materials*, 48–49 (2004) 42–103.

19. S. Prasana, D. Sandip, D. Sitaram, K.T. Ashok, One dimensional GaN nanostructures: Growth kinetics and applications, *Nanoscience and Nanotechnology-Asia*, 1 (2011) 140–170.

20. C. Adelmann, N. Gogneau, E. Sarigiannidou, J.L. Rouvière, B. Daudin, GaN islanding by spontaneous rearrangement of a strained two-dimensional layer on (0001) AlN, *Applied Physics Letters*, 81 (2002) 3064–3066.

21. M. Kumar, T.N. Bhat, M.K. Rajpalke, B. Roul, P. Misra, L.M. Kukreja, N. Sinha, A.T. Kalghatgi, S.B. Krupanidhi, Self-assembled flower-like nanostructures of InN and GaN grown by plasma-assisted molecular beam epitaxy, *Bulletin of Materials Science*, 33 (2010) 221–226.

22. S. Fernández-Garrido, J. Grandal, E. Calleja, M.A. Sánchez-García, D. López-Romero, A growth diagram for plasma-assisted molecular beam epitaxy of GaN nanocolumns on Si(111), *Journal of Applied Physics*, 106 (2009) 126102.

23. T.-T. Kang, X. Liu, R.Q. Zhang, W.G. Hu, G. Cong, F.-A. Zhao, Q. Zhu, InN nanoflowers grown by metal organic chemical vapor deposition, *Applied Physics Letters*, 89 (2006) 071113.

24. R.K. Debnath, R.Meijers, T. Richter, T. Stoica, R. Calarco, H. Lüth, Mechanism of molecular beam epitaxy growth of GaN nanowires on Si(111), *Applied Physics Letters*, 90 (2007) 123117.

25. L. Yu, Z. Hu, Y. Ma, K. Huo, Y. Chen, H. Sang, W. Lin, Y. Lu, Evolution of aluminum nitride nanostructures from nanoflower to thin film on silicon substrate by direct nitridation of aluminum precursor, *Diamond and Related Materials*, 16 (2007) 1636–1642.

26. S. Dhamodaran, D.S. Chander, J. Ramkumar, Antireflective and hydrophobic surface of self-organized GaN nano-flowers, *Applied Surface Science*, 257 (2011) 9612–9615.

27. N. Aggarwal, S. Krishna, A. Sharma, L. Goswami, D. Kumar, S. Husale, G. Gupta, A highly responsive self-driven UV photodetector using GaN nanoflowers, *Advanced Electronic Materials*, 3 (2017) 1700036.

28. W. Zhao, L. Wang, J. Wang, Z. Hao, Y. Luo, Edge dislocation induced self-assembly of InGaN nano-flower on GaN by metal organic vapor phase epitaxy, *Journal of Applied Physics*, 110 (2011) 014311.

29. M.J. Shin, M.J. Kim, H.S. Jeon, H.S. Ahn, S.N. Yi, D.H. Ha, Y. Huh, B.H. Kim, S.H. Park, Evolution of GaN nanoflowers from AlN–SiO2 grains on a silicon substrate by chemical vapor reaction, *Vacuum*, 86 (2011) 201–205.

30. G.H. Gilmer, M.H. Grabow, Models of thin film growth modes, *JOM*, 39 (1987) 19–23.

31. S. Raghavan, X. Weng, E. Dickey, J.M. Redwing, Effect of AlN interlayers on growth stress in GaN layers deposited on (111) Si, *Applied Physics Letters*, 87 (2005) 142101.

32. M.A. Moram, M.E. Vickers, X-ray diffraction of III-nitrides, *Reports on Progress in Physics*, 72 (2009) 036502.

33. B. Heying, R. Averbeck, L.F. Chen, E. Haus, H. Riechert, J.S. Speck, Control of GaN surface morphologies using plasma-assisted molecular beam epitaxy, *Journal of Applied Physics*, 88 (2000) 1855–1860.

34. N. Aggarwal, S.T.C. Krishna, L. Goswami, M. Mishra, G. Gupta, K.K. Maurya, S. Singh, N. Dilawar, M. Kaur, Extenuation of stress and defects in GaN films grown on a metal–organic chemical vapor deposition-GaN/c-sapphire substrate by plasma-assisted molecular beam epitaxy, *Crystal Growth and Design*, 15 (2015) 2144–2150.

35. Z.C. Feng, W. Wang, S.J. Chua, P.X. Zhang, K.P.J. Williams, G.D. Pitt, Raman scattering properties of GaN thin films grown on sapphire under visible and ultraviolet excitation, *Journal of Raman Spectroscopy*, 32 (2001) 840–846.

36. S.-W. Feng, Y.-Y. Chen, C.-M. Lai, L.-W. Tu, J. Han, Anisotropic strain relaxation and the resulting degree of polarization by one- and two-step growth in nonpolar a-plane GaN grown on r-sapphire substrate, *Journal of Applied Physics*, 114 (2013) 233103.

37. D. Won, X. Weng, Z.Y. Al Balushi, J.M. Redwing, Influence of growth stress on the surface morphology of N-polar GaN films grown on vicinal C-face SiC substrates, *Applied Physics Letters*, 103 (2013) 241908.

38. S. Strite, H. Morkoç, GaN, AlN, and InN: A review, *Journal of Vacuum Science and Technology B: Microelectronics and Nanometer Structures Processing, Measurement, and Phenomena*, 10 (1992) 1237–1266.

39. W.C. Ke, C.P. Fu, C.C. Huang, C.S. Ku, L. Lee, C.Y. Chen, W.C. Tsai, W.K. Chen, M.C. Lee, W.C. Chou, W.J. Lin, Y.C. Cheng, Optical properties and carrier dynamics of self-assembled GaN/Al 0.11 Ga 0.89 N quantum dots, *Nanotechnology*, 17 (2006) 2609.

40. S.P. Young, W.K. Tae, R.A. Taylor, Abnormal photoluminescence properties of GaN nanorods grown on Si(111) by molecular-beam epitaxy, *Nanotechnology*, 19 (2008) 475402.

41. M.A. Reshchikov, H. Morkoç, Luminescence properties of defects in GaN, *Journal of Applied Physics*, 97 (2005) 061301.

42. M.A. Khan, J.N. Kuznia, D.T. Olson, J.M. Van Hove, M. Blasingame, L.F. Reitz, High-responsivity photoconductive ultraviolet sensors based on insulating single-crystal GaN epilayers, *Applied Physics Letters*, 60 (1992) 2917–2919.

43. A. Gundimeda, S. Krishna, N. Aggarwal, A. Sharma, N.D. Sharma, K.K. Maurya, S. Husale, G. Gupta, Fabrication of non-polar GaN based highly responsive and fast UV photodetector, *Applied Physics Letters*, 110 (2017) 103507.

44. B.D. Boruah, A. Misra, Energy-efficient hydrogenated zinc oxide nanoflakes for high-performance self-powered ultraviolet photodetector, *ACS Applied Materials and Interfaces*, 8 (2016) 18182–18188.

45. L. Sang, M. Liao, M. Sumiya, A comprehensive review of semiconductor ultraviolet photodetectors: From thin film to one-dimensional nanostructures, *Sensors*, 13 (2013) 10482–10518.

46. E. Monroy, M. Hamilton, D. Walker, P. Kung, F.J. Sánchez, M. Razeghi, High-quality visible-blind AlGaN p-i-n photodiodes, *Applied Physics Letters*, 74 (1999) 1171–1173.

47. Z.M. Zhao, R.L. Jiang, P. Chen, D.J. Xi, Z.Y. Luo, R. Zhang, B. Shen, Z.Z. Chen, Y.D. Zheng, Metal–semiconductor–metal GaN ultraviolet photodetectors on Si(111), *Applied Physics Letters*, 77 (2000) 444–446.

48. N. Biyikli, T. Kartaloglu, O. Aytur, I. Kimukin, E. Ozbay, High-speed visible-blind GaN-based indium–tin–oxide Schottky photodiodes, *Applied Physics Letters*, 79 (2001) 2838–2840.

49. Z. Alaie, S. Mohammad Nejad, M.H. Yousefi, Recent advances in ultraviolet photodetectors, *Materials Science in Semiconductor Processing*, 29 (2015) 16–55.

50. M. Peng, Y. Liu, A. Yu, Y. Zhang, C. Liu, J. Liu, W. Wu, K. Zhang, X. Shi, J. Kou, J. Zhai, Z.L. Wang, Flexible self-powered GaN ultraviolet photoswitch with piezo-phototronic effect enhanced on/off ratio, *ACS Nano*, 10 (2016) 1572–1579.

51. Y.-Z. Chiou, GaN ultraviolet photodetectors with transparent titanium tungsten and tungsten electrodes, *Journal of The Electrochemical Society*, 152 (2005) G639–G642.

52. Y. Su, S.-J. Chang, C. Chen, J.F. Chen, G.C. Chi, J.-K. Sheu, W.-C. Lai, J.M. Tsai, GaN metal-semiconductor-metal ultraviolet sensors with various contact electrodes, *IEEE Sensors Journal*, 2 (2002) 366–371.

53. A.G. Reddy, N. Aggarwal, S. Krishna T. C, M. Singh, R. Rakshit, G. Gupta, Correlation of current–voltage–temperature analysis with deep level defects in epitaxial GaN films, *Applied Physics Letters*, 106 (2015) 233501.

54. J.A. Garrido, E. Monroy, I. Izpura, E. Muñoz, Photoconductive gain modelling of GaN photodetectors, *Semiconductor Science and Technology*, 13 (1998) 563.

55. R.-S. Chen, H.-Y. Chen, C.-Y. Lu, K.-H. Chen, C.-P. Chen, L.-C. Chen, Y.-J. Yang, Ultrahigh photocurrent gain in m-axial GaN nanowires, *Applied Physics Letters*, 91 (2007) 223106.

56. L. Wang Zhong, Progress in piezotronics and piezo-phototronics, *Advanced Materials*, 24 (2012) 4632–4646.

57. O. Katz, V. Garber, B. Meyler, G. Bahir, J. Salzman, Gain mechanism in GaN Schottky ultraviolet detectors, *Applied Physics Letters*, 79 (2001) 1417–1419.

58. J.L. Freeouf, T.N. Jackson, S.E. Laux, J.M. Woodall, Effective barrier heights of mixed phase contacts: Size effects, *Applied Physics Letters*, 40 (1982) 634–636.

59. X. Sun, D. Li, Z. Li, H. Song, H. Jiang, Y. Chen, G. Miao, Z. Zhang, High spectral response of self-driven GaN-based detectors by controlling the contact barrier height, *Scientific Reports*, 5 (2015) 16819.

60. T. Zhai, X. Fang, M. Liao, X. Xu, H. Zeng, B. Yoshio, D. Golberg, A comprehensive review of one-dimensional metal-oxide nanostructure photodetectors, *Sensors*, 9 (2009) 6504–6529.

61. M.S. Son, S.I. Im, Y.S. Park, C.M. Park, T.W. Kang, K.H. Yoo, Ultraviolet photodetector based on single GaN nanorod p–n junctions, *Materials Science and Engineering: C*, 26 (2006) 886–888.

62. B. Andres de Luna, T. Maria, J. Gwenole, R. Lorenzo, J. François Henri, C. Shu-Ting, L. Yuan-Ting, T. Po-Han, T. Li-Wei, Visible-blind photodetector based on p–i–n junction GaN nanowire ensembles, *Nanotechnology*, 21 (2010) 315201.

The Future of Nanoelectronics

9.1 Energy Efficient and Sustainable Nanoelectronics 9-1
  Moore's Law and Scaling of CMOS Transistors • Market Drivers, Innovations, and Trends • Scaling Logic: MOSFET Electrostatic Integrity • Memory Scaling • Towards Zero Intrinsic Variability through New Fabrication Paradigms
9.2 More Moore and More than Moore Meeting for 3D into Zero Power Systems... 9-24
References 9-27

Simon Deleonibus
Université Grenoble Alpes

9.1 Energy Efficient and Sustainable Nanoelectronics

9.1.1 Moore's Law and Scaling of CMOS Transistors

It is very important to distinguish Moore's law from individual device and geometrical scaling. Moore's law started from an economic viewpoint by noticing an historical trend in the increase of the number of devices "crammed on a chip" [1] by a factor of 2 every year. The interest for such a trend had an obvious advantage to reduce the cost per function. Moore mentioned the fact that the trend would continue in the future (Figure 9.1). Actually, bipolar transistors were the active devices at that time. Microelectronics was not betting on metal oxide semiconductor field effect transistors (MOSFETs) at all. The rise of MOSFET popularity, tackling bipolar domination, occurred by 1970 [2], with a prediction that progress would slow down by the end of the 1970s! Actually, no geometrical scaling rule was clearly set forth before Dennard et al. published their paper [3] on scaling, thanks to the possibility to self-align source and drain doping by ion implantation.

Today, we can observe that the historical trend has been followed naturally until the beginning of the 1990s and was then agreed to be sustainable, when the first coordinated road map appeared, by the U.S. National and the International Technology Roadmap for Semiconductors (ITRS). As a matter of fact, the equipment suppliers requested from integrated devices manufacturers to "sit around a table" and discuss about an objective, argumented, and realistic road map [4,5] (Figure 9.1). It took a while to select the good benchmarkers, as microelectronics kept gaining maturity. The main guideline was agreed to be the geometrical scaling of Complementary Metal Oxide Semiconductor (CMOS) transistors and memories, based on MOSFET technology [6]. Thanks to its success, dynamic memories and

microprocessors overcame 1 billion transistors level as early as 1995 [7] and 2006 [8], respectively. CMOS scaling is a technical rule listing geometrical constraints to make sure that power consumption will be limited and performance will be maximized. In order to address the major categories of products that the industry would be delivering, the ITRS identified three main families of devices: high performance (HP), low operating power (LOP), and low standby power (LSTP) architectures were defined. Among the large number of active devices described in literature, three domains were defined depending on their maturity, pervasiveness, and specificity to applications or aptitude to scaling. The Microelectronics Development for European Applications (European MEDEA) initiative first defined the domains More Moore and More than Moore, which respectively distinguished CMOS (linearly scalable) and the devices that were more application oriented and susceptible to improve the diversification of integrated circuits. Those devices were sensors and actuators, memories, Radio Frequency (RF) devices and passives, power/high voltage devices, bioelectronic devices, etc. [4] (Figure 9.2). Besides the mainstream devices, physicists and engineers have been looking into various types of devices that could challenge CMOS, either in their ultimate configuration or by using a state variable different from electric charge. The so-called Beyond CMOS or Beyond Moore domain lists and discusses the features of these devices that have been benchmarked systematically year after year to measure their progress or development. Some of them could join a transition phase before being considered for the road map. For example, that was the case for Fin-shaped Field Effect Transistors (FinFET) or multigate devices at the very beginning of their introduction in the road map. Other devices such as Magnetic Random Access Memories (MRAMs) (in particular, spin transfer torque RAM (STTRAM)) followed the same track.

The evolution at each step is not only due to dimensional scaling. Admittedly, higher resolution lithography plays a

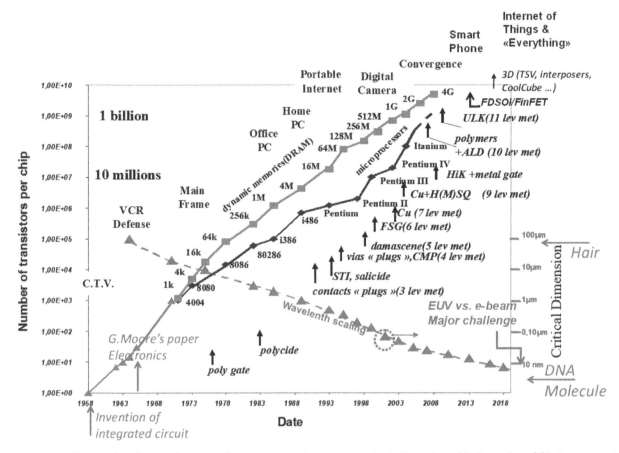

FIGURE 9.1 Historical evolution of integrated circuits since their invention in 1958 to date. (Updated from [5].) Process modules appear in black next to arrows on the right hand side, and the main end applications appear as gray bullets on the left hand side.

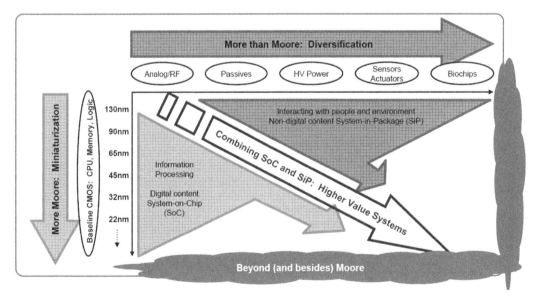

FIGURE 9.2 More Moore, More than Moore, and beyond CMOS (or beyond Moore). (By courtesy of [4].)

major role but is not sufficient to make a new generation of chip to effectively work with acceptable speed and power consumption specifications. At each generation, new process modules have been introduced to make geometrical scaling effective and workable (Figure 9.1) [5]. Innovation introduced in our technologies is a necessary condition to enable new applications and invent new usages [5] that bring a strong qualitative leap: this is illustrated by the gray bullets in Figure 9.1. It is obvious today that it would have been impossible to make a smartphone in the 1970s when the maturity and complexity of integrated circuits technology could only allow professionals to play with the first office microcomputers! Today, we are expecting to increase connectivity and grow in a future world where potentially everything will be addressable or actuated/controlled via networks, thanks to secure and reliable software. This will be a revolution in our way of life, which is being named "Internet of Things" (IoT) or even more emphasized as "Internet of Everything" (IoE) (Figure 9.1). The introduction of heterogeneous integration is necessary due to the fact that interfacing things of various and different natures will request the association of components that are not necessarily feasible today by means of a single mainstream technology. Actually, sensors, RF components, digital electronics, or stand-alone memories do not use the same technology and are not necessarily integrated in a unique process flow. Historically, there has been a strong trend to cointegrate different components on a single chip or in a single package. Even if it were possible technically speaking, it might have been impossible for economic reasons at a given time. This is one of the main dilemma and challenges that microelectronics has been facing historically and will continue to deal with. As soon as the technology can offer an innovation allowing a higher integration capability, it becomes possible to introduce a new economically viable heterogeneous solution. At the end, cointegration will win in many ways. In an IoE era, cointegration of heterogeneous technologies and 3D integration will become mature enough to be considered as a commodity!

9.1.2 Market Drivers, Innovations, and Trends

CMOS Technology as a Driver

In the early microelectronics history, the challenge between the advent of MOSFET tackling bipolar technology was debated. The demonstration of the first silicon-based MOSFET in 1960 [6], 30 years after the invention of field effect devices [9], was certainly a major step in the development of future integrated circuits. The integrity and reliability of silicon dioxide (SiO_2) was questioned as a clue to make such a technology workable and available. MOSFET was an appealing solution to envisage high input impedance-based devices that were bringing in low power consumption-actuated devices. The possibility to avoid the use of the base current of bipolar transistors to actuate a switch became very popular: this would enable low-power and high-density logic circuits based on MOSFETs. The fabrication process was potentially much less expensive than for bipolar. The superior integration density of logic circuits was rapidly shown as an important asset for the success of MOSFET [2] (Figure 9.3a), especially for a benchmark function such as a latch, which had a perfect match for static memory applications. Finally, bipolar devices association to CMOS technology, giving birth to BiCMOS, demonstrated improved abilities compared with CMOS for analog or high-frequency applications, thanks to the higher minority carrier mobility and superior noise figures due to their bulk conduction nature in bipolar devices. The increase

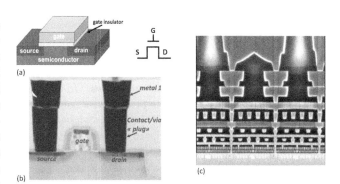

FIGURE 9.3 (a) MOSFET physical scheme (left) and electrical symbol (right), (b) cross section by transmission electron microscopy of MOSFET. Contacts to source and drain are plugged with high step coverage metal deposition and planarization. The same principle is used since the introduction of the 500 nm node and still will be at the 10 nm node. In this example, the refill metal is tungsten. Metal 1 is the first interconnect level, and (c) advanced node chips can stack up to 11 metal copper levels by using the same principle, since the 180 nm node is down to the 10 nm node. Damascene process uses the same refill and planarization sequence as contact/via. (By courtesy of CEA, LETI, and [8].)

of the number of components per chip was eased by the introduction of polysilicon gate MOSFET and its compatibility with self-aligned shallow source and drain implantation [3], which enabled a higher control of short channel effects. Such a breakthrough was a good opportunity to maximize the component density on a chip and envisage increasing memory density with device scaling.

By the mid 1990s, the increase of memory capacity by itself happened to be insufficient to increase the logic circuit bandwidth. Interconnect performance was a bottleneck to achieve ultra high-density and high-speed logic circuits. The number of interconnect levels increased, but the scaling of interconnect wiring suffered from aluminum resistivity increase and parasitic capacitance increase. A dual strategy resulted in (i) the increase of the number of interconnect levels; (ii) change of aluminum to copper as interconnect material. This was conditioned by the use of the plugging effect, thanks to the high step coverage of Chemical Vapor Deposition (CVD) deposited metals [10,11] from the first contact level (Figure 9.3b), followed by the planarization of the insulator topography and conductors by chemical mechanical polishing (Figure 9.3b,c), introduced in 1997 [12].

Memories as Drivers and Hierachy in Information Processing

The needs for memory have kept on increasing in the frame of von Neumann type of computing chips. Thanks to the scaling of 1T-1C dynamic random access memories (DRAMs) [13], the memory market soon became a commodity market and a good benchmark for technology performance (Figure 9.4a,b). Not-And(NAND) architectured Flash memories have become the technology driver instead of logic and DRAMs: the basic memory cell has been sized down by a factor of 10 in 5 years

(Figure 9.4a). By analyzing finely the way computing processes use memory, a huge number of different requirements result in terms of data storage capacity, speed, and power consumption. Intermittent storage needs high speed, but not necessarily high retention capabilities. On the other hand, medium- to long-term code or data storage needs as high retention as possible but not necessarily high speed. In the first case, volatile memories fit to the needs, whereas in the second case, nonvolatile memories (NVMs) are requested (Figure 9.4b). In all cases, energy efficiency and thus the minimization of power consumption are sought.

Volatile memories can either be directly achieved in the CMOS process flow (embedded) or in a specific process for large capacity needs (stand alone). In the first case, static random access memories (SRAM) do not need specific materials or mask step, besides cell transistor threshold voltage adjust. The requirement for the stability will need robust input capacitance: a memory cell using transistor latches has become the most popular and stable way to store information in a six-transistor SRAM cell, as long as the supply voltage is maintained: the flip-flop structure has been used since the late 1960s [14]. If stability is less, a requirement with a strong need in capacity, DRAM can be used with the disadvantage that data refresh is needed. In this case, the memory cell can be compact by using one transistor for addressing and one capacitor for storage [13]. However, the power consumption is ten times larger than for an SRAM, while the memory cell footprint will be about ten times less for DRAM when compared with SRAM (see Section 9.1.4). DRAM is dedicated to high-capacity temporary storage, while SRAM is rather recommended as an embedded working memory.

The increase of embedded software usage and high capacity storage pushed the necessity for higher capacity NVM. For NVM, long-term storage (usually 10 years) is desired, even though the power supply is off. At the same time, the largest number of write/erase endurance cycles as possible is sought. Flash memories took the lead as a market technology driver, especially because the NAND architecture gives it an unbeatable $4F^2$ size (Figure 9.4a). The automotive, leisure, and mobile phone markets were emblematic of huge expectations for large capacity NVM. As expected, memory hierarchy needs to be constantly redefined from a system perspective (Figure 9.4b) to cope with the technology evolution. Nowadays, more and more personalization of application requests increased the capacity of embedded NVMs. The needs are different from storage-type memories, which more or less need to be of the one or several times programmable read only memory (PROM) type, for example, in the case of picture storage. Today, two-dimensional (2D) linear scaling of NAND Flash Memories has run out of steam because of the stress-induced leakage current (SILC) limitation to data retention and endurance as well as weak short channel effect immunity. This is why several companies developed 3D architectures for stand-alone mass storage of the 1 TB type using gate all-around transistors, while embedded memories do not request such a high density. As a consequence, several memory families are being distinguished to respond to the applications and market needs, whereas the microelectronics community has tried to keep a sole way to supply memories in a universal manner: nowadays, it is very difficult to face all the challenges posed by different applications with a single solution. With the increasing complexity of advanced systems, it has become necessary to fill in the gap between flash NVMs and dynamic memories with a new category of high-density embedded devices with high endurance capability: the so-called storage class memories (SCM) [15] would fulfill the needs (Figure 9.4b), either for storage (S type) or working purposes (M type).

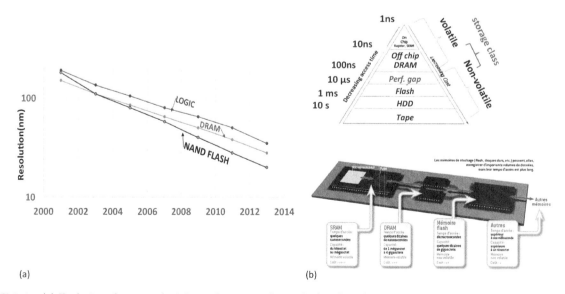

(a)   (b)

FIGURE 9.4 (a) Evolution of requested minimum feature resolution for benchmark products between 2000 and 2013 and (b) memory hierarchy from a system perspective (By courtesy of CEA, LETI).

Pushing the Limits or Introducing Innovative Approaches?

After many discussions, the ITRS finally sorted the different types of devices and their specific utility and maturity.

First of all, a difference was made between devices that were seen as being integratable during the next 15 years and the emerging research devices for which maturity was still a question. The integratable devices appeared as serving logic or memory applications. The different partners defined three categories of logic devices mentioned earlier (HP, LOP, and LSTP), depending on the products they would develop in the future. Desktop computers or equipment, supplied by a power grid, would request HP devices. Intermittently supplied portable devices, such as laptops, would be designed with LOP devices, whereas autonomous and mobile equipment, such as mobile phones, rather demand for LSTP devices. Very simply, memories were sorted into two categories: volatile (DRAMs, SRAMs) and NVMs.

Another limit, appearing on the scaling route, is the reduction of the number of carriers involved in a switching or memory charge storage process (Figure 9.5a,b) ([16], see also p. 268 in [5]). In Figure 9.5a, one can see that the number of carriers in the on state is reduced drastically to a few tens for logic devices gate length smaller than 22 nm. We reach the same situation for high-density memories at the 28 nm node. This feature is at the origin of random fluctuations of the electrical characteristics, such as threshold voltage, which could vary stochastically in the case of charge-based memories by trapping/detrapping mechanisms having an increasing weight on variability.

Emerging devices were finally defined by different criteria (Figure 9.6) depending on the state variable they are using, any specific material bringing in novel properties, the device architecture, the data representation, and the system architecture that they would need to express their figures of merit optimally.

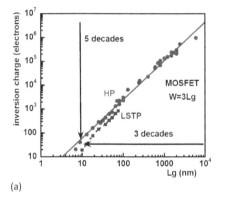

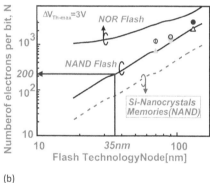

(a)   (b)

FIGURE 9.5 Number of electrons reduced with scaling: (a) per switch in logic devices [4] and (b) stored per bit in NVMs (By courtesy of CEA, LETI).

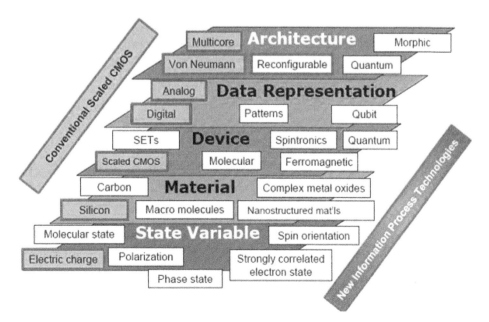

FIGURE 9.6 Taxonomy for emerging research devices. (By courtesy of ITRS [4].)

9.1.3 Scaling Logic: MOSFET Electrostatic Integrity

Dimensional scaling [3,17] was seen from the very beginning as a good way to increase MOSFET integration density while improving the circuit delay time and keeping its power density unchanged together with an improved power dissipation per function, thanks to a supply voltage scaling. In such an approach, the major features of MOSFETs (t_{ox}, L, W) are changed in three dimensions (Figure 9.7). This strategy cannot be applicable without the use of ion implantation to self-align shallow source and drain on the gate and adjust the threshold voltage, thanks to a semiretrograde channel profile. It was possible, provided that polysilicon can be used as a gate material compatible with SiO_2, capable of being processed at high temperature to activate the implanted dopants. By using such a flexible technology, increased performance and reduced short channel effects could be obtained. One of the conditions set by Dennard was that the linear scaling of gate oxide would be possible without any yield/damage or reliability issues.

Major Breakthrough Modules

Several breakthroughs were necessary to continue the "happy scaling" period, during which MOSFET scaling seemed to be obvious, and technology was considered to be

Channel length	K
Voltage	U
Gate oxide	K
Junction depth	K
Electric field	U/K^2
Channel doping	U/K
Parasitic capacitance	**K(ACox,ACj)**
Current (vel. sat.)	$U^2/K(U)$
Delay (vel. sat.)	$K^2/U(K)$
Power (vel. sat.)	$U^3/K(U^2)$
Speed.Power product	KU^2

FIGURE 9.7 CMOS-dimensional scaling rules and deducted metrics. (Adapted and modified from [3] and [17].)

a commodity or even a black box for design! The introduction of metal gate/high dielectric constant dielectric (HiK) stack to replace the polysilicon (Poly)/SiO_2 gate stack, biaxial, and uniaxial strain and fully depleted (FD) multigate channels related to random dopant fluctuations (RDF) allowed to continue linear scaling.

From SiO_2 Scaling to Metal Gate/HiK Stack

The scaling of SiO_2 thermally grown on the Si substrate had much success until 2000–2010 when a 1.2 nm SiO_2 applied to the 45–32 nm nodes was needed. Iwai and his team demonstrated that a tunnel oxide of 1.5 nm was still usable for HP small geometry transistors [18] by 1994. In parallel, many teams prepared the advent of HiK insulators likely to replace SiO_2 in the 1 nm equivalent oxide thickness (EOT) range. The first issue came from the band structure and bandgap values of the candidate HiK materials (Figure 9.8a). However, replacing SiO_2 by a metallic oxide posed a huge problem of compatibility to polysilicon gate because of a possible reaction between the metal and Si, which resulted in a pinning of the Fermi level on the gate side. This is why polysilicon gate had to be replaced by a metal gate (Figure 9.8b) that could match the band engineering necessary to scale the supply voltage, as suggested by Dennard's scaling. The first demonstration of high K and metal gate using a totally self-aligned replacement gate process integration [19] on small transistors was achieved in 2002 [20], as an anticipation on bulk 75 nm gate length MOSFETs (Figure 9.9a) and has been scaled down for several nodes on revised architectures (Figure 9.9b). Intel demonstrated the dual-gate version [21], in 2007, inspired by the same type of architecture as in [19,20].

MOSFET's short channel effect results from inversion charge sharing between the different electrodes (source, drain, gate, and background electrodes) as well as drain-induced barrier lowering (DIBL). Its control with scaling requests the increase of channel doping. However, this strategy has a drawback on the transport properties that

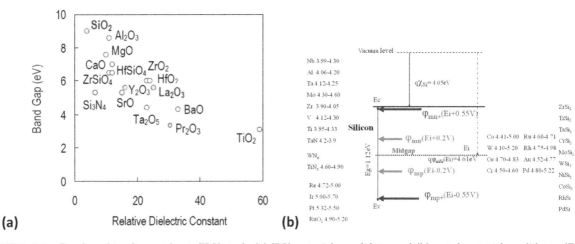

FIGURE 9.8 Benchmarking for metal gate/HiK stack: (a) HiK materials candidates and (b) metal material candidates. (By courtesy of CEA, LETI, and modified from [5].)

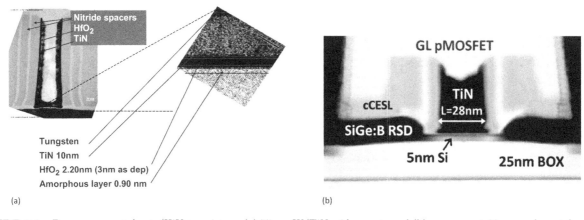

FIGURE 9.9 Damascene metal gate/HiK transistors: (a) 75 nm W/TiN midgap gate and (b) more recent 28 nm node metal gate/HiK pMOS transistor with raised source and drain on ultrathin silicon on insulator (SOI). (By courtesy of CEA, LETI.)

are being degraded at very short channel lengths due to increased Coulomb interactions of electrons with dopants as well as increased electric fields. In bulk transistors, retrograde profiles limit the mobility degradation by minimizing surface channel doping while maximizing backgroundplane doping [22–24]. A similar effect can be obtained on short channels by using pocket implants to enhance their background doping [22–24], while minimizing source and extension lateral diffusion of dopants.

From Threshold Voltage Fluctuations to Undoped Channels

Nevertheless, device scaling reduces the device volume, and the number of dopants in a channel fluctuates randomly just as an ion implanter introduces them in the device channel area [5]. Many routes have been followed, at the sub 100 nm level, to master the dopants profile and concentration fluctuations, such as the use of rapid thermal processing in different flavors or heavy ion implantation [22–24]. Still, approaching the sub 50 nm node, such a strategy became unsufficient due to severe dopant fluctuations. Let us suppose a distribution of ionized doping species that follows a Poisson law on axis x. The distribution standard deviation will be given by [5,25,26]

$$\sigma = \sqrt{\frac{n}{x_0^3}}$$

where n is the average dopant concentration and x_0 is the unit dimension of the considered volume. It is very simple to understand that σ increases with increasing n and decreasing x_0. RDF became a severe issue for devices with gate lengths lower than 50 nm [27]. As an example, threshold voltage would be affected directly as it depends on the flat band voltage MOS heterostructure: fluctuations comparable to the threshold voltage values would be obtained [5]. Such a situation is not acceptable to design a circuit. Thus, nondoped channels become interesting to control threshold voltage at such a level. On top of RDF, in small geometries, other fluctuations show a first-order impact on the electrical characteristics of devices: line edge roughness (LER) due to

the combination of lithography performance and the underlying layer roughness, gate granularity, workfunction variation, dielectric thickness fluctuations, and interface trapped charge due to device degradation [27]. A large number of results have been published, and simulation tools have been developed [27,28] on decanometer level device variability for the past two decades.

At the sub 50 nm node level, the physical fluctuations can introduce up to 100% threshold voltage variation if the threshold voltage value is in the range of 0.2 V: for HP devices, especially, the design of complex circuits would simply be impossible.

Because random dopant distribution introduces a major hurdle, then the device architecture that could be done without extrinsic channel doping gains much interest. However, the threshold voltage adjust for system on chip or microcontrollers is then questioned, because these circuits request several threshold voltage values to differentiate high speed from low power parts as well as SRAM design. FD channel devices are of much interest: the most popular architectures are FD silicon on insulator (FDSOI) [29–31] and FinFET [32–34] (Figure 9.10). At its initial stage, FinFET

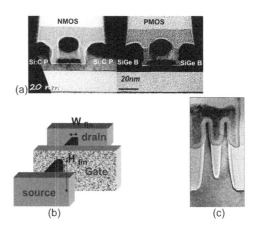

FIGURE 9.10 (a) An FDSOI (by courtesy of CEA, LETI), (b) a schematic representation of FinFET on SOI, and (c) FinFET on bulk. (Adapted from [34].)

was proposed with an SOI-based version (Figure 9.10b). More recently, a bulk version of FinFET (Figure 9.10c) was proposed. However, this device requests an antipunch through implant as a backgroundplane. It poses as well the issue of random dopant fluctuations [35]. The difference between FDSOI and FinFET channels is essentially due to the physical channel geometrical control. FDSOI channel thickness is adjusted by a planar thin silicon thickness control, whereas FinFET vertical channel structure is controlled by the precision of lithography. However, the requirement in terms of final control needs to be the same. A technology based on the transfer of spacers can drastically reduce the nominal geometry control. Finally,

FDSOI on ultrathin buried oxide (UTBO$_x$) (Figure 9.11a,b) demonstrates by far the best electrostatic integrity and device pair analog matching parameters [35,36].

Thanks to the control of SOI thickness, scalability of FDSOI devices is foreseen down to the sub 7 nm node [29,30] (Figure 9.12a). Beyond the 7 nm node, gate all-around devices with nanowire cylindrical channels will be necessary [29,30] (Figure 9.12b), thanks to their very high electrostatic integrity [37–39], through their immunity to short channel effects (charge sharing and DIBL). A good screening on device architecture can be achieved by comparing their natural lengths [37–39]. The natural length λ is defined by the relations given in Table 9.1 for each type of thin-film

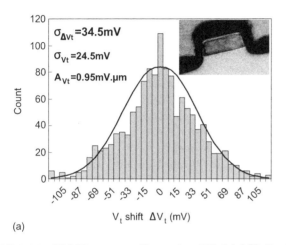

(a)

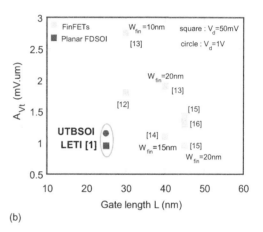

(b)

FIGURE 9.11 FDSOI parameter fluctuations [35]: (a) ΔVt distribution for short and narrow devices and (b) Benchmarking transistor pairs analog Vt mismatching coefficient AVT [35] $\sigma_{Vt} = \frac{A_{Vt}}{\sqrt{WL}}$ with $\sigma_{Vt} = \frac{\sigma_{\Delta Vt}}{\sqrt{2}}$ ΔVt is the threshold voltage deviation between pairs. (By courtesy of CEA, LETI.)

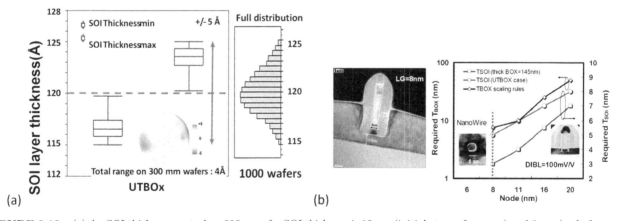

(a)   (b)

FIGURE 9.12 (a) An SOI thickness control on 300 m wafer SOI thickness is 12 nm (initial stage of process) and 6 nm (end of process) [30] and (b) LG = 8 nm device with design space for scalability of SOI devices [29,30] (By courtesy of CEA, LETI.)

TABLE 9.1 Characteristic Scale Length Expressions for Various Thin-Film Device Architectures Calculated From 2D Poisson Equation (p. 60 in [5])

Device Architecture	Surface Conduction Scale Length	Volume Conduction Scale Length
FDSOI single gate	$\lambda = \sqrt{\frac{\varepsilon_{Si}}{\varepsilon_{OX}} t_{Si} t_{OX}}$	$\lambda = \sqrt{\frac{\varepsilon_{Si}}{\varepsilon_{OX}} t_{Si} \left(t_{OX} + \frac{\varepsilon_{OX}}{\varepsilon_{Si}} \frac{t_{Si}}{2} \right)}$
Double gate	$\lambda = \sqrt{\frac{\varepsilon_{Si}}{\varepsilon_{OX}} \frac{t_{Si}}{2} t_{OX}}$	$\lambda = \sqrt{\frac{\varepsilon_{Si}}{\varepsilon_{OX}} \frac{t_{Si}}{2} \left(t_{OX} + \frac{\varepsilon_{OX}}{\varepsilon_{Si}} \frac{t_{Si}}{4} \right)}$
Cylindrical channel		$\lambda = \sqrt{\frac{\varepsilon_{Si}}{\varepsilon_{OX}} \frac{t_{Si}}{4} \left(\frac{t_{Si}}{2} \ln\left(1 + \frac{2t_{OX}}{t_{Si}}\right) + \frac{\varepsilon_{OX}}{\varepsilon_{Si}} \frac{t_{Si}}{4} \right)}$

device architecture: subthreshold volume conduction is obtained for low-doped channels, while subthreshold surface conduction corresponds to heavily doped channels, where T_{Si}, T_{ox}, ε_{ox}, ε_{Si} are the silicon thickness, gate insulator thickness, gate insulator dielectric constant, and Si dielectric constant, respectively.

The threshold roll-off, the DIBL, and the subthreshold swing degradation scale are roughly given as $\exp(-L/(2\lambda))$ and the minimum channel length with acceptable short channel effects (DIBL < 100 mV/V) is approximately 5λ [5]. From Table 9.1, we can conclude the superior short channel immunity of cylindrical channel architectures, also named as gate all-around nanowires, when compared with all other architectures (as long as $t_{ox} < t_{Si}$).

Alternative Materials, Strained Silicon, or Architectures Based on Nanowires and Nanosheets

Research on device architecture usually brings questions of different nature:

- Is the semiconductor material used as channel or for active devices the right choice?
  Can we replace silicon by any other material?
- Can scaling meet the ultimate limit of devices based on single atoms or molecules or atomic sheets or atomic wires?
- How many electrons/charges can we use to transfer or store information reliably?
- Is the electron charge the everlasting state variable?

3D Materials. The choice of the active device material could increase the internal electric field at any junction necessary to obtain nonlinear characteristics used in present electron devices [5]. As a consequence, the transport properties of charge transfer devices can be degraded.

Generally speaking, the choice of a semiconducting material necessary to build a device cannot rely on a unique parameter. We need to look into a minimum set of parameters (Figure 9.13), which will make design, integration technology, test and commercialization possible [40]. Moreover, application-specific aspects should be taken into account in pervading applications such as analog, RF, high temperature, radiation hardening, etc. As an example, we cannot rely only on the value of low field mobility to evaluate device performance: the supply voltage conditions and the band structure of conducting channels will determine the electric field conditions. Consequently, the drift velocity might linearly depend on electric field or be equivalent to saturation velocity, depending on the applied voltages. For sub 100 nm channels, under 1.5 V supply voltage, saturation velocity v_{sat} rather than low field mobility μ_n (or μ_p) is the right feature to benchmark for carrier velocity estimation, due to the average electric field values ($>10^4$ V/cm) (Figure 9.13). The transport properties will also be evaluated using self-heating effects at high current density. So, the thermal conductivity of the material s_{th} should be seriously considered, especially at high operating temperatures. The number of carriers will as well depend on the intrinsic concentration n_i at high temperature and for nondoped materials. Finally, the short channel effects (charge sharing and DIBL) and parasitic capacitances will depend on the relative dielectric constant (Rel.K). As an example, Si has the lowest electron and low field mobility values. However, v_{sat} in Si is higher than in Ge or GaAs/InGaAs/InSb. Moreover, these materials have higher Rel.K than Si: they will have a weaker immunity to short channel effects (electrostatic integrity) and higher conducting channel parasitic capacitance. Consequently, these materials will show worse electron transport properties than Si, in a MOSFET operating in the saturation regime. If we consider the immunity to self-heating, Si is much better than the mentioned

Material	μ_n (cm²V⁻¹s⁻¹)	μ_p (cm²V⁻¹s⁻¹)	s_{th}(W/m/K)	Rel. K	Eg(eV)	v_{sat}(10⁷cm/s)	n_i(cm⁻³) (m*_e m*_h /m²)T^{3/2} exp(-Eg/2kT)
Si	1400	500	141	11.9	1.12	0.86	2x10¹⁰
Ge	3900	1900	59.9	16	0.66	0.60	2x10¹³
GaAs	8500	400	55	12.9	1.42	0.72	2.1x10⁶
InGa$_{0.47}$As$_{0.53}$	12 000	300	5	13.9	0.74	0.6	6X10¹¹
InSb	77000	850	1.8	16.9	0.17	5,0 @77K	2x10¹⁶
C-Diamond sp3	2200	1800	2000	5.7	5.47	2,7	10⁻²⁷
Graphene (CNT) sp2	10⁴-10⁵	10⁴-10⁵	1000	5.7	Semi-metal	4	*1x10¹²cm⁻²* *(1x10¹⁵)*

FIGURE 9.13 Comparison of the physical properties of column IV and compound III–V materials (in bulk and nanoobject versions for carbon-based materials).

materials. Among the mentioned III–V materials, only InSb has a v_{sat} (at 77 K) higher than Si. However, its thermal conductivity is much poorer than in Si, which handicaps InSb. Moreover, the ratio between low field mobilities μ_n/μ_p is widely in favor of silicon over all III–Vs and Ge: this gives a major advantage to Si for ratio CMOS designs. To this respect, carbon-based materials have a huge advantage because μ_n and μ_p values are very close or equal. These materials also have very high saturation velocities, much higher values of thermal conductivity (record for diamond), and twice smaller Rel.K than Si. So, these carbon-based materials would be serious competitors of Si. However, the extreme values of bandgap (5.47 eV for diamond and 0 for graphene) will impact their ability to supply carriers, in the first case, or to contain leakage currents, on the other extreme side. Furthermore, extrinsic doping and establish of contacts in these materials are still a question. As a first conclusion, silicon is by far a very good choice to design high-density circuits with very small features. Silicon is also a reliable material, because it is very easily oxidized: sub nm (few monolayers) very high-quality SiO_2 can be grown from a high-quality Si material on large 300 or 450 mm diameter wafers [41]. A huge amount of knowhow is now available on the SiO_2/Si interface quality and its major role played on the HiK/Si channel interface [5,40].

Effect of Strain on Transport Properties. The effects of strain on the band structure of semiconductors and transport properties of MOSFET have been studied extensively. Two ways have been adopted and should be distinguished: biaxial and uniaxial strains. In the case of biaxial strain, a global strain is applied by growing a thin epitaxial layer on top of a substrate that does not match perfectly by its crystal lattice constant. Depending on the difference of lattice parameter values between the bottom and upper layers, the strain induced in the epitaxial layer will be compressive (negative difference) or tensile (positive difference). They induce band offset and splitting, giving to charge carriers an access to lower effective mass subbands. Biaxially strained channel MOSFETs have been extensively studied on Si, pseudomorphic Si:Ge and

Si:Ge:C alloys [42–46] grown on bulk silicon that could also be bonded on SOI substrates. Strain induces degeneracy lift and band splitting. As an example, in the case of tensile strain obtained by a growth of Si:C on Si, electron mobility increases, thanks to the access of electrons to the $\Delta(2)$ valley (Figure 9.14). If $SiGe_x$ is grown on a Si substrate, then compressive strain is obtained by the contribution of low effective mass holes valley to transport. Unfortunately, high transverse electric fields in the channel tend to compensate the effect of strain in small transistors. At the micro/nanometric level, a MOSFET is strain added by the surrounding thin films which serve its isolation (*Inter-Level Dielectric* (ILD), gate stack, etc.) or process integration (spacers, contact etch stop layer (CESL), source and drain metallization, contacts structure, etc.) (Figure 9.15a,b). Local strain can compensate the loss of mobility in channels smaller than 100 nm, and still effective at the sub 20 nm level [21,28,46] (Figures 9.9b and 9.15a,b). Strain has a huge effect on ultranarrow and short device geometries, such as FinFET or SOI nanowire-like based devices [46–48]: the strain can be reversed from large to small depending on the environment. Strategies have been developed to selectively strain n channel and p channel devices and end up with n channel and p channel devices delivering the same current: such a feature has a dramatic impact on the design layout optimization [47,49–51] (Figure 9.16b).

Nanowires and Nanosheets 2D Materials. Scaling of channel volume by thinning the channel area (FinFET or FDSOI) poses the question of ultimate thickness of such a channel to one atomic or molecular layer. The shift to Gate All Around (GAA) nanowires can relax the constraint on the channel geometry by a factor of 2, whilst maintaining the same electrostatic integrity. On the other side, their scaling to the size of atomic strings could become a concern. In all cases, the semiconductor cannot be considered as a 3D isotropic material anymore. In the first case at the extreme limit, we would need to consider the material as a 2D media. A nanowire would ultimately have to be investigated as a 1D material.

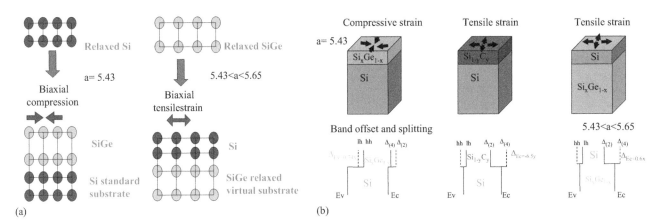

FIGURE 9.14 Strain induced by epitaxial growth: (a) examples of compressive and tensile strain obtained by growth on a relaxed substrate and (b) degeneracy lifting by strain and splitting of bands into light and heavy carrier subbands. (By courtesy of CEA, LETI.)

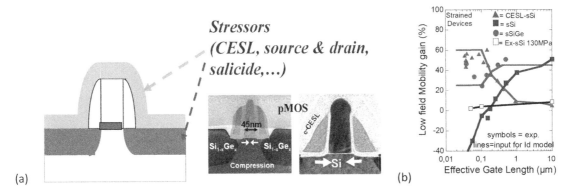

FIGURE 9.15 Local strain on MOSFETs: (a) examples of the way local strain is applied: CESL: contact etch stop layer and (b) compensation of mobility gain loss with scaling by local strain on p-MOSFET strained channels. (By courtesy of CEA, LETI.)

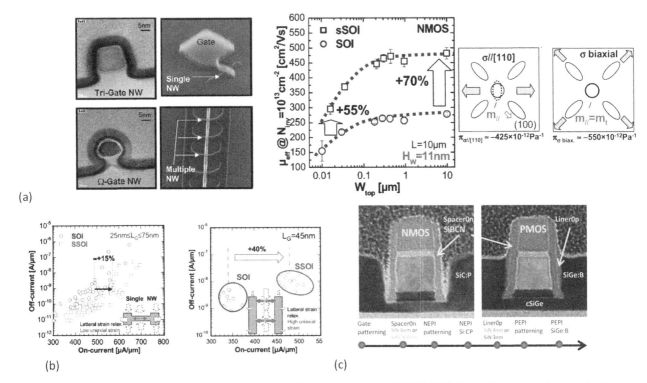

FIGURE 9.16 (a) nMOSFET mobility improves with global tensile strained SOI (sSOI), but enhancement is degraded in narrow active areas, still significant due to local uniaxial tensile strain, (b) enhancement of transport properties depends on layout density and gate length, and (c) effectiveness of strain depends on the combination of global biaxial and local uniaxial contributions on transistor channel. (By courtesy of CEA, LETI.)

Nanowires: The main questions about nanowires turn around their process integration and variability. Nanowires can be achieved in various ways, either by a natural material growth/deposition (bottom-up approach) [40,52,53] (Figure 9.17a) or by using a finely processed initial substrate (top-down approach) (Figure 9.17b,c) [40,54]. As already mentioned in Section 9.1.3.1.3.2, degradation of transport properties can be observed in narrow devices: the loss of mobility can be compensated by wrapping the channel by a straining layer [55] (Figure 9.17b). By using a Si/SiGe superlattice initial substrate, defect-free stacked nanowires can be obtained [30,54,56]: transistor layout can be much more flexible than with FinFET and can cope with digital as well as analog design requirements [56]. Large current values per footprint can be reached by stacking nanowires [56]. In the 'bottom-up' case, their placement and alignment to underlying patterns is the biggest challenge. Many attempts have been carried out to deposit and place/position carbon nanotubes (CNTs) [57–59] (Figure 9.17d), for example, by selective growth on gold, copper, ferritin, or iron catalysts, provided that single-walled and multiwalled nanotubes have been selected. However, no trivial and efficient solution has been proposed yet, which would be eager to build a chip with a large number of components [60] in a reproducible way on large Si initially processed wafers. The most critical issue comes from establishing ohmic contacts to the

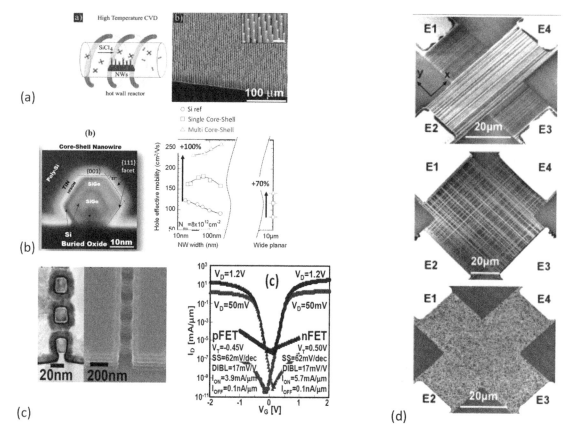

FIGURE 9.17   (a) Bottom-up approach [53], (b) core–shell structure [55], (c) stacked nanowires by top-down approach [54,56], and (d) printed single-walled carbon nanotubes in different layouts: aligned arrays (top), cross bars (middle), random networks (bottom) [57].

CNTs. Furthermore, selecting their size and scaling them would request for a GAA transistor architecture. Intrinsic cutoff frequencies of 153 GHz [61] have been obtained on 100 nm gate length MOSFETs. Other devices and applications such as nanoelectromechanical systems (NEMS), near field Atomic Force Microscopy (AFM) Ntips, field emission electrodes, integrated electronic devices on flexible substrates, hydrogen storage, and composites have been demonstrated [5,60].

Rather than trying to position CNTs on a chip, the integration of graphene, based on the same carbon molecule, that demonstrates 2D transport properties is another alternative. As a matter of fact, graphene [62,63] or CNTs demonstrate very high values of low field mobilities for both electrons and holes. However, the measurement and de-embedding of high-frequency figures of merit, on MOSFETs smaller than 100 nm gate lengths, do not demonstrate superior values of cutoff frequencies but rather poor maximum oscillation frequencies [64]. The low value of drain conductance translates into lack of saturation of the current voltage characteristics and explains poor maximum oscillation frequencies. On another side, their low bandgap values, which can be modulated by the channel width, do not give them any advantage, for low-power applications, compared with silicon [64].

2D Materials, Nanosheets: Under the light of different debates and choices on transistor architectures, the question

of ultimately scaled thin channels received much interest. There has been a great emphasis on single atomic sheet materials showing 2D conduction. Many candidates have been studied and still are in the literature, such as hexagonally arranged 2D atomic thick sheets [40,62,63]: graphene, hexagonal Boron Nitride (hBN), silicene, germanene, layered oxides and metal di- and trichalcogenides received much attention, i.e., MoS_2 [65], $MoSe_2$, WSe_2, WS_2, Bi_2Se_3, Bi_2Te_3, etc. [66]. The last category in the list attracts interest because their bandgap is much higher than the bandgap of graphene and possibly qualifies them to be used for low-power applications, as being much closer to silicon bandgap. Silicene [67] and germanene [68] are also studied but suffer from chemical stability. This topic will be reviewed more precisely in Section 9.1.5.

Scaling Supply Voltage to Sub 0.5 V: Opportunity for Tunneling Field Effect Transistors

Drastic supply voltage scaling will be necessary with dimensional scaling to the sub 5 nm scale of logic and memory devices to meet the energy efficiency challenges at the system level. At the same time, new device architecture and a huge effort in variability efficiency will be requested to make it viable.

The most promising candidate compatible with CMOS architecture is the tunnel field effect transistor (TFET). The transport mechanisms on which TFETs are based is

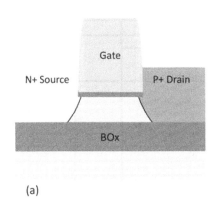
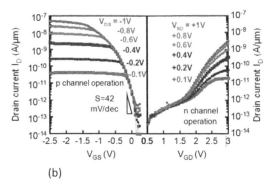

(a) (b)

FIGURE 9.18 The TFET: (a) an SOI-based structure, (b) subthreshold slope lower than 60 mV/dec is demonstrated on SOI pTFET cointegrated with FDSOI CMOS. The GeOI pTFET gives I_{on} increased by 2,700 compared with the SOI version. (By courtesy of CEA, LETI, and modified from [70].)

direct band-to-band tunneling (DBTBT): its evidence will request thin-film body channels free of any bulk leakage currents originating from short channel effects [69]. The high electrostatic integrity request for TFET achievement appeals strongly to the use of GAA nanowires or at least FD channel architectures: in this frame, $SiGe_xOI$-based TFETs demonstrated high I_{on}/I_{off} capabilities [70] with a less than 60 mV/dec subthreshold swing. (Figure 9.18). Low effective mass materials are theoretically the best candidates to increase the output on current, thanks to the higher tunneling probability of charge carriers. That is why III–V materials have been given much attention [71]: TFET switches demonstrating 21 mV/dec subthreshold swinging at room temperature at low Vg have been reported. Thanks to the DBTBT governing the transport mechanism, TFET-based architectures could potentially beat CMOS at supply voltages [70,72] lower than 0.5 V. Their use in a circuit design will request a low process variability. Actually, the operation under DBTBT makes TFETs much more sensitive than CMOS to device geometric variations, and the existence of possible extrinsic charge originate from interface states, traps, defects, environment-related stress, or induced roughness. These latter perturbations do impact the output current and degrade the steepness of their Id(Vg) characteristics [71,73], especially under low supply voltages. The necessity to use multigated nanowires to make a TFET channel increases the strength of the low variability requirement.

9.1.4 Memory Scaling

By the end of 1960s, charge-based memory architectures were proposed either using floating gate, discrete trap sites in an insulator or a capacitor. In the following, we will exclude the read only memories (ROM) or onetime programmable ROMs (OTP) that follow the scaling rules either of CMOS or NVMs, respectively.

Conventional Scaling Hits the Limit

In all cases of today's flash NVMs, one transistor is enough to define a memory cell: the feature size of $4–10F^2$ is the main difference when compared with SRAM or DRAM sizes that stand at $150F^2$ and $12–30F^2$, respectively (Figure 9.19).

The scaling of SRAM follows the rules of CMOS, as they come naturally with the CMOS process. One major hurdle to SRAM scaling is its static noise margin, which becomes harder because input capacitance values are reduced.

In the case of DRAM, scaling challenges come from the access transistor leakage clamping and capacitor footprint reduction [13], while maintaining a minimum charge to make sure enough sensing signal margin and data retention time can be obtained. The access transistor is of the LSTP type. Practically, the sensing signal should have the minimum value of

$$\frac{C_s}{CBL + C_s}\frac{VCC}{2} \geq 150\,mV$$

C_s is the storage capacitance, CBL is the bit line parasitic capacitance, and VCC is the memory array voltage. C_s is consequently maintained at values higher than 25fF/cell. As a consequence, the capacitor footprint scaling imposes to increase its aspect ratio or introduce a new dielectric of higher dielectric constant. Introducing a vertical transistor under the capacitor [74], requests to revise process integration which consequently and reduces the footprint to $4F^2$ (Figure 9.20).

In the case of NVM, the stress-induced leakage through the transistor tunnel dielectric is responsible for data retention and endurance degradation. First of all, in the case of a polysilicon floating gate, a coupling dielectric has been introduced between the control gate and floating gate: an oxide(SiO_2) nitride(S_3N_4) oxide(SiO_2) stack, so-called Oxide Nitride Oxide (ONO), is used for such a purpose (Figure 9.21a). In order to keep enough margins to soft read and soft programming immunity, whilst injecting enough tunnel current in the floating gate, an optimal capacitive coupling ratio is defined:

$$\gamma = \frac{CONO}{C_{tunnel} + CONO} \approx 0.5\,to\,0.6$$

Until the 1980s, UV-erasable NVMs could not face the challenge of high-density massive storage that could be achieved only by hard disk drives (HDD). A much faster solution

		Baseline Technologies					Prototypical technologies [A]		
		DRAM		SRAM [C]	Flash		FeRAM	STT-MRAM	PCM
		Stand-alone [A]	Embedded [C]		NOR Embedded [C]	NAND Stand-alone [A]			
Storage Mechanism		Charge on a capacitor		Inter-locked state of logic gates	Charge trapped in floating gate or in gate insulator		Remnant polarization on a ferroelectric capacitor	Magnetization of ferromagnetic layer	Reversibly changing amorphous and crystalline phases
Cell Elements		1T1C		6T	1T		1T1C	1(2)T1R	1T(D)1R
Feature size F, nm	2013	36	65	45	45	16	180	65	45
	2026	9	20	10	25	>10	65	16	8
Cell Area	2013	6F²	(12-30)F²	140 F²	10 F²	4 F²	22F²	20F²	4F²
	2026	4F²	(12-50)F²	140 F²	10 F²	4 F²	12F²	8F²	4F²
Read Time	2013	<10 ns	2 ns	0.2 ns	15 ns	0.1ms	40 ns [G]	35 ns [J]	12 ns [K]
	2026	<10 ns	1 ns	70 ps	8 ns	0.1ms	<20 ns [H]	<10 ns	< 10 ns
W/E Time	2013	<10 ns	2 ns	0.2 ns	1µs/10ms	1/0.1 ms	65 ns [G]	35 ns [J]	100 ns [K]
	2026	<10 ns	1 ns	70 ps	1µs/10ms	1/0.1 ms	<10 ns[H]	<1 ns	<50 ns
Retention Time	2013	64 ms	4 ms	[D]	10 y	10 y	10 y	>10 y	>10 y
	2026	64 ms	1 ms	[D]	10 y	10 y	10 y	>10 y	>10 y
Write Cycles	2013	>1E16	>1E16	>1E16	1E5	1E5	1E14	>1E12	1E9
	2026	>1E16	>1E16	>1E16	1E5	1E5	>1E15	>1E15	1E9
Write Operating Voltage (V)	2013	2,5	2,5	1	8-10	15-20	1.3-3.3	1,8	3 [K]
	2026	1,5	1,5	0,7	8	15	0.7-1.5	<1	<3
Read Operating Voltage (V)	2013	1,8	1,7	1	4,5	4,5	1.3-3.3	1,8	1,2
	2026	1,5	1,5	0,7	4,5	4,5	0.7-1.5	<1	<1
Array Write Energy (J/bit)	2013	1,00E-13	1,00E-13	unavailable	2,00E-10	1,00E-11	unavailable	unavailable	unavailable
	2026	1,00E-13	1,00E-13	unavailable	2,00E-10	1,00E-12	unavailable	unavailable	unavailable
Single Cell Write Energy (J/bit)	2013	4E-15 [B]	5,00E-15	5,00E-16	1E-10 [E]	4E-16 [F]	3E-14 [I]	2.5E-12 [A]	6E-12 [L]
	2026	2E-15 [B]	2,00E-15	3,00E-17	1E-10 [E]	1E-16 [F]	7E-15 [I]	1.5E-13 [A]	~1E-15 [M]

FIGURE 9.19 Current baseline and prototypical memory technologies: ITRS 2013 status and 2026 projections. (By courtesy of [4].)

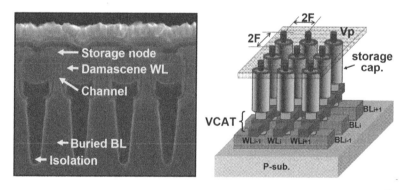

FIGURE 9.20 SEM cross section of vertical channel access transistor and a schematic diagram of 4F² DRAM cell array. (Modified from [74].)

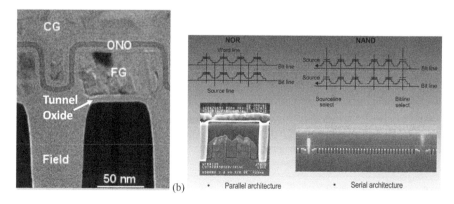

FIGURE 9.21 Floating gate flash memories: (a) Transmission Electron Microscopy (TEM) cross section of 60 nm NAND flash memory floating gate cell along word line and (b) TEM cross section of NOR and NAND flash memory arrays along bit line.

was needed to achieve electrically erasable programmable ROMs (EEPROM). Flash memories were proposed in 1984 by Toshiba [75] to face the challenge of massive data storage; erasing was achieved collectively by memory sectors, while keeping a Not-OR(NOR) type circuit architecture requesting bit-by-bit access and programming. Automatically, each erasing process would need reprogramming by sectors. Due to the higher density obtained and low write energy per bit, speed was not an issue for the user.

Several types of flash memory cells are used today, depending mainly on the application and needed capacity for the memory array. (i) NOR gates use hot electron injection in bit–by-bit programming/erasing and reading; (ii) NAND-type cells using Fowler Nordheim tunnel injection and serial reading. The NAND structure was introduced by Toshiba in 1987 [76]. In the NOR type, each cell is contacted to a bit line and a word line. In the NAND type, reading is achieved by sectors: there is no need to contact each cell individually. NOR flash cells have thus a larger footprint than NAND flash memory cells (Figure 9.21b).

The scaling of NAND flash NVM devices needs to face the immunity to severe short channel effects, because of the thick tunnel oxide being a hurdle to scale down the floating gate transistor. Scaling down its thickness will as well request to scale the ONO dielectric to keep constant capacitive coupling ratio and will also generate higher SILC, whenever the tunnel oxide thickness reaches 4–5 nm. In a dense array, at the sub 40 nm level, the ONO thickness becomes comparable to the bit line-to-bit line spacing and the cell-to-cell parasitic coupling increases. Different alternative paradigms have been proposed to correct this limitation to continue the increase of NVM capacity.

Nanofloating Gates to Help Conventional NVM Scaling?

Reducing the probability to lose carriers from the floating gate after defect creation by the already mentioned SILC can be achieved by reducing the floating gate dimension or by band engineering of the storage sites. Silicon nanocrystal floating gates (Si NC) were proposed [5,16,77,78] (Figure 9.22a) to reduce the SILC issue and revamp the

scaling capabilities. However, the reduction of SILC was accompanied by the reduction of CTunnel, which increases γ that could be compensated by an increase of CONO. In parallel, the CONO needs to be reduced from the floating gate sidewall coupling for integration purposes. The replacement of ONO by a high K stack (with K values higher than $K_{Si_3N_4} = 7$) was a possible solution (Figure 9.22b). Combined to Si NC, this solution seemed to be a good way to continue the scaling of floating gate architectures. However, the placement of Si NC in a deterministic way is an issue whenever <22 nm devices need to be integrated. Practically, the scaling of floating gate devices are limited by the reduction of the number of stored electrons whenever the device geometry is reduced: the dispersion of retention time can reach a factor of 2 with a number of stored electrons of 100, which is the case at the sub 22 nm level.

As shown in Figure 9.4a (Section 9.1.2.2), NVMs have become a major technological driving force in today's market. The increase in mass storage needs in the different market sectors, including professional, automotive, and consumer electronic applications (digital cameras, smartphones, automotive, etc.), has drastically boosted the development of NVMs. Different needs for stand-alone high capacity and embedded memories for system on chip (SOC) come along the way and thus the different specifications in terms of features and cost.

3D Integration for Mass Storage

For mass storage, 3D integration [79–81] is already an adopted approach that avoids using linear scaling while increasing memory capacity. Band engineering was necessary to reintroduce a direct Silicon-Oxide-Nitride-Oxide-Silicon (SONOS) structure with a p+ control gate, thanks to heavily doped polysilicon or metal (Tantalum Nitride Oxide Silicon - TaNOS) to ensure sufficient data retention, either in a vertical channel scheme [79], stacked SOI [80], or Thin Film Transistors (TFT) [81]. The vertical transistor architecture can easily integrate a gate all-around structure that improves the short channel effect immunity of NVM.

The solution proposed in [79] has been applied to mass storage NVM: up to 512 GB with 3 bit/cell has

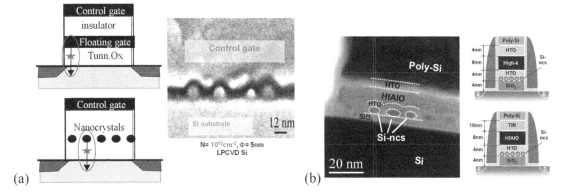

FIGURE 9.22 (a) SILC limits floating gate NVM scaling. Nanocrystals act as nanofloating gates to allow further scaling and (b) Si NCs and HiK coupling insulators increase coupling ratio in NVM cell. (Modified from [78] and by courtesy of CEA, LETI.)

been demonstrated [82]. Another approach by stacking partitioned 2D design has demonstrated up to 1 TB with 4 bit/cell [83]. In each case, specific design architecture and process steps are needed, but limit their integration as embedded in CMOS SOC circuits.

Alternative Architectures to Floating Gate Cells

Resistive devices such as resistive RAMs (ReRAMs) or phase-change memories (PCMs) are often mentioned as good candidates for embedded solutions, because they require materials already available in most fabs and, thanks to the recent progress, as well for massive storage applications (Figure 9.23). Usually, they request an access device and a storage node, being a resistor that can take two highly contrasted values defining two states (Figure 9.24).

Because of their low voltage-switching capabilities, these devices can be mixed and cointegrated with logic CMOS

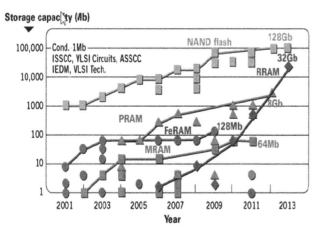

FIGURE 9.23 Evolution of different types of flash memory ReRAMs. have a great potential due to the availability for process integration. (By courtesy of Yole Dev.)

beyond data storage purposes [84–86]. They enable new architectures for drastic reduction of power consumption, latency, and to design reconfigurable, programmable, or neuromorphic architecture circuits [84–86]. Using such memories, technologies in combination with logic might request zero intrinsic variability solutions that conduct filaments (metallic or oxygen vacancies) [84–86] or deterministically placed crystalline clusters [87] that might be paving the way to.

They show high-resistance contrast states that might define at least two states and sometimes multilevels. In PCMs, chalcogenide materials, such as $GeSe_xTe_y$ [88], can be switched from an amorphous (high resistivity) to a crystalline (low resistivity) state, by thermally heating the material sandwiched between two electrodes. Thermal cross talk and stability during the packaging process as well as power consumption are often mentioned as major challenges for the integration of these materials: N and C doping improve the thermal stability [89]. Memory multilevels can be defined by gradually varying the set current elevating the sandwiched material to temperatures higher than the recrystallization temperature. High endurance ($>10^8$ cycles) can be obtained on carbon-doped PCMs, possibly qualifying them also as embedded NVM type [90].

Magnetic RAMs and, more recently, STTRAM use spin-polarized current flow through a tunnel barrier sandwiched between two magnetic films (examples in [91–94])—also called spin valve—being a pinned or free layer, which have parallel and antiparallel magnetizations. When the electrons flow through the pinned layer, the minority electrons get scattered and the majority electrons pass through to the free layer. When the polarized electrons reach the free layer, the spin angular momentum exerts a torque on the magnetization of the layer that is oriented antiparallel to the pinned layer. Reading the memory state is achieved through the giant magnetoresistive resistance effect [95].

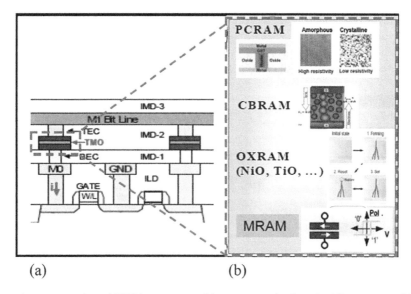

(a)   (b)

FIGURE 9.24 One transistor one resistor (1T1R) memory architectures are back end of line integratable. (By courtesy of CEA, LETI.)

The relative spin orientations of both layers will define two different states (ON and OFF). The high endurance (>10^{12} cycles) and very low access time obtained with STTRAM can still not qualify the memory architecture to replace DRAM, SRAM, or NVM due to the $9F^2$ footprint [96], and to the lack of memory window for sufficient read margin, sensitivity to noise, and write error rate. Power consumption, thermal stability, and cost are still showstoppers to replace established SRAM and NVM. A 4 GB STTRAM has been demonstrated by using a perpendicular magnetic tunnel junction architecture [96]. Magnetic tunnel junctions have been proposed to design 3D serial-type domain wall memory (DWM), the so-called Racetrack [97], as the universal memory and, consequently, a direct competitor to flash NVM and DRAM. The key difference between DWM and other NVMs (STTRAM, ReRAM, and PCM) is that accessing DWM requires shifting operation since many magnetic domains share one read/write port for achieving high density. By comparison, in other NVMs, each storage

element has its own access path, and thus, an entire block (e.g., 64 B) can be simultaneously accessed. Still, it is much slower than volatile memories (SRAM and DRAM), even though no or negligible standby leakage is expected. Several teams worldwide are carrying out efforts to mix tunnel junctions with CMOS ([98], see also pp. 389–436 in [28]), thanks to their compatibility in terms of cointegration for Field Programmable Gate Arrays (FPGA), analog, and neuromorphic applications.

Many materials show resistance changes either by filament formation by electrochemical, valence change, and thermochemical effects or contact electrode interface change (Table 9.2). Their switching and conduction mechanisms are of unipolar or bipolar type (Figure 9.25). Due to their potential low cost, the most popular studied architectures are based on metallic migration (conductive bridge RAM, CBRAM) or oxygen vacancy diffusion (oxygen vacancy RAM, OxRAM) from one of the two electrodes (Figures 9.24 and 9.26). These devices show memristor-like characteristics

TABLE 9.2 Electrical Performance of Different Types of Memristors

		A. Electrochemical Metallization Bridge	B. Metal Oxide: Bipolar Filament	C. Metal Oxide: Unipolar Filament	D. Metal Oxide Bipolar Interface Effects
Storage mechanism		Electrochemical filament formation	Valence change filament formation	Thermochemical effect filament formation	Change in interface
Feature size F	Best projected	<5 nm	1 nm	<30 nm	<10 nm
	Demonstrated	<10 nm (CuGeSe)	<3 nm (HfO$_x$)	30 nm	40 nm
Cell area (2D)	Best projected	$4F^2$	$4F^2$	$4F^2$	$4F^2$
	Demonstrated	$4F^2$	$4F^2$	$4F^2$	$4F^2$
Read current	Best projected	Not available	Not available	Not available	Not available
	Demonstrated	Not available	~100 pA/~10 nA	1 nA/20 nA	Not available
Write/Erase time	Best projected	<1 ns	<100 ps	Not available	10 ns
	Demonstrated	<1 ns	100 ps	10 ns (W), 5 ns (E)	<100 ns
Retention time	Best projected	>10 years	>10^5 years @ 85°C	>10 years	>10 years
	Demonstrated	1,000 h @ 200°C or 1 h @ 400°C	55.55 h @ 300 °C	1,008 h @ 150°C	~3 h @ 200°C
Write cycles	Best projected	>10^{10}	10^{12}	>10^7	>10^7
	Demonstrated	10^{10}	10^{12}	10^7	10^7
Write operating voltage (V)	Best projected	<0.15 V	<1 V	2 V	Not available
	Demonstrated	<0.2 V	1–3 V	1 V (write)/1.5 V (erase)	2
Read operating voltage (V)	Best projected	<0.02 V	0.1 V	0.3 V	0.1
	Demonstrated	0.02 V	0.1–0.2 V	0.3 V	0.5
Write energy (J/bit)	Best projected	<200 fJ	<200 fJ	<100 pJ	1 fJ
	Demonstrated	200 fJ	230 fJ	100 pJ	1.23 fJ

Source: By courtesy of [4].

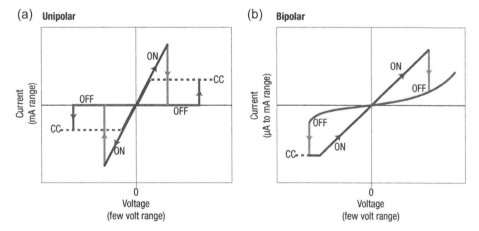

FIGURE 9.25 Two basic operation modes of memristive devices. (a) Unipolar (or nonpolar) and (b) bipolar. CC denotes compliance current controlled by an external circuit. (By courtesy of [4].)

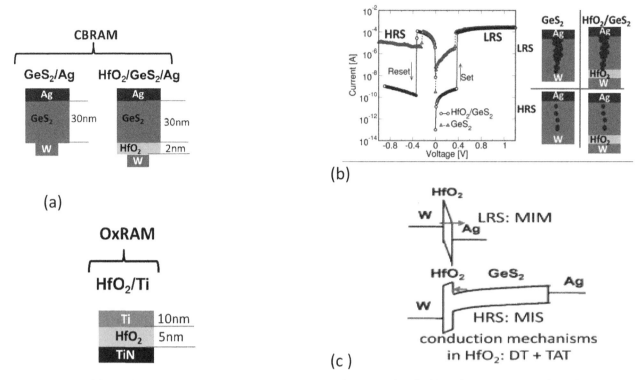

FIGURE 9.26 (a) ReRAM's integrate 1T1R. Two types are described: CBRAM Ag/GeS₂/W with self-rectifying heterostructure Ag/GeS₂/HfO₂/W and OxRAM Ti/HfO₂/TiN, (b) reduction of leakage current in SET and RESET states by insertion of HfO₂ on I(V) characteristics of CBRAM. Phenomenological model and (c) band structure, showing the behavior of OxRAM as a Metal Insulator Metal (MIM) structure in the low-resistance state and CBRAM as an MIS heterostructure in the high-resistance state. (By courtesy of CEA, LETI.)

[99] of the bipolar type and can switch at very low voltages (Table 9.2 and Figure 9.25). Data retention compatible with nonvolatile applications ($>10^4$ s at 200°C) and endurance cycles higher than 10^{11}, compatible with 30 ns switching time and programming window $R_{off}/R_{on} >10^3$, have been obtained on OxRAM [100] with W/Al₂O₃/TaOₓ/doped ZrOₓ/Ru resistors. In this stack, TaOₓ acts as an oxygen vacancy supplier, ZrOₓ is an oxygen-rich exchange layer and Al₂O₃ a reaction retardation layer between the bottom electrode and the vacancy-rich layer. CBRAM devices use Cu or Ag electrodes to generate an ionic current into the electrolyte sandwiched in between two electrodes when switched on: the minimum R_{on} is much lower than for an OXRAM and the programming window lower. However, the CBRAM system Ag/GeS/W, in which Ag+ ions flow into GeS, demonstrates a much higher programming window by insertion of a HfO₂ layer (Figure 9.26). HfO₂ inserts a potential barrier in between GeS and W [84]. The Ag/GeS/HfO₂/W stack behaves as a Metal Insulator Semiconductor (MIS) heterostructure and demonstrates a much higher R_{off}/R_{on} ratio ($>10^6$), even though a lower endurance than for the OXRAM Ti/HfO₂/TiN Ref. [85]. In the case of OxRAM, Ti acts as an oxygen-gettering layer that strongly reacts with HfO₂ and generates oxygen vacancies at the contact with HfO₂ [101] (Figure 9.27a). The low SET and RESET modes voltages are compatible with the low power supply of

CMOS, but might represent an issue for disturb immunity [102,103]. In the same spirit as for CBRAMs, self-rectifying cells have been attempted on OxRAMs using a TaOₓ/ZrOₓ barrier [100], which acts as a nonlinear element stacked, without the need to add a transistor for the select operation. Other examples have also demonstrated the interest of a self-rectifying heterostructure inside the RRAM stack (PrₓCa₁₋ₓ,MnO₃, and TiO₂/Ta₂O₅), thanks to high nonlinearity in the on state of bipolar devices. TiN/Hf/HfOₓ/TiN OxRAMs have been identified with a strong scaling potential to less than 1 nm feature [104,105].

RRAMs can offer new possibilities to be embedded or distributed in CMOS [84,102] (Figure 9.28a), competing directly with split gate architectures [106]. Their codesign with CMOS offers low-cost opportunities to implement nonvolatile SRAM without any battery backup, reduces CMOS power consumption by intermittent data storage (latency reduction and "close to data" computing), and neuromorphic [85,86] (Figure 9.28b) and programmable circuits such as FPGA [30,84].

Memories need to be essentially defined by a storage device associated to an access (or select) device. Since the late 1960s, we have been lucky to be able to scale a floating gate transistor in which the select device and storage node are recently stacked into NAND flash transistors having a 4F² footprint. With the paradigm shift in

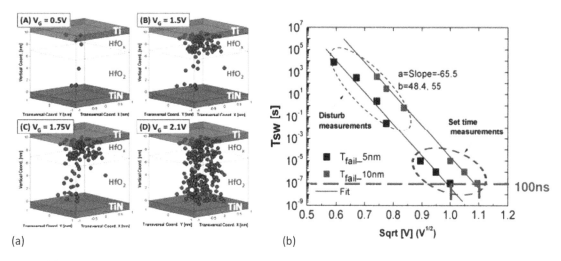

(a)   (b)

FIGURE 9.27 (a) The different stages of 3D defect distribution (O vacancies) in TiN/HfO$_2$/Ti-resistive and switch-resistive RAM. (A) Preexisting defect distribution. (B) Defects are initially generated only in the HfO$_x$ layer close to the Ti layer. (C) After complete breakdown of the HfO$_x$ layer, defects are generated in the stoichiometric portion of the oxide. (d) Full conductive filament is formed. (Modified from [101].) (b) Low voltage switching offers advantageous compatibility with logic devices cointegration; however, it might reduce programming window and affect read disturb (TiN/HFO$_2$ 5 and 10 nm/Ti). (Modified from [102,103], by courtesy of CEA, LETI.)

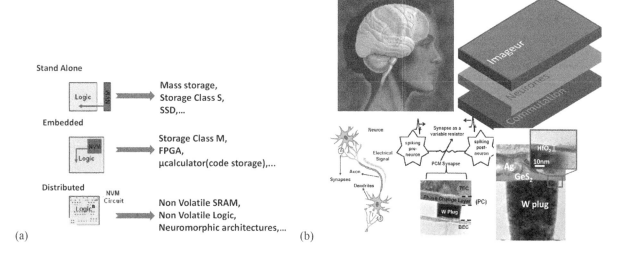

(a)   (b)

FIGURE 9.28 (a) Embedded nonvolatile resistive memories (ReRAMs) to be cointegrated with logic, beyond data storage purposes, to reduce power consumption, latency, and design reconfigurable programmable architectures. (b) Illustration of biological synapse and concept using ReRAM as a synapse in neural circuits, either by Phase Change RAM (PCRAM) or CBRAM (Ag/GeS$_2$) as synapse for ultradense neuromorphic systems: applications to complex visual (retina) and auditory (cochlea) cognitive processing applications. (By courtesy of CEA, LETI.)

the introduction of RRAMs, it was challenging to reduce a 2D 1T1R to the same footprint as NAND Flash. ReRAM indeed has a great potential to be scaled into crossbar arrays, including the resistor element and the selector provided that enough signal can be supplied by the selector element. Many research projects have been and are still being carried out worldwide exploring the different alternatives summarized in Figure 9.29a,b [4]. The crossbar structure (Figure 9.29c) has been designated in the past [4] as having the highest integration density and a first approach to stackable 3D layout [107]. Such an approach is efficient to stack the RRAM with its selector, which would be the best suited to

the application. ReRAM has gone out of infancy, and scaled capacity chips as high as 32 GB have been demonstrated (the memory cross point cell included the ReRAM and a diode selector) [108].

9.1.5 Towards Zero Intrinsic Variability through New Fabrication Paradigms

At the sub 8 nm level, low dimensionality and ultimate effects need to be considered. As a matter of fact, the number of electrons required to switch a MOS transistor or program a charge-based memory cell will be in the

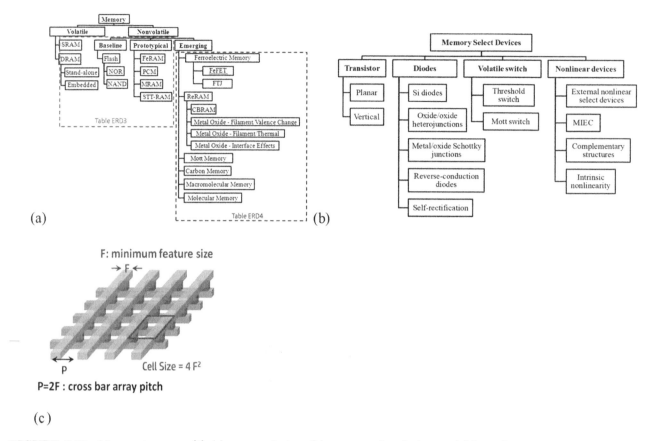

(a)   (b)

(c)

FIGURE 9.29   Memory taxonomy [4]: (a) memory devices, (b) memory select devices, and (c) crossbar structure.

range of ten to less (see Figure 9.5 in Section 9.1.2.3). This limitation is one more reason why variability treatment should be revised. In this range, the device size approaches molecular dimension: quantum confinement and stochastic phenomena need to be taken into account much more at first order. Few to single electrons phenomena, such as Coulomb blockade, appear at temperatures close to room temperature in MOSFETs. The resulting large channel conductance fluctuations can be exploited for multivalued logic applications [109]. The selection of electronic carriers (electrons or holes) spin has been suggested to build quantum computing circuits [110]. A single electron transistor can, as well, serve as a highly resolvant ionized single dopant sensor (Figure 9.30) [109,111]. The doping technique adopted to adjust the threshold voltage of CMOS transistors is ion implantation: classically, precisely selected ionized impurities are randomly distributed by an accelerated ion beam into a target, which is the material to dope. Stochastic dopant diffusion may occur in the intentionally undoped channel of a MOSFET from heavily doped source and drain extensions. At a tunneling distance from the doped layer, trap-assisted tunneling [73] could be at the origin of conduction paths. The stochastic nature of the dopant distribution represents a major source of variability for these devices. Deterministic doping techniques have been proposed by several research teams [112–114] (see also p. 3 in [40]) aiming at the control and placement of single dopant atoms (Figure 9.31). These techniques draw the

pathway to obtain the zero intrinsic variability of the number of dopants. Novel functions, classically requesting several devices, could be designed in one single device: a single electron pump circuit has been evidenced to be functional by the presence of two active dopants in a MOSFET channel [115,30]. At the same level of scaling, resistive memories will suffer from the variability of conduction paths formation, whether they are formed through oxygen vacancies or conductive bridges. Deterministic crystalline clusters placement [87] (Figure 9.31d), in materials such as chalcogenides, has been suggested to approach the near zero variability or deterministic position of conduction paths. A process integration approach has proven to reduce bit error rate in high- and low-resistance states, thanks to high-dose Si-localized implantation, in HfO_2-based RRAM [116].

Reaching the molecular size as the distance between electrodes or the volume of active body gives emphasis to 1D or 2D materials. The integration of 2D materials is envisaged as a response to the quantum confinement issue, resulting from the body thickness scaling due to steep increase of 3D material bandgap. Single atomic sheet-based materials generally demonstrate 2D transport properties. One atom-thick, hexagonally arranged 2D sheets are good candidates for new applications, thanks to their optical and mechanical properties [117,118] (Figure 9.32a). Typical examples are graphene, hBN, silicene, germanene, layered oxides and chalcogenides (MoS_2, WSe_2, Bi_2Se_3, Bi_2Te_3), and materials made from isolated atomic planes that result in so-called

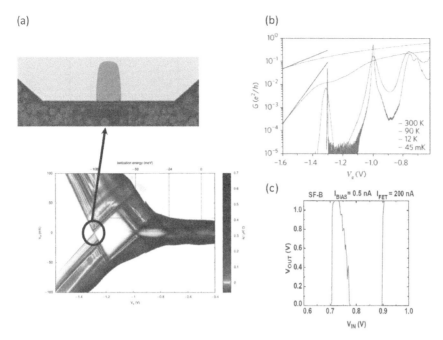

FIGURE 9.30 FETs/SOI evidence single electron on single atom interaction. (a) Application to detect single dopant in small MOSFETs, (b) tunnel barriers demonstrate Coulomb blockade conductance fluctuations [111]. Zero intrinsic variability and new applications [111] will emerge, and (c) multilevel logic architectures are possible by combination with CMOS [109,111]. (By courtesy of CEA, LETI, modified from [30].)

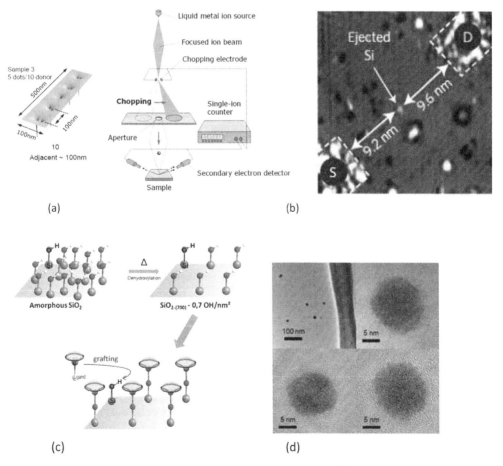

FIGURE 9.31 Zero intrinsic variability deterministic doping can be approached by (a) single ion implantation, (b) Scanning Tunneling Microscope (STM) based lithography and CVD, (c) chemical grafting, and (d) control of nanocluster placement in chalcogenides (GeSe). (By courtesy of CEA, LETI, adapted from [40].)

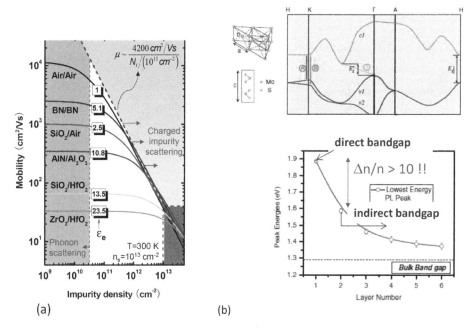

FIGURE 9.32 Use of 2D material properties request zero intrinsic variability: (a) transport properties are very sensitive to substrate and top layer encapsulation. Case of MoS_2 [118]: (b) band structure and optical and electronic properties strongly depend on the precise number of atomic monolayers [120]. (Modified from [21], by courtesy [118,119].)

van der Walls heterostructures [66] The strong dependence of their physical properties on the number of monolayers [118,119] appeals to control zero intrinsic variability in their fabrication, to keep the remarkable properties of these materials (Figure 9.32b).

In logic devices, doping by ion implantation for V_T adjustment has been suppressed with the adoption of FD channels, but is still used to dope source and drain extensions. We already mentioned (Section 9.1.3.1.2 and Figure 9.30a–c) that dopants could stochastically diffuse to the channel area from the source and drain extensions. The extension doping profile inside the channel will be added by a pseudoroughness located at the edge of the

extensions. The resulting localized states [30] can act as centers for assisted tunneling or as barriers. For sub 10 nm channel devices, such an issue is a source of variability, if it is not controlled deterministically. It thus makes sense to address dopantless contacts and dual heterostructured extensions as interesting options realized in nonalloyed contact areas [30]. Proper band alignment is needed to avoid rectifying high-resistance Schottky contact [120,121], which would, anyway, request the use of highly doped shallow extensions to become quasi-ohmic. Nonalloyed contact band engineering, employing interfacial oxides [121–123] between the metal interconnect and the extension area (Figure 9.33), becomes a major research topic.

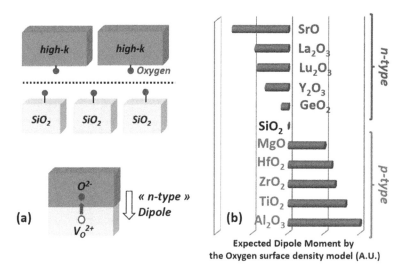

FIGURE 9.33 Nonalloyed contacts using heterostructures bandgap engineering and matching would ultimately reduce variability in transistor access resistance. Model underlines contribution of oxygen to interface dipole. (Modified from [30,121].)

Patterning is at the center of concerns when variability is discussed: it is credited as a main cause of variability of the critical dimensions in a circuit, strongly depending on the combined performances of lithography and etching. Classical optical lithography uses a mask that is exposed to monochromatic light, which is transferred to a photoresist either through (ultraviolet (UV)) or by reflection (extreme ultraviolet (EUV)) on it. Electron beam lithography does not require a mask and can reach resolutions as low as 2 nm [124], obtained thanks to an aberration- corrected scanning transmission electron microscope and compatible with a 5 nm half-pitch in hydrogen silsesquioxane resist. The resolution limits of this technique, by measuring the point-spread function at 200 keV, was also analyzed to be far less than EUV capabilities. However, patterning is not collective as it is in the case of optical lithography because patterns are realized by fine e beams. Multibeams have been proposed to circumvent the throughput issue of Gaussian-shaped single e beams [125].

At the sub 5 nm level, the existence of patterning solutions that can lead to zero intrinsic variability will be questioned: as a matter of fact, the device size to be reached represents the size of a few molecules. Solutions to achieve such a goal can be obtained from polymer chemistry. Direct Self-Assembly (DSA) using block co-polymers (BCP) with high χ resists (Figure 9.34) [126] is the most popular candidate. The Flory Huggins parameter χ takes account of the energy of interdispersing polymer and solvent molecules [127]. By mix and matching this technique to optical or multibeam lithography, phase separation of polymers from a composite resist can be obtained: monodisperse high density patterns far less than 10 nm have been proven (Figure 9.34).

Scaling of Cu interconnect is also facing difficulties related to resistivity highly nonlinear increase with Cu lines less than 20 nm wide [128] (Figure 9.35). The granular morphology of copper will increase the material resistivity because the grain size searched to decrease resistitivity becomes comparable to the pattern size! At the same time, electromigration becomes more of a restriction due to grain boundaries resistance and material to barrier interfaces increasing weight. Increasing the material thickness can be seen as a repairing solution but will increase parasitic capacitance, thus negatively impacting the interconnect delay [30,129].

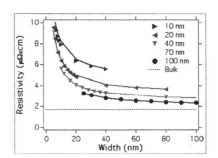

FIGURE 9.35 Copper resistivity as a function of metal line width. (By courtesy of [128].)

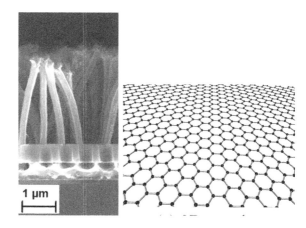

FIGURE 9.36 CNT bundles and graphene interconnects. (By courtesy of CEA, LETI.)

New paradigms are needed to meet the goal of drastic zero intrinsic variability and getting rid of electromigration or at least reducing it significantly as compared to copper results. To face this challenge, Carbon based materials have been proposed for Post Cu interconnect: CNTs bundles [130] (Figure 9.36) or graphene nanoribbons associated to DSA [30] have been demonstrated and could offer monodisperse wiring capabilities. Grafting Cu [131] or Au [132] on DNA molecules has already been proposed to achieve single atom wide metallic strings. Novel possibilities to build future 2D or 3D nanosystems thanks to the use of DNA as a template [133,134] or its functionalization and linkage (origamis) [135] are explored today.

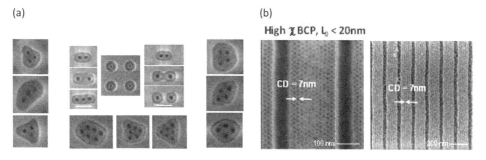

FIGURE 9.34 Patterning by direct self assembly using block co-polymers demonstrates monodisperse patterns: (a) new layout strategies and (b) 7 nm resolution. (By courtesy of CEA, LETI, see also [126].)

9.2 More Moore and More than Moore Meeting for 3D into Zero Power Systems

Since the 1960s, innovative functional designs have been possible thanks to the linear down scaling of electron devices, especially MOSFETs, and thus integrated circuits density increase. The progress was mainly due to the scaling of lithography wavelength and the constant introduction of new process modules (see Section 9.1.1). Today, dimensional scaling is a question driven by the cost of optical lithography: the cost and lack of availability of EUV lithography tools encouraged the search for alternatives, such as multiple patterning [136], maskless solutions using, for example, spacer-based image transfer or multielectron beams [125,137].

As mentioned at the beginning of this chapter, a whole question remains on whether novel functions can be introduced without MOSFET device scaling or not. In parallel with micro and nanoelectronics development, different types of devices, such as microelectromechanical systems (MEMS), have been developed, which were feasible by using the silicon integrated circuit technology: this effect, defined in the 1960s, is named pervasiveness [138]. Nanoelectronics can take new added values by the cointegration of various devices with CMOS and trace the new route with diversification: the new progress laws will be driven by applications in the field of energy, bio and healthcare, mobility and communication, IoT, and big data.

As an example, NEMS obtained from top-down fabricated nanowires and nanoribbons can be used as resonating cantilevers with increased mass sensing capabilities (Figure 9.37a,b) [139,140]. Sensitivity of a few zeptogram (10^{-21} g) [140] has been demonstrated for a few 10 nm size nanowires. Ultrasmall signals are delivered by nanometer-sized wires: their practical use poses new challenges. Their cointegration with CMOSFETs results

in huge advantages; thanks to the reduction of interconnect distance between the readout circuit and the sensor (Figure 9.37a). When compared to stand-alone NEMS, noise figures (Figure 9.37b), compactness, readout speed, and power consumption are substantially improved.

The use of NEMS as mechanical switches could compete with MOSFETs or TFETs, thanks to very sharp electrical characteristics, thus approaching the ideal switch properties. Stitching of electrodes can be a problem depending on the electrode material that usually introduces an extra series resistance and capacitance, introducing hysteresis and loss of performance in the switch characteristics [140]. The appearance of pull-in and pull-out voltages defines the nonreversibility of electrical characteristics [141]. Adiabatic gate architectures have been proposed [142] to replace CMOS devices for ultralow power logic: they could reduce power consumption by about a factor of 1,000 when compared with CMOS, provided that low contact resistance is achieved. In order to obtain such a property, flexible and highly conducting electrodes are needed: the integration of materials such as graphene [143,144] (Figure 9.38) has been reported with difference between pull-in and pull-out voltages values of about 2 V.

An alternative paradigm that could offer increased number of CMOS as well as diversification devices on a chip is the monolithic or sequential 3D [145] integration. Monolithic 3D [145] (Figure 9.39) offers unique opportunities to partition CMOS design, which consequently improves layout density, circuit speed, and power consumption, thanks to shorter interconnect distances between devices. Integration of new materials at low process temperature [145], memories [146], sensors, actuators, imagers, etc. enhances the circuit's figures of merit such as increased bandwidth and reduced power consumption by in-memory immersed computing or quantum computing [146–149]. Such an integration approach can lead to HP-embedded intelligence sensing [30,147,150]. Dopant activation at upper levels

(a)   (b)

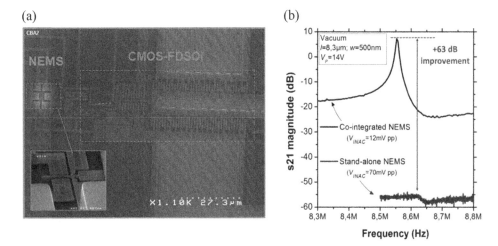

FIGURE 9.37 Mixing nanowires by top-down approach for sensing and readout by CMOS [140]: (a) cointegration of CMOS readout circuit with NEMS for ultrahigh sensitivity gas sensing and (b) comparison of signal-to-background and signal-to-noise ratios of CMOS cointegrated and stand-alone NEMS. Mass resolution sensed is down to ten nanograms of hydrocarbons. (By courtesy of CEA, LETI.)

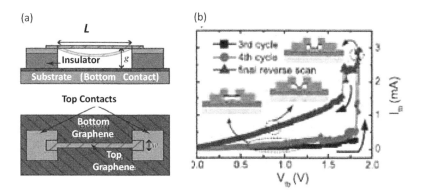

FIGURE 9.38 Single sheet graphene switches are candidates as nanoscale switches for adiabatic computing: (a) an ideal design with two graphene electrodes [143] and (b) graphene-suspended electrode demonstrates less than 2 V pull-in and pull-out voltages and high drivability [144] in contact with gold fixed electrode.

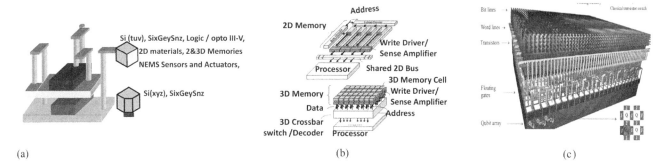

FIGURE 9.39 (a) Monolithic 3D integration [145] makes heterogeneous cointegration of materials and sensors possible; thanks to cold end processing at midprocess: 50% improved area and energy delay product can be obtained on partitioned Si CMOS FPGAs [146]. High-density embedded intelligence (sensors, actuators, etc.) is accessible, (b) high density or 3D memories (DRAM, RERAM) [147] embedded with digital circuit can deliver more bandwidth, reduce latency, introduce neuromorphic and programmable architectures [30,129,147], and (c) alternative computing paradigms from in-memory-immersed to massively parallel (Q computing, neuromorphic computing, etc.) increase bandwidth and energy efficiency (up to ×1,000). (By courtesy of CEA, LETI and partially from [148].)

preserving device characteristics at the first level is the main challenge for sequential 3D: a CoolCube solution, based on the bonding/stacking of ultrathin SOI, solid-phase epitaxy, and nanosecond laser annealing [151], has been proposed: device layout density is kept, without sacrificing their performance.

The sequential 3D approach opens up new opportunities to build a technology/design platform based on UTBO$_x$, low power and HP nanoelectronics logic scalable devices (More Moore) that can be cointegrated with diversification-type devices (More than Moore) to access new application opportunities [30,129,150] (Figure 9.40). It also gives the possibilities to hybridize, with aforementioned objects, new materials, new process integration paradigms, innovative functions from the beyond CMOS domain, with possible maximization of density and drastic reduction of power consumption. Whether this represents a second life for Moore's law or a different route is not necessarily the most interesting question. We are for sure entering a new era in which the possibilities of new fields are widened, while keeping societal needs awareness.

New application development needs to meet time to market needs. Instead of developing a completely new

design, different chips can be packaged together, either by stacking through a flip-chip process or side by side on the same board. Of course, the stacked chips might deliver the best performance system if cost is not a major hurdle. This so-called approach of 3D integration [152–155] (including multichip modules) can be distinguished from sequential/monolithic 3D and has proven to "cram more and more components in a package" (RF, MEMS, passive components, etc.). Different chips can thus be "taken from the shelf" and stacked to make a new product, without expecting the next technology node. Hybrid and heterogeneous integration is, by this way, made accessible at reasonable cost, with benefits to performance and power consumption at the system level. Today, with an emphasized trend in the future, the increasing demand for nomadic products requesting heterogeneous integration for many societal needs, such as healthcare, energy, communication, IoT, etc., will request the integration of energy sources, storage, and harvesting as well as communication capabilities. Energy harvesting sources [30] (Figure 9.41), thin-film batteries [156], and communication capabilities [154] could maximize energy efficiency to zero power consumption from a grid. This is an important consideration, given the fact

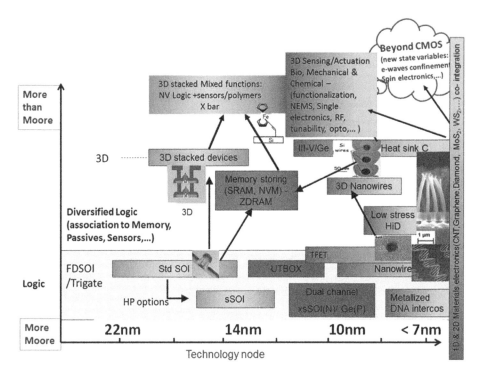

FIGURE 9.40   Advanced SOI-based road map for cointegration of More Moore and More than Moore devices [30,129]. (By courtesy of CEA, LETI.)

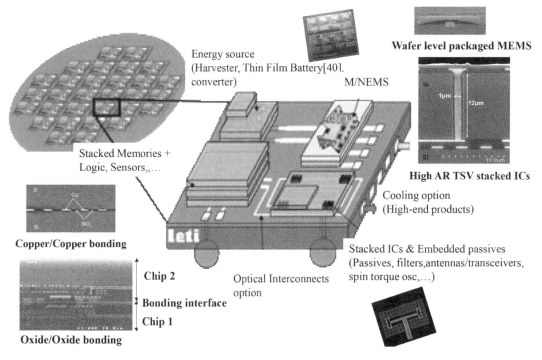

FIGURE 9.41   Parallel 3D integration, defined as a flexible set of tools, could profit by collective fabrication on a silicon interposer, which takes into account packaging at wafer level. The choice of organic, glass, or silicon substrate depends on applications by challenging performance, thermal losses, compatibility to the environment, reliability, and cost. (By courtesy of CEA, LETI.)

that nomadic and autonomous systems (mobile devices, e-assistants, robots, etc.) will be the next drivers in the IoE market. New process modules, such as through silicon vias (TSV), oxide/oxide and metal/metal bonding, and wafer-level packaging, are needed to make the 3D integration practical and viable (Figure 9.42) [30,129,152–156]. The issue of energy harvesting is certainly a serious question to take into account in future road maps. The complexity and diversity of possible applications makes the choice of energy harvesting policy even more difficult and will

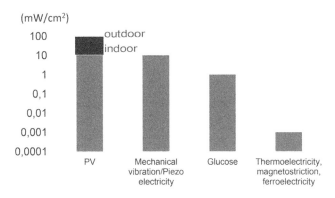

FIGURE 9.42 Energy harvesting is a major topic for autonomous system integration. The choice of the energy source is very much dependent on the application. (Adapted partially from [157].)

be a major technology research subject in the following decades. An autonomous power source needs three basic functions: energy harvesting, energy storage, and power management. Thin-film batteries [85] have made a lot of progress recently: energy storage normalized to the weight of electrolyte can be maximized: typical values of 250 µAh\cm^2 µm can be reached with LiPON. Major efforts are made towards flexible and biocompatible packaging solutions. The choice of the energy source harvested from the environment (Figure 9.42) ([129] and the data partially from [157]) depends mainly on the application: photovoltaics is by far the most abundant way to harvest amounts of energy that could be usable for many usages. The problem becomes very complex when biocompatibility and environment friendliness are desired. As an example, sources based on the enzymatic transformation of glucose are considered as serious candidates: storage densities of 596 Ah/kg, first-order magnitude higher than lithium-based batteries, have been reported [158]. Depending on the application specificities, 3D system design and their packaging will constantly be challenged by power dissipation and reliability, particularly affected by the mastering of temperature distributions [30,152,153].

References

1. G.E. Moore, Cramming more components onto integrated circuits, *Electronics*, vol 38, no 8, 1965.
2. G.E. Moore, Process in digital integrated electronics, *IEDM Technical Digest*, pp. 11–13, Washington, DC, 1975.
3. R.H. Dennard, F.H. Gaensslen, H.N. Yu, V.L. Rideout, E. Bassous, A.R. Le Blanc, Design of ion-implanted MOSFET's with very small physical dimensions, *IEEE Jouranl of Solid State Circuits*, vol 9, no 5, pp. 256–268, 1974.
4. www.itrs2.net/itrs-reports.html.
5. S. Deleonibus, Ed., *Electronic Device Architectures for the Nano-CMOS Era: From Ultimate CMOS Scaling to Beyond CMOS Devices*, Pan Stanford Publishing, Singapore, 2009.
6. D. Kahng, M.M. Atalla, Silicon-silicon dioxide field induced surface devices, *IRE-IEEE Solid-State Device Research Conference, Carnegie Institute of Technology*, Pittsburgh, PA, 1960.
7. M. Horiguchi, T. Sakata, T. Sekiguchi, S. Ueda, H. Tanaka, E. Yamasaki, Y. Nakagome, M. Aoki, I. Kaga, M. Ohkura, R. Nagai, F. Murai, T. Tanaka, S. Iijima, N. Yokoyama, Y. Gotoh, K. Shoji, T. Kisu, H. Yamashita, T. Nishida, E. Takeda, An experimental 220 MHz 1 GB DRAM, *Proceeding of ISSCC*, pp. 252–253, San Francisco (CA), 1995.
8. www.intel.com/pressroom/archive/releases/2006/2006 0718comp.htm.
9. J.E. Lilienfeld, Method and apparatus for controlling electric currents, US Patent 1,745,175, 1930.
10. S. Deleonibus, G. Dubois, Planarized contacts for integrated circuits obtained by conformal deposition and planarization, Patent: Filed April 13 1984 France National File N: 84.05906; International extension. April 10 1985(US File N: 721,779; Patent n: 4592802).
11. A. Bergemont, S. Deleonibus, B. Guillaumot, M. Laurens, F. Martin, G. Guegan, A high performance CMOS process for submicron 16 Mbit EPROM, *Proceedings of IEDM 1989*, Washington, DC, pp. 591–594, December 1989.
12. P. Singer, Making the move to dual damascene processing, *Semiconductor International*, vol 20, pp. 79–82, 1997.
13. R.H. Dennard, Field effect transistor memory, US Patent 3,387,286, Filed: July 14, 1967, Published: June 4, 1968.
14. L. Terman, MOSFET memory circuits, *Proceeding of the IEEE*, vol 59, pp. 1044–1058, 1971.
15. R.F. Freitas, W.W. Wilcke, Storage-class memory: The next storage system technology, *IBM Journal of Research and Development*, vol 52, no 4/5, pp. 449–464, 2008.
16. G. Molas, M. Bocquet, J. Buckley, J.P. Colonna, L. Masarotto, H. Grampeix, F. Martin, V. Vidal, A. Toffoli, P. Brianceau, L. Vermande, P. Scheiblin, M. Gély, A. M. Papon, G. Auvert, L. Perniola, C. Licitra, T. Veyron, N. Rochat, C. Bongiorno, S. Lombardo, B. De Salvo, S. Deleonibus, Thorough investigation of Si-nanocrystal memories with high-k interpoly dielectrics for sub-45nm node Flash NAND applications, *IEEE IEDM Technical Digest*, Washington, DC, p. 453, 2007.
17. G. Baccarani, M.R. Wordeman, R.H. Dennard, Generalized scaling theory and its application to a 1/4 micrometer MOSFET design, *IEEE Transaction on Electron Devices*, vol 31, no 4, pp. 452–462, 1984.
18. H.S. Momose, M. Ono, T. Yoshitomi, S. Nakamura, M. Saito, H. Iwai, Tunneling gate oxide approach to ultra high current drive in small geometry

MOSFET, *IEDM Proceedings*, pp. 593–596, San Francisco (CA), 1994.

19. S. Deleonibus, F. Martin, CEA, LETI, totally self aligned, planarized metal gate/HiK stack by gate last principle, European Patent #: 96.15436; US Patent #: 6,346,450, 1996.

20. B. Guillaumot, X. Garros, F. Lime, K. Oshima, B. Tavel, J.A. Chroboczek, P. Masson, R. Truche, A.-M. Papon, F. Martin, J.F. Damlencourt, S. Maitrejean, M. Rivoire, C. Leroux, S. Cristoloveanu, G. Ghibaudo, J.L. Autran, T. Skotnicki, S. Deleonibus, A 75 nm damascene metal gate and high- k integration for advanced CMOS devices, *IEDM Technical Digest*, San Francisco, CA, pp. 355–358, December 2002.

21. K. Mistry, C. Allen, C. Auth, B. Beattie, D. Bergstrom, M. Bost, M. Brazier, M. Buehler, A. Cappellani, R. Chau, C.-H. Choi, G. Ding, K. Fischer, T. Ghani, R. Grover, W. Han, D. Hanken, M. Hattendorf, J. He, J. Hicks, R. Huessner, D. Ingerly, P. Jain, R. James, L. Jong, S. Joshi, C. Kenyon, K. Kuhn, K. Lee, H. Liu, J. Maiz, B. McIntyre, P. Moon, J. Neirynck, S. Pae, C. Parker, D. Parsons, C. Prasad, L. Pipes, M. Prince, P. Ranade, T. Reynolds, J. Sandford, L. Shifren, J. Sebastian, J. Seiple, D. Simon, S. Sivakumar, P. Smith, C. Thomas, T. Troeger, P. Vandervoorn, S. Williams, K. Zawadzki, A 45 nm logic technology with High-k+metal gate transistors, strained silicon, 9 Cu interconnect layers, 193 nm dry patterning, and 100% Pb-free packaging, *IEDM Technical Digest Papers*, pp. 247–250, Washington (DC), 2007.

22. S. Deleonibus, C. Caillat, G. Guegan, M. Heitzmann, M.-E. Nier, S. Tedesco, B. Dal Zotto, F. Martin, P. Mur, A.-M. Papon, G. Lecarval, S. Biswas, A 20 nm physical gate length NMOSFET featuring 1.2 nm gate oxide shallow iimplanted source and drains and BF2 pockets, *IEEE Electron Devices Letters*, vol 21, pp. 173–175, 2000.

23. C. Caillat, S. Deleonibus G. Guegan, S. Tedesco, B. Dal'Zotto, M. Heitzmann, F. Martin, P. Mur, B. Marchand, F. Balestra, A 65 nm physical gate length NMOSFETs with heavy ion implanted pockets and highly reliable 2 nm-thick gate oxide for 1.5 V operation, *Symposium on VLSI Technology. Digest of Technical Papers*, Kyoto, Japan, pp. 89–90, 1999.

24. G. Bertrand, S. Deleonibus, B. Previtali, G. Guegan, X. Jehl, M. Sanquer, F. Balestra, Towards the limits of conventional MOSFETs: Case of sub 30 nm NMOS devices, *Solid-State Electronics*, vol 48, pp. 505–509, 2004.

25. W. Shockley, Problems related to pn junction in silicon, *Solid State Electronics*, vol 2, pp. 35–67, 1961.

26. R.W. Keyes, Fundamental limits in digital processing, *Proceedings of the IEEE*, vol 69, no 2, pp. 267–278, 1981.

27. A. Asenov, Simulation of statistical variability in nano MOSFETs, *Symposium on VLSI Technology. Digest of Technical Papers*, pp. 86–87, Kyoto (Japan), 2007.

28. S. Deleonibus, Ed., *Intelligent Integrated Systems: Devices, Technologies and Architectures*, Pan Stanford Series on Intelligent Nanosystems, vol 1, Pan Stanford Publishing, Singapore, 2014.

29. O. Faynot, F. Andrieu, O. Weber, C. Fenouillet-Béranger, P. Perreau, J. Mazurier, T. Benoist, O. Rozeau, T. Poiroux, M. Vinet, L. Grenouillet, J.-P. Noel, N. Posseme, S. Barnola, F. Martin, C. Lapeyre, M. Cassé, X. Garros, M-A. Jaud, O. Thomas, G. Cibrario, L. Tosti, L. Brévard, Tabone, P. Gaud, S. Barraud, T. Ernst and S. Deleonibus, Planar fully depleted SOI technology: A powerful architecture for the 20 nm node and beyond, *IEEE International Electron Devices Meeting, Technical Digest Papers*, Washington, DC, pp. 3.2.1–3.2.4, 2010.

30. S. Deleonibus, O. Faynot, T. Ernst, M. Vinet, P. Batude, F. Andrieu, O. Weber, D. Cooper, F. Bertin, H. Moriceau, L. DiCioccio, T. Signamarcheix, M. Sanquer, X. Jehl, O. Cueto, H. Fanet, F. Martin, H. Okuno, F. Nemouchi, G. Poupon, Y. Lamy, D. Gasparutto, X. Baillin, L. Duraffourg, J. Arcamone, L. Perniola, B. De Salvo, E. Vianello, L. Hutin, C. Poulain, E. Beigne, R. Tiron, L. Pain, S. Tedesco, S. Barnola, N. Posseme, C. Le Royer, A. Villalon, R. Salot, Future challenges and opportunities for heterogeneous process technology. Towards the thin films, zero intrinsic variability devices, zero power era, *International Electron Devices Meeting. IEDM Technical Digest Paper*, Washington DC, pp. 9.2.1–9.2.4, 2014. doi: 10.1109/IEDM.2014.7047015.

31. O. Weber, E. Josse, F. Andrieu, A. Cros, E. Richard, P. Perreau, E. Baylac, N. Degors, C. Gallon, E. Perrin, S. Chhun, E. Petitprez, S. Delmedico, J. Simon, G. Druais, S. Lasserre, J. Mazurier, N. Guillot, E. Bernard, R. Bianchini, L. Parmigiani, X. Gerard, C. Pribat, O. Gourhant, F. Abbate, C. Gaumer, V. Beugin, P. Gouraud, P. Maury, S. Lagrasta, D. Barge, N. Loubet, R. Beneyton, D. Benoit, S. Zoll, J. Chapon, L. Babaud, M. Bidaud, M. Gregoire, C. Monget, B. Le-Gratiet, P. Brun, M. Mellier, A. Pofelski, L.R. Clement, R. Bingert, S. Puget, J.-F. Kruck, D. Hoguet, P. Scheer, T. Poiroux, J.-P. Manceau, M. Rafik, D. Rideau, M.-A. Jaud, J. Lacord, F. Monsieur, L. Grenouillet, M. Vinet, Q. Liu, B. Doris, M. Celik, S.P. Fetterolf, O. Faynot, M. Haond, A 14 nm FDSOI technology for high speed and energy efficient applications, *International Electron Devices Meeting. IEDM Technical Digest Papers*, 2014, Honolulu, HI. doi: 10.1109/VLSIT.2014.6894343.

32. D. Hisamoto, W.-C. Lee, J. Kedzierski, E. Anderson, H. Takeuchi, K. Asano, T.-J. King, J. Bokor, C. Hu, A folded-channel MOSFET for deep–sub-tenth

micron era, *International Electron Devices Meeting, Technical Digest*, San Francisco, CA, pp. 1031–1034, 1998.

33. B. Yu, L. Chang, S. Ahmed, H. Wang, S. Bell, C.-Y. Yang, C. Tabery, C. Hu, T.-J. King, J. Bokor, M.-R. Lin, D. Kyser, FinFET scaling to 10 nm gate length, *International Electron Devices Meeting, Technical Digest*, San Francisco, CA, pp. 251–254, 2002.

34. www.intel.com/content/dam/www/public/us/en/ documents/presentation/advancing-moores-law.

35. O. Weber, O. Faynot, F. Andrieu, C. Buj-Dufournet, F. Allain, P. Scheiblin, J. Foucher, N. Daval, D. Lafond, L. Tosti, L. Brevard, O. Rozeau, C. Fenouillet-Beranger, M. Marin, F. Boeuf, D. Delprat, K. Bourdelle, B.-Y. Nguyen, S. Deleonibus, High immunity to threshold voltage variability in undoped ultra-thin FDSOI MOSFETs and its physical understanding, *International Electron Devices Meeting, Technical Digest Papers*, San Francisco, CA, pp. 245–248, 2008.

36. M.J.M. Pelgrom, A.C.J. Duinmaijer, A.P.G. Welbers, Matching properties of MOS transistors, *IEEE Journal of Solid-State Circuits*, vol 24, pp. 1433–1439, 1989.

37. R.-H. Yan, A. Ourmazd, K.F. Lee, Scaling the Si MOSFET: From bulk to SOI to bulk, *IEEE Transactions on Electron Devices*, vol 39, no 7, pp. 1704–1710, 1992.

38. K. Suzuki, T. Tanaka, Y. Tosaka, H. Horie, Y. Arimoto scaling theory for double-gate SOI MOSFET's, *IEEE Transactions on Electron Devices*, vol 40, no 12, pp. 2326–2329, 1993.

39. C.P. Auth, J.D. Plummer, Scaling theory for cylindrical, fully-depleted, surrounding-gate MOSFET's, *IEEE Electron Device Letters*, pp. 74–76, 1997.

40. S. Deleonibus, Ed., *Integrated Nanodevice and Nanosystem Fabrication: Materials, Techniques and New Opportunities*, Pan Stanford Series on Intelligent Nanosystems, vol 2, Pan Stanford Publishing, Singapore, 2017.

41. Z. Lu, S. Kimbel, Growth of 450 mm diameter semiconductor grade silicon crystals, *Journal of Crystal Growth*, vol 318, pp. 193–195, 2011.

42. E.A. Fitzgerald, Y.H. Xie, M.L. Green, D. Brasen, A.R. Kortan, J. Michel, Y.J. Mii, B. Weir, Totally relaxed Ge_xSi_{1-x} layers with low threading dislocation densities grown on Si substrates, *Journal of Applied Physics*, vol 59, no 7, pp. 811–813, 1991.

43. J. Welser, J.L. Hoyt, J.F. Gibbons, NMOS and PMOS transistors fabricated in strained silicon/relaxed silicon-germanium structures, *IEDM*, pp. 1000–1002, San Francisco (CA), 1992.

44. K. Eberl, S.S. Iyer, S. Zollner, J.C. Tsang, F.K. LeGoues, Growth and strain compensation effects in the ternary $Si_{1-x-y}Ge_xC_y$ alloy system, *Applied Physics Letters* vol 60, p. 3033, 1992. doi: org/10.1063/1.106774.

45. T. Ernst, J.M. Hartmann, V. Loup, F. Ducroquet, P. Dollfus, G. Guegan, D. Lafond, P. Holliger, B. Prévitali, A. Toffoli, S. Deleonibus, Fabrication of a novel strained SiGe:C-channel planar 55 nm nMOSFET for high-performance CMOS, *Proceedings of VLSI Technology Symposium*, Honolulu, HI, pp. 92–93, June 2002.

46. F. Andrieu, T. Ernst, F. Lime, F. Rochette, K. Romanjek, S. Barraud, C. Ravit, F. Boeuf, M. Jurczak, M. Casse, O. Weber, L. Brevard, G. Reimbold, G. Ghibaudo, S. Deleonibus, Experimental and comparative investigation of low and high field transport in substrate- and process-induced strained nanoscaled MOSFETs, *Symposium on VLSI Technology Digest of Technical Papers*, Kyoto, Japan, pp. 176–177, 2005.

47. S. Morvan, F. Andrieu, M. Cassé, O. Weber, N. Xu, P. Perreau, J.M. Hartmann, J.C. Barbé, J. Mazurier, P. Nguyen, C. Fenouillet-Béranger, C. Tabone, L. Tosti, L. Brévard, A. Toffoli, F. Allain, D. Lafond, B.Y. Nguyen#, G. Ghibaudo, F. Boeuf, O. Faynot, T. Poiroux, Efficiency of mechanical stressors in planar FDSOI n and p MOSFETs down to 14 nm gate length, *Symposium on VLSI Technology*, Honolulu, HI, pp. 11–12, 2012.

48. R. Coquand, M. Cassé, S. Barraud, P. Leroux, D. Cooper, C. Vizioz, C. Comboroure, P. Perreau, V. Maffini-Alvaro, C. Tabone, L. Tosti, F. Allain, S. Barnola, V. Delaye, F. Aussenac, G. Reimbold, G. Ghibaudo, D. Munteanu, S. Monfray, F. Boeuf, O. Faynot, T. Poiroux, Strain-induced performance enhancement of tri-gate and omega-gate nanowire FETs scaled down to 10 nm width, *Symposium on VLSI Technology*, Honolulu, HI, pp. 13–14, 2012.

49. Q. Liu, M. Vinet, J. Gimbert, N. Loubet, R. Wacquez, L. Grenouillet, Y. Le Tiec, A. Khakifirooz, T. Nagumo, K. Cheng, H. Kothari, D. Chanemougame, F. Chafik, S. Guillaumet, J. Kuss, F. Allibert, G. Tsutsui, J. Li, P. Morin, S. Mehta, R. Johnson, L.F. Edge, S. Ponoth, T. Levin, S. Kanakasabapathy, B. Haran, H. Bu, J.-L. Bataillon, O. Weber, O. Faynot, E. Josse, M. Haond, W. Kleemeier, M. Khare, T. Skotnicki, S. Luning, B. Doris, M. Celik, R. Sampson, High performance UTBB FDSOI devices featuring 20 nm gate length for 14 nm node and beyond, *IEDM Technical Digest Papers*, San Francisco, CA, pp. 228–231, 2013.

50. O. Weber, E. Josse, F. Andrieu, A. Cros, E. Richard, P. Perreau, E. Baylac, N. Degors, C. Gallon, E. Perrin, S. Chhun, E. Petitprez, S. Delmedico, J. Simon, G. Druais, S. Lasserre, J. Mazurier, N. Guillot, E. Bernard, R. Bianchini, L. Parmigiani, X. Gerard, C. Pribat, O. Gourhant, F. Abbate, C. Gaumer, V. Beugin, P. Gouraud, P. Maury, S. Lagrasta, D. Barge, N. Loubet, R. Beneyton, D.

Benoit, S. Zoll, J. Chapon, L. Babaud, M. Bidaud, M. Gregoire, C. Monget, B. Le-Gratiet, P. Brun, M. Mellier, A. Pofelski, L.R. Clement, R. Bingert, S. Puget, J.-F. Kruck, D. Hoguet, P. Scheer, T. Poiroux, J.-P. Manceau, M. Rafik, D. Rideau, M.-A. Jaud, J. Lacord, F. Monsieur, L. Grenouillet, M. Vinet, Q. Liu, B. Doris, M. Celik, S.P. Fetterolf, O. Faynot, M. Haond, A 14 nm FDSOI technology for high speed and energy efficient applications, *Symposium on VLSI Technology (VLSI-Technology): Digest of Technical Papers*, Honolulu, HI, 2014. doi: 10.1109/VLSIT.2014.6894343.

51. A. Bonnevialle, C. Le Royer, Y. Morand, S. Reboh, C. Plantier, N. Rambal, J.-P. Pédini, S. Kerdiles, P. Besson, J.-M. Hartmann, D. Marseilhan, B. Mathieu, R. Berthelon, M. Cassé, F. Andrieu, D. Rouchon, O. Weber, F. Bœuf, M. Haond, A. Claverie, M. Vinet, Smart solutions for efficient dual strain integration for future FDSOI generations, *Symposium on VLSI Technology (VLSI-Technology): Digest of Technical Papers*, Honolulu, HI, 2016.

52. Z. Zhong, D. Wang, Yi Cui, M.W. Bockrath, C.M. Lieber, Nanowire crossbar arrays as address decoders for integrated nanosystems. *Science*, vol 302, pp. 1377–1379, 2003.

53. B.M. Kayes, M.A. Filler, M.C. Putnam, M.D. Kelzenberg, N.S. Lewis, H.A. Atwater, Growth of vertically aligned Si wire arrays over large areas (>1 cm^2)(>1 cm^2) with Au and Cu catalysts, *Applied Physics Letters*, vol 91, p. 103110, 2007.

54. T. Ernst, C. Dupré, C. Isheden, E. Bernard, R. Ritzenthaler, V. Maffini-Alvaro, J.C. Barbé, F. De Crecy, A. Toffoli, C. Vizioz, S. Borel, F. Andrieu, V. Delaye, D. Lafond, G. Rabillé, J.M. Hartmann, M. Rivoire, B. Guillaumot, A. Suhm, P. Rivallin, O. Faynot, G. Ghibaudo, S. Deleonibus, Novel 3D integration process for highly scalable nano-beam stacked-channels GAA (NBG) FinFETs with HfO$_2$/TiN gate stack, *IEDM Technical Digest*, San Francisco, CA, pp. 1–4, 2006.

55. S. Barraud, J.M. Hartmann, V. Maffini-Alvaro, L. Tosti, V. Delaye, D. Lafond, Top-down fabrication of epitaxial SiGe/Si multi-(core/shell) p-FET nanowire transistors, *IEEE Transactions on Electron Devices*, vol 61, no 4, pp. 953–956, 2014.

56. C. Dupré, A. Hubert, S. Bécu, M. Jublot, V. Maffini-Alvaro, C. Vizioz, F. Aussenac, C. Arvet, S. Barnola, J.-M. Hartmann, G. Garnier, A. Allain, J.P. Colonna, M. Rivoire, L. Baud, S. Pauliac, V. Loup, T. Chevolleau, P. Rivallin, B. Guillaumot, G. Ghibaudo, O. Faynot, T. Ernst, S. Deleonibus, 15-diameter 3D stacked nanowires with independent gates operation: φFET, *IEDM Technical Digest*, San Francisco, CA, pp. 1–4, 2008.

57. S.J. Kang, C. Kocabas, H.S. Kim, Q. Cao, M.A. Meitl, D.Y. Khang, J.A. Rogers, Printed multilayer superstructures of aligned single-walled carbon nanotubes for electronic applications, *Nano Letters*, vol 7, no 11, pp. 3343–3348, 2007.

58. J. Kong, H. Soh, A. Cassell, C. Quate, H. Dai, Synthesis of individual single-walled carbon nanotubes on patterned silicon wafers, *Nature*, vol 395, p. 878, 1998.

59. A. Javey, Q. Wang, A. Ural, Y. Li, H. Dai, Carbon nanotube transistor arrays for multistage complementary logic and ring oscillators. *Nanoletters*, vol 2, p. 929, 2002.

60. P. Wong, D. Akinwande, *Carbon Nanotube and Graphene Device Physics*, Cambridge University Press, Cambridge, UK, 2011.

61. M. Steiner, M. Engel, Y.M. Lin, Y. Wu, K. Jenkins, D.B. Farmer, P. Avouris high-frequency performance of scaled carbon nanotube array field-effect transistors, *Applied Physics Letters*, vol 101, no 5, p. 053123, 2012. doi: 10.1063/1.4742325.

62. K.S. Novoselov, A.K. Geim, S.V. Morozov, D. Jiang, Y. Zhang, S.V. Dubonos, I.V. Grigorieva, A.A. Firsov, Electric field effect in atomically thin carbon films, *Science*, vol 306, pp. 666–669, 2004.

63. C. Berger, Z. Song, T. Li, X. Li, A.Y. Ogbazghi, R. Feng, Z. Dai, A.N. Marchenkov, E.H. Conrad, P.N. First, W.A. de Heer, Ultrathin epitaxial graphite: 2D electron gas properties and a route toward graphene-based nanoelectronics, *The Journal of Physical Chemistry B*, vol 108, pp. 19912–19916, 2004.

64. F. Schwierz, Graphene transistors: Status, prospects and problems, *Proceedings of the IEEE*, vol 101, no 7, 1567–1584, 2013.

65. K.-K. Liu, W. Zhang, Y.-H. Lee, Y.-C. Lin, M.-T. Chang, C.-Y. Su, C.-S. Chang, H. Li, Y. Shi, H. Zhang, C.-S. Lai, L.-J. Li, Growth of large-area and highly crystalline MoS thin layers on insulating substrates, *Nano Letters*, vol 12, no 3, pp. 1538–1544, 2012.

66. A.K. Geim, I.V. Grigorieva, Van der Waals heterostructures, *Nature*, vol 499, pp. 419–425, 2013. doi: 10.1038/nature12385.

67. P. Vogt, P. De Padova, C. Quaresima, J. Avila, E. Frantzeskakis, M.C. Asensio, A. Resta, B. Ealet, G.L. Lay, Silicene: Compelling experimental evidence for graphenelike two-dimensional silicon, *Physical Review Letters*, vol 108, p. 155501, 2012.

68. D. María Eugenia, L.L. Guy, Few layer epitaxial germanene: A novel two-dimensional Dirac material. *Scientific Reports*, vol 6, p. 20714, 2016.

69. S. Gupta, P.P. Manik, R.K. Mishra, A. Nainani, M.C. Abraham, S. Lodha, Contact resistivity reduction through interfacial layer doping in metal-interfacial layer-semiconductor contacts, *Journal of Applied Physics*, vol 113, p. 234505, 2013.

70. F. Mayer, C. Le Royer, J.F. Damlencourt, K. Romanjek, F. Andrieu, C. Tabone, B. Previtali,

S. Deleonibus. Impact of SOI, Si$_{1-x}$Ge$_x$OI and GeOI substrates on CMOS compatible tunnel FET performance, *IEDM*, San Francisco, CA, 2008. doi: 10.1109/IEDM.2008.4796641.

71. K. Tomioka, M. Yoshimura, F. Ishizaka, T. Fukui, Integration of III–V nanowires on Si: From high-performance vertical FET to steep-slope switch. *IEDM Technical Digest*, Washington, DC, pp. 88–91, 2013. doi: 10.1109/IEDM.2013.6724557.

72. U.E. Avci, R. Rios, K. Kuhn, I.A. Young, Comparison of performance, switching energy and process variations for the TFET and MOSFET in Logic, *Symposium on VLSI Technology: Digest of Technical Papers*, Kyoto, Japan, pp. 124–125, June 2011.

73. A. Villalon, C. Le Royer, P. Nguyen, S. Barraud, F. Glowacki, A. Revelant, L. Selmi S. Cristoloveanu, L. Tosti, C. Vizioz, J.-M. Hartmann, N. Bernier, B. Previtali, C. Tabone, F. Allain, S. Martinie, O. Rozeau, M. Vinet, First demonstration of strained SiGe nanowires TFETs with ION beyond 700 μA/μm, *Symposium on VLSI Technology: Digest of Technical Papers*, Honolulu, HI, pp. 66–67, 2014. doi: 10.1109/VLSIT.2014.689436.

74. K.-W. Song, J.-Y. Kim, H. Kim, H.-W. Chung, H. Kim, K. Kim, H.-W. Park, H.C. Kang, S. Kim, N,-K. Tak, D. Park, W.-S. Kim, Y.-T. Lee, Y.C. Oh, G.-Y. Jin, J. Yoo, K. Oh, C. Kim, W.-S. Lee, A 31ns random cycle VCAT-based 4F^2 DRAM with enhanced cell efficiency, *Symposium on VLSI Circuits Digest of Technical Papers*, Kyoto, Japan, pp. 132–133, 2009.

75. F. Masuoka, M. Asano, H. Iwahashi, T. Komuro, S. Tanaka. A new flash E2PROM cell using triple polysilicon technology, *IEDM Technical Digest Papers*, Washington, DC, pp. 464–467, 1984.

76. F. Masuoka, M. Momodomi, Y. Iwata, R. Shirota, New ultra high density EPROM and flash EEPROM with NAND structure cell, *IEDM Technical Digest Papers*, Washington, DC, pp. 552–555, 1987.

77. B. De Salvo, C. Gerardi, R. van Schaijk, S. Lombardo, D. Corso, C. Plantamura, S. Serafino, G. Ammendola, M. van Duuren, P. Goarin, W.Y. Mei, K. van der Jeugd, T. Baron, M. Gely, P. Mur, S. Deleonibus, Performance and reliability features of advanced nonvolatile memories based on discrete traps (silicon nanocrystals, SONOS), *IEEE Transactions on Device and Materials Reliability*, vol 4, no 3, p. 377, 2004.

78. B. DeSalvo, C. Gerardi, S. Lombardo, T. Baron, L. Perniola, D. Mariolle, P. Mur, A. Toffoli, M. Gely, M.N. Semeria, S. Deleonibus, G. Ammendola, V. Ancarani, M. Melanotte, R. Bez, L. Baldi, D. Corso, I. Crupi, R.A. Puglisi, G. Nicotra, E. Rimini, F. Mazen, G. Ghibaudo, G. Pananakakis, C. Monzio Compagnoni, D. Ielmini, A. Spinelli, A. Lacaita, Y.M. Wan, K. van der Jeugd, How far will silicon nanocrystals push the scaling limits of NVMs technologies? *IEEE IEDM Technical Digest*, Washington, DC, p. 597, 2003.

79. H. Tanaka, M. Kido, K. Yahashi, M. Oomura, R. Katsumata, M. Kito, Y. Fukuzumi, M. Sato, Y. Nagata, Y. Matsuoka, Y. Iwata, H. Aochi, A. Nitayama, Bit cost scalable technology with punch and plug process for ultra high density Flash memory, *IEEE Symposium on VLSI Technology Digest of Technical Papers*, pp. 14–15, 2007. doi: 10.1109/VLSIT.2007.4339708.

80. S.-M. Jung, J. Jang, W. Cho, H. Cho, J. Jeong, Y. Chang, J. Kim, Y. Rah, Y. Son, J. Park, M.-S. Song, K.-H. Kim, J.-S. Lim, K. Kim, Three dimensionally stacked NAND Flash memory technology using stacking single crystal Si layers on ILD and TANOS structure for beyond 30 nm node. *IEDM*, San Francisco, CA, 2006. doi: 10.1109/IEDM.2006.346902.

81. E.-K. Lai, H.-T. Lue, Y.-H. Hsiao, J.-Y. Hsieh, C.-P. Lu, S.-Y. Wang, L.-W. Yang, T. Yang, K.-C. Chen, J. Gong, K.-Y. Hsieh, R. Liu, C.-Y. Lu, A multi-layer stackable Thin-Film Transistor (TFT) NAND-type Flash memory, *IEDM*, San Francisco, CA, 2006. doi: 10.1109/IEDM.2006.346903.

82. H. Maejima, K. Kanda, S. Fujimura, T. Takagiwa, S. Ozawa, J. Sato, Y. Shindo, M. Sato, N. Kanagawa, J. Musha, S. Inoue, K. Sakurai, N. Morozumi, R. Fukuda, Y. Shimizu, T. Hashimoto, X. Li, Y. Shimizu, K. Abe, T. Yasufuku, T. Minamoto, H. Yoshihara, T. Yamashita, K. Satou, T. Sugimoto, F. Kono, M. Abe, T. Hashiguchi, M. Kojima, Y. Suematsu, T. Shimizu, A. Imamoto, N. Kobayashi, M. Miakashi, K. Yamaguchi, S. Bushnaq, H. Haibi, M. Ogawa, Y. Ochi, K. Kubota, T. Wakui, D. He, W. Wang, H. Minagawa, T. Nishiuchi, H. Nguyen, K.-H. Kim, K. Cheah, Y. Koh, F. Lu, V. Ramachandra, S. Rajendra, S. Choi, K. Payak, N. Raghunathan, S. Georgakis, H. Sugawara, S. Lee, T. Futatsuyama, K. Hosono, N. Shibata, T. Hisada, T. Kaneko, H. Nakamura, A 512 Gb 3b/cell 3D Flash memory on a 96-word-line-layer technology, *IEEE International Solid-State Circuits Conference: Digest of Technical Papers*, San Francisco, CA, pp. 336–337, 2018.

83. S. Lee, C. Kim, M. Kim, S.-M Joe, J. Jang, S. Kim, K. Lee, J. Kim, J. Park, H.-J. Lee, M. Kim, S. Lee, S.G. Lee, J. Bang, D. Shin, H. Jang, D. Lee, N. Kim, J. Jo, J. Park, S. Park, Y. Rho, Y. Park, H.-J. Kim, C.A. Lee, C. Yu, Y. Min, M. Kim, K. Kim, S. Moon, H. Kim, Y. Choi, Y.H. Ryu, J. Choi, M. Lee, J. Kim, G.S. Choo, J.-D. Lim, D.-S. Byeon, K. Song, K.-T. Park, K.-H. Kyung, A 1Tb 4b/cell 64-stacked-WL 3D NAND Flash memory with 12MB/s program throughput, *IEEE International Solid-State Circuits Conference: Digest of Technical Papers*, San Francisco, CA, pp. 340–342, 2018.

84. E. Vianello, O. Thomas, G. Molas, O. Turkyilmaz, N. Jovanovic, D. Garbin, G. Palma, M. Alayan, C. Nguyen, J. Coignus, B. Giraud, T. Benoist, M. Reyboz, A. Toffoli, C. Charpin, F. Clermidy, L. Perniola, Resistive memories for ultra-low-power embedded computing design, *IEDM Technical Digest*, San Francisco, CA, pp. 6.3.1–6.3.4, 2014. doi: 10.1109/IEDM.2014.7046995.

85. M. Suri, O. Bichler, D. Querlioz, G. Palma, E. Vianello, D. Vuillaume, C. Gamrat, B. DeSalvo, CBRAM devices as binary synapses for low-power stochastic neuromorphic systems: Auditory (Cochlea) and Visual (Retina) cognitive processing applications, *IEDM Technical Digest*, San Francisco, CA, pp. 10.3.1–10.3.4, 2012. doi: 10.1109/IEDM.2012.6479017.

86. M. Suri, O. Bichler, D. Querlioz, O. Cueto, L. Perniola, V. Sousa, D. Vuillaume, C. Gamrat, B. De Salvo, Phase change memory as synapse for ultra-dense neuromorphic systems: Application to complex visual pattern extraction, *IEDM Technical Digest*, Washington, DC, pp. 4.4.1–4.4.4, 2011. doi: 10.1109/IEDM.2011.6131488.

87. G.E. Ghezzi, R. Morel, A. Brenac, N. Boudet, M. Audier, F. Fillot, S. Maitrejean, F. Hippert, Crystallization of $Ge_2Sb_2Te_5$ nanometric phase change material clusters made by gas-phase condensation, *Applied Physics Letters*, vol 101, p. 233113, 2012.

88. S.R. Ovshinsky, Reversible electrical switching phenomena in disordered structures, *Physical Review Letters*, vol 21 no 20, 1968.

89. G. Navarro, M. Coué, A. Kiouseloglou, P. Noé, F. Fillot, V. Delaye, A. Persico, A. Roule, M. Bernard, C. Sabbione, D. Blachier, V. Sousa, L. Perniola, S. Maitrejean, A. Cabrini, G. Torelli, P. Zuliani, R. Annunziata, E. Palumbo, M. Borghi, G. Reimbold, B. De Salvo, Trade-off between SET and data retention performance thanks to innovative materials for phase-change memory, *IEDM Technical Digest Papers*, San Francisco, CA, pp. 21.5.1–21.5.4, 2013.

90. J.H. Park, S.W. Kim, J.H. Kim, D.H. Ko, Z. Wu, D. Ahn, D.H. Ahn, J.M. Lee, S.B. Kang, S.Y. Choi, Enhancement of a cyclic endurance of phase change memory by application of a high-density $C_{15}(Ge_{21}Sb_{36}Te_{43})$ film, *AIP Advances* vol 6, p. 025013, 2016. doi: 10.1063/1.4942110.

91. L. Berger, Emission of spin waves by a magnetic multilayer traversed by a current, *Physical Review B*, vol 54, p. 9353, 1999.

92. J.A. Katine, F.J. Albert, R.A. Buhrman, E.B. Myers, D.C. Ralph, Current-driven magnetization reversal and spin-wave excitations in Co/Cu/Co pillars, *Physical Review Letters*, vol 84, p. 3149, 2000.

93. K.-T. Nam, S.C. Oh, J.E. Lee, J.H. Jeong, I.G. Baek, E.K. Yim, J.S. Zhao, S.O. Park, H.S. Kim, U-In Chung, J.T. Moon, Switching properties in spin transper torque MRAM with sub-50nm MTJ

94. I.L. Prejbeanu, W. Kula, K. Ounadjela, R.C. Sousa, O. Redon, B. Dieny, J.P. Nozieres, Thermally assisted switching in exchange- biased storage layer magnetic tunnel junctions, *IEEE Transactions on Magnetics*, vol 40, no 4, 2004.

95. M. Baibich, J.M. Broto, A. Fert, F. Nguyen Van Dau, F. Petroff, P. Etienne, G. Creuzet, A. Friederch, J. Chazelas, Giant magnetoresistance of (001)Fe/(001)Cr magnetic superlattices, *Physical Review Letters*, vol 61, no 247, 1988.

96. S.-W. Chung, T. Kishi, J.W. Park, M. Yoshikawa, K.S. Park, T. Nagase, K. Sunouchi, H. Kanaya, G.C. Kim, K. Noma, M.S. Lee, A. Yamamoto, K.M. Rho, K. Tsuchida, S.J. Chung, J.Y. Yi, H.S. Kim, Y.S. Chun, H. Oyamatsu, and S.J. Hong, 4Gbit density STT-MRAM using perpendicular MTJ realized with compact cell structure, *IEDM Technical Digest Papers*, San Francisco, CA, pp. 659–662, 2016.

97. S.S.P. Parkin, M. Hayashi, L. Thomas, Magnetic domain-wall racetrack memory, *Science*, vol 320, no 5873, pp. 190–194, 2008. doi: 10.1126/science.1145799.

98. S. Matsunaga, J. Hayakawa, S. Ikeda, K. Miura, H. Hasegawa, T. Endoh, H. Ohno, T. Hanyu, Fabrication of a nonvolatile full adder based on logic-in-memory architecture using magnetic tunnel junctions, *Applied Physics Express*, vol 1, p. 091301, 2008.

99. L.O. Chua, Memristor: The missing circuit element. *IEEE Transactions on Circuits Theory*, vol 18, no 5, pp. 507–519, 1971. doi:10.1109/TCT.1971.1083337.

100. Y.-B. Kim, S.R. Lee, D. Lee, C.B. Lee, M. Chang, J.H. Hur, M.-J. Lee, G.-S. Park, C.J. Kim, U.-I Chung, I.-K. Yoo, K. Kim, Bi-layered RRAM with unlimited endurance and extremely uniform switching, *Symposium on VLSI Technology: Digest of Technical Papers*, Honolulu, HI, pp. 52–53, 2011.

101. C. Cagli, J. Buckley, V. Jousseaume, T. Cabout, A. Salaun, H. Grampeix, J.F. Nodin, H. Feldis, A. Persico, J. Cluzel, P. Lorenzi, L. Massari, R. Rao, F. Irrera, F. Aussenac, C. Carabasse, M. Coue, P. Calka, E. Martinez, L. Perniola, P. Blaise, Z. Fang, Y.H. Yu, G. Ghibaudo, D. Deleruyelle, M. Bocquet, C. Müller, A. Padovani, O. Pirrotta, L. Vandelli, L. Larcher, G. Reimbold, B. de Salvo, Experimental and theoretical study of electrode effects in HfO_2 based RRAM, *IEDM Technical Digest Papers*, pp. 658–661, 2011.

102. E. Vianello, O. Thomas, M. Harrand, S. Onkaraiah, T. Cabout, B. Traoré, T. Diokh, H. Oucheikh, L. Perniola, G. Molas, P. Blaise, J.F. Nodin, E. Jalaguier, B. De Salvo, Back-end 3D integration of HfO_2-based RRAMs for low-voltage advanced IC digital design, *International Conference on IC*

Design and Technology (ICICDT) Technical Digest, Pavia, Italy, pp. 235–238, 2013.

103. T. Diokh, E. Le-Roux, S. Jeannot, C. Cagli, V. Jousseaume, J.-F. Nodin, M. Gros-Jean, C. Gaumer, M. Mellier, J. Cluzel, C. Carabasse, P. Candelier, B. De Salvo, Study of resistive random access memory based on $TiN/TaO_x/TiN$ integrated into a 65 nm advanced complementary metal oxide semiconductor technology, *Thin Solid Films*, vol 533, no 4, pp. 24–28, 2013.

104. B. Govoreanu, G.S. Kar, Y-Y. Chen, V. Paraschiv, S. Kubicek, A. Fantini, I.P. Radu, L. Goux, S. Clima, R. Degraeve, N. Jossart, O. Richard, T. Vandeweyer, K. Seo, P. Hendrickx, G. Pourtois, H. Bender, L. Altimime, D.J. Wouters, J.A. Kittl, M. Jurczak, 10×10 nm^2 Hf/HfO$_x$ crossbar resistive RAM with excellent performance, reliability and low-energy operation, *IEDM Technical Digest Papers*, San Francisco, CA, pp. 31.6.1–31.6.4, 2011.

105. G.S. Kar, A. Fantini, Y-Y. Chen, V. Paraschiv, B. Govoreanu, H. Hody, N. Jossart, H. Tielens, S. Brus, O. Richard, T. Vandeweyer, D.J. Wouters, L. Altimime, M. Jurczak, Process-improved RRAM cell performance and reliability and paving the way for manufacturability and scalability for high density memory application, *Symposium on VLSI Technology: Digest of Technical Papers*, Kyoto, Japan, pp. 157–158, 2012.

106. L. Masoero, G. Molas, F. Brun, M. Gély, J.P. Colonna, V. Della Marca, O. Cueto, E. Nowak, A. De Luca, P. Brianceau, C. Charpin, R. Kies, A. Toffoli, D. Lafond, V. Delaye, F. Aussenac, C. Carabasse, S. Pauliac, C. Comboroure, G. Ghibaudo, S. Deleonibus, B. De Salvo, Scalability of split-gate charge trap memories down to 20nm for low-power embedded memories, *IEDM Technical Digest Papers*, pp. 215–218, 2011.

107. X. Ma, D.B. Strukov, J.H. Lee, K.K. Likharev, Afterlife for silicon: CMOL circuit architectures, *Proceedings of IEEE-Nano'05*, Nagoya, Japan, pp. 175–178, 2005.

108. T.-Y. Liu, T.H. Yan, R. Scheuerlein, Y. Chen, J. Koon, Y. Lee, G. Balakrishnan, G. Yee, H. Zhang, A. Yap, J. Ouyang, T. Sasaki, S. Addepalli, A. Al-Shamma, C.-Y. Chen, M. Gupta, G. Hilton, S. Joshi, A. Kathuria, V. Lai, D. Masiwal, M. Matsumoto, A. Nigam, A. Pai, J. Pakhale, C.H. Siau, X. Wu, R. Yin, L. Peng, J.Y. Kang, S. Huynh, H. Wang, N. Nagel, Y. Tanaka, M. Higashitani, T. Minvielle, C. Gorla, T. Tsukamoto, T. Yamaguchi, M. Okajima, T. Okamura, S. Takase, T. Hara, H. Inoue, L. Fasoli, M. Mofidi, R. Shrivastava, K. Quader, A 130.7 mm^2 2-layer 32Gb ReRAM memory device in 24nm technology, *ISSCC Technical Digest Papers*, San Francisco, CA, pp. 210–211, 2013.

109. M. Vinet, V. Deshpande, X. Jehl, R. Wacquez, S. Barraud, M. Sanquer, R. Coquand, O. Cueto, B. Roche, B. Voisin, M. Pierre, L. Grenouillet, C. Vizioz, L. Tosti, B. Previtali, P. Perreau, T. Poiroux, O. Faynot, FDSOI nanowires: An opportunity for hybrid circuit with field effect and single electron transistors, *IEDM Technical Digest*, Washington, DC, pp. 26.4.1–26.4.4, 2013. doi: 10.1109/IEDM.2013.6724697.

110. S. De Franceschi, L. Hutin, R. Maurand, L. Bourdet, H. Bohuslavskyi, A. Corna, D. Kotekar-Patil, S. Barraud, X. Jehl, Y.-M. Niquet, M. Sanquer, M. Vinet, SOI technology for quantum information processing, *IEDM Technical Digest Papers*, pp. 13.4.1–13.4.4, San Francisco (CA), 2016.

111. R. Wacquez, M. Vinet, M. Pierre, B. Roche, X. Jehl, O. Cueto, J. Verduijn, G.C. Tettamanzi, S. Rogge, V. Deshpande, B. Previtali, C. Vizioz, S. Pauliac-Vaujour, C. Comboroure, N. Bove, O. Faynot, M. Sanquer, Single dopant impact on electrical characteristics of SOI NMOSFETs with effective length down to 10nm, *Symposium on VLSI Technology: Digest of Technical Papers*, Honolulu, HI, pp. 153–154, 2010. doi: 10.1109/VLSIT.2010.55 56224.

112. T. Shinada, M. Hori, F. Guagliardo, G. Ferrari, A. Komatubara, K. Kumagai, T. Tanii, T. Endo, Y. Ono, E. Prati, Quantum transport in deterministically implanted single-donors in Si FETs, *IEDM Technical Digest*, Washington, DC, pp. 30.4.1–30.4.4, 2011. doi: 10.1109/IEDM.2011.6131644.

113. M. Fuechsle, S. Mahapatra, F.A. Zwanenburg, M. Friesen, M.A. Eriksson, M.Y. Simmons, Spectroscopy of few-electron single-crystal silicon quantum dots, *Nature Nanotechnology*, vol 5, pp. 502–505, 2010.

114. L. Mathey, T. Alphazan, M. Valla, L. Veyre, H. Fontaine, V. Enyedi, K. Yckache, M. Danielou, S. Kerdiles, J. Guerrero, J.-P. Barnes, M. Veillerot, N. Chevalier, D. Mariolle, F. Bertin, C. Durand, M. Berthe, J. Dendooven, F. Martin, C. Thieuleux, B. Grandidier, C. Coperet, Functionalization of silica nanoparticles and native silicon oxide with tailored boron-molecular precursors for efficient and predictive p-doping of silicon, *The Journal of Physical Chemistry C*, 2015. doi: 10.1021/acs.jpcc.5b03408.

115. B. Roche, R.-P. Riwar, B. Voisin, E. Dupont-Ferrier, R. Wacquez, M. Vinet, M. Sanquer, J. Splettstoesser, X. Jehl, A two-atom electron pump, *Nature Communications*, vol 4, no 1581, 2013. doi: 10.1038/ncomms2544.

116. M. Barlas, A. Grossi, L. Grenouillet, E. Vianello, E. Nolot, N. Vaxelaire, P. Blaise, B. Traoré, J. Coignus, F. Perrin, R. Crochemore, F. Mazen, L. Lachal, S. Pauliac, C. Pellissier, S. Bernasconi, S. Chevalliez, J.F. Nodin, L. Perniola, E. Nowak, Improvement of HfO$_2$ based RRAM array performances by local Si implantation, *IEDM Technical Digest Papers*, pp. 14.6.1–14.6.4, San Francisco (CA), 2017.

117. B. Radisavljevic, A. Radenovic, J. Brivio, V. Giacometti, A. Kis, Single-layer MoS$_2$ transistors, *Nature Nanotechnology*, vol 6, pp. 147–150, 2011.

118. N. Ma, D. Jena, Charge scattering and mobility in atomically thin semiconductors, *Physical Review X*, vol 4, p. 011043, 2014.

119. K.F. Mak, C. Lee, J. Hone, J. Shan, T.F. Heinz, Atomically thin MoS$_2$: A new direct-gap semiconductor, *Physical Review Letters*, vol 105, p. 136805, 2010.

120. L. Hutin. M. Vinet, T. Poiroux, C. Le Royer, B. Previtali, C. Vizioz, D. Lafond, Y. Morand, M. Rivoire, F. Nemouchi, V. Carron, T. Billon, S. Deleonibus, O. Faynot, Dual metallic source and drain integration on planar single and double gate SOI CMOS down to 20 nm: Performance and scalability assessment, *Technical Digest*, Baltimore, MD, pp. 45–48, 2009. doi: 10.1109/IEDM.2009.5424425.

121. L. Hutin, O. Rozeau, V. Carron, J.-M. Hartmann, L. Grenouillet, J. Borrel, F. Nemouchi, S. Barraud, C. Le Royer, Y. Morand, C. Plantier, P. Batude, C. Fenouillet-Beranger, H. Boutry, T. Ernst, M. Vinet, Junction technology outlook for sub-28nm FDSOI CMOS, *Proceedings of IWJT*, Shanghai, PR China, pp. 149–154, 2014. doi: 10.1109/IWJT.2014.6842050.

122. K. Kita, A. Toriumi, Intrinsic origin of electric dipoles formed at high-k/SiO$_2$ interface, *IEDM Technical Digest*, San Francisco, CA, pp. 29–32, 2008. doi: 10.1109/IEDM.2008.4796605.

123. S. Gupta, P.P. Manik, R.K. Mishra, A. Nainani, M.C. Abraham, S. Lodha, Contact resistivity reduction through interfacial layer doping in metal-interfacial layer-semiconductor contacts, *Journal of Applied Physics*, vol 113, p. 234505, 2013.

124. V.R. Manfrinato, L. Zhang, D. Su, H. Duan, R.G. Hobbs, E.A. Stach, K.K. Berggren, Resolution limits of electron-beam lithography toward the atomic scale, *Nano Letters*, vol 13, no 4, pp. 1555–1558, 2013. doi: 10.1021/nl304715p.

125. L. Pain, S.V. Tedesco, B. Icard, M. Martin, C. Constancias, B.J. Kampherbeek, IMAGINE: An open consortium to boost maskless lithography take off: First assessment results on MAPPER technology, *SPIE Advanced Lithography Symposium Paper*, San Jose, CA, p. 7970, 2011.

126. R. Tiron, A. Gharbi, M. Argoud, X. Chevalier, J. Belledent, P. Pimmenta Barros, C. Navarro, G. Cunge, S. Barnola, L. Pain, M. Asai, C. Pieczulewski, The potential of block copolymer's directed self-assembly for contact hole shrink and contact multiplication, *Proceedings of SPIE, Alternative Lithographic Technologies*, San Jose, CA, vol 868012. doi: 10.1117/12.2011477.

127. H. Feng, X. Lu, W. Wang, N.-G. Kang, J.W. Mays, Block copolymers: Synthesis, self-assembly, and applications, *Polymers*, vol 9, no 10, p. 494, 2017. doi: 10.3390/polym9100494.

128. R.L. Graham, G.B. Alers, T. Mountsier, N. Shamma, S. Dhuey, S. Cabrini, R.H. Geiss, D.T. Read, S. Peddeti, Resistivity dominated by surface scattering in sub-50 nm Cu wires, *Applied Physics Letters*, vol 96, p. 042116, 2010.

129. S. Deleonibus, The Energy and Variability Efficient Era (E.V.E.) is ahead of us, *IEEE Journal of the Electron Devices Society*, vol 4, no 5, pp. 236–245, 2016.

130. J. Dijon, H. Okuno, M. Fayolle, T. Vo, J. Pontcharra, D. Acquaviva, D. Bouvet, A.M. Ionescu, C.S. Esconjauregui, B. Capraro, E. Quesnel, J. Robertson, Ultra-high density carbon nanotubes on Al-Cu for advanced vias, *IEDM Technical Digest*, San Francisco, CA, pp. 760–763, 2010. doi: 10.1109/IEDM.2010.5703470.

131. K. Tanaka, A. Tengeiji, T. Kato, N. Toyama, M. Shionoya, A discrete self-assembled metal array in artificial DNA, *Science*, vol 299, no 5610, pp. 1212–1213, 2003. doi: 10.1126/science.1080587.

132. M.G. Warner, J.-E. Hutchinson, Linear assembly of nanoparticles electrostatically organized on DNA scaffolds, *Nature Materials*, vol 2, pp. 272–277, 2003.

133. G. Clavé, G. Chatelain, A. Filoramo, D. Gasparutto, C. Saint-Pierre, E. Le Cam, O. Piétrement, V. Guérineau, S. Campidelli, Synthesis of a multi branched porphyrin-oligonucleotide scaffold for the construction of DNA-based nano-architectures, *Organic and Biomolecular Chemistry*, vol 12, p. 2778, 2014.

134. C. Brun, P.-H. Elchinger, G. Nonglaton, C. Tidiane-Diagne, R. Tiron, A. Thuaire, D. Gasparutto, X. Baillin, Metallic conductive nanowires elaborated by PVD metal deposition on suspended DNA bundles, *Small*, vol 13, p. 1700956, 2017. doi: 10.1002/smll.201700956.

135. A. Kumar, J.H. Hwang, S. Kumlar, J.M. Nam, Tuning and assembling metal nanostructures with DNA, *Chem Commun*, vol 49, p. 2597, 2013.

136. J. Hwang, J. Seo, Y. Lee, S. Park, J. Leem, J. Kim, T. Hong, S. Jeong, K. Lee, H. Heo, H. Lee, P. Jang, K. Park, M. Lee, S. Baik, J. Kim, H. Kkang, M. Jang, J. Lee, G. Cho, J. Lee, B. Lee, H. Jang, S. Park, J. Kim, S. Lee, S. Aritome, S. Hong, S. Park, A middle-1X nm NAND Flash memory cell (M1X-NAND) with highly manufacturable integration technologies, *IEDM Technical Digest*, pp. 199–202, Washington (DC), 2011.

137. L. Lattard, Maskless lithography for volume manufacturing, *SEMICON Europa*, Grenoble, France, October 7–9, 2014; L. Pain, S. Tedesco, C. Constancias, Direct write lithography: The global solution for R&D and manufacturing, *Comptes Rendus de Physique*, Académie des Sciences, vol 7, no 8, pp. 910–923, Elsevier, Paris, France.

138. J.-P. Dauvin, Il faut sauver le soldat Kondratief, In: *La Micro-Nanoélectronique, Enjeux et Mutations*, Ed. CNRS Editions, BIPE, Paris, pp. 20–25, 2009, citing the report Le transistor: Perspectivess à long terme d'une innovation majeure, Paris, 1966.

139. T. Ernst, L. Duraffourg, C. Dupre, E. Bernard, P. Andreucci, S. Becu, E. Ollier, A. Hubert, C. Halte, J. Buckley, O. Thomas, G. Delapierre, S. Deleonibus, B. de Salvo, P. Robert, O. Faynot, Novel Si-based nanowire devices: Will they serve ultimate MOSFETs scaling or ultimate hybrid integration? *IEDM Technical Digest Papers*, San Francisco, CA, 2008. doi: 10.1109/IEDM.2008.4796804.

140. J. Arcamone, J. Philippe, G. Arndt, C. Dupre, M. Savoye, S. Hentz, T. Ernst, E. Colinet, L. Duraffourg, E. Ollier, Nanosystems monolithically integrated with CMOS: Emerging applications and technologies, *IEDM Technical Digest Papers*, pp. 22.1.1–22.1.4, 2014. doi: 10.1109/IEDM.2014.7047098.

141. T. Jae King, L. Hutin, I-R. Chen, R. Nathanael, Y. Chen, E. Alon, Recent progress and challenges for relay logic switch technology, *Presented at the 2012 Symposium on VLSI Technology*, Honolulu, HI, June 2012.

142. S. Houri, G. Billiot, M. Belleville, A. Valentian, H. Fanet, Limits of CMOS technology and interest of NEMS relays for adiabatic logic applications, *IEEE Transactions on Circuits and Systems*, 62, no 6, pp. 1546–1554, 2015.

143. K.M. Milaninia, M.A. Baldo, A.Reina, J. Kon, All graphene electromechanical switch fabricated by chemical vapor deposition, *Applied Physics Letters*, vol 95, p. 183105, 2009. doi: 10.1063/1.3259415.

144. J. Sun, W. Wang, M. Muruganathan, H. Mizuta, Low pull-in voltage graphene electromechanical switch fabricated with a polymer sacrificial spacer, *Applied Physics Letters*, vol 105, no 3, p. 033103, 2014.

145. P. Batude, M. Vinet, B. Previtali, C. Tabone, C. Xu, J. Mazurier, O. Weber, F. Andrieu, L. Tosti, L. Brevard, B. Sklenard, P. Coudrain, S. Bobba, H. Ben Jamaa, P.-E. Gaillardon, A. Pouydebasque, O. Thomas, C. Le Royer, J.-M. Hartmann, L. Sanchez, L. Baud, V. Carron, L. Clavelier, G. De Micheli, S. Deleonibus, O. Faynot, T. Poiroux, Advances, challenges and opportunities in 3D CMOS sequential integration, *IEDM Technical Digest Papers*, Washington, DC, pp. 7.3.1–7.3.4, 2011. doi: 10.1109/IEDM.2011.6131506.

146. O. Turkyilmaz, G. Cibrario, O. Rozeau, P. Batude, F. Clermidy. 3d FPGA using high density interconnect monolithic integration. *In Design, Automation and Test in Europe Conference and Exhibition (DATE)*, Dresden, Germany, 2014. doi: 10.7873/DATE. 2014.351.123.

147. K. Abe, M.P. Tendulkar, J.R. Jameson, P.B. Griffin, K. Nomura, S. Fujita, Y. Nishi, Ultra-high bandwidth memory with 3D-stacked emerging memory cells, *ICICDT Technical Digest Papers*, Austin, TX, pp. 203–206, 2008. doi: 10.1109/ICICDT.2008.4567279.

148. M. Veldhorst, H.G.J. Eenink, C.H. Yang, A.S. Dzurak, Silicon CMOS architecture for a spin-based quantum computer, *Nature Communications*, vol 8, no 1, p. 1766, 2017.

149. M.M. Shulaker, T.F. Wu, A. Pal, L. Zhao, Y. Nishi, K. Saraswat, H.-S. Philip Wong, S. Mitra, Monolithic 3D integration of logic and memory: Carbon nanotube FETs, resistive RAM, and silicon FETs, *IEDM Technical Digest Papers*, San Francisco, CA, pp. 27.4.1–27.4.4, 2014.

150. P. Batude, C. Fenouillet-Beranger, L. Pasini, V. Lu, F. Deprat, L. Brunet, B. Sklenard, F. Piegas-Luce, M. Casse, B. Mathieu, O. Billoint, G. Cibrario, O. Turkyilmaz, H. Sarhan, S. Thuries, L. Hutin, S. Sollier, J. Widiez, L. Hortemel, C. Tabone, M.-P. Samson, B. Previtali, N. Rambal, F. Ponthenier, J. Mazurier, R. Beneyton, M. Bidaud, E. Josse, E. Petitprez, O. Rozeau, M. Rivoire, C. Euvard-Colnat, A. Seignard, F. Fournel, L. Benaissa, P. Coudrain, P. Leduc, J.-M. Hartmann, P. Besson, S. Kerdiles, C. Bout, F. Nemouchi, A. Royer, C. Agraffeil, G. Ghibaudo, T. Signamarcheix, M. Haond, F. Clermidy, O. Faynot, M. Vinet, 3DVLSI with Cool-Cube process: An alternative path to scaling, *In Symposium on VLSI Technology: Digest of Technical Papers*, Kyoto, Japan, pp. 40–41, 2015. doi: 10.1109/VLSIT.2015.7223698.

151. P. Batude, L. Brunet, C. Fenouillet-Beranger, F. Andrieu, J.-P. Colinge, D. Lattard, E. Vianello, S. Thuries, O. Billoint, P. Vivet, C. Santos, B. Mathieu, B. Sklenard, C.-M.V. Lu, J. Micout, F. Deprat, E. Avelar Mercado, F. Ponthenier, N. Rambal, M.-P. Samson, M. Cassé, S. Hentz, J. Arcamone, G. Sicard, L. Hutin, L. Pasini, A. Ayres, O. Rozeau, R. Berthelon, F. Nemouchi, P. Rodriguez, J.-B. Pin, D. Larmagnac, A. Duboust, V. Ripoche, S. Barraud, N. Allouti, S. Barnola, C. Vizioz, J.-M. Hartmann, S. Kerdiles, P. Acosta Alba, S. Beaurepaire, V. Beugin, F. Fournel, P. Besson, V. Loup, R. Gassilloud, F. Martin, X. Garros, F. Mazen, B. Previtali, C. Euvrard-Colnat, V. Balan, C. Comboroure, M. Zussy, Mazzocchi, O. Faynot, M. Vinet, 3D sequential integration: Application-driven technological achievements and guidelines, *IEDM Technical Digest Papers*, pp. 3.1.1–3.1.4, San Francisco (CA), 2017.

152. M. Koyanagi, Y. Nakagawa, K.-W. Lee1, T. Nakamura, Y. Yamada, K. Inamura, K.-T. Park, H. Kurino, Neuromorphic vision chip fabricated using three-dimensional integration technology, *Proceedings of ISSCC,* San Francisco, CA, p. 17.2, 2001.

153. G. Poupon, Y. Lamy, A. Rouzaud, New era for packaging and coming challenges for interposers,

Proceedings of Pan Pacific Microelectronics Symposium, Hawaii, HI, p. 19, 2014.

154. O. El Bouayadi, Y. Lamy, L. Dussopt, A high-impedance surface antenna on silicon interposer for 3D integrated mmW transceivers, *European Microwave Conference (EuMC)*, Rome, Italy, October 5–10, 2014.

155. L. Di Cioccio, F. Baudin, P.Gergaud, V. Delaye, P.-H. Jouneau, F. Rieutord, T. Signamarcheix, Modeling and integration phenomena of metal-metal direct bonding technology, *ECS Transaction*, vol 64, no 5, pp. 339–355, 2014.

156. V.P. Phan, B. Pecquenard, F. Le Cras, High-performance all-solid-state cells fabricated with silicon electrodes, *Advanced Functional Materials*, vol 22, pp. 2580–2584, 2012. doi: 10.1002/adfm.201200104.

157. Guardian Angel, Guardian Angel project final report July 2012. Available: www.ga-project.eu/.

158. Z. Zhu, T.K. Tam, F. Sun, C. You, Y.-H. Percival Zhang, A-high-energy-density-sugar-biobattery-based-on-a- synthetic-enzymatic-pathway, *Nature Communications*, vol 5, p. 3026, 2004. doi: 10.1038/ncomms4026.

10

Semimetal Electronics: Quantum Confinement and Surface Chemistry as Design Tools

Alfonso Sanchez-Soares
EOLAS Designs

Christian König and Conor
O'Donnell
Tyndall National Institute

Jean-Pierre Colinge
CEA, LETI

James C. Greer
University of Nottingham Ningbo China

10.1 Introduction... 10-1
  Nanopatterned Materials: Novel Electronic Properties at the Nanoscale •
  Semimetals at the Nanoscale: A New Class of Semiconductors • New
  Opportunities for Device Designs: Challenges in Scaling Traditional Designs
10.2 Quantum-Size Effects in Semimetals 10-4
  Semimetal-to-Semiconductor Transition: The Quantum Confinement Effect •
  Surface Effects: Tuning Confinement Effects through Surface Chemistry
10.3 Engineering Monomaterial Electronic Devices: Novel
  Schottky-Type Device Designs 10-8
  Monomaterial Schottky-Type Junctions • Confinement-Modulated Designs •
  Surface Chemistry: Electrostatic Effects of Surface Passivants
10.4 Conclusions and Outlook................................. 10-13
References 10-13

10.1 Introduction

Modern electronic devices commonly used every day around the globe are capable of processing impressive amounts of information almost instantly. At the core of their capabilities lies their ability to perform vast numbers of mathematical operations in a fraction of a second. Under several layers of software and hardware, electronic devices' mathematical abilities lie in core components, including central processing units (CPUs) and memory devices (such as random access memory (RAM)), which are composed at their core by large arrays of a single basic component: the transistor. Nowadays, commercial CPUs incorporate up to 20 billion of these building blocks into the same chip, arranged into digital logic gates capable of performing basic mathematical operations.

The type of transistor commonly found in logic and memory devices, the field-effect transistor (FET), is a semiconductor device in which current flow between two of its terminals (commonly referred to as *source* and *drain*) through the region in between them–the *channel*–is modulated by a potential applied on a third terminal that electrostatically influences the channel: the *gate*. Figure 10.1 shows a schematic indicating the basic anatomy of a metal-oxide-semiconductor FET (MOSFET) commonly found in logic devices.

Since the birth of electronics and for decades to come, the transistor was embodied as a planar device in which a channel region is formed on a silicon substrate by introducing a doping profile that alternates n-type and p-type dopants (*i.e.* electron-rich or electron-deficient regions)

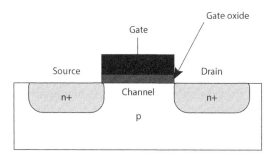

FIGURE 10.1 A schematic of an n-type MOSFET typically found in digital circuits. The source and drain regions are heavily doped n-type (n^+), while the substrate is doped p-type (p). By applying a potential at the gate electrode, an n-type channel forms, allowing the flow of electrons between the source and the drain.

along the intended conduction path, thus defining the channel as the region between two back-to-back p-n junctions. The gate electrode sitting on top of the channel–electrically insulated by an oxide barrier–then modulates current flow between the source and drain electrodes near the surface of the channel by electrostatically shifting its electronic levels and thus varying band alignments between the channel, source, and drain to either promote electron flow (*ON* state) or impede it (*OFF* state). Such basic switches can be arranged to design logic gates capable of implementing Boolean functions; logic gates can then be arranged into higher-level digital circuits that can perform familiar operations such as addition and subtraction. In an effort to increase the processing power in integrated circuits while reducing their cost per function and power consumption, research in the field over the past 50 years has focused on increasing the amount of digital logic gates present in chips by miniaturization of transistor designs. Advances in semiconductor processing over the past decades have led to a reduction of typical feature sizes present in commercial transistor designs by orders of magnitude: from micrometers to nanometers.

This dramatic reduction in size has brought several challenges as a result of a collection of effects detrimental to device operation, collectively known as *short-channel effects* that have become increasingly important [1]. As designs are scaled down, electric fields in different regions of transistors increasingly influence each other, resulting in degraded performance in devices with channel lengths below approximately 1μm. Other effects that are exacerbated with scaling and hamper device operation include increasing leakage currents between the gate electrode and substrate due to quantum-mechanical tunneling currents through thinner gate oxides, and reliability issues arising from accumulated material damage due to acceleration of charge carriers in increasingly large electric fields, as supply voltages are kept relatively constant. As channel lengths drop below 100 nm, the electrical characteristics of traditional transistor designs begin to deviate significantly from those of larger devices, as all these effects come into play and limit their performance. The introduction of increasingly complex processing techniques incorporating new materials and modified architectures has allowed mitigation of these effects and maintained suitable electrical characteristics up to channel lengths below 20 nm.

Aside from the aforementioned short-channel effects, miniaturization below approximately 10 nm brings into play other fundamental effects arising due to the nature of matter and its discreteness, which call for novel transistor designs employing new architectures. At feature sizes on the order of tens of atoms across, the discrete nature of matter becomes relevant as the use of dopants in regions composed of relatively small numbers of atoms introduces variability issues that render traditional designs unsuitable for commercial production. Additionally, confinement of electrons into regions with dimensions comparable to their de Broglie wavelength [2] results in significantly modified electronic structure and charge-transport related properties. Although the effects of *nanopatterning* hamper the operation of traditional designs, new possibilities arise as electron confinement brings about new ways of engineering material properties that may be exploited in innovative device concepts.

10.1.1 Nanopatterned Materials: Novel Electronic Properties at the Nanoscale

Advances in fabrication and characterization techniques over the past decades have allowed the realization of material structures with characteristic dimensions as low as a few nanometers. In such small structures, confinement of electrons to regions with characteristic lengths below their de

Broglie wavelength and the concomitant high density of interfaces result in effects that can severely alter how materials behave: patterning materials such that the distance between interfaces confining charge carriers becomes comparable to the their associated effective wavelength, quantization of their energy spectrum results in striking changes to their electrical and optical properties; thus, opening the possibility of engineering structures with different properties than those of larger pieces of materials with the same chemical composition. In addition, surface effects that can typically be ignored when describing the electrical characteristics of traditional electronic devices begin to play a key role in nanoscale device physics. The presence of interfaces between materials induces effects that typically penetrate a few nanometers into their structures and can be ignored for the most part when describing the properties of bulk materials, but the close proximity of interfaces and large surface-to-volume ratio inherent to nanostructures translates into an increasingly strong dependence of device properties on those of its interfaces.

The effects of nanostructuring materials have been extensively studied over the last few decades, with unusual behavior springing from confinement effects often reported in the literature and increasing amounts of applications being proposed regularly. In the following, we shall provide a brief historical overview of *quantum-size* or *quantum confinement* effects in inorganic semiconductors and semimetals, and mention only a handful of their many applications relevant to optoelectronic devices.

Experimental evidence of quantum-size effects have been reported in the literature since the 1980s. In one of the first of such studies, the absorption spectra of crystalline semiconductor clusters with radii below 30 nm grown on silicate glasses was found to exhibit a strong dependence on cluster size; the measured optical bandgap was observed to shift towards energies as much as 33% larger with decreasing cluster size [3]. In this case, the measured structures were *zero-dimensional*: in these type of nanostructures–also referred to as *quantum dots, nanoparticles*, or *nanocrystals*–electrons are confined in all spatial directions and thus exhibit the most prominent quantum-size effects. The size-induced discretization of electronic levels and consequent tunability of their optical properties have made quantum dots particularly studied for applications in photonics, as their light absorption and emission spectra can be customized by controlling their shape and dimensions.

An analogous effect was observed in the optical properties of silicon–the semiconductor industry's material of choice–when the photoluminescence spectra of nanoscale wire arrays fabricated via electrochemical and chemical dissolution were measured to *blueshift* (*i.e.* shift toward higher energies) with decreasing wire cross section [4]. In these *one-dimensional* structures–often referred to as *quantum wires* or *nanowires* (NW)–electron movement is restricted along two spatial dimensions, and thus observed quantum-size effects are generally less prominent than in quantum dots.

Since their geometrical shape includes one extended dimension along which electrical current can flow, NWs have been extensively studied over the past two decades for applications in electronic devices and presently comprise one of the most promising successors of traditional planar technology [1]. Although the first mechanism for synthesizing semiconductor NWs was elucidated in the 1960s [5], it wasn't until the late 1990s that it became a research interest for the electronics community [6]. Since then, the properties of semiconductor NWs have been extensively studied, and their electronic, mechanical, thermal, and optical properties have been found to exhibit a strong size dependence in the nanoscale regime. For silicon NWs (SiNWs) with approximately cylindrical cross section, the electronic bandgap–a key material property for optoelectronic applications–is known to increase for diameters below 10 nm from its bulk value of 1.1 eV to as large as 3.5 eV for NWs with diameters around 1 nm. This translates into the possibility of tuning silicon's optical spectrum from infrared all the way to ultraviolet radiation merely by geometric effects.

In *two-dimensional* nanostructures, confinement is introduced in one spatial dimension, allowing electrons to move along the remaining two unconstrained ones. While the impact of confinement in these *thin films* or *quantum wells* is generally lower than in other types of nanostructures discussed previously, exploitation of quantum-size effects for optoelectronic devices was first proposed in this type of nanostructures: in the 1970s, while the field of integrated optics was gaining momentum, research in using thin films as optical waveguides led to elucidation of an application that later became known as *quantum well lasers* [7,8]. Figure 10.2 shows the band profile associated with thin-film heterostructures typically employed in semiconductor lasers, illustrating this type of device name's etymology: arranging alternating layers of two materials with different bandgaps $E_{G,1}$ and $E_{G,2}$ effectively creates a band profile in which electrons (holes) located in thin regions composed of the material with the lower bandgap are confined to *wells* at the conduction (valence) band edge. The distribution of discretized energy levels in these wells can then be tuned by varying chemical compositions and layer thickness, allowing great control over the optical properties of these heterostructures. Quantum well lasers were extensively investigated in the coming years, eventually becoming one of the most popular and versatile laser designs with applications in several mass-produced devices such as fiber optic communication links, optical disc players, and laser printers.

During the 1960s and in parallel with the first reports of bandgap *widening* in semiconductor thin films, observation of quantum-size effects in semimetals was first reported in the literature. The electric and magnetic properties of bismuth thin films were measured to qualitatively deviate from their bulk behavior for film thicknesses below 40 nm, indicating a transition of their electronic properties of films towards semiconducting (SC) behavior and thus introducing a novel class of nanoscale-exclusive semiconductors with potential applications in the field of nanoelectronics that would emerge a few decades later [9].

10.1.2 Semimetals at the Nanoscale: A New Class of Semiconductors

Semimetals are materials with an electronic structure somewhat in between that of metals and semiconductors. In metals, the electronic structure is characterized by a partially filled CB; that is, the Fermi level lies inside at least one band, thus providing electrons in the vicinity with a large number of states for electrical conduction. In contrast to metals, semiconductors and insulators have well-defined VB and CBs separated by an energy bandgap. Electrons occupying states at the top of the VB thus have no available states for conduction, unless they are promoted to the bottom of the CB, requiring relatively large energies on the order of electron volts. Therefore, semiconductors allow control of current flow in gated devices by shifting band alignments between device regions through application of electrostatic potentials with magnitudes on the order of volts. The electronic structure of semimetals closely resembles that of semiconductors in that VB and CB can be clearly identified, but exhibit a small overlap between them instead of being separated by a bandgap. Although the list of elemental semimetals is limited and includes arsenic, antimony, bismuth, and some allotropes of tin and carbon, the number of reported alloys and low-dimensional materials with SM properties is ever increasing. Some examples include mercury telluride, tin–germanium alloys, and some transition metal dichalcogenides (TMDs) such as tungsten ditelluride [10]. Figure 10.3 schematically illustrates the characteristic band structure of metals, semiconductors/insulators, and semimetals.

In a similar fashion to bandgap widening observed in semiconductors, nanopatterning semimetals results in a reduction of the VB and CB overlap and subsequent bandgap opening, such that their electronic properties transition into those of a semiconductor. As previously mentioned, the transition was first measured in bismuth thin films–where the critical film thickness for SC behavior has been established

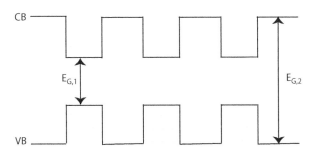

FIGURE 10.2 Band profile of a quantum well heterostructure showing the conduction band (CB) and valence band (VB) edges. Alternating layers of two materials with bulk bandgaps $E_{G,1}$ and $E_{G,2}$ creates a band profile in which charge carriers are confined to thin material regions, significantly altering electronic and optical properties.

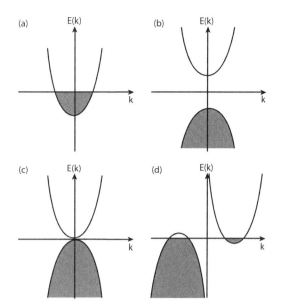

FIGURE 10.3 Characteristic band dispersion of different types of materials. (a) In metals, the Fermi level crosses at least one (partially filled) band. (b) In semiconductors, the Fermi level lies within a clear gap between completely filled (valence) bands and completely empty (conduction) bands. Similarly, semimetals exhibit VB and CB that may show *zero bandgap* (c) or a *negative bandgap* (d).

to be around 30–60 nm [11,12]–and has since been reported for other types of nanostructures and semimetals, such as antimony, tin, bismuth–antimony alloys, and TMDs such as molybdenum telluride [13–16]. As feature sizes in modern electronic components have already reached dimensions below those at which transitions have been reported to occur and, given the increasing difficulties in downscaling silicon-based technology further, the new class of materials that nanopatterned semimetals comprise is currently being investigated for applications in nanoelectronic component designs.

10.1.3 New Opportunities for Device Designs: Challenges in Scaling Traditional Designs

Given that the degree of miniaturization of electronic components has already reached dimensions below 10 nm and efforts in continuing downscaling of silicon-based technology are running into fundamental limits in terms of material properties and suitability of the doping paradigm, we may be approaching the end of traditional designs and possibly the dominance of silicon in most electronic applications. In the following sections, we provide a theoretical overview of quantum-size effects observed in the electronic properties of semiconductors and semimetals which are particularly relevant to charge transport and briefly discuss how surface effects can further influence their electronic structure in significant ways. Finally, we provide some examples of how quantum-size and surface effects induced in SM nanostructures can be exploited for electronic

component designs that overcome some of the difficulties encountered by further miniaturization of designs presently in production.

10.2 Quantum-Size Effects in Semimetals

10.2.1 Semimetal-to-Semiconductor Transition: The Quantum Confinement Effect

In previous generations of silicon microelectronic devices, charge carriers would flow through a relatively large transistor channel with dimensions of the order of micrometers. From a practical point of view, the transport of electrons and holes can be described accurately by the band structure of the bulk material in such devices. In this regime, all relevant quantum mechanical aspects of the system can then be absorbed into a single quantity per band–the *effective mass*–given by the dependence of the charge carrier energy on crystal momentum near the band's edge.

As discussed, several issues arise upon the reduction of the size of transistors down to nanometer length scales. We shall focus here on changes in the material properties rather than on issues with device operation. To understand the underlying principle, we make use of a simple model employing several assumptions. First of all, we assume that the material is crystalline with a lattice that does not change with respect to the bulk material despite its finite size. Strictly speaking, this is not always the case, but this approximation is nevertheless useful to understand the underlying principles. Second, it is useful to think in terms of independent noninteracting electrons for the sake of simple equations. We shall thus adopt the Schrödinger equation describing a single electron in a mean-field potential V as

$$\left[-\frac{\hbar^2}{2m_e}\left(\frac{\partial^2}{\partial x^2}+\frac{\partial^2}{\partial y^2}+\frac{\partial^2}{\partial z^2}\right)+V\left(x,y,z\right)\right]\Psi\left(x,y,z\right)$$
$$= E\Psi\left(x,y,z\right). \tag{10.1}$$

Given that the band structure of semiconductors can be locally approximated as parabolas around the energies at which charge transport relevant to device operation takes place (*i.e.* around band edges), we employ the effective mass approximation whereby we assume a parabolic band dispersion as

$$E(\vec{k}) = E_0 + \frac{\hbar^2 \vec{k}^2}{2m^*}, \tag{10.2}$$

where E_0 is the band offset with respect to some origin of energies, $\hbar$ is the reduced Planck constant, and the band curvature is described by its effective mass m^*.

Onset of Quantum-Size Effects

The name *quantum size* already implies that the electrons and holes in our model devices are located in a thin film or a narrow wire. To clarify what *thin* means in this context, it is beneficial to recall some of the findings from the early days of quantum mechanics. The wave-particle duality describes that the behavior of particles–such as photons or electrons–can, depending on the situation, be best described as particles with a certain mass and momentum, or as a wave with an energy-dependent wavelength. As proposed in 1924 by de Broglie [2], this principle can be applied to electrons, where the following relation holds:

$$\lambda = h/p , \tag{10.3}$$

where the wavelength λ is determined by the particle momentum p and Planck's constant h. We know from diffraction experiments that the wave character of a particle becomes important at length scales of the order of λ. We can therefore expect the onset of quantum effects in films or wires to occur at length scales of confinement of approximately the wavelength of charge carriers in the channel [17].

In a semiconductor device, charge carriers usually are excited thermally into a conducting state. Thus, in the effective mass approximation, we can deduce the particle momentum p from setting the kinetic energy E equal to the thermal energy $k_B T$ (k_B being Boltzmann's constant and T the temperature). Our Eq. (10.3) then becomes

$$\lambda = h/p = \frac{h}{\sqrt{2m^*E}} = \frac{h}{\sqrt{2m^*k_B T}} . \tag{10.4}$$

Here it becomes already clear that the bulk material properties (e.g. the effective mass m^*) determine the influence of the quantum effects. Let us consider, for example, an electron moving along the [100] direction in a silicon crystal where the effective mass associated to the bottom of the CB is $m^* = 0.98 \cdot m_e$. At room temperature, the thermal energy is approximately 25meV, therefore

$$\lambda_{RT}^{\mathrm{Si[100]}} \approx \frac{h}{\sqrt{2 \cdot (0.98 \cdot m_e) \cdot 25\mathrm{meV}}} \approx 7.8\,\mathrm{nm}. \tag{10.5}$$

We can learn from this simple example that the wavelength of electrons in silicon can be as large as feature sizes in transistors comprising modern digital integrated circuits and hence quantum-size effects are becoming increasingly relevant to the design of electronic components. Note that most materials exhibit anisotropic band curvatures, and thus the effective mass depends on the crystallographic direction considered. If the electron travels along the [110] direction in a silicon crystal, its effective mass m^* is $0.19 \cdot m_e$ and the corresponding wavelength $\lambda_{RT}^{\mathrm{Si[110]}} \approx 1/\sqrt{5} \cdot 7.8\,\mathrm{nm} \approx 17.2\,\mathrm{nm}$ is even larger. Historically, quantum-size effects were first observed in materials with lower effective mass values; in bismuth–the first semimetal in which such effects were reported–the onset of size effects at room temperature occurs at dimensions as large as $\lambda_{RT}^{\mathrm{Bi}} \approx 70\,\mathrm{nm}$.

It is important to note from Eq. (10.4) that the temperature is also relevant for the occurrence of quantum-size effects. Because of a longer associated particle wavelength, the onset of confinement effects can be observed at larger dimensions with decreasing temperature.

Increase of the Bandgap

Now that we know under which conditions to expect the onset of quantum-size effects, we shall now consider their consequences. If we fabricate a thin film, the charge carriers are confined in one direction of space; likewise, confinement in two dimensions can be achieved within a NW. While the charge carriers can move freely as Bloch waves in the bulk material, their motion is constrained along certain directions in nanostructures. Considering a single particle in a free-standing film, the Schrödinger equation (10.1) allows us to separate the wave function $\Psi(x, y, z)$ into a product $\psi(x, y)\psi_c(z)$, where $\psi_c(z)$ describes the behavior in the confinement direction independently of $\psi(x, y)$.

Just like in the textbook example of a particle in a box, $\psi_c(z)$ forms standing waves between the confining potential barriers that represent the edges of the material. The confining potential then forces electrons to remain in the box and their wave function to vanish at the surface of the material. For the following considerations, we define the film thickness along z to be L, drawing a parallel to the particle-in-a-box textbook example. The wave vectors and corresponding energies are quantized accordingly (see for example, [18, Chapter 4]):

$$k = \frac{\pi}{L} \cdot n, \tag{10.6}$$

$$\Delta E_n = \frac{\hbar^2 k^2}{2m^*} \propto \frac{n^2}{L^2 m^*}, \tag{10.7}$$

with $n \in \mathbb{N}$.

For our previous example of a silicon thin film confined along [100], we can estimate the following shift in energy for the bottom of the CB and top of the VB as

$$\Delta E_n^{\mathrm{CB}} = \frac{\hbar^2 k^2}{2 \cdot 0.98 \cdot m_e} \approx 384\mathrm{meV} \cdot \frac{n^2}{(L/1\,\mathrm{nm})^2}, \tag{10.8}$$

$$\Delta E_n^{\mathrm{VB}} = \frac{\hbar^2 k^2}{2 \cdot (-0.49) \cdot m_e} \approx -767\mathrm{meV} \cdot \frac{n^2}{(L/1\,\mathrm{nm})^2}. \tag{10.9}$$

The bandgap may then be estimated within our model as the bandgap in the bulk material plus confinement contributions of both band edges with $n = 1$ (*i.e.* separation between first *subband* in the VB and CB) as

$$E_g(L) = E_{g,\mathrm{bulk}} + \left(\Delta E_1^{\mathrm{CB}} - \Delta E_1^{\mathrm{VB}}\right) \approx 1.1\,\mathrm{eV} + \frac{1.15\,\mathrm{eV}}{(L/1\,\mathrm{nm})^2}, \tag{10.10}$$

where effective mass values employed in shift estimates correspond to the bottom of the CB and top of the VB, and their opposite signs reflect their opposite curvatures in momentum space. Note additionally that the relevant effective mass values employed in calculations correspond to crystallographic directions along which charge carriers are confined. For systems with two-dimensional confinement (*i.e.* NWs), we accordingly get two contributions to each band of the form in Eq. (10.7), resulting in an expression analogous to Eq. (10.10). Figure 10.4a shows the CB

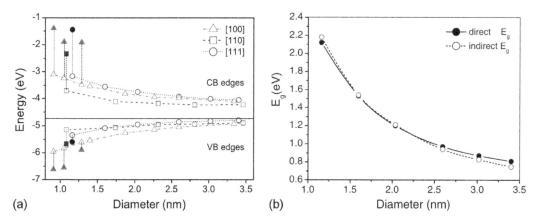

FIGURE 10.4 Band-edge shifts in SiNWs as a function of diameter. (a) Density functional theory results for the CB and VB edges in NWs with different orientations. With decreasing wire diameter, the CB edge is pushed towards higher energies while states at the VB edge are pushed towards lower energies. The solid data points show results employing more accurate many-body corrections. (b) Corresponding direct and indirect bandgap for NWs oriented along [111]. Reprinted figures with permission from Jia-An Yan, Li Yang, and M. Y. Chou, *Physical Review B* (76), 115319 (2007). Copyright (2007) by the American Physical Society.

and VB edges for SiNWs grown along different crystallographic orientations, as calculated from first principles [19]. As expected, the electron states at the bottom of the CB are shifted towards higher energies with decreasing NW diameter while hole states at the bottom of the VB shift towards lower energies. Figure 10.4b shows the bandgap widening of both direct and indirect bandgaps calculated for SiNWs grown along [111]. Since the band associated with the indirect bandgap exhibits a lighter effective mass, its confinement-induced shift is more pronounced than for the band associated with the direct bandgap (as per Eq. (10.7), $\Delta E \propto m^{*-1}$), and we observe a crossover from indirect to direct bandgap at NW diameters below ≈ 2 nm.

Despite the simplicity of the effective mass model employed in our calculations thus far, it qualitatively describes size-induced effects on semiconductor bandgaps. Although more complex effects need to be accounted for to accurately describe real systems, a similar expression to Eq. (10.10), which is commonly employed to fit results from both experiments and more sophisticated theoretical calculations, is given by [18, Chapter 5]

$$E_g(L) = E_{g,\text{bulk}} + C/L^\alpha \;, \qquad (10.11)$$

where C and α are fitting parameters and the case of $\alpha = 2$ being equivalent to Eq. (10.10). For SiNWs with varying crystallographic orientations, fitting to bandgaps computed from first principle calculations yield values of α between 1.4 and 1.8 [19].

Application to Semimetals

We have seen that the bandgap of semiconductors is modified by confinement effects in nanostructures; the smaller the confinement length and the more spatial directions along which charge carrier movement is restricted, the larger the bandgap. In a similar manner, the principle of quantum confinement also applies to materials that do not exhibit a

bandgap at all in their bulk form, *i.e.* metals and semimetals. While the effects of quantum confinement are detrimental for silicon devices in highly integrated digital circuits, transistor designs based on semimetals can take advantage of these effects. Therefore, semimetal devices offer not just an alternative technology to bulk semiconductors but resolve some of the crucial problems currently encountered by the semiconductor industry.

In Section 10.1.2, we discussed that the electronic structure of semimetals is somewhere in between those of metals and semiconductors. Semimetals are characterized by having relatively few states around the Fermi level due to small overlaps between their CB and VB. For confined semimetals, significant changes in material properties can be expected as the effect of quantum confinement can shift their bands around the Fermi level and reduce their overlap to the point of inducing a bandgap in the material, *i.e.* confining semimetals can lead to a semimetal-to-semiconductor transition.

Some electronic component designs exploiting the effects of quantum confinement in nanostructured semimetals, such as bismuth and α-tin, have already been proposed [20–24]. The predicted and measured bandgaps are much larger than the thermal energy at room temperature for both materials when their characteristic dimensions drop below 10 nm, making these types of structures suitable for applications in digital integrated circuits. Some of these designs are discussed in Section 10.3.

Formation of Subbands

In the discussion earlier, we have only considered the band edges, *i.e.* the CB minimum and VB maximum, which shift *up* and *down* in energy, respectively. Taking also into account the cases with $n > 1$ in Eq. (10.7), we find that as a direct consequence an infinite number of so-called subbands appear in the band structure. Their confinement-induced shift increases rapidly with increasing subband index as $\Delta E_n \propto n^2$, and therefore, their effect increases when moving away from band edges. In a NW, we expect a subband for each combination of the independent quantum numbers n and n' corresponding to the two confined directions.

In Figure 10.5, we show schematically how subbands form above the CB edge of a thin film and their position with respect to a reference energy. Without confinement ($n = 0$), the band edge lies below the reference energy. However, the lowest conduction state that is possible with confinement ($n \geq 1$) is already located above the reference energy. We show two additional subbands corresponding to $n = 2$ and $n = 3$. This simple picture illustrates the existence of additional subbands, which is a general characteristic of the electronic structure of confined materials.

Band Folding: Electron and Hole Pockets

In the previous sections we have discussed, the principle of quantum confinement and how it can change the magnitude of the bandgap in the material. We have thus far considered the standing waves induced in the confinement direction but not the effects on bands in reciprocal space. We shall now have a closer look at the underlying band structure along the transport direction. What kind of band structure can we expect, assuming we already know the bulk band structure? Where are the charge carrier pockets that contribute to the current located in reciprocal space?

In the context of current flowing through a device, the position of the CB minimum and VB maximum are the most important features in the material's band structure. These electron and hole pockets are populated with mobile charge carriers by thermal excitation and determine the transport properties relevant for device operation. For typical bulk semiconductors, the electron and hole pocket positions in

k-space are well known. The curvature of the bands at these points is parametrized by an effective mass that can be used to describe electrical transport through the material within a simple model.

If the charge carriers are confined in a thin film or wire, they may not move in every direction, and the electron and hole pockets of the bulk material may be located along directions that have been confined and are thus not accessible in the nanostructure. As a general rule of thumb, their positions in the bulk band structure are folded back to the transport direction [18, Chapter 2]. This means that, in a thin film, the three-dimensional band structure is projected into the two-dimensional plane defined by nonconfined directions; a material's bulk band structure is similarly projected to a one-dimensional dispersion in NWs. A few illustrative examples for the band folding in confined silicon are given in Figure 10.6.

To sum up, we have found that the electronic properties of a confined material can be very different from their bulk counterpart. The general features of band structures like the position of charge carrier pockets in momentum space may be modified, and the symmetry and character of states at band edges may change. The quantum mechanical nature of electrons changes the energy spacing between the VB and CB, and states within bands discretize with the formation of *subbands*. Although the models presented here are not sufficient to accurately describe the band structure of a real material, they illustrate the main relevant concepts. More sophisticated theoretical frameworks such as density functional theory (DFT) and many-body theories like the GW approximation are state-of-the-art methods to calculate the electronic structure of both bulk and nanostructured materials, and take into account interactions between electrons that were completely neglected in the simple model presented here.

10.2.2 Surface Effects: Tuning Confinement Effects through Surface Chemistry

As discussed in the previous section, the crystal orientation along which confinement occurs has a crucial influence on a nanostructure's electronic structure. However, it alone does not determine the bandgap of a real nanostructure with a given geometry. As the surface-to-volume ratio increases rapidly with decreasing material size, surface effects become increasingly important. In particular, effects that influence the shape of the confining potential near surfaces such as the crystallographic orientation of exposed surface facets and the nature of chemical bonds at a nanostructure's surface have been found to be an important factor in determining the electronic properties, such as the bandgap magnitude and electron affinity [21,25,26]. For [110]-oriented α-tin NWs with a diameter of 1.5 nm, SM and SC behavior was predicted with bandgaps ranging between 0.70 and 2.05 eV, depending on the atomic species bonding to the NW surface [25]. Variations in bandgap magnitude at constant crystallographic orientation and cross-sectional dimensions were

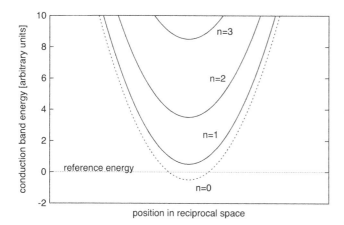

FIGURE 10.5 Subbands emerge as a result of confinement in a thin film. In this example, the CB edge without confinement, $n = 0$, lies below a reference energy. Quantum confinement increases the energy of the lowest electronic states that correspond to $n = 1$ in Eq. (10.7). This can push the CB edge above the reference energy. Higher quantum numbers give rise to additional subbands with increasingly shifted energies.

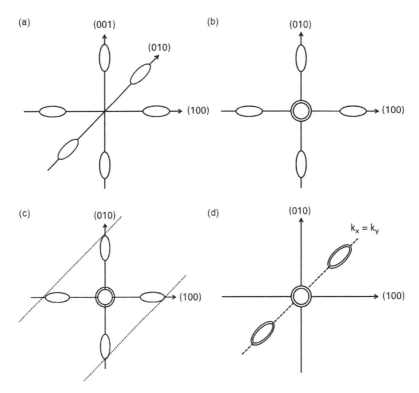

FIGURE 10.6 Band folding example. (a) Electron pockets in bulk silicon. (b) Electron pockets in Si(001); electrons can no longer move along [001] in reciprocal space. Therefore, the CB minima are projected onto the remaining two-dimensional k-space, where they lie at the Brillouin zone center (Γ). (c) Further confinement of the film that results in a NW with [110] orientation. In the wire (d), only one extended direction remains, onto which the electron pockets are projected. Reprinted figures with permission from Jean-Pierre Colinge, and James C. Greer, *Nanowire transistors: Physics of devices and materials in one dimension.* Copyright (2016) by Cambridge University Press.

ascribed to surface chemical effects that were found to be the same order of magnitude as quantum confinement effects for nanostructures with sufficiently large surface-to-volume ratio.

The electronegativity of atoms or molecules passivating the surface determines how much charge is transferred from the material to the surface as a result of chemical bonding. An accumulation of charge at the surface induces a depletion of charge to a few nanometers into the material itself. For the semimetal NW example mentioned earlier, this situation was found to be similar to a cylindrical capacitor: the charge on the surface lowers the potential energy for electrons in the interior of the wire and thus increases the wire's electron affinity. In a similar study carried out for SiNWs, the bandgap was reported to increase with the strength of the bonds between surface passivants and NW, as stronger bonds were found to reduce the surface-like character of band edges [25].

Surface chemistry thus comprises an additional tool with which the electronic structure of nanostructures can be significantly manipulated. Although the mechanism by which such effects occur can be complicated and will in general depend on the nature of interactions between the nanostructured material and surface passivant, trends reported in the literature seem to be consistent: terminations that withdraw more charge from the nanostructure tend

to lower the magnitude of its bandgap. In this manner, surface chemistry can be employed to *counteract* the effects of confinement on the bandgap. Additionally, and as will be discussed in the next section, this opens the possibility of designing electronic components by inducing regions with significantly different electronic properties by varying surface treatments along an otherwise homogeneous piece of material.

10.3 Engineering Monomaterial Electronic Devices: Novel Schottky-Type Device Designs

10.3.1 Monomaterial Schottky-Type Junctions

One interesting consequence of quantum-size phenomena in semimetals discussed in the previous sections is the possibility of designing nanostructures composed of a single material with a path along which the electronic character can vary between SM and SC. This allows the realization of Schottky-like barriers at the interface between SM and SC regions, thus allowing the design of nanoelectronic components such as Schottky-barrier (SB) diodes and Schottky-barrier FETs (SB-FETs) in which current flows through a chemically homogeneous structure: a *monomaterial* Schottky-type device.

The physics of electronic transport through a metal–semiconductor interface have been long established and discussed in the literature [27]. Figure 10.7 shows a band diagram corresponding to a conventional metal–semiconductor interface; the difference in barrier heights for electrons flowing in each direction results in an asymmetry in the junction's current–voltage characteristics and hence current rectification. In such a junction, the barrier for electrons flowing from the semiconductor's CB into the metal is lower than the SB height Φ_B for electrons flowing in the opposite direction. The resulting asymmetrical transfer curve is characteristic of a diode.

There are a number of parameters that play a role in determining the properties of a SB and thus of any related device designs. The Schottky–Mott rule provides a first approximation to the properties of a metal–semiconductor junction, in which the properties of the barrier are largely determined by the difference between the metal's work function and the semiconductor's electron affinity:

$$\Phi_B \approx \Phi_{\text{Metal}} - \chi_{\text{SC}} \qquad (10.12)$$

In reality, however, different semiconductors follow the Schottky–Mott rule to varying degrees. The properties of the metal–semiconductor interface, such as defect density and chemical interactions at the heterogeneous junction, result in the barrier height being somewhat independent of the metal work function in what is known as *Fermi level pinning*. The properties of the potential barrier at the monomaterial semimetal–semiconductor interface considered in this section can vary significantly from traditional SBs, as the height of the barrier does not depend solely on the chemical nature of the materials involved, since variations on the confinement potential induced by the geometrical shape of the device play a key role. Moreover, chemically homogeneous junctions may exhibit lower density of interface defects and thus less Fermi level pinning. This, in combination with the fact that the use of SB design eliminates the need for dopants and results in designs relying on junctions that are

potentially easier to realize experimentally by avoiding two key difficulties associated with the fabrication of sub-10-nm devices: integration of different nanostructured materials and variability issues arising from random dopant fluctuations. Additionally, the fact that there is no bonding between different atomic species in the case of chemically homogeneous junctions results in less scattering at the interface due to the absence of a chemically induced interface dipole and may potentially lead to devices with better electrical characteristics.

In the monomaterial interfaces discussed in the following sections, the shape of the confining potential comprises the key property determining the electronic structure of each material region and of the interfaces between them. Similar to the case of nanostructured semiconductors, the electronic structure shows a large dependence on confinement dimensions and surface effects, such as exposed crystallographic planes at surfaces and chemical details of their passivation. In addition, other structural effects such as strain have been reported to allow further control over the electronic properties of sub-10-nm nanostructures based on semimetals [28].

The crystallographic orientation along which NWs based on semimetals are grown has been shown to significantly impact the magnitude of their bandgap. In Ref. [20], the bandgap of NWs based on SM α-tin grown along $\langle 100 \rangle$, $\langle 110 \rangle$, and $\langle 111 \rangle$ crystallographic orientations was computed with DFT for various cross-sectional dimensions. Figure 10.8 shows the transition from SM to SC behavior to occur at different diameters for NWs grown along different orientations, with differences in bandgap magnitudes as large as an electron volt across wires grown along varying crystallographic directions with similar cross-sectional dimensions. This result highlights the versatility of nanostructured semimetals and how critical the choice of orientation and surface properties can be when designing nanoelectronic components.

In the following sections, we shall discuss how to engineer useful interfaces for nanoelectronic devices by variations in the confining potential achieved through geometry modulation and surface chemistry effects. Representative proof-of-concept Schottky-type devices reported in the literature are presented.

10.3.2 Confinement-Modulated Designs

The first way of engineering a Schottky-type band profile we shall discuss is solely based on modulating the confining potential along the length of a chemically homogeneous nanostructure through geometrical effects. As discussed earlier, semimetals patterned below certain critical dimensions–which can depend on several factors such as the bulk semimetal employed, number of confined directions, and surface properties–exhibit SC properties. Thus by patterning adjacent sections of a semimetal above and below such critical dimensions, a semimetal–semiconductor Schottky-like potential barrier is induced. As previously mentioned, electronic transport through such an interface

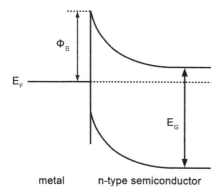

FIGURE 10.7 Band diagram of an n-type semiconductor SB. E_F is the Fermi level, E_G is the semiconductor bandgap, and Φ_B is the SB height.

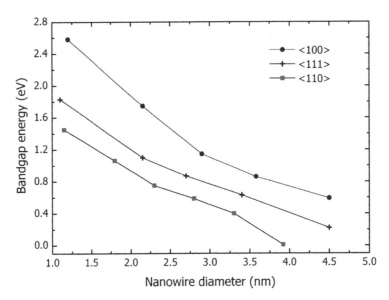

FIGURE 10.8 Bandgap of α-Sn NWs as a function of NW diameter for various crystallographic directions. (Reprinted with permission from G. Fagas, J.-P. Colinge, and J. C. Greer, A proposed confinement modulated gap nanowire transistor based on a metal (tin), *Nano letters*, vol. 12, no. 5, pp. 2222–2227, 2012. Copyright 2017 American Chemical Society.)

exhibits asymmetric properties and may be employed in the design of rectifier diodes. Furthermore, by patterning a *thin* SC region in between two *wide* SM regions, two back-to-back Schottky-like barriers are induced in the material, thus realizing a band profile similar to that of SB-FETs. Figure 10.9 shows a schematic of such a geometry and the induced band profile across the two junctions between regions wider and thinner than critical dimensions.

In conventional SB-FETs, an SC channel lies in between source and drain electrodes comprised of a metal; similar

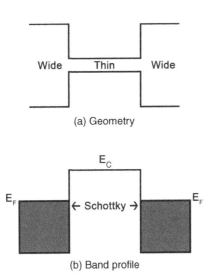

FIGURE 10.9 (a) Geometry and (b) the corresponding band profile for a confinement-modulated gap transistor (CMGT). The *thick-thin-thick* geometry profile alternates between sizes above and below critical dimensions, inducing two back-to-back Schottky-like barriers suitable for the design of an SB-FET. E_F indicates the position of the Fermi level in SM regions and E_C the CB edge of the SC region.

to MOSFETs, an oxide layer electrically insulates the SC channel from a gate electrode that modulates current flow between source and drain electrodes through electrostatically shifting electronic energy levels in the channel [29]. A similar device design realized through geometrically induced variations in the confining potential was first demonstrated through theoretical ab initio simulations [20]. In their study, variations in cross-sectional dimensions were employed to define an SC channel along an SM α-tin NW; the thinner SC channel is insulated from the gate electrode by an oxide layer in a gate-all-around (GAA) design, as illustrated in Figure 10.10.

The electrical characteristics of this semimetal-based CMGT are shown in Figure 10.11, where conventional transistor-like behavior can be observed, demonstrating the applicability of semimetals in nanoelectronic component designs. Figure 10.11b shows the switching characteristics of this design as simulated for two different source-drain bias V_{DS}; the fact that the gate threshold voltage is maintained for both high and low source-drain bias operation shows remarkable robustness against drain-induced barrier lowering even at sub-5-nm channel lengths, demonstrating some of the advantages of the increased electrostatic control inherent to NW-based GAA designs with regard to short-channel effects.

The switching efficiency of a transistor design is measured through its *subthreshold swing* (SS), defined as the gate voltage required to change the drain current by an order of magnitude when in the *subthreshold region* (*i.e.* region between gate voltages of 0.0 and 0.2 V in Figure 10.11b, in which drain current increases exponentially with increasing gate voltage). For transistor designs operating under the usual MOSFET paradigm, there is a lower limit of 60 mV/dec imposed by thermodynamics at room temperature [30], with typical state-of-the-art devices

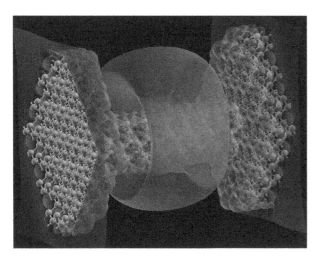

FIGURE 10.10 Atomistic illustration of a CMGT design based on SM α-tin. The wider sections of the NW are above critical dimensions (SM), while the narrow region in between is below the critical dimensions (SC). The ring around the channel indicates an isopotential surface from the application of gate bias in a GAA configuration. (Reprinted with permission from G. Fagas, J.-P. Colinge, and J. C. Greer, A proposed confinement modulated gap nanowire transistor based on a metal (tin), *Nano letters*, vol. 12, no. 5, pp. 2222–2227, 2012. Copyright 2017 American Chemical Society.)

exhibiting SS values around 70 mV/dec. The reported SS value for theoretical simulations of this CMGT design is 72.6 mV/dec, indicating performance on par with modern devices but at lower dimensions.

10.3.3 Surface Chemistry: Electrostatic Effects of Surface Passivants

The formation of semimetal–semiconductor Schottky-like junctions in nanostructures can also be achieved through the exploitation of surface chemistry effects. At a crystal surface, undercoordinated atoms result in *dangling bonds*, which must be passivated by chemically bonding to surface terminations. The large surface-to-volume ratio inherent to nanostructures results in a substantial dependence of their electronic structure on surface phenomena, such as surface chemistry. Since the influence of surface effects usually extends a few nanometers into materials, it is possible to modify the electronic structure throughout the cross section of sub-10-nm nanostructures by varying surface chemical treatments. Similar to effects reported on semiconductor NWs [25], a study on the effect of surface chemistry on the electronic structure of α−tin NWs reported their magnitude to be comparable with that of confinement effects. In that study, the large influence of surface chemistry on SM NWs' electronic properties, such as their bandgap magnitude, was ascribed to electrostatic effects related to charge transfer at surfaces resulting from chemical bonding; passivating surfaces with increasingly electronegative terminations was found to induce charge rearrangements throughout the NW's cross section, modifying the confining potentials and

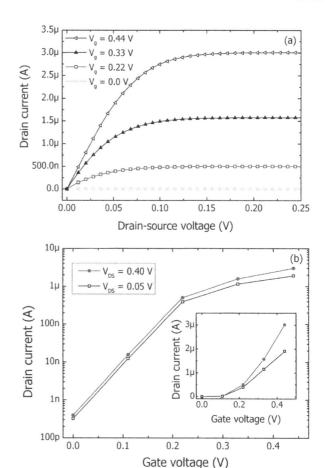

FIGURE 10.11 (a) Output characteristics of a CMGT, at varying gate voltages and (b) I_D vs. gate voltage V_{GS} in the subthreshold region, on a log scale. The inset shows a linear scale. (Reprinted with permission from G. Fagas, J.-P. Colinge, and J. C. Greer, A proposed confinement modulated gap nanowire transistor based on a metal (tin), *Nano letters*, vol. 12, no. 5, pp. 2222–2227, 2012. Copyright 2017 American Chemical Society.)

electron affinities such that NWs passivated with varying chemical species may exhibit significantly different critical dimensions [21].

A semimetal NW rectifier design based on this effect was theoretically demonstrated in a subsequent study [22]. A semimetal–semiconductor Schottky-like junction was induced along the length of a SM NW of homogeneous cross-sectional dimensions by varying surface passivants, such that one section of the wire is below critical dimensions (*i.e.* SC) while the adjacent section is above critical dimensions and thus SM. Figure 10.12 illustrates the device geometry and local density of states (LDOS) along the length of the NW when in equilibrium: the left portion of the NW is passivated with fluorine (a highly electronegative species) and remains SM, while the right portion is passivated with hydrogen and exhibits a bandgap of around 0.7 eV.

By switching surface terminations, a Schottky-like barrier is induced, as shown in Figure 10.12b, where the vertical dashed line indicates the physical position where the passivant switch occurs; the LDOS profile around the

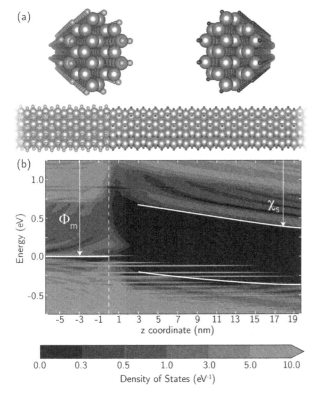

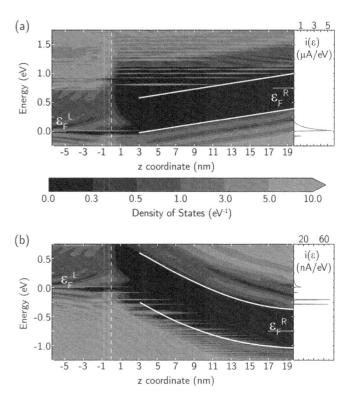

FIGURE 10.12 (a) Atomistic visualization of a chemically and geometrically homogeneous α-tin NW in which a Schottky-like junction is induced by switching surface passivants between fluorine (left) and hydrogen (right) and (b) Contour plot of the predicted LDOS profile across the junction. (Reprinted with permission from A. Sanchez-Soares and J. C. Greer, Electronic structure tuning via surface modication in semimetallic nanowires, *Phys. Rev. B*, vol. 94, p. 23544, 2016. Copyright 2017 American Chemical Society.)

FIGURE 10.13 Contour plot of the predicted LDOS of a surface-induced semimetal NW rectifier under (a) a forward bias of +0.75 V and (b) a reverse bias of −0.75 V. (Reprinted with permission from A. Sanchez-Soares and J. C. Greer, Electronic structure tuning via surface modication in semimetallic nanowires, *Phys. Rev. B*, vol. 94, p. 23544, 2016. Copyright 2017 American Chemical Society.)

junction shows that the electronic structure does not sharply transition between SM and SC even when the passivant switch is abrupt, with notable extensions of the SM region's states into the SC region's bandgap energy range, akin to *metal-induced gap states* (MIGS) observed in metal–semiconductor junctions.

Figure 10.13 shows the LDOS and energy-resolved current when the device is under forward and reverse bias. When the device is under forward bias the SM region's Fermi level ε_F^L lies below the SC region's Fermi level ε_F^R, similar to conventional SB diodes. The conduction mechanism, however, qualitatively differs from that of conventional diodes; the conventional diode designs typically employ doped semiconductors, whereas the semimetal-based device's SC region is intrinsic and thus exhibits a very low electron population in its CB. As shown in the energy-resolved current plot in Figure 10.13a, conduction in the *ON* state occurs at energies near the SM region's Fermi level (ε_F^L), as the SC region's VB edge aligns with it, and electrons can flow into the SM region's unoccupied CB edge states in a mechanism sometimes described as *hole injection* into the SC region.

When the device is under reverse bias, the SM region's Fermi level (ε_F^L) is situated at higher energies than that of

the SC region's (ε_F^R), as depicted in Figure 10.13b. Under these conditions, the SC region's occupied states align with fully occupied states in the SM region and no current can flow in this energy range. Conversely, occupied states in the SM region align with either occupied VB states in the SC region or with its bandgap. Thus, electronic transport in reverse bias can only occur when electrons from the SM region tunnel through the SC region's bandgap and into its CB, as depicted by the energy-resolved current shown in Figure 10.13b. This qualitative difference in electronic transport mechanisms translates into the asymmetry in transfer characteristics observed in Figure 10.14a: the fact that electrons have to tunnel through the bandgap when in reverse bias (*OFF* state) translates into a current that is orders of magnitude lower than the current observed when under forward bias, as evidenced by the different scales employed when plotting the corresponding energy-resolved currents in Figure 10.13.

Figure 10.14 shows the transfer characteristics of the device, where negative values of applied voltage correspond to reverse bias, and positive values to forward bias. For reverse bias conditions, very low currents are reported, as reverse bias transport is limited by electron tunneling through the SC region's bandgap. When forward bias is applied, a *step-like* structure can be observed in

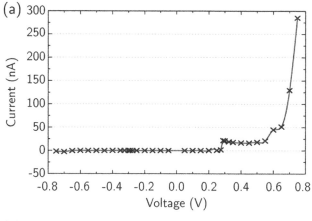

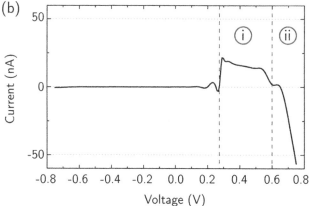

FIGURE 10.14 (a) Current–voltage characteristics of a surface-induced semimetal NW rectifier junction at 300 K and (b) the difference between current–voltage characteristics at 300 and 0 K. (Reprinted with permission from A. Sanchez-Soares and J. C. Greer, Electronic structure tuning via surface modication in semimetallic nanowires, *Phys. Rev. B*, vol. 94, p. 23544, 2016. Copyright 2017 American Chemical Society.)

Figure 10.14a around $V = 0.3$ V, as a flat condition is achieved in the SC region when its VB edge aligns with the SM region's Fermi level and electrons can begin to flow. In the $V \in [0.3, 0.6]$ V region, the device begins to turn *ON*, as unoccupied states in the SM region begin to align with the SC region's occupied VB. To further illustrate the nature of this conduction mechanism, Figure 10.14b shows the difference in computed currents at electronic temperatures of 300 and 0 K: as both region's VB edges align, more current can flow at larger temperatures as the SM region's VB exhibits partially unoccupied states due to thermally excited electrons being promoted to the bottom of the conduction band, accessible due to the region's negligible bandgap. These unoccupied states in the SM region's VB represent open channels for electrons coming from the SC region to flow into. In contrast, all states at the SM region's VB are fully occupied at zero temperature and thus unavailable for conduction. With increasing applied bias, a larger portion of the *ON* state conduction begins to occur through the SM region's conduction band, and for $V > 0.6$ V, we observe a crossover in Figure 10.14b, as a fully unoccupied

CB becomes more favorable and thus the current is greater at lower temperatures.

This proof-of-concept device design illustrates the relevance of surface effects in nanostructures and how its interplay with quantum-size effects can be exploited in device engineering. Furthermore, the results of this theoretical study offer insights into the conduction mechanisms expected for a SM Schottky-type device that does not rely on the use of dopants, and hence offers an alternative to silicon-based technology at feature sizes below 10 nm.

10.4   Conclusions and Outlook

In this chapter, we have briefly discussed some of the challenges currently faced by silicon-based nanoelectronic device designs, and how some effects detrimental to their performance can be exploited in semimetal-based designs. Quantum-size effects, found to result in an undesirable increase in power consumption in silicon-based technology, can be used for tailoring the properties of semimetal-based nanostructures, allowing for a wide range of properties out of chemically homogeneous designs, including a novel class of nanoscale semiconductors with small bandgap magnitudes and thus potentially useful in the design of low-power electronics. The large effects induced by the close proximity of surfaces and interfaces in nanostructures allow further customization of the potential confining electrons to such nanostructures and, hence, their electronic properties. Still in the infancy of the field of semimetal nanoelectronics, the processing of new materials currently represents one of the biggest difficulties. Theoretical studies show promising predictions, and a recent experimental study has already demonstrated the associated principle by showing current rectification in SM thin films [24]. In their study, SC sub-10-nm bismuth thin films were measured to have remarkably high electron mobility values up to two orders of magnitude larger than silicon thin films, showing a great potential for applications in high-speed electronics. In spite of the presently encountered fabrication difficulties, potential advantages, including the use of chemically homogeneous materials in different device regions, favorable transport-related material properties, the possibility to fine-tune properties through geometric and surface effects–on top of some of the usual design tools such as strain engineering–and, in particular, the associated potential to design components without the need to employ the doping paradigm, have attracted significant efforts in electronic component research, which may result in the birth of semimetal electronics.

References

1. International technology roadmap for semiconductors, www.itrs2.net/itrs-reports.html (accessed May 2018).
2. L. de Broglie, Recherches sur la théorie des quanta, *Ann. Phys.*, vol. 10, pp. 22–128, 1925.

3. A. Ekimov, A. Efros, and A. Onushchenko, Quantum size effect in semiconductor microcrystals, *Solid State Commun.*, vol. 56, pp. 921–924, 1985.

4. L. T. Canham, Silicon quantum wire array fabrication by electrochemical and chemical dissolution of wafers, *Appl. Phys. Lett.*, vol. 57, pp. 1046–1048, 1990.

5. R. S. Wagner and W. C. Ellis, Vapor-liquid-solid mechanism of single crystal growth, *Appl. Phys. Lett.*, vol. 4, pp. 89–90, 1964.

6. A. M. Morales and C. M. Lieber, A laser ablation method for the synthesis of crystalline semiconductor nanowires, *Science*, vol. 279, pp. 208–211, 1998.

7. L. Esaki and R. Tsu, Superlattice and negative differential conductivity in semiconductors, *IBM J. Res. Dev.*, vol. 14, pp. 61–65, 1970.

8. P. S. Zory, ed., *Quantum Well Lasers (Quantum Electronics–Principles and Applications)*. Academic Press, Boston, MA, 1993.

9. Y. F. Ogrin, V. N. Lutskii, and M. I. Elinson, Observation of quantum size effects in thin bismuth films, *ZhETF Pisma Redaktsiiu*, vol. 3, p. 114, 1966.

10. C.-H. Lee, E. C. Silva, L. Calderin, M. A. T. Nguyen, M. J. Hollander, B. Bersch, T. E. Mallouk, and J. A. Robinson, Tungsten ditelluride: A layered semimetal, *Sci. Rep.*, vol. 5, p. 10013, 2015.

11. C. A. Hoffman, J. R. Meyer, F. J. Bartoli, A. D. Venere, X. J. Yi, C. L. Hou, H. C. Wang, J. B. Ketterson, and G. K. Wong, Semimetal-to-semiconductor transition in bismuth thin films, *Phys. Rev. B*, vol. 48, pp. 11431–11434, 1993.

12. X. Sun, Z. Zhang, and M. S. Dresselhaus, Theoretical modeling of thermoelectricity in Bi nanowires, *Appl. Phys. Lett.*, vol. 74, pp. 4005–4007, 1999.

13. X. Yi, H. Wang, S. Chen, J. Lai, M. He, S. Wang, and G. K. Wong, Quantum size effect in antimony thin films and its application, *Infrared Phys. Technol.*, vol. 46, pp. 263–266, 2005.

14. L. L. Wang, X. C. Ma, S. H. Ji, Y. S. Fu, Q. T. Shen, J. F. Jia, K. F. Kelly, and Q. K. Xue, Epitaxial growth and quantum well states study of Sn thin films on Sn induced Si(111)-$(2\sqrt{3} \times 2\sqrt{3})$ $R30°$ surface, *Phys. Rev. B*, vol. 77, p. 205410, 2008.

15. S. Tang and M. S. Dresselhaus, Electronic phases, band gaps, and band overlaps of bismuth antimony nanowires, *Phys. Rev. B*, vol. 89, p. 045424, 2014.

16. R. Sankar, G. N. Rao, I. P. Muthuselvam, C. Butler, N. Kumar, G. S. Murugan, C. Shekhar, T.-R. Chang, C.-Y. Wen, C.-W. Chen, W.-L. Lee, M.-T. Lin, H.-T. Jeng, C. Felser, and F. C. Chou, Polymorphic layered MoTe₂ from semiconductor, topological insulator, to weyl semimetal, *Chem. Mater.*, vol. 29, pp. 699–707, 2017.

17. V. K. Arora, Quantum well wires: Electrical and optical properties, *J. Phys. C: Solid State Phys.*, vol. 18, no. 15, p. 3011, 1985.

18. J.-P. Colinge and J. C. Greer, *Nanowire Transistors: Physics of Devices and Materials in One Dimension*. Cambridge University Press, Cambridge, UK, 1st ed., 2016.

19. J.-A. Yan, L. Yang, and M. Y. Chou, Size and orientation dependence in the electronic properties of silicon nanowires, *Phys. Rev. B*, vol. 76, p. 115319, 2007.

20. L. Ansari, G. Fagas, J.-P. Colinge, and J. C. Greer, A proposed confinement modulated gap nanowire transistor based on a metal (tin), *Nano Lett.*, vol. 12, no. 5, pp. 2222–2227, 2012.

21. A. Sanchez-Soares, C. O'Donnell, and J. C. Greer, Electronic structure tuning via surface modification in semimetallic nanowires, *Phys. Rev. B*, vol. 94, p. 23544, 2016.

22. A. Sanchez-Soares and J. C. Greer, A semimetal nanowire rectifier: Balancing quantum confinement and surface electronegativity, *Nano Lett.*, vol. 16, no. 12, pp. 76397644, 2016.

23. L. Ansari, F. Gity, and J. C. Greer, Electronic and structural properties of rhombohedral [111] and [110] oriented ultra-thin bismuth nanowires, *J. Phys. Condens. Matter*, vol. 29, no. 6, p. 065301, 2017.

24. F. Gity, L. Ansari, M. Lanius, P. Schüffelgen, G. Mussler, D. Grützmacher, and J. C. Greer, Reinventing solid state electronics: Harnessing quantum confinement in bismuth thin films, *Appl. Phys. Lett.*, vol. 110, no. 9, p. 093111, 2017.

25. P. W. Leu, B. Shan, and K. Cho, Surface chemical control of the electronic structure of silicon nanowires: Density functional calculations, *Phys. Rev. B*, vol. 73, p. 195320, 2006.

26. M. Nolan, S. O'Callaghan, G. Fagas, J. C. Greer, and T. Frauenheim, Silicon nanowire band gap modification, *Nano Lett.*, vol. 7, no. 1, pp. 3438, 2007.

27. R. T. Tung, The physics and chemistry of the Schottky barrier height, *Appl. Phys. Rev.*, vol. 1, no. 1, p. 011304, 2014.

28. L. Ansari, G. Fagas, and J. C. Greer, Strain induced effects on electronic structure of semi-metallic and semiconducting tin nanowires, *Appl. Phys. Lett.*, vol. 105, no. 12, p. 123105, 2014.

29. N. Dasgupta and A. Dasgupta, *Semiconductor Devices: Modelling and Technology*. PHI Learning Pvt. Ltd., New Delhi, 2004.

30. K. P. Cheung, On the 60 mV/dec @ 300 K limit for MOSFET subthreshold swing, *In 2010 International Symposium on VLSI Technology Systems and Applications (VLSI-TSA)*, Piscataway, NJ, pp. 72–73, IEEE, 2010.

11

Neuromorphic Nanoelectronics

11.1 Nanoscale Memristance... 11-2
11.2 Neuromorphic Computing 11-4
11.3 Thin Films and 2D Materials... 11-5
11.4 Nanoparticles 11-5
11.5 Networks of Neuromorphic Devices.................................... 11-6
11.6 Conclusion 11-7
Acknowledgment 11-7
References 11-7

Jessamyn A. Fairfield
National University of Ireland Galway

The nanoscale is where quantum effects emerge in the measurable behavior of materials, leading to exciting opportunities for electronic devices, as well as new challenges in material sensitivity and reliability. For example, quantum dots possess unique size-dependent optical properties that stem from quantum confinement of charge carriers; however, this same confinement leads to poor conduction through films of quantum dots, and the high number of surface trap states results in photoconductive behavior, which is heavily dependent on device measurement history.[1] In general, the high proportion of surface atoms to volume atoms in nanomaterials leads to a large number of surface dangling bonds compared with bulk materials, which can both enable new gas sensor technologies and restrict the environments in which nanoelectronic devices can operate.[2−4] The stochastic nature of electronic conduction at the nanoscale has also led to reliability issues when trying to replicate Boolean logic for computation, although noise-assisted computation has also been demonstrated,[5] making use of this inherent feature of nanoscale conduction.

Many of the properties of nanomaterials that were initially considered challenging have re-emerged as strengths in the proper application. The most fascinating example of this is in the case of memory-dependent conductivity or memristance. Current flow can often cause small changes in nanoscale materials, by moving atoms or even by changing the electron distribution between mobile states and trapped states. As a result, nanomaterial researchers frequently observe contradictory conduction data and had struggled to find the source of it, across many classes of nanomaterial.[6−9] But memristors as a fourth circuit element were predicted theoretically in 1971, as a consequence of the mathematical symmetries observed in the rules that governed the existing three circuit elements: resistors, capacitors, and inductors,[10] as shown in Figure 11.1.[11] However, a physical device exhibiting hysteresis was first described as a

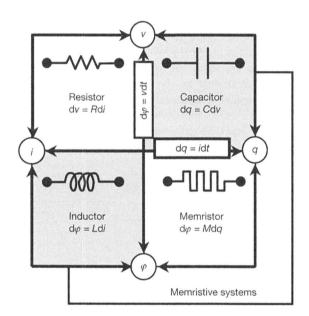

FIGURE 11.1 Mathematical relations between the four basic circuit elements, including the memristor. (Copyright 2008 Nature Publishing Group.[11])

memristor in 2008, for a 5 nm thin film of TiO_2 sandwiched between conducting electrodes.[11] In retrospect, there had been a long history of observed memristive hysteresis in various materials dating back as far as 1939.[12] Other memory-based circuit elements have since been proposed,[13] but the memristor remains the simplest to realize experimentally, and in fact, memristance seems to be a feature of nanoscale electronics due to the conduction mechanisms and energy landscape prevalent at the nanoscale.

Memristors challenge the very nature of computing as it has developed, based on the architecture of transistors acting as binary switches. However, the computing architecture of the brain provides another possible model—rather than digital logic, the brain runs on analog connections between neurons and synapses, which are made more or less conductive by stimulation, leading to the repeated firing of the synapse. This is the physical process underpinning learning at the neurological level, and is generally described as a change in "synaptic weight," which over time can make synapses biochemically easier (or more difficult) to fire.[14] This efficient approach to computation leads to the brain's relatively low power consumption (20 W) and fault tolerance deriving from plasticity.[15]

"Neuromorphic" device design aims to emulate the function of neurons and the synaptic connections between them, by reproducing memory effects or timing dependence. Electronics built from these devices avoid the so-called "von Neumann bottleneck" in computing, where processing and memory are in different locations, and the speed of passing information between them limits the maximum speed of the computer.[16] Neuromorphic devices are also able to route around breaks or faults in their electrical connection, and adapt to damage, as well as having much lower power consumption compared with standard computing architectures.[17,18] This chapter will cover the basics of memristance, the physical mechanisms at the nanoscale that underlie this approach, as well as an overview of neuromorphic behavior that has been demonstrated in nanoscale thin films, colloidal nanomaterial devices, and networks of memristive nanoscale devices. This area of research is currently very active, driven by a desire to use nanostructured materials to understand how we can build better

computers, improve our knowledge of neural electronic circuitry, and perhaps better understand the brain itself.

11.1   Nanoscale Memristance

Memristive electronic response is commonly observed at the nanoscale, but in such a variety of forms that it might more accurately be described as a family of electronic responses. The first nanoscale thin film memristor described in the seminal Hewlett Packard paper[11] displayed what is now called continuum memristance, where increasing voltage stimulus drives increased conductance in a continuous, reversible manner. The underlying mechanism proposed was the field-driven migration of dopants, oxygen vacancies in the case of TiO_2, which create a higher-conductivity region in the material. On reversal of the applied voltage, the oxygen vacancies migrate in the opposite direction, shrinking the conductive region and lowering the conductivity of the overall device. This process is shown for an ideal memristor in Figure 11.2.

However, not all memristive behavior is continuous, and sometimes, conductivity modification can take the form of switching between two discrete conductive states. This "resistive switching" behavior has been observed in many transition metal oxides and, generally, requires a forming step at a higher voltage to initiate the jump to a low resistance state (LRS).[19] After this forming step, the device can be cycled between its high resistance state (HRS)

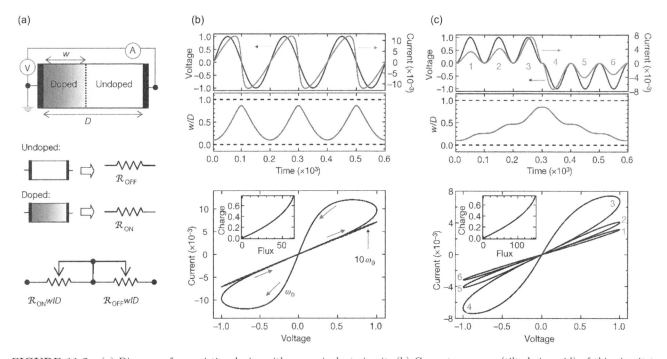

FIGURE 11.2   (a) Diagram of memristive device with an equivalent circuit. (b) Current response (tilted sinusoid) of this circuit to applied voltage (straight sinusoid) as a function of time. (c) Current response to bipolar applied voltage, showing the reversibility of the conductance change. The numbers 1–6 indicate separate I–V sweeps that are also indicated in the current—voltage graphs. The frequency-dependent Lissajous figure is now considered a typical memristive current–voltage feature. (Copyright 2008 Nature Publishing Group.[11])

and the LRS at a lower voltage a large number of times, with on/off ratios as high as 10^5.[20] Resistive switching is a special case of memristance, which is digital in nature, because of a difference in mechanism compared with continuum memristance.[21,22] Resistive switching is generally divided into two major subtypes: electrochemical metallization (ECM) and valence change mechanism (VCM). In ECM, metal cations become mobile in the resistive switching material, whereas in VCM, oxygen anions or vacancies are the mobile species[23]; both forms of resistive switching are also impacted by electronic effects, which are inherently stochastic in nature.[24] In both cases, these migrating ions lead to the formation of a conducting filament through the insulating oxide, which requires a high field to form, but which can then be partially broken and reformed over many cycles, leading to the observed HRS and LRS. At the microscopic level, Landau–Zener tunneling of electrons leads to a phase transition takes place in the oxide, leading to the formation of hot conductive filaments that can be broken by Joule heating.[25,26] So-called "atomic switches"

operate on a similar principle to resistive switches, but with the conducting filament forming through air instead of oxide.[27,28]

Conducting filaments can be imaged directly during formation and dissolution, as shown in Figure 11.3. The release and reincorporation of oxygen to form conducting filaments have been observed directly using in situ transmission electron microscopy (TEM), leading to resistance change.[30] Filament formation has also been observed in various materials with TEM,[29,31–33] leading to filament evolution with repeated cycles that can contribute to unreliability.[34] Variability in memristive devices can be observed even after forming, stemming from subfilamentary networks of oxygen vacancies that have been observed.[35] Filament thickness can also affect the type of conduction observed in filamentary resistive switch materials.[36] Indeed, this enables multilevel memory in resistive switching materials by modulating filament geometry,[37] as well as in continuum memristors by modulating oxygen vacancy distribution.[38] Resistive switching reliability is also affected

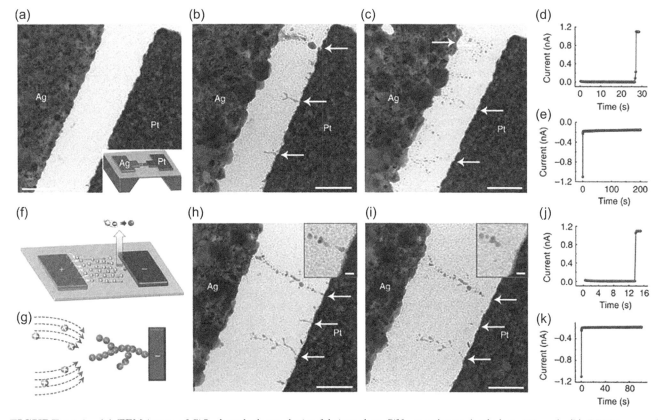

FIGURE 11.3 (a) TEM image of SiO_2-based planar device fabricated on SiN_x membrane (scale bar: 200 nm). (b) TEM image of the same device after forming, with filaments indicated by arrows (scale bar, 200 nm). (c) TEM image of the same device after erasing (scale bar: 200 nm). (d) I–t curve for forming that led to (b), with 8 V applied. (e) I–t curve for erasing process that led to (c), with −10 V applied. (f) Schematic of filament growth showing transport and reduction of Ag cations, and (g) dendrite structure formation. (h) TEM of another device with larger electrode spacing after forming (scale bar: 200 nm). Inset: zoomed-in image of the filament is highlighted by the upper arrow (scale bar: 20 nm). (i) TEM image of same device as in (h) after erasing. Filament dissolution takes place at the inert electrode interface (scale bar: 200 nm). Inset: same region as inset in (h) after erasing (scale bar: 20 nm). (j) I–t curve during forming that led to (h), with 10 V applied. (k) I–t curve during erasing that led to (i), with −10 V applied. (Copyright Nature Publishing Group 2012.[29])

by the maximum voltage applied to the device.[39] Methods explored thus far to improve filament control, such as creating the filament with a voltage applied by an atomic force microscope tip,[40] are not industrially viable.

Quantized conductance has been observed in multiple resistive switch materials[23,36,41−44]; however, this is correlated with decreased stability and increased noise.[45] Random telegraph noise is observed in some resistive switching filaments, depending on compliance current during forming. Characteristics of random telegraph noise stem from Joule heating and Poole–Frenkel emission in the resistive switching material.[46] Noise can also be increased by irradiation of memristive nanodevices and used to enhance computation efficiency.[47]

Continuum resistance and resistive switching are not necessarily mutually exclusive, and in fact, both electromigration and filament formation have been observed in the same material (TiO_2) under different conditions.[48] Networks of resistive switching devices can display continuum memristance as well as semidigital memristance at larger scales, as demonstrated with Ni and Ni/Ag nanowire networks and shown in Figure 11.4.[20,49] Each of these behaviors relies on nanoscale ionic movement, and can be modulated by the sensitivity of nanoscale surfaces, and as has been repeatedly shown across many materials, memristance is a feature of conduction at the nanoscale. However, the challenges of emulating traditional architectures for computing, with switch-based Boolean architecture, are clear when one considers the stochastic and responsive nature of nanoscale materials. What other paradigms for computing might be possible that exploit this inherent electronic feature of nanoscale materials?

11.2   Neuromorphic Computing

An alternate model for computation is offered by the brain itself, the original computer. Electrical connections in the brain, initially semirandomly distributed, are strengthened and weakened by experience, leading to continuous modification of overall network connectivity to improve capability and survivability of the organism possessing

the brain. In recent years following the establishment of memristance as a common nanoscale conduction feature, researchers have turned to the so-called "neuromorphic" device design, aiming to emulate specific features of neuronal and synaptic function. The term neuromorphic was popularized in the 1980s by Carver Mead to describe attempts to emulate the electronic structure of the brain and associated sensory organs using very large scale integrated (VLSI) technology.[17] But more recently, neuromorphic device researchers have begun to demonstrate some of the following behaviors in inorganic devices:

1. Spike timing-dependent plasticity (STDP): STDP is a change in synaptic weight due to the timing difference between the presynaptic and postsynaptic pulses (see Figure 11.5).[50]

2. Short-term plasticity (STP): Changes in synaptic weight that occur over timescales of minutes or shorter, assisting in computation in neural circuits.[51] Sometimes also called short-term memory.

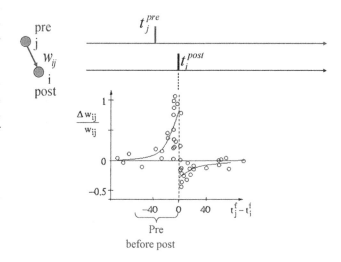

FIGURE 11.5 STDP observed in hippocampal neurons, in the now seminal paper by Bi and Poo. (Copyright Society for Neuroscience 1998.[50])

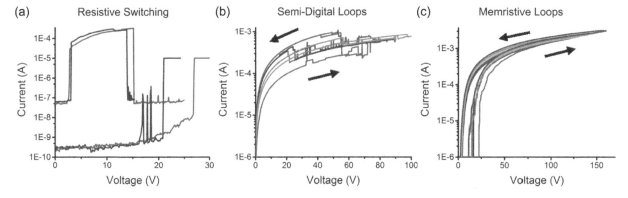

FIGURE 11.4 Current–voltage data showing (a) resistive switching, (b) semidigital loops, and (c) memristive hysteresis loops in nanowire networks of the same Ni and Ag nanowires but with different device sizes. (Copyright American Chemical Society 2016.[49])

3. Long-term plasticity (LTP): Changes in synaptic weight that last for hours or longer, critical to learning and memory.[51] LTP is sometimes also called long-term memory.

4. Pair-pulsed depression and facilitation (PPD, PPF): Decrease or increase in synaptic weight due to time spacing of two action potentials[52] leads to STP changes.

5. Hodgkin–Huxley neuronal model: A set of differential equations modeling the propagation of the action potential through the neuron over time based on ion channels, ion pumps, and electrochemical gradients.[53–55]

6. Integrate-and-fire neuronal model: A model of synaptic firing where the neuron is characterized by its membrane potential, which receives weighted inhibitory or excitatory input pulses that are subject to decay over time, the sum of which determines whether or not the neuron will fire.[55–57]

7. Winner-takes-all neuronal model: A nonlinear response model for networked neurons, in which they compete for conductivity with the first successful connection winning the full synaptic weight.[58]

These behaviors have been exhibited in a broad range of nanoscale devices, which will be examined in detail in the following sections. While more advanced biological functions will require networked devices, many researchers begin neuromorphic design by emulating STDP, the classic biological definition of which is shown in Figure 11.5.

STDP emerges naturally from memristive systems,[59,60] and both continuous and resistive switching memristors can be used to implement STDP.[61] The same device can demonstrate LTP and STP, as shown with electrolyte-gated WO_3 memristors.[62] Silicon oxide devices have been used to demonstrate both the Hodgkin–Huxley conductance model and the leaky integrate-and-fire model for neuronal behavior.[63] Field-driven oxygen migration results in the formation of localized conducting filaments in this material, even if the initial structure is amorphous.[64]

Inhibiting neural signals as well as stimulating them is an important part of true synaptic behavior. This has been less studied in the inorganic device space, but devices showing both excitatory and inhibitory behavior have been demonstrated at the active interface of black phosphorus and tin selenide.[65] In the following sections, an overview of various nanoscale device geometries is given, with a review of existing research demonstrating neuromorphic functions in these nanoscale devices.

11.3 Thin Films and 2D Materials

Nanoscale films in a sandwich device geometry was used for the first experimental demonstration of memristance, and hence it may come as no surprise that thin film devices were some of the first examples of neuromorphic function. Thin film memristors can have nanosecond switching between HRS and LRS, with a persistence of months.[66] Moreover, the memristive behavior observed is highly tunable: asymmetry in electrode composition can lead to multiple types of observed resistive switching behavior,[67] due to both chemical and mechanical factors in the filament formation.[68] The counterelectrode material can have a strong influence on the type of memristive behavior observed because it affects the electrode potentials and filament formation, as shown by Lübben et al.[69] Thin film geometries can also include a selector to decrease leakage current.[70] For filamentary conductors, the size and shape of the filament affect the conductance leading to STP and LTP, as demonstrated in a Ag/Ag_2S ECM device.[71]

STDP was an early neuromorphic behavior that has been often replicated, for example, in a HfO_x/AlO_x stack.[72] Neuromorphic devices made from two transistors and one resistor (2T1R) have been shown, with thin films of HfO_x, to demonstrate STDP.[73] Artificial synapses made from one transistor and one variable resistor can emulate STDP, as well as long-term potentiation and depression.[74,75] Second-order memristance can be observed when resistance change is controlled by filament width as well as vacancy conductance, allowing memristors to emulate the dynamics of Ca^{2+} in neurons.[76] TiO_2 thin film memristors showing STDP can also be used to emulate leaky integrate-and-fire neurons with a random spike train input.[77,78]

Thin film phase change materials have also emerged as a promising avenue for neuromorphic device design, with STDP, LTP, and LTD demonstrated in nanoscale films of GeSbTe.[16,79] The change in conduction defines a "memory window" for the device, as shown in Figure 11.6. This intrinsic memristance results from trapping at defect states in the material, which are modified by phase change, leading to ultrafast memory functions.[80]

Neuromorphic functions are also beginning to be demonstrated in two-dimensional (2D materials) comprised of a single sheet of atoms. In the canonical 2D material, graphene, electrochemical modification of graphene using Li^+ as a dopant leads to STDP, PPF, and other neuromorphic functions.[81] MoS_2, a 2D material with a multiatom basis, can be used as a resistive switching material,[82] and has demonstrated photosensitive switching, which could enable multi-input associative memory.[83] Additional inputs other than electronics have been shown with neural-like computing in nanoscale magnetic tunnel junctions.[84,85]

11.4 Nanoparticles

Colloidal nanoparticles, as well as nanoparticles incorporated in a matrix of another material, can also be incorporated into electronic devices that exhibit memristance, and can hence demonstrate neuromorphic behavior. Surface functionalization and coatings are of high importance in these materials, due to the high proportion of surface

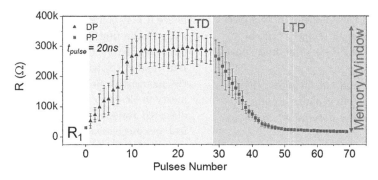

FIGURE 11.6 Long-term depression and potentiation demonstrated in a thin film phase change device, showing a change in conductance that is persistent. (Copyright Wiley 2018.[79])

atoms compared with bulk atoms. The first demonstration of STDP in an inorganic device made use of nanoparticles of Ag in a Si matrix in the first inorganic demonstration of STDP.[86] Potentiation and long-term memory have also been demonstrated in nanorod/polymer hybrid devices.[87] Synaptic behaviors such as STP and LTP have been demonstrated in nanoparticle/polymer hybrids in field effect transistor (FET) configurations with filaments formed from mobile silver ions, shown in Figure 11.7.[88]

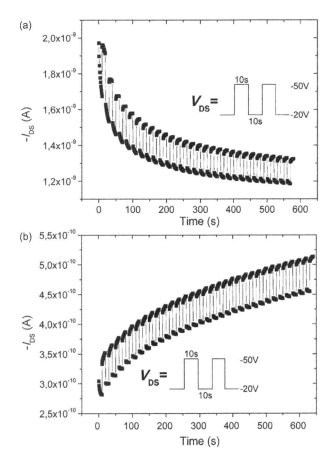

FIGURE 11.7 Potentiation and depression in a nanoparticle/organic FET, with pulse train following an initial pulse of (a) +50 V or (b) −50 V. (Copyright Wiley-VCH Verlag GmbH 2010.[88])

STDP and multilevel memory have also been demonstrated in TiO_2 nanowires,[38] following the same mechanism but a different device geometry compared with thin film TiO_2 devices.

Nanoparticles also offer routes to incorporate nonelectronic physical signals that can be transduced to electrical memristance by modifying conductivity behavior.[89] Artificial synapses that incorporate optical signals as well as electronic signals have been demonstrated in perovskite quantum dots.[90] Neuromorphic TiO_2 single nanowires, as well as ZnO nanowire devices, have shown optoelectronic STDP emulating associative memory.[87,91]

11.5 Networks of Neuromorphic Devices

Given that the human brain contains approximately 10^{12} neurons and 10^{15} synapses,[18] true neuromorphic electronics must be operable in networks that scale to a high level. While much of the research done so far has focused on the individual device level, the neuromorphic device community is now beginning to ask how to begin this scaling: how can we build an elephant from a toe?

Networks of neuromorphic devices have also been used to emulate higher order biological functions, such as networked memristors used to implement a Hopfield neural network with associative memory.[92] STDP has been shown using networks of memristors and memristive FETs,[93] and used to build a self-learning visual cortex.[94] Networks of TiO_2 memristors have been shown to emulate a winner-takes-all neural network, enabling reversible learning.[95] Networks of 125 synapses as a neural network were used to demonstrate unsupervised learning for character recognition.[96] Networks of phase change memristors have also been shown to provide high fidelity character recognition.[97] However, new algorithms for spike sorting will be needed to process neural spike data from many parallel devices.[98]

Field-programmable gate arrays can also be used to implement spiking neuromorphic devices demonstrating STDP,[99] and more advanced pattern recognition could be on the way for nanoscale memristors if results from VLSI chips shown in Figure 11.8 are an indication.

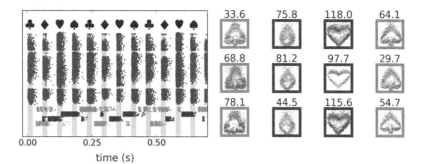

FIGURE 11.8 Implementation of reliable pattern recognition of playing card suits by a convolutional neural network implemented in VLSI artificial neurons. (Copyright IEEE Publishing 2017.[100])

VLSI neuromorphic networks present another promising technology to implement, integrate, and fire neurons,[101,102] and can be used for pattern classification,[103] visual[104] and auditory processing,[105] collocating sensory and learning functions. Networks of neuromorphic devices have been shown to replicate all two-input logic operations with one step (or two steps in the case of XOR), showing the flexibility of the neuromorphic approach.[106] Bioinspired computing is an obvious application of neuromorphic nanodevices, with low power consumption and good pattern recognition.[107]

11.6   Conclusion

Memristance at the nanoscale is enabling entirely new types of electronic devices, which can emulate many types of synaptic behavior. Devices based on nanoscale thin films and colloidal nanomaterials have emulated a wide variety of neuromorphic functions, including spike timing-dependent plasticity, pair pulse depression and facilitation, short and long-term plasticity, and several different models of synaptic firing. However, neuromorphic functionality is not strictly limited to nanoscale devices, and has been demonstrated in organic electrochemical transistors where the functional material is poly(3,4-ethylenedioxythiophene) doped with either polystyrene sulfonate[108–110] or polytetrahydrofuran.[111] Organic memristors have also been used to implement Boolean logic in networks with metal-oxide-semiconductor artificial neurons.[112] However, polymer devices suffer from lower ON/OFF ratios and worse reliability, and progress will need to be made in these areas to rival inorganic devices.[113] Scaling up from single devices, networks of neuromorphic devices are beginning to be used to replicate higher order pattern recognition and logic functions, demonstrating the beginning of neuromorphic computation. A major future application of this work will undoubtedly be using neuromorphic devices for novel neural interfaces. Neuromorphic metal-oxide memristors have been used to encode and compress neuronal spike signals,[114] and neurons have been remotely connected to VLSI neurons to create biohybrid networks.[115] Neurons have also been grown with nanoparticle/polymer electrolyte-gated transistors.[116] Despite the major challenges of signal loss and reliability that have yet to be fully addressed, the progress in the development of neuromorphic nanoelectronics in the last decade has been impressive, and the applications of this work toward computing and neuroscience seem nearly limitless.

Acknowledgment

The author acknowledges COST Action 1401: Memristors— Devices, Circuits, and Applications (MemoCIS) for many helpful workshops and discussions.

References

1. Willis, L. J.; Fairfield, J. A.; Dadosh, T.; Fischbein, M. D.; Drndic, M. Controlling nanogap quantum dot photoconductivity through optoelectronic trap manipulation. *Nano Lett.* **2009**, *9* (12), 4191–4197.

2. Charlie Johnson, A. T.; Staii, C.; Chen, M.; Khamis, S.; Johnson, R.; Klein, M. L.; Gelperin, A. DNA-decorated carbon nanotubes for chemical sensing. *Phys. Status Solidi B* **2006**, *243*, 3252–3256.

3. Sysoev, V. V; Schneider, T.; Goschnick, J.; Kiselev, I.; Habicht, W.; Hahn, H.; Strelcov, E.; Kolmakov, A. Percolating SnO$_2$ nanowire network as a stable gas sensor: Direct comparison of long-term performance versus SnO$_2$ nanoparticle films. *Sensors Actuators B Chem.* **2009**, *139* (2), 699–703.

4. Ponzoni, A.; Comini, E.; Sberveglieri, G.; Zhou, J.; Deng, S. Z.; Xu, N. S.; Ding, Y.; Wang, Z. L. Ultrasensitive and highly selective gas sensors using three-dimensional tungsten oxide nanowire networks. *Appl. Phys. Lett.* **2006**, *88* (20), 203101–203103.

5. Guerra, D. N.; Bulsara, A. R.; Ditto, W. L.; Sinha, S.; Murali, K.; Mohanty, P. A noise-assisted reprogrammable nanomechanical logic gate. *Nano Lett.* **2010**, *10* (4), 1168–1171.

6. Ginger, D. S.; Greenham, N. C. Charge injection and transport in films of CdSe nanocrystals. *J. Appl. Phys.* **2000**, *87* (3), 1361–1368.

7. Drndic, M.; Jarosz, M. V; Morgan, N. Y.; Kastner, M. A.; Bawendi, M. G. Transport properties of annealed CdSe colloidal nanocrystal solids. *J. Appl. Phys.* **2002**, *92* (12), 7498–7503.

8. Porter, V. J.; Geyer, S.; Halpert, J. E.; Kastner, M. A.; Bawendi, M. G. Photoconduction in annealed and chemically treated CdSe/ZnS inorganic nanocrystal films. *J. Phys. Chem. C* **2008**, *112* (7), 2308–2316.

9. Leatherdale, C. A.; Kagan, C. R.; Morgan, N. Y.; Empedocles, S. A.; Kastner, M. A.; Bawendi, M. G. Photoconductivity in CdSe quantum dot solids. *Phys. Rev. B* **2000**, *62* (4), 2669.

10. Chua, L. Memristor-the missing circuit element. *IEEE Trans. Circuit Theory* **1971**, *18* (5), 507–519.

11. Strukov, D. B.; Snider, G. S.; Stewart, D. R.; Williams, R. S. The missing memristor found. *Nature* **2008**, *453* (7191), 80–83.

12. Bruce, J. H.; Hickling, A. Electrical conduction of commercial boron crystals. *Trans. Faraday Soc.* **1939**, *35*, 1436–1439.

13. Di Ventra, M.; Pershin, Y. V.; Chua, L. O. Putting memory into circuit elements: Memristors, memcapacitors, and meminductors [point of view]. *Proc. IEEE* **2009**, *97*, 1717.

14. Kim, S.; Choi, B.; Lim, M.; Yoon, J.; Lee, J.; Kim, H.-D.; Choi, S.-J. Pattern recognition using carbon nanotube synaptic transistors with an adjustable weight update protocol. *ACS Nano* **2017**, *11*, 2814–2822.

15. Kuzum, D.; Yu, S.; Philip Wong, H. S. Synaptic electronics: Materials, devices and applications. *Nanotechnology* **2013**, *24* (38), 382001.

16. Wright, C. D.; Hosseini, P.; Diosdado, J. A. V. Beyond von-neumann computing with nanoscale phase-change memory devices. *Adv. Funct. Mater.* **2013**, *23*, 2248–2254.

17. Mead, C. Neuromorphic electronic systems. *Proc. IEEE* **1990**, *78*, 1629–1636.

18. Rajendran, B.; Liu, Y.; Seo, J. S.; Gopalakrishnan, K.; Chang, L.; Friedman, D. J.; Ritter, M. B. Specifications of nanoscale devices and circuits for neuromorphic computational systems. *IEEE Trans. Electron Devices* **2013**, *60* (1), 246–253.

19. Waser, R.; Aono, M. Nanoionics-based resistive switching memories. *Nat. Mater.* **2007**, *6* (11), 833–840.

20. Bellew, A. T.; Bell, A. P.; McCarthy, E. K.; Fairfield, J. A.; Boland, J. J. Programmability of nanowire networks. *Nanoscale* **2014**, *6* (16), 1–3.

21. Chua, L. Resistance switching memories are memristors. *Appl. Phys. A Mater. Sci. Process.* **2011**, *102*, 765–783.

22. Ielmini, D. Resistive switching memories based on metal oxides: Mechanisms, reliability and scaling. *Semicond. Sci. Technol.* **2016**, *31* (6), 063002.

23. Li, Y.; Long, S.; Liu, Y.; Hu, C.; Teng, J.; Liu, Q.; Lv, H.; Suñé, J.; Liu, M. Conductance quantization in resistive random access memory. *Nanoscale Res. Lett.* **2015**, *10*, 420.

24. Yang, Y.; Lu, W. Nanoscale resistive switching devices: Mechanisms and modeling. *Nanoscale* **2013**, *5* (21), 10076.

25. Li, J.; Aron, C.; Kotliar, G.; Han, J. E. Microscopic theory of resistive switching in ordered insulators: Electronic versus thermal mechanisms. *Nano Lett.* **2017**, *17* (5), 2994–2998.

26. Kim, S.; Choi, S.; Lu, W. Comprehensive physical model of dynamic resistive switching in an oxide memristor. *ACS Nano* **2014**, *8* (3), 2369–2376.

27. Lutz, C.; Hasegawa, T.; Chikyow, T.; Hasegawa, T.; Hill, J. P.; Ariga, K.; Aono, M.; Yamada, H.; Deranlot, C.; Mathur, N. D.; et al. Ag_2S atomic switch-based 'Tug of War' for decision making. *Nanoscale* **2016**, *8* (29), 14031–14036.

28. Hasegawa, T.; Ohno, T.; Terabe, K.; Tsuruoka, T.; Nakayama, T.; Gimzewski, J. K.; Aono, M. Learning abilities achieved by a single solid-state atomic switch. *Adv. Mater.* **2010**, *22* (16), 1831–1834.

29. Yang, Y.; Gao, P.; Gaba, S.; Chang, T.; Pan, X.; Lu, W. Observation of conducting filament growth in nanoscale resistive memories. *Nat. Commun.* **2012**, *3*, 732.

30. Cooper, D.; Baeumer, C.; Bernier, N.; Marchewka, A.; La Torre, C.; Dunin-Borkowski, R. E.; Menzel, S.; Waser, R.; Dittmann, R. Anomalous resistance hysteresis in oxide ReRAM: Oxygen evolution and reincorporation revealed by in situ TEM. *Adv. Mater.* **2017**, 1700212.

31. Liu, Q.; Sun, J.; Lv, H.; Long, S.; Yin, K.; Wan, N.; Li, Y.; Sun, L.; Liu, M. Real-time observation on dynamic growth/dissolution of conductive filaments in oxide-electrolyte-based ReRAM. *Adv. Mater.* **2012**, *24* (14), 1844–1849.

32. Midya, R.; Wang, Z.; Zhang, J.; Savel'ev, S. E.; Li, C.; Rao, M.; Jang, M. H.; Joshi, S.; Jiang, H.; Lin, P.; et al. Anatomy of Ag/Hafnia-based selectors with 10^{10} nonlinearity. *Adv. Mater.* **2017**, *29* (12), 1604457.

33. Yang, Y.; Gao, P.; Li, L.; Pan, X.; Tappertzhofen, S.; Choi, S.; Waser, R.; Valov, I.; Lu, W. D. Electrochemical dynamics of nanoscale metallic inclusions in dielectrics. *Nat. Commun.* **2014**, *5*, 377–383.

34. Lv, H.; Xu, X.; Liu, H.; Liu, R.; Liu, Q.; Banerjee, W.; Sun, H.; Long, S.; Li, L.; Liu, M. Evolution of conductive filament and its impact on reliability issues in oxide-electrolyte based resistive random access memory. *Sci. Rep.* **2015**, *5*, 7764.

35. Baeumer, C.; Valenta, R.; Schmitz, C.; Locatelli, A.; Menteş, T. O.; Rogers, S. P.; Sala, A.; Raab, N.; Nemsak, S.; Shim, M.; et al. Subfilamentary networks cause cycle-to-cycle variability in memristive devices. *ACS Nano* **2017**, *11* (7), 6921–6929.

36. Oliver, S. M.; Fairfield, J. A.; Bellew, A. T.; Lee, S.; Champlain, J. G.; Ruppalt, L. B.; Boland, J. J.; Vora, P. M. Quantum point contacts and resistive switching in Ni/NiO nanowire junctions. *Appl. Phys. Lett.* **2016**, *109* (20), 203101.

37. Nardi, F.; Larentis, S.; Balatti, S.; Gilmer, D. C.; Ielmini, D. Resistive switching by voltage-driven ion migration in bipolar RRAM: Part I: Experimental study. *IEEE Trans. Electron Devices* **2012**, *59* (9), 2461–2467.

38. O'Kelly, C.; Fairfield, J. A.; Boland, J. J. A single nanoscale junction with programmable multilevel memory. *ACS Nano* **2014**, *8* (11), 11724–11729.

39. Balatti, S.; Ambrogio, S.; Wang, Z.; Sills, S.; Calderoni, A.; Ramaswamy, N.; Ielmini, D. Voltage-controlled cycling endurance of HfO<Sub><Italic>X</Italic></Sub>-based resistive-switching memory. *IEEE Trans. Electron Devices* **2015**, *62* (10), 3365–3372.

40. Crouch, G. M.; Han, D.; Fullerton-Shirey, S. K.; Go, D. B.; Bohn, P. W. Addressable direct-write nanoscale filament formation and dissolution by nanoparticle-mediated bipolar electrochemistry. *ACS Nano* **2017**, *11* (5), 4976–4984.

41. Mehonic, A.; Vrajitoarea, A.; Cueff, S.; Hudziak, S.; Howe, H.; Labbé, C.; Rizk, R.; Pepper, M.; Kenyon, A. J. Quantum conductance in silicon oxide resistive memory devices. *Sci. Rep.* **2013**, *3*, 2708.

42. Sun, Y.; Wen, D. Conductance quantization in nonvolatile resistive switching memory based on the polymer composite of zinc oxide nanoparticles. *J. Phys. Chem. C* **2018**, *122* (19), 10582.

43. Fadaly, E. M. T.; Zhang, H.; Conesa-Boj, S.; Car, D.; Gül, Ö.; Plissard, S. R.; Op het Veld, R. L. M.; Kölling, S.; Kouwenhoven, L. P.; Bakkers, E. P. A. M. Observation of conductance quantization in InSb nanowire networks. *Nano Lett.* **2017**, *11*, 6511–6515.

44. Zhu, X.; Su, W.; Liu, Y.; Hu, B.; Pan, L.; Lu, W.; Zhang, J.; Li, R. W. Observation of conductance quantization in oxide-based resistive switching memory. *Adv. Mater.* **2012**, *24* (29), 3941–3946.

45. Yi, W.; Savel'ev, S. E.; Medeiros-Ribeiro, G.; Miao, F.; Zhang, M.-X.; Yang, J. J.; Bratkovsky, A. M.; Williams, R. S. Quantized conductance coincides with state instability and excess noise in tantalum oxide memristors. *Nat. Commun.* **2016**, *7*, 11142.

46. Ambrogio, S.; Balatti, S.; Cubeta, A.; Calderoni, A.; Ramaswamy, N.; Ielmini, D. Statistical fluctuations in HfOx resistive-switching memory: Part II: Random telegraph noise. *IEEE Trans. Electron Devices* **2014**, *61* (8), 2920–2927.

47. Mikhaylov, A. N.; Gryaznov, E. G.; Belov, A. I.; Korolev, D. S.; Sharapov, A. N.; Guseinov, D. V.; Tetelbaum, D. I.; Tikhov, S. V.; Malekhonova, N. V.; Bobrov, A. I.; et al. Field- and irradiation-induced phenomena in memristive nanomaterials. *Phys. Status Solidi* **2016**, *13* (10–12), 870–881.

48. Tang, K.; Meng, A. C.; Hui, F.; Shi, Y.; Petach, T.; Hitzman, C.; Koh, A. L.; Goldhaber-Gordon, D.; Lanza, M.; McIntyre, P. C. Distinguishing oxygen vacancy electromigration and conductive filament formation in TiO_2 resistance switching using liquid electrolyte contacts. *Nano Lett.* **2017**, *17* (7), 4390–4399.

49. Fairfield, J. A.; Rocha, C. G.; O'Callaghan, C.; Ferreira, M. S.; Boland, J. J. Co-percolation to tune conductive behaviour in dynamical metallic nanowire networks. *Nanoscale* **2016**, *8* (43), 18516–18523.

50. Bi, G. Q.; Poo, M. M. Synaptic modifications in cultured hippocampal neurons: Dependence on spike timing, synaptic strength, and postsynaptic cell type. *J. Neurosci.* **1998**, *18*, 10464–10472.

51. Abbott, L. F.; Regehr, W. G. Synaptic computation. *Nature* **2004**, *431*, 796–803.

52. Jackman, S. L.; Regehr, W. G. The mechanisms and functions of synaptic facilitation. *Neuron* **2017**, *94*, 826–839.

53. Hodgkin, A. L.; Huxley, A. F. Action potentials recorded from inside a nerve fibre. *Nature* **1939**, *144*, 710–711.

54. Hodgkin, A. L.; Huxley, A. F. Resting and action potentials in single nerve fibres. *J. Physiol.* **1945**, *104* (2), 176–195.

55. Gerstner, W. Spiking neuron models. *Encycl. Neurosci.* **2010**, 277–280. doi: 10.1016/B978-008045046-9.01405-4.

56. Burkitt, A. N. A review of the integrate-and-fire neuron model: I. Homogeneous synaptic input. *Biol. Cybern.* **2006**, *95* (1), 1–19.

57. Burkitt, A. N. A review of the integrate-and-fire neuron model: II. Inhomogeneous synaptic input and network properties. *Biol. Cybern.* **2006**, *95* (2), 97–112.

58. Oster, M.; Wang, Y.; Douglas, R.; Liu, S. C. Quantification of a spike-based winner-take-all VLSI network. *IEEE Trans. Circuits Syst. I Regul. Pap.* **2008**, *55*, 3160–3169.

59. Linares-Barranco, B.; Serrano-Gotarredona, T. Memristance can explain spike-time-dependent-plasticity in neural synapses. *Nature* **2009**, *14*, 32.

60. Serrano-Gotarredona, T.; Masquelier, T.; Prodromakis, T.; Indiveri, G.; Linares-Barranco, B. STDP and STDP variations with memristors for spiking neuromorphic learning systems. *Front. Neurosci.* **2013**, *7* (2), 1–15.

61. Serrano-Gotarredona, T.; Masquelier, T.; Prodromakis, T.; Indiveri, G.; Linares-Barranco, B. STDP and STDP variations with memristors for spiking neuromorphic learning systems. *Front. Neurosci.* **2013**, *7*, 2.

62. Yang, J.-T.; Ge, C.; Du, J.-Y.; Huang, H.-Y.; He, M.; Wang, C.; Lu, H.-B.; Yang, G.-Z.; Jin, K.-J. Artificial synapses emulated by an electrolyte-gated tungsten-oxide transistor. *Adv. Mater.* **2018**, *30* (34), 1801548.

63. Mehonic, A.; Kenyon, A. J. Emulating the electrical activity of the neuron using a silicon oxide RRAM cell. *Front. Neurosci.* **2016**, *10*, 57.

64. Mehonic, A.; Buckwell, M.; Montesi, L.; Munde, M. S.; Gao, D.; Hudziak, S.; Chater, R. J.; Fearn, S.; McPhail, D.; Bosman, M.; et al. Silica: nanoscale transformations in metastable, amorphous, silicon-rich silica. *Adv Mater.* **2016**, *28*, 7549.

65. Tian, H.; Cao, X.; Xie, Y.; Yan, X.; Kostelec, A.; DiMarzio, D.; Chang, C.; Zhao, L.-D.; Wu, W.; Tice, J.; et al. Emulating bilingual synaptic response using a junction-based artificial synaptic device. *ACS Nano* **2017**, *11* (7), 7156–7163.

66. Beck, A.; Bednorz, J. G.; Gerber, C.; Rossel, C.; Widmer, D. Reproducible switching effect in thin oxide films for memory applications. *Appl. Phys. Lett.* **2000**, *77* (1), 139.

67. Brivio, S.; Frascaroli, J.; Spiga, S. Role of metal-oxide interfaces in the multiple resistance switching regimes of Pt/HfO¡2¡/Inf¿/TiN devices. *Appl. Phys. Lett.* **2015**, *107* (2), 023504.

68. Ambrogio, S.; Balatti, S.; Choi, S.; Ielmini, D. Impact of the mechanical stress on switching characteristics of electrochemical resistive memory. *Adv. Mater.* **2014**, *26* (23), 3885–3892.

69. Lübben, M.; Menzel, S.; Park, S. G.; Yang, M.; Waser, R.; Valov, I. SET kinetics of electrochemical metallization cells: Influence of counter-electrodes in SiO_2/Ag based systems. *Nanotechnology* **2017**, *28* (13), 135205.

70. Choi, B. J.; Zhang, J.; Norris, K.; Gibson, G.; Kim, K. M.; Jackson, W.; Zhang, M.-X. M.; Li, Z.; Yang, J. J.; Williams, R. S. Trilayer tunnel selectors for memristor memory cells. *Adv. Mater.* **2016**, *28* (2), 356–362.

71. La Barbera, S.; Vuillaume, D.; Alibart, F. Filamentary switching: Synaptic plasticity through device volatility. *ACS Nano* **2015**, *9* (1), 941–949.

72. Yu, S.; Wu, Y.; Jeyasingh, R.; Kuzum, D.; Wong, H. S. P. An electronic synapse device based on metal oxide resistive switching memory for neuromorphic computation. *IEEE Trans. Electron Devices* **2011**, *58* (8), 2729–2737.

73. Wang, Z.; Ambrogio, S.; Balatti, S.; Ielmini, D. A 2-transistor/1-resistor artificial synapse capable of communication and stochastic learning in neuromorphic systems. *Front. Neurosci.* **2015**, *8*, 438.

74. Ambrogio, S.; Balatti, S.; Nardi, F.; Facchinetti, S.; Ielmini, D. Spike-timing dependent plasticity in a transistor-selected resistive switching memory. *Nanotechnology* **2013**, *24* (38), 384012.

75. Ambrogio, S.; Balatti, S.; Milo, V.; Carboni, R.; Wang, Z.-Q.; Calderoni, A.; Ramaswamy, N.; Ielmini, D. Neuromorphic learning and recognition with one-transistor-one-resistor synapses and bistable metal oxide RRAM. *IEEE Trans. Electron Devices* **2016**, *63* (4), 1508–1515.

76. Kim, S.; Du, C.; Sheridan, P.; Ma, W.; Choi, S.; Lu, W. D. Experimental demonstration of a second-order memristor and its ability to biorealistically implement synaptic plasticity. *Nano Lett.* **2015**, *15* (3), 2203–2211.

77. Prezioso, M.; Merrikh Bayat, F.; Hoskins, B.; Likharev, K.; Strukov, D. Self-adaptive spike-time-dependent plasticity of metal-oxide memristors. *Sci. Rep.* **2016**, *6*, 21331.

78. Seo, K.; Kim, I.; Jung, S.; Jo, M.; Park, S.; Park, J.; Shin, J.; Biju, K. P.; Kong, J.; Lee, K.; et al. Analog memory and spike-timing-dependent plasticity characteristics of a nanoscale titanium oxide bilayer resistive switching device. *Nanotechnology* **2011**, *22* (25), 254023.

79. La Barbera, S.; Ly, D. R. B.; Navarro, G.; Castellani, N.; Cueto, O.; Bourgeois, G.; De Salvo, B.; Nowak, E.; Querlioz, D.; Vianello, E. Narrow heater bottom electrode-based phase change memory as a bidirectional artificial synapse. *Adv. Electron. Mater.* **2018**, *4*, 1800223.

80. Li, Y.; Zhong, Y.; Xu, L.; Zhang, J.; Xu, X.; Sun, H.; Miao, X. Ultrafast synaptic events in a chalcogenide memristor. *Sci. Rep.* **2013**, *3*, 1619.

81. Sharbati, M. T.; Du, Y.; Torres, J.; Ardolino, N. D.; Yun, M.; Xiong, F. Low-power, electrochemically tunable graphene synapses for neuromorphic computing. *Adv. Mater*, **2018**, *30* (36), 1870273.

82. Rehman, M. M.; Siddiqui, G. U.; Gul, J. Z.; Kim, S.-W.; Lim, J. H.; Choi, K. H. Resistive switching in all-printed, flexible and hybrid MoS2-PVA nanocomposite based memristive device fabricated by reverse offset. *Sci. Rep.* **2016**, *6* (1), 36195.

83. Wang, W.; Panin, G. N.; Fu, X.; Zhang, L.; Ilanchezhiyan, P.; Pelenovich, V. O.; Fu, D.; Kang, T. W. MoS2 memristor with photoresistive switching. *Sci. Rep.* **2016**, *6* (1), 31224.

84. Mizrahi, A.; Hirtzlin, T.; Fukushima, A.; Kubota, H.; Yuasa, S.; Grollier, J.; Querlioz, D. Neural-like computing with populations of superparamagnetic basis functions. *Nat. Commun.* **2018**, *9* (1), 1533.

85. Vodenicarevic, D.; Locatelli, N.; Mizrahi, A.; Friedman, J. S.; Vincent, A. F.; Romera, M.; Fukushima, A.; Yakushiji, K.; Kubota, H.; Yuasa, S.; et al. Low-energy truly random number generation with superparamagnetic tunnel junctions for unconventional computing. *Phys. Rev. Appl.* **2017**, *8*, 054045.

86. Jo, S. H.; Chang, T.; Ebong, I.; Bhadviya, B. B.; Mazumder, P.; Lu, W. Nanoscale memristor device as synapse in neuromorphic systems. *Nano Lett.* **2010**, *10* (4), 1297–1301.

87. Jaafar, A. H.; Gray, R. J.; Verrelli, E.; O'Neill, M.; Kelly, S. M.; Kemp, N. T. Reversible optical switching memristors with tunable STDP synaptic plasticity: A route to hierarchical control in artificial intelligent systems. *Nanoscale* **2017**, *9* (43), 17091–17098.

88. Alibart, F.; Pieutin, S.; Guérin, D.; Novembre, C.; Lenfant, S.; Lmimouni, K.; Gamrat, C.; Vuillaume, D. An organic nanoparticle transistor behaving as a biological spiking synapse. *Adv. Funct. Mater.* **2010**, *20* (2), 330–337.

89. Fairfield, J. A.; Dadosh, T.; Drndic, M. Characterization of memory and measurement history in photoconductivity of nanocrystal arrays. *Appl. Phys. Lett.* **2010**, *97* (14), 143112–143113.

90. Wang, Y.; Lv, Z.; Chen, J.; Wang, Z.; Zhou, Y.; Zhou, L.; Chen, X.; Han, S.-T. Photonic synapses based on inorganic perovskite quantum dots for neuromorphic computing. *Adv. Mater.* **2018**, *30*, 1802883.

91. O'Kelly, C. J.; Fairfield, J. A.; McCloskey, D.; Manning, H. G.; Donegan, J. F.; Boland, J. J. Associative enhancement of time correlated response to heterogeneous stimuli in a neuromorphic nanowire device. *Adv. Electron. Mater.* **2016**, *2* (6), 1500458.

92. Hu, S. G.; Liu, Y.; Liu, Z.; Chen, T. P.; Wang, J. J.; Yu, Q.; Deng, L. J.; Yin, Y.; Hosaka, S. Associative memory realized by a reconfigurable memristive hopfield neural network. *Nat. Commun.* **2015**, *6*, 7522.

93. Afifi, A.; Ayatollahi, A.; Raissi, F. STDP implementation using memristive nanodevice in CMOS-nano neuromorphic networks. *IEICE Electron. Express* **2009**, *6* (3), 148–153.

94. Linares-Barranco, B.; Serrano-Gotarredona, T.; Camunas-Mesa, L. A.; Perez-Carrasco, J. A.; Zamarreno-Ramos, C.; Masquelier, T. On spike-timing-dependent-plasticity, memristive devices, and building a self-learning visual cortex. *Front. Neurosci.* **2011**, *5*, 26.

95. Serb, A.; Bill, J.; Khiat, A.; Berdan, R.; Legenstein, R.; Prodromakis, T. Unsupervised learning in probabilistic neural networks with multi-state metal-oxide memristive synapses. *Nat. Commun.* **2016**, *7*, 12611.

96. Covi, E.; Brivio, S.; Serb, A.; Prodromakis, T.; Fanciulli, M.; Spiga, S. Analog memristive synapse in spiking networks implementing unsupervised learning. *Front. Neurosci.* **2016**, *10*, 482.

97. Ambrogio, S.; Ciocchini, N.; Laudato, M.; Milo, V.; Pirovano, A.; Fantini, P.; Ielmini, D. Unsupervised learning by spike timing dependent plasticity in Phase Change Memory (PCM) synapses. *Front. Neurosci.* **2016**, *10*, 56.

98. Rey, H. G.; Pedreira, C.; Quian Quiroga, R. Past, Present and future of spike sorting techniques. *Brain Res. Bull.* **2015**, *119* (Pt B), 106–117.

99. Cassidy, A. S.; Georgiou, J.; Andreou, A. G. Design of silicon brains in the nano-CMOS era: Spiking neurons, learning synapses and neural architecture optimization. *Neural Networks* **2013**, *45*, 4–26.

100. Moradi, S.; Qiao, N.; Stefanini, F.; Indiveri, G. A scalable multicore architecture with heterogeneous memory structures for Dynamic Neuromorphic Asynchronous Processors (DYNAPs). *IEEE Trans. Biomed. Circuits Syst.* **2018**, *12*, 106–122.

101. Chicca, E.; Stefanini, F.; Bartolozzi, C.; Indiveri, G. Neuromorphic electronic circuits for building autonomous cognitive systems. *Proc. IEEE* **2014**, *102* (9), 1367–1388.

102. Indiveri, G.; Linares-Barranco, B.; Hamilton, T. J.; van Schaik, A.; Etienne-Cummings, R.; Delbruck, T.; Liu, S. C.; Dudek, P.; Häfliger, P.; Renaud, S.; et al. Neuromorphic silicon neuron circuits. *Front. Neurosci.* **2011** *5*, 73.

103. Mitra, S.; Fusi, S.; Indiveri, G. Real-time classification of complex patterns using spike-based learning in neuromorphic VLSI. *IEEE Trans. Biomed. Circuits Syst.* **2009**, *3* (1), 32–42.

104. Neftci, E.; Binas, J.; Rutishauser, U.; Chicca, E.; Indiveri, G.; Douglas, R. J. Synthesizing cognition in neuromorphic electronic systems. *Proc. Natl. Acad. Sci. U. S. A.* **2013**, *110* (37), E3468–E3476.

105. Sheik, S.; Coath, M.; Indiveri, G.; Denham, S. L.; Wennekers, T.; Chicca, E. Emergent auditory feature tuning in a real-time neuromorphic VLSI system. *Front. Neurosci.* **2012**, *6*, 17.

106. Sun, Z.; Ambrosi, E.; Bricalli, A.; Ielmini, D. Logic computing with stateful neural networks of resistive switches. *Adv. Mater.* **2018**, *30*, e1802554.

107. Aono, M.; Naruse, M.; Kim, S.-J.; Wakabayashi, M.; Hori, H.; Ohtsu, M.; Hara, M. Amoeba-inspired nanoarchitectonic computing: Solving intractable computational problems using nanoscale photoexcitation transfer dynamics. *Langmuir* **2013**, *29* (24), 7557–7564.

108. Gkoupidenis, P.; Schaefer, N.; Garlan, B.; Malliaras, G. G. Neuromorphic functions in PEDOT: PSS organic electrochemical transistors. *Adv. Mater.* **2015**, *27* (44), 7176–7180.

109. Gkoupidenis, P.; Koutsouras, D. A.; Lonjaret, T.; Fairfield, J. A.; Malliaras, G. G. Orientation selectivity in a multi-gated organic electrochemical transistor. *Sci. Rep.* **2016**, *6*, 27007.

110. Gkoupidenis, P.; Koutsouras, D. A.; Malliaras, G. G. Neuromorphic device architectures with global connectivity through electrolyte gating. *Nat. Commun.* **2017**, *8*, 15448.

111. Gkoupidenis, P.; Schaefer, N.; Strakosas, X.; Fairfield, J. A.; Malliaras, G. G. Synaptic plasticity functions in an organic electrochemical transistor. *Appl. Phys. Lett.* **2015**, *107* (26), 263302.

112. Emelyanov, A. V.; Lapkin, D. A.; Demin, V. A.; Erokhin, V. V.; Battistoni, S.; Baldi, G.;

Dimonte, A.; Korovin, A. N.; Iannotta, S.; Kashkarov, P. K.; et al. First steps towards the realization of a double layer perceptron based on organic memristive devices. *AIP Adv.* **2016**, *6* (11), 111301.

113. Wang, L.; Yang, C.; Wen, J.; Gai, S.; Peng, Y. Overview of emerging memristor families from resistive memristor to spintronic memristor. *J. Mater. Sci. Mater. Electron.* **2015**, *26* (7), 4618–4628.

114. Gupta, I.; Serb, A.; Khiat, A.; Zeitler, R.; Vassanelli, S.; Prodromakis, T. Real-time encoding and compression of neuronal spikes by metal-oxide memristors. *Nat. Commun.* **2016**, *7*, 12805.

115. Serb, A.; Corna, A.; George, R.; Khiat, A.; Rocchi, F.; Reato, M.; Maschietto, M.; Mayr, C.; Indiveri, G.; Vassanelli, S.; et al. A geographically distributed bio-hybrid neural network with memristive plasticity. *arXiv preprint arXiv* **2017**, *1709*, 04179.

116. Desbief, S.; di Lauro, M.; Casalini, S.; Guerin, D.; Tortorella, S.; Barbalinardo, M.; Kyndiah, A.; Murgia, M.; Cramer, T.; Biscarini, F.; et al. Electrolyte-gated organic synapse transistor interfaced with neurons. *Org. Electron.* **2016**, *38*, 21–28.

Single Electronics: Modeling and Simulation Techniques

B. S. Pês
IFPR

E. Oroski
UTFPR

J. G. Guimarães
UFSC

12.1 Introduction.. **12**-1
12.2 Fundamental Concepts **12**-2
    Tunneling • Tunnel Junctions • Single-Electron Transistor
12.3 Simulation Methods.. **12**-3
    Master Equation of Single-Electron Tunneling • Monte Carlo Method
12.4 SET: A Revision of Proposed Models **12**-6
    Analytical Models for SET • Macromodels for SET
12.5 Conclusion **12**-11
References **12**-12

12.1 Introduction

In 1965, Gordon Moore enunciated that the miniaturization rate of electronic devices would double every 2 years. Since then, several efforts were made to keep the validity of the so-called Moore's Law. However, at the end of 1990s, performance issues due to shrinking the characteristic dimension of transistors emerged. Companies as such as Intel and IBM invested in new materials and fabrication techniques to overcome those problems. An example of a new technique is stretching silicon lattice by layering it in another crystal (Hain 2015).

Current microprocessors have transistors with gate terminals of about 14 nm. This significant advance is due to research efforts driven by the promise of faster and more efficient circuits. The results of such efforts are innovative technologies that kept Moore's law alive for 50 years. Today, the miniaturization rate doubles every 2.5 years, approximately (Toumey2016).

Despite all efforts, Metal-Oxide Semiconductor (MOS) technology will eventually stop providing performance improvements (Garner, Kloster, Astwood, Mosley & Palanduz 2007). As MOS devices approach its shrinking limit, new technologies are needed to maintain the progress of electronics industry. Nanoelectronics emerge as a promising alternative.

Nanoelectronics provide advantages that go beyond scalability. Nanoelectronic devices are ruled by quantum physics (Goser, Pacha, Kanstein & Rossman 1997). This feature allows current to flow through tunneling events, a quantum phenomenon enables individual electron transport (Likharev 1999). It is common to highlight this particularity of nanoelectronic devices by referring it as Single-Electron (SE) devices. Charge transport through tunneling impacts power consumption, scale, and switching capacity (Paul 2002).

Unique features make the Single-Electron Transistor (SET) as one of the most studied SE devices (Hasaneen, Wahab & Ahmed 2011). Though promising, SET application in practical circuits still faces many challenges. Extreme sensibility to charge fluctuations makes SET susceptible to impurities, which may lead the transistor to conduct in a blockade region (Abutaleb 2015). Besides, SET has low voltage gain (Lientschnig, Weymann & Hadley 2003). Though operation at room temperature is still challenging, some research groups reported the successful observation of Coulomb blockade in temperatures up to 350 K (Shin et al. 2010, 2011, Lavieville, Barraud, Corna, Jehl, Sanquer & Vinet 2015).

Hybrid implementations of SET-MOS circuits are increasingly common in literature (Abutaleb 2015, Jain, Singh, Ghosh & Sarkar 2014, Jana, Singh, Sing & Sarkar 2013). Circuit simulators and accurate models are fundamental for analyzing both hybrid and pure nanoelectronic approaches.

According to Ljung (1999), inferring models from observation and studying their properties is really what science is all about (Ljung 1999). In many scenarios, simply performing the experiment is not possible or feasible. In these cases, the behavior of the system can be analyzed through numerical simulation. In this context, modeling can be defined as the process of creating mathematical models for real systems. Furthermore, in practical aspects, modeling is important for analysis and forecast of the system behavior, especially, in research areas.

Virtually, every component and system one can think of can be described by a mathematical model (Billings 2013).

The same applies to Nanoelectronics, circuit or component features can be adapted and tested by simulation, before physical implementation. In design theory, it is very important to achieve a mathematical model for a system to be designed. The component sizing can be done in an easier way if the fundamental mathematical relations are known.

There are several SET models available in literature. Three methods are commonly applied when developing these models: Monte Carlo, Master Equation of single-electron tunneling, and Macromodeling for Simulated Program with Integrated Circuits Emphasis (SPICE). This chapter is dedicated to present and discuss each one of these methods.

12.2 Fundamental Concepts

To study SET simulation methods and modeling techniques, first, it is important to remember the basic concepts regarding SET operation. Thus, this section will present a summarized description of SET operation, starting by the charge transport mechanism: single-electron tunneling.

12.2.1 Tunneling

In 1923, Louis de Broglie suggested that particles should have oscillatory properties. De Broglie was referring to properties observed with light experiments: photoelectric effect, which suggested a particle nature, and interference pattern, which suggested a wave nature.

In 1926, Clinton Davisson and Lester Germer observed electron diffraction patterns. The conclusion was that when the wavelength, λ, is small enough in comparison to the distance in which the energy changes, quantum particles behave in a classic way (Hanson 2008).

In general terms, tunneling is a quantum phenomenon in which particles are allowed to cross classically forbidden regions. That is, regions with potential energy greater than the total particle energy. Figure 12.1 depicts the idea.

It is worth mentioning that charge transport in nanostructures involves a great amount of concepts and phenomena that were not treated in this section. A good reference concerning the theme is (Ferry, Goodnick & Bird 2009).

Next section is dedicated to present the tunnel junction, the fundamental unit of a SET.

12.2.2 Tunnel Junctions

SE devices are all about the control over electron flow. For that control to be possible, it is necessary to have a device through which the electrons can flow in a discrete manner. This device is known as tunnel junction, and it is depicted in Figure 12.2.

A tunnel junction is characterized by two macroscopic parameters: (i) its tunneling resistance, R_T and (ii) its capacitance, C_j. In this way, a tunnel junction may be seen as a capacitor with a dielectric layer that is thin enough so that electrons may tunnel through it: a leaky capacitor (Grabert & Devoret 1992). Figure 12.3 illustrates the idea.

To ensure that tunneling is the dominant mechanism of charge transport, the tunnel junction parameters must fulfill the conditions expressed by (12.1) and (12.2) (Grabert & Devoret 1992).

$$R_T \geq \frac{h}{e^2}, \tag{12.1}$$

$$C_j \ll \frac{e^2}{k_B T}, \tag{12.2}$$

in which h stands for the Planck constant, e is the elementary charge, k_B is the Boltzmann constant, and T is the temperature in Kelvin. In Eq. (12.1), $\frac{h}{e^2}$ is also known as the quantum resistance, R_Q.

12.2.3 Single-Electron Transistor

A SET consists of a series connection of two tunnel junctions. Since the conducting region between the junctions is isolated by insulator layers, it is usual to call it an island. The electrostatic potential of the island is controlled via a gate terminal that is capacitively coupled to the island (Heinzel 2007). Figure 12.4 illustrates a SET with one gate. It is possible to add a second gate to the island, which is a common practice in applications such as decimal adders (Fahmy & Ismail 1997).

Consider the circuit shown in Figure 12.5. Let V_{DS} be the bias voltage and V_{GS} the voltage applied in the gate terminal. If V_{DS} is kept constant and V_{GS} is increased monotonically, the current across the transistor will present a

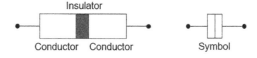

FIGURE 12.2 A schematic and symbol of a tunnel junction.

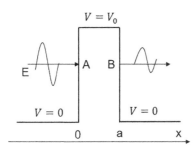

FIGURE 12.1 Tunneling of a particle with total energy, E, across a barrier with potential $V_0 > E$. A and B are the classical turning points.

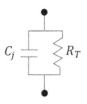

FIGURE 12.3 Tunnel junction as a leaky capacitor.

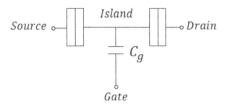

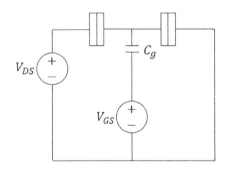

FIGURE 12.4 A SET. The island is the conducting region between the source and drain terminals, and charge transport through the island is via tunneling. C_g stands for the gate capacitance.

FIGURE 12.5 A SET with bias voltage equal to V_{DS} and gate voltage, V_{GS}.

feature known as Coulomb oscillations. This is the characteristic dynamic to be reproduced in SET models, where Figure 12.6 depicts Coulomb oscillations.

There is a theory commonly used to simplify the analysis of SET behavior, the orthodox theory of single-electron tunneling. It was first proposed by Kulik and Shekter (1975) and, later, generalized by Averin and Likharev (1991). Models developed for the transistor are usually based on this theory, since it provides a good accuracy under specific circumstances. The main postulates are as follows:

- The electron energy quantization inside the conductor is ignored, i.e., the electron energy spectrum is considered continuous. This assumption is able to give an adequate description if the charging energy, E_c, is greater than both quantum and thermal fluctuations;
- The time, τ_t, taken by electron tunneling through the barrier is considered negligible when compared

with other time scales. For practical devices, $\tau_t \approx 10^{-15}$ s. Thus, this is a reasonable consideration.

- Rare events, such as cotunneling (when more than one electron tunnels simultaneously), are ignored. The tunneling resistance, R_T, must be considerably larger than the quantum resistance, R_Q, for this assumption to be valid.

The main result of the orthodox theory of single-electron tunneling is to describe tunneling as a random event. Thus, tunneling has a rate of occurrence, i.e., a ratio between probability and time. The tunneling rate, Γ, depends only on the reduction in the system-free electrostatic energy, ΔF, caused by the tunneling event (Likharev 1999). Equation (12.3) describes the tunneling rate for single-electron systems.

$$\Gamma = \frac{-\Delta F}{e^2 R_T [1 - e^{\frac{\Delta F}{k_B T}}]}, \qquad (12.3)$$

in which ΔF is the variation of the system-free electrostatic energy, R_T is the tunneling resistance, k_B stands for the Boltzmann constant, and T is the temperature in Kelvin.

The orthodox theory postulates and Eq. (12.3) are widely applied in SE devices simulation and modeling. Keeping these concepts in mind, the main methods of simulation are discussed in the next section.

12.3 Simulation Methods

Due to low power consumption, high switching capacity, and low occupied area, SE devices are ideal for implementing circuits in Very Large Scale Integration (VLSI) (Radwan, Marzouk, Rehan & Abdel-Fattah 2015). For the physical implementation of the circuit to be possible, accurate and fast simulation tools are needed (Wasshuber 2001). In this scenario, several simulation softwares for nanoelectronic devices were proposed. Among the most famous, one can mention SIMulation Of Nanostructures (SIMON) (Wasshuber, Ksoina & Selberherr 1997), KOrea Single Electron Circuit simulator for spice (KOSEC) (Yu, Jung, Hwang & Doyeol 2000), Single Electron NanoElectronic Circuit Analyser (SENECA) (Fonseca, Korotkov & Likharev 1995), MOnte carlo Single Electron Simulator (MOSES) (Chen 1996), Single Electron Circuit Simulator (SECS)

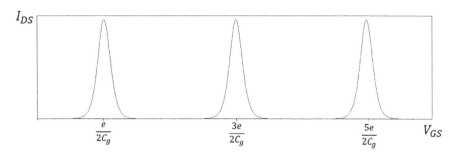

FIGURE 12.6 Electrons are able to tunnel across the island for the following multiples of gate voltage: $\frac{e}{2C_g}$, $\frac{3e}{2C_g}$, $\frac{5e}{2C_g}$, ..., $\frac{(2n+1)e}{2C_g}$ (Hanson 2008).

(Zardalidis & Karafyllidis 2008), and Single Electron Multiscale Simulation (SEMSIM) (Allec, Knobel & Shang 2008).

This section is dedicated to present two of the most used methods for simulating single-electron circuits: the Master Equation of Single-Electron Tunneling and the Monte Carlo method.

12.3.1 Master Equation of Single-Electron Tunneling

The orthodox theory of single-electron tunneling, Section 12.2.3, is able to calculate the tunneling rate for one electron through a potential barrier. In a single-electron circuit, however, it is of interest to establish current through the tunnel junction or through the island in a SET. That is, a succession of tunneling events is needed.

The Master Equation is used to calculate the occupation probabilities for each island state. An island state is defined by the number of electrons in the island, n. The following assumptions are made in the Master Equation method:

- The tunneling rate depends only on the current state of the system;
- The states are discrete.

Consider a Markov chain with three states, as shown in Figure 12.7. The number of possible transitions grows exponentially with the number of states (Papoulis & Pillai 2002). For that reason, a common simplification in the Master Equation algorithm is to treat the single-electron device as a birth–death process. That is, transitions are only allowed between neighboring states.

From this simplification, the Master Equation of single-electron tunneling is written as Eq. (12.4) (Wasshuber 2001). The terms Γ_{ij} and Γ_{ji} express the tunneling rate from state j to i and vice versa. $P_i(t)$ is the time-dependent occupation probability of state i.

$$\frac{\partial P_i(t)}{\partial t} = \sum_{j \neq i}[\Gamma_{ij}P_j(t) - \Gamma_{ji}P_i(t)]. \qquad (12.4)$$

Figure 12.8 illustrates the birth–death process described by Eq. (12.4). It is worth mentioning that considering transitions only between neighboring states is the same as assuming only one electron tunneling at a time. In other words, cotunneling events are inherently ignored (Hasaneen et al. 2011).

Equation (12.4) is the general stochastic description of a single-electron system (Wasshuber 2001). It may be written in matrix form as Eq. (12.5) or Eq. (12.6).

$$\begin{bmatrix} \frac{\partial P_1(t)}{\partial t} \\ \frac{\partial P_2(t)}{\partial t} \\ \vdots \\ \frac{\partial P_n(t)}{\partial t} \end{bmatrix} = \begin{bmatrix} -\sum_{i \neq 1}\Gamma_{i1} & \Gamma_{12} & \cdots & \Gamma_{1n} \\ \Gamma_{21} & -\sum_{i \neq 2}\Gamma_{i2} & \cdots & \Gamma_{2n} \\ \vdots & \vdots & \vdots & \vdots \\ \Gamma_{n1} & \Gamma_{n2} & \cdots & -\sum_{i \neq n}\Gamma_{in} \end{bmatrix} \begin{bmatrix} P_1(t) \\ P_2(t) \\ \vdots \\ P_n(t) \end{bmatrix}, \qquad (12.5)$$

$$\dot{\boldsymbol{p}} = \boldsymbol{\Gamma}\boldsymbol{p}. \qquad (12.6)$$

The first assumption is known as the Markov property or memoryless property (Papoulis & Pillai 2002). For systems with that feature, future states are only affected by the current state of the system (Haverkort 1998).

In Eq. (12.6), the matrix $\boldsymbol{\Gamma}_{n \times n}$ is the transition matrix, whose elements are the tunneling rates. The dimension n describes the number of states considered and $\boldsymbol{p}$ is the vector of state probabilities, given by

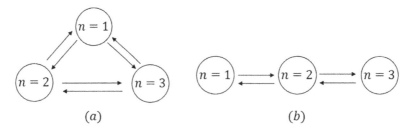

(a)   *(b)*

FIGURE 12.7   (a) A Markov chain with three states and six possible transitions. (b) A birth–death process with three states and four possible transitions.

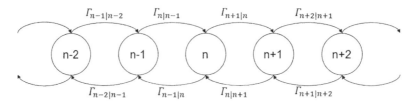

FIGURE 12.8   Possible transitions in the birth–death process are described by Eq. (12.4). $\Gamma_{n-1|n-2}$ is the tunneling rate from state $n-2$ to $n-1$, where n stands for the number of electrons in the island.

$$\boldsymbol{p} = [P_1(t), P_2(t), ..., P_n(t)]^T. \qquad (12.7)$$

For the stationary case, $\dot{\boldsymbol{p}}$ in Eq. (12.6) is a null vector. Thus, $\boldsymbol{0} = \boldsymbol{\Gamma}\boldsymbol{p}$ is a system of linear equations (Wasshuber 2001). The kernel of $\boldsymbol{\Gamma}_{n \times n}$ is the solution for the stationary case. For the transient case, the system of equations may be solved by the formal integration of Eq. (12.6), which will result in

$$\boldsymbol{p}(t) = e^{\boldsymbol{\Gamma}t}\boldsymbol{p}(0), \qquad (12.8)$$

in which the exponential term may be approximated by the Taylor series:

$$e^{\boldsymbol{\Gamma}t} = \sum_{k=0}^{\infty} \frac{(\boldsymbol{\Gamma}t)^k}{k!}. \qquad (12.9)$$

The orthodox theory provides the calculation of the tunneling rates, as expressed in Eq. (12.3). There are three widely used ways for calculation of the energy change, ΔF: Hadley and Lientschnig method (Hadley, Lientschnig & Lai 2002), the critic voltage method (Geerligs, Anderegg, Holweg & Mooiji 1990), and using the Gibbs free energy, also known as thermodynamic approach. The first one is the most used when simulating single-electron devices. A detailed explanation of any of those methods may be found in reference (Wasshuber 2001).

After computing the state probabilities and the tunneling rates, the current can be calculated by Eq. (12.10).

$$I = e \sum_{n=-\infty}^{n=+\infty} P(n)\Gamma_X(n), \qquad (12.10)$$

in which I is the current flowing through the device, e is the elementary charge, $P(n)$ is the occupation probability of state n, and $\Gamma_X(n)$ is the total tunneling rate for the state n in the tunnel junction X, which is given by Eq. (12.11).

$$\Gamma_X(n) = \Gamma_X^+(n) - \Gamma_X^-(n), \qquad (12.11)$$

being $\Gamma_X^+(n)$ the tunneling rate through the junction from left to right and $\Gamma_X^-(n)$, from right to left. Figure 12.9 depicts the junction and the corresponding tunneling rates.

A single-electron circuit has a discrete and infinite number of states, since the number of electrons in excess is unbound (Hasaneen et al. 2011). Thus, the first step in solving Eq. (12.6) is determining the number of states to be considered. There is a trade-off between accuracy and computational cost in this choice. The higher the number of states considered, the more accurate and slower is the simulation.

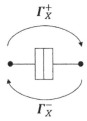

FIGURE 12.9 Tunneling rates in a tunnel junction X.

A common way to start the calculation is to consider an initial state, generally $n = 0$, from which tunneling rates for all possible transitions are calculated. These rates are then used as an initial estimate for the state probabilities. Another common practice is to establish an inferior limit for the rate, $\Gamma_{\min}$. States with respective rates $\Gamma < \Gamma_{\min}$ are ignored, since they have low probability of being achieved. This is an iterative process and, for each iteration, new states may be added (Wasshuber 2001).

Figure 12.10 illustrates a flowchart with the steps to be followed in the Master Equation simulation method.

12.3.2 Monte Carlo Method

The Monte Carlo method was presented to the scientific community by the mathematicians Stanislaw Ulam and Nicholas Metropolis in 1949 (Metropolis & Ulam 1949). Named in allusion to the casino, the Monte Carlo method applies random sampling to obtain numeric results. The basic idea is to use distributions of random numbers to get samples that approximate the system behavior (Anderson 1986).

SIMON, one of the most accurate nanoelectronic simulators applies the Monte Carlo method (Radwan et al.

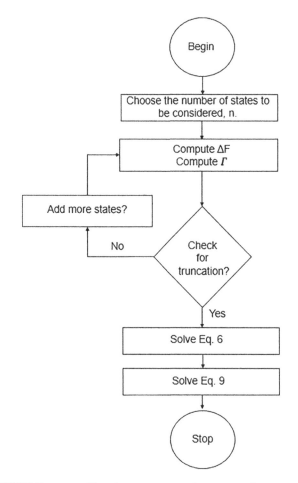

FIGURE 12.10 Flowchart presenting the steps in obtaining the current through a single-electron device using the Master Equation of single-electron tunneling.

2015). SIMON was developed by Cristoph Wasshuber in 2001 (Wasshuber 2001). Results obtained using SIMON are often used as benchmark in the development of new models and simulation techniques for SE devices.

To start the simulation, SIMON treats the tunneling events as Poisson processes, which are characterized by the exponential distribution of time between events, as expressed in Eq. (12.12).

$$P_0(\tau) = e^{-\Gamma \tau}, \qquad (12.12)$$

in which Γ is the tunneling rate, calculated with Eq. (12.3).

Monte Carlo demands a generator of random numbers uniformly distributed in the interval $[0,1]$. In this case, to achieve random numbers with $P_0(\tau)$ distribution, starting from uniformly distributed numbers, one can use the inverse of the distribution $P_0(\tau)$ (Eckhardt 1987):

$$\tau = \frac{-\ln(r)}{\Gamma}, \qquad (12.13)$$

with r being a random number with uniform distribution in the interval $[0,1]$.

The simulation through the Monte Carlo method starts with a list, in which all tunneling events and its rates are tabled. The tunneling times, τ_i, are computed for all events according to Eq. (12.13). The event with smaller τ occurs first and, hence, it is the winner. The current is calculated directly through the ratio between the charge and time interval. This procedure is repeated until the precision limits are achieved. Figure 12.11 shows, in a flowchart way, the steps in the Monte Carlo method, used by SIMON.

Simulations based on Monte Carlo are not the best options to compute rare events (Wasshuber et al. 1997). In nanoelectronics, this fact directly affects cotunneling simulation. Especially, for circuits with a large number of tunnel junctions in which there are parts of the circuit in conduction and parts in blockade, the simulation would be impaired. In this case, cotunneling events would only have a chance to beat normal tunneling for an elevated number of simulated events.

Another drawback of the Monte Carlo method concerns the contribution of each event for the computation of the current. If an event never wins the simulation, it has no contribution in the current. This could generate imprecision in simulation results. A way, described in (Wasshuber 2001), to overcome this drawback is to consider the contribution of all events in the final current computation. The contributions are weighted by the transition probability. The probability of a transition from state k to state j is given by Eq. (12.14).

$$P_{jk} = \frac{\Gamma_{jk}}{\sum_i \Gamma_{ik}}, \qquad (12.14)$$

in which P_{jk} is the transition probability from state k to state j and Γ_{jk} is the tunneling rate from state k to state j.

Regarding the charge transport, this modification in the current computation is equivalent to distribute the charge among all tunneling events. It is worth to emphasize that

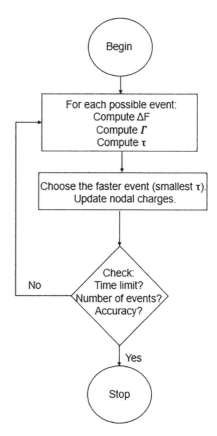

FIGURE 12.11 Flowchart describing the basic steps of the Monte Carlo method.

this is a modification that concerns only the final current calculation. The change in the state space is a discrete process in which electrons hop from one state to the other.

12.4 SET: A Revision of Proposed Models

As already exposed, unique features make SET one of the most studied SE devices. Among the drawbacks that prevent SET application in practical circuits, one might mention the low voltage gain and the high output impedance.

SET voltage gain is defined by the ratio between gate and junction capacitance (Kirihara, Nakazato & Wagner 1999). Since these capacitances tend to have similar values, the voltage gain is close to unit. The high output impedance is due to the tunneling resistance.

A way to overcome these difficulties is to combine SET with MOS transistors. Several proposals of hybrid circuits have already been made (Ohata, Toriumi & Uchida 1997, Mahapatra et al. 2003, Gerousis & Grepiotis 2013, Jana et al. 2013). For those circuits, however, SET bias voltage might assume high values that may result in a suppression of Coulomb oscillations (Radwan et al. 2015).

With the goal of studying SET performance, accurate models are needed. In this section, a comprehensive review concerning previously proposed models is made. The proposals may be divided into two groups: analytical and macromodels. The first ones are based on the Master Equation method, presented in Section 12.3.1. Macromodels use conventional electronic devices and are explained in Section 12.4.2.

12.4.1 Analytical Models for SET

Analytical models for SETs are based on the orthodox theory, Section 12.2.3, and in the solution of Eq. (12.6) for the stationary case. As discussed in Section 12.3.1, the first step in solving the Master Equation is determining the number of states to be considered. This number will affect the model accuracy and the model computational cost. In this way, one might consider that analytical models of SET aim to find an optimal number of states, capable of providing accuracy and low computational cost.

Master Equation on a SET

Figure 12.12 depicts a SET and the tunneling rates of interest. There are four possible events on a SET: a charge, e, can tunnel from source to the island (Γ_S^+), from island to source (Γ_S^-), from island to drain (Γ_D^+), and from drain to island (Γ_D^-).

For the stationary case, Eq. (12.6) can be written, for a SET, as Eq. (12.15) (Hasaneen et al. 2011).

$$
\begin{aligned}
0 = &... + \Gamma_{n-1,n-2}P(n-2) + (\Gamma_{n,n-1} + \Gamma_{n-2,n-1})P(n-1) \\
&+ (\Gamma_{n+1,n} + \Gamma_{n-1,n})P(n) + +(\Gamma_{n+2,n+1} + \Gamma_{n,n+1}) \\
&\times P(n+1) + \Gamma_{n+1,n+2}P(n+2) + \cdots,
\end{aligned}
\tag{12.15}
$$

in which $\Gamma_{n,n+1}$ is the tunneling rate from $n+1$ to n, and $P(n+1)$ is the occupation probability of state $n+1$. n is the number of electrons in the island, which characterizes each state.

The tunneling rates are computed according to the orthodox theory, using Eq. (12.3). The current in the SET may be calculated, considering either the source or the drain junction. Equation (12.16) expresses the current, considering the drain junction.

$$
I_{\text{SET}} = e \sum_n P(n)(\Gamma_D^+ - \Gamma_D^-).
\tag{12.16}
$$

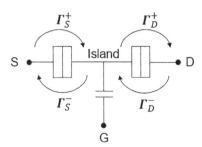

FIGURE 12.12 A diagram showing the tunneling rates on a SET. The letters S, G, and D stand for the terminals source, gate, and drain, respectively.

Previously Proposed Analytical Models

The development of analytical models for SETs is due to the need for studying the device features and limitations. Numerical simulations, based on the Monte Carlo method, are not efficient for large circuits (Uchida, Matsuzawa, Koga, Ohba, Takagi & Toriumi 2000).

The first analytical models were proposed in the beginning of the 1990s (Glazman & Shekhter 1989, Beenakker 1991). These models aimed to reproduce intrinsic SET features, such as the Coulomb blockade. The models of Glazman and Beenakker considered $V_{DS} = 0$, which condition simplifies the calculation of state probabilities. However, $V_{DS} \neq 0$ in most practical applications (Uchida et al. 2000). The higher the bias, i.e. V_{DS} voltage, the more states must be considered in the Master Equation for the model to present an acceptable accuracy.

Uchida et al. proposed the first analytical SET model in which $V_{DS} \neq 0$. This model considered symmetric devices. That is, the tunnel junctions of source and drain have the same parameters. The authors considered only two charge states: $n = 0$ and $n = 1$. This model was validated for bias voltages respecting Eq. (12.17).

$$
|V_{DS}| \leq \frac{e}{C_{\text{island}}},
\tag{12.17}
$$

$$
C_{\text{island}} = C_D + C_S + C_{g1} + C_{g2}
\tag{12.18}
$$

in which e is the elementary charge, C_D, C_S, C_{g1}, and C_{g2} are the capacitances of drain and source tunnel junctions and of the gate terminals, respectively. For Uchida model, $C_{g2} = 0$ F, since the authors considered SET with only one gate terminal.

Uchida validated the model functionality using I_{DS}–V_{GS} curves obtained with CAMSET, Circuit Analysis program including a Model of Single Electron Tunneling, as benchmark. CAMSET (Kirihara et al. 1999) is a single-electron simulator that applies the Master Equation method, varying the number of states in accordance with the temperature and V_{DS}. The results presented good agreement with CAMSET curves for temperatures up to 20 K and a maximum bias voltage of 20 mV.

Wang and Porod (2001) proposed an extension of Uchida model. This new model could be applied to describe asymmetric devices and took background charge effect in consideration. Background charges can be induced by impurities resulting from manufacturing processes. Since SET is very sensitive, these charges may cause errors in the transistor operation, leading it to conduct in a blockade region. Background charges represent one major challenge in the practical application of SET in integrated circuits (Lientschnig et al. 2003, Abutaleb 2015).

In Wang and Porod model, the current is treated as a combined process, in which the electron first tunnels through source junction and, then, through drain junction. Thus, considering Figure 12.12, the current is given by

$$
I_{DS} = e \frac{\Gamma_S^+ \Gamma_D^+}{\Gamma_S^+ + \Gamma_D^+}.
\tag{12.19}
$$

The authors used SIMON $I_{DS}-V_{GS}$ curves as benchmark and concluded that their model provided good accuracy only when the island had one electron in excess. For other conditions of charge, Wang and Porod model underestimates the current. When more than one electron is involved in conduction, the authors suggested that an interpolation process may be used to estimate current.

One of the most famous analytical models for SET is known as the MIB model, named after the authors (Mahapatra, Ionescu & Banerjee 2002). MIB model considers unidirectional current flow. That is, in Figure 12.12, $\Gamma_S^- = \Gamma_D^- = 0$. The total current is computed in two separate amounts: I_{dsu} and I_{dsth}. I_{dsu} is the portion of the total current due to tunneling events and I_{dsth} is due to the thermal energy. This division of total current characterizes MIB as a quasi-analytical model. Also, one might say that this model is not based on the orthodox theory of single-electron tunneling, since tunnel events are considered temperature independent.

The first version of MIB was able to provide a good agreement with SIMON results for temperatures up to 5 K. The authors developed an improved version of MIB (Mahapatra et al. 2002), which computed the current in accordance with the orthodox theory. This second version was used for analyzing power dissipation in logic circuits.

Lientschnig et al. (2003) presented a widely used SET model, which applies a recursion relation to compute the state probabilities. The relation is expressed in Eq. (12.20). The use of this recursion relation makes simulation more efficient, since the computational effort is reduced.

$$P(n) = P(n-1)\left(\frac{\Gamma_S^+(n-1) + \Gamma_D^-(n-1)}{\Gamma_S^-(n) + \Gamma_D^+(n)}\right). \quad (12.20)$$

$$\sum_n P(n) = 1. \quad (12.21)$$

Lientschnig et al. also applied a calculation of the most probable state of an island:

$$n_{\text{opt}} = \frac{-(Q_0 + C_D V_D + C_S V_S + C_{g1} V_{g1} + C_{g2} V_{g2})}{e}$$
$$+ \frac{C_{\text{island}}}{e} \frac{V_D R_S + V_S R_D}{R_S + R_D}, \quad (12.22)$$

in which n_{opt} stands for the optimal number of electrons in the island, Q_0 is the background charge, C_D, C_S, C_{g1}, and C_{g2} are drain, source, first and second gate capacitances, respectively. V_{g1}, V_{g2}, V_D, and V_S are the voltages applied to gate, drain, and source terminals. C_{island} is given by Eq. (12.18), R_S and R_D are the tunneling resistances of source and drain junctions.

In order to achieve room temperature operation, the authors considered 11 charge states around n_{opt}.

Subsequent models applied Lientschnig recursion relation and sought to extend the interval of V_{DS} in which the models could be applied (Inokawa & Takahashi 2003, Mahapatra, Vaish, Wasshuber, Banerjee & Ionescu 2004).

Pruvost et al. developed a model that is able to adapt the number of states needed in accordance with the simulation temperature and the V_{DS} value (Pruvost, Mizuta & Oda 2008). To do that, the authors used the stability plot of SET, i.e., the $V_{DS} \times V_{GS}$ characteristic. The interval of states to be considered is $[-N, N]$, where N is given by (12.23).

$$N = \left\lfloor \frac{|V_{DS}|}{\frac{2e}{C_{island}}} \right\rfloor + 1, \quad (12.23)$$

in which $\lfloor x \rfloor$ is the floor function, which gives the next integer less or equal to x.

The authors compared their model to the models of Uchida (Uchida et al. 2000) and the third version of MIB (Mahapatra et al. 2004). Pruvost model was the only one to present Coulomb blockade for $T = 300$ K.

Hasaneen et al. (2011) proposed a model that used ten states to present accurate results up to 77 K. Radwan et al. (2015) developed an adaptive model, which changes the number of states with the temperature and V_{DS}. Radwan et al. used the Mean Squared Error (MSE) as a metric to show that their model provided the best fit to SIMON results.

12.4.2 Macromodels for SET

Aiming a more efficient simulation, macromodels were developed to reduce the computational cost of techniques like Master Equation and Monte Carlo. Generally, macromodels are based on conventional electronic components, such as resistors, capacitors, and diodes. Thus, the first step in studying macromodeling is to explore the dynamics of these components. The devices used in SET macromodeling are briefly described later.

Conventional Electronic Components

The simple and most passive component, in the set of macromodels, is the resistor. It can have its voltage, V_r, and its current, I_r, related by its resistance, R, as expressed in Eq. (12.24).

$$V_r = RI_r. \quad (12.24)$$

Other important passive component is the capacitor, and its dynamic can be described by the following equation:

$$I_c = C\frac{dVc}{dt}, \quad (12.25)$$

in which I_c, V_c, and C are the current, voltage, and capacitance of the capacitor, respectively.

Regarding active components, i.e., devices capable of providing energy to the system, voltage sources can be described by the following equation:

$$V_s = V - r_i I_s, \quad (12.26)$$

in which V_s and I_s are the voltage and current in the voltage source, respectively, and r_i stands for internal losses.

Diodes are also used in SET macromodels. The main behavior of a diode, the Volt–Ampere characteristic, is described by the Shockley equation, expressed as Eq. (12.27).

$$I = I_0 \left(e^{V/\eta V_T} - 1\right), \qquad (12.27)$$

in which V is the voltage between the diode terminals, I_0 is the reverse saturation current, η depends on the material, and V_T is a temperature-dependent constant.

Previously Proposed Macromodels

In 1999, Yu, Hwang & Ahn (1999) proposed a macromodel for a SET, which would later be an inspiration for several authors. This model built the theoretical basis for macro-modeling of a SET. The first step was to determine under what conditions SE devices could be simulated in SPICE. Circuit simulators such as SPICE describe the behavior of a circuit from its topology. Hence, regarding transistors, there are two basic assumptions:

1. Once the transistor parameters are established, they are valid for all circuits;

2. The transistor $I \times V$ curve is affected only by voltage variation in neighboring devices. The inter-action between nonadjacent devices is negligible.

For SETs, the second assumption is generally not valid. The terminal current of a SET is a function of the average states of the island. Thus, when two or more SETs are connected, the charge state of the island in one SET is affected by the other(s). In this context, an isolated SET and a SET placed in a circuit (with two or more SETs) can present different currents (Yu et al. 1999).

The solution proposed by Yu consists in assuming that the interconnection capacitance is large enough, so the transistors may be treated singly. To find the lower value from which the second assumption holds, the authors made

simulations with a single-electron inverter in SPICE and KOSEC. Through the comparison of the results, the authors were able to conclude that if the condition expressed in Eq. (12.28) was satisfied, the SET could be simulated in SPICE with acceptable accuracy.

$$C_i > 6,25C_j, \qquad (12.28)$$

in which C_i is the interconnection capacitance and C_j is the capacitance of the tunnel junction.

Yu et al. used diodes, resistors, and voltage sources to build the first macromodel for a SET. The idea behind this model is to mimic Coulomb oscillations by forcing a sinusoidal variation in its parameters, see the $\cos(.)$ factor in Eq. (12.29). On the other hand, Coulomb blockade is achieved with the diodes, as V_{GS} varies monotonically. Figure 12.13 shows four of the main macromodels proposed for SETs.

In Figure 12.13a, the resistance R_G must be large enough to inhibit current flow between Gate, G, and Source, S, terminals. The resistances R_1 and R_2 must follow the dynamics expressed in Eqs. (12.29) and (12.30).

$$R_1(V_{GS}) = P_{r1} + P_{r2}\,\cos(P_f V_{GS}), \qquad (12.29)$$

$$R_2(V_{GS}) = R_3(V_{GS}) = \frac{V_p}{P_{i2} - \frac{2\,V_p}{R_1(V_{GS})}}, \qquad (12.30)$$

in which the values of the parameters P_{r1}, P_{r2}, P_f, V_p, and P_i can be found in Table 12.1.

TABLE 12.1 Values for Curve Fitting Parameters Used by Yu et al.

Parameter	P_i	P_{r1}	P_{r2}	P_f	V_p
Value	0.2×10^{-9}	300×10^6	100×10^6	40	0.02

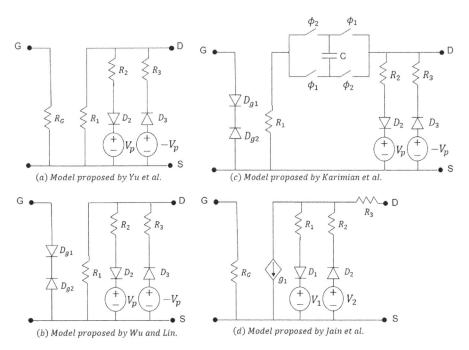

FIGURE 12.13 SET macromodels. (a) The first macromodel of a SET was proposed by Yu et al. (1999). (b) Wu and Lin contribution aimed to improve the model behavior in the blockade region. (Wu & Lin 2003). (c) Karimian et al. model is able to estimate output delay (Karimian, Dousti, Pouyan, & Faez 2009). (d) Jain et al. macromodel. (Jain, Ghosh, Singh, & Sarkar 2015)

It is important to mention that parameter V_p corresponds to the absolute value of voltage sources V_p in Figure 12.13. In other words, it represents the values from which there will be conduction between source, S, and drain, D, terminals. P_f is given by Eq. (12.31).

$$P_f = \frac{2C_G}{e}, \qquad (12.31)$$

with C_G being the gate capacitance and e the elementary charge.

To validate the functionality of their model, Yu et al. made simulations of a single-electron inverter and a NOR logic gate. The simulations were made in SPICE, and KOSEC results were used as benchmark. Yu model presented good agreement with the benchmark for temperatures up to 30 K, and SPICE simulation was 10^4 times faster than KOSEC.

In 2003, Wu and Lin presented their contribution to SET macromodeling. Their model is depicted in Figure 12.13b. The idea of replacing R_G with two diodes, D_g1 and D_g2, is justified by a performance issue presented by Yu model and observed by Wu and Lin: the relation between the current I_{DS} and the bias voltage V_{DS}. Experimental results, as well as SIMON results, show an exponential dependence of I_{DS} on V_{DS} in the blockade region (Likharev 1999). However, in Yu et al. model, this dependence was linear. As expressed in Eq. (12.27), the insertion of the diodes provides the required exponential relation.

R_1 and R_2 were also altered, as shown in Eqs. (12.32) and (12.33).

$$R1(V_{GS}, V_{DS}) = P_{r1} + P_{r2}[\cos(P_f\pi V_{GS}) + 1]2^{V_p - V_{DS}x}, \qquad (12.32)$$

$$R2(V_{GS}, V_{DS}) = R3(V_{GS}, V_{DS}) = \frac{V_p}{P_i - \frac{2V_p}{R1(V_{GS},V_{DS})}}, \qquad (12.33)$$

in which x is a function of temperature, P_f is given by (12.31), and P_{r1} and P_{r2} are expressed by Eqs. (12.34) and (12.35).

$$P_{r1} = 4R_T, \qquad (12.34)$$

$$P_{r2} = 1.33R_T, \qquad (12.35)$$

in which R_T is the tunneling resistance.

The reverse diode, D_{g2}, reduces the leakage current between Gate and Source to the value expressed by the reverse saturation current, I_0, in the diode D_{g2}. The second diode, D_{g1}, turns the voltage relation (V_{GS}, I_{DS}) exponential, as a diode is governed by an exponential relation, as expressed by Eq. (12.27).

Wu and Lin model presented $I_{DS} \times V_{GS}$ and $I_{DS} \times V_{DS}$ curves in better agreement with the benchmark, SIMON results, than Yu model (Wu & Lin 2003).

An important drawback of the first macromodel, proposed by Yu et al. in 1999, was the inability to treat the nonstationary analysis. That is, condition (12.28) expresses the minimum value of the interconnect capacitance that enables SPICE to perform steady case simulations (Yu, Hwang & Ahn 2005).

Yu himself proposed a new model, in 2005, which was able to treat the transient case. This model combined macromodeling and Master Equation. Also, the interconnection capacitance should satisfy Eq. (12.36), for the transient case. The structure of this model can be seen in Figure 12.14.

$$C_L > 64C_j. \qquad (12.36)$$

The parameters of the model depicted in Figure 12.14 are described by Eqs. (12.37) and (12.38).

$$\begin{bmatrix} J_{DS} \\ J_{GS} \\ J_{BS} \end{bmatrix} = \begin{bmatrix} g_{ds} & g_{dmg} & g_{dmb} \\ g_{gmd} & g_{gs} & g_{gmb} \\ g_{bmd} & g_{bmg} & g_{bs} \end{bmatrix} \begin{bmatrix} V_{DS} \\ V_{GS} \\ V_{BS} \end{bmatrix} - \begin{bmatrix} I_D \\ I_G \\ I_B \end{bmatrix}, \qquad (12.37)$$

in which J_{DS}, J_{GS}, and J_{BS} are equivalent current sources, connected between drain and source, gate and source, and backgate and source, respectively. I_D, I_G, and I_B stand for drain, gate, and backgate currents, respectively.

$$\begin{bmatrix} g_{ds} & g_{dmg} & g_{dmb} \\ g_{gmd} & g_{gs} & g_{gmb} \\ g_{bmd} & g_{bmg} & g_{bs} \end{bmatrix} = \begin{bmatrix} \frac{\partial I_D}{\partial V_{DS}} & \frac{\partial I_D}{\partial V_{GS}} & \frac{\partial I_D}{\partial V_{BS}} \\ \frac{\partial I_G}{\partial V_{DS}} & \frac{\partial I_G}{\partial V_{GS}} & \frac{\partial I_G}{\partial V_{BS}} \\ \frac{\partial I_B}{\partial V_{DS}} & \frac{\partial I_B}{\partial V_{GS}} & \frac{\partial I_B}{\partial V_{BS}} \end{bmatrix}, \qquad (12.38)$$

in which, V_{DS} stands for the drain–source voltage, V_{GS} for the gate–source voltage, and V_{BS} for the backgate–source voltage.

The improved model by Yu is based on a linearized equivalent circuit, its first model depicted in Figure 12.13a, and the solution of the Master Equation is attained by the programming capabilities of the simulator, SmartSpice (Yu et al. 2005). The results presented by the improved model show a $V_{GS} \times I_{DS}$ curve more accurate than the one provided by Yu's first model, than by Wu model as well. However, the results were by far less accurate than applications of Master Equation and Monte Carlo Method. Moreover, SET transient analysis is only possible when condition (12.36) is

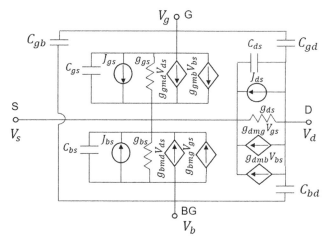

FIGURE 12.14 SET macromodel proposed by Yu, combining Master Equation and macromodeling techniques, in which G stands for Gate, BG for BackGate, S for the Source, and D for the Drain.(Yu et al. 2005).

satisfied, which is not realistic. Another relevant drawback in the last model was the lack of accuracy in the time delay between input, V_{GS}, and output signals, I_{DS}. In order to overcome some of these issues, Karimian presented a new macromodel, depicted in Figure 12.13c.

The SET model proposed by Karimian et al. (2009) follows the line given by Yu and Wu. However, this model had an additional capacitive set, aiming to match the model time delay and the time delay found in real SETs.

The idea behind Karimian model was to use a switched capacitor in order to compute the time delay between input and output signals. The structure formed by the capacitor C and the switches ϕ_1 and ϕ_2, in Figure 12.13c, compose a quantizer block, as named by the authors. The charge in the capacitor executes the quantization. The switches are controlled by two clock signals, depicted in Figure 12.15.

If the input, V_{GS}, and output, I_{DS}, signals do not present variation during the period T, the quantizer block can be approximated by an equivalent resistance $\frac{T}{4C}$ (Karimian et al. 2009).

In the same way as for Yu model, in the Karimian model, the resistances R_1, R_2, and R_3 are functions of V_{GS} and I_{DS}. They are given by Eqs. (12.39) and (12.40).

$$R_1(V_{GS}, V_{DS}) = P_{r2}\left[\cos\left(P_f \pi V_{GS}\right) + Y\right] 2^{V_p - V_{DS}} \; x, \tag{12.39}$$

$$R_2(V_{GS}, V_{DS}) = R_3(V_{GS}, V_{DS}) = \frac{V_p}{P_{i2} - \frac{2V_p}{R_1(V_{GS}, V_{DS}) + P_{r1}}}, \tag{12.40}$$

in which Y is defined as a positive parameter and x is a temperature function. The parameters P_{r2}, P_f, V_p, and P_i are same as the ones in Wu and Lin model. However, the parameter P_{r1} is a function of the period T (commutation of the switches ϕ_1 and ϕ_2), and it is given by Eq. (12.41).

$$P_{R1} = \frac{T}{4C}, \tag{12.41}$$

with C being the capacitance inserted in Karimian model.

The results provided by Karimian et al. presented more accuracy, when compared with the model of Wu and Lin.

The benchmark was the $I_{DS} \times V_{GS}$ curve obtained with SIMON (Karimian et al. 2009).

In a more recent scenario, Jain et al. (2015), created a model based on the original structure proposed by Yu et al. The main idea was improving the $I \times V$ characteristic of the SET, in the Coulomb blockade region. As mentioned before, the original Yu model failed to describe SET behavior in this region. Furthermore, Wu model presents the correct characteristic only for a fixed value of V_{GS}. In this scenario, Jain et al. proposed changes in the equivalent circuit as well as in equations that define its parameters. The model obtained can be seen in Figure 12.13d.

The values of the equivalent circuit components are given by Eqs. (12.42), (12.43), and (12.44).

$$R_1(V_{GS}) = \frac{\frac{V_p}{P_{I2} - \frac{2V_p}{P_{r1} + P_{r2} \; cos(P_f \pi V_{GS})}}}{K_1}, \tag{12.42}$$

$$R_2(V_{GS}) = \frac{\frac{V_p}{P_{I2} - \frac{2V_p}{P_{r1} + P_{r2} \; cos(P_f \pi V_{GS})}}}{K_2}, \tag{12.43}$$

$$g_1(V_{GS}) = \frac{P_{r3} \; sen(\pi V_{GS})}{K_3}, \tag{12.44}$$

in which K_1, K_2, and K_3 are scale factors, $P_{r2} = 220 \times 10^6$, $P_{r3} = 0.5 \times 10^{-9}$, and $R_G = 100 \; G\Omega$. The parameters P_f, P_i, and P_{r1} e V_p have the same values used by Yu, detailed in Table 12.1.

The factors K_1 and K_2 allow variation of the calculation interval for the drain current I_D, without changing the characteristic of the curve. The factor K_3 controls the vertical displacement of the curve $I_{DS} \times V_{DS}$ for different values of V_{GS}.

Jain et al. model presented more accurate results than Yu et al. and Wu and Lin reaching a $I_{DS} \times V_{GS}$ curve closer from the benchmark, which was also SIMON results (Jain et al. 2015).

In general, macromodels are less accurate than probabilistic models, as Monte Carlo and Master Equation. However, their computational efficiency motivated many researchers to find more accurate and efficient macromodels.

12.5 Conclusion

This chapter presented a detailed explanation of the two methods applied in SE device simulations: Master Equation and Monte Carlo. Also, a comprehensive review of the most recent proposed models for the SET were carried out.

It is expected that the reader is capable to apply the probabilistic methods in simulation platforms and to reproduce the results obtained by the authors of the macromodels.

It is worth to recapitulate the main reasons why modeling SE devices is necessary. Traditional electronic devices may, in a near future, stop providing advances and performance improvements due to the miniaturization tendency. Thus, new paradigms are needed to overcome the challenges posed

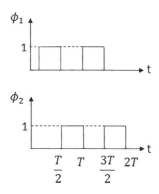

FIGURE 12.15 Commutation sequence for the switches ϕ_1 and ϕ_2 for the SET macromodel proposed by Karimian et al.(2009). T is the period of the signal.

by scale issues. To perform studies that are able to provide insights concerning promising technologies, accurate models are essential.

Regarding previously proposed models, it can be said that these are either analytical or macromodels. The works published so far point out the need to improve accuracy and reduce the simulation time. Therefore, studying modeling techniques and understanding its pros and cons is fundamental. The lack of consolidated manufacturing techniques for SE devices is also challenging. This fact is especially difficult to compare model results with experimental results. Hence, the prevalent use of SIMON results was used as a benchmark.

References

Abutaleb, M. M. (2015), A new static differential design style for hybrid SET-CMOS logic circuits, *Journal of Computational Electronics* **14**, 329–340.

Allec, N., Knobel, R. G. & Shang, L. (2008), SEMSIM: Adaptive multiscale simulation for single-electron devices, *IEEE Transaction Nanotechnology* **7(3)**, 351–354.

Anderson, H. L. (1986), Metropolis, Monte Carlo and the MANIAC, *Los Alamos Science* **14**, 96–108.

Averin, D. V. & Likharev, K. K. (1991), Single-electronics: Correlated transfer of single electrons and cooper pairs in small tunnel junctions. In: B. L. Altshuler, P. A. Lee & R. A. Webb, Eds, *Mesoscopic Phenomena in Solids*, Elsevier, Amsterdam, pp. 173–271.

Beenakker, C. W. J. (1991), Theory of Coulomb-Blockade oscillations in the conductance of a quantum dot, *Physical Review B* **44**, 1646–1656.

Billings, S. A. (2013), *Nonlinear System Identification: NARMAX Methods in the Time, Frequency, and Spatio—Temporal Domains*, John Wiley & Sons, London.

Chen, R. H. (1996), MOSES: A general Monte Carlo simulator for single electron circuits, *The Eletrochemical Society* **96(2)**, 576.

Eckhardt, R. (1987), Stan Ulam, John Von Neumann and the Monte Carlo Method, *Los Alamos Science* **Special Issue 1987**, 131–143.

Fahmy, H. & Ismail, K. (1997), Analysis of a single-electron decimal adder, *Applied Physics Letters* **70**, 2613–2615.

Ferry, D., Goodnick, S. & Bird, J. (2009), *Transport in Nanostructures*, Cambridge University Press, Cambridge.

Fonseca, L. R. C., Korotkov, A. N. & Likharev, K. (1995), A numerical study of the dynamics and statistics of single electron systems, *Journal of Applied Physics* **78(5)**, 3238–3251.

Garner, C. M., Kloster, G., Astwood, G., Mosley, L. & Palanduz, A. V. (2007), Challenges for dielectric materials in future integrated circuit technology, *Microelectronics Reliability* **47**, 937–943.

Geerligs, L. J., Anderegg, V. F., Holweg, P. A. M. & Mooiji, J. E. (1990), Frequency-locked turnstile for single electrons, *Physical Review Letters* **64(22)**, 2691–2694.

Gerousis, C. & Grepiotis, A. (2013) , Reconfigurable gate array architecture for logic functions in tunneling transistors technology, *Microelectronics Journal* **44(8)**, 706–711.

Glazman, L. I. & Shekhter, R. I. (1989), Coulomb oscillations of the conductance in a laterally confined heterostructure, *Journal of Physics: Condensed Matter* **1(33)**, 5811–5815.

Goser, K. F., Pacha, C., Kanstein, A. & Rossman, M. (1997), Aspects of systems and circuits for nanoelectronics, *Proceedings of the IEEE* **85(4)**, 558–573.

Grabert, H. & Devoret, M. H. (1992) , *Single Charge Tunneling: Coulomb Blockade Phenomena in Nanostructures*, Plenum Press, New york.

Hadley, P., Lientschnig, G. & Lai, M. (2002), Single-electron transistors, *Proceedings of the 29th International Symposium on Compound Semiconductors*, Lausanne, Switzerland, pp. 125–132.

Hain, L. (2015), More from Moore, *Nature* **520**, 408.

Hanson, G. W. (2008), *Fundamentals in Nanoelectronics*, Prentice Hall, Upper Saddle River, NJ.

Hasaneen, E. A. M., Wahab, M. A. A. & Ahmed, M. G. (2011), Exact analytical model of single electron transistor for pratical IC design, *Microelectronics Reliability* **51**, 733–745.

Haverkort, B. R. (1998), *Performance of Computer Communication Systems: A Model-Based Approach*, 1st edn, Wiley, New York.

Heinzel, T. (2007), *Mesoscopic Electronics in Solid State Nanostructures*, 3rd edn, Wiley-VCH, New York.

Inokawa, H. & Takahashi, Y. (2003), A compact analytical model for asymmetric single-electron tunneling transistors, *IEEE Transactions on Electronic Devices* **50**, 455–461.

Jain, A., Ghosh, A., Singh, N. B. & Sarkar, S. K. (2015), A new SPICE macro model of single electron transistor for efficient simulation of single-electronics circuits, *Analog Integrated Circuits and Signal Processing* **82**, 653–662.

Jain, A., Singh, N. B., Ghosh, A. & Sarkar, S. K. (2014), Stability and reliability analysis of hybrid CMOS-SET circuits: A new approach, *Journal of Computational and Theoretical Nanoscience* **11(12)**, 2519–2524.

Jana, A., Singh, N. B., Sing, J. K. & Sarkar, S. K. (2013), Design and simulation of hybrid CMOS-SET circuits, *Microelectronics Reliability* **53(4)**, 592–599.

Karimian, M., Dousti, M., Pouyan, M. & Faez, R. (2009), A new SPICE macro-model for simulation of single electron circuits, *International Conference on Microelectronics*, Marrakech, Morocco, pp. 228–231.

Kirihara, M., Nakazato, K. & Wagner, M. (1999), Hybrid circuit simulator including a model for single electron tunneling devices, *Japanese Journal of Applied Physics* **38**, 2028–2032.

Kulik, I. O. & Shekhter, R. I. (1975), Kinetic phenomena and charge discreteness effects in granulated media, *Soviet Physics JETP* **41**, 308–316.

Lavieville, R., Barraud, S., Corna, A., Jehl, X., Sanquer, M. & Vinet, M. (2015), A 350 K operating silicon nanowire single electron/hole transistors scaled down to 3.4 nm diameter and 10 nm gate length, *EUROSOI-ULIS 2015*. Bologna, Italy.

Lientschnig, G., Weymann, I. & Hadley, P. (2003), Simulating hybrid circuits of single-electron transistors and field-effect transistors, *Japanese Journal of Applied Physics* **42**, 6467–6472.

Likharev, K. K. (1999), Single-electron devices and their applications, *Proceedings of the IEEE* **87**, 606–632.

Ljung, L. (1999), *System Identification: Theory for the User*, Prentice Hall PTR, Upper Saddle River, NJ.

Mahapatra, S., Ionescu, A. M. & Banerjee, K. (2002), A quasi-analytical SET model for few electron circuit simulation, *IEEE Electronic Device Letters* **23**, 366–368.

Mahapatra, S., Ionescu, A. M., Banerjee, K. & Declercq, M. J. (2002), Modelling and analysis of power dissipation in single electron logic, *IEDM*, San Francisco, CA pp. 323–326.

Mahapatra, S., Pott, V., Ecoffey, S., Schmid, A., Wasshuber, C., Tringe, J. W., Leblebici, Y., Declercq, M., Banerjee, K. & Ionescu, A. M. (2003), SETMOS: A novel true hybrid SET-MOS high current Coulomb Blockade oscillation cell for future nanoscale analog ICs, *IEDM Technical Digest*, Washington, DC, pp. 703–706.

Mahapatra, S., Vaish, V., Wasshuber, C., Banerjee, K. & Ionescu, A. M. (2004), Analytical modeling of single-electron transistor for hybrid CMOS-SET analog IC design, *IEEE Transactions on Electronic Devices* **51(11)**, 1772–1782.

Metropolis, N. & Ulam, S. (1949), The Monte Carlo method, *Journal of the American Statistical Association* **44**, 335–341.

Ohata, A., Toriumi, A. & Uchida, K. (1997), Coulomb Blockade effects in edge quantum wire SOI MOSFETs, *Japanese Journal of Applied Physics* **36**, 1686–1689.

Papoulis, A. & Pillai, S. U. (2002), *Probability, Random Variables and Stochastic Processes*, 4th edn, McGraw Hill, New York.

Paul, D. J. (2002), Nanoelectronics, *Encyclopedia of Physical Science and Technology* **10**, 285–301.

Pruvost, B., Mizuta, H. & Oda, S. (2008), Voltage-limitation free analytical single-electron transistor model incorporating the effects of spin-degenerate discrete energy states, *Journal of Applied Physics* **103**, 1–10.

Radwan, M. S., Marzouk, E. A., Rehan, S. E. & Abdel-Fattah, A. I. (2015), A computationally efficient model of single electron transistor, *Microelectronics Journal* **46**, 301–309.

Shin, S. J., Jung, C. S., Park, B. J., Yoon, T. K., Lee, J., Kim, S. J., Choi, J. B., Takahashiand, Y. & Hasko, D. G. (2010), Si-based ultrasmall multiswitching single-electron transistor operating at room-temperature, *Applied Physics Letters* **97(10)**, 103101–103103.

Shin, S. J., Lee, J., Kang, H. J., Choi, J. B., Yang, S. R. E., Takahashiand, Y. & Hasko, D. G. (2011), Room-temperature charge stability modulated by quantum effects in a nanoscale silicon island, *Nano Letters* **11(4)**, 1591–1597.

Toumey, C. (2016), Less is Moore, *Nature Nanotechnology* **11**, 2–3.

Uchida, K., Matsuzawa, K., Koga, J., Ohba, R., Takagi, S. & Toriumi, A. (2000), Analytical single-electron transistor (SET) model for design and analysis of realistic (SET) circuits, *Japanes Journal of Applied Physics* **39**, 2321–2324.

Wang, X. & Porod, W. (2001), Analytical I-V model for single-electron transistors, *VLSI Design* **13**, 189–192.

Wasshuber, C. (2001), *Computational Single-Electronics*, 1st edn, Springer-Verlag Wien, New York.

Wasshuber, C., Ksoina, H. & Selberherr, S. (1997), SIMON: A simulator for single-electron tunnel devices and circuits, *IEEE Transactions on Computer-Aided Design* **16**, 937–944.

Wu, Y. & Lin, S. (2003), An improved single-electron transistor model for SPICE application, *Nanotechnology* **3**, 321–324.

Yu, Y., Jung, H. O., Hwang, S. & Doyeol, A. (2000), Implementation of single electron circuit simulation by SPICE: KOSECSPICE, *Proceedings of Asia Pacific Workshop on Fundamental and Application of Advanced Semiconductor Device*, Nagasaki, Japan, pp. 85–90.

Yu, Y. S., Hwang, S. W. & Ahn, D. (1999), Macromodeling of single-electron transistors for efficient circuit simulation, *IEEE Transactions on Electronic Devices* **46**, 1667–1671.

Yu, Y. S., Hwang, S. W. & Ahn, D. (2005), Transient modelling of single-electron transistors for efficient circuit simulation by SPICE, *IEE Proceedings of Circuits, Devices and Systems* **152**, 691–696.

Zardalidis, G. & Karafyllidis, I. G. (2008), SECS: A new single-electron-circuit simuator, *IEEE Transactions on Circuits and Systems* **55-9**, 2774–2784.

Single Electron Transport and Possible Quantum Computing in 2D Materials

13.1 Introduction of the Family of 2D Materials 13-1
    Graphene and hBN • 2H-Transition Metal Dichalcogenides • 1T′-Transition
    Metal Dichalcogenides

13.2 Theoretical Background on Quantum Transport 13-10
    Single QD • Double Quantum Dot • Andreev Reflections in Ballistic S-N-S
    JJs

13.3 Single-Electron Transport in Graphene 13-18
    Graphene SQDs on SiO_2/Si Substrates • Coulomb Blockade at Zero Field •
    Electron–Hole Crossover in a Perpendicular Magnetic Field • Spin States in
    In-Plane Magnetic Field • Charge Relaxation Time • Graphene Double
    Quantum Dots on SiO_2/Si Substrates • Coulomb Blockade and
    Magnetotransport • Charge Pumping • Graphene Quantum Dots on hBN

13.4 Single-Electron Transport
    in 2H-TMDs 13-29

13.5 2D Material-Based JJ 13-32
    Graphene-Based JJ • TMD-Based JJ

13.6 Quantum Spin Hall Edge States in 1T′-TMDs.......................... 13-38

13.7 Summary 13-41

References 13-41

K. L. Chiu

National Sun Yat-Sen University

13.1 Introduction of the Family of 2D Materials

Since the 1960s, the density of components on silicon chips has doubled approximately every 18 months, following a trend known as Moore's law after Intel's cofounder Gordon Moore, who predicted the phenomenon. Silicon-based transistor manufacturing has now reached the sub-10nm scale, heralding the limit of Moore's law and stimulating the development of alternative switching technologies and host materials for processing and storing bits of information. Quantum bits, or "qubits," are at the heart of quantum computing, an entirely different paradigm in which information is encoded using the superposition states of individual quanta. Ideally, the charge and spin degrees of freedom of a single electron trapped in quantum dots (QDs) are nature's candidates of qubits for use in quantum computing operations. To reach this goal, tremendous efforts have been dedicated to study the transport properties of QDs made from semiconductors, such as GaAs and silicon, and, more recently, graphene and other two-dimensional (2D) materials [1–6]. One important parameter in quantum computing is the quality factor defined by $Q = T_2^*/t_G$, where T_2^* is the decoherence time and t_G is the gate operation time. A good quantum computing system requires a long decoherence time and a short gate operation time, thus many calculations can be performed before the information is lost. Although the high mobility (clean) and light effective mass (allowing for wider gate separation, hence relatively easy fabrication) in GaAs-based QDs have enabled the rapid development of spin qubits [7], the strong nuclear field limits the spin decoherence time ($T_2^* \approx 10 - 100$ ns), making this material less ideal for upscaling. The QDs fabricated in isotopically purified silicon (^{28}Si) do not suffer from the nuclear field and have shown a sufficiently long spin decoherence time ($T_2^* \approx 0.12$ ms) [8,9], but the number of entanglement is hindered by the fabrication difficulty (shorter gate separation required) resulted from the heavy effective mass in silicon. While research on these materials is ongoing, 2D materials such as graphene and transition metal dichalcogenides (TMDs) have attracted considerable attention over the past few years because of their novel electronic properties [10,11]. Graphene is expected to be a robust material for spintronics owing to its weak spin–orbit and hyperfine interactions. Over the last decade, attempts to confine and manipulate single charges in graphene quantum dots (GQDs) have been widely studied and reported, as noted in several review articles [12–16]. However, early studies of GQDs on SiO_2 have indicated an absence of spin-related phenomena, such as spin blockade and the Kondo effect. To reduce the substrate disorder, which is one of the major sources of fast spin relaxation, recent efforts have been focused on GQDs on atomically flat substrates [e.g., hexagonal boron nitride (hBN)]. Nevertheless, the edge disorder may still play a role; hence, no significant differences compared with studies on SiO_2 have been reported either. Other 2D materials, such as 2H-TMDs, exhibit direct bandgap in monolayer form and are promising for switch applications due to the high current on/off rates in their transistors. In addition, the absence of inversion symmetry and the existence of strong spin–orbit coupling (SOC) in monolayer 2H-TMDs allow the charge carriers to be simultaneously valley- and spin-polarized, providing more degrees of freedom that can be controlled as qubits. On the other hand, 1T′ phase TMDs possess a completely different band structure compared with their 2H phase family. They are semimetal in bulk but become 2D topological insulators (TI) in the monolayer form. These insulators hold promise for hosting Majorana zero modes, which are topologically protected and form the core element for performing fault-tolerant quantum computing at the hardware level. In this chapter, we aim to provide an overview of experimental studies that are relevant to the development of various qubits in 2D materials. We supply the entry-level knowledge for this field by first introducing the fundamental properties

of various 2D materials and nanostructures followed by a selection of experimental studies. We discuss the transport properties of graphene nanodevices fabricated on both SiO_2 and hBN substrates at low temperatures and under high magnetic fields. Our primary focus is the single-electron tunneling regime in transport. In the second part, our focus will be directed to 2H-TMD nanostructures. We review recent developments in the fabrication and understanding of the electronic properties of these 2D nanostructures, including MoS_2 nanoribbons, WSe_2 single QDs (SQDs), and MoS_2 double QDs(DQDs). In the third part, we extend our discussion to 2D material-based Josephson junctions (JJs) and their potential applications in quantum computing. Finally, we review the quantum spin Hall (QSH) edge states observed in monolayer 1T′-TMDs, which combined with superconductor, could be useful for probing Majorana zero modes. In the summary, we outline how future work should pursue the development of various qubits in 2D materials.

13.1.1   Graphene and hBN

Graphene is a single layer of carbon atoms packed tightly in a honeycomb lattice as shown in Figure 13.1a. An early

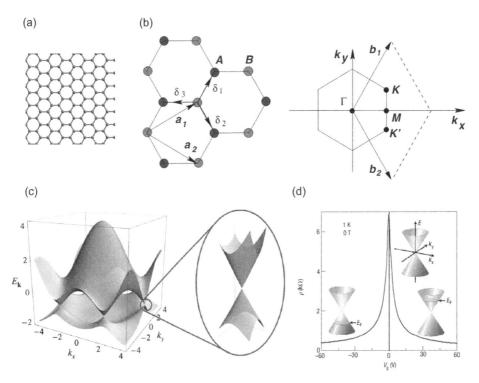

FIGURE 13.1   (a) Graphene is a honeycomb lattice of carbon atoms. (b) Honeycomb lattice and its Brillouin zone. Left: lattice structure of graphene, made out of two interpenetrating triangular lattices ($\mathbf{a}_1$ and $\mathbf{a}_2$ are the lattice unit vectors, and $\vec{\delta_i}$, $i =1, 2, 3$ are the nearest neighbor vectors). Right: corresponding Brillouin zone. The Dirac cones are located at the K and K' points. (c) The band structure of graphene is calculated using a tight-binding model. The zoom in shows the conical dispersion relation around the Dirac point. (d) The ambipolar electric field effect of graphene. V_g is the back-gate voltage and ρ is resistivity. By varying V_g, one can shift the Fermi energy level and, therefore, determine the type of carriers (either electrons or holes) in graphene. ((b,c) Adapted with permission from Ref. [18]. Copyright 2009 American Physical Society. (d) Adapted with permission from Ref. [19]. Copyright 2007 Nature Publishing Group.)

study on few-layer graphene can be tracked back to 1948 by G. Ruess and F. Vogt, in which they occasionally observed extremely thin graphitic flakes in transmission electron microscope (TEM) images. However, no one isolated single layer graphene until 2004, when the physicists at the University of Manchester first isolated and spotted graphene on a chosen SiO_2 substrate [17]. The first line of enquiry stems from graphene's unique gapless bandstructure. The unit cell of graphene consists of two carbon atoms, labeled as A and B sublattices, and can be described by the two lattice vectors $\mathbf{a}_1$ and $\mathbf{a}_2$, as shown in Figure 13.1b (left panel). They include an angle of 60° and have a length of $|\mathbf{a}_1| = |\mathbf{a}_2| = \sqrt{3}a_0 - 2.461\ \mathring{A}$, where a_0 is the carbon–carbon bond length ($a_0 = 1.42\ \mathring{A}$). The lattice vectors can be determined as $\mathbf{a}_1 = \frac{a_0}{2}(3, \sqrt{3})$ and $\mathbf{a}_2 = \frac{a_0}{2}(3, -\sqrt{3})$, and the reciprocal lattice is described by $\mathbf{b}_1 = \frac{2\pi}{3a_0}(1, \sqrt{3})$ and $\mathbf{b}_2 = \frac{2\pi}{3a_0}(1, -\sqrt{3})$, as shown in the right panel of Figure 13.1b. The lattice has high-symmetry points Γ, K, and M, where $K = \left(\frac{2\pi}{3a_0}, \frac{2\pi}{3\sqrt{3}a_0}\right)$ and $K' = \left(\frac{2\pi}{3a_0}, -\frac{2\pi}{3\sqrt{3}a_0}\right)$ are two points at the corners of the hexagonal Brillouin zone [18]. Around the K point, a tight-binding calculation for the bandstructure of this lattice yields a 2D Dirac-like Hamiltonian $\hat{H}_K$ for massless fermions (and around the K' point, the Hamiltonian is simply $\hat{H}_{K'} = \hat{H}_K^T$):

$$\hat{H}_K\psi(\mathbf{r}) = \hbar v_F \begin{pmatrix} 0 & k_x - ik_y \\ k_x + ik_y & 0 \end{pmatrix} \psi(\mathbf{r})$$
$$= -i\hbar v_F \begin{pmatrix} 0 & \frac{\partial}{\partial x} - i\frac{\partial}{\partial y} \\ \frac{\partial}{\partial x} + i\frac{\partial}{\partial y} & 0 \end{pmatrix} \psi(\mathbf{r})$$
$$= -i\hbar v_F \vec{\sigma}\nabla\psi(\mathbf{r}) = v_F\vec{\sigma}\cdot\vec{p}\,\psi(\mathbf{r}) = E\psi(\mathbf{r}) \quad (13.1)$$

where v_F is the Fermi velocity, $\vec{\sigma} = (\sigma_x, \sigma_y)$ is the 2D Pauli matrix, and $\psi(\mathbf{r})$ is the two-component electron wavefunction. This Hamiltonian gives rise to the most important aspect of graphene's energy dispersion, $E = \hbar v_F k$, which is a linear energy–momentum relationship at the edge of the Brillouin zone as shown in Figure 13.1c. The two-component vector part of the wavefunction, which corresponds to the A or B sublattices, is the so-called pseudospin degree of freedom, since it resembles the two-component real spin vector. The Pauli matrices σ_x and σ_y combined with the direction of the momentum leads to the definition of a chirality in graphene ($h = \vec{\sigma}\cdot\vec{p}/2\,|\vec{p}|$), meaning that the wavefunction component of A or B sublattice is polarized with regard to the direction of motion of electrons [18]. The existence of the K and K' points (as a result of graphene's hexagonal structure), where the Dirac cones for electrons and holes touch each other in momentum space (Figure 13.1b,c), is sometimes referred as isospin, and gives rise to a valley degeneracy $g_\nu = 2$ for graphene. The linear dispersion along with the presence of potential disorder leads to a maximum resistivity in the limit of vanishing carrier density (or the so-called Dirac point), as shown in Figure 13.1d. To change the Fermi level, and hence the charge carrier density, voltage needs to be applied to a nearby gate capacitively coupled to the graphene, which in

the case of Figure 13.1d is a back-gate-doped Si substrate that is isolated from the graphene by a SiO_2 insulator layer.

There are rich physics originated from the Dirac nature of the fermions in graphene, such as its electronic, optical, and mechanical properties [18,20,21]. Here, we introduce an important phenomenon in graphene transport, which is relevant to the subjects to be discussed in this chapter: the extreme quantum Hall effect (QHE) can be observed even at room temperature [19]. Because the low-energy fermions in graphene are massless, it is obvious that for graphene we cannot apply the results valid for standard semiconductor 2D electron gas (2DEG) systems. Charge carriers in a standard 2DEG have an effective mass, which is related to the parabolic dispersion relation of conduction(valence)band *via* $E = E_c + \frac{\hbar^2 k^2}{2m_e^*}\left(E = E_v - \frac{\hbar^2 k^2}{2m_h^*}\right)$, where $E_{c(v)}$ is the conduction (valence) band minimum (maximum) and $m_{e(h)}^*$ is the effective mass for the electrons (holes). The band dispersion leads to a constant density of state (DOS) of $\frac{m_{e(h)}^*}{\pi\hbar^2}$ for the conduction (valence) band region. In a perpendicular magnetic field, the DOS of electrons in a 2DEG system is quantized at discrete energies given by:

$$E_n = \pm\hbar\omega_c(n + 1/2), \quad (13.2)$$

which is the so-called Landau Level (LL) energy, with n the integer number and $\omega_c = eB/m_e^*$ the cyclotron frequency, as sketched in Figure 13.2a. The resulting Hall plateaus of a 2DEG lie at the conductivity values as follows:

$$\sigma_{xy} = \nu e^2/h, \quad (13.3)$$

where ν is the filling factor and takes only integer values, as illustrated in Figure 13.2b. For the QHE in graphene, the 2D massless Dirac equation must be solved in the presence of a perpendicular magnetic field B to find the LL energy E_n [18,22]. Thus, the Hamiltonian for graphene now reads:

$$v_F\vec{\sigma}\cdot\left(\vec{p} + \frac{e\vec{A}}{c}\right)\psi(\mathbf{r}) = E\psi(\mathbf{r}), \quad (13.4)$$

where momentum $\vec{p}$ in Eq. (13.1) has been replaced by $\vec{p} + \frac{e\vec{A}}{c}$, and $\vec{A}$ s the in-plane vector potential generating the perpendicular magnetic field $\vec{B} = B\hat{z}$. The solution of this equation gives rise to the eigenenergy of each LL for monolayer graphene:

$$E_n = \text{sgn}(n)v_F\sqrt{2e\hbar B\,|n|} \quad (13.5)$$

with the LL index $n = 0, \pm1, \pm2$, etc., and $\text{sgn}(n)$ stands for the sign of n. Unlike 2DEG, there will be a LL at zero energy ($n = 0$) separating the positive and negative LLs, and their energies are proportional to $\sqrt{B}$ (instead of B in 2DEG), as sketched in Figure 13.2c. In addition, the resulting Hall conductivity for monolayer graphene is given by

$$\sigma_{xy} = 4e^2/h\left(n + \frac{1}{2}\right) = \nu e^2/h, \quad (13.6)$$

where n is an integer and the factor 4 is due to the double valley and double spin degeneracy [18,19,24]. Note the filling

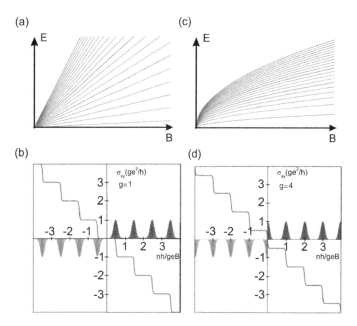

FIGURE 13.2 (a) A schematic illustration of LL for a standard 2DEG system. (b) Illustration of the integer QHE for a 2DEG system. (c) Same as (a) but for graphene. (d) Same as (b) but for graphene. The QHE plateau σ_{xy} lies at half integers of $4e^2/h$. ((b,d) Adapted with permission from Ref. [23]. Copyright 2006 Nature Publishing Group.)

factor now reads: $\nu \equiv 4N/N_\phi = 4(n+1/2) = \pm 2, \pm 6, \pm 10$ etc., where N is the total electron occupancy and N_ϕ is the magnetic flux divided by the flux quantum h/e. This result differs from the conventional QHE found in GaAs heterostructure 2DEGs (Figure 13.2b) and is a hallmark of Dirac fermion in monolayer graphene. The quantization of σ_{xy} has been observed experimentally [19], and a sketch of the data is illustrated in Figure 13.2d. The lowest LL in the conduction band and the highest LL in the valence band merge and contribute equally to the joint level at $E = 0$, resulting in the half-odd-integer QHE. The factor $1/2$ in Eq. (13.6) is due to the additional Berry phase π that the electrons, due to their chiral nature, acquire when completing a cyclotron trajectory [25,26]. The observation of the QHE at room temperature is also a consequence of the Dirac nature of the fermions in graphene. Because in graphene E_n is proportion to $v_F\sqrt{B\,|n|}$ (where $v_F=10^6$ m/s is the Fermi velocity), and at low energy, the energy spacing $\Delta E_n \equiv E_{n+1} - E_n$ between LLs can be rather large. For example, for fields of the order of $B = 10$ T, the cyclotron energy in a GaAs 2DEG system is of the order of 10 K; however, the same field in graphene gives rise to the cyclotron energy of the order of 1,000 K, that is, two orders of magnitude larger.

Having briefly introduced graphene, we extend our discussion to hBN, which is isostructural to graphene but has boron and nitrogen atoms on the A and B sublattices, as shown in Figure 13.3a. Due to the different onsite energy of A and B sublattices, the tight-binding calculation shows that hBN is an insulator with a large bandgap of around 6 eV [10, 27,31–33]. Traditionally, hBN has been used as a lubricant or a charge leakage barrier layer in electronic equipments [27]. More importantly, recent studies have shown the use

of hBN thin films as a dielectric layer for gating or as a flat substrate for graphene transistor to improve the electronic transport quality of devices by a factor of ten (or more), compared with the case of graphene on SiO_2 substrates [29,30,34,35]. The high quality of graphene/hBN heterostructures originates from the atomic-level smooth surface of hBN that can suppress surface ripples in graphene. Scanning tunneling microscopy (STM) topographic images (Figure 13.3b) show that the surface roughness of graphene on hBN is greatly decreased compared with that of graphene on SiO_2 substrates. While graphene on SiO_2 exhibits charge puddles with diameters of 10 ∼ 30 nm, the sizes of charge puddles in graphene on hBN are roughly one order of magnitude larger. The enhanced high mobility of graphene on hBN (up to 10^6 cm^{2}V s reported [36]) has enabled the studies of many-body physics and phase coherent transport that cannot be accessed in low-mobility samples, such as the observation of fractional QHE and supercurrent in the quantum Hall regime [37,38].

Due to the similarity in lattice structure, when graphene is stacked on hBN with a small twist angle ($\leq 5°$), it can form a superlattice (called the moiré pattern, as shown in Figure 13.3c) with a wavelength ranging from a few to 14 nm [29,30,35,39]. The superlattice with a relatively large wavelength compared with the bond length of carbon atom introduces additional minibands in graphene's band structure [30]. Figure 13.3d shows typical transfer curves for three graphene/hBN stacks with different moiré wavelengths, in which two extra Dirac peaks, situated symmetrically about the charge neutrality point ($V_g = 0$ V), are observed in all devices. These newly appeared Dirac peaks result from the superlattice minibands, which are away from the original Dirac point of graphene, as shown in the inset of

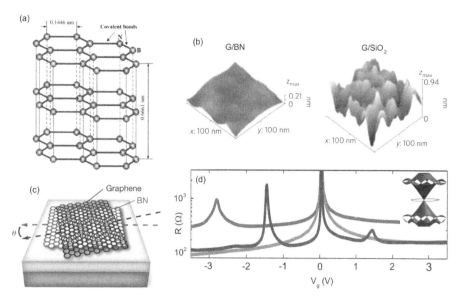

FIGURE 13.3 (a) Structure of the multilayered hBN. (b) STM topographic images of monolayer graphene on hBN (left) and SiO$_2$(right) showing the underlying surface corrugations. (c) Schematic of the moiré pattern formed from a graphene/hBN stack. The moiré wavelength varies with the twist angle θ. (d) Resistance as a function of gate voltage measured from three graphene/hBN stacks (with different moiré wavelengths), showing two extra Dirac peaks as a result of superlattice minibands. Inset shows the band diagram of graphene on hBN. ((a) Adapted with permission from Ref. [27]. Copyright 2013 American Chemical Society. (b) Adapted with permission from Ref. [28]. Copyright 2011 Nature Publishing Group. (c) Adapted with permission from Ref. [29]. Copyright 2013 Nature Publishing Group. (d) Adapted with permission from Ref. [30]. Copyright 2013 American Association for the Advancement of Science.)

Figure 13.3d. Such hybrid band structures lend novel transport features to graphene; for example, the observation of the Hofstadter Butterfly spectrum in high magnetic fields [29,30,35].

13.1.2 2H-Transition Metal Dichalcogenides

2D materials with a hexagonal lattice structure (such as graphene or TMDs with 2H phase) possess a valley of energy–momentum dispersion at the corner of the hexagonal Brillouin zone. In graphene, this dispersion at the K and $-K$ points gives rise to a valley degeneracy (note that in this and the subsequent sections we use the notation $-K$ to replace K' for simplicity). The situation is different in 2H-TMDs because of the absence of inversion symmetry, which allows the valley degree of freedom to be accessed independently (valleytronics), although it is still degenerate in energy. 2H-TMDs are semiconductors and have hexagonal lattices of MX$_2$, where M is a transition metal element from group VI (Mo or W) and X is a chalcogen atom (S, Se or Te), as illustrated in Figure 13.4a. Unlike graphene and hBN, the lattice structure of such a TMD consists of hexagons M and X, with the M atom being coordinated by the six neighboring X atoms in a trigonal prismatic geometry, as shown in Figure 13.4b. A key aspect of semiconducting TMD is the effect exerted by the number of layers on the electronic band structure. Figure 13.4c shows the calculated band structure of a 2H-TMD (MoS$_2$), which exhibits a crossover from an indirect gap in the bulk form to a direct gap in the monolayer form as a result of

a decreasing interlayer interaction. The photoluminescence (PL) from monolayer MoS$_2$ has shown the quantum yield to be two orders of magnitude larger than that from the multilayer material, providing evidence of such a crossover in the bandgap [40,41]. In the monolayer limit, the conduction and valence band edges are at the $\pm K$ points and are predominantly formed by the partially filled d-orbitals of the M atoms and have the following forms:

$$|\phi_c\rangle = |d_{z^2}\rangle \tag{13.7}$$

$$|\phi_v^\tau\rangle = \frac{1}{\sqrt{2}}\left(\left|d_{x^2-y^2}\right\rangle + i\tau\left|d_{xy}\right\rangle\right), \tag{13.8}$$

where d_{z^2}, $d_{x^2-y^2}$ and d_{xy} are the d-orbitals of the M atom, the subscript $c(v)$ indicates the conduction (valence) band, and $\tau = \pm 1$ is the valley index. At the valley points ($\pm K$), a two-band $k \cdot p$ Hamiltonian that takes the form of the massive Dirac fermion model can be used to describe the dispersion at the conduction and valence band edges [44]:

$$H = at(\tau k_x \sigma_x + k_y \sigma_y) + \frac{\Delta}{2}\sigma_z - \lambda\tau\frac{\sigma_z - 1}{2}\hat{S}_z, \tag{13.9}$$

where σ denotes the Pauli matrices for the two basic functions given in Eq. (13.7) and (13.8), a is the lattice constant, t is the effective nearest neighbor hopping integral, and Δ is the bandgap. The last term in Eq. (13.9) represents the SOC, where 2λ is the spin splitting at the top of the valence band and $\hat{S}_z$ is the Pauli matrix for spin. The spin splitting is due to the strong spin–orbit interaction arising from the d-orbitals of the heavy metal atoms. The conduction band-edge state consists of d_{z^2} orbitals and remains almost spin-degenerate at the $\pm K$ points, whereas the valence-band-edge

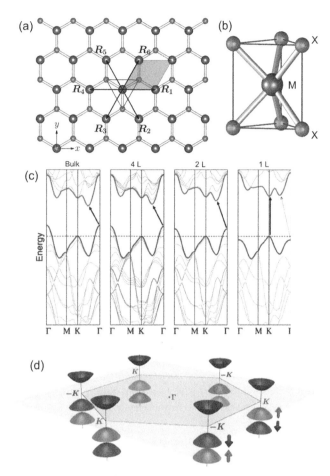

FIGURE 13.4 (a) Top view of monolayer 2H-MX$_2$. Larger balls represent M atoms, and smaller balls represent X atoms. The shadowed diamond region shows the 2D unit cell with lattice constant a. $\mathbf{R}_1$ - $\mathbf{R}_6$ denote the M-M nearest neighbors. (b) A schematic illustration for the structure of trigonal prismatic coordination, corresponding to a side view of the triangle in (a). (c) Energy dispersion in bulk, quadrilayer (4L), bilayer (2L), and monolayer (1L) 2H-MoS$_2$, from left to right, showing the transition from an indirect bandgap to a direct bandgap. (d) A schematic illustration of the band structure at the band edges located at the edges of the Brillouin zone. ((a,b) Adapted with permission from Ref. [42]. Copyright 2013 American Physical Society. (c) Adapted with permission from Ref. [43]. Copyright 2015 Royal Society of Chemistry. (d) Adapted with permission from Ref. [44]. Copyright 2012 American Physical Society.)

state shows a pronounced split. A schematic illustration of the band dispersion at the edges of the hexagonal Brillouin zone is shown in Figure 13.4d. Note that the spin splitting at the different valleys is opposite because the K and $-K$ valleys are related to one another by time-reversal symmetry (TRS).

Because of the large valley separation in momentum space, the valley index is expected to be robust against scattering by smooth deformations and long-wavelength phonons. To manipulate such a valley degree of freedom for valleytronic applications, measurable physical quantities

that distinguish the $\pm K$ valleys are required. The Berry curvature ($\boldsymbol{\Omega}$) and the orbital magnetic moment ($\mathbf{m}$) are two physical quantities for $\pm K$ valleys to have opposite values. The Berry curvature is defined as a gauge field tensor derived from the Berry vector potential $\mathbf{A}_n(\mathbf{R})$ through the relation $\boldsymbol{\Omega}_n(\mathbf{R}) = \nabla_{\mathbf{R}} \times \mathbf{A}_n(\mathbf{R})$, where n is the energy band index (in the case of 2H-TMDs and at the $\pm K$ points, n is either the conduction or valence band), and $\mathbf{R}$ is the parameter to be varied in a physical system (in the case below, $\mathbf{R}$ is the wavevector $\mathbf{k}$) [45]. The Berry curvature can be written as a summation over the eigenstates as follows [46]:

$$\boldsymbol{\Omega}_{\mathbf{n}}(\mathbf{k}) = i\frac{\hbar^2}{m^2}\sum_{i \neq n}\frac{\mathbf{P}_{n,i}(\mathbf{k}) \times \mathbf{P}_{i,n}(\mathbf{k})}{\left[E_n^0(\mathbf{k}) - E_i^0(\mathbf{k})\right]^2} \qquad (13.10)$$

Here, $\mathbf{P}_{n,i}(\mathbf{k}) \equiv \langle u_n(\mathbf{k})|\hat{\mathbf{p}}|u_i(\mathbf{k})\rangle$ is the interband matrix element of the canonical momentum operator $\hat{\mathbf{p}}$, where $u(\mathbf{k})$ is the periodic part of the Bloch wavefunction, and $E_{n(i)}^0(\mathbf{k})$ denotes the energy dispersion of the $n(i)$-th band. Upon substituting the eigenfunctions of Eq. (13.9) into Eq. (13.10), the Berry curvature in the conduction band is given by

$$\boldsymbol{\Omega}_{\mathbf{c}}(\mathbf{k}) = -\tau\frac{2a^2 t^2 \Delta'}{(4a^2 t^2 k^2 + \Delta'^2)^{3/2}} \qquad (13.11)$$

where τ is the valley index and $\Delta' \equiv \Delta - \tau S_z\lambda$ is the spin-dependent bandgap. Note that the Berry curvature has opposite signs in opposite valleys, and this also occurs in the conduction and valence bands [$\boldsymbol{\Omega}_{\mathbf{v}}(\mathbf{k}) = -\boldsymbol{\Omega}_{\mathbf{c}}(\mathbf{k})$]. Here, we write the equations of motion for Bloch electrons under the influence of the Berry curvature and applied electric and magnetic fields [45]:

$$\dot{\mathbf{r}} = \frac{1}{\hbar}\frac{\partial E_n(\mathbf{k})}{\partial \mathbf{k}} - \dot{\mathbf{k}} \times \boldsymbol{\Omega}_{\mathbf{n}}(\mathbf{k}) \qquad (13.12)$$

$$\hbar\dot{\mathbf{k}} = -e\mathbf{E} - e\dot{\mathbf{r}} \times \mathbf{B} \qquad (13.13)$$

It can be seen that in the presence of an in-plane electric field, carriers with different valley indices will acquire opposite velocities in the transverse direction because of the opposite signs of their Berry curvatures, leading to the so-called valley Hall effect, as illustrated in Figure 13.5a,b. Here, we note that this result is valid not only for monolayer 2H-TMDs but also for thin films with an odd number of layers, because odd numbers of layers also exhibit inversion symmetry breaking, which is a necessary condition for the $\pm K$ valleys to exhibit valley contrast in the Berry curvature.

The valley contrast in 2H-TMDs can also reflect on the optical interband transitions from the top of the spin-split valenceband to the bottom of the conduction band at the $\pm K$ points. The coupling strength with optical fields of $\sigma_{\pm}$ circular polarization is given by $\boldsymbol{P}_{\pm}(\mathbf{k}) \equiv \boldsymbol{P}_x(\mathbf{k}) \pm i\boldsymbol{P}_y(\mathbf{k})$, where $\boldsymbol{P}_{\alpha}(\mathbf{k}) \equiv m_0\langle u_c(\mathbf{k})|\frac{1}{\hbar}\frac{\partial \hat{H}}{\partial k_{\alpha}}|u_v(\mathbf{k})\rangle$ is the interband matrix element of the canonical momentum operator ($u_{c(v)}(\mathbf{k})$ is the Bloch function for the conduction (valence) band, and m_0 is the free electron mass). For transitions near the $\pm K$ points and for a reasonable approximation of $\Delta' \gg atk$ (see the parameters in Ref. [44]), this expression has the following form [44]:

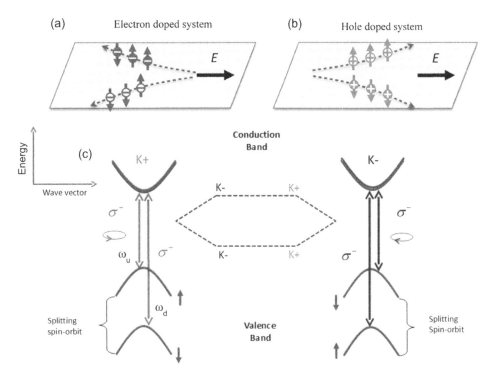

FIGURE 13.5 Schematic illustration of the valley Hall effect in (a) the electron-doped regime and (b) the hole-doped regime. The electrons and holes in the $+K$ valley are denoted by white "$-$" and "$+$" symbols in dark circles and their counterparts in the $-K$ valley are denoted by an inverted color scheme. (c) Illustrations of the valley- and spin-dependent optical transition selection rules. ((a,b) Adapted with permission from Ref. [44]. Copyright 2012 American Physical Society.)

$$|\boldsymbol{P}_{\pm}(\mathbf{k})|^2 = \frac{m_0^2 a^2 t^2}{\hbar^2}(1 \pm \tau)^2 \qquad (13.14)$$

It is evident that the coupling strength between circularly polarized light and interband transitions is valley dependent; $\boldsymbol{P}_{+}(\mathbf{k})$ has a nonzero value in the $+K$ valley, as does $\boldsymbol{P}_{-}(\mathbf{k})$ in the $-K$ valley. This valley-dependent optical selection rule is illustrated in Figure 13.5c, where a $\sigma_{+(-)}$ circularly polarized optical field exclusively couples with the interband transitions at the $+(-)K$ valley. Note that the spin is selectively excited through this valley-dependent optical selection rule, and consequently, the spin index becomes locked with the valley index at the band edges. For example, an optical field with σ_+ circular polarization and a frequency of $\omega_d(\omega_u)$ can generate spin-up (spin-down) electrons and spin-down (spin-up) holes in the $+K$ valley, whereas the excitation in the $-K$ valley is precisely the time-reversed counterpart of the above [44].

13.1.3 1T'-Transition Metal Dichalcogenides

The TMD family has three typical phases, including 2H, 1T, and 1T', as shown in Figure 13.6a–c, respectively. In contrast to 2H structure, the M atoms in the 1T structure are octahedrally coordinated with the nearby six X atoms, resulting in ABC stacking with the $P\bar{3}m1$ space group, as shown in Figure 13.6b. 1T-MX$_2$ have very different electronic properties compared with the semiconducting 2H structures. 1T-TMDs are metallic (Fermi level lying in the middle of degenerate $d_{xy,yz,xz}$ single band) and are often unstable in ambient condition, which usually leads to a spontaneous lattice distortion and a doubling periodicity in the x-direction [47]. Eventually, they form a 2×1 superlattice structure, i.e., the 1T' structure, consisting of one-dimensional (1D) zigzag chains along the y direction, as shown in Figure 13.6c. The lattice distortion from the 1T phase to the 1T' phase induces band inversion and causes 1T'-TMDs to become topologically nontrivial [48]. Figure 13.7a schematically illustrates this topological phase transition in 1T'-WTe$_2$ [47]. The bulk band starts with a topological trivial phase and then evolves into a nontrivial phase where the energy of the original valence band (darker gray) is higher than that of the original conduction band (lighter gray), resulting in an inverted band crossing at a momentum point along the Γ-Y direction. Finally, a strong SOC lifts the degeneracy and opens up a bulk bandgap as shown in the rightmost panel of Figure 13.7a. The actual calculated electronic band structure of 1T'-MX$_2$ (here taking MoS$_2$ as an example) using many-body perturbation theory is shown in Figure 13.7b. As can be seen, the band of 1T'-MoS$_2$ shows a gap (E_g) of about 0.08 eV, located at $\Lambda = \pm(0,0.146)A^{-1}$. The conduction and valence bands display a camelback shape near Γ point and present a large inverted gap (2δ) of about 0.6 eV. To better understand the nature of the inverted bands near Γ, a low-energy $\boldsymbol{k} \cdot \boldsymbol{p}$ Hamiltonian for 1T'-MX$_2$, in which the valence band mainly consists of d-orbitals of M atoms (d_{yz} and d_{xy}) and the conduction band mainly consists of p_y-orbitals of X atoms, is written as [48]

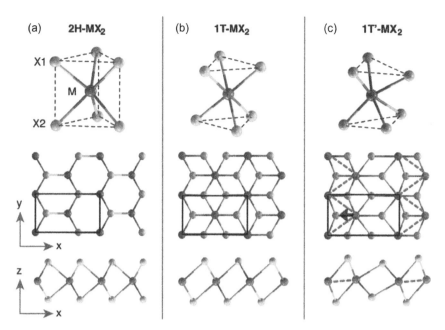

FIGURE 13.6 Three atomistic structures of monolayer TMDs MX_2, where M stands for (Mo,W) and X stands for (S, Se, Te). (a) 2H-MX_2 structure where M atoms are trigonal-prismatically coordinated by six X atoms. (b) 1T-MX2 structure where M atoms are octahedrally coordinated with the nearby six X atoms. (c) 1T′-MX2, distorted 1T-MX2, where the distorted M atoms form 1D zigzag chains indicated by the dashed lines. The unit cell is indicated by gray rectangles. (Figure adapted with permission from Ref. [48]. Copyright 2014 American Association for the Advancement of Science.)

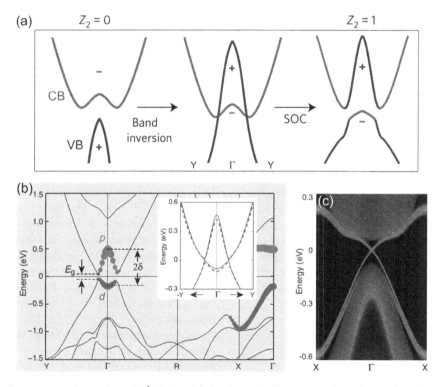

FIGURE 13.7 Band structures of monolayer 1T′-TMDs. (a) A schematic diagram to show the band evolution from a topologically trivial phase, to a nontrivial phase, and then to a bandgap opening due to the SOC. (b) First-principles calculated band structures for monolayer 1T′-MoS_2. E_g and 2δ represent the fundamental gap and inverted gap, respectively. The dots labeled by p and d indicate the major orbital characters in the top valence band and bottom conduction band. The inset compares band structures with (dashed line) and without (solid line) SOC. (c) Calculated edge DOSs of monolayer 1T′-MoS_2. ((a) Adapted with permission from Ref. [47]. Copyright 2017 Nature Publishing Group. (b,c) Adapted with permission from Ref. [48]. Copyright 2014 American Association for the Advancement of Science.)

$$H = \begin{pmatrix} E_p(k_x, k_y) & 0 & -i\nu_1\hbar k_x & \nu_2\hbar k_y \\ 0 & E_p(k_x, k_y) & \nu_2\hbar k_y & -i\nu_1\hbar k_x \\ i\nu_1\hbar k_x & \nu_2\hbar k_y & E_d(k_x, k_y) & 0 \\ \nu_2\hbar k_y & i\nu_1\hbar k_x & 0 & E_d(k_x, k_y) \end{pmatrix},$$

(13.15)

where $E_p = -\delta - \frac{\hbar^2 k_x^2}{2m_x^p} - \frac{\hbar^2 k_y^2}{2m_y^p}$, and $E_d = \delta + \frac{\hbar^2 k_x^2}{2m_x^d} + \frac{\hbar^2 k_y^2}{2m_y^d}$. Here $\delta < 0$ corresponds to the $d - p$ band inversion ($E_p > E_d$ near Γ, see Figure 13.7b). Note that the band inversion arises from the formation of quasi-1D M chains in the 1T′ structure, which lowers the metal d orbital below chalcogenide p orbital with respect to the original 1T structure, leading to the band inversion at Γ point. By fitting with first-principles band structure in Figure 13.7b, parameters in Eq. (13.15), such as δ, m_x^p, m_y^p, m_x^d, and m_y^d, can be estimated [48]. Since the 1T′ structure has inversion symmetry, the Z_2 band topology can be determined by the parity of valence bands at four time-reversal invariant momenta (TRIM), Γ, X, Y and R [11,49]. Apart from 1T′-MoS$_2$, Qian et al. have calculated the band structures and TRIM of other five 1T′-MX$_2$, including MoSe$_2$, MoTe$_2$, WS$_2$, WSe$_2$, and WTe$_2$ [48]. Their results suggest that all 1T′-TMDs have Z$_2$ nontrivial band topology resulting from the above $p-d$ band inversion, with inverted bandgaps at Γ of 1.04, 0.36, 0.28, 0.94, and 1.17 eV, respectively. 1T′-MoSe$_2$, WS$_2$, and WSe$_2$ have fundamental gaps of 0.11, 0.12, and 0.12 eV, respectively; while 1T′-MoTe$_2$ and WTe$_2$ are semimetals due to the increase of valence band maximum at the Γ point (although recent experiments indicated that monolayer 1T′-WTe$_2$ is actually a 2D TI, see Section 13.6).

The topological phase (Z$_2$ = 1) in monolayer 1T′-MX$_2$ makes them a 2D TI, which carries helical edge states that are protected from elastic backscattering by TRS.

Figure 13.7c shows the calculated edge DOS of 1T′-MoS$_2$ (similar results are found for other 1T′-MX$_2$) using iterative Green's function and many-body GW theory [48]. The edge states present a Dirac-like (linear) dispersion located inside the bulk bandgap at Γ, with a high Fermi velocity of $\approx 1 \times 10^5$ m/s. These edge states, also known as QSH edge states, have the special "spin-filter" property in which upward and downward spins propagate in opposite directions, leading to a phenomenon called spin–momentum locking. Further investigations also indicate that the decay length (from the edge to bulk) of these helical edge states to be as short as 5 nm (50 nm in HgTe quantum wells [50]), which can greatly reduce scattering with bulk states and hence increase the transport lifetime [48]. Most interestingly, theory has predicted that the topological phase, and hence the existence of helical edge states within the bandgap, can be controlled by gating (vertical electric field) in monolayer 1T′-TMDs [48]. This tunable topological phase arises from the vertically well-separated planes between chalcogenide's p and metal's d orbitals, which allow a vertical electric field to modify the inverted band. Figure 13.8a displays the first-principles calculated bulk band structures of 1T′-MoS$_2$ under different vertical electric fields from 0 to 0.2 V/A, while Figure 13.8b shows the corresponding edge DOSs along X-Γ-X. The electric field breaks the inversion symmetry and introduces Rashba spin splitting of the original doubly degenerate bands near the fundamental gap E_g (see middle panel of Figure 13.8a). As the field increases, E_g first decreases to zero at a critical field strength of 0.142 V/A and then reopens (see the rightmost panel in Figure 13.8a). This gap-closing transition induces a topology change to a trivial phase, leading to the destruction of helical edge states, as shown in Figure 13.8b. In addition to the vertical

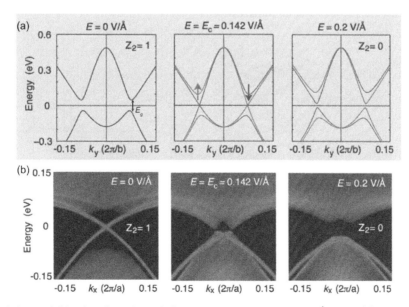

FIGURE 13.8 Vertical electric field induced topological phase transition in monolayer 1T′-MoS$_2$. (a) Bulk band structure of monolayer 1T′-MoS$_2$ (along Y-Γ-Y) under different vertical electric field, showing a bandgap closing and reopening around $k_y = 0.08$ 2π/b. (b) The corresponding edge DOSs under different electric fields. (Figure adapted with permission from Ref. [48]. Copyright 2014 American Association for the Advancement of Science.)

electrical field, Qian et al. also reported that a few percent of in-plane elastic strain can change monolayer $1T'$-$MoTe_2$ and WTe_2 from semimetals to small-gap QSH insulators by lifting the band overlap. The gate-tunable topological phases are viable for designing all electric-field controlled topological devices and could be useful for probing Majorana zero modes [51]. The quantized conductance of the QSH edge states in $1T'$-TMDs has been observed in several literature, which will be reviewed in Section 13.6. However, the topological phase transition induced by vertical gating has not been reported up to date.

In this section, we introduced the fundamental properties of various 2D materials that are to be discussed in the rest of the chapter. Combined with the quantum transport physics presented in the next section, these discussions will serve as a basis for our examination of the experimental studies in the subsequent sections.

13.2 Theoretical Background on Quantum Transport

13.2.1 Single QD

A QD is an artificially structured system that can be filled with only a few electrons or holes [7]. The charged carriers in such a system are generally confined in a submicron area, and the confinement potential in all directions is so strong that it gives rise to quantized energy levels that can be observed at low temperatures. The electronic properties of QDs are dominated by several effects [7]. First, the Coulomb repulsion between electrons on the dot leads to an energy cost called charging energy $E_C = e^2/C$, where C is the total capacitance of the dot, for adding an extra electron to the dot. Because of this charging energy, the tunneling of electrons to or from the reservoirs can be suppressed

at low temperatures (when $E_C > k_B T$), which leads to a phenomenon called Coulomb blockade. Second, the tunnel barrier resistance R_t, which describes the coupling of the dot to both the source and drain reservoirs, has to be sufficiently opaque, such that the electrons are located either in the source, in the drain, or on the dot. The minimal R_t can be estimated using the uncertainty principle, $\Delta E \cdot \Delta t > h$. From $\Delta E = e^2/C$, and $\Delta t = R_t C$, the condition $R_t > h/e^2$ for R_t can be found. This means that the energy uncertainty corresponding to the tunneling time cannot be greater than the charging energy; otherwise, it would lead to uncertainty in the number of carriers occupying the dot. Third, if the confinement in all three directions is strong enough for electrons residing on the dot to form quantized energy levels E_n (often denoted as single-particle energy), the energy spacing $\Delta E = E_n - E_{n-1}$ can be observed on top of charging energy if $\Delta E > k_B T$. Because of this discrete energy spectrum E_n, QDs behave in many ways as artificial atoms. Figure 13.9a shows an example of a QD formed in a GaAs/AlGaAs 2DEG system, where the dot is defined by a gate-depleted area and is tunnel coupled to the reservoir on each side. Varying the voltages on the surface gates enables several important parameters, such as the number of electrons and the tunnel barrier resistance, to be finely tuned. To understand the dynamics of an SQD, a constant interaction model has been proposed [7] and is illustrated in Figure 13.9b. The model is based on two assumptions. First, the Coulomb interactions among electrons in the dot, between electrons in the dot, and those in the environment are parameterized by a single, constant capacitance C. This capacitance is the sum of capacitance between the dot and the source C_S, the drain C_D and the gate C_G: $C = C_S + C_D + C_G$. The second assumption is that the single-particle energy spectrum E_n is independent of the Coulomb interaction and, therefore, of the number of electrons in the dot. Using this model, the total

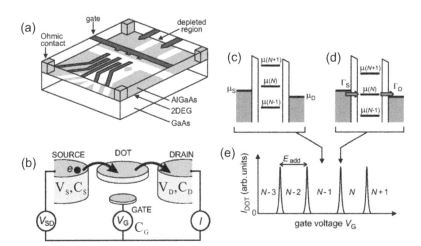

FIGURE 13.9 (a) A schematic view of a lateral QD device defined by metal surface electrodes on a GaAs/AlGaAs 2DEG system. (b) Electrical network diagram of anSQD. (c) The electrochemical potential energies for a single dot with transport blocked due to Coulomb blockade. (d) With one of the electrochemical potentials lying within the bias window, transport through the dot is then permitted. (e) A schematic plot of the current through a single dot against gate voltage showing periodic oscillations with the dot charge. (Adapted with permission from Ref. [7]. Copyright 2007 American Physical Society.)

energy of a single dot with N electrons in the ground state (GS) is given by [7]

$$U(N) = \frac{\left(-|e|(N - N_0) + C_S V_S + C_D V_D + C_G V_G\right)^2}{2C}$$
$$+ \sum_{n=1}^{N} E_n \qquad (13.16)$$

where $-|e|$ is the electron charge, N_0 is the charge on the QD due to the positive background charge of the donors, and V_S, V_D, and V_G are the voltages of the source, drain, and gate, respectively. The last term is a sum over the occupied single-particle energy levels E_n, which depend on the characteristics of the confinement potential.

The electrochemical potential of the dot $\mu(N)$ is defined as the energy needed to add the N-th electron to a dot with $N - 1$ occupied electrons [7]:

$$\mu(N) = U(N) - U(N - 1)$$
$$= \left(N - N_0 - \frac{1}{2}\right) E_C - \frac{E_C}{e}(C_S V_S + C_D V_D$$
$$+ C_G V_G) + E_N \qquad (13.17)$$

where

$$E_C = e^2/C \qquad (13.18)$$

is the charging energy. The addition energy is then given by the energy difference between two successive electrochemical potentials:

$$E_{\text{add}}(N) = \mu(N + 1) - \mu(N) = E_C + \Delta E \qquad (13.19)$$

where $\Delta E = E_{N+1} - E_N$ is the single-particle energy spacing and is independent of the electron number on the dot (the second assumption).

When the temperature is low enough ($k_B T \ll \Delta E, E_C$), the transport through the QD depends on whether the dot electrochemical potentials align with bias window, which is defined as the spacing between the electrochemical potentials of the source and drain, i.e., $-eV_{\text{SD}} \equiv \mu_S - \mu_D = -eV_S - (-eV_D)$. In the low-bias regime where $-eV_{\text{SD}} < E_C$,

electron tunneling can only happen when the dot electrochemical potential lies in a small bias window, such that $\mu_D < \mu(N) < \mu_S$, as shown in Figure 13.9d. When the electrochemical potential is outside the bias window, the transport is blocked and no current flows through the dot, which is the Coulomb blockade regime, as shown in Figure 13.9c. When a gate V_G constantly tunes the electrochemical potential of the QD, an on–off current can be observed as peaks with constant spacing (E_{add}) between each other, as shown in Figure 13.9e. Each current forbidden regime corresponds to a different electron number on the dot, so in this way, the number of electrons on the dot can be varied.

In the high-bias regime where $-eV_{\text{SD}} > \Delta E$ and/or $-eV_{\text{SD}} > E_{\text{add}}$, more dot levels are allowed to lie within the bias window and give rise to multiple tunneling paths, as shown in Figure 13.10a,b. Depending on how wide the bias window is, the transition can involve a GS and its ES as shown in Figure 13.10a, or in an even wider window ($-eV_{\text{SD}} > E_{\text{add}}$), it can couple to two successive GSs as shown in Figure 13.10b. From Eq. (13.17), the electrochemical potential is a function of V_S, V_D, and V_G. Since $\mu_{S(D)} = -eV_{S(D)}$, if we measure the conductance of dot as a function of bias eV_{SD} and gate voltage V_G, a spectrum called "Coulomb diamond" is formed, as shown in Figure 13.10c. Since larger biases require a wider spacing in gate voltage for dot levels being pulled out of the window, the V-shape feature can be expected. In Figure 13.10c, along the left (right) edge of the black V-shape following the slope at $\frac{-|e|C_G}{C - C_S}$ $\left(\frac{|e|C_G}{C_S}\right)$, the level of the N-electron GS is aligned with the source (drain) level while the bias window is becoming wider. The black V-shape shows the transition between the N-electron GS and $N + 1$-electron GS, and defines the regimes of blockade (outside the V-shape) and tunneling (within the V-shape). The left(lighter gray) and right(darker gray)V-shapes shown in Figure 13.10c correspond to two different transitions between the dot states thatare the N-electon ES to $N + 1$-electron GS ($ES(N) \rightarrow GS(N + 1)$) and the N-electon GS to $N + 1$-electron ES ($GS(N) \rightarrow ES(N + 1)$). Since the ES energy $ES(N)$ and $ES(N + 1)$ are separated from the GSs $GS(N)$ and $GS(N + 1)$ by $\Delta E(N)$ and

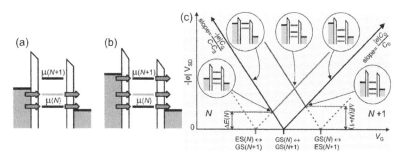

FIGURE 13.10 Schematic diagrams of the electrochemical potential levels of a QD in the high-bias regime. (a) V_{SD} exceeds ΔE so that the electron transport couples to an excited state (ES) (the gray level). (b) V_{SD} exceeds the addition energy so that the electron transport couples to two successive GS levels. (c) Differential conductance $dI_{\text{DOT}}/dV_{\text{SD}}$ through a QD as a function of gate V_G and bias voltage $-|e|V_{\text{SD}}$. The insets show different configurations of the dot level with respect to the lead potential in the $V_{\text{SD}} - V_G$ plane. (Adapted with permission from Ref. [7]. Copyright 2007 American Physical Society.)

$\Delta E(N + 1)$, respectively (see Figure 13.10c), Coulomb diamond measurements are very useful for studying the ES spectroscopy in a QD system. The insets shown in Figure 13.10c represent a different configuration of dot levels with respect to the source–drain level. Note that the $ES(N) \rightarrow GS(N + 1)$ and $GS(N) \rightarrow ES(N + 1)$ transitions are forbidden outside the black V-shape as $ES(N)$ and $ES(N + 1)$ states only exist when the $GS(N) \rightarrow GS(N + 1)$ transition is within the bias window. Finally, the dimension of the Coulomb diamond (current-suppressed region) in the bias direction is a direct measure of E_{add} or the charging energy E_C, because beyond the edge of the diamond the bias window is greater than E_{add} and transport is no longer blocked.

Here, we discuss the effect exerted by a magnetic field on the single-particle energy of QDs. The energy spectrum of a 2DEG QD in the presence of a magnetic field is typically solved using a single-particle approximation with a parabolic confinement potential [52,53]. Such a spectrum is called the Fock–Darwin diagram which describes how 0D levels evolve with respect to an applied perpendicular magnetic field. The symmetric parabolic potential can be approximated as $U(x, y) = \frac{m^*}{2}\omega_0^2(x^2 + y^2)$, where m^* is the effective mass and ω_0^2 denotes the strength of the confinement potential. Thus, the Hamiltonian of an electron in the dot can be written as follows:

$$H = \frac{1}{2m^*}(\mathbf{p} + e\mathbf{A})^2 + \frac{m^*}{2}\omega_0^2(x^2 + y^2) \quad (13.20)$$

If we choose the symmetric gauge for the vector potential $\mathbf{A} = (-By/2,\ Bx/2,\ 0)$, then the energy spectrum of the Hamiltonian can be solved as follows:

$$E_{n_+, n_-} = (n_+ + 1)\hbar\Omega + \frac{1}{2}\hbar\omega_c n_- \quad (13.21)$$

with $\Omega^2 \equiv \omega_0^2 + \frac{\omega_c^2}{4}$ where $\omega_c = \frac{|eB|}{m^*}$ is the cyclotron frequency, and with quantum numbers $n_\pm = n_x \pm n_y$, where $n_x, n_y = 0, 1, 2, ...,$ etc. This spectrum is plotted in Figure 13.11a. For $B = 0$, the spectrum has a constant level spacing and is simply the spectrum of the 2D harmonic oscillator. In the high-field limit, the spectrum goes over into that of the LLs

(see Figure 13.2a), with the confinement effects of the dot playing an ever-decreasing role.

In a GQD, the Fock–Darwin spectrum is notably different from that in the 2DEG case owing to the existence of an LL at zero energy, which does not shift in energy with increasing magnetic field [54,55]. Together with quantum confinement, the unique linear band dispersion of graphene results in an electron–hole crossover in GQD's magnetotransport [3,56]. To solve the Fock–Darwin spectrum for a GQD, we start from a free Dirac equation with a circular confinement potential $V(r)$ and include a perpendicular magnetic field, where the symmetric gauge $A = \frac{B}{2}(-y, x, 0) = \frac{B}{2}(-r\sin\phi, r\cos\phi, 0)$ for the vector potential is used (ϕ is the polar angle). Thus, the Hamiltonian now reads (ignoring spin) [54]:

$$H = \nu_F(\mathbf{p} + e\mathbf{A}) \cdot \vec{\sigma} + \tau V(r)\sigma_z, \quad (13.22)$$

where $\vec{\sigma} = (\sigma_x, \sigma_y)$ represents Pauli's matrices and $\tau = \pm 1$ is the valley index for $\pm K$. Note that the quantum confinement effect is introduced in the Hamiltonian *via* a mass-related potential $V(r)$ coupling to the σ_z Pauli matrix. We let the mass in the dot to be zero, i.e., $V(r) = 0$ for $r < R$, but let it tend toward infinity at the edge of the dot, i.e., $V(r) = \infty$ for $r > R$. In this way, charge carriers are confined inside the QD that has a radius of R. This leads to a boundary condition, which yields the simple relation that $\frac{\psi_2}{\psi_1} = \tau i \exp[i\phi]$ for circular confinement [54], where $\psi = (\psi_1, \psi_2)$ is the eigenfunction of the Hamiltonian. Hence, in the following, we can set $V(r) = 0$; thus, the energy E is related to the wavevector k via $E = \hbar\nu_F k$, and we can determine k using the boundary condition. Following Ref. [54], the implicit equation for determining the wavevector k (and therefore, the energy E) that satisfies the boundary condition is given by

$$\left(1 - \tau \frac{kl_B}{R/l_B}\right)L\left(\frac{k^2l_B^2}{2} - (m + 1), m, \frac{R^2}{2l_B^2}\right)$$
$$+ L\left(\frac{k^2l_B^2}{2} - (m + 2), m + 1, \frac{R^2}{2l_B^2}\right) = 0, \quad (13.23)$$

where $l_B = \sqrt{\hbar/eB}$ is the magnetic length and m is the angular momentum quantum number. The functions

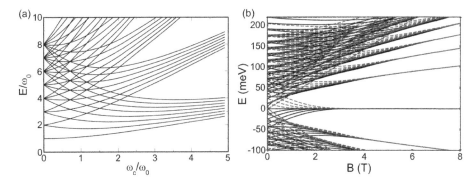

FIGURE 13.11   (a) Fock–Darwin spectrum of a 2DEG symmetric QD up to a quantum number of $n = 7$. (b) Energy spectrum of a graphene QD with $R = 70$ nm for $m = -4, ..., 4$ and $n = 1, ..., 6$. The energy states for $\tau = +1$ are drawn as solid black lines, and those for $\tau = -1$ are drawn as dashed lines. The dashed line enclosed areas highlight the continuously crossing of different states at low energy, as will be discussed in Section 13.3.1.2. ((b) Adapted with permission from Ref. [54]. Copyright 2008 American Physical Society.)

$L(a, b, x)$ are generalized Laguerre polynomials, which are oscillatory functions. Hence, there are an infinite number of wavevectors k_n for a given B, m, and τ that fulfill Eq. (13.23). This condition defines the radial quantum number n, from which the energy spectrum $E(n, m, \tau) = \hbar \nu_F k_n$ can be plotted, as shown in Figure 13.11b for a QD of radius R = 70 nm. Note that $-E(n, m, \tau) = E(n, m, -\tau)$, which gives rise to the electron–hole symmetry in the spectrum. We discuss Eq. (13.23) under two particular limits. For $B \to 0$, Eq. (13.23) can be written as follows:

$$\tau J_m(kR) = J_{m+1}(kR), \tag{13.24}$$

where J_m is a Bessel function. This relation yields the single-particle energy spectrum and can be used to estimate the energy of the ESs on a graphene dot with N confined charge carriers $[\Delta(N) = \hbar \nu_F / (d\sqrt{N})$, where d is an effective dot diameter; see Ref. [1,57]]. In addition, there is no state at zero energy under zero magnetic field, which leads to an energy gap separating the states of negative and positive energies. By contrast, at high field, where $R/l_B \to \infty$, Eq. (13.23) gives rise to the following:

$$E_m = \hbar \nu_F k_m = \pm \nu_F \sqrt{2e\hbar B(m+1)} \tag{13.25}$$

which are the LLs for graphene. Therefore, as the B-field increases, there will be a transition governed by the parameter R/l_B, from a regime in which the confinement plays an important role ($R \leq l_B$) to the LL regime ($R \geq l_B$).

Note that the resonances on both sides of the electron–hole crossover have opposite slopes and merge into the zeroth LL. An experimental observation of this effect would constitute a clear identification of this crossover, as will be presented in Section 13.3.1.2.

13.2.2 Double Quantum Dot

When two single dots are placed in series and separately connected to a source and drain reservoir, a DQD with a network of source–dot–dot–drain is formed. To apply the constant interaction model in such a system [58], a schematic diagram of its equivalent electrical network is shown in Figure 13.12a. In this model, the dots QD1 (QD2) are capacitively coupled to their nearest plunger-gate PG1 (PG2) via a capacitance C_{g1} (C_{g2}); however, they are also coupled to the further gate PG2 (PG1) through the cross capacitance C_{g21} (C_{g12}). The dots themselves also couple to each other through an interdot capacitance C_m and to the source and drain reservoir through C_S and C_D individually. The voltages applied to plunger-gate 1, plunger-gate 2, source, and drain are denoted by V_{PG1}, V_{PG2}, V_S, and V_D, respectively, as shown in Figure 13.12a. The charge and its equivalent voltage on QD1 (QD2) are denoted by $Q_{1(2)}$ and $V_{1(2)}$, also shown in Figure 13.12a. Based on this model, the charge at each dot is given by the vector $\mathbf{Q} = \mathbf{C}\mathbf{V}$, where $\mathbf{C}$ is the capacitance matrix, $\mathbf{Q} = (Q_1, Q_2)$ is the vector of charges, and $\mathbf{V} = (V_1, V_2)$ is the vector of electrostatic potentials. Therefore the components of $\mathbf{Q}$ are given by [58]

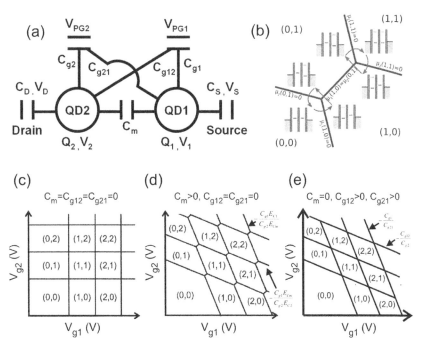

FIGURE 13.12 (a) Electrostatic network model for a DQD considering the cross-capacitance coupling. (b) Charge evolution during conduction at triple points. The outer numbers in brackets give the stable electron number for that region. The insets shown are the energy levels of the stable regions at points close to the triple point. (c)–(e) Charge-stability diagrams for a DQD with (c) no interdot and cross-capacitance coupling ($C_{g21} = C_{g21} = E_{Cm} = 0$), (d) intermediate interdot coupling but no cross-capacitance coupling ($C_{g21} = C_{g12} = 0$, $E_{Cm} \neq 0$), and (e) no interdot coupling but intermediate cross-capacitance coupling ($C_{g21} \neq 0$, $C_{g12} \neq 0$, $E_{Cm} = 0$). ((c,d) Adapted with permission from Ref. [58]. Copyright 2003 American Physical Society.)

$$\begin{pmatrix} Q_1 + C_S V_S + C_{g1} V_{g1} + C_{g21} V_{g2} \\ Q_2 + C_D V_D + C_{g2} V_{g2} + C_{g12} V_{g1} \end{pmatrix}$$
$$= \begin{pmatrix} C_1 & -C_m \\ -C_m & C_2 \end{pmatrix} \begin{pmatrix} V_1 \\ V_2 \end{pmatrix}, \qquad (13.26)$$

where $C_{1(2)} = C_{S(D)} + C_{g1(2)} + C_{g21(12)} + C_m$ is the total capacitance of dot 1(2). Making the substitution $Q_{1(2)} = -N_{1(2)}e$, and taking $V_S = V_D = 0$ [in the low-bias regime and $N_{1(2)}$ is the electron number in dot 1(2)], Eq. (13.26) then reads [58]

$$\begin{pmatrix} V_1 \\ V_2 \end{pmatrix} = \frac{1}{C_1 C_2 - C_m^2} \begin{pmatrix} C_2 & C_m \\ C_m & C_1 \end{pmatrix}$$
$$\times \begin{pmatrix} -N_1 e + C_{g1} V_{g1} + C_{g21} V_{g2} \\ -N_2 e + C_{g2} V_{g2} + C_{g12} V_{g1} \end{pmatrix} \qquad (13.27)$$

So the total electrostatic energy of such a system is given by [7],

$$U(N_1, N_2) = \frac{1}{2} \mathbf{Q} \cdot \mathbf{V}$$
$$= \frac{1}{2} N_1^2 E_{C1} + N_1 N_2 E_{Cm} + \frac{1}{2} N_2^2 E_{C2}$$
$$- \frac{1}{2e} C_{g1} V_{g1} N_1 E_{C1} - \frac{1}{2e} C_{g21} V_{g2} N_1 E_{C1}$$
$$- \frac{1}{2e} C_{g2} V_{g2} N_1 E_{Cm} - \frac{1}{2e} C_{g12} V_{g1} N_1 E_{Cm}$$
$$- \frac{1}{2e} C_{g1} V_{g1} N_2 E_{Cm} - \frac{1}{2e} C_{g21} V_{g2} N_2 E_{Cm}$$
$$- \frac{1}{2e} C_{g2} V_{g2} N_2 E_{C2} - \frac{1}{2e} C_{g12} V_{g1} N_2 E_{C2}$$
$$\qquad (13.28)$$

where the charging energies for the dots E_{C1} and E_{C2} and the coupling energy E_{Cm} are given by

$$E_{C1} = \frac{e^2}{C_1} \frac{1}{1 - \frac{C_m^2}{C_1 C_2}} \qquad (13.29)$$

$$E_{C2} = \frac{e^2}{C_2} \frac{1}{1 - \frac{C_m^2}{C_1 C_2}} \qquad (13.30)$$

$$E_{Cm} = \frac{e^2}{C_m} \frac{1}{\frac{C_1 C_2}{C_m^2} - 1} \qquad (13.31)$$

The electrostatic potentials for the dots are then given by [7],

$$\mu_1(N_1, N_2) = U(N_1, N_2) - U(N_1 - 1, N_2)$$
$$= (N_1 - 1) E_{C1} + N_2 E_{Cm} - \frac{1}{2e} (C_{g1} V_{g1} E_{C1}$$
$$+ C_{g2} V_{g2} E_{Cm} + C_{g21} V_{g2} E_{C1} + C_{g12} V_{g1} E_{Cm})$$
$$\qquad (13.32)$$

$$\mu_2(N_1, N_2) = U(N_1, N_2) - U(N_1, N_2 - 1)$$
$$= (N_2 - 1) E_{C2} + N_1 E_{Cm} - \frac{1}{2e} (C_{g1} V_{g1} E_{Cm}$$
$$+ C_{g2} V_{g2} E_{C2} + C_{g12} V_{g1} E_{C2} + C_{g21} V_{g2} E_{Cm})$$
$$\qquad (13.33)$$

The physical meaning of each term, for example, $(N_1 - 1)E_{C1}$ and $N_2 E_{Cm}$, stand for the Coulomb statistic energy increases on dots 1 and 2 when the N_1-th electron is added to dot 1. The term $C_{g1} V_{g1} E_{C1}$ in Eq. (13.32) is the direct coupling energy between PG1 and QD1, while $C_{g2} V_{g2} E_{Cm}$ is the indirect coupling energy between PG2 and QD1 in which PG2 couples to QD2 first and then QD2 influences QD1 through interdot coupling. The last two terms in Eq. (13.32) show the cross-coupling effect, where $C_{g21} V_{g2} E_{C1}$ is the cross-coupling energy between PG2 and QD1 and $C_{g12} V_{g1} E_{Cm}$ is the indirect cross-coupling energy that PG1 couples to QD2 first and QD2 influences QD1 through interdot coupling. At low temperature ($k_B T < e^2/C$), the electrical transport through DQD is only possible in the case where the energy levels in both dots are aligned with the source–drain bias window, and this gives rise to the charge-stability diagram shown in Figure 13.12b. The outer numbers in brackets (N_1, N_2) are the stable electron numbers residing in the dot for that region, and the condition for electron transport is met whenever three charge states meet at one point (the so-called triple point). The arrows in Figure 13.12b circling each triple point mark the route around the stability diagram that the system takes as electrons shuttle through. The counterclockwise path follows the sequence of charge state $(N_1, N_2) \to (N_1 + 1, N_2) \to (N_1, N_2 + 1) \to (N_1, N_2)$, corresponding to moving an electron to the right. The clockwise path follows the sequence of charge state $(N_1 + 1, N_2 + 1) \to (N_1 + 1, N_2) \to (N_1, N_2 + 1) \to (N_1 + 1, N_2 + 1)$, corresponding to moving a hole to the left. We here try to find a specific slope for $\mu_{1(2)}$ in the V_{g1}-V_{g2} plane, along which $\mu_{1(2)}$ will remain constant for a given (N_1, N_2). We make the second row of Eqs. (13.32) and (13.33) = 0, which gives

$$V_{g1} (C_{g1} E_{C1} + C_{g12} E_{Cm}) = -V_{g2} (C_{g2} E_{Cm} + C_{g21} E_{C1})$$
$$\Rightarrow \frac{V_{g2}}{V_{g1}} = -\left(\frac{C_{g1} E_{C1} + C_{g12} E_{Cm}}{C_{g2} E_{Cm} + C_{g21} E_{C1}} \right), (for\ \mu_1) \quad (13.34)$$

$$V_{g2} (C_{g2} E_{C2} + C_{g21} E_{Cm}) = -V_{g1} (C_{g1} E_{Cm} + C_{g12} E_{C2})$$
$$\Rightarrow \frac{V_{g2}}{V_{g1}} = -\left(\frac{C_{g1} E_{Cm} + C_{g12} E_{C2}}{C_{g2} E_{C2} + C_{g21} E_{Cm}} \right), (for\ \mu_2) \quad (13.35)$$

We discuss the stability diagram for a DQD with three different coupling regimes.

1. **No interdot and cross-capacitance coupling**
   If we do not consider the cross-capacitance and interdot coupling (i.e., $C_{g12} = C_{g21} = E_{Cm} = 0$), so that PG1 only influences QD1 and PG2 only influences QD2, Eq. (13.34) and Eq. (13.35) now read

$$\frac{V_{g2}}{V_{g1}} = -\infty, (for\ \mu_1) \qquad (13.36)$$

$$\frac{V_{g2}}{V_{g1}} = 0, (for\ \mu_2) \qquad (13.37)$$

The resulting stability diagram is shown as Figure 13.12c, where the lines for $\mu_{1(2)}$ to stay constant appear as vertical (horizontal) lines.

2. **Finite interdot but no cross-capacitance coupling**

As the interdot coupling or the cross-capacitance coupling opens, the gate PG1(2) has the ability to influence QD2(1). We first consider the case that interdot coupling is finite but the cross-capacitance coupling is weak, so $C_{g21} = C_{g12} = 0$, $E_{Cm} \neq 0$. In such a case, the only way that PG1(2) influences dot 2(1) is to influence dot 1(2) first and through interdot capacitance to tune the other dot indirectly. So now Eqs. (13.34) and (13.35) read

$$\frac{V_{g2}}{V_{g1}} = -\left(\frac{C_{g1} E_{C1}}{C_{g2} E_{Cm}} \right), \text{(for } \mu_1) \qquad (13.38)$$

$$\frac{V_{g2}}{V_{g1}} = -\left(\frac{C_{g1} E_{Cm}}{C_{g2} E_{C2}} \right), \text{(for } \mu_2) \qquad (13.39)$$

The resulting stability diagram is shown as Figure 13.12d. Instead of appearing as vertical (horizontal) lines, now $\mu_{1(2)}$ has a slope that is determined by the strength of the interdot coupling E_{Cm}. The larger E_{Cm} is, the more $\mu_{1(2)}$ deviates from a vertical(horizontal) line.

3. **No interdot but finite cross-capacitance coupling**

Finally, in the case of no interdot coupling but with cross-capacitance coupling, i.e., $C_{g12} \neq 0$, $C_{g21} \neq 0$, $E_{Cm} = 0$, Eqs. (13.34) and (13.35) read

$$\frac{V_{g2}}{V_{g1}} = -\frac{C_{g1}}{C_{g21}}, \text{(for } \mu_1) \qquad (13.40)$$

$$\frac{V_{g2}}{V_{g1}} = -\frac{C_{g12}}{C_{g2}}, \text{(for } \mu_2) \qquad (13.41)$$

and the resulting stability diagram is shown in Figure 13.12e where the slopes are now determined by the ratio between direct capacitance $C_{g1(2)}$ and cross-capacitance $C_{g21(12)}$.

Usually a double-dot system has a finite interdot and weak cross-capacitance coupling strength, so the charge-stability diagram is made up of hexagonal regions of a fixed charge, as shown in Figure 13.12d (also an enlarged illustration in Figure 13.13a). The dimensions of the hexagonal regions, as indicated in Figure 13.13a, are given by [58]:

$$\Delta V_{g1} = e/C_{g1} \qquad (13.42)$$

$$\Delta V_{g2} = e/C_{g2} \qquad (13.43)$$

$$\Delta V_{g1}^m = \Delta V_{g1} \frac{C_m}{C_2} \qquad (13.44)$$

$$\Delta V_{g2}^m = \Delta V_{g2} \frac{C_m}{C_1} \qquad (13.45)$$

In the high-bias regime, the triple points evolve into bias-dependent triangular regions where the two dot levels lie

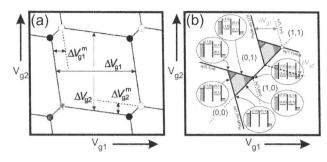

FIGURE 13.13 (a) An enlarged stability diagram of Figure 13.12d, with the dimensions for one hexagonal region. (b) Schematic for transport through the double dot at the triple points in the high-bias regime. The electrochemical potentials at different points of the triangular regions are also indicated. ((a) Adapted with permission from Ref. [58]. Copyright 2003 American Physical Society. (b) Adapted with permission from Ref. [7]. Copyright 2007 American Physical Society.)

within the bias window as shown in Figure 13.13b. The dimensions of the triangles are related with the applied bias via [58]:

$$\alpha_1 \delta V_{g1} = \frac{C_{g1}}{C_1} e\delta V_{g1} = |eV_{\mathrm{SD}}| \qquad (13.46)$$

$$\alpha_2 \delta V_{g2} = \frac{C_{g2}}{C_2} e\delta V_{g2} = |eV_{\mathrm{SD}}| \qquad (13.47)$$

Here $\alpha_{1(2)}$ is the conversion factor between gate voltage and energy, which could be extracted from the dimension of the bias triangle. Therefore, the charging energy of the dots and the interdot coupling energy can be found:

$$E_{C1} = e^2/C_1 = \frac{\alpha_1 e}{C_{g1}} = \alpha_1 \Delta V_{g1} \qquad (13.48)$$

$$E_{C2} = e^2/C_2 = \frac{\alpha_1 e}{C_{g2}} = \alpha_2 \Delta V_{g2} \qquad (13.49)$$

$$E_{Cm} = \alpha_1 \Delta V_{g1}^m = \alpha_2 \Delta V_{g2}^m \qquad (13.50)$$

A phenomenon closely related to the manipulation of the dot levels in DQDs is charge pumping. Charge pumping refers to a quantized number n of electrons transferred from the source to the drain at a driven frequency f, leading to a total current $I \equiv nef$ even if zero bias is applied (e is the elementary charge). Such quantized charge transport was first demonstrated in single-electron turnstile devices in which an external RF signal was applied to linear arrays of tunnel junctions. By doing so, electrons could be clocked through each tunnel junction one at a time by exploiting the Coulomb blockade effect [59,60]. When an RF signal is applied to the plunger-gates (instead of barriers) of a DQD device, it is also possible to generate an accurate and frequency-dependent quantized current through the device. The AC voltages on both plunger-gates with a phase difference between them drives the DQD into different charge states around the triple point. The schematic diagram to illustrate such a pumping mechanism is shown in Figure 13.14a. Assuming that the voltages applied on the plunger-gates are AC sinusoidal waves with a phase difference of 90°, it effectively

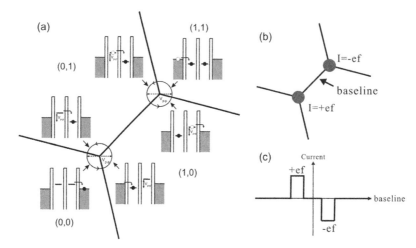

FIGURE 13.14 (a) The pump loop (counterclockwise close trajectory) and its corresponding charge configuration obtained by modulating the gate voltages by two out-of-phase radio-frequency (RF) signals induce one electron to go around the circuit. $V_{p-p}/2$ is the RF amplitude. (b) A schematic for the resulting current map from (a). If the pump is working, one should observe a quantized current $I = \pm ef$ at the location of two nearby triple points. (c) A schematic for the linecut along the baseline in (b). (Reproduced from K. L. Chiu[56].)

forms a circular pump loop in the stability diagram. The radius of the circle is determined by the amplitude of the sinusoidal wave ($V_{p-p}/2$). When the circular route passes through three charge states around the triple point, it corresponds to shuttling a charge carrier from source reservoir to drain reservoir and generating a current. If the AC amplitude is small enough for the pump loop to just enclose a triple point and the frequency is large enough to produce a measurable pumping current, a current $I = ef$ will follow even when zero source–drain voltage is applied. Depending on the type of triple point that the pumping circle encloses, it generates a different direction of current; i.e., positive current for the electron-transport-type triple point and negative current for the hole-transport-type triple point. So if the pumping is successful the current recorded around two nearby triple points will present a circular shape with equal values but with different signs as shown in Figure 13.14b. This effect can be seen in a linecut along the baseline of triple points, where the current appears as two plateaus as shown in Figure 13.14c. The experimentally observed quantized pumped current in graphene double dot will be presented in Section 13.3.2.2.

13.2.3   Andreev Reflections in Ballistic S-N-S JJs

Having discussed the fundamental physics related to transport in QDs, we introduce another important topic in this chapter—the proximity effect in JJs. This section will serve as a basis for understanding the superconducting physics in 2D material-based JJs, as will be discussed in Section 13.5. To start with the proximity effect, we use the following defining line [61]: If a normal metal N is deposited on top of a superconductor S, and if the electrical contact between the two is good, Cooper pairs can leak from S to N. In such a way, the normal metal acquires some superconducting-like

properties at a low temperature. This proximity effect is a well-known phenomenon in superconductivity for over 50 years, and is still attracting enormous interest owing to its rich physics underneath. The key mechanism responsible for the proximity effect, the Andreev reflection, offers phase correlations in a system without interacting electrons at mesoscopic scales, is the main topic to be introduced below. Andreev reflection is a microscopic process that happens in an S-N-S junction, in which single particles in the normal region cannot enter the superconductor and therefore experience a special type of reflection at each S-N interface. This process results in Andreev bound states (ABS), which are capable of carrying superconducting current across the normal region. Thus, one can say Andreev reflection and the proximity effect are intimately connected and not two distinct phenomena. In the following, we adopt the discussions in Ref. [62] to illustrate the process of Andreev reflections in an S-N-S junction (Figure 13.15). Assuming an incident electron with energy E_k and spin σ ($E_k - E_F \leq \Delta$, where E_F is the Fermi energy of the normal metal and Δ is the energy gap of the superconductor) is moving toward the N/S_1 interface. The incoming electron would grab an electron with a spin and momentum that is opposite to its own, thereby forming a Cooper pair that can propagate freely into the superconductor S_1. In order to conserve the momentum, spin, and charge, this process leaves behind an empty electronic state (hole) with the opposite spin $-\sigma$ and wavevector $-k$ as shown in the Figure 13.15 by the dashed light gray arrow. The bounced hole follows the time-reversed trace of the incoming electron and eventually hit the N/S_2 interface, where another Andreev reflection takes place. The hole will pass through the N/S_2 interface with another hole excitation that pairs with it (the Cooper pair of hole-like). A Cooper pair of electrons in superconductor S_2 is thus annihilated, resulting in an electron with momentum k and spin σ (identical to the electron we started with),

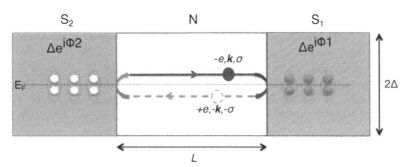

FIGURE 13.15 An S-N-S junction and Andreev roundtrips responsible for transporting Cooper pairs across a ballistic normal metal weaklink. (Reproduced with permission from Ref. [62].)

ejected into the normal metal and completes a roundtrip. In completing each roundtrip, one Cooper pair in the left superconductor is annihilated, while another one in the right superconductor is created, leading to a transfer of Cooper pair from left to right. This microscopic process gives rise to the transport of supercurrent across the S-N-S junction, in which current flows without dissipation from one superconductor to another by passing through a normal metal that is not inherently superconducting. Andreev reflections in an S-N-S junction can lead to a series of bound states, known as ABS, whose energy strongly depends on the length of the junction and the phase acquired in each roundtrip. To understand the mechanism of ABS, let us consider the phase an electron would acquire through the Andreev reflection that converts the incoming electron into a hole at the N/S_1 interface:

$$\phi_{e \to h} = \phi_1 + \arccos\left(\frac{E}{\Delta}\right), \qquad (13.51)$$

where ϕ_1 denotes the phase of the superconductor S_1 and E is the excitation energy of electron measured with respect to E_F ($E = E_k - E_F$). The first phase term arises from the requirement that particles absorbed by the superconductor must be in phase with the macroscopic wave function that describes the condensate. The second phase term $arccos(\frac{E}{\Delta})$ comes from the reflection probability amplitude, which depends on the relative strength of the excitation energy E and the barrier Δ [63]. Similarly, the phase a hole acquired through the Andreev reflection at the N/S_2 interface can be written as

$$\phi_{h \to e} = -\phi_2 + \arccos\left(\frac{E}{\Delta}\right) \qquad (13.52)$$

In addition to the phase associated with Andreev reflections, we also need to consider the dynamic phase $\phi_{e(h)} = \int k_{e(h)} \cdot d\vec{l}$ that electron (hole) acquired when traveling between the superconductors. In a 1D ballistic case (no scattering), the phase accumulated by an electron traveling from S_2 to S_1 can be simply written as

$$\phi_e = k_e \cdot L = \sqrt{\frac{2m}{\hbar^2}(E + E_F)} \cdot L, \qquad (13.53)$$

where L is the length of the normal metal region. Similarly, the dynamic phase a hole acquired when traveling from S_1 to S_2 is

$$\phi_h = k_h \cdot (-L) = \sqrt{\frac{2m}{\hbar^2}(E_F - E)} \cdot (-L) \qquad (13.54)$$

As $E \ll E_F$, the total dynamic phase can be approximated as

$$\phi_e + \phi_h = k_F L \frac{E}{E_F} \qquad (13.55)$$

Therefore, the ABS energy is determined by requiring the total phase accumulated in the Andreev process (roundtrip) to be multiples of 2π (i.e., $\phi_{e \to h} + \phi_{h \to e} + \phi_e + \phi_h = 2\pi n$), which can be written as follows [64]:

$$2 \arccos\left(\frac{E}{\Delta}\right) \pm (\phi_1 - \phi_2) + k_F L \frac{E}{E_F} = 2\pi n, \qquad (13.56)$$

where n is an integer and $\pm$ accounts for the two possible directions the roundtrip can take. Note that in the above expressions, we have limited our discussion to ballistic normal metal, meaning that no scattering event (which usually introduces extra phases) takes place within the roundtrip. As can be seen now, the ABS energy depends on the phase difference between the two superconductors and the junction length. The term $k_F L \frac{E}{E_F}$ provides a criteria to estimate the short- and long-junction limits. The junction is "short" if this term is negligible, so that the phase depends almost completely on the Andreev reflection and is insensitive to the dynamic phase associated with geometry. The short-junction limit corresponds to a condition $L \ll \xi$, where $\xi = \sqrt{\hbar D/\Delta}$ is the superconducting coherence length, with $D = \nu_F l_e/2$ the Einstein diffusion coefficient, ν_F the Fermi velocity, and l_e the mean free path in normal metal [65]. In the short-junction limit, the bound state energy can be solved from Eq. (13.56) as [66]

$$E_{L \ll \xi}^{\pm}(\Delta\phi) = \pm\Delta\cos\left(\frac{\Delta\phi}{2}\right), \quad -\pi \leq \Delta\phi \leq \pi \qquad (13.57)$$

where $\Delta\phi = \phi_1 - \phi_2$ is the phase difference between two superconductors. Note that only a pair of ABS exists within the superconducting gap Δ. In the opposite limit, where L is much larger than the phase coherencelength ξ, the phase acquired in each roundtrip is dominated by the dynamic phase associated with L and the ABS energy in this long-junction limit can be written as [66]

$$E_{L \gg \xi}^{\pm}(\Delta\phi) = \frac{\hbar\nu_F}{L}\left[\left(n - \frac{1}{2}\right)\pi \mp \frac{\Delta\phi}{2}\right], \quad -\pi \leq \Delta\phi \leq \pi$$

$$(13.58)$$

Due to the weaker quantum confinement in a long junction when compared with that in a short junction, the energy spacing between bound states decreases with L, and thus multiple bound states (denoted by the index n) can be accommodated within the gap. Here, we emphasize again that the above formulas are based on the assumption of a ballistic (scattering-free) normal metal in an S-N-S junction. However, the general results presented here (especially the oscillation behavior with phase $\Delta\phi$) still qualitatively cover the experimental results presented in this chapter. For ABS considering the scattering from defects, more information can be found in Ref. [67]. The phase-dependent ABS energy in Eqs. (13.57) and (13.58) provides a way to study the ABS spectrum in devices capable of varying magnetic flux, as will be discussed in Section 13.5.

In summary, we have introduced the relevant physics useful to understand the transport properties in QDs and JJs. In the subsequent sections, we will review a series of experimental studies relevant to developing qubits in 2D materials. Following the development of spin qubits, we will discuss single-electron transport properties of various graphene nanostructures in Section 13.3, while the same properties of 2H-TMDs nanostructures will be reviewed in Section 13.4. For potential use in superconducting qubits, we investigate the Josephson effects of 2D material-based S-N-S junctions in Section 13.5. In Section 13.6, we provide recent studies on QSH edge states in 1T′-TMDs and discuss their potential applications in probing Majorana zero modes.

13.3 Single-Electron Transport in Graphene

In this section, we will review the early development of graphene nanostructures fabricated on SiO_2/Si substrates. After briefly introducing graphene nanoribbons and their function as tunnel barriers, we will focus mainly on GQDs and their transport properties.

Although graphene is a superb conductor thatoffers advantages in terms of sensing and analog electronics, its gapless bandstructure hinders its use in logic circuit

applications. Owing to the absence of a bandgap, the current in graphene cannot be completely turned off, leading to low on/off ratios that are insufficient for switches [19]. Engineering bandgaps in graphene is thus a major challenge that must be addressed to enable the use of graphene-based transistors in digital electronics. First-principle calculations predict that cutting graphene into 1D nanoribbons can open up a scalable bandgap $E_g = \alpha/w$, where w is the nanoribbon width and α is in the range of 0.2–1.5 eV·nm, depending on the model and the crystallographic orientation of the edges [12,68]. Similar results are also obtained from tight-binding calculations [69,70]. A graphene nanoribbon (GNR) can have two possible types of edge terminations, namely armchair and zigzag edges, as shown in Figure 13.16a. These two edge types correspond to different boundary conditions from which the energy band dispersion can be found. The tight-binding calculated energy band structures for armchair GNRs (of two different ribbon widths) and zigzag GNRs are shown in Figure 13.16b–d, where N denotes the number of dimer (carbon-site pair) lines (for the armchair ribbons) or the number of zigzag lines (for the zigzag ribbons). The band dispersion for an armchair nanoribbon with $N = 3m - 2$ dimers exhibits a bandgap (semiconducting), whereas for an armchair nanoribbon with $N = 3m - 1$ dimers, the dispersion is metallic (m is an integer). For semiconducting ribbons, the direct gap decreases with increasing ribbon width and tends toward zero in the limit of very large N. Zigzag nanoribbons always exhibit a metallic behavior (Figure 13.16d) regardless of how the width (N) is varied. The predicted existence of bandgaps in GNRs has motivated an experimental effort to establish whether nanostructuring graphene is a feasible route for preparing graphene-based switches [71–78]. GNRs can be fabricated by means of O_2 plasma etching using physical masks [71–75], unzipping carbon nanotubes [76–78], gas-phase etching [79], or functionalization [80,81]. Such devices have been tested for their transport properties at various temperatures, and the general results will be discussed below.

Figure 13.17a shows the conductance of an O_2 plasma-etched GNR (Figure 13.17a) as a function of the voltage applied to the back-gate. This back-gate

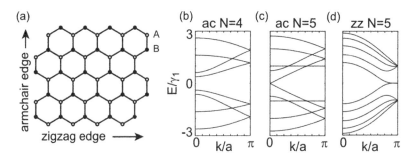

FIGURE 13.16 (a) Lattice of a zigzag (armchair) graphene nanoribbon by extension in the x-(y-) direction. (b–d) Tight-binding calculations of the nanoribbon subbands for (b) an $N = 4$ semiconducting armchair (ac) nanoribbon, (c) an $N = 5$ metallic armchair (ac) nanoribbon, and (d) an $N = 5$ metallic zigzag (zz) nanoribbon. N denotes the number of the dimer (carbon-site pair) lines for the armchair ribbon and the number of zigzag lines for the zigzag ribbon, respectively. (Adapted with permission from Ref. [14]. Copyright 2012 IOP Publishing Ltd.)

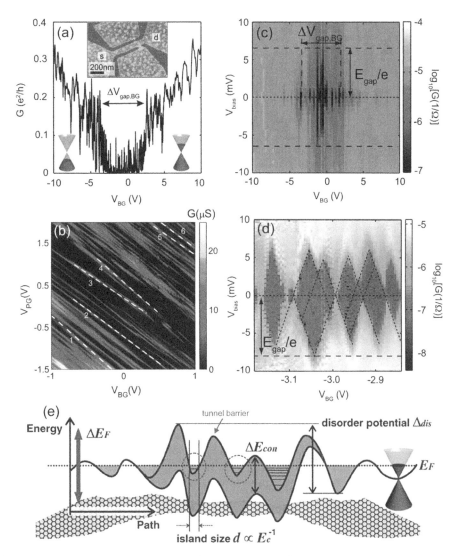

FIGURE 13.17 (a) Conductance through a nanoribbon (shown in the inset) as a function of the back-gate voltage V_{BG}, recorded under an applied bias voltage of $V_{bias} = 300$ μV and at a temperature of $T = 2$ K. Inset: Atomic force microscopy (AFM) image of a graphene nanoribbon ($w = 85$ nm, $l = 500$ nm) etched using O_2 plasma. (b) Conductance as a function of V_{BG} and the plunger-gate voltage V_{PG} of another GNR device, showing a variation in the slopes of the Coulomb resonances (indicated by dashed lines and numbers). (c) Logarithmic conductance as a function of V_{bias} and V_{BG} at $T = 2$ K for the device shown in (a), with indication of the extent of the transport gap $\Delta V_{gap,BG}$ in the back-gate direction and bias-gap E_{gap} in the bias direction. (d) Zoom in of the region of suppressed conductance depicted in (c). (e) A schematic illustration of the formation of localized states induced by disorders. Δ_{dis} characterizes the strength of the charge-neutrality point fluctuation, and ΔE_{con} is a confinement gap induced by local constriction. ΔE_F denotes the Fermi energy spacing that the transport gap has to overcome. ((a,c,d) Adapted with permission from Ref. [73]. (b) Reproduced from K. L. Chiu [56]. (e) Adapted with permission from Ref. [82]. Copyright 2009 American Physical Society.)

sweep shows a typical V-shape, with a region around 0 V separating the hole- from electron-transport regime, where the conductance is strongly suppressed. In contrast to the prediction of energy gaps in clean GNRs (i.e., without considering bulk disorder and edge roughness), where transport should be completely pinched-off, this gap exhibits a large number of conductance peaks reminiscent of Coulomb blockade resonances in QDs. The nature of these resonances can be interrogated by varying the potential of GNR. Figure 13.17b shows the conductance as a function of both back-gate and plunger-gate (an in-plane

gate close to the GNR) voltages within the transport gap. The conductance resonances exhibiting a range of relative lever arms indicated by dashed lines are present over a wide range of V_{BG} and V_{PG} voltages. One explanation for this behavior draws on its similarity to a series of charge islands (or QDs), each coupled to the plunger-gate through different capacitive coupling strength, assuming the lever arm of the back-gate to the charge islands is nearly constant all over the GNR. More information about such localized states in the GNR can be gleaned by the Coulomb diamond measurements (see Section 13.2.1), in which the differential conductance as a function of back-gate voltage and source-drain bias is recorded, as shown in Figure 13.17c. Within this picture, the extent in bias voltage of the diamond-shaped regions of suppressed current (see E_{gap}/e in Figure 13.17c and its zoomin in Figure 13.17d) is a direct indication of the charging energy of the dots (see Section 13.2.1), which fluctuates strongly with V_{BG} and extends to ≈ 8.5 meV. The overlapping diamonds in Figure 13.17d resembles the behavior of a QD network [83], supporting the notion that multiple QDs form along the GNR. In addition, the gap in Fermi energy ΔE_F corresponding to the transport gap $\Delta V_{\text{gap},BG}$ can be estimated using $\Delta E_F \approx \hbar\nu_F\sqrt{2\pi C_g \Delta V_{\text{gap},BG}/|e|}$, where C_g is the back-gate capacitance per area and ν_F is the Fermi velocity in graphene [14,82]. This leads to an energy gap $\Delta E_F \approx 110 - 340$ meV, which is significantly larger than the observed E_{gap} (8.5 meV) and the bandgaps ΔE_{con} (≤ 50 meV) estimated from calculations of a GNR with width $W = 45$ nm [14].

A schematic model shown in Figure 13.17e is able to qualitatively explain the findings described above [82]. This model consists of a combination of quantum confinement energy gap ΔE_{con} (the intrinsic bandgap of a clean GNR) and strong bulk and edge-induced disorder potential fluctuation Δ_{dis}. The confinement energy ΔE_{con} alone can neither explain the observed energy scale ΔE_F nor the dots formation in the GNR. However, superimposing a fluctuation in the disorder potential (Δ_{dis}) can result in tunnel barriers separating different localized states (i.e., puddles or QDs), as shown in in Figure 13.17e. Therefore, transport in such a system is described by a percolation between the puddles (in Figure 13.17e, the dashed circles indicate the puddles, whereas the gray arrow indicates the tunnel barrier). Within this model, ΔE_F depends on both the confinement energy gap and the disorder potential fluctuation, and can be approximated using the relation $\Delta E_F = \Delta_{\text{dis}} + \Delta E_{\text{con}}$. Δ_{dis} can be estimated from the bulk carrier density fluctuations Δn (due to substrate disorder) using $\Delta_{\text{dis}} = \hbar\nu_F\sqrt{4\pi\Delta n}$, where $\Delta n \approx \pm 2 \times 10^{11}$ is extracted from Ref. [84]. This in turns gives $\Delta E_F = \hbar\nu_F\sqrt{4\pi\Delta n} + \Delta E_{\text{con}} \approx 126$ meV [14], which is comparable to the experimental value (110 – 340 meV). The energy gap in the bias direction (E_{gap}) is not directly related with the magnitude of the disorder potential but rather with its spatial variation. When the Fermi energy (or said V_{BG}) lies in the center of the transport gap, the smaller localized states are more likely to form, giving rise to the larger charging energies (larger Coulomb diamonds). By contrast, when the Fermi energy is tuned away from the charge-neutrality point, the size of the relevant diamonds generally gets smaller due to the merging of individual puddles.

Although the localized states in GNRs pose additional complications, their tunability in resistances still allows them to be used as tunnel barriers for transport in GQDs. While a large number of studies on GNRs have been reported in the field, in this section, we will focus primarily on GQDs in which GNRs are used as tunnel barriers. Further

discussion of the transport properties of GNRs can be found in Ref. [85].

13.3.1 Graphene SQDs on SiO$_2$/Si Substrates

Owing to the expected long spin relaxation time, GQDs are considered to be a viable candidate for preparing spin qubits and spintronic devices [86]. Over the past decade, GQDs have proven to be a useful platform for confining and manipulating single electrons [2,3,56,87–91]. In this section, we will review a few relevant transport experiments performed on graphene single quantum dots (GSQDs) fabricated on SiO$_2$/Si substrates. These include the Coulomb blockade at zero field, Fock–Darwin spectrum, spin states, and charge relaxation dynamics, as will be discussed below.

13.3.2 Coulomb Blockade at Zero Field

GQDs can be formed by etching isolated islands connected to source and drain graphene reservoirs via nanoconstrictions that are resistive enough to act as tunnel barriers [2,3,87]. An example of such a device is shown in Figure 13.18a, in which in-plane graphene side and plunger-gates (SG1, SG2, and PG) are used to locally tune the potential of the tunnel barriers and the 50 nm diameter dot, while the doped-silicon back-gate (BG) is used to adjust the overall Fermi level. Another way to define a GQD is to induce a bandgap in bilayer graphene by applying an electric field perpendicular to the layers; in this way, charges are confined in an island defined by top-gate geometry [92,93]. Such a structure can be seen in Figure 13.18b, where a bilayer graphene is suspended between two Cr/Au electrodes and sits below suspended local top gates that are used to break interlayer symmetry. Graphene QDs can also be formed from the disorder potential [94,95], strain engineering [96], and gated GNRs [89], in all of which Coulomb blockade can be observed.

Fig. 13.18(c) shows the back-gate sweep (conductance as a function of back-gate voltage) of the device shown in Figure 13.18a. The measurement shows a transport gap ranging from $0 \leq V_{\text{BG}} \leq 10$ V, in which current is suppressed except for multiple sharp Coulomb resonances, separating hole- from electron-transport regime. The transport gap resulting from the GNR tunnel barriers can be lifted using the side-gate voltage. Figure 13.19a shows the current measurements of another GQD (diameter ≈ 180 nm) as a function of its side-gate voltages V_{SG1} and V_{SG2} at a fixed back-gate voltage within the transport gap. There is a cross-like region of suppressed current separating four large conductance regions, which correspond to different doping configurations of the constrictions, labeled as NN, NP, PP, and PN at the corners of the diagram, respectively. For example, keeping $V_{SG1} = -20$ V constant and sweeping V_{SG2} from -20 V to $+20$ V keeps constriction 1 in the p-doped regime, whereas constriction 2 is tuned from p-doped to n-doped (PP to PN transition). In order to observe single electron transport, it is necessary to operate in a region of gate space

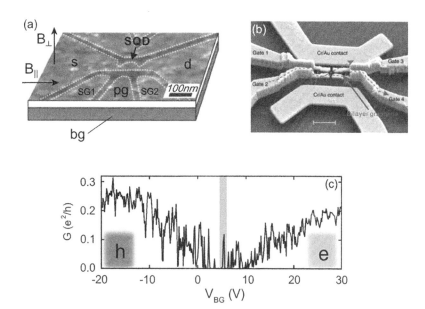

FIGURE 13.18 (a) AFM image of a graphene SQD (≈ 50 nm wide and ≈ 80 nm long) etched by O_2 plasma. (b) Scanning Electron Micrograph (SEM) of a suspended bilayer GQD device. Bilayer graphene (indicated by the arrow) is suspended between two electrodes below local top gates. Scale bar, 1 μm. (c) Source–drain conductance as a function of back-gate voltage V_{BG} at bias $V_b = 4$ mV measured from the device shown in (a). ((a,c) Adapted with permission from Ref. [3]. Copyright 2009 American Physical Society. (b) Adapted with permission from Ref. [92]. Copyright 2012 Nature Publishing Group.)

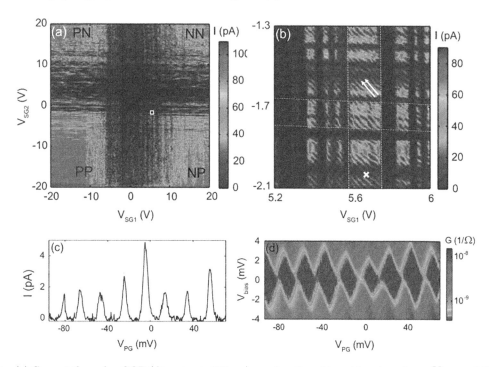

FIGURE 13.19 (a) Current through a GQD (diameter ≈ 180 nm) as a function of two side-gate voltages V_{SG1} and V_{SG2}. (b) Current as a function of V_{SG1} and V_{SG2} in the range indicated by the white square in (a). (c) Current as a function of plunger-gate voltage V_{PG} at $V_{SG1} = 5.67$ V and $V_{SG2} = -2.033$ V [the white cross in (b)]. (d) Coulomb diamonds associated with the Coulomb resonances in (c). (Adapted with permission from Ref. [13].)

where both tunnel barriers are resistive (i.e., within the center of the cross-like current suppressed regime). Figure 13.19b shows the case with the Fermi energy located at the edge of the transport gap for both constrictions (marked by the white square in Figure 13.19a). The measurement shows

broaden vertical and horizontal resonances (white and gray dashed lines in Figure 13.19b), which correspond to resonant transmission through the localized states in the left and right constrictions, tuned with the respective side-gate. The fact that those lines are almost perfectly vertical and horizontal indicates that the side-gate only influences its adjacent constriction. A closer inspection of Figure 13.19b shows a series of diagonal lines (indicated by arrows), which correspond to the Coulomb blockade resonances from the central QD, where both side-gates are expected to have a similar lever arm. These 0D Coulomb resonances can be unambiguously resolved as a series of well-defined and regular peaks, as shown in Figure 13.19c, by sweeping a plunger-gate voltage V_{PG} with side-gates fixed at $V_{SG1} = 5.67$ V and $V_{SG2} = 2.03$ V (the white cross in Figure 13.19b). A Coulomb diamond measurement of these resonances further confirms their origin. A charging energy $E_C \approx 3.2$ meV is extracted from the vertical extent of the Coulomb diamonds shown in Figure 13.19d, in reasonable agreement with the dot diameter if the disc plate capacitance model $E_C = e^2/8\epsilon\epsilon_0 r$, where r is the radius of the QD, is used [13]. In the following sections, we discuss how these Coulomb blockade peaks evolve with the applied perpendicular and in-plane magnetic fields.

13.3.3 Electron–Hole Crossover in a Perpendicular Magnetic Field

In Section 13.2.1, we have shown the calculated Fock–Darwin spectrum of a GQD. Here, we consider a more practical case where a charging energy is included in the spectrum. Figure 13.20a shows a tight-binding simulated Fock–Darwin spectrum of a 50×80 nm GQD, where a constant charging energy $E_C = 18$ meV has been added to each single-particle level spacing (≈ 4 meV in

average). Several key features seen from the spectrum are summarized in the following. At low B-field, the 0D levels fluctuate but stay at roughly the same energy, as can be seen in the regime I of Figure 13.20a. This fluctuation of the Coulomb blockade resonances at low B is due to the continuous crossing of different unfilled states at low energy, as seen in Figure 13.11b (light gray dashed line highlighted regimes). This situation changes when the second lowest LL (LL_1) is full, at which point the levels show a kink (regime II) indicating that the electrons (or holes) start to condense into the lowest LL (i.e., LL_0 at energy E_0), and the B-field onset of this kink increases with increasing number of particles in the QD. Beyond this B-field, the levels tend to move towards the charge-neutrality point (regime III), meaning the hole levels move to higher energies while the electron levels move to lower energies. At large enough B-field, eventually, the levels stop moving and stay roughly at the same energy again (regime IV), indicating the full condensation of electrons/holes into the lowest LL. The Fock–Darwin spectrum of the GQD in Figure 13.18a has been studied experimentally by tracking the position of Coulomb peaks under the influence of perpendicular magnetic fields, as shown in Figure 13.20b. Comparing the numerical simulation and the experimental data (Fig. 13.20a,b), one can find the same qualitative trend of states running toward the center (E_0). The arrows in Figure 13.20b indicate the kinks beyond which all the levels start to fall into the lowest LL. These kinks in the magnetic-field dependence of Coulomb resonances can be used to identify the few-carrier regime in GQDs. The opposite energy shift for electrons and holes in the Fock–Darwin spectrum also provides a method to estimate the charge neutrality point in GQDs [56], but the precise first electron-to-hole transition is difficult to identify. This can be attributed to the formation of localized states near the Dirac point, which exhibit a weak magnetic-field dependence that alters the spectrum. It is also worth noting that the parasitic magnetic resonances in the tunnel barrier GNRs can also alter the magneto-transport in the GQD [56], which complicates a direct comparison with the simulated Fock–Darwin spectrum.

13.3.4 Spin States in In-Plane Magnetic Field

Perpendicular magnetic fields strongly affect the component of the electron wavefunctions in QD, resulting in the Fock–Darwin spectrum. In-plane magnetic fields, on the other hand, leave the orbital component unaffected, making it possible to explore Zeeman splitting of QD states [88,97,98]. It is critical to perfectly align the sample plane to the magnetic field to reduce the perpendicular components, which can be technically difficult. However, this problem can be minimized if one can analyze spin pairs, i.e., two subsequently filled electrons occupying the same orbital state with opposite spin orientation. In this case, the orbital contributions can be significantly reduced by subtracting the positions of individual peaks sharing the same orbital shift in

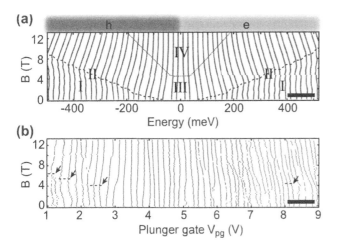

FIGURE 13.20 (a) The Fock–Darwin spectrum of a 50×80 nm GQD calculated by assuming a constant charging energy and spin degenerate states. The dashed line (regime II) indicates a filling factor $\nu=2$ above which all eigenstates continuously evolve into the zero-energy LL. The captions I–IV denote different regimes mentioned in the content. (b) Coulomb peak position as a function of perpendicular magnetic field, measured from the device shown in Figure 13.18a. The arrows indicate the filling factor $\nu=2$ kinks. (Adapted with permission from Ref. [3]. Copyright 2009 American Physical Society.)

a perpendicular magnetic field. Potential spin pairs can be identified by tracking the evolution of two subsequent Coulomb peaks with increasing perpendicular magnetic field, as shown in Figure 13.21a. For example, the lowest two peaks (B1 and B2) and the following two (B3 and B4) are identified as potential spin pairs due to their similar peak evolution. Figure 13.21b shows a measurement of the same peaks in Figure 13.21a, but with increasing in-plane magnetic fields after the sample is carefully rotated into an orientation parallel to the applied B-field. The peaks show a small energy shift with in-plane B-field, indicating that the orbital effect is negligible. In order to analyze the movement of the peaks in detail, Figure 13.21c shows the fit of the data selected from Figure 13.21a and (b), in which two adjacent peaks (a spin pair) are plotted with suitable offsets in V_{pg} such that pairs coincide at $B = 0$ T. As can be seen from the left panel of Figure 13.21c, the orbital states of each pair have approximately the same $B_\perp$ dependence, hence, spurious orbital contributions (from slight misalignment) to the peak spacing in $B_\parallel$ are limited, resulting in a resolvable Zeeman splitting (the right panel of Figure 13.21c). The energy scale of the Zeeman splitting for the spin pairs in Figure 13.21c and for two additional peak spacings (A3–A4 and A5–A4, not shown in Figure 13.21c), are plotted in

Figure 13.21d. The spin differences between three successive spin GSs take the integer values $\Delta^{(2)} = 0, \pm 1, ...$[e.g., for two successive states, the spin difference can be 1/2 $(-1/2)$ for adding a spin-up (spin-down) electron or 3/2 $(-3/2)$ for adding a spin-up (spin-down) electron while flipping another spin from down (up) to up (down)]. Therefore, apart from the slight deviation of B2–B1, all spin pairs in Figure 13.21d follow the relation $\Delta E^Z = \Delta^{(2)} g \mu_B B$ and a g-factor value of approximately 2 can be extracted. The study of Zeeman splitting on spin pairs enables the extraction of the spin-filling sequence in a GQD, which follows an order of $\downarrow\uparrow\uparrow\downarrow\downarrow\uparrow\uparrow\downarrow$ (data not shown) [88]. It is deviated from a sequence of $\uparrow\downarrow\uparrow\downarrow$ observed in the low carrier regime of carbon nanotube QDs [99,100]. This phenomenon has been attributed to the exchange interaction between the charge carriers in graphene, which is comparable to the single-particle energy spacing in GQDs and can therefore lead to a ground-state spin polarization [88]. The spin states in GQDs can in principle be considered as a candidate of spin qubits. However, the spin-related transport in graphene has shown to suffer from extrinsic perturbations [101–103]. We will address this issue again in Section 13.3.3, where transport properties of GQDs on less disordered substrate will be discussed.

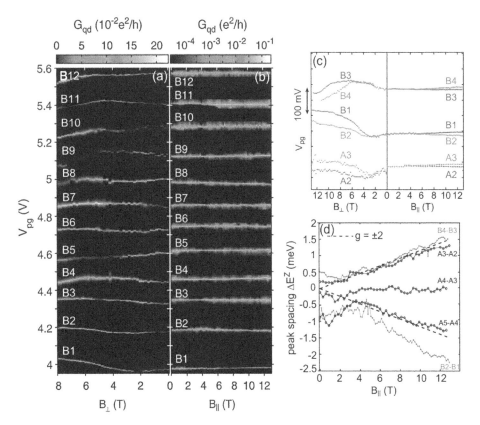

FIGURE 13.21 (a) Coulomb peaks as a function of perpendicular magnetic field recorded at V_b=100 μV, measured from the device shown in Figure 13.18a. (b) The same Coulomb peaks in (a) but measured in parallel (in-plane) magnetic field. (c) Comparing the evolution of three peak pairs in perpendicular (left) and parallel (right) magnetic field. The peak positions are extracted by fitting the data in (a) and (b), and are offset in V_{PG} voltage such that the pairs coincide at $B = 0$ T. (d) Peak spacing as a function of in-plane magnetic field for the three pairs in (c). The dashed lines represent the Zeeman splitting $\Delta E^Z = \pm |g| \mu_B B$ for a g-factor $|g|$=2. (Adapted with permission from Ref. [88]. Copyright 2010 American Physical Society.)

13.3.5   Charge Relaxation Time

Pulsed gating, in which a RF voltage is applied to the gates, is a powerful tool to manipulate electron spin and to study the spin relaxation time in 2DEG QD systems [7]. In this section, we will describe how pulse gating can be used to investigate the charge relaxation dynamics of ESs in GSQDs [87,104]. In these measurements, a rectangular pulse V_{pp} with a duration T (Figure 13.22a) is applied on top of a DC voltage (V_{PG}) to the plunger-gate located in the vicinity of the GQD. If the frequency of the pulse is low ($2/T \leq \Gamma_R$, Γ_L, where $\Gamma_{R(L)}$ is the tunneling rates of the right (left) barrier), the square-wave modulation of the gate voltage results simply in the splitting of the Coulomb resonance into two peaks. Figure 13.22a shows such a behavior when pulses with increasing amplitude (from bottom to top) are applied to the plunger-gate. These peaks (labeled A and B in Figure 13.22a) result from the QD GS entering the biaswindow at two different values of V_{PG}, one for the lower pulselevel (A) and one for the upper one (B). This situation changes dramatically at higher frequencies ($2/T \geq \Gamma_R$, Γ_L), as shown in the bottom panel of Figure 13.22b, where the splitting is broadened due to the reduced electron tunneling probability set by T (black dashed line), and a number of additional peaks appear due to transient transport through the ESs of GQD (light gray dashed line). Each of these additional resonances corresponds to a situation in which the QD levels are pushed well outside the bias window in the first half of the pulse (Figure 13.22b, top left panel),

and then brought into a position where transport can occur only through the ESs in the second one (Figure 13.22b, top middle panel). When the ES lies within the bias window, an electron occupying the GS, either because of tunneling from the leads or relaxation from the ES, will block the current. Therefore, the additional resonances can be resolved in the DC-current measurements only if the frequency of the pulse is higher than the characteristic rate γ of the blocking processes. As both tunneling and ES relaxation lead to the occupation of the GS, γ is approximately given by $\gamma \approx \Gamma + 1/\tau$, where Γ is the tunneling rate from lead to dot and τ is the intrinsic relaxation time of the ES. Since the lowest frequency at which signatures of transport through ESs emerge provides an upper bound for γ, Γ can be determined by fitting thepeak current through the dot. This in turn gives a lower bound $\tau \geq 78$ ns for the charge relaxation time of the GQD ESs [87].

The ES relaxation timescale is related to the lifetime of charge excitations, which is limited by electron–phonon interactions. The main potential source that induces the charge relaxation in supported graphene is through coupling to the longitudinal-acoustic (LA) phonon *via* deformation potential (due to an area change of the unit cell) [87,105,106]. The fact that the observed timescale is a factor 5–10 larger than what has been reported in III-V QDs [107–109] indicates that the electron–phonon interaction in sp^2-bound carbons is relatively weak, which is likely due to the absence of piezoelectric phonons in graphene [87].

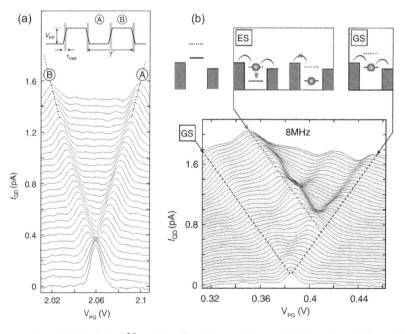

FIGURE 13.22   (a) Current through the dot at V_{SD}=1.5 mV while applying a 100-kHz pulse. Different lines (from bottom to top) correspond to V_{pp} being varied from 0 to 1.4 V in steps of 50 mV. Inset: Sketch of the pulse scheme employed in the measurements presented in this figure. Low and high pulse levels are labeled A and B, and T is the period of the pulse. (b) Top panel: A schematic of transport via GS, ES, and, on the left, of a possible initialization stage. Bottom panel: Measurement similar to the ones shown in (a), but with a higher frequency of 8 MHz. V_{pp} is varied from 0 to 2 V in steps of 25 mV (from bottom to top). (Adapted with permission from Ref. [87]. Copyright 2013 Nature Publishing Group.)

13.3.6 Graphene Double Quantum Dots on SiO$_2$/Si Substrates

Graphene DQDs (GDQDs) are formed when two graphene islands are located close enough such that they are capacitively coupled to each other and individually coupled to the adjacent gates. DQDs in a wide range of semiconductors are a model system for investigating the spin dynamics of electrons [7,58,110–112]. For example, spin-to-charge conversion using the Pauli spin blockade phenomenon and measurements of spin decoherence time were pioneered in GaAs and later realized in carbon nanotube and silicon DQDs [7,8,113–117]. Graphene has been predicted to be particularly suitable for preparing spin-based qubits because of its weak spin–orbit interaction and hyperfine effect [86], which should lead to a long spin decoherence time (T_2^*). However, although the energy levels in GQD have shown the ability to distinguish spin (see Section 13.3.1.3), spin-related transport phenomena such as Kondoeffect [118] and spin blockade have thus far not been observed [89,90,119–124]. Although attempts to probe the spin dynamics in such a system have failed, the control of confined charges in GDQDs can still be achieved. These include gate-tunable interdot coupling [89,90,122,125] and charge pumping [91], which are discussed in the following sections.

13.3.7 Coulomb Blockade and Magnetotransport

GDQDs can be fabricated lithographically by O$_2$ plasma etching out of a graphene flake or by defining the potential landscape using top gates on an etched GNR [89,90,119–122]. Figure 13.23a shows an AFM image of an etched GDQD device on SiO$_2$/Si substrate. Two plunger-gates $V_{GR(GL)}$ are used to tune the energy levels in QD$_{R(L)}$, while three side gates ($V_{CL,GC,CR}$) are used to tune the tunnel barriers. Figure 13.23b shows the current through the device as a function of V_{GR} and V_{GL} at V_b= 500 μV, in which a honeycomb-like charge stability pattern typical for a DQD device can be seen. In this low-bias regime, transmission is only possible within small areas (known as triple points) in the stability diagram, where the levels of two dots are aligned with a small bias window. When the applied bias is large, the current flow is possible over a wider range in gate space, resulting in current measured in the bias-dependent triangle-shaped regions (known as bias triangles), as shown in Figure 13.23c. The dimensions of bias triangle allow the determination of the conversion factors between gate voltage and energy. The charging energies for the left dot E_C^L=$\alpha_L \cdot \Delta V_{GL}$=13.2 meV and for the right dot E_C^R=$\alpha_R \cdot \Delta V_{GR}$=13.6 meV are obtained using the voltage-energy conversion factor $\alpha_{L(R)}$=$eV_b/\delta V_{GL(GR)}$, which can be extracted from the bias triangles shown in Figure 13.23c (see Section 13.2.2). The interdot coupling energy can also be determined from the splitting of the triangles (Figure 13.23c): E_C^m=$\alpha_L \cdot \Delta V_{GL}^m$=$\alpha_R \cdot \Delta V_{GR}^m$=2.2 meV (see Section 13.2.2). It is possible to modulate the interdot coupling strength by changing the voltage applied to the central gate, i.e., V_{GC}. The inset in Figure 13.23d shows examples of two charge stability diagrams recorded with exactly the same parameters, except for the voltage applied to the central

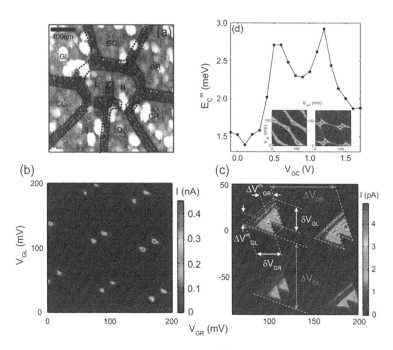

FIGURE 13.23 (a) AFM image of a GDQD etched by O$_2$ plasma. (b) Current through the GDQD in (a) as a function of V_{GR} and V_{GL} measured at a low bias voltage $V_b = 500$ μV showing the triple points. (c) The same as (b) but at a higher bias voltage $V_b = 6$ mV shows the bias triangles. (d) Mutual capacitive coupling between the two dots as a function of central plunger-gate V_{GC}. All the data points correspond to the same triple point. Inset: Current as a function of V_{GR} and V_{GL} for two different central plunger-gate voltages $V_{GC} = 1.9$ V (left) and $V_{GC} = 0$ V (right). (Adapted with permission from Ref. [120]. Copyright 2010 European Physical Society.)

plunger-gate. This is also shown in Figure 13.23d, where the interdot coupling energy E_C^m extracted from the data is plotted as a function of V_{GC}. The oscillating behavior has been also reported in three different GDQD devices and was attributed to resonances induced by disorder states either in the middle GNR (connecting two dots) or in the graphene gate itself [89,122,125]. Since large gate-voltage ranges are used, the capacitive coupling of the gates to the disorder states can add or subtract charges discretely to these localized states, thus altering the entire environment abruptly and unpredictably. Consequently, the wavefunction in DQD needs to reconstruct itself, leading to the nonmonotonic changes in the interdot coupling strength with gate voltage.

When a perpendicular magnetic field is applied to a large graphene dot in which substrate disorder plays an important role (meaning that the size of the QD is greater than the size of the disorder-induced charge puddles; see Figure 13.3b), it is possible to induce charge redistribution due to the merging of charge puddles in the dot. Charge stability diagrams of DQDs reveal a wealth of information about their charging energy, interdot coupling, and cross-gate coupling strength, making them an ideal way to probe charge rearrangements in QDs. Figure 13.24a shows the evolution of the charge stability diagram of a large GDQD (200 nm in diameter) for applied perpendicular magnetic fields ranging from 4 to 10 T [126]. The field-dependent changes in the dimensions of the honeycomb (highlighted by the dotted hexagonal outlines) indicate the variations in the capacitances C_{g1} and C_{g2} and

thus the changes in the charging energies of both dots [see Eqs. (13.42), (13.43), (13.48) and (13.49)]. The QD charging energies vary from $E_{C1}\approx3$ meV and $E_{C2}\approx6$ meV at $B=4$ T to $E_{C1}\approx2.2$ meV and $E_{C2}\approx3.5$ meV at 10 T. Note that the subscript 1(2) in $C_{g1(2)}$ and $E_{C1(2)}$ denotes the gate-dot capacitance and the charging energy for $QD_{1(2)}$. These results suggest that the "effective sizes" of both dots increase at high B-fields, which is reflected by the decrease in charging energies. A schematic model shown in Figure 13.24b–d serves as a qualitative explanation for this observation [126]. Consider a varying background potential V in a model QD, as shown in Figure 13.24b, where V fluctuates from positive (V>0) to negative (V<0), passing through $V=0$. If V varies slowly, in each region of a large dot, the energy bands will approximately correspond to the shifted energy bands of 2D graphene with the Fermi energy set to zero. A bandgap is introduced to represent the quantum confinement effects of the dot, such that in the $V=0$ region, the DOS is very low or 0, whereas in the $V\ll0$ ($V\gg0$) regions, it gives rise to the electron (hole) puddles with a high DOS, as shown in Figure 13.24c. The DOS in the dot changes dramatically at high B-fields, where the lowest LL (LL_0) is well developed, with the consequent closing of the bandgap. Thus, in the $V=0$ region, the DOS is expected to increase, resulting in the development of nonchiral channels connecting the puddles (the light gray region in Figure 13.24d), whereas in the $V\ll0$ ($V\gg0$) regions, the DOS decreases due to the more energetically separated LLs in high B-fields. At the same time, the other LLs begin developing together with the chiral magnetic edge channel, as indicated in Figure 13.24d by the lighter gray (darker gray) arrows for the electron (hole) puddles. Since in this regime the DOS decreases in the bulk of the puddles while it increases at their edges, electron transport through the dot is not confined to a particular puddle but can be delocalized in the dot by flowing through both the chiral edge channels (arrows on the edge of QD) and nonchiral channels (channels whthin the QD). In this sense, the current is delocalized in the dot, and charge rearrangement can be observed compared with the case of low B-fields.

13.3.8 Charge Pumping

Charge pumping, which refers to a device that can shuttle n electrons per cyclic variation of control parameters to give the quantized current $I \equiv nef$, provides an exquisite way to link the electrical current to the elementary charge e and frequency f [59,60,127]. Such quantized charge transport can be realized when out-of-phase RF signals are applied to the plunger gates of a DQD, as discussed in Section 13.2.2 [91,128,129]. Figure 13.25a shows a schematic of the measurement circuit and AFM image of a GDQD device used for charge pumping. The AC voltages $V_{RF}(t)$ on both plunger gates with a phase difference φ between them drives the DQD into different charge states around the triple point. When $\varphi = 90$, it effectively forms a circular pump loop through three charge states in the stability diagram: (i) loading an electron from source reservoir into the left dot,

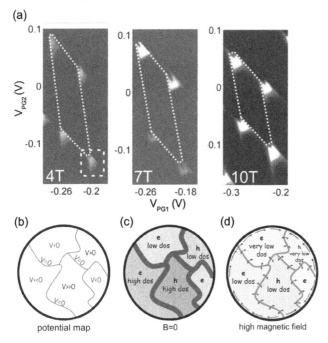

FIGURE 13.24 (a) The evolution of the charge stability diagram of a large GDQD under the influence of perpendicular magnetic field from 4 to 10 T, measured at $V_b = -1$ mV. (b) Example of a potential distribution in a large disordered QD. (c,d) Expected DOS distributions in the dot at zero magnetic field and high magnetic field, respectively. (Adapted with permission from Ref. [126]. Copyright 2015 American Physical Society.)

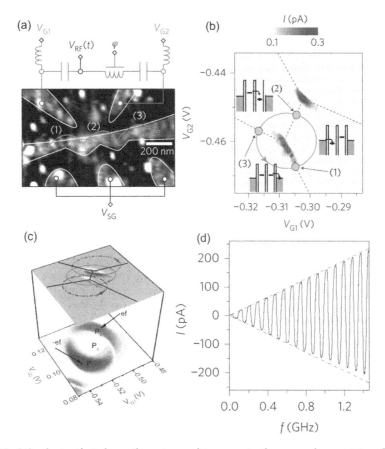

FIGURE 13.25 (a) AFM of the device that shows the gates used to generate the pumped current in a GDQD device. An oscillating voltage $V_{\text{RF}}(t)$ is added to the DC voltages V_{G1} and V_{G2}. A phase difference φ is added to V_{RF} before being added to one of the gates, which describes a circular trajectory shown in (b). (b) Source–drain current as a function of V_{G1} and V_{G2} with an applied bias ≤ 1 µV. The trajectory (circle with arrow) that encircles a triple point, passing through the sequence of transitions (1)→(2)→(3), as indicated in both (a) and (b). The insets denote different configuration of QD's energy level. (c) Plot showing a direct comparison between the DC (top) and AC (bottom) current behavior with $f = 12$ MHz and $P = -25$ dBm. Regions P_+, P_- and P_0 refer to the positive, negative, and zero pumped current, respectively. (d) Pumped current as a function of frequency at a power of $P = -15$ dBm. (Adapted with permission from Ref. [91]. Copyright 2013 Nature Publishing Group.)

(ii) electron transfer from left dot to the right dot, and (iii) unloading an electron from the right dot to the drain reservoir, as shown in Figure 13.25a,b. When a cycle is complete, a single charge has been transferred from source to drain reservoir and establishes a current. The frequency f of V_{RF} determines the value of the quantized pumped current $I = ef$, and the amplitude of V_{RF} determines the size of area in gate space where pumped current is generated. Depending on the type of triple point that the pumping circle encloses, it generates a different direction of current. Thus, the current recorded around two nearby triple points will present a circular shape with equal values but different signs. Figure 13.25c shows a direct comparison of the locations in gate space around a pair of triple points without RF (top) and with RF (bottom) voltages applied to the plunger gates. If the pump loop only encloses one triple point (bottom and top loop), it results in a flat regions, labeled P_+ and P_-, with a quantized pumped current $P_{+,-} = \pm ef$ in the stability diagram. However, when the pump loop encloses a pair of triple points (middle loop), it leads to repeatedly increasing

and decreasing the occupancy of each QD without any net transfer of electrons from source to drain. Thus, there is a central region (labeled P_0) where $I \approx 0$, giving rise to the crescent shape of pumped current as shown in Figure 13.25c. Unambiguous confirmation of quantized charge pumping is shown in Figure 13.25d, which plots the pumped current as a function of f with the DC gate voltages fixed at the center of the P_+ region. The oscillatory behavior is introduced because of a frequency-dependent phase shift in the RF circuit. The pumped current follows the quantized value $I = \pm ef$ over a range of frequencies up to gigahertz, an order of magnitude faster than the traditional metallic pump [130]. The pumping frequency in graphene is characterized by the RC time constant of the tunnel barriers, where R and C are the effective resistance and capacitance of GNRs. The 2D nature of graphene leads to a small C and results in a large pump frequency set by the tunnel rate of tunnel barriers (GNRs) [91].

We have reviewed the transport properties of graphene single dots and double dots fabricated on SiO$_2$ substrates.

Additional relevant reviews of research on GQDs on SiO_2 substrates can be found in Ref. [12–15]. Next, we will review GQDs fabricated on hBN and discuss how the transport properties change with the reduced influence of substrate disorder.

13.3.9 Graphene Quantum Dots on hBN

While the behavior of graphene nanostructures fabricated on SiO_2 is clearly influenced by localized states, it remains an open question whether they originate predominantly from substrate disorder or edge roughness. Here, we review the studies of GQDs fabricated on hBN substrate. These devices, with reduced substrate disorder potential, are expected to enable the influence of substrate and edge disorder to be studied separately.

GQDs with different diameters ranging from 100 to 300 nm have been fabricated on hBN substrates for transport characterization, as noted in a few literature [131,132]. The sizes of the dots are close to the order of the expected size of charge puddles in bulk graphene on hBN (≈ 100 nm in diameter) [28], so the substrate disorder is expected to play a less important role. Figure 13.26a shows the schematic illustration of such a device (top panel) and the AFM of an etched GQD on hBN with a diameter of 180 nm (bottom panel). The QD levels are tuned by a plunger gate (PG) while two side gate (SGR and SGL) are used to tune the resistance of the tunnel barrier GNRs. In a regime where the two barriers are pinched-off, the current I_{SD} as a function of plunger gate voltage, as shown in Figure 13.26b, confirms that the QD

is operating in the Coulomb blockade regime. For a more detailed comparison between GQDs resting on hBN and SiO_2, the distribution of the Coulomb-peak spacing ΔV_{PG}, i.e., the spacing between two subsequent Coulomb peaks, are statistically studied among dots with different sizes fabricated both on hBN and SiO_2 substrates. The normalized Coulomb peak spacings $\Delta V_{PG}/\overline{\Delta V_{PG}}$ for GQDs on hBN are reported as histograms in Figure 13.26c, for QD diameter $d=110$ nm (left panel), $d=180$ nm (middle panel) and $d=300$ nm (right panel), respectively. The same type of measurements are also performed for GQDs on SiO_2, and the results are summarized in Figure 13.26d, where the standard deviation of the normalized peak spacing distribution (σ) as a function of QD diameter is presented for both hBN and SiO_2 substrates. A clear difference can be seen between these two cases. The standard deviation for GQDs on hBN shows a clear decreasing dependence from 0.16 for the dot with $d=110$ nm to 0.05 for the dot with $d=300$ nm, while in the case of GQDs on SiO_2, it is independent of d. The standard deviation σ, which can be considered as the strength of peak-spacing fluctuations, may result from (i) the fluctuations of single particle level spacing Δ, (ii) fluctuations of the charging energy E_C (i.e., fluctuations in the size of the dot), or (iii) fluctuations of the lever arm α (i.e., the position of the dot). The single-particle level spacing in GQD is $\Delta(N)=\hbar v_F/(d\sqrt{N})$, where N is the number of charge carriers on the dot and v_F is the Fermi velocity [1]. If N is the only variable, the single particle level spacing $\Delta(N)$ gives an upper limit on the order of 0.03 to σ for $N=600$ (the number of peaks studied), and should be independent of the dot size and substrate. This is not in agreement with

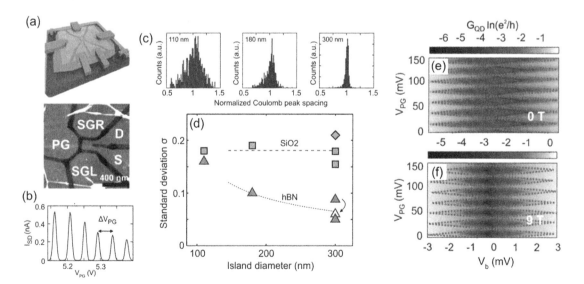

FIGURE 13.26 (a) Top panel: A schematic illustration of a graphene SET on hBN. Bottom panel: AFM of an etched GQD on hBN with a diameter of 180 nm. (b) Source–drain current I_{SD} as a function of V_{PG} for the device shown in (a). (c) Normalized peak-spacing distribution for GQDs on hBN with diameters of $d=110$ nm (left panel), $d=180$ nm (middle panel), and $d=300$ nm (right panel). (d) Summary plot of the standard deviation σ of the normalized peak-spacing distribution for different sized GQD on a SiO_2 (rectangular data points) and hBN (triangular data points) substrates. (e) Coulomb diamond measurement of a GQD ($d=300$ nm) on hBN at a perpendicular B-field of 0 T. (f) The same measurement as (e) but at $B=9$ T. ((a–d) Adapted with permission from Ref. [131]. Copyright 2013 American Institute of Physics. (e,f) Adapted with permission from Ref. [132]. Copyright 2013 John Wiley and Sons.)

the data shown in Figure 13.26d and leads to the assumption that the remaining two sources are responsible for the variability in peak spacing. The standard deviation for GQDs on hBN can be represented as $\sigma \approx \sigma^{hBN} + \sigma^{edge}/d \approx 0.01 + 16/d$ (nm), where σ^{hBN} represents the substrate-induced disorder (independent of dot size) and σ^{edge} represents the edge-induced disorder (scale with size as the edge-to-bulk ratio changes). Note that both values are obtained from the fit of the dotted line in Figure 13.26d. By contrast, the standard deviation for GQDs on SiO$_2$ is independent of dot size and reads $\sigma^{SiO_2} \approx 0.18$. This suggests that the potential landscape in the dot on SiO$_2$ is dominated by substrate-induced disorder, while contributions due to edge roughness, which are expected to scale with the size of the sample, play a minor role. These σ values also lead to the conclusions that (i) the substrate-induced disorder in GQDs on hBN is reduced by roughly a factor 10 when compared with SiO$_2$ (σ^{SiO_2}=0.18 to σ^{hBN}=0.01), (ii) edge roughness is the dominating source of disorder for GQDs with diameters less than 100 nm.

The reduced substrate disorder of GQDs on hBN can also reflect on magnetotransport. If the magnetic length of the electrons on GQDs is on the order of the disorder potential length scale, the electrons can accumulate in different charge puddles, leading to charge redistribution in the dot, thus changing the charging energy (as discussed in 3.2.1). However, as a result of the reduced bulk disorder, this effect is assumed not to occur for GQDs on hBN substrates. Figure 13.26e,f shows the comparison of Coulomb diamond measurements of a $d = 300$ nm GQD on hBN at $B = 0$ T and $B = 9$ T. It can be seen by similar Coulomb diamonds (with a charging energy $E_C \approx 3$ meV) for both magnetic fields, that the QD is stable and welldefined at 9 T, supporting the notion that the effective size of GQD is not affected athigh magnetic fields.

In summary, we have reviewed the transport properties of GQDs fabricated on both SiO$_2$ and hBN substrates. We conclude the main observations with the summary of references given in Table 13.1. Here we note that, although the reduced substrate disorders for GQDs on hBN, can in principle suppress some possible sources for fast spin relaxation, the spin-related transport phenomena (such as spin blockade) are still unreported [89,90,119–123]. It has been reported that the spin relaxation time in monolayer graphene ranges from 100 ps to 2 ns, significantly shorter than theoretically predicted [101–103,133,134]. Two mechanisms have been proposed to explain this observation. One involves local magnetic moments, which enhance spin relaxation through the resonant scattering of electrons off magnetic moments. Adatoms, organic molecules, vacancies, or spin-active edges are the possible sources of such local magnetic moments [135]. The other mechanism is related to the interplay between the spin and pseudospin quantum degrees of freedom when the disorder does not induce valley mixing [136]. Since graphene constrictions have been widely used as tunnel barriers in most GQDs reported thus far, resonant scattering of electrons off spin-active GNR

TABLE 13.1 References for the Main Observations

References	Main Observations of GSQDs
[1]	Observation of ESs
[2]	Charge detection
[3,56]	Fock–Darwin spectrum in the few-electron and many-electron regimes
[88]	Zeeman splitting of spin states
[92]	First suspended GQDs
[142]	GQD defined by AFM cutting
[92,93,143]	Bilayer GQDs defined by top gates
[87,104]	High-frequency gate manipulation on GQDs
[96]	GQDs defined *via* strain engineering
References	Main Observations of GDQDs (GDQDs)
[89,90,120]	Observation of ESs
[90]	Zeeman splitting
[90,144]	Bilayer graphene double dot
[89]	GDQDs defined by gated GNRs
[145]	Electron–phonon coupling
[122]	Metal gate tuning
[91]	Charge pumping
[126]	Charge redistribution in magnetic fields
[124]	RF sensing of the number of charges
References	Main Observations of GQDs on hBN
[132]	Size-dependent mean Coulomb peak spacing fluctuation
[131]	Magnetotransport

edges can be inevitable, leading to enhanced spin relaxation that lifts the spin blockade. In fact, GQDs fabricated from lithographic etching are all expected to possess edge roughness and suffer from unwanted edge scattering. One possible solution to this problem is to use an electrical-field-induced bandgap in bilayer graphene to define GQDs [92,93]. However, the small induced energy gap ($\approx$200 meV [137]) may limit the available energy range for QD operation. The other approach is to use the tip-induced deformation to define edge-free GQD. When graphene is deformed, the strain in the membrane can induce a local pseudomagnetic field, which has been reported to be as high as a real magnetic field of 300 T [138–140]. This strain-generated pseudomagnetic field can introduce strong quantum confinement to electrons, from which a QD can be formed. A few STM studies have been reported, but a novel method to probe spin dynamics in such a system is still lacking [96,141].

13.4 Single-Electron Transport in 2H-TMDs

The existence of bandgaps close to the wavelengths of visible light has earned 2H-TMD nanostructures considerable attention in optical studies [147–149]. However, it is also the sufficiently large bandgaps that distinguish 2H-TMDs from graphene and allows their nanostructures to be defined using electrical gating, as is commonly done in GaAs 2DEG systems. In this manner, the edge roughness created during the lithographic etching process, which is a common case in GQD fabrications, can be avoided. In this section, we will introduce a series of nanostructures defined from 2H-TMDs, including MoS$_2$ nanoribbons, WSe$_2$ SQDs, and MoS$_2$ DQDs [6,146,150,151]. In the former case, the nanoribbons were fabricated using Reactive-Ion Etching (RIE) (Figure 13.27a), whereas in the latter, the QDs were defined using metal top gates (Figure 13.27b). Both devices were fabricated on SiO$_2$/Si substrates. The Coulomb

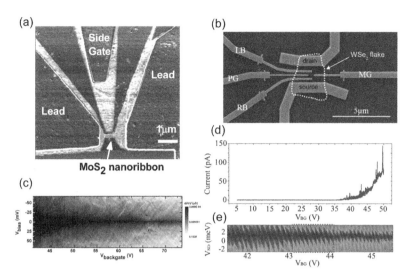

FIGURE 13.27 (a) AFM image of the MoS_2 nanoribbon. (b) SEM image of the WSe_2 QD device. The WSe_2 flake (4.5 nm in thickness) is highlighted by the white dotted line. The WSe_2 flake was directly contacted by the source/drain electrodes, and was separated from the four top gates (LB, PG, RB, and MG) by a 40 nm layer of Al_2O_3 grown by atomic layer deposition (ALD). (c) Differential conductance of the MoS_2 nanoribbon as a function of DC bias voltage V_{bias} and back-gate voltage $V_{backgate}$, with the side gate floated. (d) Source–drain current flow through the WSe_2 device as a function of back-gate voltage (V_{BG}). (e) Coulomb diamond measurements for WSe_2 QD from $V_{BG} = 41.5$ V to $V_{BG} = 45.5$ V, with all other top-gate voltages fixed at -2 V. (a,c) Adapted with permission from Ref. [6]. (b,d,e) Adapted with permission from Ref. [146]. Copyright 2015 Royal Society of Chemistry.)

diamond measurement for a MoS_2 nanoribbon is shown in Figure 13.27c. The existence of diamonds confirms the presence of small localized states (or QDs) in the nanoribbon. Moreover, the fact that larger diamonds are formed in the middle of the transport gap, whereas smaller diamonds are located away from the gap, indicate that the size of the localized state is strongly dependent on the Fermi energy, as is also observed for GNRs on SiO_2/Si substrates. This finding suggests that, in both cases, the potential in the nanoribbons can be described as a superposition of the substrate disorder potential and the confinement-induced energy gap, as already discussed in Section 13.3. The back-gate sweep for the WSe_2 device is shown in Figure 13.27d and exhibits the characteristic behavior of an n-doped semiconductor with a transport gap for $V_{BG} \leq 35$ V. The Coulomb blockade regime for a single dot can be achieved by tuning the WSe_2 flake into the conducting regime ($V_{BG} > 35$ V) while keeping the area below the top gates in an insulating state ($V_{MG} = V_{PG} = V_{LB} = V_{RB} = -2$ V), such that the Coulomb diamonds can be measured as a function of V_{BG}, as shown in Figure 13.27e. In this study, the charging energy E_C was estimated to be approximately 2 meV, which corresponds to a QD radius of $r = 260$ nm (the plate capacitance model $E_C = e^2/8\epsilon\epsilon_0 r$ is used, where ϵ is the relative permittivity of WSe_2) and is in reasonable agreement with the area defined by the top gates. Figure 13.28a shows the optical micrograph of a DQD defined in MoS_2 using the same gating technique. MoS_2 and h-BN flakes were exfoliated and transferred onto the local bottom gates (labeled "UM," "LB," "LP," "DM," "RP," and "RB" in the inset of Figure 13.28a), which are prepatterned on SiO_2/Si substrate. After depositing the Ti/Au source and drain contacts (labeled "S" and "D"),

another h-BN flake was transferred onto the whole structure to prevent the oxidation of MoS_2. By keeping the n-doped MoS_2 conductive ($V_{BG} = 30$ V) while pinching off the two middle gates ($V_{LP} = V_{RP} = 0$ V, $V_{UM} = -1.8$ V, and $V_{DM} = -1$ V), a conductance map of the plunger gates V_{RB} and V_{LB} demonstrates a honeycomb-like pattern (Figure 13.28b), which is typical for DQD's charge stability diagram. Upon applying a suitable bias voltage, the triple points evolve into bias triangles, as shown in Figure 13.28c, from which various parameters could be extracted (see Section 13.2.2). The estimated charging energies for the left and right dots are $E_C^L = 4.8$ meV and $E_C^R = 4.7$ meV, while the interdot coupling energy is extracted to be $E_C^m = 1.2$ meV. Furthermore, the estimated dot radius was ≈ 68 nm for both dots. These energy factors extracted from the gate-defined QDs are comparable with those obtained from the etched GQDs with similar size. One thing this DQD resembles the 2DEG QDs rather than GQDs is the capability of monotonic tuning on the interdot coupling, which allows a smooth transition from a single-dot state to a double-dot state in the same device. Such an effect is shown in Figure 13.28d, where the interdot coupling strength was monotonically tuned by changing the middle gate voltage V_{DM}, and the charge stability diagram for SQD and DQD was also observed accordingly (Figure 13.28d). Although the 0D behaviors were demonstrated, no magnetic-field dependence of the Coulomb resonances was reported in the above works.

Despite a lack of experimental studies on the topic, here, we briefly discuss how the Coulomb resonances of 2H-TMD QDs should evolve with perpendicular magnetic fields [152]. The Hamiltonian for a QD in a 2D semiconducting TMD defined by electrostatic gates under the influence of a

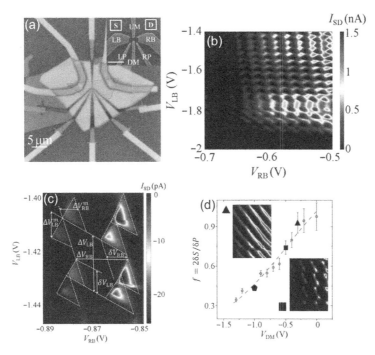

FIGURE 13.28 Transport through a gate-defined MoS$_2$ DQD. (a) Optical microscopy image of the device with a scale bar of 5 μm. The area enclosed in the light gray dashed lines indicates the location of the sandwiched MoS$_2$ flake. The inset with a scale bar of 500 nm shows an SEM image of the bottom gate structure taken before the stacks were transferred. The bottom gates are formed by a 5-nm-thick Pd. (b) DC current through the DQD vs. V_{LB} and V_{RB} (plunger gates LB and RB), for $V_{BG} = 30$ V, $V_{LP} = V_{RP} = 0$ V, $V_{UM} = -1.8$ V, $V_{DM} = -1$ V, and bias voltage $V_{SD} = 3$ mV. (c) Charge stability diagram at $V_{BG} = 30$ V, $V_{LP} = V_{RP} = 0$ V, $V_{UM} = -2.1$ V, $V_{DM} = -1.2$ V, and a bias voltage at $V_{SD} = -2$ mV. The triple points expand into bias triangles, which allow the estimation of the lever arm between the gates and the dots and the charging energy of both dots. (d) $f = 2\delta S/\delta P$ as a function of V_{DM}, where δS is the diagonal splitting measured between vertices in a triangle and δP is the distance between triple points. f essentially represents the interdot coupling energy and shows a decreasing tendency when the middle gate V_{DM} becomes more negative, demonstrating a monotonic change of f by tuning V_{DM}. The insets indicate charge stability diagram at different values of V_{DM}, whereas other parameters are kept the same. The left inset corresponds to an SQD state, while the right one is associated with a DQD state. (Adapted with permission from Ref. [150]. Copyright 2017 American Association for the Advancement of Science.)

perpendicular magnetic field can be written as follows (note that the model assumes that the system is n-doped and that the Hamiltonian describes the conduction band at the K (K') point) [152]:

$$H_{dot} = H_{el}^{\tau,s} + H_{SO}^{intr} + H_{vl}^{\tau} + H_{sp} + V_{dot}$$

$$= \frac{\hbar^2 \hat{q}_+ \hat{q}_-}{2m_{eff}^{\tau,s}} + \frac{1+\tau}{2} sgn(B_z)\hbar\omega_c^{\tau,s} + \tau\Delta_{CB}s_z$$

$$+ \frac{\tau}{2} g_{vl}\mu_B B_z + \frac{1}{2}\mu_B g_{sp}^{\perp} s_z B_z + V_{dot} \quad (13.59)$$

The wave numbers $q_{\pm} = q_x \pm iq_y$ are measured from the K and K' valley points of the TMD; therefore, the band dispersion is parabolic and isotropic (note that at zero B-field, $q_+ q_- = q_x^2 + q_y^2$). Here, the cyclotron energy $\hbar\omega_c^{\tau,s} = e|B_z|/m_{eff}^{\tau,s}$, where B_z is the applied magnetic field in z-axis, and $m_{eff}^{\tau,s}$ denotes the effective masses for bands with different valley ($\tau = \pm 1$) and spin ($s = \pm 1$) indices. $H_{SO}^{intr} = \tau\Delta_{CB}s_z$ denotes the intrinsic SOC in the TMD, where s_z is the spin Pauli matrix and Δ_{CB} determines the coupling strength. The next term, $H_{vl}^{\tau} = \frac{\tau}{2} g_{vl}\mu_B B_z$, breaks the valley symmetry of the LLs (g_{vl} is the valley g-factor

and μ_B is the Bohr magneton) and describes how the valley states move in the magnetic field. Finally, $g_{sp}^{\perp} = g_e + g_{so}^{\perp}$ is the total g-factor, where g_e is the free-electron g-factor and $g_{so}^{\perp}$ is the out-of-plane effective spin g-factor addressing the SOC. V_{dot} is the confinement potential for a QD of radius R_d and describes the hard-wall boundary conditions: $V_{dot}(r) = 0$ for $r \leq R_d$ and $V_{dot}(r) = \infty$ for $r \geq R_d$. To solve the band dispersion for this Hamiltonian, we follow the route for solving the Fock–Darwin spectrum of the GQD, as depicted in Section 13.2.1. We set $V_{dot} = 0$ and find the eigenvalue and eigenfunction of Eq. (13.59), and the bound-state solutions of the QD can then be determined from the condition that the wave function must vanish at $r = R_d$. For further details on this formalism, refer to Ref. [152].

The numerically calculated spectra for QDs of $R_d = 40$ nm in MoS$_2$ and WS$_2$ are shown in Figure 13.29a,b. Note that the parameters $m_{eff}^{\tau,s}$, g_{vl} and $g_{sp}^{\perp}$ used in the simulations for MoS$_2$ and WS$_2$ are different [152]. As can be seen in Figure 13.29a, this spectrum mimics the one that we derived for a 2DEG QD (Figure 13.11a) because of the quadratic dispersion in the model. At zero magnetic field, states with an angular momentum of $\pm l$ within the same

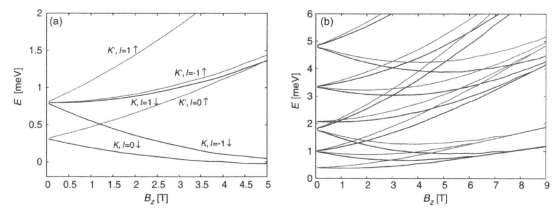

FIGURE 13.29 (a) Fock–Darwin spectrum of an MoS_2 QD of radius $R_d = 40$ nm. Labels show the valley (K or K'), the orbital quantum number l, and the spin state ($\uparrow$ or $\downarrow$) for each level. (b) Fock–Darwin spectrum of a WS_2 QD of radius $R_d = 40$ nm. Black (gray) lines represent the spin $\uparrow(\downarrow)$ states from the $K(K')$ valley. (Adapted with permission from Ref. [152]. Copyright 2014 American Physical Society.)

valley are degenerate, due to the effective TRS acting within each valley. For a finite magnetic field, all levels are both valley and spin split, which is different from Figure 13.11a, where spin is not considered. For large magnetic fields, when $l_B \leq R_d$, the dot levels merge into LLs, as shown in Figure 13.11a,b. From Figure 13.29a, one can see that the spin and valley degrees of freedom are locked together (meaning that spin $\downarrow$ ($\uparrow$) electrons only reside in the K (K') valley), suggesting that TMD QDs can be used as simultaneous valley and spin filters for single electrons. The spectrum for WS_2, as shown in Figure 13.29b, is similar to that for MoS_2, but the level spacing at zero field is larger because of the smaller effective mass in WS_2 (the mean level spacing can be approximated as $\delta = \frac{2\pi\hbar^2}{m_{\text{eff}}A}$, where A is the area of the dot [152]). By contrast, at a finite B-field, the splitting between states belonging to different valleys (and also spins, as spin and valley are locked) is significantly larger for the MoS_2 than for the WS_2 because of the different signs of Δ_{CB} and, consequently, the different spin polarizations of the lowest levels in the two materials. The valley and spin pairs could, in principle, serve as valley or spin qubits, but the crossing between levels with different quantum numbers l in finite magnetic fields may add complication to its realization. One way to circumvent this problem may be to use the lowest Kramer pairs [$|l = 0, K', \uparrow\rangle$ and $|l = 0, K, \downarrow\rangle$; see Figure 13.29a] at a low B-field as a combined spin-valley qubit. Note that the above results apply not only for monolayer 2H-TMDs but also for 2H-TMDs with an odd number of layers, in which the inversion symmetry is broken.

After reviewing the relevant experimental studies related to developing spin qubits in 2D materials, in the next section, we will discuss the prospect of using 2D materials as an element in superconducting qubits.

13.5 2D Material-Based JJ

JJ is a core element in superconducting based quantum computing. When JJ is connected in parallel to a capacitor,

under the condition that Josephson energy is much larger than the charging energy ($E_J \gg E_C$), the eigenenergy of the quantum circuit forms anharmonic bound states with unequal energy spacing. The two lowest bound states of the quantum circuit can be used as a two-level system for qubit operation, and such qubit is generally referred as a transmon qubit. The anharmonic nature of the bound states arises from the nonlinear phase relation in JJ's inductance (L_J), which is given by $V = L_J dI/dt$ and can be written as

$$L_J = \frac{\Phi_0}{2\pi I_C \cos\varphi}, \qquad (13.60)$$

where $I = I_C \sin\varphi$, I_C is the critical current of JJ, and φ is the phase difference across the junction, $V = \frac{\Phi_0}{2\pi}\frac{d\varphi}{dt}$, and Φ_0 is the magnetic flux quanta. The Josephson energy E_J is given by $E_J = \hbar I_C/2e$, and the corresponding energy spacing between the lowest two bound states can be approximated as (when $E_J \gg E_C$) [153]:

$$E_{01} = \sqrt{8E_C E_J(V_g)}, \qquad (13.61)$$

The critical current of an S-I-S (S stands for superconductor and I stands for insulator) JJ is usually determined by its geometry, which is fixed after fabrications. Thus, the Josephson energy, hence the qubit frequency ($f_Q = E_{01}/h$), is fixed correspondingly. Although a configuration of multiple JJs can allow some tunability with magnetic flux [155,156], it usually requires complex fabrications that can be difficult for upscaling. On the other hand, an S-N-S JJ, in which N stands for a semiconductor, allows the critical current to be mediated by the DOS of the middle semiconductor. By this manner, the qubit frequency can be readily tunable by an electric gate close to the semiconductor, leading to a gate-tunable transmon qubit, also known as gatemon. An example of a gatemon qubit is shown in Figure 13.30, where an InAs nanowire is contacted by two Al electrodes to form an Al-InAs-Al junction. The JJ is formed from a molecular beam epitaxy-grown InAs nanowire, 75 nm in diameter, with an *in situ* grown 30 nm

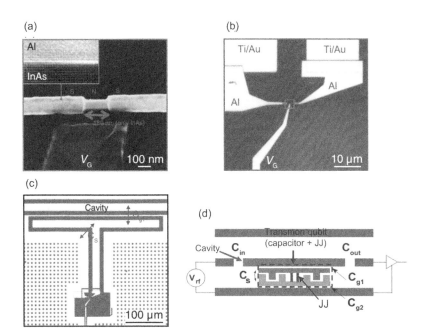

FIGURE 13.30 InAs nanowire-based superconducting gatemon qubit. (a) SEM of the Al-InAs-Al JJ. A segment (180 nm) of the epitaxial Al shell is etched to create a semiconducting weak link. Inset shows a TEMof the epitaxial InAs/Al interface. (b) Optical micrograph of the gatemon device. The square indicates the region shown in (a). (c) Optical micrograph showing the gatemon defined between the T-shaped island and the surrounding ground plane. The circuit (JJ + capacitor) is coupled to a transmission line cavity *via* C_{g1}. (d) A schematic of the entire circuit for readout and control of qubit. ((a–c) Adapted with permission from Ref. [153]. Copyright 2015 American Physical Society. (d) Adapted with permission from Ref. [154].)

thick Al shell. By wet etching away a 180 nm segment of the Al shell, followed by a deposition of a gate 100 nm away from it, a gateable junction is fabricated as shown in Figure 13.30a,b. The JJ is bridging between a T-shape Al island and the surrounding Al ground plane, leading to a junction shunted by a capacitor C_S (see the arrow next to C_S in Figure 13.30c), which effectively forms a transmon qubit (JJ + capacitor). Figure 13.30d illustrates the measurement scheme for the transmon qubit. The on-chip cavity is capacitively coupled to the qubit through C_{g1} (also indicated in Figure 13.30c), and the C_{in} (C_{out}) defines the capacitive coupling between the incoming (outgoing) transmission line and the cavity. When the incoming RF signal resonates with the qubit ($f_{RF} = f_Q$), the transmitted signal will undergo a significant loss, from which qubit frequency can be determined. The readout of qubit state is operated in a dispersive regime, where the resonant frequency of the cavity will depend on the qubit state. For readers who are interested in the relevant techniques, please refer to [157].

It is natural to consider 2D materials as a good intermediate medium in an S-N-S junction owing to their 2D (allowing different JJ sizes) and gateable nature. In order to do so, a gateable JJ based on 2D materials must be demonstrated. We will review a selection of experimental studies in the subsequent sections.

13.5.1 Graphene-Based JJ

When a graphene flake is in contact with a superconductor, it acquires superconducting properties because of the proximity effect, a well-known phenomenon that can be described as the leakage of Cooper pairs of electrons from a superconductor (S) into a normal-type conductor (N). The proximity effect takes place only if the S-N interface is highly transparent to electrons, which makes graphene a good candidate as it can form Schottky barrier-free contacts with metals. Graphene-based JJs have been widely studied over the past decade, from the early demonstration of bipolar supercurrent [158] to the more recent studies on Andreev reflection [65,159,160], current phase relation [161], edge-mode superconductivity [162,163], and supercurrents in the quantum Hall regime [38]. The inset in Figure 13.31a illustrates an example of a dual-gated graphene JJ, where graphene is contacted by two superconducting electrodes (Al) to form an S-G-S junction. The graphene flake can also be electrically gated through a global Si back-gate and a local Ti/Au finger-like top gate using a piece of hBN as a dielectric layer. The graphene device exhibits ohmic behavior with a normal state resistance, R_N, in the range from 0.2 to 1 kΩ within the gate-tunable range. Figure 13.31b shows R_N measured as a function of top gate V_{TG} and back-gate V_{BG} by applying a current excitation much larger than the critical Josephson current of the device (hence, no Josephson effect is expected). Since the top gate in the S-G-S junction can tune the carrier density in the locally gated region, the polarity of the entire device can be determined by four distinct regions in the 2D resistance map, i.e., P-N-P, N-P-N, P-P'-P, and N-N'-N, as indicated in Figure 13.31b. In particular, the P-N interface in graphene can be highly transparent for charge carriers (as the chiral Dirac

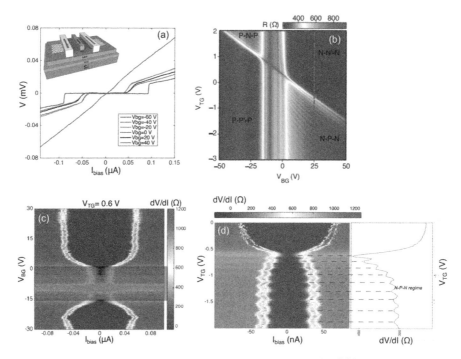

FIGURE 13.31   Electrical characterization on a dual-gated S-graphene-S junction. (a) I-V characteristic of the junction measured at different back-gate voltage showing the gate-tunable critical current. The inset shows a 3-D schematic of the dual-gated S-G-S junction. (b) 2-D normal state resistance map measured as a function of V_{TG} and V_{BG}. Note that Fabry–Prot oscillations are observed in N-P-N and P-N-P configurations. (c) Differential resistance dV/dI as a function of back-gate voltage V_{BG}. Top-gate voltage V_{TG} is set at 0.6 V. (d) Fabry–Prot oscillation of Josephson effect, where the critical current I_C oscillates outofphase with the normal state resistance R_N. (Adapted with permission from Ref. [62].)

fermion can pass the potential barrier *via* Klein tunneling), while their trajectories could resemble that of refracted light at the interface of metamaterials with negative refractive index. Therefore, in analogy to wave optics, the locally gated region can be regarded as a Fabry–Prot cavity for the charge carriers and result in the interference pattern as observed in the P-N-P and N-P-N regions in Figure 13.31b. This interference pattern can be correlated with the supercurrent as will be discussed below.

When the temperature is below the critical temperature of the superconductor (T_C=1.1 K for Al), the pseudo four-probe current bias measurement, in which the voltage across the graphene is measured while sweeping the bias current through the same source–drain electrodes, was performed at different back-gate voltages, as shown in Figure 13.31a. The proximity effect in graphene manifests itself through the appearance of a dissipationless supercurrent, i.e., no voltage drop accompanies the finite current flow, and the switching from superconductive to dissipative conduction occurs when I approaches a critical current, I_C, leading to an abrupt appearance of finite voltage. Similar measurements can also be performed by a standard AC+DC technique, in which a small AC modulation is added on top of the DC bias current, and the resulting dV/dI was detected by a lock-in amplifier. The I_c, defined by the boundary of the zero dV/dI region, displays a strong gate-dependence as shown in Figure 13.31c. The gate-dependence of I_C can be attributed to the $I_C R_N$ relation in an S-N-S junction [164]. The product of critical

current and normal state resistance for an S-N-S junction can be roughly characterized by $I_C R_N \approx \Delta_0/e$, where Δ_0 is the superconducting energy gap of the contact electrodes (note that the exact form of the formula may vary depending on whether the junction is in the diffusive or ballistic regime [165,166]). Since the 2D and semiconducting nature of the graphene allows its carrier density, hence the normal state resistance, to be controlled by the voltage applied to the back-gate, the critical current also shows a strong gate-dependence. As expected, I_C nearly diminishes at the Dirac point ($V_{BG} = -13$ V), where R_N is at its maximum (see Figure 13.31b), as can be observed in Figure 13.31c. The strong correlation between I_C and R_N can be further investigated in Figure 13.31d, where the left panel displays the dV/dI vs. I_{bias} as a function of V_{TG} in the Fabry–Pérot oscillation regime, and the right panel plots normal resistance in the same regime [linecut along the dashed line in Figure 13.31b] for comparison. It can be clearly seen that I_c, defined by the boundary of the zero dV/dI region, oscillates outofphase with the Fabry–Prot oscillation of normal state resistance R_N, in agreement with the $I_C R_N$ relation mentioned above. The observed I_C varies from 10 nA at $V_{BG} = -13$ V to 80 nA at $V_{BG} = -60$ V, demonstrating a good gate-tunability readily for use in graphene-based gatemon qubits.

In Section 13.2.3, we have discussed how Andreev reflection, and hence the corresponding ABS, can be related with proximity effect. To manifest the phase dependence of ABS energy, here we introduce a SQUID-like graphene JJ, in which the phase difference across the weak link can be well-controlled by applying an external magnetic field [65]. The top and side views of this SQUID-like S-G-S junction are shown in Figure 13.32a,b. A monolayer graphene sheet is encapsulated between two hBN sheets, where the bottom one is 15 nm thick and the top one is one atom thick (0.3 nm). The bottom hBN isolates graphene from a graphite local back-gate, which enables the electrostatic control of the Fermi energy of graphene. On top of the top hBN sits a 150-nm-wide metallic probe made of thin graphite. The use of a graphite probe (instead of metal) is to limit the doping in the underneath graphene and allow the low carrier density regime to be accessed. The graphene sheet, with a width W = 2 µm, is connected to two superconducting Al electrodes, with a spacing $L = 500$ nm. The extension of the electrode is patterned in a loop that enables the control of phase difference $\phi = 2\pi\Phi/\Phi_0$ across the graphene weak link by applying a magnetic flux $\Phi = B \cdot$ Area through the loop (Φ_0 is flux quantum). In such a way, one could study the phase dependence of ABS energy discussed in Section 13.2.3. The tunneling measurement to study the DOS of proximitized graphene was performed by applying a voltage modulation $dV = 10$ µV to the graphite probe and measuring the differential conductance dI/dV through the Al lead using a standard lock-in technique. Figure 13.32c shows dI/dV measured as a function of DC bias voltage V_b and magnetic field B for a device with the geometry shown in Figure 13.32b. The observed oscillation within the induced superconducting gap in graphene (-160 µV $\leq V_b \leq 160$ µV) strongly indicates the phase dependence of ABS in the short-junction limit, where only a pair of ABS is modulated by phase [65]. However, it is worth noting that there exists nonvanishing DOS inside the gap at all fields, suggesting there are ABS whose energies span the full spectrum within the gap. These

ABS can be regarded effectively as contributed by a set of long-junction modes, which are less sensitive to phase modulation compared with the short-junction modes [65]. Their origin may be due to the weak coupling between graphene and the superconductor or impurity scattering within the normal region. The oscillation in Figure 13.32c possesses a periodicity $\delta B = 360$ µT, which corresponds to one flux quantum Φ_0 threading the loop of enclosed area $A = 5.7$ µm^2 (taking into account the Meissner effect). The corresponding phase $\phi = 2\pi A(B - B_0)/\Phi_0$ is shown on the top axis of Figure 13.32c, where $B_0 = 250$ µT is an offset magnetic field in this experiment. The gap and side peaks are most pronounced at $\phi = 0$ and get reduced when the phase is swept towards $\phi = \pi$, as expected from Eq. (13.57).

13.5.2 TMD-Based JJ

Apart from graphene, the S-N-S junctions made of other layered materials are relatively limited. Although there are works using superconducting layered materials as a component, they are used to provide superconductivity not as an intermediate layer in an S-N-S junction [167,168]. Owing to the intrinsic bandgap in semiconducting 2H-TMDs, the critical current in S-TMD-S junctions is expected to be highly tunable. However, it is also this semiconducting nature that leads to a difficulty to make good ohmic contacts on them, as high Schottky barriers are usually present in the metal/TMD interfaces. The resistive Schottky barrier tends to weaken the induced superconductivity, while the transport in 2H-TMDs is usually diffusive owing to low mobility, both effects result in a very poor weak link for the supercurrent to be observed. On the other hand, the T$_d$-phase TMDs (which is three-dimensional (3D) stack of 1T'-phase monolayers), such as T$_d$-WTe$_2$ and T$_d$-MoTe$_2$, are highly

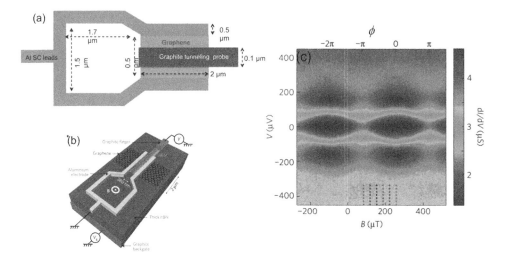

FIGURE 13.32 ABS oscillations in a SQUID-like graphene JJ (S-G-S). (a) Top view of the SQUID-like S-G-S junction, showing the relevant dimensions of the device. (b) A schematic side view of the SQUID-like S-G-S junction. An encapsulated graphene flake is connected to two superconducting electrodes. (c) A 2D map of dI/dV measured on the device with the geometry shown in (b), showing the ABS oscillations within the induced superconducting gap in graphene. ((a) Adapted with permission from Ref. [62]. (b,c) Adapted with permission from Ref. [65]. Copyright 2017 Nature Publishing Group.)

conducting semimetal and could be readily used as a weak link to transmit supercurrent. In this section, we will introduce experiments that have been performed to study the Josephson effects in junctions made of 2H-MoS$_2$ [169] and T$_d$-WTe$_2$ [170].

Fig. 13.33 (a) shows the optical image of the vertical S-MoS$_2$-S junction, where a monolayer 2H-MoS$_2$ is sandwiched between two MoRe electrodes. The use of superconducting MoRe ($T_c \approx 10$ K) as electrodes is to provide Schottky barrier-free contact to MoS$_2$ as a result of their work function match. The successfully induced superconductivity in the monolayer weak link can be seen in Figure 13.33b, in which a critical current I_C of 2.5 μA was observed in the V-I_b curve. The differential conductance measured as a function of bias voltage was also performed (Figure 13.33b), which shows symmetric peaks at ±2.6 mV marking the quasiparticle gaps. In addition, subgap conductance peaks indicating multiple Andreev reflections (MAR) were also observed symmetrically in voltage axis, suggesting that the transport is in the diffusive regime despite the short junction [171]. The positions of the MAR peaks in energy are given by $eV_n = 2\Delta/n$, where n is a positive integer [171], as indicated by the arrows and number in the inset of Figure 13.33b. Linear fitting of MAR peaks gives the superconducting gap of $\Delta = 1.3$ meV, which is expected from MoRe thin film [169]. The thickness dependence of the vertical junction reveals more information on the Josephson coupling between the top and bottom MoRe electrodes. Figure 13.33c–eshows the measured V-I_b (main panels) and dI_b/dV-V (insets) curves for three devices consisting of 2, 3, and 4 layers of MoS$_2$, respectively. As can be seen in Figure 13.33b–e, the critical current decreases with increase in thethickness, with high critical currents for the single and bilayer devices and several orders of magnitude lower values for the three- and four-layer junctions. The observed MAR in monolayer and bilayer junctions (Figure 13.33b,c, see arrows) indicates a highly transparent and metallic weak link [63], possibly due to the doping of MoS$_2$ from direct contact with the MoRe electrodes. On the other hand, the trilayer and four-layer devices show a tunneling behavior through an undoped semiconductor, which is supported by the formation of a reduced quasiparticle gap, as shown in the inset of Figure 13.33d,e. These results indicated that the uncoupled layers (layers that are not directly in contact with the MoRe electrodes) in the three- and four-layer devices provide a tunnel barrier, which reduces the critical current densities and results in more well-defined quasiparticle transport gaps.

Weyl semimetals are 3D phases of matter with gapless electronic excitations that are protected by topology and symmetry [172]. The Weyl fermions, which are used to describe the low-energy excitations in Weyl semimetals, can be viewed as a 3D analog of graphene, i.e., dispersing linearly along all the three momentum directions across the Weyl points (WPs). The WPs always appear in pairs

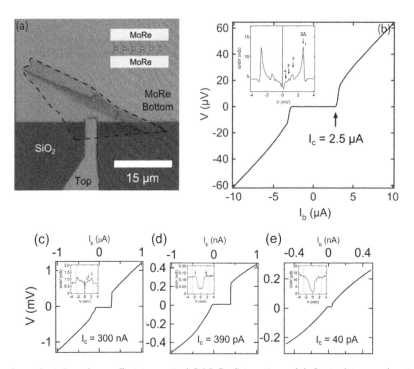

FIGURE 13.33 Layer-dependent Josephson effect in vertical S-MoS$_2$-S junctions. (a) Optical image showing a monolayer 2H-MoS$_2$ is sandwiched between two MoRe electrodes to form a vertical junction. (b) V-I_b curve of the monolayer MoS$_2$ junction presenting a critical current of 2.5 μA. The inset shows the differential conductance (dI_b/dV) as a function of V. The arrows with integer number mark the conductance peaks arising from Andreev reflections. (c–e) Current bias sweeps at $T = 1.2$ K for vertical junctions have two to four layers of MoS$_2$, respectively. Insets show the differential conductance (dI_b/dV) vs. V for each junction, taken at 30 mK. (Figure adapted with permission from Ref. [169]. Copyright 2016 IOP Publishing Ltd.)

with opposite chirality, and there exists open curve-like Fermi surfaces with nondegenerate spin texture, also known as surface Fermi arcs, connecting them [172]. The Weyl semimetal can be further classified into two types based onwhether the Lorentz symmetry is respected: type I hosts point-like Fermi surfaces and its WPwith respect to the Lorentz symmetry, whereas type II breaks the Lorentz symmetry and contains electron and hole pockets in its Fermi surfaces [173,174]. Recently, type-II Weyl semimetals have been proposed to exist in the orthorhombic T_d-phase TMDs (such as WTe_2 and $MoTe_2$ [175,176]), which is later supported by several experimental studies [177,178]. As an analogy to the Dirac fermions in graphene, the Weyl fermion carriers in T_d-TMDs are expected to transmit electrical currents effectively, and therefore, could provide a good weak link in JJs. Figure 13.34a illustrates a lateral S-N-S junction made of superconducting indium leads and T_d-WTe_2. The device is fabricated by weakly pressing a WTe_2 single crystal (≈ 0.5 mm $\times$ 100 μm $\times$ 0.5 μm dimensions) onto the indium lead pattern (100 nm thick) so that planar In-WTe_2 junctions [5 (W) $\times$ 5 (L) μm^2] are formed

at the bottom surface of the WTe_2 crystal. The In-WTe_2-In junctions were measured by a quasi-four-probe technique (as illustrated in Figure 13.34a) at a temperature around 1.4 K, which is well below the critical temperature of indium ($T_c \approx$ 3.4 K). The measured *V-I* curve between two 5 μm spaced contacts S1 and S2 shows a supercurrent with $\pm I_C \approx 4$ mA (darker gray curve in Figure 13.34b), while that between the 80 μm separated S1 and S3 leads always presents a finite resistance (lighter gray curve in Figure 13.34b). The observed supercurrent between S1 and S2 indicates an unprecedented long junction 5 μm $\gg \xi_{In}$, where $\xi_{In} \approx 300$ nm is the indium phase coherence length. The temperature and *B*-field dependence of critical current I_C were performed on the junction to reveal more information aboutJosephson effect. To analyze I_C (*B,T*) behavior, dV/dI vs. *I* was measured by adding an additional AC modulation (100 nA, 10 kHz) on top of the DC current, followed by an AC part of *V* ($\approx dV/dI$) detected by a lock-in amplifier. The right inset in Figure 13.34b shows an example of the measured dV/dI-*I* curve, from which I_C can be determined by the edge of the gap. The obtained I_C (*B,T*) was summarized

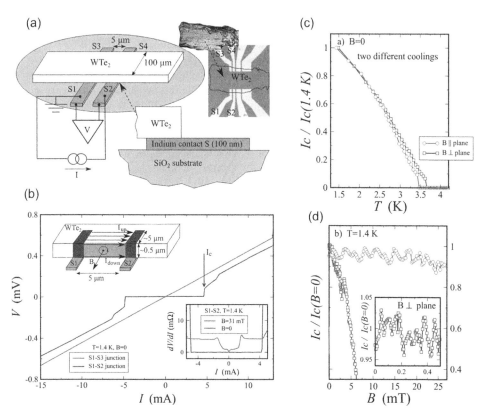

FIGURE 13.34 Josephson effects in a lateral In-WTe_2-In junction. (a) Sketch of the sample with indium contacts to the bottom surface of a T_d-WTe_2 crystal (not to scale). Right inset shows the top-view image of the indium leads and WTe_2 crystal. (b) *V-I* characteristics in two different experimental configurations in zero magnetic field at $T = 1.4$ K. The darker gray curve is obtained for 5-μm-long In-WTe_2-In junction between the superconducting leads S1 and S2, as depicted in (a). In contrast, the resistance is always finite between 80 μm separated S1 and S3 indium leads, see the lighter gray curve. Right inset: dV/dI-*I* characteristics for the S1-WTe_2-S2 junction at $T = 1.4$ K, obtained in zero field (the darker gray curve) and for the critical field $B_c = 31$ mT (the lighter gray one). Left inset: A schematic diagram of a double-slit SQUID geometry, realized by Weyl surface states in WTe_2 semimetal. (c, d) Suppression of the critical current I_C by temperature (c) and perpendicular (darker gray) and parallel (lighter gray) magnetic fields (d). (Figure adapted with permission from Ref. [170].)

in Figure 13.34c,d. The temperature dependence $I_C(T)$ in Figure 13.34c does not present an exponential decay that is expected for long S-N-S junctions ($L \gg \xi_{In}$). Instead, the $I_C(T)$ dependence is even slower than the linear function of T, which is usually the case for a short-junction limit [64]. The slow temperature dependence of $I_C(T)$ in this long device may be associated with the topological Fermi arc surface states in Weyl semimetal [170]. Because of the topological protection, they can efficiently transfer the Josephson current that leads to a slow temperature dependence. The field dependence of $I_C(B)$ also reveals a scenario of surface transport, which can respond differently to magnetic field orientation, as shown in Figure 13.34d. A strong suppression of $I_C(B)$ is observed when the applied field is perpendicular to the junction plane (as expected for a standard Josephson effect), while $I_C(B)$ diminishes very slowly with the parallel magnetic field. For both orientations, $I_C(B)$ oscillates with B within 5 percent of I_C magnitude, with a period of $\Delta B = 2$ mT for the parallel field and $\Delta B = 0.1$ mT for the perpendicular one. The observed $I_C(B)$ suppression in parallel magnetic fields resembles a double-slit SQUID behavior, indicating a possible surface transport as illustrated in the left inset of Figure 13.34b. Since the thickness of WTe$_2$ (500 nm) is comparable with indium phase coherence length (300 nm), it is likely that the regions of proximity-induced superconductivity couples two opposite sample surfaces near the In leads (dark gray regions in the inset of Figure 13.34b) and essentially forms a SQUID configuration. Therefore, parallel magnetic field induces the magnetic flux threading the loop area enclosed by the surfaces and results in the oscillation of $I_C(B_{\parallel})$. Since $A \cdot \Delta B \approx \Phi_0$, $\Delta B = 2$ mT gives $A \approx 10^{-8}$ cm^2, which corresponds to 300 nm sample thickness for a 5 µm long junction and is in good agreement with the device dimensions. The perpendicular field dependence of $I_C(B_{\perp})$ reflects a homogeneous supercurrent distribution within the surface state in the top and bottom S-N-S junctions [170]. The observed period of $\Delta B = 0.1$ mT corresponds to an effective junction area of $A \approx 2 \times 10^{-7}$ cm^2 ($\approx$5 µm $\times$ 5 µm), which again corresponds to the sample dimensions.

T$_d$-TMDs have more exotic topological properties in the monolayer form, which will be introduced in the next section.

13.6 Quantum Spin Hall Edge States in 1T′-TMDs

Majorana fermions (MFs) or Majorana bound states (MBS) in condensed matter systems are a special type of excitation that is predicted to exist in 1D or 2D p-wave superconductors [51,179,180]. They obey non-Abelian (noncommutative) exchange statistics, which effectively means that particle exchange with different routes will lead to different end states. This is the essential property for utilizing them as an element for quantum computing. By braiding a few MFs (computing) and annihilating them (readout), one could encode bits of information [181]. What

stands out from conventional quantum computing (based on a two-level system) is that Majorana-based qubit is expected to have a very long decoherence time. Another way to say this is that Majorana states are topologically protected, i.e., as long as they are well separated from each other, any local perturbation will not be able to affect them simultaneously. This nonlocal property distinguishes them from qubits made of two-level systems, which often suffer from local couplings to charges, phonons, or magnetic moments that induce decoherence. Engineering MFs in a condensed matter system requires several important ingredients. They are believed to exist in the vortices of a 2D p-wave superconductor or in an s-wave superconductor proximatized 2D TI or semiconducting nanowire with a large g-factor and strong SOC [51,180]. The last has been realized experimentally in several InSb and InAs nanowires since 2012 [182–184]. In the former case, the spin-polarized edge states in TIs provide the spin nondegenerate states while the s-wave superconductors provide Cooper pairing strength, the combination of the two essentially forms a "spinless" superconductor that is capable of hosting MFs [51]. Experimental search for MBS in 3D TI platforms has been focused on using STM to probe the center of vortices formed on the surfaces of proximatized 3D TIs [185,186]. As for 2D TIs, although possible Majorana-related physics have been revealed in HgTe-based JJs [187,188], definite experimental evidence for MBS in such a system remains elusive to date. 1T′-TMDs as nature 2D TIs with large bulk gap and narrow edge states (compared with HgTe quantum wells, see Section 13.1.3) thus provide another interesting platform to probe MBS.

In Section 13.1.3, we have briefly introduced the theoretical background for 1T′-TMDs (monolayer form of T$_d$-TMDs), whose band structures indicate that they are 2D TIs with QSH edge states. Quantum transport through a 2D TI should exhibit the following characteristics [189]: (i) helical edge modes protected by TRS, characterized by an edge conductance that is approximately the quantum value of e^2/h per edge; (ii) saturation to the conductance quantum in the short-edge limit; and (iii) suppression of conductance quantization upon application of a magnetic field, owing to the breakdown of TRS. In the following, we will present an experimental evidence that, at low temperatures, monolayer 1T′-WTe$_2$ exhibits an insulating bulk and conducting edge and describes the properties of edge conduction, including its dependence on gate voltage, magnetic field, temperature, and contact separation. All the properties eventually indicate that 1T′-WTe$_2$ meet the above criteria for being a 2D TI. Figure 13.35a shows the two-terminal differential conductance measurements of an encapsulated monolayer 1T′-WTe$_2$ device, which has a row of bottom Pd contacts along one edge (see the inset in Figure 13.35a) and a graphite top gate V_g to tune the Fermi energy of the whole flake. On cooling from 300 K, the monolayer shows a metallic behavior but develops a strong V_g dependence with a wide minimum near $V_g = 0$. However, below $T = 100$ K, the minimum stops dropping and instead broadens into a plateau of conductance (at $\approx$16 µS) reminiscent of

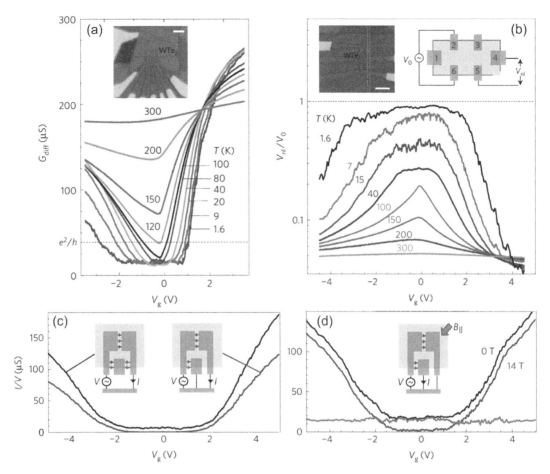

FIGURE 13.35 (a) Temperature dependence of the characteristics for Pd contacts on an edge of a monolayer 1T′-WTe$_2$ ($L = 0.24$ μm, $W = 3.3$ μm) device. Inset shows the optical image of a monolayer 1T′-WTe$_2$ device. Scale bar, 5 μm. (b) Nonlocal measurements on a device shown in the left inset. The measurement scheme is illustrated in the right inset, where a voltage V_0 ($V_0 = 100$ μV at 11.3 Hz) is applied across contacts 2 and 6, and avoltage drop V_{nl} between contacts 4 and 5 is detected. (c) Measurements on a device with pincer-shaped contacts overlapping one straight edge (the contact separation along the edge is 0.22 μm; the pincer spacing is 0.28 μm). Measurements were performed at $T = 10$ K. (d) Effect of in-plane magnetic field $B_{||} = 14$ T on I/V between adjacent contacts ($T = 10$ K). The trace near the bottom with a value of around 15 is the magnitude of the decrease. (Figure adapted with permission from Ref. [190]. Copyright 2017 Nature Publishing Group.)

quantized edge state conductance. We will show below that the plateau is due to edge conduction remaining when the bulk becomes insulating below $T \approx 100$ K. Figure 13.35b shows the nonlocal measurements that were used to detect edge conduction in another monolayer 1T′-WTe$_2$ device (left inset of Figure 13.35b). In the measurement setup shown in the right inset of Figure 13.35b, a small excitation V_0 between contacts 2 and 6 on opposite edges was applied, and the nonlocal voltage V_{nl} induced between contacts 4 and 5 (which are far out of the normal current path between contacts 2 and 6) was detected. At low T and small V_g, V_{nl}/V_0 increases, suggesting that, in this regime, most of the current propagates along the edge. At higher T or larger V_g, V_{nl}/V_0 falls off as more current takes the direct path through the bulk. Although this measurement indicates that the current follows the edge, in this geometry, it is difficult to quantitatively separate edge contributions form bulk conduction. To address this, another geometry thatemploys a series of pincer-shaped contacts overlapping one straight

edge of a monolayer flake is adopted and shown schematically in the insets of Figure 13.35c,d. The gate-dependence of conductance between a pair of pincers (black trace in Figure 13.35c) behaves similarly to a pair of adjacent contacts as shown in Figure 13.35a. However, if the smaller rectangular contact interposed between them is grounded (right inset in Figure 13.35c), so that any current flowing near the edge is shorted out, the measured I/V between pincers is suppressed nearly to zero around $V_g = 0$ (bottom trace in Figure 13.35c). This indicates that, around $V_g = 0$, most of the current flows near the edge while the bulk is insulating. In contrast, with the middle contact still grounded but V_g larger than about ± 2 V, the appearance of nonzero I/V indicates the conduction through the 2D bulk (directly across the gap between the pincers).

The effect of an in-plane B-field on edge conduction is shown in Figure 13.35d, where the black trace is a measurement at $B_{||} = 0$ T between two adjacent contacts (see the configuration shown in the inset), and the dark gray trace is the same measurement done with an in-plane field of $B_\parallel = 14$ T. As can be seen in the dark gray trace, near $V_g = 0$, where the bulk is insulating, I/V drops nearly to zero, implying that the edge conduction is strongly suppressed by the magnetic field due to the breakdown of TRS. In addition, the magnitude of the drop, plotted in light gray in Figure 13.35d, is similar at all V_g, indicating that the edge makes a roughly constant contribution to the conductance, independent of gate voltage and bulk conductivity.

The temperature, gate, and magnetic-field dependence on the edge conduction have been checked. However, we note that the edge conductance never reaches the expected quantum value of $e^2/h = 38.7$ μS, as can be seen in Figure 13.35a. This could be due to the imperfect transmission between the metal contacts and the edge or the backscattering from multiple magnetic impurities, which all prevent observation of the intrinsic edge conductance [190]. To address this, a study adopting another device geometry is shown in Figure 13.36a, in which the device consists of eight contact electrodes (four on each side), a top graphite gate, and a series of in-channel local bottom gates with width L_c varying from 50 to 900 nm. The monolayer flake is carefully selected to have a long strip shape (a few μm wide and about 10μm long) and is fully encapsulated between two hBN flakes. The goal of the design was to minimize the effect of contact resistance and to enable a length-dependence study

on a single device. Figure 13.36b shows a typical measurement of the four-terminal conductance across all local gates (≈ 8 μm long) as a function of top-gate voltage, V_{tg}. Like previous study (Figure 13.35a), a finite conductance plateau develops around $V_{tg} = 0$ V, indicating a regime of edge conduction. To study the length-dependence of edge conduction, a short transport channel with length L_c was selectively defined by a local gate voltage V_c, whereas the rest of the flake is highly doped by V_{tg} to ensure good contact to the electrodes (see the left inset of Figure 13.36c). Figure 13.36c shows the 2D resistance map as a function of V_{tg} and V_c for a local gate with $L_c = 100$ nm. The step in conductance (color contrast) indicates a transition from a bulk-metallic state (doped) to a bulk-insulating state (undoped) within the locally gated region, from which the offset resistance (resistance change from the value in the highly doped limit) can be defined as $\Delta R = R(V_c) - R(V_c = -1V)$. The saturated value of ΔR at large negative V_c, as shown in Figure 13.36d, thus measures the resistance of the undoped channel, which can only originate from the edges because the monolayer interior is insulating. Notably, ΔR for different channel length (100-nm channel and the 60- and 70-nm channels) all saturate at $h/2e^2$. Given that the sample has two edges, the observed conductance per edge is therefore e^2/h, pointing to helical edge modes as the source of the conductance and confirming the abovementioned criteria (i)

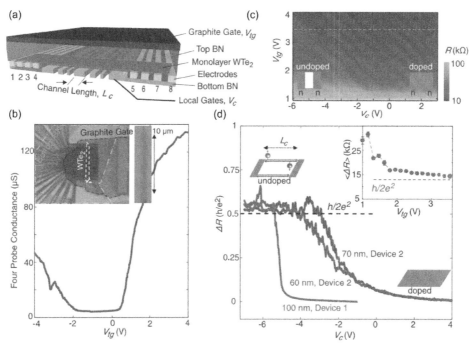

FIGURE 13.36 Length-dependence study on edge conduction. (a) A schematic of the device structure. (b) Four-terminal conductance measurement at 4 K of the device (see inset) as a function of V_{tg} across all the local gates, which are floating. Inset shows the optical image of the device and the corresponding monolayer WTe$_2$ flake before fabrication. (c) 2D map of the resistance tuned by V_{tg} and the 100-nm-wide local gate V_c at 4 K. Two regions are separated by a step in the resistance distinguishing the doped and undoped local channel, as depicted by the inset schematics. (d) ΔR vs.V_c for the 100-nm-wide gate on the device shown in (b) at $V_{tg} = 3.5$ V [white dashed line in (c)] and the 60- and 70-nm-wide gates on another device at $V_{tg} = 4.1$ V (taken at 5 K). Inset shows the average step height ΔR, extracted from (c), as a function of V_{tg}, showing a clear saturation toward $h/2e^2$ for large V_{tg}. (Figure adapted with permission from Ref. [189]. Copyright 2018 American Association for the Advancement of Science.)

and (ii). The observed QSH edge states in 1T′-TMDs can be readily integrated with s-wave superconductors for testing Majorana-related physics, such as the 4π-periodicity current phase relation [191] and doubling of the voltage spacing for Shapiro steps [187].

13.7 Summary

In summary, we have provided the theoretical background and a number of experimental studies relevant to developing various qubits in 2D materials. The single-electron transport properties of graphene nanodevices (GNRs, GSQDs, and GDQDs) fabricated on SiO_2 and hBN substrates were both reviewed. GSQDs fabricated on SiO_2 and hBN show a distinct difference in their Coulomb blockade peak-spacing fluctuations, indicating that edge roughness is the dominant source of disorder for QDs with diameters of less than 100 nm. For 2H-TMD nanostructures, despite Coulomb blockade being demonstrated in both SQDs and DQDs, B-field dependence of the dot levels is still lacking, presumably due to the intrinsic high disorder in the materials. Nevertheless, we discussed their Fock–Darwin spectra for the potential use in spin and valley qubits. We also reviewed the transport properties of JJs made of various 2D materials. The $I_C R_N$ correlation and current-phase relation in graphene-based lateral junctions have been studied, while the layer dependence of critical current was demonstrated in vertical junctions made of 2H-MoS_2. In addition, Weyl semimetal T_d-WTe_2-based junctions have shown an unusual temperature dependence of critical current, suggesting the existence of topologically protected surface states. Finally, we reviewed recent experimental studies showing that monolayer 1T′-WTe_2 is a natural 2D TI with QSH edge states thatare protected by TRS.

The absence of spin blockade in GDQDs [89,90,119–123] and the fact that the measured spin relaxation times in 2D graphene flakes are shorter than expected [101–103,133,134] suggest that there are extrinsic effects that govern the spin relaxation dynamics in graphene. These effects may be related to the scattering of electrons of magnetic impurities originating from carbon atom vacancies or graphene edge roughness [16]. Since graphene constrictions have been used as tunnel barriers in most GQDs reported thus far, alteration of the electron spin by the constriction edge during transport could be inevitable. Possible solutions to circumvent this issue could be the following: (i) using an electrical-field-induced bandgap in bilayer graphene to define GQDs [92,93,192,193]; (ii) functionalization of graphene, such as fluorinated graphene (FG), to simultaneously define the quantum confinement and passivate the graphene edges [80,81]; and (iii) choosing magnetic inert materials, such as hBN, as a tunnel barrier to build vertical graphene nanodevices [194,195]. In the first two cases, nanodevices defined in gated bilayer and functionalized graphene have been achieved, but no DQD structures (useful for T_1 and T_2^* measurements) have been reported. Vertical tunneling [196] to GQDs using hBN as a tunnel barrier may also serve as a promising solution for minimizing the edge effect from graphene constrictions. Despite the edge of GQDs still exists, the atomically flat and thin hBN as tunnel barriers should contain less disorders to perturb spin during the electron tunnelling. It is also possible to probe spin relaxation time in vertical QDs using pulse gating techniques [7]. However, spin qubits made in vertical geometry are difficult for upscaling, as currently, they are mostly made in lateral configurations in GaAs and Si systems [7,8]. The large bandgaps of few-layer 2H-TMDs can be advantageous for defining QDs *via* electrical gating. But the disorder nature of the material might lead to a strong decoherence that shorten T_1 and T_2^*. Nevertheless, spin blockade or spin relaxation time measurements have not been reported from this material either.

On the other hand, the graphene-based JJs have shown promising prospects for use in superconducting qubits owing to its ballistic nature and gate-tunable critical current [197,198]. The lateral 2H-TMDs JJs have not been reported so far, perhaps due to their diffusive nature that is difficult to preserve the induced superconductivity. On the contrary, T_d phase TMD-based junctions have shown great conductivity for sustaining large Josephson current, but they are not gateable except for the few-layer form (≤3 layers). The few-layer T_d-TMDs are reported to oxidize very quickly in air [189,190], which may add difficulty for making T_d-TMDs-based gatemons. Apart from graphene, experimental reports on other 2D material-based gatemon qubits are still lacking to date.

The pursuit of MFs in various solid-state systems has been a hot topic over the past decade, from the early tunneling experiments on semiconducting nanowires [182–184] to the more recent studies on proximatized TIs [185–188] and quantum anomalous Hall insulator [199]. Later studies of STM experiments further confirm their location at the end of a chain of magnetic atoms [200]. However, whether these unconventional quasiparticles obey non-Abelian exchange statistics will not be known until braiding and readout are performed. More sophisticated devices, such as QD chains and T-junctions [201–207], that allow exchange of particles are needed for further investigations. 1T′-TMDs with the proposed gate-tunable topological phases are particular interesting to test these theoretical proposals and other potential computing schemes [208].

References

1. S. Schnez, *Applied Physics Letters* **94**, 012107 (2009), doi:10.1063/1.3064128.
2. J. Guttinger, *Applied Physics Letters* **93**, 212102 (2008), doi:10.1063/1.3036419.
3. J. Guttinger, C. Stampfer, F. Libisch, T. Frey, J. Burgdorfer, T. Ihn, and K. Ensslin, *Physical Review Letters* **103**, 046810 (2009), doi:10.1103/PhysRevLett.103.046810.

4. S. Cho, D. Kim, P. Syers, N. P. Butch, J. Paglione, and M. S. Fuhrer, *Nano Letters* **12**, 469 (2012), doi:10.1021/nl203851g.

5. S. S. Hong, Y. Zhang, J. J. Cha, X.-L. Qi, and Y. Cui, *Nano Letters* **14**, 2815 (2014), doi:10.1021/nl500822g.

6. Y. Li and N. Mason, ArXiv e-prints (2013),

7. R. Hanson, L. P. Kouwenhoven, J. R. Petta, S. Tarucha, and L. M. K. Vandersypen, *Reviews of Modern Physics* **79**, 1217 (2007), doi:10.1103/RevModPhys.79.1217.

8. F. A. Zwanenburg, A. S. Dzurak, A. Morello, M. Y. Simmons, L. C. L. Hollenberg, G. Klimeck, S. Rogge, S. N. Coppersmith, and M. A. Eriksson, *Reviews of Modern Physics* **85**, 961 (2013), doi:10.1103/RevModPhys.85.961.

9. M. Veldhorst, C. H. Yang, J. C. C. Hwang, W. Huang, J. P. Dehollain, J. T. Muhonen, S. Simmons, A. Laucht, F. E. Hudson, K. M. Itoh, et al., *Nature* **526**, 410 (2015), doi:10.1038/nature15263.

10. A. K. Geim and I. V. Grigorieva, *Nature* **499**, 419 (2013), ISSN 0028-0836, doi:10.1038/nature12385.

11. K.-L. Chiu and Y. Xu, *Physics Reports* **669**, 1 (2017a), ISSN 0370-1573, single-electron Transport in Graphene-like Nanostructures, www.sciencedirect.com/science/article/pii/S0370157316303933.

12. C. Stampfer, S. Fringes, J. Guttinger, F. Molitor, C. Volk, B. Terrs, J. Dauber, S. Engels, S. Schnez, A. Jacobsen, et al., **6**, 271 (2011), ISSN 2095-0462, doi:10.1007/s11467-011-0182-3.

13. F. Molitor, J. Guttinger, C. Stampfer, S. Druscher, A. Jacobsen, T. Ihn, and K. Ensslin, *Journal of Physics: Condensed Matter* **23**, 243201 (2011), ISSN 0953-8984, http://stacks.iop.org/0953-8984/23/i=24/a=243201.

14. J. Guttinger, F. Molitor, C. Stampfer, S. Schnez, A. Jacobsen, S. Droscher, T. Ihn, and K. Ensslin, *Reports on Progress in Physics* **75**, 126502 (2012), ISSN 0034-4885, http://stacks.iop.org/0034-4885/75/i=12/a=126502.

15. C. Neumann, C. Volk, S. Engels, and C. Stampfer, *Nanotechnology* **24**, 444001 (2013), ISSN 0957-4484, http://stacks.iop.org/0957-4484/24/i=44/a=444001.

16. K.-L. Chiu and Y. Xu, *Physics Reports* **669**, 1 (2017b), ISSN 0370-1573, single-electron Transport in Graphene-like Nanostructures, www.sciencedirect.com/science/article/pii/S0370157316303933.

17. K. S. Novoselov, A. K. Geim, S. V. Morozov, D. Jiang, Y. Zhang, S. V. Dubonos, I. V. Grigorieva, and A. A. Firsov, *Science* **306**, 666 (2004), www.sciencemag.org/content/306/5696/666.abstract.

18. A. H. Castro Neto, F. Guinea, N. M. R. Peres, K. S. Novoselov, and A. K. Geim, *Reviews of Modern Physics* **81**, 109 (2009), doi:10.1103/RevModPhys.81.109.

19. A. K. Geim and K. S. Novoselov, *Nature Materials* **6**, 183 (2007), ISSN 1476-1122, doi:10.1038/nmat1849.

20. D. Basov, M. Fogler, A. Lanzara, F. Wang, and Y. Zhang, *Reviews of Modern Physics* **86**, 959 (2014), doi:10.1103/RevModPhys.86.959.

21. B. Amorim, A. Cortijo, F. de Juan, A. Grushin, F. Guinea, A. Gutirrez-Rubio, H. Ochoa, V. Parente, R. Roldn, P. San-Jose, et al., *Physics Reports* **617**, 1 (2016), ISSN 0370-1573, www.sciencedirect.com/science/article/pii/S0370157315005402.

22. N. M. R. Peres, F. Guinea, and A. H. Castro Neto, *Physical Review B* **73**, 125411 (2006), doi:10.1103/PhysRevB.73.125411.

23. K. S. Novoselov, E. McCann, S. V. Morozov, V. I. Falko, M. I. Katsnelson, U. Zeitler, D. Jiang, F. Schedin, and A. K. Geim, *Nature Physics* **2**, 177 (2006), ISSN 1745-2473, doi:10.1038/nphys245.

24. S. Das Sarma, S. Adam, E. H. Hwang, and E. Rossi, *Reviews of Modern Physics* **83**, 407 (2011), doi:10.1103/RevModPhys.83.407.

25. I. A. Lukanchuk, I. A.nchuk and Y. Kopelevich, *Physical Review Letters* **93**, 166402 (2004), doi:10.1103/PhysRevLett.93.166402.

26. J. Xue, ArXiv e-prints (2013),

27. M. Xu, T. Liang, M. Shi, and H. Chen, *Chemical Reviews* **113**, 3766 (2013), ISSN 0009-2665, doi:10.1021/cr300263a.

28. J. Xue, J. Sanchez-Yamagishi, D. Bulmash, P. Jacquod, A. Deshpande, K. Watanabe, T. Taniguchi, P. Jarillo-Herrero, and B. J. LeRoy, *Nature Materials* **10**, 282 (2011), ISSN 1476-1122, doi:10.1038/nmat2968.

29. C. R. Dean, L. Wang, P. Maher, C. Forsythe, F. Ghahari, Y. Gao, J. Katoch, M. Ishigami, P. Moon, M. Koshino, et al., *Nature* **497**, 598 (2013), ISSN 0028-0836, doi:10.1038/nature12186.

30. B. Hunt, J. D. Sanchez-Yamagishi, A. F. Young, M. Yankowitz, B. J. LeRoy, K. Watanabe, T. Taniguchi, P. Moon, M. Koshino, P. Jarillo-Herrero, et al., *Science* **340**, 1427 (2013), www.sciencemag.org/content/340/6139/1427.abstract.

31. D. Golberg, Y. Bando, Y. Huang, T. Terao, M. Mitome, C. Tang, and C. Zhi, *ACS Nano* **4**, 2979 (2010), ISSN 1936-0851, doi:10.1021/nn1006495.

32. H. Wang, F. Liu, W. Fu, Z. Fang, W. Zhou, and Z. Liu, *Nanoscale* **6**, 12250 (2014), ISSN 2040-3364, doi:10.1039/C4NR03435J.

33. A. C. Ferrari, F. Bonaccorso, V. Fal'ko, K. S. Novoselov, S. Roche, P. Boggild, S. Borini, F. H. L. Koppens, V. Palermo, N. Pugno, et al., *Nanoscale* **7**, 4598 (2015), ISSN 2040-3364, doi:10.1039/C4NR01600A.

34. C. R. Dean, A. F. Young, I. Meric, C. Lee, L. Wang, S. Sorgenfrei, K. Watanabe, T. Taniguchi, P. Kim, K. L. Shepard, et al., *Nature*

Nanotechnology **5**, 722 (2010), ISSN 1748-3387, doi:10.1038/nnano.2010.172.

35. L. A. Ponomarenko, R. V. Gorbachev, G. L. Yu, D. C. Elias, R. Jalil, A. A. Patel, A. Mishchenko, A. S. Mayorov, C. R. Woods, J. R. Wallbank, et al., *Nature* **497**, 594 (2013), ISSN 0028-0836, doi:10.1038/nature12187.

36. F. Amet, A. J. Bestwick, J. R. Williams, L. Balicas, K. Watanabe, T. Taniguchi, and D. Goldhaber-Gordon, *Nature Communications* **6**, (2015), doi:10.1038/ncomms6838.

37. A. Kou, B. E. Feldman, A. J. Levin, B. I. Halperin, K. Watanabe, T. Taniguchi, and A. Yacoby, *Science* **345**, 55 (2014), ISSN 0036-8075, http://science.sciencemag.org/content/345/6192/55.

38. F. Amet, C. T. Ke, I. V. Borzenets, J. Wang, K. Watanabe, T. Taniguchi, R. S. Deacon, M. Yamamoto, Y. Bomze, S. Tarucha, et al., *Science* **352**, 966 (2016), ISSN 0036-8075, http://science.sciencemag.org/content/352/6288/966.

39. M. Yankowitz, J. Xue, D. Cormode, J. D. Sanchez-Yamagishi, K. Watanabe, T. Taniguchi, P. Jarillo-Herrero, P. Jacquod, and B. J. LeRoy, *Nature Physics* **8**, 382 (2012), ISSN 1745-2473, doi:10.1038/nphys2272.

40. A. Splendiani, L. Sun, Y. Zhang, T. Li, J. Kim, C.-Y. Chim, G. Galli, and F. Wang, *Nano Letters* **10**, 1271 (2010), ISSN 1530-6984, doi:10.1021/nl903868w.

41. K. F. Mak, C. Lee, J. Hone, J. Shan, and T. F. Heinz, *Physical Review Letters* **105**, 136805 (2010), doi:10.1103/PhysRevLett.105.136805.

42. G.-B. Liu, W.-Y. Shan, Y. Yao, W. Yao, and D. Xiao, *Physical Review B* **88**, 085433 (2013), doi:10.1103/PhysRevB.88.085433.

43. X. Duan, C. Wang, A. Pan, R. Yu, and X. Duan, *Chemical Society Reviews* **44**, 8859 (2015), doi:10.1039/C5CS00507H.

44. D. Xiao, G.-B. Liu, W. Feng, X. Xu, and W. Yao, *Physical Review Letters* **108**, 196802 (2012), doi:10.1103/PhysRevLett.108.196802.

45. D. Xiao, M.-C. Chang, and Q. Niu, *Reviews of Modern Physics* **82**, 1959 (2010), doi:10.1103/RevModPhys.82.1959.

46. X. Xu, W. Yao, D. Xiao, and T. F. Heinz, *Nature Physics* **10**, 343 (2014), ISSN 1745-2473, doi:10.1038/nphys2942.

47. S. Tang, C. Zhang, D. Wong, Z. Pedramrazi, H.-Z. Tsai, C. Jia, B. Moritz, M. Claassen, H. Ryu, S. Kahn, et al., *Nature Physics* **13**, 683 (2017), doi:10.1038/nphys4174.

48. X. Qian, J. Liu, L. Fu, and J. Li, *Science* **346**, 1344 (2014), ISSN 0036-8075, http://science.sciencemag.org/content/346/6215/1344.

49. L. Fu and C. L. Kane, *Physical Review B* **76**, 045302 (2007), doi:10.1103/PhysRevB.76.045302.

50. M. Konig, S. Wiedmann, C. Brne, A. Roth, H. Buhmann, L. W. Molenkamp, X.-L. Qi, and S.-C. Zhang, *Science* **318**, 766 (2007), www.sciencemag.org/content/318/5851/766.abstract.

51. J. Alicea, *Reports on Progress in Physics* **75**, 076501 (2012), http://stacks.iop.org/0034-4885/75/i=7/a=076501.

52. V. Fock, *Pysics* **47**, 446 (1928), ISSN 0044-3328, doi:10.1007/BF01390750.

53. L. P. Kouwenhoven, D. G. Austing, and S. Tarucha, *Reports on Progress in Physics* **64**, 701 (2001), ISSN 0034-4885, http://stacks.iop.org/0034-4885/64/i=6/a=201.

54. S. Schnez, K. Ensslin, M. Sigrist, and T. Ihn, *Physical Review B* **78**, 195427 (2008), doi:10.1103/PhysRevB.78.195427.

55. P. Recher, J. Nilsson, G. Burkard, and B. Trauzettel, *Physical Review B* **79**, 085407 (2009), doi:10.1103/PhysRevB.79.085407.

56. K. L. Chiu, M. R. Connolly, A. Cresti, C. Chua, S. J. Chorley, F. Sfigakis, S. Milana, A. C. Ferrari, J. P. Griffiths, G. A. C. Jones, et al., *Physical Review B* **85**, 205452 (2012), doi:10.1103/PhysRevB.85.205452.

57. L. A. Ponomarenko, F. Schedin, M. I. Katsnelson, R. Yang, E. W. Hill, K. S. Novoselov, and A. K. Geim, *Science* **320**, 356 (2008), www.sciencemag.org/content/320/5874/356.abstract.

58. W. G. van der Wiel, S. De Franceschi, J. M. Elzerman, T. Fujisawa, S. Tarucha, and L. P. Kouwenhoven, *Reviews of Modern Physics* **75**, 1 (2002), doi:10.1103/RevModPhys.75.1.

59. L. J. Geerligs, V. F. Anderegg, P. A. M. Holweg, J. E. Mooij, H. Pothier, D. Esteve, C. Urbina, and M. H. Devoret, *Physical Review Letters* **64**, 2691 (1990), doi:10.1103/PhysRevLett.64.2691.

60. L. P. Kouwenhoven, A. T. Johnson, N. C. van der Vaart, C. J. P. M. Harmans, and C. T. Foxon, *Physical Review Letters* **67**, 1626 (1991), doi:10.1103/PhysRevLett.67.1626.

61. T. M. Klapwijk, *Journal of Superconductivity* **17**, 593 (2004a), ISSN 1572-9605, doi:10.1007/s10948-004-0773-0.

62. I.-J. Wang, Doctoral dissertation, Harvard University (2016), https://dash.harvard.edu/handle/1/26718763.

63. G. E. Blonder, M. Tinkham, and T. M. Klapwijk, *Physical Review B* **25**, 4515 (1982), doi:10.1103/PhysRevB.25.4515.

64. I. O. Kulik, *Soviet Journal of Experimental and Theoretical Physics* **30**, 944 (1969).

65. L. Bretheau, J. I.-J. Wang, R. Pisoni, K. Watanabe, T. Taniguchi, and P. Jarillo-Herrero, *Nature Physics* **13**, 756 (2017), doi:10.1038/nphys4110.

66. T. M. Klapwijk, *Journal of Superconductivity* **17**, 593 (2004b), ISSN 1572-9605, doi:10.1007/s10948-004-0773-0.

67. P. F. Bagwell, *Physical Review B* **46**, 12573 (1992), doi:10.1103/PhysRevB.46.12573.

68. Y.-M. Lin, V. Perebeinos, Z. Chen, and P. Avouris, *Physical Review B* **78**, 161409 (2008), doi:10.1103/PhysRevB.78.161409.

69. K. Nakada, M. Fujita, G. Dresselhaus, and M. S. Dresselhaus, *Physical Review B* **54**, 17954 (1996), doi:10.1103/PhysRevB.54.17954.

70. K. Wakabayashi, M. Fujita, H. Ajiki, and M. Sigrist, *Physical Review B* **59**, 8271 (1999), doi:10.1103/PhysRevB.59.8271.

71. M. Y. Han, B. Ozyilmaz, Y. Zhang, and P. Kim, *Physical Review Letters* **98**, 206805 (2007), doi:10.1103/PhysRevLett.98.206805.

72. K. Todd, H.-T. Chou, S. Amasha, and D. Goldhaber-Gordon, *Nano Letters* **9**, 416 (2008), ISSN 1530-6984, doi:10.1021/nl803291b.

73. F. Molitor, A. Jacobsen, C. Stampfer, J. Guttinger, J.ttinger, T. Ihn, and K. Ensslin, *Physical Review B* **79**, 075426 (2009), doi:10.1103/PhysRevB.79.075426.

74. J. Bai, R. Cheng, F. Xiu, L. Liao, M. Wang, A. Shailos, K. L. Wang, Y. Huang, and X. Duan, *Nature Nanotechnology* **5**, 655 (2010), ISSN 1748-3387, doi:10.1038/nnano.2010.154.

75. M. R. Connolly, K. L. Chiu, A. Lombardo, A. Fasoli, A. C. Ferrari, D. Anderson, G. A. C. Jones, and C. G. Smith, *Physical Review B* **83**, 115441 (2011), doi:10.1103/PhysRevB.83.115441.

76. X. Wang, Y. Ouyang, L. Jiao, H. Wang, L. Xie, J. Wu, J. Guo, and H. Dai, *Nature Nanotechnology* **6**, 563 (2011), ISSN 1748-3387, doi:10.1038/nnano.2011.138.

77. L. Jiao, X. Wang, G. Diankov, H. Wang, and H. Dai, *Nature Nanotechnology* **5**, 321 (2010), ISSN 1748-3387, doi:10.1038/nnano.2010.54.

78. D. Wei, L. Xie, K. K. Lee, Z. Hu, S. Tan, W. Chen, C. H. Sow, K. Chen, Y. Liu, and A. T. S. Wee, *Nature Communications* **4**, 1374 (2013a), doi:10.1038/ncomms2366.

79. X. Wang and H. Dai, *Nature Chemistry* **2**, 661 (2010), ISSN 1755-4330, doi:10.1038/nchem.719.

80. F. Withers, T. H. Bointon, M. Dubois, S. Russo, and M. F. Craciun, *Nano Letters* **11**, 3912 (2011), doi:10.1021/nl2020697.

81. W.-K. Lee, J. T. Robinson, D. Gunlycke, R. R. Stine, C. R. Tamanaha, W. P. King, and P. E. Sheehan, *Nano Letters* **11**, 5461 (2011), doi:10.1021/nl203225w.

82. C. Stampfer, J. Guttinger, S. Hellmuller, F. Molitor, K. Ensslin, and T. Ihn, *Physical Review Letters* **102**, 056403 (2009), doi:10.1103/PhysRevLett.102.056403.

83. A. Dorn, T. Ihn, K. Ensslin, W. Wegscheider, and M. Bichler, *Physical Review B* **70**, 205306 (2004), doi:10.1103/PhysRevB.70.205306.

84. J. Martin, N. Akerman, G. Ulbricht, T. Lohmann, J. H. Smet, K. von Klitzing, and A. Yacoby, *Nature Physics* **4**, 144 (2008), ISSN 1745-2473, doi:10.1038/nphys781.

85. D. Bischoff, A. Varlet, P. Simonet, M. Eich, H. C. Overweg, T. Ihn, and K. Ensslin, *Applied Physics Reviews* **2**, 031301 (2015a), http://scitation. aip.org/content/aip/journal/apr2/2/3/10.1063/ 1.4926448.

86. B. Trauzettel, D. V. Bulaev, D. Loss, and G. Burkard, *Nature Physics* **3**, 192 (2007), ISSN 1745-2473, doi:10.1038/nphys544.

87. C. Volk, C. Neumann, S. Kazarski, S. Fringes, S. Engels, F. Haupt, A. Muller, T.ller, and C. Stampfer, *Nature Communications* **4**, 1753 (2013), doi:10.1038/ncomms2738.

88. J. Guttinger, T. Frey, C. Stampfer, T. Ihn, and K. Ensslin, *Physical Review Letters* **105**, 116801 (2010), doi:10.1103/PhysRevLett.105.116801.

89. X. L. Liu, D. Hug, and L. M. K. Vandersypen, *Nano Letters* **10**, 1623 (2010), ISSN 1530-6984, doi:10.1021/nl9040912.

90. C. Volk, S. Fringes, B. Terres, J. Dauber, S. Engels, S. Trellenkamp, and C. Stampfer, *Nano Letters* **11**, 3581 (2011), ISSN 1530-6984, doi:10.1021/nl201295s.

91. M. R. Connolly, K. L. Chiu, S. P. Giblin, M. Kataoka, J. D. Fletcher, C. Chua, J. P. Griffiths, G. A. C. Jones, V. I. Fal'ko, and C. G. S. T. J. B. M. Janssen, *Nature Nanotechnology* **8**, 417 (2013), ISSN 1748-3387, doi:10.1038/nnano.2013.73.

92. M. T. Allen, J. Martin, and A. Yacoby, *Nature Communications* **3**, 934 (2012), doi:10.1038/ncomms1945.

93. A. S. M. Goossens, S. C. M. Driessen, T. A. Baart, K. Watanabe, T. Taniguchi, and L. M. K. Vandersypen, *Nano Letters* **12**, 4656 (2012), ISSN 1530-6984, doi:10.1021/nl301986q.

94. Y. Zhang, V. W. Brar, C. Girit, A. Zettl, and M. F. Crommie, *Nature Physics* **5**, 722 (2009a), ISSN 1745-2473, doi:10.1038/nphys1365.

95. F. Amet, J. R. Williams, A. G. F. Garcia, M. Yankowitz, K. Watanabe, T. Taniguchi, and D. Goldhaber-Gordon, *Physical Review B* **85**, 073405 (2012), doi:10.1103/PhysRevB.85.073405.

96. N. N. Klimov, S. Jung, S. Zhu, T. Li, C. A. Wright, S. D. Solares, D. B. Newell, N. B. Zhitenev, and J. A. Stroscio, *Science* **336**, 1557 (2012), www.sciencemag.org/content/336/ 6088/1557.abstract.

97. J. A. Folk, C. M. Marcus, R. Berkovits, I. L. Kurland, I. L. Aleiner, and B. L. Altshuler, *Physica Scripta* **2001**, 26 (2001), ISSN 1402-4896, http://stacks.iop. org/1402-4896/2001/i=T90/a=003.

98. S. Lindemann, T. Ihn, T. Heinzel, W. Zwerger, K. Ensslin, K. Maranowski, and A. C. Gossard, *Physical Review B* **66**, 195314 (2002), doi:10.1103/PhysRevB.66.195314.

99. M. R. Buitelaar, A. Bachtold, T. Nussbaumer, M. Iqbal, and C. Schnenberger, *Physical Review Letters* **88**, 156801 (2002), doi:10.1103/PhysRevLett.88.156801.

100. D. H. Cobden and J. Nygard, *Physical Review Letters* **89**, 046803 (2002), doi:10.1103/PhysRevLett.89.046803.

101. N. Tombros, C. Jozsa, M. Popinciuc, H. T. Jonkman, and B. J. van Wees, *Nature* **448**, 571 (2007), ISSN 0028-0836, doi:10.1038/nature06037.

102. W. Han, K. Pi, K. M. McCreary, Y. Li, J. J. I. Wong, A. G. Swartz, and R. K. Kawakami, *Physical Review Letters* **105**, 167202 (2010), doi:10.1103/PhysRevLett.105.167202.

103. W. Han and R. K. Kawakami, *Physical Review Letters* **107**, 047207 (2011), doi:10.1103/PhysRevLett.107.047207.

104. S. Droscher, J. Guttinger, T. Mathis, B. Batlogg, T. Ihn, and K. Ensslin, *Applied Physics Letters* **101**, 043107 (2012), http://scitation.aip.org/content/aip/journal/apl/101/4/10.1063/1.4737937.

105. T. Ando, *Journal of the Physical Society of Japan* **74**, 777 (2005), ISSN 0031-9015, doi:10.1143/JPSJ.74.777.

106. K. Kaasbjerg, K. S. Thygesen, and K. W. Jacobsen, *Physical Review B* **85**, 165440 (2012), doi:10.1103/PhysRevB.85.165440.

107. T. Fujisawa, Y. Tokura, and Y. Hirayama, *Physical Review B* **63**, 081304 (2001), doi:10.1103/PhysRevB.63.081304.

108. T. Fujisawa, D. G. Austing, Y. Tokura, Y. Hirayama, and S. Tarucha, *Nature* **419**, 278 (2002a), ISSN 0028-0836, doi:10.1038/nature00976.

109. T. Fujisawa, D. G. Austing, Y. Tokura, Y. Hirayama, and S. Tarucha, *Physical Review Letters* **88**, 236802 (2002b), doi:10.1103/PhysRevLett.88.236802.

110. A. Pfund, I. Shorubalko, R. Leturcq, and K. Ensslin, *Applied Physics Letters* **89**, 252106 (2006), http://scitation.aip.org/content/aip/journal/apl/89/25/10.1063/1.2409625.

111. M. D. Schroer, K. D. Petersson, M. Jung, and J. R. Petta, *Physical Review Letters* **107**, 176811 (2011), doi:10.1103/PhysRevLett.107.176811.

112. S. Pecker, F. Kuemmeth, A. Secchi, M. Rontani, D. C. Ralph, P. L. McEuen, and S. Ilani, *Nature Physics* **9**, 576 (2013), ISSN 1745-2473, doi:10.1038/nphys2692.

113. K. Ono, D. G. Austing, Y. Tokura, and S. Tarucha, *Science* **297**, 1313 (2002), www.sciencemag.org/content/297/5585/1313.abstract.

114. A. C. Johnson, J. R. Petta, J. M. Taylor, A. Yacoby, M. D. Lukin, C. M. Marcus, M. P. Hanson, and A. C. Gossard, *Nature* **435**, 925 (2005a), ISSN 0028-0836, doi:10.1038/nature03815.

115. A. C. Johnson, J. R. Petta, C. M. Marcus, M. P. Hanson, and A. C. Gossard, *Physical Review B* **72**, 165308 (2005b), doi:10.1103/PhysRevB.72.165308.

116. K. D. Petersson, C. G. Smith, D. Anderson, P. Atkinson, G. A. C. Jones, and D. A. Ritchie, *Nano Letters* **10**, 2789 (2010), ISSN 1530-6984, doi:10.1021/nl100663w.

117. S. J. Chorley, G. Giavaras, J. Wabnig, G. A. C. Jones, C. G. Smith, G. A. D. Briggs, and M. R. Buitelaar, *Physical Review Letters* **106**, 206801 (2011), doi:10.1103/PhysRevLett.106.206801.

118. D. Goldhaber-Gordon, H. Shtrikman, D. Mahalu, D. Abusch-Magder, U. Meirav, and M. A. Kastner, *Nature* **391**, 156 (1998), ISSN 0028-0836, doi:10.1038/34373.

119. S. Moriyama, D. Tsuya, E. Watanabe, S. Uji, M. Shimizu, T. Mori, T. Yamaguchi, and K. Ishibashi, *Nano Letters* **9**, 2891 (2009), ISSN 1530-6984, doi:10.1021/nl9011535.

120. F. Molitor, H. Knowles, S. Droscher, U. Gasser, T. Choi, P. Roulleau, J. Guttinger, A. Jacobsen, C. Stampfer, K. Ensslin, et al., *EPL (Europhysics Letters)* **89**, 67005 (2010), http://stacks.iop.org/0295-5075/89/i=6/a=67005.

121. L.-J. Wang, H.-O. Li, T. Tu, G. Cao, C. Zhou, X.-J. Hao, Z. Su, M. Xiao, G.-C. Guo, A. M. Chang, et al., *Applied Physics Letters* **100**, 022106 (2012), http://scitation.aip.org/content/aip/journal/apl/100/2/10.1063/1.3676083.

122. D. Wei, H.-O. Li, G. Cao, G. Luo, Z.-X. Zheng, T. Tu, M. Xiao, G.-C. Guo, H.-W. Jiang, and G.-P. Guo, *Scientific Reports* **3**, (2013b), doi:10.1038/srep03175.

123. K. L. Chiu, M. R. Connolly, A. Cresti, J. P. Griffiths, G. A. C. Jones, and C. G. Smith, ArXiv e-prints (2015a),

124. G.-W. Deng, D. Wei, J. Johansson, M.-L. Zhang, S.-X. Li, H.-O. Li, G. Cao, M. Xiao, T. Tu, G.-C. Guo, et al., *Physical Review Letters* **115**, 126804 (2015), doi:10.1103/PhysRevLett.115.126804.

125. F. Molitor, *Applied Physics Letters* **94**, 222107 (2009), doi:10.1063/1.3148367.

126. K. L. Chiu, M. R. Connolly, A. Cresti, J. P. Griffiths, G. A. C. Jones, and C. G. Smith, *Physical Review B* **92**, 155408 (2015b), doi:10.1103/PhysRevB.92.155408.

127. H. Pothier, P. Lafarge, C. Urbina, D. Esteve, and M. H. Devoret, *EPL (Europhysics Letters)* **17**, 249 (1992), http://stacks.iop.org/0295-5075/17/i=3/a=011.

128. A. Fuhrer, C. Fasth, and L. Samuelson, *Applied Physics Letters* **91**, 052109 (2007), http://scitation.aip.org/content/aip/journal/apl/91/5/10.1063/1.2767197.

129. S. J. Chorley, J. Frake, C. G. Smith, G. A. C. Jones, and M. R. Buitelaar, *Applied Physics Letters* **100**, 143104 (2012), http://scitation.aip.org/content/aip/journal/apl/100/14/10.1063/1.3700967.

130. M. W. Keller, J. M. Martinis, N. M. Zimmerman, and A. H. Steinbach, *Applied Physics Letters* **69**, 1804 (1996), http://scitation.aip.org/content/aip/journal/apl/69/12/10.1063/1.117492.

131. S. Engels, A. Epping, C. Volk, S. Korte, B. Voigt-lander, K. Watanabe, T. Taniguchi, S. Trellenkamp, and C. Stampfer, *Applied Physics Letters* **103**, 073113 (2013), http://scitation.aip.org/content/aip/journal/apl/103/7/10.1063/1.4818627.

132. A. Epping, S. Engels, C. Volk, K. Watanabe, T. Taniguchi, S. Trellenkamp, and C. Stampfer, *Physica Status Solidi B* **250**, 2692 (2013), ISSN 1521-3951, doi:10.1002/pssb.201300295.

133. T. Maassen, J. J. van den Berg, N. IJbema, F. Fromm, T. Seyller, R. Yakimova, and B. J. van Wees, *Nano Letters* **12**, 1498 (2012), ISSN 1530-6984, doi:10.1021/nl2042497.

134. M. Drogeler, F. Volmer, M. Wolter, B. Terres, K. Watanabe, T. Taniguchi, G. Gntherodt, C. Stampfer, and B. Beschoten, *Nano Letters* **14**, 6050 (2014), ISSN 1530-6984, doi:10.1021/nl501278c.

135. D. Kochan, M. Gmitra, and J. Fabian, *Physical Review Letters* **112**, 116602 (2014), doi:10.1103/PhysRevLett.112.116602.

136. D. V. Tuan, F. Ortmann, D. Soriano, S. O. Valenzuela, and S. Roche, *Nature Physics* **10**, 857 (2014), ISSN 1745-2473, doi:10.1038/nphys3083.

137. Y. Zhang, T.-T. Tang, C. Girit, Z. Hao, M. C. Martin, A. Zettl, M. F. Crommie, Y. R. Shen, and F. Wang, *Nature* **459**, 820 (2009b), ISSN 0028-0836, doi:10.1038/nature08105.

138. N. Levy, S. A. Burke, K. L. Meaker, M. Panlasigui, A. Zettl, F. Guinea, A. H. C. Neto, and M. F. Crommie, *Science* **329**, 544 (2010), www.sciencemag.org/content/329/5991/544.abstract.

139. A. A. Pacheco Sanjuan, Z. Wang, H. P. Imani, M. Vanevic, and S. Barraza-Lopez, *Physical Review B* **89**, 121403 (2014), doi:10.1103/PhysRevB.89.121403.

140. S. Barraza-Lopez, A. A. Pacheco Sanjuan, Z. Wang, and M. Vanevic, *Solid State Communications* **166**, 70 (2013), ISSN 0038-1098, www.sciencedirect.com/science/article/pii/S0038109813002287.

141. N. M. Freitag, T. Reisch, L. A. Chizhova, P. Nemes-Incze, C. Holl, C. R. Woods, R. V. Gorbachev, Y. Cao, A. K. Geim, K. S. Novoselov, et al., *ArXiv e-prints* (2017),

142. R. K. Puddy, C. J. Chua, and M. R. Buitelaar, *Applied Physics Letters* **103**, 183117 (2013), doi:10.1063/1.4828663.

143. A. Muller, B. Kaestner, F. Hohls, T. Weimann, K. Pierz, and H. W. Schumacher, *Journal of Applied Physics* **115**, 233710 (2014), http://scitation.aip.org/content/aip/journal/jap/115/23/10.1063/1.4884617.

144. S. Fringes, C. Volk, B. Terres, J. Dauber, S. Engels, S. Trellenkamp, and C. Stampfer, *Physica Status Solidi C* **9**, 169 (2012), ISSN 1610-1642, doi:10.1002/pssc.201100340.

145. P. Roulleau, S. Baer, T. Choi, F. Molitor, J. Guttinger, T. Muller, S. Droscher, K. Ensslin, and T. Ihn, *Nature Communications* **2**, 239 (2011), doi:10.1038/ncomms1241.

146. X.-X. Song, D. Liu, V. Mosallanejad, J. You, T.-Y. Han, D.-T. Chen, H.-O. Li, G. Cao, M. Xiao, G.-C. Guo, et al., *Nanoscale* **7**, 16867 (2015), doi:10.1039/C5NR04961J.

147. D. Gopalakrishnan, D. Damien, and M. M. Shaijumon, *ACS Nano* **8**, 5297 (2014), ISSN 1936-0851, doi:10.1021/nn501479e.

148. H. Lin, C. Wang, J. Wu, Z. Xu, Y. Huang, and C. Zhang, *New Journal of Chemistry* **39**, 8492 (2015), ISSN 1144-0546, doi:10.1039/C5NJ01698C.

149. Y. Yan, C. Zhang, W. Gu, C. Ding, X. Li, and Y. Xian, *The Journal of Physical Chemistry C* pp. - (2016), ISSN 1932-7447, doi:10.1021/acs.jpcc.6b01868.

150. Z.-Z. Zhang, X.-X. Song, G. Luo, G.-W. Deng, V. Mosallanejad, T. Taniguchi, K. Watanabe, H.-O. Li, G. Cao, G.-C. Guo, et al., *Science Advances* **3** (2017), http://advances.sciencemag.org/content/3/10/e1701699.

151. R. Pisoni, Z. Lei, P. Back, M. Eich, H. Overweg, Y. Lee, K. Watanabe, T. Taniguchi, T. Ihn, and K. Ensslin, *Applied Physics Letters* **112**, 123101 (2018), ISSN 0003-6951, doi:10.1063/1.5021113.

152. A. Kormanyos, V. Zolyomi, N. D. Drummond, and G. Burkard, *Physical Review X* **4**, 011034 (2014), doi:10.1103/PhysRevX.4.011034.

153. T. W. Larsen, K. D. Petersson, F. Kuemmeth, T. S. Jespersen, P. Krogstrup, J. Nygård, and C. M. Marcus, *Physical Review Letters* **115**, 127001 (2015), doi:10.1103/PhysRevLett.115.127001.

154. J. W.D. Oliver. Editor: D. P. DiVincenzo, *Quantum Information Processing: Lecture Notes of the 44th IFF Spring School* (2013), http://equs.mit.edu/wp-content/uploads/2016/11/SC_qubits_Oliver_IFF_Spring_School_20140330.pdf.

155. J. E. Mooij, T. P. Orlando, L. Levitov, L. Tian, C. H. van der Wal, and S. Lloyd, *Science* **285**, 1036 (1999), ISSN 0036-8075, http://science.sciencemag.org/content/285/5430/1036.

156. T. P. Orlando, J. E. Mooij, L. Tian, C. H. van der Wal, L. S. Levitov, S. Lloyd, and J. J. Mazo, *Physical Review B* **60**, 15398 (1999), doi:10.1103/PhysRevB.60.15398.

157. M. D. Reed, Ph.D. thesis, Yale University (2014).

158. H. B. Heersche, P. Jarillo-Herrero, J. B. Oostinga, L. M. K. Vandersypen, and A. F. Morpurgo, *Nature* **446**, 56 (2007), doi:10.1038/nature05555.

159. T. Dirks, T. L. Hughes, S. Lal, B. Uchoa, Y.-F. Chen, C. Chialvo, P. M. Goldbart, and N. Mason, *Nature Physics* **7**, 386 (2011), doi:10.1038/nphys1911.

160. D. K. Efetov, L. Wang, C. Handschin, K. B. Efetov, J. Shuang, R. Cava, T. Taniguchi, K. Watanabe,

J. Hone, C. R. Dean, et al., *Nature Physics* **12**, 328 (2015), doi:10.1038/nphys3583.

161. G. Nanda, J. L. Aguilera-Servin, P. Rakyta, A. Kormnyos, R. Kleiner, D. Koelle, K. Watanabe, T. Taniguchi, L. M. K. Vandersypen, and S. Goswami, *Nano Letters* **17**, 3396 (2017), ISSN 1530-6984, doi:10.1021/acs.nanolett.7b00097.

162. M. T. Allen, O. Shtanko, I. C. Fulga, A. R. Akhmerov, K. Watanabe, T. Taniguchi, P. Jarillo-Herrero, L. S. Levitov, and A. Yacoby, *Nature Physics* **12**, 128 (2015), doi:10.1038/nphys3534.

163. G.-H. Lee, K.-F. Huang, D. K. Efetov, D. S. Wei, S. Hart, T. Taniguchi, K. Watanabe, A. Yacoby, and P. Kim, *Nature Physics* **13**, 693 (2017), doi:10.1038/nphys4084.

164. K. K. Likharev, *Reviews of Modern Physics* **51**, 101 (1979), doi:10.1103/RevModPhys.51.101.

165. Y.-J. Doh, J. A. van Dam, A. L. Roest, E. P. A. M. Bakkers, L. P. Kouwenhoven, and S. De Franceschi, *Science* **309**, 272 (2005), ISSN 0036-8075, http://science.sciencemag.org/content/309/5732/272.

166. M. Titov and C. W. J. Beenakker, *Physical Review B* **74**, 041401 (2006), doi:10.1103/PhysRevB.74.041401.

167. N. Yabuki, R. Moriya, M. Arai, Y. Sata, S. Morikawa, S. Masubuchi, and T. Machida, *Nature Communications* **7**, 10616 (2016), doi:10.1038/ncomms10616.

168. M. Kim, G.-H. Park, J. Lee, J. H. Lee, J. Park, H. Lee, G.-H. Lee, and H.-J. Lee, *Nano Letters* **17**, 6125 (2017), ISSN 1530-6984, doi:10.1021/acs.nanolett.7b02707.

169. J. O. Island, G. A. Steele, H. S. J. van der Zant, and A. Castellanos-Gomez, *2D Materials* **3**, 031002 (2016), http://stacks.iop.org/2053-1583/3/i=3/a=031002.

170. O. O. Shvetsov, A. Kononov, A. V. Timonina, N. N. Kolesnikov, and E. V. Deviatov, *ArXiv e-prints* (2018),

171. M. Octavio, M. Tinkham, G. E. Blonder, and T. M. Klapwijk, *Physical Review B* **27**, 6739 (1983), doi:10.1103/PhysRevB.27.6739.

172. N. P. Armitage, E. J. Mele, and A. Vishwanath, *Reviews of Modern Physics* **90**, 015001 (2018), doi:10.1103/RevModPhys.90.015001.

173. H. Weng, C. Fang, Z. Fang, B. A. Bernevig, and X. Dai, *Physical Review X* **5**, 011029 (2015), doi:10.1103/PhysRevX.5.011029.

174. D. Bulmash, C.-X. Liu, and X.-L. Qi, *Physical Review B* **89**, 081106 (2014), doi:10.1103/PhysRevB.89.081106.

175. A. A. Soluyanov, D. Gresch, Z. Wang, Q. Wu, M. Troyer, X. Dai, and B. A. Bernevig, *Nature* **527**, 495 (2015), doi:10.1038/nature15768.

176. T.-R. Chang, S.-Y. Xu, G. Chang, C.-C. Lee, S.-M. Huang, B. Wang, G. Bian, H. Zheng, D. S. Sanchez, I. Belopolski, et al., *Nature Communications* **7**, 10639 (2016), doi:10.1038/ncomms10639.

177. J. Jiang, Z. Liu, Y. Sun, H. Yang, C. Rajamathi, Y. Qi, L. Yang, C. Chen, H. Peng, C.-C. Hwang, et al., *Nature Communications* **8**, 13973 (2017), doi:10.1038/ncomms13973.

178. P. Li, Y. Wen, X. He, Q. Zhang, C. Xia, Z.-M. Yu, S. A. Yang, Z. Zhu, H. N. Alshareef, and X.-X. Zhang, *Nature Communications* **8**, 2150 (2017), ISSN 2041-1723, doi:10.1038/s41467-017-02237-1.

179. A. Y. Kitaev, *Physics-Uspekhi* **44**, 131 (2001), http://stacks.iop.org/1063-7869/44/i=10S/a=S29.

180. M. Leijnse and K. Flensberg, *Semiconductor Science and Technology* **27**, 124003 (2012), http://stacks.iop.org/0268-1242/27/i=12/a=124003.

181. C. Nayak, S. H. Simon, A. Stern, M. Freedman, and S. Das Sarma, *Reviews of Modern Physics* **80**, 1083 (2008), doi:10.1103/RevModPhys.80.1083.

182. V. Mourik, K. Zuo, S. M. Frolov, S. R. Plissard, E. P. A. M. Bakkers, and L. P. Kouwenhoven, *Science* **336**, 1003 (2012), ISSN 0036-8075, http://science.sciencemag.org/content/336/6084/1003.

183. S. M. Albrecht, A. P. Higginbotham, M. Madsen, F. Kuemmeth, T. S. Jespersen, J. Nygrd, P. Krogstrup, and C. M. Marcus, *Nature* **531**, 206 (2016), doi:10.1038/nature17162.

184. M. T. Deng, S. Vaitiekenas, E. B. Hansen, J. Danon, M. Leijnse, K. Flensberg, J. Nygård, P. Krogstrup, and C. M. Marcus, *Science* **354**, 1557 (2016), ISSN 0036-8075, http://science.sciencemag.org/content/354/6319/1557.

185. J.-P. Xu, M.-X. Wang, Z. L. Liu, J.-F. Ge, X. Yang, C. Liu, Z. A. Xu, D. Guan, C. L. Gao, D. Qian, et al., *Physical Review Letters* **114**, 017001 (2015), doi:10.1103/PhysRevLett.114.017001.

186. H.-H. Sun, K.-W. Zhang, L.-H. Hu, C. Li, G.-Y. Wang, H.-Y. Ma, Z.-A. Xu, C.-L. Gao, D.-D. Guan, Y.-Y. Li, et al., *Physical Review Letters* **116**, 257003 (2016), doi:10.1103/PhysRevLett.116.257003.

187. E. Bocquillon, R. S. Deacon, J. Wiedenmann, P. Leubner, T. M. Klapwijk, C. Brne, K. Ishibashi, H. Buhmann, and L. W. Molenkamp, *Nature Nanotechnology* **12**, 137 (2016), doi:10.1038/nnano.2016.159.

188. J. Wiedenmann, E. Bocquillon, R. S. Deacon, S. Hartinger, O. Herrmann, T. M. Klapwijk, L. Maier, C. Ames, C. Brne, C. Gould, et al., *Nature Communications* **7**, 10303 (2016), doi:10.1038/ncomms10303.

189. S. Wu, V. Fatemi, Q. D. Gibson, K. Watanabe, T. Taniguchi, R. J. Cava, and P. Jarillo-Herrero, *Science* **359**, 76 (2018), ISSN 0036-8075, http://science.sciencemag.org/content/359/6371/76.

190. Z. Fei, T. Palomaki, S. Wu, W. Zhao, X. Cai, B. Sun, P. Nguyen, J. Finney, X. Xu, and D. H. Cobden, *Nature Physics* **13**, 677 (2017), doi:10.1038/nphys4091.

191. L. Fu and C. L. Kane, *Physical Review B* **79**, 161408 (2009), doi:10.1103/PhysRevB.79.161408.

192. H. Overweg, P. Rickhaus, M. Eich, Y. Lee, R. Pisoni, K. Watanabe, T. Taniguchi, T. Ihn, and K. Ensslin, *New Journal of Physics* **20**, 013013 (2018a), ISSN 1367-2630, http://stacks.iop.org/1367-2630/20/i=1/a=013013.

193. H. Overweg, H. Eggimann, X. Chen, S. Slizovskiy, M. Eich, R. Pisoni, Y. Lee, P. Rickhaus, K. Watanabe, T. Taniguchi, et al., *Nano Letters* **18**, 553 (2018b), ISSN 1530-6984, doi:10.1021/acs.nanolett.7b04666.

194. D. Bischoff, M. Eich, A. Varlet, P. Simonet, T. Ihn, and K. Ensslin, *Physical Review B* **91**, 115441 (2015b), doi:10.1103/PhysRevB.91.115441.

195. D. Bischoff, M. Eich, O. Zilberberg, C. Rssler, T. Ihn, and K. Ensslin, *Nano Letters* **15**, 6003 (2015c), ISSN 1530-6984, doi:10.1021/acs.nanolett.5b02167.

196. R. C. Ashoori, H. L. Stormer, J. S. Weiner, L. N. Pfeiffer, K. W. Baldwin, and K. W. West, *Physical Review Letters* **71**, 613 (1993), doi:10.1103/PhysRevLett.71.613.

197. J. G. Kroll, W. Uilhoorn, D. de Jong, F. Borsoi, K. van der Enden, S. Goswami, M. Cassidy, and L. P. Kouwenhoven, "Magnetic field compatible circuit quantum electrodynamics with graphene Josephson junctions" (2017), https://www.nature.com/articles/s41467-018-07124-x.

198. J. Wang, D. R. Legrain, L. Bretheau, F. Yan, M. Kjaergaard, D. Kim, J. Yoder, G. Samach, D. Campbell, P. Krantz, et al., "Coherent control of a hybrid superconducting circuit made with graphene-based van der Waals heterostructures" (2018), https://www.nature.com/articles/s41565-018-0329-2.

199. Q. L. He, L. Pan, A. L. Stern, E. C. Burks, X. Che, G. Yin, J. Wang, B. Lian, Q. Zhou, E. S. Choi, et al., *Science* **357**, 294 (2017), ISSN 0036-8075, http://science.sciencemag.org/content/357/6348/294.

200. S. Nadj-Perge, I. K. Drozdov, J. Li, H. Chen, S. Jeon, J. Seo, A. H. MacDonald, B. A. Bernevig, and A. Yazdani, *Science* **346**, 602 (2014), ISSN 0036-8075, http://science.sciencemag.org/content/346/6209/602.

201. B. van Heck, A. R. Akhmerov, F. Hassler, M. Burrello, and C. W. J. Beenakker, *New Journal of Physics* **14**, 035019 (2012), http://stacks.iop.org/1367-2630/14/i=3/a=035019.

202. I. C. Fulga, A. Haim, A. R. Akhmerov, and Y. Oreg, *New Journal of Physics* **15**, 045020 (2013), http://stacks.iop.org/1367-2630/15/i=4/a=045020.

203. J. Alicea, Y. Oreg, G. Refael, F. von Oppen, and M. P. A. Fisher, *Nature Physics* **7**, 412 (2011), doi:10.1038/nphys1915.

204. J. D. Sau, D. J. Clarke, and S. Tewari, *Physical Review B* **84**, 094505 (2011), doi:10.1103/PhysRevB.84.094505.

205. T. Hyart, B. van Heck, I. C. Fulga, M. Burrello, A. R. Akhmerov, and C. W. J. Beenakker, *Physical Review B* **88**, 035121 (2013), doi:10.1103/PhysRevB.88.035121.

206. K. Flensberg, *Physical Review Letters* **106**, 090503 (2011), doi:10.1103/PhysRevLett.106.090503.

207. C. Schrade, A. A. Zyuzin, J. Klinovaja, and D. Loss, *Physical Review Letters* **115**, 237001 (2015), doi:10.1103/PhysRevLett.115.237001.

208. S. Hoffman, C. Schrade, J. Klinovaja, and D. Loss, *Physical Review B* **94**, 045316 (2016), doi:10.1103/PhysRevB.94.045316.

Assembly of Plasmonic Nanoparticles

Eric H. Hill
The University of Texas at Austin

Christoph Hanske
CIC biomaGUNE

Cyrille Hamon
University of Paris-Sud

Yuebing Zheng
The University of Texas at Austin

14.1 Introduction... 14-1
14.2 Standard Assembly Methods 14-1
    Forces Involved in Colloidal Self-Assembly • Self-Assembly via Solvent
    Evaporation • Template-Directed Assembly • External Field-Directed
    Assembly • Directed Self-Assembly by Molecular Interactions
14.3 NP Assemblies... 14-7
    Self-Assembled Plasmonic Structures in Dispersion • Assembly of 1D NP
    Chains • Two-Dimensional NP Assemblies • Three-Dimensional NP
    Assemblies
14.4 Concluding Remarks.. 14-18
Acknowledgments 14-18
References 14-18

14.1 Introduction

Colloidal plasmonic nanoparticles (NPs) have attracted a great deal of research attention in many fields, ranging from biosensors to photonic materials.[1] The so-called plasmonic properties of these NPs originate from localized surface plasmon resonance (LSPR), occurring when electrons in a metal NP collectively oscillate in resonance with the frequency of incident light.[2] This has important implications in both near and far fields. The local electromagnetic field near the nanostructure surface can be several orders of magnitude higher than the incident field at the LSPR wavelength, and has been utilized for various plasmon-enhanced optical phenomena, such as surface-enhanced Raman scattering (SERS),[3] metal-enhanced fluorescence,[4] or plasmonic photocatalysis.[5] Besides, the absorption and scattering coefficients of plasmonic NPs are several orders of magnitude higher than those of the conventional organic dyes, and therefore, they exhibit strong optical properties in a wide range of electromagnetic spectrum.[6] The visible LSPR is particularly interesting for colorimetric sensing, such as plasmonic enzyme-linked immunosorbent assay.[7] Moreover, the plasmonic response can be tuned by directed self-assembly in which the close spatial location of at least two NPs induces new optical properties by means of plasmon hybridization.[8] Given the almost infinite orientations and structures achievable from a single colloidal suspension, a large range of plasmonic responses can be targeted. In other words, the assembly of NPs can produce a large variety of new plasmonic nanostructures with a minimal synthetic effort. Herein, we review relevant examples of the assembly of plasmonic NPs with a focus on the mechanisms involved and recent promising nanostructures.

The topics covered herein can be investigated neither from the magnifying glass of the physicist nor chemist, but should instead be seen from an interdisciplinary perspective. On the one hand, plasmonic NPs have interesting physical properties that differ from the bulk, which are not limited to optical properties (e.g., melting temperature, heat generation, carrier recombination). On the other hand, NPs are prepared by wet chemical approaches and stabilized by a soft corona, giving properties like solvation, colloidal stability, and often tunable interaction. Depending on the techniques used to assemble NPs, other interfacial interactions should be considered for wetting, capillary flows, and so forth. The structural diversity of NP assembly is illustrated as in Figure 14.1.

The synthetic approaches to prepare NPs will not be covered, but we refer the readers to recent reviews.[2,10−12] Although nature doesn't build impermeable walls, this manuscript will be compartmentalized for the sake of clarity. We begin by introducing the interactions between NPs and the typical forces involved. Next, we review the standard assembly methods to obtain colloidal nanomaterials in suspension or solid form. The third and final section presents an overview of different 1D, 2D, and 3D plasmonic nanostructures that have been constructed.

14.2 Standard Assembly Methods

14.2.1 Forces Involved in Colloidal Self-Assembly

The first self-assembled structures were observed by drying an NP suspension on transmission electron microscopy (TEM) grids, and coincided with the discovery of

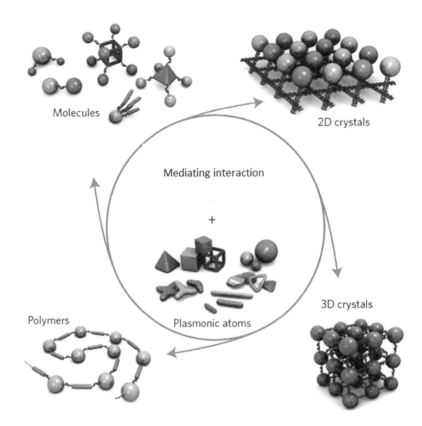

FIGURE 14.1 Structural diversity in the assembly of plasmonic NPs. The terminology adopted here refers to an analogy between NPs and atoms. Starting from plasmonic NPs "atoms," one can target many structures such as NP clusters (molecules), one-dimensional (1D) assemblies (polymers), two-dimensional (2D) crystals, and three-dimensional (3D) crystals. (Adapted from Tan et al.[9] Copyright 2011 Springer Nature.)

monodisperse colloidal synthesis. Although monodispersity is a key element to obtain higher-order structures,[13] it is not the keystone, as well-defined assemblies have been observed from polydisperse NPs[14] or binary mixtures.[15] As can be seen in the following sections, assembled structures can be obtained by a mosaic of techniques that involve evaporation of an NP dispersion, tuning of the NP surface chemistry, or external fields. We start with the introduction of the main components of NP interactions that are present in most cases. Understanding these interactions is central for the predictive design of innovative structures.

After synthesis, plasmonic NPs are stabilized by a soft corona, with its inner part interacting with the NP surface and the outer part contributing to surface properties (e.g., solvent affinity, charge, and steric hindrance). A great amount of work has been dedicated to understanding the forces mediating the interaction of micron-sized colloids (e.g., polymers, silica). The most representative model is probably the Derjarguin, Landau, Verwey, and Overbeek model, or simply DLVO, which describes colloidal stability as a balance of repulsive and attractive forces.[16−18] Repulsive interactions include electrostatic forces between two similarly charged species and steric hindrance generating elastic and osmotic forces. Electrostatic forces are encountered in both polar and nonpolar solvents, though charges dissociate more easily in a media with high dielectric

constant. Elastic and osmotic forces enter into action when the soft corona of two particles begin to overlap and the detailed local interaction can be described by the Flory–Krigbaum theory.[19,20]

In short, depending on the affinity of the ligand shell with the solvent, either interpenetration of the soft corona or exclusion can be favored, which is the basis of self-assembling strategies mediated by hydrophobic interactions described in Section 14.3.1. Attractive forces include van der Waals forces that originate from transient fluctuations in the distribution of electrons. van der Waals forces increase dramatically as the interparticle distance is reduced, but can be overcome by the elastic deformation of the soft corona. Two scenarios can be considered: repulsive forces overcome attractive forces (stable scenario), or the inverse case, in which the particles flocculate as soon as they interact (nonstable scenario). In the latter case, it is rare to obtain an assembly, and particles are often fused. Thus, it is important to fine-tune the interaction potentials to obtain a higher-order assembly. If NP attraction remains comparable to $k_B T$,[21] nanocrystals are allowed to sample multiple sites on the superlattice surface before settling. It is worth stressing that the DLVO model imperfectly describes the interaction of NPs but can serve as a first approximation.

The reduction of particle dimensions from the micrometer scale to the nanoscale leads to a more complex phenomena.

For instance, NPs may be commensurate with globular proteins and can show similar self-organization[19,22,23] or even lead to hybrid nanomaterials composed of two subunits.[24] The use of plasmonic NPs as building blocks led to the discovery of "plasmonic molecules" showing plasmon hybridization schemes, which can be viewed through the lens of molecular systems.[8,25] Another striking example is the ability of nanoassemblies to exhibit chirality due to the interaction of achiral plasmonic NPs in a chiral configuration.[26,27] We don't intend to give an exhaustive list of examples here, but it is important to realize that concepts in nanotechnology have been inspired from chemistry and biology in many ways.[28−30] Such discrepancies between NPs and micron-sized colloids can be partially addressed by considering the increased influence of the soft corona in NP interactions.[31,32] Currently, much effort is devoted to build predictive models that take into account the fluctuation of ionic species around the NPs.[19,33−35] One must always keep in mind the nonadditivity of NP interactions that leads to such intriguing effects.[19,36,37]

Another important interaction to leverage NP assemblies is known as depletion interaction. Depletion interaction occurs when the colloids of interest are in the presence of depletants, which are frequently polymers or micelles but can also be colloids of another morphology.[38−40] The driving force is the maximization of the free volume available for the depletants to increase the overall entropy of the system. This interaction has been successfully applied to purify polydisperse dispersions of gold NPs by adjusting the concentration of the surfactant in solution.[41−43] It can also be employed to form well-defined plasmonic assemblies and supercrystals.[44,45]

14.2.2 Self-Assembly via Solvent Evaporation

One of the primary ways to form assemblies of plasmonic NPs is via a simple evaporation of a dispersion of NPs on a surface. This process follows the familiar pattern of a dried spot of spilled coffee, where the evaporation of coffee deposits suspended particulate matter in a ring pattern along the perimeter of the former drop. In fact, such ring deposits are a common effect from drying of drops of dispersed solids of all sizes, even down to the nanoscale. Early work by Deegan and coworkers showed that capillary flow was the cause of the formation of this ring deposit, allowing the prediction and control of the deposition without knowing the exact nature of the liquid, solid, or substrate.[46] The mechanism of the ring deposition is that the edges of the droplet become pinned to the surface, and capillary flow outward from the center of the droplet delivers the suspended particles to the drop edges.

Later studies showed that this effect involves the suppression of Marangoni flow (a flow driven by surface tension gradients) and that Marangoni flow present in organic solvents reverses the deposition to leave material in the center of the droplet.[47] A short time later, it was shown that the morphology of the suspended particles can affect the coffee ring effect. In a study by Yunker et al., it was shown that, by using ellipsoidal particles instead of spherical particles, the coffee-ring effect was not observed.[48] The strong interparticle capillary interactions between the ellipsoidal particles lead to the formation of assemblies at the air–water interface, which prevent them from reaching the drop edge. This approach is highly used in plasmonic NP assembly. Curiously, compared to coffee, whiskey droplets yield much more uniform dried films. Kim et al. proposed that surface-active macromolecules and continuous mixing during the evaporation of an ethanol/water mixture causes an unexpected behavior.[49] Their studies of binary model dispersions containing different combinations of low molecular surfactant and polyethylene oxide suggested that both interaction of substrate-adsorbed polymer chains with the particles and different Marangoni flows were responsible for the improved macroscopic homogeneity of the final films.

14.2.3 Template-Directed Assembly

To obtain surface-bound colloidal assemblies with precisely defined dimensions, prestructured target substrates are commonly employed. Thereby, local contrast in adhesiveness, wettability, topography, or combinations of these guide the spatial distribution of particles.[50] The necessary patterns can be prepared by techniques like e-beam writing, microcontact printing, photolithography, or wrinkling.

The affinity of colloids toward a surface can be controlled via specific interactions like complementary DNA-strand recognition or nonspecific interactions such as electrostatics or hydrophobicity.[51] A widespread method for the fabrication of microscale colloidal assemblies was introduced in 1999 by Aizenberg et al., which achieved site-selective colloidal deposition by immersion of a charge-patterned gold surface into a dispersion of anionic colloids followed by washing and drying.[52] Shortly afterwards, the method was extended by Hammond and coworkers to arbitrary surfaces and binary particle assemblies using polymer-on-polymer stamping for patterning polyelectrolyte coatings for subsequent colloidal adsorption.[53] Thereby, the particles adhere to the oppositely charged surface regions due to counterion release, and their density can be controlled to a certain degree via the ionic strength in solution governing the repulsion of adsorbed and adsorbing particles as well as site selectivity.[52,54] Plasmonic assemblies, with feature sizes comparable to the particle dimensions, were realized by nanoxerography, e-beam lithography, or employing microphase-separated block copolymer films as substrates.[55−58] In this regime, the interplay of repulsive and attractive forces could lead to crystalline layers with nonclose-packed lattices.[59] Due to the often irreversible particle attachment, however, well-ordered arrangements are hard to achieve by patterned adsorption and may require additional structure annealing, as demonstrated by Noh et al. for DNA-guided assembly of gold nanospheres.[60]

An alternative to adhesion-controlled colloidal adsorption consists of confining the particle-containing liquid to designated template regions. For aqueous dispersions, this can be achieved with chemical patterns that exhibit strong hydrophilic/hydrophobic contrasts and thus lead to local dewetting as shown in Figure 14.2a. The necessary wettability contrasts are accessible, for example, by thiol and silane chemistry combined with microcontact printing.[61,62] In the absence of hit-stick interaction between the substrates and particles,[63] close-packed arrangements are normally formed upon drying due to attractive capillary forces between the colloids, which allows a relatively straightforward construction of colloidal crystals.[64] As for

the deposition on featureless surfaces, various techniques can be used with hydrophilicity-patterned substrates, including simple droplet drying,[65] vertical drying, dip coating,[66] or deposition under blade.[67] At the (sub)micrometer scale, near-instant liquid evaporation, shape instabilities (especially for nonspherical droplets), and surface contaminations can be problematic to control with 2D wettability patterns. As a consequence, this approach is best suited for the construction of large colloidal crystals, but rarely used for colloidal assembly with submicrometer features.[68]

Equally apt for the manipulation of particle dispersions on the micrometer and submicrometer scales are topographically structured templates. Such structured templates

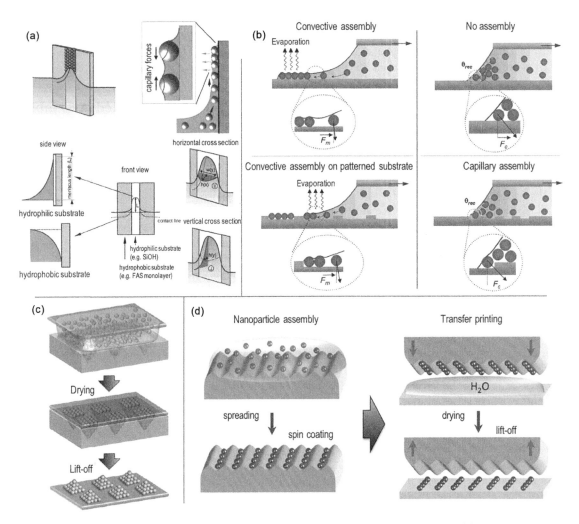

FIGURE 14.2 Template-assisted techniques for the colloidal assembly with defined dimensions. (a) Dip coating on wettability-patterned, flat surfaces. The thinning meniscus follows the hydrophilic regions in contact with the liquid, while convection and capillary forces drive particle deposition at the curved three-phase contact line. (b) Particle assembly on flat and topographically structured substrates using a sandwiched geometry with a moving blade. Although only convective assembly with a low contact angle facilitates particle deposition on flat substrate, topographically structured templates can trap the particles, leading to distinct structures depending on the deposition mechanism. (c,d) With soft templates of low-to-medium wettability, colloids can be deposited either directly on an unstructured target substrate or first be arranged and then transferred by contact printing. All of these methods yield densely packed assemblies. ((a) Adapted from Fustin et al.,[66] Copyright 2004 American Chemical Society. (b) Adapted from Malaquin et al.,[84] Copyright 2007 American Chemical Society. (c) Adapted from Hanske et al.,[85] Copyright 2017 American Chemical Society. (d) Adapted from Hanske et al.,[79] Copyright 2014 American Chemical Society.)

have been introduced for colloidal assembly by Xia and coworkers in 2001 in a process coined template-assisted self-assembly (TASA).[69] In this widespread technique for micro- and NP assembly, thin liquid films are formed over a solvophilic substrate by dip coating[70] or spin coating[71] followed by drying in confinement.[72] Thereby, meniscus geometry is a critical parameter that determines whether the particles are transported toward the target substrate at the three-phase contact line by convective and capillary forces, eventually getting trapped between the topographic features as illustrated in Figure 14.2b.[73] Consequently, arrays of close-packed or even isolated single particles are formed in the recessed cavities.[74–77] Variants with soft elastomeric templates (i.e., usually polydimethylsiloxane (PDMS) allow either direct fabrication of the particle arrangements on flat, functionalized target substrates or assembly and subsequent transfer printing to create attached or partially free-standing films, as shown in Figure 14.2c,d.[78–80] Notably, for a sufficiently high particle concentration, TASA normally leads to maximum space filling in the confined geometry, which has been exploited for the construction of colloidal crystals with uncommon lattices or controlled defect creation.[81] The true reason for the success of this method, however, stems its ability to bridge several length scales, allowing ordered assembly of microparticles as well as arrangement of particles as small as 2 nm and heteroaggregates.[82,83]

14.2.4 External Field-Directed Assembly

Colloidal assembly in electric fields is generally driven by field-induced interactions. The field-induced interactions in colloidal suspensions have been thoroughly explored for colloids that are magnetically and electrically polarizable. The mechanisms behind the polarization of such colloids in magnetic and electric fields have been reviewed extensively in the past, and thus will not be covered here.[86] In the case of electrically polarizable colloidal particles, electric fields will typically induce polarization due to the mismatch of dielectric properties of the particles with the surrounding medium.[87] Mobile charges in the electric double layer also respond to an applied electric field, further contributing to the polarization.[88] This also plays a role in thermophoretic assembly of particles. When it comes to far-field interactions, the induced field surrounding a polarized particle gives a strong anisotropic interparticle interaction in the form of a dipole. Once Brownian motion is overcome with strong-enough interactions, such particles will form dipolar chains, followed by coarsening as the chains interact laterally. In an ideal case, the coarsening proceeds until the lowest energy lattice structure is reached, depending on the particle concentration and the electric-field orientation and strength.[89–91] Weaker interaction strengths give rise to ordered 2D and 3D colloidal crystals, and the use of an alternating current (AC) field avoids unwanted electroosmotic and electrochemical effects that occur when direct current (DC) is used.

The use of electric fields for plasmonic particle assembly takes advantage of the high dielectric constant of metallic particles relative to plastics and silicates, enabling stronger field gradients to direct such particles of nanometer dimensions. As an example, Hermanson et al. used dielectrophoresis in an AC electric field to drive 30–45 nm gold NPs towards the electrodes, where their concentration and self-assembly resulted in micrometer-diameter wires (Figure 14.3A).[92] The induced assembly of particles via magnetism follows the same lines of assembly as described for electric fields and, however, requires a magnetic component to be present. Plasmonic particles with magnetic properties generally include a magnetic component such as Fe_2O_3 in the form of Janus particles, alloys, and core–shell.[93] In one example by Landfester and coworkers, the field strength was modified to achieve different "block" lengths in "polymers" of iron oxide NPs assembled under the influence of a magnetic field (Figure 14.3B).[94]

Another interesting direction for assembling plasmonic NPs has been to use optothermal effects to trap and manipulate NPs, including optically generated thermophoretic and thermoelectric fields.[95–97] This has been previously achieved with silica and polystyrene microparticles. Recent extension of this technique to plasmonic NPs means that low-power laser heating on a plasmonic substrate can be used to direct the assembly of plasmonic architectures. The optothermal control of NPs is enabled by using NPs coated with the surfactant cetyltrimethylammonium chloride (CTAC) as well as CTAC micelles suspended in the medium. Following the optical heating of the plasmonic substrate, a temperature gradient is established. In a typical example, 0.216 mW of 532 nm laser light was used to generate a temperature gradient of 12 K. The temperature gradient causes both micellar CTAC macroions and Na^+ counterions to undergo thermophoresis, migrating from the hot to the cold region of the gradient. The electric field gradient resulting from the migration of the charged species traps the positively charged plasmonic NP at the laser spot. The facile optical setup of this technique allows the use of controllable laser beam configurations through a digital micromirror device, where arbitrary assemblies of NPs can be formed (Figure 14.4A).[97]

Thermophoresis has also been used to form assemblies of plasmonic NPs. In another study by Lin and coworkers, heating a plasmonic substrate in the presence of gold nanotriangles functionalized with CTAC led to the formation of NP aggregates.[98] These aggregates could then be manipulated with the laser beam and moved from the plasmonic substrate to the glass substrate, indicating that the trapped plasmonic NP assembly is generating the thermal gradient that is, in fact, keeping it trapped (Figure 14.4B).

Another recent study by Lin and coworkers has shown that particles are directed to a bubble formed by optical heating at a plasmonic substrate.[99] The formation of strong flows and Marangoni effects upon formation of the bubble leads to the delivery of particles to the bubble surface. Once they arrive at the bubble/air interface, they are trapped by

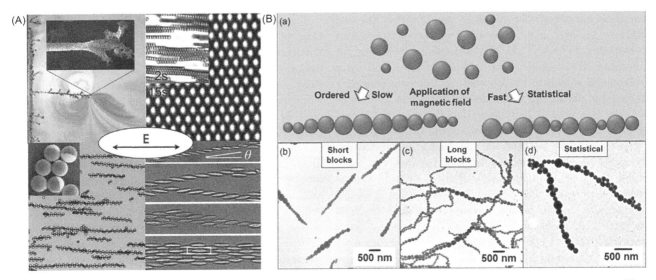

FIGURE 14.3 Directed assembly of particles in external fields. (A) Growth of microwires from gold NPs in an AC electric field.[92] (B) Controlled size-self-assembly of large and small NPs within a chain. (a) Schematics depicting the two possibilities to align a polydisperse sample of NPs. (b) TEM image of shorter chains of NPs with the tendency to form small blocks of larger and smaller particles. Often, the larger particles are found toward the middle of the chains and the smaller ones toward the ends. (c) TEM image of larger chains with large blocks of larger and smaller particles. (d) TEM image of statistically assembled NPs into chains. ((a) Adapted from Hermanson et al.,[92] Copyright 2001 AAAS and (b) Adapted from Bannwarth et al.,[94] Copyright 2015 American Chemical Society.)

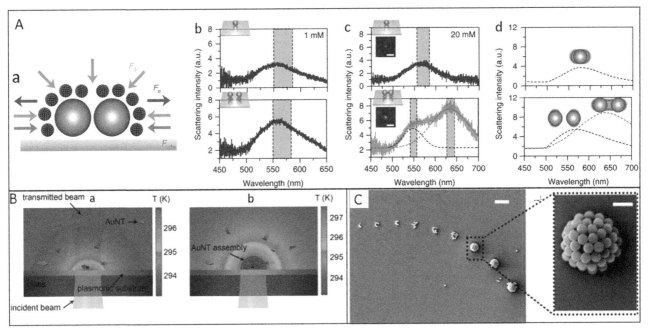

FIGURE 14.4 Optothermal methods of particle assembly. (A): (a) Interaction forces between two trapped NPs. (b,c) Scattering spectra of a single gold nanosphere (top) and two gold nanospheres (bottom) in 1 mM (b) and 20 mM (c) CTAC solutions. The rectangles represent the peak distributions recorded in multiple experiments. (d) Simulated scattering spectra of a single gold nanosphere (top) and two gold nanospheres (bottom) in 20 mM CTAC solution. The left and right dashed curves in (d) represent the longitudinal and transverse plasmon modes, respectively. (B) Simulated temperature distribution at the interface between plasmonic substrate and particle solution in a cross-sectional view: (a) before the formation of a gold triangle assembly, and (b) after the formation of a gold triangle assembly. (C) Scanning electron micrograph of a series of patterned 540 nm polystyrene (PS) beads generated at increasing laser power densities from 0.56 to 1.12 mW/μm^2. (inset) High-magnification image of the 3D hollow structure formed at a laser power density of 0.97 mW/μm^2. ((A) Adapted from Lin et al.,[97] Copyright 2018 Springer Nature. (B) Adapted from Lin et al.,[98] Copyright 2016 American Chemical Society. (C) Adapted from Lin et al.,[99] Copyright 2016 American Chemical Society.)

the surface tension at the interface. Upon collapse of the bubble, the particles follow the contact line and adhere with the solid interface, resulting in printing of an aggregate of the particles that were attracted to the bubble. The resulting structures formed from these particle aggregates can be both a collapsed dome or a hollow shell, depending on the bubble and particle size (Figure 14.4C).[99]

14.2.5 Directed Self-Assembly by Molecular Interactions

Although surfactants can serve to control the drying behavior of colloidal dispersions, ligands typically employed in nanocrystal syntheses like cetyltrimethylammonium bromide (CTAB) are rarely ideal particle surface functionalizations for pursuing ordered assemblies. First, their weak attachment and small dimensions can cause stability problems under the conditions needed for many assembly approaches. This is especially problematic for concentrated dispersions of metallic particles with diameters of several tens of nanometers, due to strong van der Waals interactions. Second, excess surfactant normally needed for stabilization of bare colloids interferes with assembly and patterning processes based on locally controlled wettability and further promotes ring stain formation. This can severely limit precision and scalability. Third, control over interparticle distance or particle orientation is often desired. Therefore, it has become a standard procedure to replace surfactants with either covalently bonded thiolated molecules or macromolecular ligands that strongly attach through multiple physical bonds and give tunable (electro)steric stabilization.

Useful ligands for particle assembly from aqueous media are hydrophilic polymers (e.g., thiolated polyethylene glycol, PEG-SH),[85,100] polyelectrolytes (e.g., polyacrylic acid, PAA or polystyrene sulfonate, PSS),[101] or proteins (e.g., bovine serum albumin, BSA),[79,102] whereas controlled deposition from apolar solvents and mixtures requires coating with hydrophobic thiols or polymers (e.g., polystyrene-SH).[103] Amphiphilic polymer coatings (such as polyvinylpyrrolidone, PVP) and mixed ligand shells (e.g., of PEG-SH and 1-dodecanethiol, DDT) further allow phase transfer and interfacial NP assembly enabling, for example, the use of the Langmuir–Blodgett technique.[104] Overall, control of surface chemistry provides unmatched flexibility when it comes to self-assembly of NPs under various stimuli that cause chemical transformations to lead to self-assembly. Stimuli such as light, temperature, and pH allow for flexible fabrication strategies that can utilize a mix of different stimuli to achieve the desired assembly constituents or geometries. Some examples of the diverse arrays of NP assemblies formed by directed assembly via molecular interactions are shown in Figure 14.5.

The construction of freely dispersed colloidal assemblies relies mainly on controlled aggregation via manipulation of particle surface chemistry. Biomolecules are often used for this purpose. A typical example of binding interactions from biology is that of DNA.[116] Single-stranded DNA (ssDNA)

and RNA are able to recognize their complementary pairs, with which they bind to form the double-stranded structure that is the basis for all life. In 1996, the use of DNA hybridization as a mechanism for NP assembly was established by Mirkin et al. In their classic study, two batches of 13 nm gold NPs were functionalized with thiolated strands of two different, noncomplementary strands of ssDNA.[117] Upon addition of a DNA linker strand with the complements of ssDNA molecules bound to both batches of NPs, the hybridization of the linker strand with the NP-bound ssDNAs leads to the oligomerization of gold NPs. This simple example set the stage for more complicated DNA-based NP linkage strategies. In a classic example by Aldaye and Sleiman, thiolated DNA is attached to gold NPs, and the monofunctionalized particles are isolated. Gold NPs with complementary DNA strands are linked together, resulting in the formation of different configurations of 3- and 4-NP aggregates, depending on particle size and relative orientation (Figure 14.6).[118] In addition, dynamic control of the assembly can be achieved through writing/erasing of the linked NP through the introduction of an "eraser" strand that takes up the entire 60 base pairs of the linkage region, compared with 40 taken by the NP linker strand. Apart from DNA, protein coatings, dithiols, and amino acids like cysteine are frequently used to link plasmonic particles and create dispersible plasmonic assemblies.[119,120] Also, there have been reports of successful particle assembly in confinement or by direct adsorption on an adhesive template.[121,122]

14.3 NP Assemblies

14.3.1 Self-Assembled Plasmonic Structures in Dispersion

The colloidal control over self-assembly can be realized by tuning surface chemistry of NPs and the properties of the solvent. An important contribution to the field by Nie, Rubenstein, and coworkers showed that long chains and closed rings of gold nanorod aggregates could be formed by changing the solvent mix with the three solvents, i.e., dimethylformamide, tetrahydrofuran, and water (Figure 14.7).[113] In this case, the ends of nanorods were functionalized with polystyrene, and the sides were functionalized with CTAB, giving a amphiphilic surface chemistry to the nanorods.

Liz-Marzan group has made numerous contributions to the field of plasmonic NP synthesis and assembly over several decades. One example is in the colloidal assembly of plasmonic NPs controlled by hydrophobic interactions.[123] In one study, they found that the functionalization of gold NPs by polystyrene allowed their self-assembly in a tetrahydrofuran (THF)–water mixture, where the following addition of a polymeric surfactant (PS-b-PAA) could suppress further aggregation, allowing the study of aggregation kinetics.[123] The colloidal aggregates maintained a 3D geometry related to the packing of NPs, and they collapsed into disk-like structures when dried (Figure 14.8).

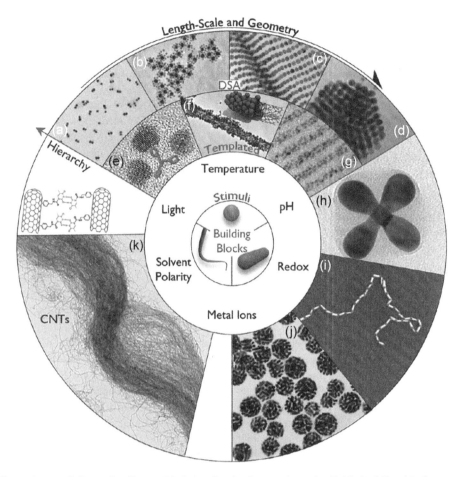

FIGURE 14.5 The universe of directed self-assembly by molecular interactions. Available building blocks are classified in terms of aspect ratio (AR), ranging from AR = 1 (isotropic, mostly spheres (a–g)) through 1 < AR < 15 (rod-like (h–j)) to AR > 15 (wires (k)). Self-assembly of gold NPs on templates (macromolecules[105] (e); carbon nanotubes[106] (f); block copolymers[107] (g)) can lead to geometries that are complementary to those of the templates. In template-free self-assembly, stimuli-responsive molecules adsorbed on NP surfaces induce controlled aggregation. Temperature-sensitive DNA can induce particle clustering[108,109] (a,b) via H-bonding or induce formation of 2D superlattices,[110] (c) whereas light-sensitive azobenzenes induce formation of colloidal crystals,[111] (d) via molecular dipole–dipole interactions. Hydrophobic interactions can induce the assembly of nanorods into low-symmetry clusters,[112] (h) chain-like structures,[113] (i) or spherical objects,[113] (j) by tuning the volume fraction of different solvents. Similarly, highly anisotropic carbon nanotubes undergo assembly into bundles via H-bonding-induced changes in the medium polarity[114] (k). Finally, hierarchical self-assembly can be achieved using preformed assemblies as building blocks. (Reprinted from Grzelczak et al.[115], Copyright 2010 American Chemical Society.)

In further research studying the influence of hydrophobic interactions on self-assembly of plasmonic NPs, Liz-Marzán et al. turned their focus from isotropic to anisotropic particle morphologies. Gold nanodumbbells or dumbbell-shaped gold NPs were functionalized with polystyrene on the lobes and with CTAB at the center neck region.[112] This led to two interesting findings related to inter-particle interactions directly resulting from the function-alization of the two regions and the steric hindrance involved in stacking dumbbell-shaped particles efficiently. Once transferred from DMF to DMF with 10% water, the polystyrene-functionalized lobes of dumbbells associated strongly with one another, leading to side-by-side aggre-gates of particles. The encapsulation of the nanodumbbells in a PS-PAA block copolymer leads to the assembly of nanodumbbells in a cross-like configuration, which can then be suspended in water (Figure 14.9a). In addition, the

chirality of the assembly of the two crossed nanodumbbells gives them a strong chiroptical plasmonic response.[124] These types of solvent-mediated hydrophobic interactions have been applied to a variety of particle morpholo-gies to create novel and interesting colloidal assemblies (Figure 14.9).[125]

14.3.2 Assembly of 1D NP Chains

The optical properties of close-packed colloidal chains have received much attention from the scientific community. Modeling the optical properties of 1D chains showed that the dominant spectral feature, a strongly scattering super-radiant mode, typically red-shifts upon particle addition until reaching a gap-size-dependent "infinite chain limit" around ten colloidal repeat units.[126] Perhaps the most comprehensive experimental studies of the

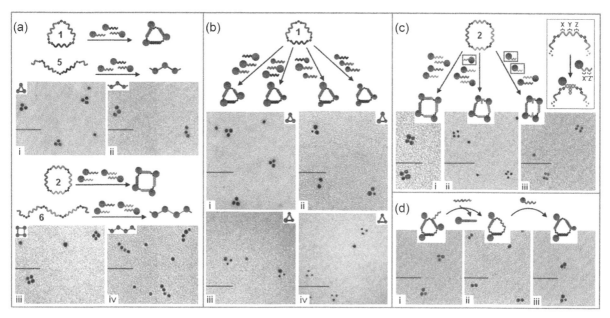

FIGURE 14.6 (a) 1 and 2 organize gold particles into triangles and squares; 5 and 6 result in open linear assemblies of three and four particles. (b) 1 generates triangles of (i) three large (15 nm), (ii) two large/one small (5 nm), (iii) one large/two small, and (iv) three small particles. (c) 2 assembles four gold particles into (i) squares (15 nm particles), (ii) trapezoids, and (iii) rectangles (5 nm). Inset: use of a loop shortens the template's arm. (d) Write/erase function with 1 by (i) writing three gold particles (15 nm) into triangles, (ii) removal of a specific particle using an eraser strand, and (iii) rewriting with a 5 nm particle. Scale bars: 50 nm. (Reprinted from Aldaye et al.[118], Copyright 2007 American Chemical Society.)

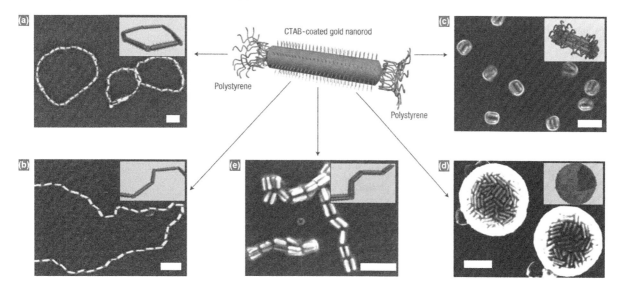

FIGURE 14.7 Self-assembly of polymer-tethered gold nanorods in selective solvents. An amphiphilic gold nanorod carrying a double layer of CTAB along the longitudinal side and polystyrene molecules grafted to both ends. (a–e) Scanning electron micrographs of the self-assembled nanorod structures: rings (a) and chains (b) self-assembled in the dimethyl formamide/water mixture at water contents of 6 and 20 wt%, respectively, side-to-side aggregated bundles of nanorods (c) and nanospheres (d) self-organized in the tetrahydrofuran/water mixture at water contents of 6 and 20 wt%, respectively, and bundled nanorod chains obtained in the ternary dimethyl formamide/tetrahydrofuran/water mixture at a weight ratio of liquids 42.5:42.5:15 (e). The scale bars are 100 nm. The insets show the corresponding schematic diagrams of nanorod assemblies. (Reprinted from Nie et al.[113], Copyright 2007 Springer Nature.)

collective plasmon modes of strongly coupled chains were conducted by Link and coworkers via polarization-sensitive extinction spectroscopy.[74] Gold nanospheres were assembled on structured electron-beam lithographic resist layers, and the resulting close-packed chains were characterized by single-particle optical spectroscopy. In line with the theoretical predictions displayed in Figure 14.10a–c, the wavelength of the extinction maximum red-shifted with increasing particle number toward a chain-morphology-dependent limit. Further investigations

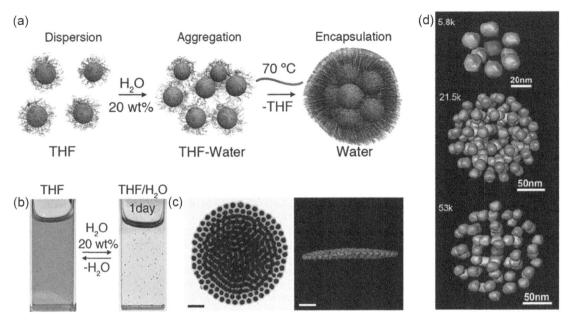

FIGURE 14.8 Colloidal self-assembly of polystyrene-coated gold NPs. (a) Colloidal dispersion of polystyrene-coated gold NPs in THF; The aggregation starts upon addition of a nonsolvent (water, 20 wt%); The addition of polymeric surfactant (PS-b-PAA) at any time during aggregation suppresses further aggregation; mild thermal treatment leads to expelling THF and induces rearrangement of the particles inside the hydrophobic core, while the polyacrylic shell ensures stability of the clusters in water. (b) Visible changes of the solution containing Au18@PS509 upon aggregation by hydrophobic interactions. (c) Before addition of PS-b-PAA, the clusters form flat, disk-like structures upon drying, as confirmed by 3D electron tomography reconstruction. (d) 3D electron tomography reconstructions of clusters comprising Au18 with progressively increasing lengths of the PS chains. Note that interparticle distances increase with increasing PS chain length. (Reprinted from Grzelczak et al.[123], Copyright 2012 American Chemical Society.)

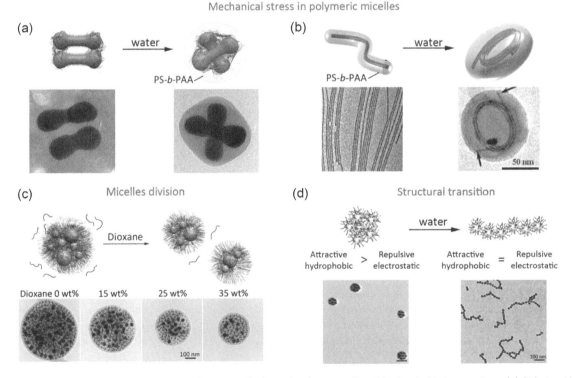

FIGURE 14.9 Structural transformations at the nanoscale through solvent-mediated hydrophobic interactions. (a) Side-to-side dimers of gold nanodumbbells transform into a cross-like conformation inside polymeric micelles. (b) Coiling of thin gold nanowires through encapsulation in polymeric micelles and subsequent increase in water content. (c) Division of plasmonic micelles with increasing amount of dioxane. (d) Structural transition from globular to chain-like assemblies through increasing the amount of water. (Reprinted from Grzelczak et al.[125], Copyright 2014 American Chemical Society.)

of single particle, dimer and quadrimer chains revealed that the chain spectra depend little on particle size dispersity and disorder, but are influenced strongly by the type of repeat unit due to retardation effects.[127]

Besides tailoring the polarization-dependent scattering and absorption properties, linear NP arrangements allow plasmonic waveguiding via near-field coupling. This concept was first presented in 1998 by Aussenegg and coworkers and was soon afterwards corroborated experimentally by Atwater and coworkers with lithographically written structures.[128–130] Such particle-based waveguides enable energy transport below the diffraction limit as well as sharp bends without excessive radiation leakage, but the achievable attenuation length depends sensitively on the interparticle distance and choice of the optical guiding mode. As displayed in Figure 14.10d–m, addressing the dark subradiant modes of close-packed, self-assembled NP

chains via asymmetric excitation is highly efficient for plasmonic energy transport over distances of several microns as well as around right-angled bends.[131] The radiative losses are comparatively low, but the effective propagation length was shown with arrays of free-standing particle wires to depend sensitively on the width of the employed particle chains.[132]

As an economic alternative to templates fabricated by nanolithography, wrinkled surfaces were suggested by Genzer et al. in 2005.[133] In fact, microscopic wrinkles, which occur when structures consisting of at least two layers with mismatching mechanical properties are subjected to compressive strain, are ubiquitous in nature, an example being the cuticular folds of plant surfaces.[134] For a thin, hard layer on an elastic substrate, the sum of bending and stretching energies is often minimized by the formation of highly regular surface corrugations.[135,136] The most

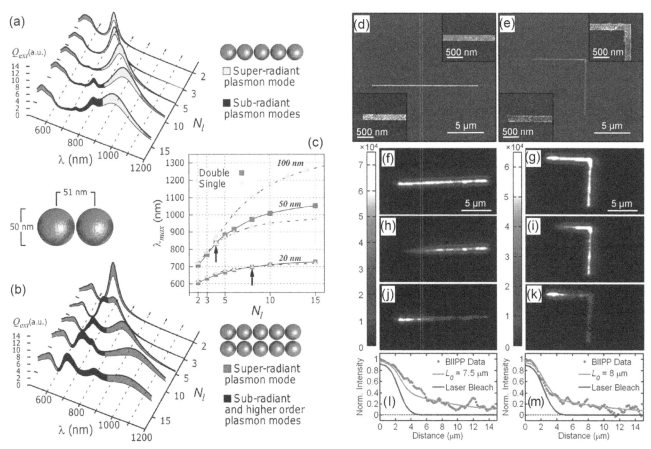

FIGURE 14.10 Optical properties of strongly coupled plasmonic NP chains. (a,b) Spectra of single particle and dimer chains modeled for gold nanospheres with a gap size of 1 nm and light polarization parallel to the long chain axis. The super-radiant mode shifts upon addition of repeat units N_l (i.e., single particles for a, and dimers for b) until reaching a maximum of around 10. Starting at three repeat units, bright subradiant modes appear. (c) The peak wavelength $\lambda_{\max}$ also depends on the particle size. (d–m) Plasmonic waveguiding in straight and right-angled silver particle chains mapped by bleach-imaged plasmon propagation. The chains shown in the scanning electron micrographs (d,e) were coated with a fluorescent dye and locally excited at the left end with the resonance wavelength of a dark, subradiant mode for photobleaching. Fluorescence images before (f,g) and after (h,i) irradiation and the difference images (j,k) illustrate the energy transport by near-field coupling. Evaluation of the local intensity (l,m) yields a propagation length of 8 μm. ((a–c) Adapted from Slaughter et al.,[74] Copyright 2012 American Chemical Society. (d–m) Adapted from Solis et al.,[131] Copyright 2013 American Chemical Society.)

common material for the fabrication of artificial wrinkles is PDMS, whose surface is easily converted into a rigid, glass-like layer via dry or wet chemical oxidation. First observed in 1982 by Martin et al., manifold strategies for wrinkling of PDMS were developed after the seminal works of Whitesides and coworkers presented in 1998.[137-142] For particle assembly, permanent-oriented corrugations as illustrated in Figure 14.11 are most relevant.[143] According to a recent study by Bayley et al., such wrinkles can be fabricated with a periodicity as small as 140 nm by plasma oxidation of prestretched PDMS slabs and subsequent relaxation.[144]

Such assemblies can be applied, for instance, as large-scale SERS substrates consisting of aligned gold or silver NPs. As shown by Pazos-Pérez et al. for close-packed chains of gold nanospheres, high SERS activity can be achieved due to hot spot formation in the nanometer-sized interparticle gaps,[145] which was also reported for assemblies of spherical silver NPs.[146] Müller et al., on the other hand, assembled gold nanostars coated with thick poly(N-isopropylacrylamide) (pNIPAM) shells. The spacer layers prevent quenching of the inherent particle hot spots, which occurs for direct contact of sharp tips. The polymer shells further served to catch and detect hydrophobic analyte molecules from the gas phase.[147] Regarding the correlation of local and macroscopic optical properties of strongly coupled plasmonic chains, the works of Fery and coworkers on the printing transfer of protein-coated gold nanospheres assembled in wrinkles onto polyethylene imine-coated substrates are worth mentioning. With this system, arrays of quasi-infinite parallel chains with average interparticle distances of 1–2 nm were realized. Interestingly, the macroscopic optical response measured over ensembles of several thousand chains not only exhibited the characteristic plasmonic mode structures of separated ideal chains but also signs of peak sharpening due to the array lattice.[148] On stretchable substrates, the optical response was further tunable by mechanically varying the average chain length, which opens avenues toward mechanoplasmonic strain sensors (Figure 14.12).[149]

14.3.3 Two-Dimensional NP Assemblies

Murray and Kagan groups have developed some interesting examples of complex architectures that can be made by 2D assembly of plasmonic NPs. They refer to their structures as "metamolecules," as the red-shifting of plasmon resonances bears similarity to the effect observed by the lengthening of pi-conjugated organic molecules. In one interesting report, they use capillary forces to assemble gold NPs into polygonal topographic templates, which they had defined using electron-beam lithography.[150] In addition to observing light scattering that was highly sensitive to the angle of incident light, they found that an increasing number of gold NPs in the 2D assembly resulted in a red-shifted plasmon resonance. This work is summarized in Figure 14.13.

Apart from spatial arrangement, the morphology of building blocks can have a pronounced influence on the optical response of plasmonic clusters. As demonstrated by Halas and coworkers for planar assemblies of gold nanoshells (deposited onto silica NPs) and heteroclusters, strong magnetic and sharp Fano-like resonances can be induced through close-packed arrangements and engineered by the aggregation number (Figure 14.14). Such clusters were realized by simple droplet evaporation,[151] complementary DNA strand linking,[152] or convective assembly on topographically structured templates.[153] Their special optical properties could open interesting avenues for metamaterials or novel sensors.[154]

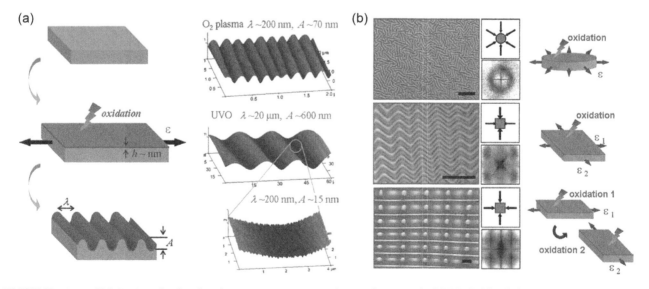

FIGURE 14.11 Fabrication of ordered surface corrugations via oxidation of prestretched PDMS. (a) While plasma treatment leads to wrinkles with periodicities of hundreds of nanometers to few microns, Ultraviolet (UV)/ozone oxidation gives access to feature dimensions in the range of tens of micrometers. (b) The obtained surface morphology can be tailored by the directionality of the applied prestrain, allowing the fabrication of one- and 2D patterns. (Adapted from Chiche et al.,[143] Copyright 2008 Royal Society of Chemistry.)

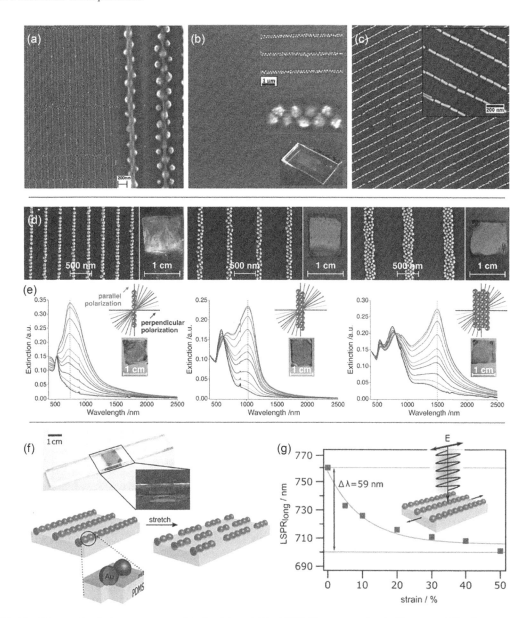

FIGURE 14.12 Plasmonic chain arrays templated with wrinkled PDMS stamps. (a–c) Plasmonic NP chain arrays prepared by wrinkle-assisted assembly. Scanning electron micrographs from left to right: Ag@pNIPAM core–shell particles, pNIPAM-coated gold nanostars, and protein-coated gold nanorods. (d) Arrays of strongly coupled plasmonic chains fabricated by spin coating of protein-functionalized gold nanospheres into wrinkled PDMS stamps followed by subsequent wet transfer printing onto adhesive, flat substrates. (e) Corresponding UV-Vis-Near-Infrared (NIR) extinction spectra of the close-packed single particle, dimer, and quadrumer chain arrays showing the strong structure and polarization dependence of the macroscopic optical properties. (f) Illustration of a plasmonic strain sensor consisting of gold NP chains on a stretchable substrate. (g) Dependence of the longitudinal resonance wavelength of single-particle chains on the applied strain. ((a) Adapted from Müller et al.,[78] Copyright 2012 Royal Society of Chemistry. (b) Adapted from Mueller et al.,[147] Copyright 2012 American Chemical Society. (c) Adapted from Tebbe et al.,[102] Copyright 2015 Royal Society of Chemistry. (d,e) Adapted from Hanske et al.,[79] Copyright 2014 American Chemical Society. (f,g) Adapted from Steiner et al.,[149] Copyright 2017 American Chemical Society.)

14.3.4 Three-Dimensional NP Assemblies

For the fabrication of 3D colloidal supercrystals, building block quality matters even more than for 1D and 2D assemblies. The last decade witnessed huge progress in the synthesis of shape- and size-monodisperse plasmonic particles, which translated directly into more regular self-assembled structures. Probably the most basic plasmonic supercrystal consists of gold nanospheres arranged in a face-centered cubic (FCC) lattice. Recently, Liz-Marzán and coworkers presented a straightforward approach to preparing macroscale arrays of micrometer-sized supercrystals.[85] A scalable synthesis of monodisperse gold nanospheres by seeded growth, etching, and subsequent

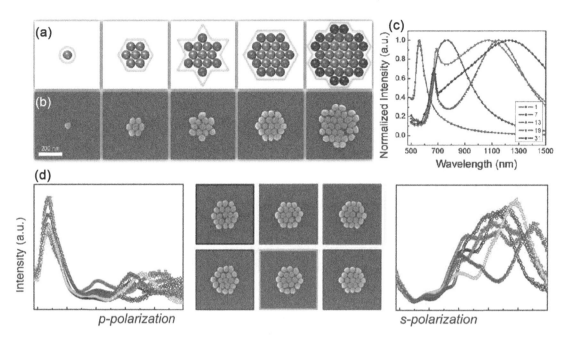

FIGURE 14.13 2D gold metamolecules reported by Murray, Kagan, and coworkers. (a) Schematic of plasmonic oligomer meta-molecules assembled in polygonal templates. The assemblies contain 1, 7, 13, 19, and 31 NPs. Like shading indicates NPs equidistant from the central NP, forming (very light) first-, (medium) second-, (light) third-, and (dark) fourth-nearest neighbor 2D shells. (b) Scanning electron micrographs of gold NP oligomers formed via TASA, shown after resist liftoff. (c) Simulated normalized scattering cross section for each metamolecule configuration. (d) Dark-field scattering spectra of different metamolecules (annotated by shaded outlines) collected under p- and s-polarized excitation, with corresponding electron micrographs. (Adapted from Greybush et al.,[150] Copyright 2017 American Chemical Society.)

PEGylation enabled TASA with extremely high particle concentrations and accurate mold replication. The regular arrays of FCC-supercrystals (in the shape of equilateral square pyramids) thus achieved over mm² (Figure 14.15).

In general, what we consider spherical nanocrystals are not truly spherical but adopt Wulff shapes.[156] In colloidal synthesis, the equilibrium shape cannot be spherical because many high-index facets with high specific surface free energies would be required to obtain a perfect sphere. Supercrystals obtained by solvent evaporation, surfactant diffusion, and sedimentation obey to the same laws and often mirror the shape of the constituent building blocks.[157–159] In this regard, astonishing examples reported the formation of supercrystals with crystallographic facets and even Wulff shapes (Figure 14.16).[45,160,161] In one example, Kalsin et al. coated gold and silver NPs with anionic and cationic thiols, respectively, and then charge-equivalent amounts of the two were mixed in an aqueous solution. The electrostatic interactions between the NPs led to the formation of crystals with specific crystallographic facets resembling diamond-like crystals. Interestingly, small precipitates were resuspended in solvent and recrystallized in 1:4 (v/v) Water/dimethyl sulfoxide to yield large numbers of regularly faceted crystals up to 3 μm in size (Figure 14.16a).[161]

The hierarchical organization of supercrystals can be tailored by varying the self-assembling conditions such as the solvent,[163,164] temperature,[159,164] or the capping agent.[165,166] Other ways to tailor the local organization include postprocesses such as applying a pressure gradient onto the supercrystals,[167,168] functionalizing with inorganic matrices[169,170] or reintroducing a little amount of solvent in the structure.[171] It is known that NPs with different shapes tend to self-assemble separately to maximize inter-particle interaction.[172–174] A recent study has been shown that even similar-sized NPs but with different crystallinity (i.e., monocrystalline vs. polycrystalline) assembled into separated supercrystals.[175] Importantly, the differences in nanocrystallinity between the two types of supercrystals had a great influence on their physical properties. In particular, the Young modulus, which describes the mechanical response of the material upon stress, was a decade larger for supercrystal composed of monocrystalline NPs.[176] Indeed, one current challenge in the field is to find new avenues[177] to control the supercrystal local organization and thus the optical properties.[178,179] The assemblies reported in this section so far were mostly prepared in apolar solvents, but it is worth stressing out that an appropriate postsurface modification may allow such supercrystals to be suspended in water.[180]

In colloidal synthesis, some specific facets of the nanocrystals can be favored by adjusting the synthesis conditions and yield polyhedral NPs. While assembly by polyhedral particles occurs in similar conditions to spherical NPs, the effect of particle shape becomes important and exotic superlattices have been observed (Figure 14.16c). For instance, Minkowski lattices, which define a very dense

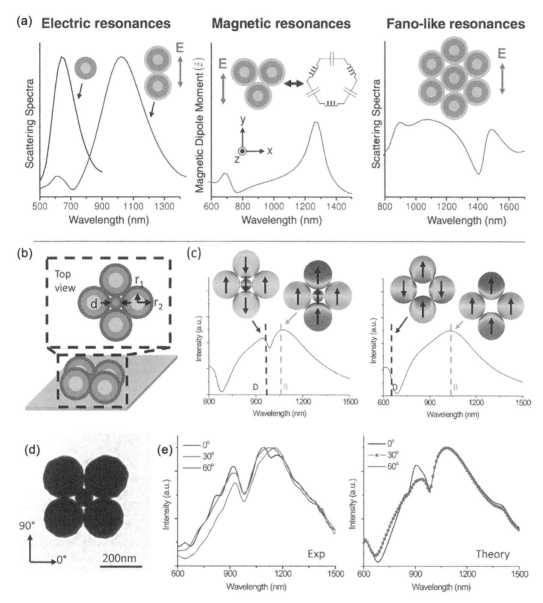

FIGURE 14.14 Optical response of complex 2D plasmonic clusters: (a) characteristic electric, magnetic, and Fano-like resonances of individual nanoshells and small clusters.[155] (b) Illustration of a heterocluster composed of four nanoshells and central solid sphere. (c) Comparison of clusters with and without central particle explaining the origin of the Fano-like resonance dip near 1,000 nm (charge distribution modeled by finite-difference time-domain method). (d) TEM of a colloidal heterocluster assembled by complementary DNA strand linking. (e) Comparison of measured and modeled cluster extinction spectra under s-polarization with varying polarization angle of the electric field.[152] ((a) Adapted from Fan et al.,[155] Copyright 2010 American Association for the Advancement of Science. (b–e) Adapted from Fan et al.,[152] Copyright 2011 American Chemical Society.)

packing, have been observed in silver[181] and gold octahedra supercrystals.[182] Other intriguing examples highlighting the effect of particle shape on tiling can be found.[183–185]

Plasmonic anisotropic NPs such as rods and plates can also form micron-sized supercrystals under controlled deposition.[184,185] The shape anisotropy of NP building blocks is of critical importance in determining their packing symmetry and assembly directionality. It has been shown that to maximize face-to-face interactions CdSe plates tend to stack into giant anisotropic superparticles.[186,187] Similar observation has been made with plasmonic plates, although such long-range organizations have not been

reached yet.[44,188] Through colloidal synthesis, beveled triangular gold nanoprisms have been prepared, which are characterized by two types of interaction surfaces, i.e., the basal plane and the beveled side.[42,189] The controlled self-assembly led to intriguing structures due to their geometrical features and the control of experimental conditions, such as the concentration of depletant and temperature (Figure 14.17a).[185]

Supercrystals made of rods have been obtained in which face-to-face interactions are also favored (Figure 3.17b). Therein, the rods were organized orthogonally with respect to the substrate, but supercrystals in which rods were

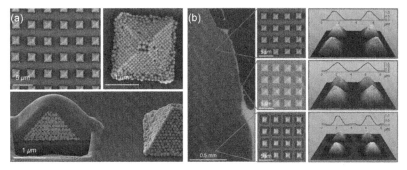

FIGURE 14.15 Arrays of micrometer-sized plasmonic supercrystals prepared by TASA with soft templates. Due to the high particle concentrations employed, near-complete space filling was achieved resulting in FCC packing with a characteristic particle layer structure. (a) Scanning electron micrographs of the main area with top views and an angled view of typical integral and cut supercrystals. (b) Characterization of regions with different assembly types via scanning electron microscopy and atomic force microscopy showing the influence of the coffee ring on array definition and supercrystal fine structure. (Adapted from Hanske et al.,[85] Copyright 2017 American Chemical Society.)

FIGURE 14.16 Example of micron-sized plasmonic supercrystals. (a) Supercrystals obtained by the controlled interaction between oppositely charged anionic gold NPs and cationic silver NPs. Their macroscopic sphalerite counterpart is displayed on the right. (b) High-resolution scanning electron micrograph of the high energy facet {221} on a gold nanocrystal superlattice plane that is vicinal to {111}. (c) Scanning electron micrographs of single polyhedral gold supercrystals. Models of supercrystals constructed from cubic, rhombic dodecahedral, and octahedral gold nanocrystals are also presented on the right. ((a) Adapted from Kalsin et al.,[161] Copyright 2006 AAAS. (B) Adapted from Goubet et al.,[157] Copyright 2014 American Chemical Society. (C) Adapted from Liao et al.[162] Copyright 2013 American Chemical Society.)

oriented horizontally have also been characterized.[191] The control of the local orientation of the building blocks is central to control of the anisotropic physical properties.[192,193] One of the key parameters is the initial concentration of NPs,[194] but the most important is the AR of the rods, which directs their orientational fate.[195–197]

As discussed previously, the Murray group has developed some interesting approaches for the self-assembly of plasmonic NPs. In addition, they have explored 3D assemblies of NPs into superlattices using thiolated dendritic ligands, or dendrons, to control the interparticle distance upon self-assembly. In addition to controlling the interparticle separation distance, they found that, by controlling the solvent evaporation rate, they could induce the gold NPs to self-organize into hexagonal close-packed superlattices.[198] Furthermore, by mixing two gold NPs of the same size, but functionalized with different sizes of dendrons, they were able to make the known crystal superlatices of $NaZn_{13}$ and $CaCu_5$ (Figure 14.18a). In a different direction, they found that, by doping silver/gold NPs into a superlattice of PbSe nanocrystals, they were able to adjust the DC conductivity of the superlattice (Figure 14.18b).[199]

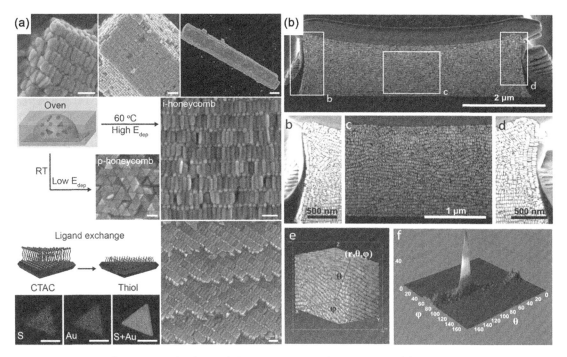

FIGURE 14.17 Micron-sized supercrystals obtained from anisotropic plasmonic NPs. (a) Beveled triangular nanoprisms and the structural diversity can be obtained by controlling the depletion attraction-driven assembly. (b) Cross-sectional views of a gold nanorod supercrystal obtained by combined templated approach and slow solvent evaporation. Below is the result of a 3D reconstruction of the supercrystal showing that the rods are oriented with an angle of 90° in respect to the substrate. ((a) Adapted from Kim et al.,[185] Copyright 2017 American Chemical Society. (b) Adapted from Hamon et al.,[190] Copyright 2016 American Chemical Society.)

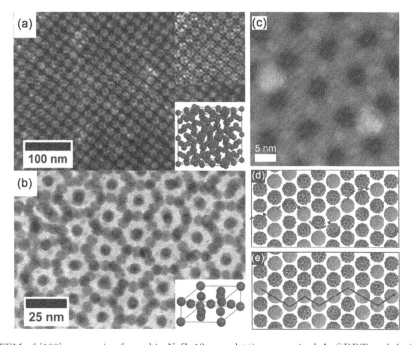

FIGURE 14.18 (a) TEM of [100] zone axis of a cubic NaZn13 superlattice comprised Au@DDT and Au@G2 blends. Inset (upper right) is a thinner region of the sample. Inset (lower right) is a diagram of the unit cell, which contains 112 gold NPs. (b) TEM of [001] zone axis of a CaCu5 superlattice formed from a mixture of Au@G1 and Au@G4 hybrids. Inset (lower right) is a drawing of the unit cell, which contains six gold NPs. (c) HAADF-STEM image of a multilayer film of Au/Ag–PbSe nanocrystals. (d,e) Schematics for conductivity in systems below (d) and above (e) the percolation threshold in a honeycomb lattice showing the hopping path for electrons in d (dashed black arrow) and the direct electron transfer between gold nanocrystals in e (filled black arrow). ((a,b) Adapted from Jishkariani et al.[198] Copyright 2015 American Chemical Society. (c–e) Adapted from Cargnello et al.,[199] Copyright 2015 Springer Nature.)

14.4   Concluding Remarks

The controlled assembly of plasmonic NPs leads to interesting and diverse properties, which result from the orientation, number, morphology, and size of the NPs in the assemblies. The tunability of collective properties in this manner enables the application of these assemblies in various areas. In recent years, the improvements in plasmonic colloidal NP synthesis have led to an explosion in the number of interesting approaches for their assembly into novel and interesting architectures of different scales and dimensions. We hope that the information presented in this chapter can help the readers in their efforts to understand, appreciate, and apply assembly of plasmonic NPs in their future research.

Acknowledgments

E.H.H. and Y.Z. acknowledge the financial support of the Beckman Young Investigator Program, the Army Research Office (W911NF-17-1-0561), the National Aeronautics and Space Administration Early Career Faculty Award (80NSSC17K0520), and the National Institute of General Medical Sciences of the National Institutes of Health (DP2GM128446).

References

1. Dreaden, E. C.; Alkilany, A. M.; Huang, X.; Murphy, C. J.; El-Sayed, M. A. The golden age: Gold nanoparticles for biomedicine. *Chem. Soc. Rev.* **2012**, *41*, 2740.
2. Cortie, M. B.; McDonagh, A. M. Synthesis and optical properties of hybrid and alloy plasmonic nanoparticles. *Chem. Rev.* **2011**, *111*, 3713–3735.
3. Schlücker, S. Surface-enhanced Raman spectroscopy: Concepts and chemical applications. *Angew. Chemie Int. Ed.* **2014**, *53*, 4756–4795.
4. Aroca, R. F. Plasmon enhanced spectroscopy. *Phys. Chem. Chem. Phys.* **2013**, *15*, 5355–5363.
5. Baffou, G.; Quidant, R. Nanoplasmonics for chemistry. *Chem. Soc. Rev.* **2014**, *43*, 3898–3907.
6. Polavarapu, L.; Pérez-Juste, J.; Xu, Q.-H.; Liz-Marzán, L. M. Optical sensing of biological, chemical and ionic species through aggregation of plasmonic nanoparticles. *J. Mater. Chem. C* **2014**, *2*, 7460–7476.
7. Roberto de la Rica, M. M. S. Plasmonic ELISA for the ultrasensitive detection of disease biomarkers with the naked eye. *Nat. Nanotechnol.* **2012**, *7*(12), 821–824.
8. Guerrero-Martínez, A.; Grzelczak, M.; Liz-Marzán, L. M. Molecular thinking for nanoplasmonic design. *ACS Nano* **2012**, *6*(5), 3655–3662.
9. Tan, S. J.; Campolongo, M. J.; Luo, D.; Cheng, W. Building plasmonic nanostructures with DNA. *Nat. Nano* **2011**, *6*, 268–276.
10. Rycenga, M.; Cobley, C. M.; Zeng, J.; Li, W.; Moran, C. H.; Zhang, Q.; Qin, D.; Xia, Y. Controlling the synthesis and assembly of silver nanostructures for plasmonic applications. *Chem. Rev.* **2011**, *111*, 3669–3712.
11. Jones, M. R.; Osberg, K. D.; Macfarlane, R. J.; Langille, M. R.; Mirkin, C. A. Templated techniques for the synthesis and assembly of plasmonic nanostructures. *Chem. Rev.* **2011**, *111*, 3736–3827.
12. Aberasturi, D. J.; Hamon, C.; Liz-Marzán, L. M. Chemical synthesis of plasmonic nanoparticles. In J. Aizpurua (ed.) *World Scientific Handbook of Metamaterials and Plasmonics*. World Scientific Publishing Co. Pte Ltd: Singapore, 2017, pp. 439–473.
13. O'Brien, M. N.; Jones, M. R.; Mirkin, C. A. The nature and implications of uniformity in the hierarchical organization of nanomaterials. *Proc. Natl. Acad. Sci. . U. S. A.* **2016**, *113*, 11717–11725.
14. Xia, Y.; Nguyen, T. D.; Yang, M.; Lee, B.; Santos, A.; Podsiadlo, P.; Tang, Z.; Glotzer, S. C.; Kotov, N. A. Self-assembly of self-limiting monodisperse supraparticles from polydisperse nanoparticles. *Nat. Nano* **2011**, *6*, 580–587.
15. Shevchenko, E. V; Talapin, D. V; Kotov, N. A.; O'Brien, S.; Murray, C. B. Structural diversity in binary nanoparticle superlattices. *Nature* **2006**, *439*, 55–59.
16. Derjaguin, B. A theory of interaction of particles in presence of electric double layers and the stability of lyophobe colloids and disperse systems. *Prog. Surf. Sci.* **1993**, *43*, 1–14.
17. Derjaguin, B. V; Landau, L. Theory of the stability of strongly charged lyophobic sols and the adhesion of strongly charged particles in solution of electrolytes. *Acta Physicochim. U.R.S.S.* **1941**, *14*, 633.
18. Verwey, E. J. W.; Overbeek, J. T. G. *Theory of the Stability of Lyophobic Colloids*. Elsevier Publishing Company Inc.: New York, 1948.
19. Boles, M. A.; Engel, M.; Talapin, D. V. Self-assembly of colloidal nanocrystals: From intricate structures to functional materials. *Chem. Rev.* **2016**, *116*, 11220–11289.
20. Napper, D. H. Steric stabilization. *J. Colloid Interface Sci.* **1977**, *58*, 390–407.
21. Goubet, N.; Richardi, J.; Albouy, P.-A.; Pileni, M.-P. Which forces control supracrystal nucleation in organic media? *Adv. Funct. Mater.* **2011**, *21*, 2693–2704.
22. Kotov, N. A. Inorganic nanoparticles as protein mimics. *Science* **2010**, *330*, 188–189.
23. Moyano, D. F.; Rotello, V. M. Nano meets biology: Structure and function at the nanoparticle interface. *Langmuir* **2011**, *27*, 10376–10385.
24. Mout, R.; Yesilbag Tonga, G.; Wang, L.-S.; Ray, M.; Roy, T.; Rotello, V. M. Programmed self-assembly of hierarchical nanostructures through

protein–nanoparticle coengineering. *ACS Nano* **2017**, *11*(4), 3456–3462.

25. Halas, N. J.; Lal, S.; Chang, W.-S.; Link, S.; Nordlander, P. Plasmons in strongly coupled metallic nanostructures. *Chem. Rev.* **2011**, *111*, 3913–3961.

26. Ma, W.; Xu, L.; de Moura, A. F.; Wu, X.; Kuang, H.; Xu, C.; Kotov, N. A. Chiral inorganic nanostructures. *Chem. Rev.* **2017**, *117*, 8041–8093.

27. Kumar, J.; Thomas, K. G.; Liz-Marzán, L. M. Nanoscale chirality in metal and semiconductor nanoparticles. *Chem. Commun. (Camb)*. **2016**, *52*, 12555–12569.

28. Tu, Y.; Peng, F.; Adawy, A.; Men, Y.; Abdelmohsen, L. K. E. A.; Wilson, D. A. Mimicking the cell: Bio-inspired functions of supramolecular assemblies. *Chem. Rev.* **2016**, *116*, 2023–2078.

29. De, S.; Klajn, R. Dissipative self-assembly driven by the consumption of chemical fuels. *Adv. Mater.* **2018**, *30*, e1706750.

30. Fratzl, P. Biomimetic materials research: What can we really learn from nature's structural materials? *J. R. Soc. Interface* **2007**, *4*, 637–642.

31. Si, K. J.; Chen, Y.; Shi, Q.; Cheng, W. Nanoparticle superlattices: The roles of soft ligands. *Adv. Sci.* **2017**, *5*, 1700179.

32. Hajiw, S.; Pansu, B.; Sadoc, J.-F. Evidence for a C14 Frank–Kasper phase in one-size gold nanoparticle superlattices. *ACS Nano* **2015**, *9*, 8116–8121.

33. Guerrero-García, G. I.; González-Mozuelos, P.; de la Cruz, M. O. Potential of mean force between identical charged nanoparticles immersed in a size-asymmetric monovalent electrolyte. *J. Chem. Phys.* **2011**, *135*, 164705.

34. Pfeiffer, C.; Rehbock, C.; Hühn, D.; Carrillo-Carrion, C.; de Aberasturi, D. J.; Merk, V.; Barcikowski, S.; Parak, W. J. Interaction of colloidal nanoparticles with their local environment: The (ionic) nanoenvironment around nanoparticles is different from bulk and determines the physico-chemical properties of the nanoparticles. *J. R. Soc. Interface* **2014**, *11*, 20130931.

35. Silvera Batista, C. A.; Larson, R. G.; Kotov, N. A. Nonadditivity of nanoparticle interactions. *Science*. **2015**, *350*, 1242477.

36. Abécassis, B. Three-dimensional self assembly of semiconducting colloidal nanocrystals: From fundamental forces to collective optical properties. *ChemPhysChem* **2016**, *17*, 618–631.

37. Min, Y.; Akbulut, M.; Kristiansen, K.; Golan, Y.; Israelachvili, J. The role of interparticle and external forces in nanoparticle assembly. *Nat. Mater.* **2008**, *7*, 527.

38. Lekkerkerker, H.N.W.;Tuinier, R. *Colloids and the Depletion Interaction*, vol. 833. Springer, Heidelberg, 2011.

39. Asakura, S.; Oosawa, F. On interaction between two bodies immersed in a solution of macromolecules. *J. Chem. Phys.* **1954**, *22*, 1255–1256.

40. Vrij, A. Polymers at interfaces and the interactions in colloidal dispersions. *Pure Appl. Chem.* **1976**, *48*, 471–483.

41. Park, K.; Koerner, H.; Vaia, R. A. Depletion-induced shape and size selection of gold nanoparticles. *Nano Lett.* **2010**, *10*, 1433–1439.

42. Scarabelli, L.; Coronado-Puchau, M.; Giner-Casares, J. J.; Langer, J.; Liz-Marzán, L. M. Monodisperse gold nanotriangles: Size control, large-scale self-assembly, and performance in surface-enhanced Raman scattering. *ACS Nano* **2014**, *8*, 5833–5842.

43. Li, Q.; Zhuo, X.; Li, S.; Ruan, Q.; Xu, Q.-H.; Wang, J. Production of monodisperse gold nanobipyramids with number percentages approaching 100% and evaluation of their plasmonic properties. *Adv. Opt. Mater.* **2015**, *3*, 801–812.

44. Young, K. L.; Jones, M. R.; Zhang, J.; Macfarlane, R. J.; Esquivel-Sirvent, R.; Nap, R. J.; Wu, J.; Schatz, G. C.; Lee, B.; Mirkin, C. A. Assembly of reconfigurable one-dimensional colloidal superlattices due to a synergy of fundamental nanoscale forces. *Proc. Natl. Acad. Sci. U. S. A.* **2012**, *109*, 2240–2245.

45. Chiu, C.-Y.; Chen, C.-K.; Chang, C.-W.; Jeng, U. S.; Tan, C.-S.; Yang, C.-W.; Chen, L.-J.; Yen, T.-J.; Huang, M. H. Surfactant-directed fabrication of supercrystals from the assembly of polyhedral Au–Pd core–shell nanocrystals and their electrical and optical properties. *J. Am. Chem. Soc.* **2015**, *137*(6), 2265–2275.

46. Deegan, R. D.; Bakajin, O.; Dupont, T. F.; Huber, G.; Nagel, S. R.; Witten, T. A. Capillary flow as the causes of ring stains from dried liquid drops. *Nature (London)* **1997**, *389*, 827–829.

47. Hu, H.; Larson, R. G. Marangoni effect reverses coffee-ring depositions. *J. Phys. Chem. B* **2006**, *110*, 7090–7094.

48. Yunker, P. J.; Still, T.; Lohr, M. A.; Yodh, A. G. Suppression of the coffee-ring effect by shape-dependent capillary interactions. *Nature* **2011**, *476*, 308–311.

49. Kim, H.; Boulogne, F.; Um, E.; Jacobi, I.; Button, E.; Stone, H. A. Controlled uniform coating from the interplay of marangoni flows and surface-adsorbed macromolecules. *Phys. Rev. Lett.* **2016**, *116*, 124501.

50. Hamon, C.; Liz-Marzan, L. M. Hierarchical assembly of plasmonic nanoparticles. *Chem. A Eur. J.* **2015**, *21*, 9956–9963.

51. Lalander, C. H.; Zheng, Y.; Dhuey, S.; Cabrini, S.; Bach, U. DNA-directed self-assembly of gold nanoparticles onto nanopatterned surfaces: Controlled placement of individual nanoparticles into regular arrays. *ACS Nano* **2010**, *4*, 6153–6161.

52. Aizenberg, J.; Braun, P. V; Wiltzius, P. Patterned colloidal deposition controlled by electrostatic and capillary forces. *Phys. Rev. Lett.* **2000**, *84*, 2997–3000.

53. Zheng, H. P.; Lee, I.; Rubner, M. F.; Hammond, P. T. Two component particle arrays on patterned polyelectrolyte multilayer templates. *Adv. Mater.* **2002**, *14*, 569–572.

54. Hanske, C.; Schneider, C.; Drechsler, M.; Wittemann, A.; Fery, A. Salt-regulated attraction and repulsion of spherical polyelectrolyte brushes towards polyelectrolyte multilayers. *Phys. Chem. Chem. Phys.* **2012**, *14*, 4196–4203.

55. Palleau, E.; Sangeetha, N. M.; Viau, G.; Marty, J.-D.; Ressier, L. Coulomb force directed single and binary assembly of nanoparticles from aqueous dispersions by AFM nanoxerography. *ACS Nano* **2011**, *5*, 4228–4235.

56. Coskun, U. C.; Mebrahtu, H.; Huang, P. B.; Huang, J.; Sebba, D.; Biasco, A.; Makarovski, A.; Lazarides, A.; LaBean, T. H.; Finkelstein, G. Single-electron transistors made by chemical patterning of silicon dioxide substrates and selective deposition of gold nanoparticles. *Appl. Phys. Lett.* **2008**, *93*, 123101.

57. Kolíbal, M.; Konečný, M.; Ligmajer, F.; Škoda, D.; Vystavěl, T.; Zlámal, J.; Varga, P.; Šikola, T. Guided assembly of gold colloidal nanoparticles on silicon substrates prepatterned by charged particle beams. *ACS Nano* **2012**, *6*, 10098–10106.

58. Nepal, D.; Onses, M. S.; Park, K.; Jespersen, M.; Thode, C. J.; Nealey, P. F.; Vaia, R. A. Control over position, orientation, and spacing of arrays of gold nanorods using chemically nanopatterned surfaces and tailored particle-particle-surface interactions. *ACS Nano* **2012**, *6*, 5693–5701.

59. Harreis, H. M.; Schmidt, M.; Lowen, H. Decoration lattices of colloids adsorbed on stripe-patterned substrates. *Phys. Rev. E* **2002**, *65*, 1–8.

60. Noh, H.; Choi, C.; Hung, A. M.; Jin, S.; Cha, J. N. Site-specific patterning of highly ordered nanocrystal superlattices through biomolecular surface confinement. *ACS Nano* **2010**, *4*, 5076–5080.

61. Brinkmann, M.; Lipowsky, R. Wetting morphologies on substrates with striped surface domains. *J. Appl. Phys.* **2002**, *92*, 4296.

62. Jeon, N. L.; Finnie, K.; Branshaw, K.; Nuzzo, R. G. Structure and stability of patterned self-assembled films of octadecyltrichlorosilane formed by contact printing. *Langmuir* **1997**, *13*, 3382–3391.

63. Ling, X. Y.; Malaquin, L.; Reinhoudt, D. N.; Wolf, H.; Huskens, J. An in situ study of the adsorption behavior of functionalized particles on self-assembled monolayers via different chemical interactions. *Langmuir* **2007**, *23*, 9990–9999.

64. Dimitrov, A. S.; Nagayama, K. Continuous convective assembling of fine particles into two-dimensional arrays on solid surfaces. *Langmuir* **1996**, *12*, 1303–1311.

65. Fan, F.; Stebe, K. J. Assembly of colloidal particles by evaporation on surfaces with patterned hydrophobicity. *Langmuir* **2004**, *20*, 3062–3067.

66. Fustin, C. A.; Glasser, G.; Spiess, H. W.; Jonas, U. Parameters influencing the templated growth of colloidal crystals on chemically patterned surfaces. *Langmuir* **2004**, *20*, 9114–9123.

67. Ressier, L.; Viallet, B.; Beduer, A.; Fabre, D.; Fabie, L.; Palleau, E.; Dague, E. Combining convective/capillary deposition and AFM oxidation lithography for close-packed directed assembly of colloids. *Langmuir* **2008**, *24*, 13254–13257.

68. Maury, P.; Escalante, M.; Reinhoudt, D. N.; Huskens, J. Directed assembly of nanoparticles onto polymer-imprinted or chemically patterned templates fabricated by nanoimprint lithography. *Adv. Mater.* **2005**, *17*, 2718–2723.

69. Yin, Y.; Lu, Y.; Gates, B.; Xia, Y. Template-assisted self-assembly: A practical route to complex aggregates of monodispersed colloids with well-defined sizes, shapes, and structures. *J. Am. Chem. Soc.* **2001**, *123*, 8718–8729.

70. Oshima, H.; Tamura, H.; Takeuchi, M.; Inomata, A.; Yanagida, Y.; Matsushita, N.; Komoriya, H.; Uzumaki, T.; Tanaka, A. Nanopattern transfer from high-density self-assembled nanosphere arrays on prepatterned substrates. *Nanotechnology* **2009**, *20*, 455303.

71. Varghese, B.; Cheong, F. C.; Sindhu, S.; Yu, T.; Lim, C.-T.; Valiyaveettil, S.; Sow, C.-H. Size selective assembly of colloidal particles on a template by directed self-assembly technique. *Langmuir* **2006**, *22*, 8248–8252.

72. Ni, S.; Leemann, J.; Wolf, H.; Isa, L. Insights into mechanisms of capillary assembly. *Faraday Discuss.* **2015**, *181*, 225–242.

73. Malaquin, L.; Kraus, T.; Schmid, H.; Delamarche, E.; Wolf, H. Controlled particle placement through convective and capillary assembly. *Langmuir* **2007**, *23*, 11513–11521.

74. Slaughter, L. S.; Willingham, B. A.; Chang, W.-S.; Chester, M. H.; Ogden, N.; Link, S. Toward plasmonic polymers. *Nano Lett.* **2012**, *12*, 3967–3972.

75. Asbahi, M.; Mehraeen, S.; Wang, F.; Yakovlev, N.; Chong, K. S. L.; Cao, J.; Tan, M. C.; Yang, J. K. W. Large area directed self-assembly of sub-10 Nm particles with single particle positioning resolution. *Nano Lett.* **2015**, *15*, 6066–6070.

76. Asbahi, M.; Mehraeen, S.; Lim, K. T. P.; Wang, F.; Cao, J.; Tan, M. C.; Yang, J. K. W. Template-induced structure transition in sub-10 Nm self-assembling nanoparticles. *Nano Lett.* **2014**, *14*, 2642–2646.

77. Flauraud, V.; Mastrangeli, M.; Bernasconi, G. D.; Butet, J.; Alexander, D. T. L.; Shahrabi, E.;

Martin, O. J. F.; Brugger, J. Nanoscale topographical control of capillary assembly of nanoparticles. *Nat. Nanotechnol.* **2016**, *12*, 73–80.

78. Müller, M.; Karg, M.; Fortini, A.; Hellweg, T.; Fery, A. Wrinkle-assisted linear assembly of hard-core/soft-shell particles: Impact of the soft shell on the local structure. *Nanoscale* **2012**, *4*, 2491–2499.

79. Hanske, C.; Tebbe, M.; Kuttner, C.; Bieber, V.; Tsukruk, V. V; Chanana, M.; Konig, T. A. F.; Fery, A. Strongly coupled plasmonic modes on macroscopic areas via template-assisted colloidal self-assembly. *Nano Lett.* **2014**, *14*, 6863–6871.

80. Ling, X. Y.; Phang, I. Y.; Schönherr, H.; Reinhoudt, D. N.; Vancso, G. J.; Huskens, J. Freestanding 3D supramolecular particle bridges: Fabrication and mechanical behavior. *Small* **2009**, *5*, 1428–1435.

81. Xia, Y.; Yin, Y.; Lu, Y.; McLellan, J. Template-assisted self-assembly of spherical colloids into complex and controllable structures. *Adv. Funct. Mater.* **2003**, *13*, 907–918.

82. Cui, Y.; Björk, M. T.; Liddle, J. A.; Sönnichsen, C.; Boussert, B.; Alivisatos, A. P. Integration of colloidal nanocrystals into lithographically patterned devices. *Nano Lett.* **2004**, *4*, 1093–1098.

83. Greybush, N. J.; Saboktakin, M.; Ye, X.; Della Giovampaola, C.; Oh, S. J.; Berry, N. E.; Engheta, N.; Murray, C. B.; Kagan, C. R. Plasmon-enhanced upconversion luminescence in single nanophosphor-nanorod heterodimers formed through template-assisted self-assembly. *ACS Nano* **2014**, *8*, 9482–9491.

84. Malaquin, L.; Kraus, T.; Schmid, H.; Delamarche, E.; Wolf, H. Controlled particle placement through convective and capillary assembly. *Langmuir* **2007**, *23*, 11513–11521.

85. Hanske, C.; González-Rubio, G.; Hamon, C.; Formentín, P.; Modin, E.; Chuvilin, A.; Guerrero-Martínez, A.; Marsal, L. F.; Liz-Marzán, L. M. Large-scale plasmonic pyramidal supercrystals via templated self-assembly of monodisperse gold nanospheres. *J. Phys. Chem. C* **2017**, *121*, 10899–10906.

86. Gast, A. P.; Zukoski, C. F. Electrorheological fluids as colloidal suspensions. *Adv. Colloid Interface Sci.* **1989**, *30*, 153–202.

87. Jackson, J. D. *Classical Electrodynamics*. John Wiley & Sons Inc., New York, *13*, 1999.

88. O'Brien, R. W.; White, L. R. Electrophoretic mobility of a spherical colloidal particle. *J. Chem. Soc. Faraday Trans. 2* **1978**, *74*, 1607.

89. Dassanayake, U.; Fraden, S.; Blaaderen, A. The structure of electrorheological fluids. *J. Chem. Phys.* **2000**, *112*, 3851.

90. Fraden, S.; Hurd, A. J.; Meyer, R. B. Electric-field-induced association of colloidal particles. *Phys. Rev. Lett.* **1989**, *63*, 2373–2376.

91. Hynninen, A. P.; Dijkstra, M. Phase diagram of dipolar hard and soft spheres: Manipulation of colloidal crystal structures by an external field. *Phys. Rev. Lett.* **2005**, *94*, 8–11.

92. Hermanson, K. D.; Lumsdon, S. O.; Williams, J. P.; Kaler, E. W.; Velev, O. D. Dielectrophoretic assembly of electrically functional microwires from nanoparticle suspensions. *Science.* **2001**, *294*, 1082–1086.

93. Yang, D.; Pang, X.; He, Y.; Wang, Y.; Chen, G.; Wang, W.; Lin, Z. Precisely size-tunable magnetic/plasmonic core/shell nanoparticles with controlled optical properties. *Angew. Chemie Int. Ed.* **2015**, *54*, 12091–12096.

94. Bannwarth, M. B.; Utech, S.; Ebert, S.; Weitz, D. A.; Crespy, D.; Landfester, K. Colloidal polymers with controlled sequence and branching constructed from magnetic field assembled nanoparticles. *ACS Nano* **2015**, *9*, 2720–2728.

95. Lin, L.; Zhang, J.; Peng, X.; Wu, Z.; Coughlan, A. C. H.; Mao, Z.; Bevan, M. A.; Zheng, Y. Opto-thermophoretic assembly of colloidal matter. *Sci. Adv.* **2017**, *3*, 1–10.

96. Lin, L.; Peng, X.; Mao, Z.; Wei, X.; Xie, C.; Zheng, Y. Interfacial-entropy-driven thermophoretic tweezers. *Lab Chip* **2017**, *17*, 3061–3070.

97. Lin, L.; Wang, M.; Peng, X.; Lissek, E. N.; Mao, Z.; Scarabelli, L.; Adkins, E.; Coskun, S.; Unalan, H. E.; Korgel, B. A.; et al. Opto-thermoelectric nanotweezers. *Nat. Photonics* **2018**, *12*, 195–201.

98. Lin, L.; Peng, X.; Wang, M.; Scarabelli, L.; Mao, Z.; Liz-Marzán, L. M.; Becker, M. F.; Zheng, Y. Light-directed reversible assembly of plasmonic nanoparticles using plasmon-enhanced thermophoresis. *ACS Nano* **2016**, *10*, 9659–9668.

99. Lin, L.; Peng, X.; Mao, Z.; Li, W.; Yogeesh, M. N.; Rajeeva, B. B.; Perillo, E. P.; Dunn, A. K.; Akinwande, D.; Zheng, Y. Bubble-pen lithography. *Nano Lett.* **2016**, *16*, 701–708.

100. Yan, B.; Thubagere, A.; Premasiri, W. R.; Ziegler, L. D.; Negro, L. D.; Reinhard, B. M. Engineered SERS substrates with multiscale signal enhancement: Nanoparticle cluster arrays. *ACS Nano* **2009**, *3*, 1190–1202.

101. Hanske, C.; Müller, M. B.; Bieber, V.; Tebbe, M.; Jessl, S.; Wittemann, A.; Fery, A. The role of substrate wettability in nanoparticle transfer from wrinkled elastomers: Fundamentals and application toward hierarchical patterning. *Langmuir* **2012**, *28*, 16745–16750.

102. Tebbe, M.; Mayer, M.; Glatz, B. A.; Hanske, C.; Probst, P. T.; Müller, M. B.; Karg, M.; Chanana, M.; König, T. A. F.; Kuttner, C.; et al. Optically anisotropic substrates *via* wrinkle-assisted convective assembly of gold nanorods on macroscopic areas. *Faraday Discuss.* **2015**, *181*, 243–260.

103. Swanglap, P.; Slaughter, L. S.; Chang, W.-S.; Willingham, B.; Khanal, B. P.; Zubarev, E. R.; Link, S. Seeing double: Coupling between substrate image charges and collective plasmon modes in self-assembled nanoparticle superstructures. *ACS Nano* **2013**, *5*, 4892–4901.

104. Giner-Casares, J. J.; Reguera, J. Directed self-assembly of inorganic nanoparticles at air/liquid interfaces. *Nanoscale* **2016**, *8*, 16589–16595.

105. Olson, M. A.; Coskun, A.; Klajn, R.; Fang, L.; Dey, S. K.; Browne, K. P.; Grzybowski, B. A.; Stoddart, J. F. Assembly of polygonal nanoparticle clusters directed by reversible noncovalent bonding interactions. *Nano Lett.* **2009**, *9*, 3185–3190.

106. Correa-Duarte, M. A.; Liz-Marzán, L. M. Carbon nanotubes as templates for one-dimensional nanoparticle assemblies. *J. Mater. Chem.* **2006**, *16*, 22–25.

107. Zhao, N.; Liu, K.; Greener, J.; Nie, Z.; Kumacheva, E. Close-packed superlattices of side-by-side assembled Au-CdSe nanorods. *Nano Lett.* **2009**, *9*, 3077–3081.

108. Maye, M. M.; Nykypanchuk, D.; Cuisinier, M.; Van Der Lelie, D.; Gang, O. Stepwise surface encoding for high-throughput assembly of nanoclusters. *Nat. Mater.* **2009**, *8*, 388–391.

109. Xu, X.; Rosi, N. L.; Wang, Y.; Huo, F.; Mirkin, C. A. Asymmetric functionalization of gold nanoparticles with oligonucleotides. *J. Am. Chem. Soc.* **2006**, *128*, 9286–9287.

110. Cheng, W.; Campolongo, M. J.; Cha, J. J.; Tan, S. J.; Umbach, C. C.; Muller, D. A.; Luo, D. Free-standing nanoparticle superlattice sheets controlled by DNA. *Nat. Mater.* **2009**, *8*, 519–525.

111. Klajn, R.; Bishop, K. J. M.; Grzybowski, B. A. Light-controlled self-assembly of reversible and irreversible nanoparticle suprastructures. *Proc. Natl. Acad. Sci. U. S. A.* **2007**, *104*, 10305–10309.

112. Grzelczak, M.; Sánchez-Iglesias, A.; Mezerji, H. H.; Bals, S.; Pérez-Juste, J.; Liz-Marzán, L. M. Steric hindrance induces crosslike self-assembly of gold nanodumbbells. *Nano Lett.* **2012**, *12*, 4380–4384.

113. Nie, Z.; Fava, D.; Kumacheva, E.; Zou, S.; Walker, G. C.; Rubinstein, M. Self-assembly of metal-polymer analogues of amphiphilic triblock copolymers. *Nat. Mater.* **2007**, *6*, 609–614.

114. Quintana, M.; Prato, M. Supramolecular aggregation of functionalized carbon nanotubes. *Chem. Commun.* **2009**, *40*, 6005.

115. Grzelczak, M.; Vermant, J.; Furst, E. M.; Liz-Marzán, L. M. Directed self-assembly of nanoparticles *ACS Nano* **2010**, *4*, 3591–3605.

116. Barrow, S. J.; Funston, A. M.; Wei, X.; Mulvaney, P. DNA-directed self-assembly and optical properties of discrete 1D, 2D and 3D plasmonic structures. *Nano Today* **2013**, *8*, 138–167.

117. Mirkin, C. A.; Letsinger, R. L.; Mucic, R. C.; Storhoff, J. J. A DNA-based method for rationally assembling nanoparticles into macroscopic materials. *Nature*, 1996, *382*, 607–609.

118. Aldaye, F. A.; Sleiman, H. F. Dynamic DNA templates for discrete gold nanoparticle assemblies: Control of geometry, modularity, write/erase and structural switching. *J. Am. Chem. Soc.* **2007**, *129*, 4130–4131.

119. Höller, R. P. M.; Dulle, M.; Thomä, S.; Mayer, M.; Steiner, A. M.; Förster, S.; Fery, A.; Kuttner, C.; Chanana, M. Protein-assisted assembly of modular 3D plasmonic raspberry-like core/satellite nanoclusters: Correlation of structure and optical properties. *ACS Nano* **2016**, *10*, 5740–5750.

120. Kumar, J.; Wei, X.; Barrow, S.; Funston, A. M.; Thomas, K. G.; Mulvaney, P. Surface plasmon coupling in end-to-end linked gold nanorod dimers and trimers. *Phys. Chem. Chem. Phys.* **2013**, *15*, 4258–4264.

121. Pazos-Pérez, N.; Wagner, C. S.; Romo-Herrera, J. M.; Liz-Marzan, L. M.; García de Abajo, F. J.; Wittemann, A.; Fery, A.; Alvarez-Puebla, R. A. Organized plasmonic clusters with high coordination number and extraordinary enhancement in Surface-Enhanced Raman Scattering (SERS). *Angew. Chemie Int. Ed.* **2012**, *51*, 12688–12693.

122. Cong, V. T.; Ganbold, E.-O.; Saha, J. K.; Jang, J.; Min, J.; Choo, J.; Kim, S.; Song, N. W.; Son, S. J.; Lee, S. B.; et al. Gold nanoparticle silica nanopeapods. *J. Am. Chem. Soc.* **2014**, *136*, 3833–3841.

123. Grzelczak, M.; Altantzis, T.; Goris, B.; Pe, J.; Sa, A.; Bals, S.; Tendeloo, G. Van; Donaldson, S. H.; Chmelka, B. F. Hydrophobic interactions modulate self-assembly of gold nanoparticles. *ACS Nano* **2012**, *6*, 11059–11065.

124. Smith, K. W.; Zhao, H.; Zhang, H.; Sánchez-Iglesias, A.; Grzelczak, M.; Wang, Y.; Chang, W. S.; Nordlander, P.; Liz-Marzán, L. M.; Link, S. Chiral and achiral nanodumbbell dimers: The effect of geometry on plasmonic properties. *ACS Nano* **2016**, *10*, 6180–6188.

125. Grzelczak, M.; Liz-Marzán, L. M. Exploiting hydrophobic interactions at the nanoscale. *J. Phys. Chem. Lett.* **2014**, *5*, 2455–2463.

126. Harris, N.; Arnold, M. D.; Blaber, M. G.; Ford, M. J. Plasmonic resonances of closely coupled gold nanosphere chains. *J. Phys. Chem. C* **2009**, *113*, 2784–2791.

127. Slaughter, L. S.; Wang, L.-Y.; Willingham, B. A.; Olson, J. M.; Swanglap, P.; Dominguez-Medina, S.; Link, S. Plasmonic polymers unraveled through single particle spectroscopy. *Nanoscale* **2014**, *6*, 11451–11461.

128. Quinten, M.; Leitner, A.; Krenn, J. R.; Aussenegg, F. R. Electromagnetic energy transport via linear chains of silver nanoparticles. *Opt. Lett.* **1998**, *23*, 1331–1333.

129. Maier, S. A.; Brongersma, M. L.; Kik, P. G.; Meltzer, S.; Requicha, A. A. G.; Atwater, H. A. Plasmonics: A route to nanoscale optical devices. *Adv. Mater.* **2001**, *13*, 1501–1505.

130. Maier, S. A.; Kik, P. G.; Atwater, H. A.; Meltzer, S.; Harel, E.; Koel, B. E.; Requicha, A. A. G. Local detection of electromagnetic energy transport below the diffraction limit in metal nanoparticle plasmon waveguides. *Nat. Mater.* **2003**, *2*, 229–232.

131. Solis, D.; Paul, A.; Olson, J.; Slaughter, L. S.; Swanglap, P.; Chang, W.-S.; Link, S. Turning the corner: Efficient energy transfer in bent plasmonic nanoparticle chain waveguides. *Nano Lett.* **2013**, *13*, 4779–4784.

132. Su, B.; Zhang, C.; Chen, S.; Zhang, X.; Chen, L.; Wu, Y.; Nie, Y.; Kan, X.; Song, Y.; Jiang, L. A general strategy for assembling nanoparticles in one dimension. *Adv. Mater.* **2014**, *26*, 2501–2507.

133. Efimenko, K.; Rackaitis, M.; Manias, E.; Vaziri, A.; Mahadevan, L.; Genzer, J. Nested self-similar wrinkling patterns in skins. *Nat. Mater.* **2005**, *4*, 293–297.

134. Kraus, T.; Brodoceanu, D.; Pazos-Pérez, N.; Fery, A. Colloidal surface assemblies: Nanotechnology meets bioinspiration. *Adv. Funct. Mater.* **2013**, *23*, 4529–4541.

135. Genzer, J.; Groenewold, J. Soft matter with hard skin: From skin wrinkles to templating and material characterization. *Soft Matter* **2006**, *2*, 310–323.

136. Jiang, H.; Khang, D.-Y.; Song, J.; Sun, Y.; Huang, Y.; Rogers, J. A. Finite deformation mechanics in buckled thin films on compliant supports. *Proc. Natl. Acad. Sci. U. S. A.* **2007**, *104*, 15607–15612.

137. Bowden, N.; Brittain, S.; Evans, A. G.; Hutchinson, J. W.; Whitesides, G. M. Spontaneous formation of ordered structures in thin films of metals supported on an elastomeric polymer. *Nature* **1998**, *393*, 146–149.

138. Bowden, N.; Huck, W. T. S.; Paul, K. E.; Whitesides, G. M. The controlled formation of ordered, sinusoidal structures by plasma oxidation of an elastomeric polymer. *Appl. Phys. Lett.* **1999**, *75*, 2557–2559.

139. Huck, W. T. S.; Bowden, N.; Onck, P.; Pardoen, T.; Hutchinson, J. W.; Whitesides, G. M. Ordering of spontaneously formed buckles on planar surfaces. *Langmuir* **2000**, *16*, 3497–3501.

140. Efimenko, K.; Wallace, W. E.; Genzer, J. Surface modification of sylgard-184 poly(dimethyl siloxane) networks by ultraviolet and ultraviolet/ozone treatment. *J. Colloid Interface Sci.* **2002**, *254*, 306–315.

141. Lu, C.; Möhwald, H.; Fery, A. A lithography-free method for directed colloidal crystal assembly based on wrinkling. *Soft Matter* **2007**, *3*, 1530–1536.

142. Watanabe, M.; Mizukami, K. Well-ordered wrinkling patterns on chemically oxidized poly(dimethylsiloxane) surfaces. *Macromolecules* **2012**, *45*, 7128–7134.

143. Chiche, A.; Stafford, C. M.; Cabral, J. T. Complex micropatterning of periodic structures on elastomeric surfaces. *Soft Matter* **2008**, *4*, 2360–2364.

144. Bayley, F. A.; Liao, J. L.; Stavrinou, P. N.; Chiche, A.; Cabral, J. T. Wavefront kinetics of plasma oxidation of polydimethylsiloxane: Limits for sub-Mm wrinkling. *Soft Matter* **2014**, *10*, 1155–1166.

145. Pazos-Pérez, N.; Ni, W.; Schweikart, A.; Alvarez-Puebla, R. A.; Fery, A.; Liz-Marzán, L. M. Highly uniform SERS substrates formed by wrinkle-confined drying of gold colloids. *Chem. Sci.* **2010**, *1*, 174–178.

146. Lee, J.; Seo, J.; Kim, D.; Shin, S.; Lee, S.; Mahata, C.; Lee, H.-S.; Min, B.-W.; Lee, T. Capillary force-induced glue-free printing of Ag nanoparticle arrays for highly sensitive SERS substrates. *ACS Appl. Mater. Interfaces* **2014**, *6*, 9053–9060.

147. Mueller, M.; Tebbe, M.; Andreeva, D. V; Karg, M.; Alvarez-Puebla, R. A.; Pazos-Pérez, N.; Fery, A. Large-area organization of pNIPAM-coated nanostars as SERS platforms for polycyclic aromatic hydrocarbons sensing in gas phase. *Langmuir* **2012**, *28*, 9168–9173.

148. Hanske, C.; Tebbe, M.; Kuttner, C.; Bieber, V.; Tsukruk, V. V; Chanana, M.; König, T. A. F.; Fery, A. Strongly coupled plasmonic modes on macroscopic areas via template-assisted colloidal self-assembly. *Nano Lett.* **2014**, *14*, 6863–6871.

149. Steiner, A. M.; Mayer, M.; Seuss, M.; Nikolov, S.; Harris, K. D.; Alexeev, A.; Kuttner, C.; Koenig, T. A. F.; Fery, A. Macroscopic strain-induced transition from quasi-infinite gold nanoparticle chains to defined plasmonic oligomers. *ACS Nano* **2017**, *11*, 8871–8880.

150. Greybush, N. J.; Liberal, I.; Malassis, L.; Kikkawa, J. M.; Engheta, N.; Murray, C. B.; Kagan, C. R. Plasmon resonances in self-assembled two-dimensional Au nanocrystal metamolecules. *ACS Nano* **2017**, *11*, 2917–2927.

151. Fan, J. A.; Wu, C.; Bao, K.; Bao, J.; Bardhan, R.; Halas, N. J.; Manoharan, V. N.; Nordlander, P.; Shvets, G.; Capasso, F. Self-assembled plasmonic nanoparticle clusters. *Science* **2010**, *328*, 1135–1138.

152. Fan, J. A.; He, Y.; Bao, K.; Wu, C.; Bao, J.; Schade, N. B.; Manoharan, V. N.; Shvets, G.; Nordlander, P.; Liu, D. R.; et al. DNA-enabled self-assembly of plasmonic nanoclusters. *Nano Lett.* **2011**, *11*, 4859–4864.

153. Fan, J. A.; Bao, K.; Sun, L.; Bao, J.; Manoharan, V. N.; Nordlander, P.; Capasso, F. Plasmonic mode engineering with templated self-assembled nanoclusters. *Nano Lett.* **2012**, *12*, 5318–5324.

154. Luk'yanchuk, B.; Zheludev, N. I.; Maier, S. A.; Halas, N. J.; Nordlander, P.; Giessen, H.; Chong, C. T.

The fano resonance in plasmonic nanostructures and metamaterials. *Nat. Mater.* **2010**, *9*, 707–715.

155. Fan, J. A.; Wu, C.; Bao, K.; Bao, J.; Bardhan, R.; Halas, N. J.; Manoharan, V. N.; Nordlander, P.; Shvets, G.; Capasso, F. Self-assembled plasmonic nanoparticle clusters. *Science* **2010**, *328*, 1135–1138.

156. Wulff, G. On the question of speed of growth and dissolution of crystal surfaces. *Krist. Miner.* **1901**, *34*, 449–530.

157. Goubet, N.; Yang, J.; Albouy, P.-A.; Pileni, M.-P. Spontaneous formation of high-index planes in gold single domain nanocrystal superlattices. *Nano Lett.* **2014**, *14*, 6632–6638.

158. Goubet, N.; Pileni, M. P. Analogy between atoms in a nanocrystal and nanocrystals in a supracrystal: Is it real or just a highly probable speculation? *J. Phys. Chem. Lett.* **2011**, *2*, 1024–1031.

159. Huang, M. H.; Thoka, S. Formation of supercrystals through self-assembly of polyhedral nanocrystals. *Nano Today* **2015**, *10*, 81–92.

160. Auyeung, E.; Li, T. I. N. G.; Senesi, A. J.; Schmucker, A. L.; Pals, B. C.; de la Cruz, M. O.; Mirkin, C. A. DNA-mediated nanoparticle crystallization into Wulff polyhedra. *Nature* **2014**, *505*, 73–77.

161. Kalsin, A. M.; Fialkowski, M.; Paszewski, M.; Smoukov, S. K.; Bishop, K. J. M.; Grzybowski, B. A. Electrostatic self-assembly of binary nanoparticle crystals with a diamond-like lattice. *Science* **2006**, *312*, 420–424.

162. Liao, C.-W.; Lin, Y.-S.; Chanda, K.; Song, Y.-F.; Huang, M. H. Formation of diverse supercrystals from self-assembly of a variety of polyhedral gold nanocrystals. *J. Am. Chem. Soc.* **2013**, *135* (7), 2684–2693.

163. Quan, Z.; Wu, D.; Zhu, J.; Evers, W. H.; Boncella, J. M.; Siebbeles, L. D. A.; Wang, Z.; Navrotsky, A.; Xu, H. Energy landscape of self-assembled superlattices of PbSe nanocrystals. *Proc. Natl. Acad. Sci. U. S. A.* **2014**, *111*, 9054–9057.

164. Courty, A.; Richardi, J.; Albouy, P.-A.; Pileni, M.-P. How to control the crystalline structure of supracrystals of 5-nm silver nanocrystals. *Chem. Mater.* **2011**, *23*, 4186–4192.

165. Mourdikoudis, S.; Çolak, A.; Arfaoui, I.; Pileni, M.-P. Hydrophilic gold supracrystals differing by the nanoparticle crystalline structure. *J. Phys. Chem. C* **2017**, *121*, 10670–10680.

166. Hamon, C.; Henriksen-Lacey, M.; La Porta, A.; Rosique, M.; Langer, J.; Scarabelli, L.; Montes, A. B. S.; González-Rubio, G.; de Pancorbo, M. M.; Liz-Marzán, L. M.; et al. Tunable nanoparticle and cell assembly using combined self-powered microfluidics and microcontact printing. *Adv. Funct. Mater.* **2016**, *26*, 8053–8061.

167. Wu, H.; Bai, F.; Sun, Z.; Haddad, R. E.; Boye, D. M.; Wang, Z.; Huang, J. Y.; Fan, H. Nanostructured gold architectures formed through high pressure-driven sintering of spherical nanoparticle arrays. *J. Am. Chem. Soc.* **2010**, *132*, 12826–12828.

168. Lequieu, J.; Córdoba, A.; Hinckley, D.; de Pablo, J. J. Mechanical response of DNA: Nanoparticle crystals to controlled deformation. *ACS Cent. Sci.* **2016**, *2*, 614–620.

169. Hamon, C.; Sanz-Ortiz, M. N.; Modin, E.; Hill, E.; Scarabelli, L.; Chuvilin, A.; Liz-Marzan, L. M. Hierarchical organization and molecular diffusion in gold nanorod/silica supercrystal nanocomposites. *Nanoscale* **2016**, *8*, 7914–7922.

170. Auyeung, E.; Morris, W.; Mondloch, J. E.; Hupp, J. T.; Farha, O. K.; Mirkin, C. A. Controlling structure and porosity in catalytic nanoparticle superlattices with DNA. *J. Am. Chem. Soc.* **2015**, *137*, 1658–1662.

171. Wan, Y.; Goubet, N.; Albouy, P.-A.; Schaeffer, N.; Pileni, M.-P. Hierarchy in Au nanocrystal ordering in a supracrystal: II. Control of interparticle distances. *Langmuir* **2013**, *29*, 13576–13581.

172. Jana, N. R. Nanorod shape separation using surfactant assisted self-assembly. *Chem. Commun.* **2003**, *9*, 1950–1951.

173. Jones, M. R.; Macfarlane, R. J.; Prigodich, A. E.; Patel, P. C.; Mirkin, C. A. Nanoparticle shape anisotropy dictates the collective behavior of surface-bound ligands. *J. Am. Chem. Soc.* **2011**, *133* (46), 18865–18869.

174. Jana, N. R. Shape effect in nanoparticle self-assembly. *Angew. Chemie Int. Ed.* **2004**, *43*, 1536–1540.

175. Goubet, N.; Portalès, H.; Yan, C.; Arfaoui, I.; Albouy, P.-A.; Mermet, A.; Pileni, M.-P. Simultaneous growths of gold colloidal crystals. *J. Am. Chem. Soc.* **2012**, *134*, 3714–3719.

176. Goubet, N.; Yan, C.; Polli, D.; Portalès, H.; Arfaoui, I.; Cerullo, G.; Pileni, M.-P. Modulating physical properties of isolated and self-assembled nanocrystals through change in nanocrystallinity. *Nano Lett.* **2013**, *13*, 504–508.

177. Gang, O.; Zhang, Y. Shaping phases by phasing shapes. *ACS Nano* **2011**, *5*, 8459–8465.

178. Ross, M. B.; Mirkin, C. A.; Schatz, G. C. Optical properties of one-, two-, and three-dimensional arrays of plasmonic nanostructures. *J. Phys. Chem. C* **2016**, *120*, 816–830.

179. Wei, J.; Schaeffer, N.; Albouy, P.-A.; Pileni, M.-P. Surface plasmon resonance properties of silver nanocrystals differing in size and coating agent ordered in 3D supracrystals. *Chem. Mater.* **2015**, *27*, 5614–5621.

180. Yang, N.; Deeb, C.; Pelouard, J.-L.; Felidj, N.; Pileni, M.-P. Water-dispersed hydrophobic Au nanocrystal assemblies with a plasmon fingerprint. *ACS Nano* **2017**, *11* (8), 7797–7806.

181. Henzie, J.; Grünwald, M.; Widmer-Cooper, A.; Geissler, P. L.; Yang, P. Self-assembly of uniform

polyhedral silver nanocrystals into densest packings and exotic superlattices. *Nat. Mater.* **2011**, *11*, 131–137.

182. Gong, J.; Newman, R. S.; Engel, M.; Zhao, M.; Bian, F.; Glotzer, S. C.; Tang, Z. Shape-dependent ordering of gold nanocrystals into large-scale superlattices. *Nat. Commun.* **2017**, *8*, 14038.

183. Ye, X.; Chen, J.; Eric Irrgang, M.; Engel, M.; Dong, A.; Glotzer, S. C.; Murray, C. B. Quasicrystalline nanocrystal superlattice with partial matching rules. *Nat. Mater.* **2017**, *16*, 214–219.

184. Scarabelli, L.; Hamon, C.; Liz-Marzán, L. M. Design and fabrication of plasmonic nanomaterials based on gold nanorod supercrystals. *Chem. Mater.* **2017**, *29*, 15–25.

185. Kim, J.; Song, X.; Ji, F.; Luo, B.; Ice, N. F.; Liu, Q.; Zhang, Q.; Chen, Q. Polymorphic assembly from beveled gold triangular nanoprisms. *Nano Lett.* **2017**, *17*, 3270–3275.

186. Jana, S.; Phan, T. N. T.; Bouet, C.; Tessier, M. D.; Davidson, P.; Dubertret, B.; Abécassis, B. Stacking and colloidal stability of CdSe nanoplatelets. *Langmuir* **2015**, *31*, 10532–10539.

187. Abécassis, B.; Tessier, M. D.; Davidson, P.; Dubertret, B. Self-assembly of CdSe nanoplatelets into giant micrometer-scale needles emitting polarized light. *Nano Lett.* **2014**, *14*, 710–715.

188. Kim, J.; Jones, M. R.; Ou, Z.; Chen, Q. In situ electron microscopy imaging and quantitative structural modulation of nanoparticle superlattices. *ACS Nano* **2016**, *10*, 9801–9808.

189. Chen, L.; Ji, F.; Xu, Y.; He, L.; Mi, Y.; Bao, F.; Sun, B.; Zhang, X.; Zhang, Q. High-yield seedless synthesis of triangular gold nanoplates through oxidative etching. *Nano Lett.* **2014**, *14*, 7201–7206.

190. Hamon, C.; Sanz-Ortiz, M. N.; Modin, E.; Hill, E.; Scarabelli, L.; Chuvilin, A.; Liz-Marzan, L. M. Hierarchical organization and molecular diffusion in gold nanorod/silica supercrystal nanocomposites. *Nanoscale* **2016**, *8* (15), 7914–7922.

191. Zhang, S.-Y.; Regulacio, M. D.; Han, M.-Y. Self-assembly of colloidal one-dimensional nanocrystals. *Chem. Soc. Rev.* **2014**, *43* (7), 2301–2323.

192. Guerrero-Martínez, A.; Pérez-Juste, J.; Carbó-Argibay, E.; Tardajos, G.; Liz-Marzán, L. M. Gemini-surfactant-directed self-assembly of monodisperse gold nanorods into standing superlattices. *Angew. Chemie Int. Ed.* **2009**, *48*, 9484–9488.

193. Hamon, C.; Novikov, S. M.; Scarabelli, L.; Solís, D. M.; Altantzis, T.; Bals, S.; Taboada, J. M.; Obelleiro, F.; Liz-Marzán, L. M. Collective plasmonic properties in few-layer gold nanorod supercrystals. *ACS Photonics* **2015**, *2*, 1482–1488.

194. Ming, T.; Kou, X.; Chen, H.; Wang, T.; Tam, H.-L.; Cheah, K.-W.; Chen, J.-Y.; Wang, J. Ordered gold nanostructure assemblies formed by droplet evaporation. *Angew. Chemie* **2008**, *120*, 9831–9836.

195. Hamon, C.; Novikov, S.; Scarabelli, L.; Basabe-Desmonts, L.; Liz-Marzán, L. M. Hierarchical self-assembly of gold nanoparticles into patterned plasmonic nanostructures. *ACS Nano* **2014**, *8*, 10694–10703.

196. Frenkel, D. Order through entropy. *Nat Mater* **2015**, *14*, 9–12.

197. Bolhuis, P. Tracing the phase boundaries of hard spherocylinders. *J. Chem. Phys.* **1997**, *106*, 666.

198. Jishkariani, D.; Diroll, B. T.; Cargnello, M.; Klein, D. R.; Hough, L. A.; Murray, C. B.; Donnio, B. Dendron-mediated engineering of interparticle separation and self-assembly in dendronized gold nanoparticles superlattices. *J. Am. Chem. Soc.* **2015**, *137*, 10728–10734.

199. Cargnello, M.; Johnston-Peck, A. C.; Diroll, B. T.; Wong, E.; Datta, B.; Damodhar, D.; Doan-Nguyen, V. V. T.; Herzing, A. A.; Kagan, C. R.; Murray, C. B. Substitutional doping in nanocrystal superlattices. *Nature* **2015**, *524*, 450–453.

Coulomb Blockade Plasmonic Switch

15.1 Background 15-1
    Coulomb Blockade • Plasmonics • Quantum-Corrected Model
15.2 Coulomb Blockade Plasmonic Switch 15-4
    Experimental Method and Results • Numerical Calculations
15.3 Summary 15-7
References 15-7

Ali Khademi, Dao Xiang, and
Reuven Gordon
University of Victoria

15.1 Background

15.1.1 Coulomb Blockade

Single-Electron Transistor

The effect of Coulomb blockade, which allows for the control of single electron, is a superlative property of the single-electron transistors (SETs). The SET can switch from the insulating to the conducting state by adding only one electron to the gate electrode, while a normal transistor requires about 1,000–10,000 electrons (Khanna 2016, Wasshuber 1997).

Figure 15.1a illustrates a SET, which consists of three electrodes: a "source" and a "drain", which are separated by an insulating gap, and an "island" in the middle of this gap, whose electron density can be tuned by a gate voltage V_G. A bias voltage V_b is applied between the source and the drain. This voltage may conduct electrons through two insulating gaps between the source–island–drain by two quantum tunneling events. Therefore, the charge of the island, which is a region containing localized electrons isolated by two tunnel junctions, can vary by the charge of one electron e (Wasshuber 1997, Grabert and Devoret 2013).

Coulomb Blockade

The energy that is required to transfer one electron onto the island is $E_C = e^2/2C$, where C is the island capacitance. Neglecting the effect of temperature, the only source of energy is coming from V_b. As long as $V_b < e/C$, no tunneling event can occur. This effect is called Coulomb blockade. As can be seen in Figure 15.1b, the Coulomb blockade effect can also be explained by energy levels of source, island, and drain in a SET (Wasshuber 1997, Grabert and Devoret 2013).

In addition to bias voltage, there are two other conditions necessary for having a Coulomb blockade (Wasshuber 1997, Grabert and Devoret 2013):

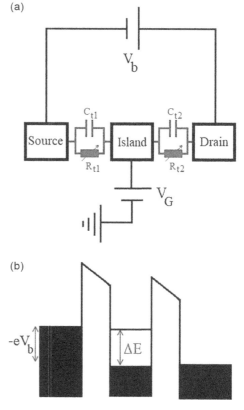

(a)

(b)

FIGURE 15.1 (a) Schematic of a SET. Each tunnel junction is modeled as a capacitor and resistor. (b) SET's energy diagram in the conducting states. Without applied bias, no electron can travel from the source to the drain through island due to Coulomb energy gap $\Delta E = \mathrm{e}^2/C$. A $V_b > e/C$ overcomes this energy gap. The dark area shows the energy states filled with electrons.

1. The charging energy should be greater than the thermal energy, which means: $e^2/2C > k_B T$, where $k_B T$ is Boltzmann's constant and T is the temperature.

2. There should be localized state of electrons on the island so that the number of electrons has negligible quantum fluctuations. In order to have this condition, the resistance of each tunnel junction R_T should exceed the resistance quantum (i.e., $R_T > h/e^2 = 25.8\,\text{k}\Omega$). This condition is derived from Heisenberg's uncertainty principle ($\Delta E \Delta t > h$), using $\Delta E = e^2/C$ and characteristic time of $\Delta t = R_T\ C$.

After Theodore Fulton and Gerald Dolan fabricated the first SET in 1987 (Fulton and Dolan 1987), Coulomb blockade has been realized experimentally on many experiments such as quantum dots (Livermore et al. 1996) or single-atom transistors (Park et al. 2002). In another example, small Pd clusters that were electrically isolated from the underlying Au substrate by an alkanethiol self-assembled monolayer (SAM) probed with scanning tunneling microscopy (STM), showing room temperature Coulomb blockade (Speets et al. 2005, Oncel et al. 2005) [See Figure 15.2].

15.1.2 Plasmonics

The history of studying the Coulomb blockade mediated quantum tunneling provided the fundamental insights into solutions for the conventional silicon devices and the future nanoelectronic circuits (Dorogi et al. 1995, Andres et al. 1996, Gittins et al. 2000, Postma et al. 2001, Cui et al. 2002, Park et al. 2002, Ray et al. 2008, Shin et al. 2011, Lee et al. 2013). The conventional test platforms of Coulomb blockade, such as STM and spectroscopy or CMOS (complementary metal-oxide-semiconductor)-compatible device structures can only probe low-frequency electronic characteristics. Plasmonics, which is the study of collective oscillations of free electrons in a metal due to their interaction with electromagnetic fields of light, can solve this issue.

The ability of quantum plasmonics field to probe the quantum properties of optical waves and their interaction

with bound oscillations of electrons at nanoscale has been explored extensively (Brongersma and Shalaev 2010, Cuche et al. 2010, Marinica et al. 2012). The ultimate limits of many applications related to plasmonics, such as harmonic generation (Kim et al. 2008, Hajisalem et al. 2014) and optical switching (MacDonald et al. 2009, Cao and Brongersma et al. 2009), were studied using the accurate platform of quantum plasmonics. Therefore, investigating the discrete electron transport characteristics by an experimental method in plasmonics, which is sensitive to the local carrier density, is intriguing. For example, metallic nanostructures can squeeze the electromagnetic waves down to the deep subwavelength scale far below that of conventional optics (Barnes et al. 2003). The subwavelength structure provides an accurate platform to study the quantum properties of light and its interaction with bound oscillations of electrons at nanoscale. In Section 15.2, we will describe this approach to observe the Coulomb blockade effect and show the transition of plasmonic gap from an insulating to conducting with bias.

15.1.3 Quantum-Corrected Model

Wentzel–Kramers–Brillouin Approximation for Quantum Tunneling

We start with the usual one-dimensional time-independent Schrödinger equation:

$$\frac{-\hbar^2}{2m}\frac{d^2\psi(x)}{dx^2} + U(x)\psi(x) = E\psi(x) \qquad (15.1)$$

It is convenient to rewrite it as

$$\frac{d^2\psi}{dx^2} = -\frac{2m\left(E - V(x)\right)}{\hbar^2}\psi = -\frac{p^2}{\hbar^2}\psi \qquad (15.2)$$

where $p = [2m(E - V(x))]^{1/2}$ is the classical formula for the momentum of an electron with total energy E moving in a one-dimensional potential $V(x)$, as shown in Figure 15.3. For the quantum tunneling, the electron energy is smaller than the potential. In general, the wave function $\psi(x)$ can be written in complex exponential form as

$$\psi(x) = A(x)e^{i\varphi(x)} \qquad (15.3)$$

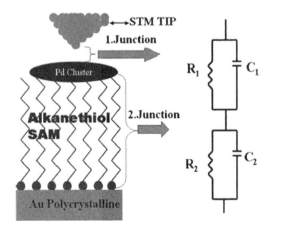

FIGURE 15.2 STM tip, Pd-cluster, SAM, and Au substrate formed double tunnel junctions. (Reproduced from Oncel et al., 2005, with the permission of AIP Publishing, doi: 10.1063/1.1996567.)

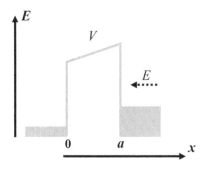

FIGURE 15.3 Electron with energy E incident on a potential barrier.

where $A(x)$ is the amplitude and $\varphi(x)$ is the phase, both of which are real functions. Substituting Eq. (15.3) into Eq. (15.2) and canceling off $e^{i\varphi}$, we get

$$A'' + 2iA'\varphi' + iA\varphi'' - A(\varphi')^2 = -\frac{p^2}{\hbar^2}A \qquad (15.4)$$

We can separate this equation into its real and imaginary parts to get two differential equations:

$$A'' - A(\varphi')^2 = -\frac{p^2}{\hbar^2}A \qquad (15.5a)$$

$$2A'\varphi' + A\varphi'' = \frac{1}{A}[A^2\varphi']' = 0 \qquad (15.5b)$$

The solution of Eq. (15.5b) is

$$A^2\varphi' = C_1^2$$
$$\Rightarrow A = \frac{C_1}{\sqrt{\varphi'}} \qquad (15.6)$$

where C_1 is a constant of integration. Equation (15.5a) cannot be solved in general since $V(x)$ could in principle be anything, but can be rewritten as

$$\frac{A''}{A} = (\varphi')^2 - \frac{p^2}{\hbar^2} \qquad (15.7)$$

Now suppose that the potential $V(x)$ varies slowly. In this case, we'd expect that the wave function $\varphi(x)$ is close to the solution for a constant potential, except that its amplitude and phase will vary slightly. The WKB approximation comes by assuming that the variation in amplitude is small enough that the derivatives of $A(x)$ are negligible compared with $A(x)$ itself. So Eq. (15.7) becomes

$$(\varphi')^2 - \frac{p^2}{\hbar^2} \approx 0 \qquad (15.8)$$

and its solution is

$$\varphi' = \pm\frac{p}{\hbar}$$
$$\Rightarrow \varphi = \pm\frac{1}{\hbar}\int p\,dx + C_2 \qquad (15.9)$$

where C_2 is a constant of integration. Substituting Eqs. (15.6) and (15.9) into Eq. (15.3), the WKB approximation for wave function is found:

$$\psi(x) = \frac{C_1\sqrt{\hbar}e^{iC_2}}{\sqrt{p}}e^{\pm\frac{i}{\hbar}\int p\,dx} = \frac{C}{\sqrt{|p|}}e^{\pm\frac{i}{\hbar}\int p\,dx} \qquad (15.10)$$

with $C = C_1\sqrt{\hbar}e^{iC_2}$. Therefore, the wave function in the whole region becomes (Griffiths 2005)

$$\psi(x) = \begin{cases} A_1e^{-ikx} + A_2e^{ikx} & x > a \\ \frac{A_3}{\sqrt{|p|}}e^{\frac{1}{\hbar}\int|p|dx} + \frac{A_4}{\sqrt{|p|}}e^{-\frac{1}{\hbar}\int|p|dx} & 0 < x < a \\ A_5e^{-ikx} & x < 0 \end{cases}$$
$$(15.11)$$

We would expect the exponential decay term (with constant A_4) to dominate in the region $0 < x < a$. If we take

the exponential decay as an estimate of the transmission probability, it is defined as

$$T = \left|\frac{A_5}{A_1}\right|^2 \approx e^{-\frac{2}{\hbar}\int_0^a|p|dx} \qquad (15.12)$$

In quantum tunnelling, $|p| = [2m(V(x) - E)]^{1/2}$ and the WKB approximation for the tunneling probability is

$$T \approx e^{-2\int_0^a\sqrt{\frac{2m}{\hbar^2}(E-V(x))}dx} \qquad (15.13)$$

This equation is used in conjunction with the quantum-corrected model (QCM) to understand the impact of modifications to tunneling on quantum plasmonics.

Introduction to QCM

The QCM has been introduced to study the quantum-induced effects in subnanometer gaps of metallic nanoparticles (Esteban et al. 2012), such as spectral redistribution of electromagnetic modes and collapse of the field enhancement. In this model, a fictitious conductive material is used to mimic the electron tunneling in the junction between two adjacent nanoparticles, and then the optical properties can be calculated within a classical electrodynamic framework, which can yield good agreement with fully quantum-mechanical calculations.

Figure 15.4 illustrates the classical and quantum descriptions for the case of two metallic spheres of radius R surrounded by vacuum with permittivity ε_0 and separated by a distance D. Figure 15.4a shows the classical description, where the interface between the different materials is assumed abrupt. The Drude model is used to characterize the local permittivity of metal ε_r by a bulk plasma frequency ω_p and damping parameter γ_p (Maier 2007):

$$\varepsilon_r = \varepsilon_\infty - \frac{\omega_p^2}{\omega^2 + i\gamma_p\omega} \qquad (15.14)$$

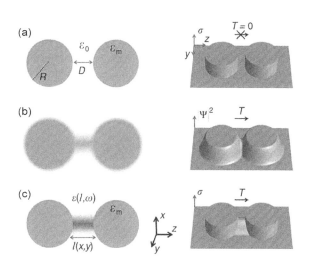

FIGURE 15.4 Comparing three different models: (a) classical electromagnetic, (b) quantum, and (c) QCM. (Reprinted by permission from Springer Nature, © 2012, Esteban et al. 2012, doi: 10.1038/ncomms1806.)

where ε_∞ explains the dielectric screening introduced by the bound valence electrons of the positive ion cores, and the second term is the contribution from free electrons. The conductivity contributed is given by Maier (2007)

$$\sigma = -i\omega\varepsilon_0(\varepsilon_r - 1) \tag{15.15}$$

The conductivity in the junction is zero in this classical formalism, which implies that the electronic density distribution is zero in vacuum and the probability of electron tunneling T between the nanoparticles is zero. Figure 15.4b shows the continuous change of the density of electrons $|\Psi|^2$ between the metal surfaces in the quantum model (QM). $|\Psi|^2$ can be considerable in the center of subnanometric gaps. Figure 15.4c shows the fictitious material with the local permittivity $\varepsilon(l, \omega)$ in the gap $l = l(x, y)$. Knowing the electron tunneling probability $T(l)$ through the potential barrier separating the two metals, $\varepsilon(l, \omega)$ can be calculated (Esteban et al. 2012):

$$\varepsilon = \varepsilon_\infty - \frac{\omega_g^2}{\omega^2 + i\gamma_g\omega} \tag{15.16}$$

Then QCM model can be modified by considering $\omega_g = \sqrt{T(E_F)}\omega_p$, where ω_g is the plasmon frequency of the fictitious material and $T(E_F)$ is the electron tunneling probability at Fermi energy (Hajisalem et al. 2014). Another modification can be using the probability of finding electron, which is the square of the normalized electron wave function, $|\psi(x)|^2$, instead of $T(E_F)$ (Khademi et al. 2018).

15.2 Coulomb Blockade Plasmonic Switch

This section investigates the Coulomb blockade effects in a plasmonic gap structure. Voltage control over the tunnel resistance of the gold nanoparticle sandwiched between two electrodes allows for direct observation of voltage-dependent changes in plasmon resonance. The influence of the bias voltage on the electron transport in the nanogap is observed optically. To control the distance between the slit edges and the nanoparticles, SAMs with three different thicknesses are prepared. The experimental findings are explained in the framework of the QCM (Xiang et al. 2017).

15.2.1 Experimental Method and Results

Nanodevice Preparation

As can be seen in Figure 15.5a, the nanodevice consists of 60 nm gold nanospheres trapped in a ~65 nm nanogap between two electrodes. First, the ~65 nm slit was fabricated in the gold slide by using focused ion beam, and then SAMs of amine-terminated alkanethiols of different lengths were prepared on the gold films' surface (referred to as "c2," "c3," "c6," for further details, see Ref. Xiang (et al. 2017)). The gold film was immersed for 18 h into the cleaned glass container of 3 mM solutions of alkanethiols, which was prepared in anhydrous ethanol, and then rinsed with anhydrous ethanol for 15 s. The concentration of the aqueous suspension of 60 nm diameter gold nanoparticles was increase to 3×10^{-5} volume fraction, using centrifugation at 1,000 rpm. The next stage was electrostatically immobilizing nanoparticles on top of SAMs using 30 min incubation with 100 µL of the colloidal stock, followed by 15 s rinsing with deionized water, and finally drying it with a slow steam of pure nitrogen gas. Surface tension pulls the particles to one edge during the drying process (Kinge et al. 2008).

The top part of Figure 15.5a shows an equivalent circuit of nanodevice, where L_{sph} and R_{sph} are effective inductance and resistance of gold nanoparticle (Engheta et al. 2005), C_1 and C_2 are capacitances of each tunnel junction, and finally, R_t is the tunnel resistance of tunnel junction 1 (Hanson 2008). Figure 15.5b demonstrates tunneling of an electron to the nanoparticle in the energy band diagram for this structure at certain bias conditions. The tunneling event occurs if the applied electric energy will be slightly larger than the Coulomb charging energy $E_C = e^2/2C_2$. Figure 15.5c shows the asymmetric shape of the device in its SEM image.

Results and Discussion

Figure 15.6a shows the setup for measuring the normalized loss spectra. To excite the gap modes, a collimated white light source was focused through a 20× microscope objective onto the sample, which was mounted on a printed board circuit and biased by the power source. Light transmitted through the sample was collected at the spectrometer.

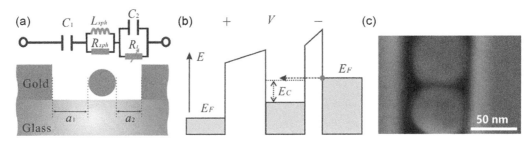

FIGURE 15.5 The nanodevice is an asymmetric double-junction structure. The schematic, energy band (the energy gap E_C comes from the charging energy), and scanning electron microscopy (SEM) image are shown in (a–c). (Reprinted with permission from Xiang et al. 2017, Copyright 2017, American Chemical Society.)

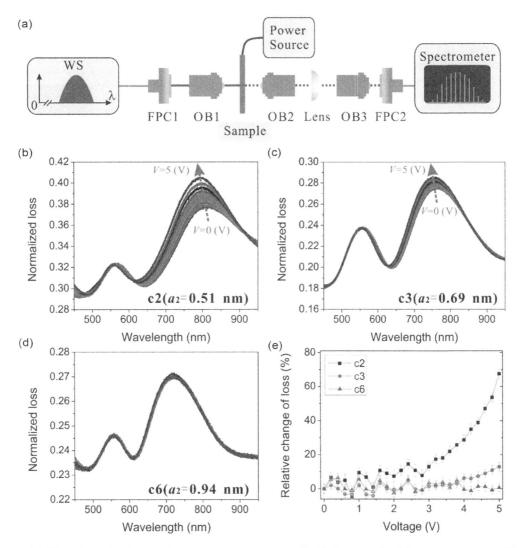

FIGURE 15.6 (a) The schematic of transmission measurement setup; (b–d) the normalized loss spectra averaged by ten repeated measurements for nanoparticles in a slit separated by SAMs: (b) cysteamine (c2), (c) 3-amino-1-propanethiol hydrochloride (c3), and (d) 6-amino-1-hexanethiol hydrochloride (c6); (e) the relative change of normalized loss at plasmonic resonance. (Reprinted with permission from Xiang et al. 2017, Copyright 2017, American Chemical Society.)

Figure 15.6b–d demonstrate the normalized loss spectra versus wavelength when varying the voltage for different gap width a_2. The nanoparticle normalized loss can be calculated by

$$\mathrm{NL} = 1 - t_\mathrm{NP}/t_\mathrm{slit} \qquad (15.17)$$

where t_NP and t_slit are the transmission spectra for the slit with and without particles, respectively.

As confirmed with linear scattering measurements in a past report (Hajisalem et al. 2014), the SAMs with varying lengths acted as the spacer layer between the nanoparticle and the electrode and were used to adjust the width a_2. The short wavelength plasmonic resonance corresponds to the wide gap, and the long wavelength one comes from the narrow gap. The large spectral change in Figure 15.6b when the SAM layer is reduced to two carbons (c2), compared with the SAM layer with three carbons (c3) and six carbons (c6), is the most striking feature of this measurement. The relative

change of normalized loss at resonance, which is shown in Figure 15.6e, can be found as

$$\frac{[\mathrm{NL}(\lambda_{r,V}) - \mathrm{NL}_{\mathrm{BG},V}] - [\mathrm{NL}(\lambda_{r,0}) - \mathrm{NL}_{\mathrm{BG},0}]}{\mathrm{NL}(\lambda_{r,0}) - \mathrm{NL}_{\mathrm{BG},0}} \qquad (15.18)$$

where $\mathrm{NL}(\lambda_r)$ is the normalized loss at resonance, $\mathrm{NL}_{\mathrm{BG}}$ is the background of normalized loss at a longer wavelength like 1,000 nm, and the indices of V and 0 indicate the measurement with and without applied voltage. At the resonance, the peak heights of normalized loss $(\mathrm{NL}(\lambda_r) - \mathrm{NL}_{\mathrm{BG}})$ are 0.042 at 0 V and 0.071 at 5 V, respectively (i.e., a 70% modulation). The threshold behavior in nonlinear way can be observed in the spectra change in c2. Furthermore, c2 has the lower threshold and stronger response than c3 and c6. Since the plasmonic resonance is very sensitive to the local carrier density in the gap, the threshold behavior can be attributed to the onset of quantum tunneling and Coulomb blockade.

15.2.2 Numerical Calculations

To numerically explore the insulator–metal transition in tunnel junction, numerical simulations were completed using the QCM. QCM was proposed to capture the effects of quantum tunneling in narrow gap nanostructures (Esteban et al. 2012, 2015, Pitarke et al. 1990). As described in Section 15.1.3, modified QCM replaces the gap material, which is a dielectric, with a factious metal of Drude model with the revised plasma frequency of $T(E_F)^{1/2}\omega_p$, where $T(E_F)$ is the transmission probability at Fermi level and ω_p is the plasma frequency of gold. Theoretical predictions of QCM for the saturation of plasmonic fields due to the quantum tunneling are consistent with more comprehensive quantum simulations (Hajisalem et al. 2017). Using the Wentzel–Kramers–Brillouin (WKB) approximation (Hanson 2008) of the wave function in the region of slowly varying potential energy, the transmission probability $T(E)$ for tunneling with the gap size a can be calculated by Eq. (15.12) with $p = \hbar\alpha\sqrt{\frac{2m}{\hbar^2}\left(e\phi - eV\frac{x}{a} + E_F - E\right)}$. As a result, $T(E)$ would be

$$T(E) \begin{cases} = \exp\left(\frac{4}{3}\alpha\frac{\sqrt{2ma}}{e\hbar V}\left[(e\phi - eV + E_F - E)^{\frac{3}{2}} - (e\phi + E_F - E)^{\frac{3}{2}}\right]\right), & eV < \phi \\ \approx \exp\left(-\frac{4}{3}\alpha\frac{\sqrt{2ma}}{e\hbar V}(e\phi + E_F - E)^{\frac{3}{2}}\right), & eV > \phi \end{cases}$$

where $\alpha = 0.65$ is the factor for electron effective mass, $\phi = 1.42\,\text{eV}$ is the barrier height (as suggested by Ref. Wang et al. (2003)), E_F is Fermi energy, m is the electron mass, V is the added voltage, and $\hbar$ is the reduced Planck constant. Using the effective Drude model, the permittivity of the tunneling region within the gap would be

$$\varepsilon = \varepsilon_\infty - \frac{\omega_g^2}{\omega(\omega + i\gamma_g)} \qquad (15.20)$$

with $\omega_g = \sqrt{T(E_F)}\omega_p$, where $\varepsilon_\infty = 2.25$ is the relative permittivity of SAMs, $\omega_p = 1.37 \times 10^{16}\,\text{rad/s}$ is the plasma frequency of gold, and $\gamma_g = 4.07 \times 10^{13}\,\text{rad/s}$, which is the scattering rate of gold, because the dominant scattering of electrons originates from the scattering inside the metal region and so the scattering rate does not alter to first approximation (Xiang et al. 2017). To estimate the accuracy of this model, it is worthwhile to use the finite-difference time-domain (FDTD) method and quantify the level of convergence (Lu and Zhou 2016):

$$\delta t(i) = \sqrt{\frac{\int (t_{\text{NP},i} - t_{\text{NP},0})^2 \, d\lambda}{\int t_{\text{NP},i}^2 \, d\lambda}} \qquad (15.21)$$

where $t_{\text{NP},0}$ is the simulated transmission spectrum for the gold nanoparticle trapped in a nanoslit with a mesh size of 0.051 nm and $t_{\text{NP},i}$ represents other mesh sizes. Plots of the near field and the level of convergence for c2 are shown in Figure 15.7 and confirm that the simulation has converged at 0.1 nm mesh size.

Using QCM, the normalized loss with bias added on the narrow gap was calculated (Figure 15.8). The

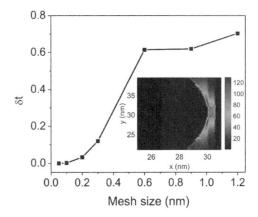

FIGURE 15.7 The numerical calculation's convergence level for c2 with mesh size. Inset is near field. (Reprinted with permission from Xiang et al. 2017, Copyright 2017, American Chemical Society.)

Coulomb blockade effect was confirmed by the nonlinear change in normalized loss, consistent with our observations.

$$eV < \phi \qquad (15.19)$$

As mentioned earlier in Section 15.1.1, there are two necessary conditions for observing Coulomb blockade (Wasshuber 1997, Grabert and Devoret 2013, Hanson 2008): (i) $E_C > k_B T \sim 25$ meV at room temperature; (ii) $R_t > h/e^2 \sim 25.8\,\text{k}\Omega$. If these conditions are satisfied in the smallest gap case of c2, it would be satisfied for c3 and c6 too, because they have smaller capacitance and larger resistance. The capacitance for c2 can be driven by the following formula (Crowley 2008):

$$C_1 \approx C_2 = C = 4\pi\varepsilon R \sum_n^\infty \frac{2\sqrt{\xi(2+\xi)}}{e^{(1+2n)\text{acosh}(1+\xi)} - 1}$$
$$\approx 3 \times 10^{-18}(F) \qquad (15.22)$$

where the ratio $\xi = a/R$ is a parameter that explains the spacing relative to the size of the conductive sphere. Thus, $E_C \sim 53\,\text{meV}$ and satisfies the condition (i). The distance-dependent dc conductivity at the junction is given by Pitarke et al. (1990)

$$\sigma_0 = \frac{a}{2\pi^2}\frac{m'}{e\hbar^3}\int_0^{E_F} T(E)dE \qquad (15.23)$$

where m' is the effective electron mass, e is the electron charge, $\hbar$ is the reduced Planck's constant, and T is the transmission probability. As a result, $R_t = a/(\pi r_s^2\sigma_0) \approx 3.8\,\text{M}\Omega$ (r_s is the Wigner–Seitz radius of gold). Therefore, the tunnel resistance satisfies the condition (ii) of Coulomb blockade.

The "turning on" of electron transport in the plasmonic structure due to the Coulomb blockade and its impact on the plasmonic resonance are observed in the numerical model. The ~70% relative change of normalized loss at the

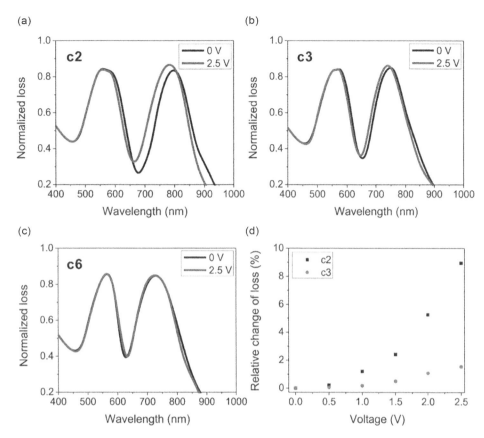

FIGURE 15.8 Simulated normalized loss for different narrow gap widths: (a) 0.51 nm (c2), (b) 0.69 nm (c3), (c) 0.94 nm (c6), and (d) comparison of the relative change of normalized loss at plasmonic resonance between c2 and c3. (Reprinted with permission from Xiang et al. 2017, Copyright 2017, American Chemical Society.)

resonances after "turning on" indicates a large change of carrier density in the subnanometer gap. The experimental findings are explained in the framework of the QCM. The time constant based on the tunnel capacitance and free-space resistance of coupled antenna $\tau = R_0 C_2 \approx 1$ (fs) demonstrates the potential of this effect for applications such as ultrafast optical switch (MacDonald et al. 2009).

15.3 Summary

As shown in this chapter, the Coulomb blockade effect can be investigated optically using plasmonic nanodevices. Combining the Coulomb blockade effect and plasmonic focusing can give very strong modification of optical properties. For example, the tiny capacitance of a Coulomb blockade plasmonic switch with subnanometer tunnel junction demonstrates the potential of this nanodevice for high-speed switching.

References

Andres, R. P., T. Bein, M. Dorogi, and S. Feng. Coulomb staircase at room temperature in a self-assembled molecular nanostructure. *Science* 272(5266), 1996: 1323–1325.

Barnes, W. L., A. Dereux, and T. W. Ebbesen. Surface plasmon subwavelength optics. *Nature* 424(6950), 2003: 824–830.

Brongersma, M.L. and V. M. Shalaev. The case for plasmonics. *Science* 328(5977), 2010: 440–441.

Cao, L. and M. L. Brongersma. Active plasmonics: Ultrafast developments. *Nature Photonics* 3(1), 2009: 12.

Crowley, J. M. Simple expressions for force and capacitance for a conductive sphere near a conductive wall. *Proceedings of the ESA Annual Meeting on Electrostatics*, Minneapolis, MN. June 2008, 1–15.

Cuche, A., O. Mollet, A. Drezet, and S. Huant. "Deterministic" quantum plasmonics. *Nano Letters* 10(11), 2010: 4566–4570.

Cui, J. B., M. Burghard, and K. Kern. Room temperature single electron transistor by local chemical modification of carbon nanotubes. *Nano Letters* 2(2), 2002: 117–120.

Dorogi, M., J. Gomez, R. Osifchin, R. P. Andres, and R. Reifenberger. Room-temperature Coulomb blockade from a self-assembled molecular nanostructure. *Physical Review B* 52(12), 1995: 9071–9077.

Engheta, N., A. Salandrino, and A. Alù. Circuit elements at optical frequencies: Nanoinductors, nanocapacitors, and nanoresistors. *Physical Review Letters* 95(9), 2005: 095504.

Esteban, R., A. G. Borisov, P. Nordlander, and J. Aizpurua. Bridging quantum and classical plasmonics with a quantum-corrected model. *Nature Communications* 3, 2012: 825.

Esteban, R., A. Zugarramurdi, P. Zhang, P. Nordlander, F. J. García-Vidal, A. G. Borisov, and J. Aizpurua. A classical treatment of optical tunneling in plasmonic gaps: Extending the quantum corrected model to practical situations. *Faraday Discussions* 178, 2015: 151–183.

Fulton, T. A. and G. J. Dolan. Observation of single-electron charging effects in small tunnel junctions. *Physical Review Letter* 59, 1987: 109–112.

Gittins, D. I., D. Bethell, D. J. Schiffrin, and R. J. Nichols. A nanometre-scale electronic switch consisting of a metal cluster and redox-addressable groups. *Nature* 408(6808), 2000: 67–69.

Grabert, H. and M. H. Devoret. *Single Charge Tunneling: Coulomb Blockade Phenomena In Nanostructures.* Springer Science & Business Media, New York, 2013.

Griffiths, D. J. *Introduction to Quantum Mechanics.* Pearson Education India, Upper Saddle River, NJ, 2005.

Hajisalem, G., M. S. Nezami, and R. Gordon. Probing the quantum tunneling limit of plasmonic enhancement by third harmonic generation. *Nano Letters* 14(11), 2014: 6651–6654.

Hajisalem, G., M. S. Nezami, and R. Gordon. Switchable metal–insulator phase transition metamaterials. *Nano Letters* 17(5), 2017: 2940–2944.

Hanson, G. W. *Fundamentals of Nanoelectronics.* Pearson/Prentice Hal, Upper Saddle River, NJ, 2008.

Khademi, A., T. Dewolf, and R. Gordon. Quantum plasmonic epsilon near zero: field enhancement and cloaking. *Optics Express* 26(12), 2018: 15656–15664.

Khanna, V. K. *Integrated Nanoelectronics: Nanoscale CMOS, Post-CMOS and Allied Nanotechnologies*, India. Springer, 2016.

Kim, S., J. Jin, Y. J. Kim, I. Y. Park, Y. Kim, and S. W. Kim. High-harmonic generation by resonant plasmon field enhancement. *Nature* 453(7196), 2008: 757–760.

Kinge, S., M. Crego-Calama, and D. N. Reinhoudt. Self-assembling nanoparticles at surfaces and interfaces. *ChemPhysChem* 9(1), 2008: 20–42.

Lee, S., Y. Lee, E. B. Song, and T. Hiramoto. Observation of single electron transport via multiple quantum states of a silicon quantum dot at room temperature. *Nano Letters* 14(1), 2013: 71–77.

Livermore, C., C. H. Crouch, R. M. Westervelt, K. L. Campman, and A. C. Gossard. The Coulomb blockade in coupled quantum dots. *Science* 274(5291), 1996: 1332–1335.

Lu, J. and H. Zhou. Finite-difference time-domain modeling of curved material interfaces by using boundary

condition equations method. *Chinese Physics B* 25(9), 2016: 090203.

MacDonald, K. F., Z. L. Sámson, M. I. Stockman, and N. I. Zheludev. Ultrafast active plasmonics. *Nature Photonics* 3(1), 2009: 55–58.

Maier, S. A. *Plasmonics: Fundamentals and Applications.* Springer Science & Business Media, New York, 2007.

Marinica, D. C., A. K. Kazansky, P. Nordlander, J. Aizpurua, and A. G. Borisov. Quantum plasmonics: Nonlinear effects in the field enhancement of a plasmonic nanoparticle dimer. *Nano Letters* 12(3), 2012: 1333–1339.

Oncel, N., A. S. Hallbäck, H. J. W. Zandvliet, E. A. Speets, B. J. Ravoo, D. N. Reinhoudt, and B. Poelsema. Coulomb blockade of small Pd clusters. *The Journal of Chemical Physics* 123, 2005: 044703.

Park, J., A. N. Pasupathy, J. I. Goldsmith, C. Chang, Y. Yaish, J. R. Petta, M. Rinkoski, J. P. Sethna, H. D. Abruña, P. L. McEuen, and D. C. Ralph. Coulomb blockade and the Kondo effect in single-atom transistors. *Nature* 417(6890), 2002: 722–725.

Pitarke, J. M., F. Flores, and P. M. Echenique. Tunneling spectroscopy: Surface geometry and interface potential effects. *Surface Science* 234(1–2), 1990: 1–16.

Postma, H. W. C., T. Teepen, Z. Yao, M. Grifoni, and C. Dekker. Carbon nanotube single-electron transistors at room temperature. *Science* 293(5527), 2001: 76–79.

Ray, V., R. Subramanian, P. Bhadrachalam, L. C. Ma, C. U. Kim, and S. J. Koh. CMOS-compatible fabrication of room-temperature single-electron devices. *Nature Nanotechnology* 3(10), 2008: 603–608.

Shin, S. J., C. S. Jeong, B. J. Park, T. K. Yoon, J. J. Lee, S. J. Kim, J. B. Choi, and D. G. Hasko. Room-temperature charge stability modulated by quantum effects in a nanoscale silicon island. *Nano Letters* 11(4), 2011: 1591–1597.

Speets, E. A., B. Dordi, B. J. Ravoo, N. Oncel, A. S. Hallbäck, H. J. W. Zandvliet, B. Poelsema, G. Rijnders, D. H. A. Blank, and D. N. Reinhoudt. Noble metal nanoparticles deposited on self-assembled monolayers by pulsed laser deposition show Coulomb blockade at room temperature. *Small* 1(4), 2005: 395–398.

Wang, W., T. Lee, and M. A. Reed. Mechanism of electron conduction in self-assembled alkanethiol monolayer devices. *Physical Review B* 68(3), 2003: 035416.

Wasshuber, C. *About Single-Electron Devices and Circuits.* PhD Dissertation, Vienna, Austria: Technischen Universität Wien, January 1997.

Xiang, D., J. Wu, and R. Gordon. Coulomb blockade plasmonic switch. *Nano Letters* 17(4), 2017: 2584–2588.

16

Amplification of Surface Plasmons

16.1 Introduction... 16-1
16.2 SPs and SPPs: General Discussion 16-2
    Mathematical Formulation
16.3 Excitation of SPP.. 16-3
    Grating Coupling • Prism Coupling/Attenuated Total Internal Reflection •
    Focused Optical Beams
16.4 Origin of Losses in Plasmonic Materials 16-4
    Scattering • Ohmic Loss
16.5 Methods for Compensating Losses.. 16-4
    Gain Created by Quantum Wells • All Semiconductor Structure •
    Incorporating Dipolar Gain Media • Fluorescent Polymer • Lead-Sulphide
    Quantum Dots • Electrical Injection • Direct Current with 2D Bragg
    Structure • Compensating Loss in Propagating SPP by Optical Gain • Thin
    Organic Gain Medium
16.6 Plasmons in a DC Electric Field... 16-9
    Analysis by Cada's Group • Analysis by Sydoruk's Group
16.7 Experimental Search... 16-11
16.8 Plasmonic-Based Lasing in Nanostructures............................. 16-11
    Spasing Process • Spaser Biological Application
16.9 Applications of SPP 16-12
    Imaging and Spectroscopy • Sensors • Nanolithography • Devices Based on
    Metal–Dielectric–Metal Stub Resonators • Nanoscale THz and Optical
    Devices
References 16-13

Gerard Tatel and M.S. Wartak
Wilfrid Laurier University

16.1 Introduction

There is a growing interest in plasmonics in recent years, on one hand motivated by the need to understand and control the behavior of light below the diffraction limit, and on the other by its potential applications in devices, sensing, giant field enhancement, negative index materials, and many others. Assuming that the appropriate materials will be found, or designed, along with practical structures, it is expected that the field of plasmonics has a potential to also revolutionize the microelectronics industry. Currently, the main issue that halts the application of plasmonics are losses in metals that affect the performance of every plasmonic structure. At present, main conventional plasmonic materials used in plasmonics are gold and silver because of their relatively small ohmic losses when compared with other metals. However, the existing losses in those materials are still one of the major problems in using those metals in plasmonics.

Nanoplasmonic which deals with plasmons is a fast-growing field as it offers the possibilities of very fast propagation speeds (up to 100 THz) and also to confine light into a subwavelength spatial dimensions. Nanoplasmonic structures are considered as promising candidates for next-generation Internet technologies, all-optical integrated photonic chips integrated with silicon.

In the present review, we summarize various methods invented to compensate or reduce losses in plasmonic materials, including introduction of optical gain and plasmons in DC electric field. The presence of losses associated with metals places a limitation on the distance over which surface plasmon–polariton modes may propagate, and these losses are associated with an imaginary part of permittivity. For several decades, it has been appreciated that loss (absorption) associated with metals places a limitation on, for example, the distance over which surface plasmon polariton (SPP) modes may propagate.

Over the years several methods to reduce these losses were analyzed, for example, by adding gain materials to plasmonic nanostructures so as to offset the losses, or by seeking alternative materials to noble metals, including doped semiconductors, graphene, tunable metal–semiconductor materials, transparent conducting oxides, heavily doped conducting polymers, and some nitrides, among others. All of these materials yield a plasmonic response owing to the free charge-carriers they contain.

In discussing those effects, one must consider localized surface plasmons (SP) as oscillations of free electrons in a metallic particle and SPP as a electromagnetic (EM) wave propagating along the interface between two media having dielectric constants of opposite signs, typically metal and dielectric. The frequency of oscillations of SP is the plasma frequency as determined for a metallic particle of a given size and of a particular shape.

SSP exists on metallic surfaces of various shapes. Mode size of SSP can be smaller than the diffraction limit [1] due to the fact that their momentum can be much smaller compared with photons.

We explain the principles of SP and SPP, provide theoretical description of those oscillations, and analyze the experiments. In terms of practical applications, we will discuss experimental observations of stimulated emission of SPP at optical frequency, where the full compensation of loss of SPP by gain was achieved. We also discuss plasmon instabilities in semiconductors with drifting electrons and the effects of plasmon loss due to carrier collisions and diffusion.

16.2   SPs and SPPs: General Discussion

Here, we will provide details of the derivation of dispersion relations SP and outline the main assumptions needed to derive dispersion relations.

Localized SPs are the oscillations of free electrons in a metal [2]. Metal can be of arbitrary shape, like ball, wire, etc. Localized SPs were discovered in single particles, in clusters, on rough surfaces, etc. SP can enhance electric fields in nanovolumes, which results, for example, in surface-enhanced Raman spectroscopy (SERS).

SPP, also known as propagating SP, is a surface EM wave accompanied by an oscillation of free electrons traveling along the interface between two media. One of these media (metal or semiconductor) should have a negative dielectric constant ε within the appropriate frequency range.

Plasmons come from the oscillations of plasma (electron gas in metal). Plasmons are considered to be quasiparticles due to the fact that they come from plasma oscillations. They are capable of existing at the interface between two materials, where the dielectric constant changes signs, here they are known as SPs. A negative value of a dielectric constant indicates that it will absorb some of the incident optical fields to a given depth, known as skin depth. The fields there have the potential to excite plasmons at the surface. SPPs are transverse magnetic excitations propagating along the surface formed by metal-dielectric interface, where its amplitude decays exponentially proportional to increasing distance into the medium. The EM field can be derived from the solution of Maxwell's equations in each corresponding medium and respective boundary conditions. The boundary conditions express the continuity of the tangential components of both electric and magnetic fields through the interface and disappear when these fields

are infinitely far from the interface. The electric field is p-polarized (TM); the magnetic vector is perpendicular to the plane defined by the direction of propagation and normal to the surface. This will be shown in a later section.

16.2.1   Mathematical Formulation

For the following discussion we assume that the system consists of two layers with z-axis perpendicular to those layers. Layers are labeled using index $i = 1, 2$ where indices 1 and 2 denote dielectric and metal regions, respectively. Propagation is assumed to be along the x-axis, which is oriented parallel to the layers.

We start with Maxwell equations in the harmonic approximation, where all time dependencies are of the form $\exp(i\omega t)$

$$\nabla \times \mathbf{E} = -i\omega\mu_0\mathbf{H} \tag{16.1}$$

$$\nabla \times \mathbf{H} = \sigma\mathbf{E} + i\omega\epsilon_0\epsilon_r\mathbf{E} \tag{16.2}$$

In the following we analyze only nonmagnetic systems with $\mu = \mu_0$. For TM mode [3] the following orientation of fields exist

$$\mathbf{E}_j = [E_x, 0, E_{z,j}]\, e^{-i\omega t} e^{i(k_x x + k_{z,j} z)} \tag{16.3}$$

$$\mathbf{H}_j = [0, H_y, 0]\, e^{-i\omega t} e^{i(k_x x + k_{z,j} z)} \tag{16.4}$$

Using the above solutions in Maxwell's equations along with boundary conditions one gets

$$\frac{\varepsilon_1}{k_{z,1}} = \frac{\varepsilon_2}{k_{z,2}} \tag{16.5}$$

$$k_x^2 + k_{z,j}^2 = \varepsilon_j k_0^2, \quad j = 1, 2 \tag{16.6}$$

$$E_x = \frac{k_{z,1}}{\omega\varepsilon_j}H_y, \quad E_{z,j} = -\frac{k_x}{\omega\varepsilon_j}H_y \tag{16.7}$$

where $k_0 = \frac{\omega}{c}$. From above one derives dispersion relations for SPP as

$$k_x^2 = k_0^2 \frac{\varepsilon_1\varepsilon_2}{\varepsilon_1 + \varepsilon_2} \tag{16.8}$$

$$k_{z,j}^2 = k_0^2 \frac{\varepsilon_j^2}{\varepsilon_1 + \varepsilon_2} \tag{16.9}$$

In the above dispersion relation the imaginary part of k_x determines losses of SPP propagating along the interface. To compensate for losses we replace dielectric (medium 1) by the material with gain, which is characterized by the complex permittivity $\varepsilon_1 = \varepsilon_1' + i\varepsilon_1''$ (negative value of ε_1'' represents gain). One can then analyze the conditions for a bound wave to propagate at the interface. After some algebra, one finds [4]

$$\varepsilon_1'' = -\frac{(\varepsilon_1')^2\,\varepsilon_2''}{|\varepsilon_2|^2} \tag{16.10}$$

The above can be related to the actual optical power gain using $\gamma = -k_0\varepsilon_1'' / (\varepsilon_1')^{1/2}$, where γ is the power gain coefficient. One obtains gain coefficient required for lossless SPP propagation

$$\gamma_0 = \frac{2\pi}{\lambda_0} \frac{\varepsilon_2''\,(\varepsilon_1')^{3/2}}{(\varepsilon_2')^2 + (\varepsilon_2'')^2} \tag{16.11}$$

For $\gamma < \gamma_0$ the SPP propagation will still be lossy, but the propagation length will increase. If the gain is large enough to compensate for both cavity and SPP losses, sustained oscillations (possible lasing) will occur.

16.3   Excitation of SPP

SPP cannot be excited by direct exposure to light beams unless the phase is synchronized. When plotting the dispersion relation of SPPs comparing light in a vacuum and in a dielectric, the SPP dispersion is entirely below that of light, implying that the propagation constant (β) must be greater than wave vector (k). Furthermore, it can be implied that SPPs cannot be excited through direct illumination of light, due to conservation of energy and momentum not being satisfied at the same moment in time. Special techniques are implemented to excite the SPPs by establishing a momentum transfer.

16.3.1   Grating Coupling

The grating pattern on a metal surface can be used to synchronize the phase. The difference between the wavevector in-plane momentum of the incoming photons ($k_x = k \sin(\theta)$) and β can be overcome. When considering a one-dimensional grating separated with lattice constant a (metal surface grating period), seen in Figure 16.1, phase syncing can take place when

$$\beta = k \sin(\theta) \pm nG \tag{16.12}$$

where G is the reciprocal lattice of the grating ($G = 2\pi/a$) and n is the number of gratings ($n = 1, 2, 3, ...$).

The excitation of SPPs detected observes a sharp dip/minimum in the reflectivity from the surface as light can be coupled with high efficiency. Significant changes to the SPP dispersion relations occur if the gratings are deep enough so that the modulation acts as a larger perturbation.

When the period is equal to half the wavelength of the SPP, an SPP bandgap is established. Two SPP standing wave solutions with equivalent wavelengths but different frequencies are formed. The upper solution (ω_+) has more energy due to the greater distance separating surface charges and larger distortion of the field.

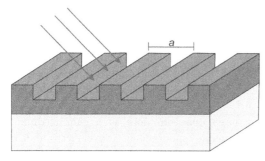

FIGURE 16.1   Grating coupling sample configuration.

16.3.2   Prism Coupling/Attenuated Total Internal Reflection

This technique utilizes coupling SPPs to the evanescent EM field formed from total internal reflection of a light beam into a specially shaped dense optical medium, specifically Krestchmann configuration (Figure 16.2) and Otto configuration (Figure 16.3). Krestchmann configuration consists of a metal film placed on a glass prism. When the incident angle from the light beam is greater than the total internal reflection, the film is illuminated. For resonant light to tunnel through the metal film and be coupled to the SPPs, a particular angle of incidence (θ) where the in-plane component of the wavevector within the prism aligns with the SPP wavevector on an air–metal interface, Eq. 13.

$$\beta = \frac{\omega}{c}\sqrt{\epsilon_p rism}\,\sin(\theta) \tag{16.13}$$

Similar to gratings, the excitation of SPPs detected observes a sharp dip/minimum in the reflectivity from the surface as light can be coupled with high efficiency. Increasing the film thickness reduces the efficiency of SPP excitation as tunneling distance increases. Between two interfaces (one prism and one metal) SPPs cannot excite, since the SPP wavevector is greater than the photon wavevector within the prism regardless of angle. To excite SPPs on the internal metal interface, another dielectric layer with a refractive index lower than the prism should be placed between the prism and metal film. Since there is an extra layer, the light photon tunneling can pass through the extra dielectric layer that will provide resonant excitation of the SPP on the inner interface. SPPs on the surface and interface can be excited in using this configuration at different angles.

When considering thick films, Krestchmann configuration characteristics cannot be applied, this is where Otto configuration is utilized. In the Otto configuration, the prism is placed close to the metal surface to enable photon tunneling through the air gap separating the prism and the metal surface. The conditions for resonance is similar to the Krestchmann configuration.

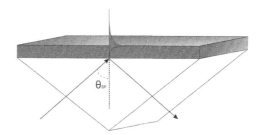

FIGURE 16.2   Kretschmann configuration.

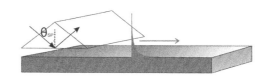

FIGURE 16.3   Otto configuration.

16.3.3 Focused Optical Beams

A microscope objective with high numerical aperture can be used to excite SPPs. This works by using an oil immersion objective touching the glass substrate of a thin metal film through a layer of index-matched immersion oil. The high numerical aperture allows a large angular spread of the beam, allowing for the angle to be greater than the critical angle for total internal refraction ($\theta > \theta_c$), which allows for phase matching to SPPs at the metal and air interface. The highly focused beam allows for localized SPP excitation within the diffraction limit.

16.4 Origin of Losses in Plasmonic Materials

As mentioned in the Introduction, a widespread problem with plasmonic integrated circuits and propagation of SPP in general are the high losses in conventional plasmonic materials like gold and silver, where the magnitude of the real part of the permittivity is very large [5]. Problematic scenarios resulting in loss are as follows: scattering due to rough metallic surfaces, propagation loss due to surrounding dielectrics/absorption of interband transitions, and ohmic losses in the metal.

16.4.1 Scattering

SPP scattering process mainly relies on size, geometrical shape, and dielectric constant of the surface. Defects occur when the surface is not perfectly smooth as anticipated. Three major results can occur due to defects [6]:

1. Scattering of SPP into SPP in another direction (SPP reflection)
2. Propagation of SPP through the defect region in the same direction as the incoming SPP (SPP transmission)
3. Scattering of SPP into light

It is noted that, although scattering can cause unwanted loss, depending on the criteria, scattering caters to many beneficial SPP applications. When analyzing scattering due to imperfections on a surface, lateral (in-plane), radial (out-of-plane), and the gradient of topographical variations are considered. This leads to complex and difficult numerical modeling for most cases (i.e., cases that are not simple and geometrical). Considering the in-plane and out-of-plane scattering processes modify the EM field distribution over the surface, which in turn affects the surface optical properties and field enhancements. Another parameter affecting loss is the distance between defects (l_d). The ratio between the distance between defects and the propagation length (L_{SP}) is critical when determining the nature of the specific scattering process. When $L_{SP} \lesssim l_d$ it is known as a single scattering regime, and when $L_{SP} \gg l_d$, it is known as a multiple scattering regime [6].

16.4.2 Ohmic Loss

Significant attenuation of EM waves due to ohmic loss (absorption by metal) is a large issue in plasmonic integrated circuits [7]. Ohmic loss is partially responsible for reducing propagation distances of SPPs, which diminishes the effectiveness of plasmonic waveguides. It is described as the imaginary part of the dielectric function of the metal and is responsible for SPP exponential decay. Ohmic losses and SPP decay can be calculated from optical constants and measured independently. Taking into account only Ohmic losses, SPP propagation length can be calculated from the imaginary part of SPP wave vector (k_x), shown in Eq. 14 [8].

$$L_\Omega = \frac{1}{2imk_x} \qquad (16.14)$$

16.5 Methods for Compensating Losses

Another candidate to acquire gain to compensate the loss of SPs is by using multiple quantum wells. This gain medium should provide a large enough amplification to realize appropriate propagation [9]. It is reported that a gain around 1,800/cm from AlGaInAs quantum wells at 1.55 μm wavelength shows that a lossless SP propagation is within current technological limits.

It is seen that through electrical pumping based on a Schottky-barrier diode shows it is able to fully compensate SPP propagation losses around 3 μ and provide net SPP gain [10]. SPP amplification through electrical injection is a complex phenomena incorporating electrical and optical processes. At 77 K (yields a small bandgap E_g^{InAs} of InAs)the Schottky-barrier height is greater than E_g^{InAs}. Since the barrier is greater, an inversion layer is established; this is where in equilibrium the Fermi level is higher than the conduction band edge of InAs, resulting in the concentration of electrons being substantially greater than holes. Through applying a positive bias voltage, the population inversion is created in InAs by injecting both electrons and holes into the active InAs area. The electrons injected from the inversion layer allow for a shift in the quasi-Fermi level towards the conduction band. At the same instance, p-InAs/p-AlAs$_{0.16}$Sb$_{0.84}$ heterojunction acts as an ohmic contact for holes which are capable of penetrating from AlAs$_{0.16}$Sb$_{0.84}$ into InAs while maintaining the hole quasi-Femi level of approximately $2k_BT$ below the valence band edge. As the energy separation between quasi-Fermi level rises over the SPP energy, the stimulated emission into SPP mode commences. The bias voltage can increase to a point where it is capable to compensate and overcome the ohmic and radiation losses of the TM$_{00}$ mode, and even amplify it.

Having nano quantum dots as a gain medium would allow for tunable spontaneous wavelength and do not suffer from excited state quenching and photobleaching. The energy levels and transition mechanism are explored by M. Stockman [11]. To begin the SP amplification by stimulated emission of radiation (spaser), the radiation from the light pumped excites an electron–hole pair in the gain medium (chromophores). The electron–hole pair then relaxes into an exciton state, with the chromophores coupled to the resonator, and the energy is transferred to the SP modes. From this, local fields are created, which then establishes a process that consists of the local field further exciting the gain medium. This creates somewhat of a feedback loop. If this loop creates a sufficient amount and the SP mode quality factor is high enough, then spasing will occur with that SP mode.

16.5.1 Gain Created by Quantum Wells

Alam et al [9] suggested to use quantum wells as the gain medium for the amplification of SPP. Their multilayer structure is shown in Figure 16.4. It consists of a metal film placed on a quantum well gain medium, consisting of five quantum well layers, each 8 nm thick and separated from each other by 16 nm thick barriers. The entire proposed structure is supported by a semi-infinite substrate (InP). Composition of quantum wells and barriers are as follows: $Al_{0.12}Ga_{0.12}In_{0.76}As$ and $Al_{0.3}Ga_{0.18}In_{0.52}As$, with 12.2 and 11.2 permittivities, respectively. The losses from substrate and barrier layers are neglected.

The proposed configuration is evaluated to find the traverse magnetic field analyzed using FEMLAB (version 3.1). The superstrate finite height enables a low-loss plasmon mode by a thin film even with the large asymmetry. This allows for less gain to be required due to there being less loss. It is noted that it is important to maintain the quality of the superstrate; if there is loss in the superstrate, this will result in a significant increase in the gain required to compensate the loss. Present day technology allows for the use of quantum wells, and a material gain of 1,800/cm from AlGaInAs quantum wells at 1.55 μm wavelength has been reported.

16.5.2 All Semiconductor Structure

Li and Ning [12] suggested an all-semiconductor structure operating in mid-infrared (MIR) wavelength range and using InAs heterostructures. These heterostructures are more optimal compared with metal-based plasmonics due to lower loss, improved confinement, and their tunability aspects. The material is selected to provide the largest plasmon frequency due to its smallest electron mass. First a prototype SPP structure is explored, including a bilayer structure with a heavily doped InAs layer and a metallic layer interfaced with a GaSb. It is proven that highly doping InAs provides an excellent plasmonic waveguide. This leads to their layer structure's schematic seen in Figure 16.5.

The system is simulated by ATLAS assuming the layers are uniformly doped at 2.5×10^{18} cm^{-3} and 5×10^{18} cm^{-3} for the layers in the plasmonic structure. Results indicate that MIR InAs-heterostructures are superior to other common semiconductors and metals for applications in plasmonic structures and metamaterials.

16.5.3 Incorporating Dipolar Gain Media

De Leon and Berini [13] suggested a structure that takes into account the nonuniformity of the gain medium due to position-dependent dipole lifetime and pump irradiance. A thin silver film is cladded on one side by a lossless dielectric and is pumped with rhodamine 6G on the opposite side, depicted in Figure 16.6. This configuration consists of 20 nm silver film extending infinitely over the x-y plane. The cladding is 25 μm of lossless dielectric material (CYTOP) on top of a semi-infinite silicon substrate.

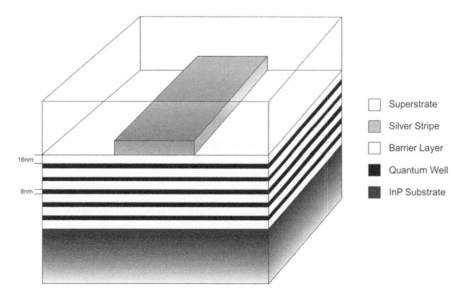

16nm

8nm

☐ Superstrate
▨ Silver Stripe
☐ Barrier Layer
■ Quantum Well
▨ InP Substrate

FIGURE 16.4 The multilayer structure proposed by [9] for compensating SP propagation loss consists of five 8 nm quantum well layers separated by 16 nm barrier layers.

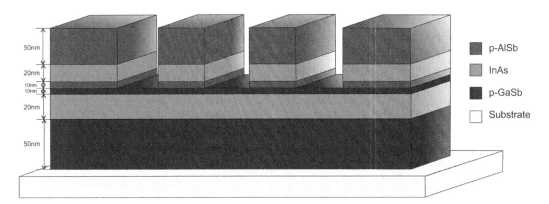

FIGURE 16.5 The all semiconductor active plasmonic structure [12] consists of an SPP source, waveguide with amplifiers, and a detector. AlSb and GaSb are p-doped, and InAs is n-doped.

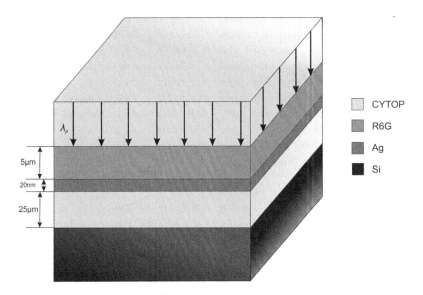

FIGURE 16.6 The multilayer structure proposed by [13].

The theoretical configuration is analyzed to describe the amplification of SPP modes in planar structures using optically pumped gain media. Simulations for long-range SPP (LRSPP) mode amplification suggest that the net amplification is possible using a reasonable amount of pump power and molecular concentration. To adequately describe SPP amplification, the pump mechanism and gain nonuniformity close to the metal surface from the position-dependent dipole lifetime are required to describe this phenomenon.

16.5.4 Fluorescent Polymer

Galther et al [14] suggested a structure that provides gain through an optically pumped layer of florescent conjugated polymer adjacent to the metal surface in a plasmonic waveguide. It contains a symmetric dielectric—metal—dielectric waveguide that supports coupled SPP modes on the top and bottom of interfaces, also known as LRSPP modes. Fluorescent polymers have a large emission cross section when compared to dyes, and they display low self-quenching due to excitons being delocalized over repetitive units of polymer. The schematic proposed is composed of a silicon substrate

with a 20 μm thick transparent polymer layer as seen in Figure 16.7. A layer of gain material (1 μm) and an extra 20 μm polymer layer are added.

The gain from the optically pumped layer of fluorescent conjugated polymer adjacent to the metal surface is analyzed using a one-dimensional solver. It is demonstrated that there is a positive net gain over macroscopic devices. Efficient energy transfer between the gain material and LRSPP mode is required for a positive net gain.

16.5.5 Lead-Sulphide Quantum Dots

Radko et al [15] suggested a schematic that uses a thin polymethylmethacrylate layer doped with lead-sulphide nanocrystals at near-infrared wavelengths. The proposed configuration, as seen in Figure 16.8, is a four-layer structure containing quartz substrate, 50 nm thick gold film, a thin layer of polymethylmethacrylate with embedded PbS quantum dots and air. The gold–polymethylmethacrylate interface allows at least one pure plasmonic mode, where they are excited by vertical illumination from periodic gold ridges focused by a continuous waveform laser beam of

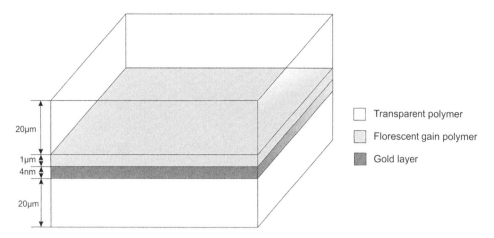

FIGURE 16.7 Proposed LRSPP waveguide with a florescent gain polymer for optical gain by [14]. A 532 nm laser is emitted to excite the gain material. The emission is obtained through a microscope objective and sent to a spectrometer.

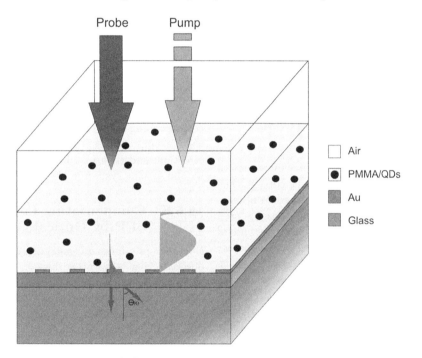

FIGURE 16.8 Experimental configuration of [15]. Excitation of the probe SPP beam (860 nm) occurs due to the grating seen in the Au layer. The intensity of the pump beam (532 nm) can be calculated.

860 nm. The gold ridges' geometry is designed to allow efficient unidirectional SPP excitation.

An analysis of experimental results of the suggested configuration shows that using a low-index dielectric (i.e., CYTOP) as a matrix medium for quantum dots are capable of increasing the SPP propagation length, thus allow compensating intrinsic SPP loss more achievable. Furthermore, a relatively high probe signal is accomplished using a schematic for efficient SPP excitation, which in turn reduces the probability of spontaneous emission.

16.5.6 Electrical Injection

Fedyanin et al [10] suggested an efficient way for amplification of SPPs, as seen in Figure 16.9. This is implemented through electronic pumping based on a single-heterostructure Schottky-barrier diode. The T-shape waveguide consists of an active p-type InAs layer with a passive $AlAs_{0.16}Sb_{0.84}$ layer. The SPP's supporting interface is formed by the active InAs layer, which serves as a Schottky contact for the creation of population inversion during electrical injection. When designing this scheme, five requirements are considered:

1. back contact is required to realize electric pumping
2. planar configuration is required for effectiveness and cost efficiency
3. single-mode guiding preference
4. bandwidth of at least 100 GB/s
5. high integration density of the plasmonic circuit

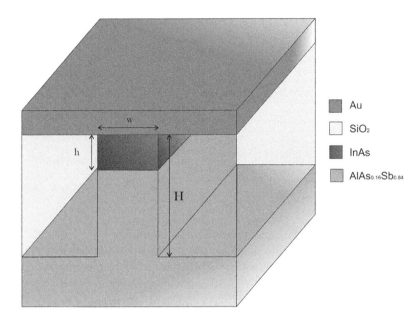

FIGURE 16.9 The T-shaped waveguide was explored [10], using an active p-type InAs layer (acceptor concentration of 2.33×10^{18} cm^{-3}) and passive p-type AlAs$_{0.16}$Sb$_{0.84}$ (acceptor concentration of 2.00×10^{18} cm^{-3}), where $w = 350$ nm, $h = 0.8$ µm, and $H = 2$ µm.

16.5.7 Direct Current with 2D Bragg Structure

Lozovskii et al. [16] proposed many experimental setups to display surface wave amplifications by direct current. The experimental setup seen in Figure 16.10 is used to observe effect amplification. It is noted that a different setup is used for numerical calculations. It is deduced that most useful systems where instability is observed are semiconductor–semiconductor or semiconductor–dielectric systems. The periodic structures of scatterers are placed on either side of the semiconductor film. Using Kretschmann or Otto configuration, one prism, and two identical light beams, two noninteractive surface waves can be excited with a specific frequency and wave vector. With two noninteractive surface waves, one will interact with DC and periodical structure of the scatterers, while the other will only interact with DC.

From their resulting data, Lozovskii et al. conclude that the effects of amplification are due to the energy transfer from plasma oscillations (DC supplied) to SPPs. With the help of Strurrock criteria, the possibility of surface wave amplification can be based on dispersion curve analysis. Strurrock criteria and dispersion curve analysis entail that, for some real frequency region, there are

corresponding imaginary values of wave vectors. During this time, asymptotes to the dispersion curves tilt at the same direction, and then instability due to convection can occur for the specific system in the frequency region.

16.5.8 Compensating Loss in Propagating SPP by Optical Gain

Noginov proposed a configuration, Figure 16.11, and observes that absorption loss in metal and propagating SPPs can be compensated through optical gain in dye-doped dielectric media adjacent to a metallic surface [17]. The setup consists of a glass prism with permittivity $\varepsilon_0 = n_0^2$, a metallic film with dielectric constant ε_1, and a layer of dielectric medium with permittivity ε_2.

16.5.9 Thin Organic Gain Medium

It is reported by Kena-Cohen et al. [18] that there is a direct observation of plasmonic gain in a structure in confined and lithographically defined SPP waveguides in various lengths. Using a thin layer of gain medium, 4-dicyanomethylene-2-methyl-6-(p-dimethylaminostyryl)-4H-pyran (DCM) dispersed in a tris(8-hydroxy-quinolinato)aluminum (Alq$_3$)

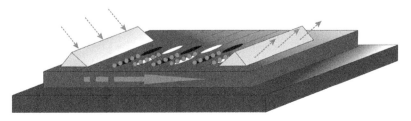

FIGURE 16.10 Configuration [16] for amplification observation.

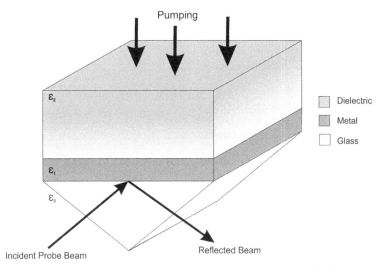

Pumping

ε_2

ε_1

ε_0

☐ Dielectric
▨ Metal
☐ Glass

Incident Probe Beam

Reflected Beam

FIGURE 16.11 Configuration exploration of the attenuated total internal reflection setup [17].

matrix; compensation of complete SPP loss is observed when it is compensated by the highest pump powers. This medium relies on Forester energy transfer from the wide optical gap host to the more narrow gap host (from Alq₃ to DCM). Throughout the entire study, it is determined that, by optically pumping this organic gain medium, there is a considerable net gain of 93 dB/mm for the best amplifier investigated.

16.6 Plasmons in a DC Electric Field

Over the years there were several attempts to analyze plasmonic effects in semiconductor plasma in a constant electric field and interacting with traveling waves [19–21]. In the summary here, we concentrate on the results obtained by the last two groups.

16.6.1 Analysis by Cada's Group

One possibility to amplify SPs is by allowing them to interact with drifting electrons in an external DC electric field. Interaction and possible amplification of EM waves with moving electrons have been discussed in classic papers [22,23]. Over the years, many groups [19,24–26] analyzed those interactions in various materials and configurations. The specific applications to plasmonics were suggested by M. Cada [27] and is extensively analyzed in [21].

Electrons moving under external DC field exchange their energies with the EM wave, and under certain conditions of phase matching, an EM wave is amplified. To facilitate the transfer of energy from electron stream to EM wave, their velocities should be close to each other. The electrons are accelerated and decelerated by the propagating EM wave while being moved under an external DC field.

To discuss the interaction we consider a two-dimensional configuration with the boundary lying along the z-axis and x-axis as the transversal one. Previous equations are rewritten to explicitly refer to a dielectric or a semiconductor.

A TM polarization is only considered, which results in the following assumptions:

$$\mathbf{H}_{S,D} = [0, H_{y,S,D}, 0] \qquad (16.15)$$

$$\mathbf{E}_{S,D} = [E_{x,S,D}, 0, E_{z,S,D}] \qquad (16.16)$$

$$\mathbf{J} = \left[J_{x,S}, 0, J_{z,S} \right] \cdot \delta_{S,D} \qquad (16.17)$$

where subscripts S, D refer to the semiconductor and dielectric materials, respectively, and

$$\delta_{S,D} = e^{i\omega t} e^{i\gamma_{S,D} x} e^{i\beta z} \qquad (16.18)$$

Here ω is the angular frequency, $\gamma_{S,D}$ and β are the transversal and longitudinal propagation constants, respectively

The remaining equations characterize electrons and are summarized below [21]

$$\nabla \times \mathbf{H}_S = \mathbf{J} + \varepsilon_S \frac{\partial \mathbf{E}_S}{\partial t}, \quad \nabla \cdot \mathbf{D}_D = 0 \qquad (16.19)$$

$$m \frac{d\mathbf{v}}{dt} = -e \left(\mathbf{E} + \mu_0 \mathbf{v} \times \mathbf{H} \right) \qquad (16.20)$$

$$\nabla \times \mathbf{H}_D = \varepsilon_D \frac{\partial \mathbf{E}_D}{\partial t}, \quad \nabla \cdot \mathbf{H}_{S,D} = 0, \quad \varepsilon_{S,D} = \varepsilon_0 \varepsilon_{\infty,d} \qquad (16.21)$$

$$\nabla \times \mathbf{E}_{S,D} = \mu_0 \frac{\partial \mathbf{H}_{S,D}}{\partial t}, \quad \nabla \cdot \mathbf{J} = -\frac{\partial \rho}{\partial t}, \quad \rho_0 = -eN \qquad (16.22)$$

$$\nabla \cdot \mathbf{D}_S = \rho_S, \quad \mathbf{J} = \rho \mathbf{v}, \quad \omega_p = e \sqrt{\frac{N}{m\varepsilon_0}} \qquad (16.23)$$

Here $\mathbf{E}, \mathbf{D}, \mathbf{H}$ are the electric, displacement, and magnetic field vectors, respectively, $\mathbf{J}, \rho, N$ are the current density, charge density, and charge current density in the semiconductor, respectively, $\mathbf{v}$ is the velocity of the moving electrons, $m = m_0 m^*$ with m^* being the effective mass, e is the electron charge, ε_0 and μ_0 are the vacuum permittivity and permeability, ε_∞ and ε_d are the relative permittivity of the semiconductor and dielectric, and ω_p is the plasma frequency.

Harmonic fields are considered and the derivatives are replaced as $\frac{\partial}{\partial t} \longrightarrow -i\omega, \frac{\partial}{\partial x} \longrightarrow i\gamma_{S,D}, \frac{\partial}{\partial z} \longrightarrow i\beta$. The velocity, charge density, and current density are assumed to be modulated around constant values, which results in the following replacements

$$\mathbf{v} \longrightarrow \mathbf{v}_0 + \mathbf{v}_S \delta_S = [v_{x,0} + v_{x,S}\delta_S, 0, v_{z,0} + v_{z,S}\delta_S] \quad (16.24)$$

$$\rho \longrightarrow \rho_0 + \rho_S \delta_S \quad (16.25)$$

$$\mathbf{J} = \rho\mathbf{v} = \mathbf{J}_0 + \mathbf{J}_S \delta_S \quad (16.26)$$

Normalize all parameters to the plasma frequency, ω_p, and the speed of light, c, like

$$\Omega = \frac{\omega}{\omega_p}, \Delta = \frac{v_d}{c}, \dots$$

After long manipulations (described in [21]), results in the following dispersion relation in a normalized form [28] are as

$$\sqrt{\varepsilon_\infty}\,(\Omega - B\Delta)\,(\Omega - B\Delta + i\Gamma)\left[(\Omega + iA\Delta_0)^2 - B^2\Delta_0^2\right]$$
$$- \Omega\Delta_0\,(\Omega + iA\Delta_0) = 0 \quad (16.27)$$

In the above Ω is the frequency, B is the complex propagation constant, Δ is the drift velocity, Γ is the electron collision loss, A is the transmission loss of the waveguide mode, Δ_0 is the velocity of a natural waveguide mode, and ε_∞ is the background permittivity of the semiconductor.

For the analysis of Eq. (16.27), the parameters summarized in Tables 16.1–16.3 were selected. Results of numerical analysis of Eq. (16.27) are shown in Figure 16.12.

16.6.2 Analysis by Sydoruk's Group

In the series of papers, Sydoruk and collaborators [20,29–34] analyzed various aspects of amplification of EM radiation in semiconductors. They concentrated on the possible amplification of semiconductor plasmas and generation of terahertz

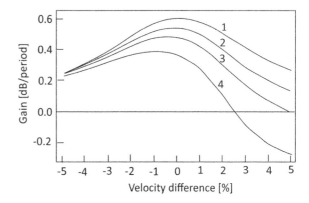

FIGURE 16.12 Gain per period as a function of velocity difference [39].

(THz) radiation. Among others, they considered traveling-wave interactions and the Dyakonov–Shur instability. They analyzed the effects of plasmon loss due to carrier collisions and diffusion.

A THz source is an important part in many applications. However, the generation of THz radiation is very difficult, and the existing sources have a number of limitations. It was recognized a long time ago that under proper conditions there can be transfer of kinetic energy from electron beam in a vacuum into EM waves for wave amplification and generation. A similar mechanism exists in semiconductors, and dc currents flowing there can transfer kinetic energy of electrons into the energy of THz plasmons, thus creating a solid-state THz source.

The system considered by Sydoruk is shown in Figure 16.13, and the computed gain is shown in Figure 16.14 It consists of a two-dimensional electron channel terminated by a conducting boundary. Plasmons propagate in the channel and electric field decays exponentially from the channel.

Sydoruk recognized the crucial role played by carrier collisions and diffusion influencing loss in solid-state plasmas. Those effects are important for the design of THz plasmonic oscillators and amplifiers. The conclusion derived by Sydoruk from his theoretical analysis is that, despite collisions, diffusion amplification of semiconductor plasmons is possible for realistic parameters corresponding to GaAs.

TABLE 16.1 Silicon Parameters Used

Symbol	Quantity	Value
ε_∞	Background permittivity	11.7
m^*	Effective electron mass	$0.26\ m_0$
α	Absorption coefficient	$10^4\ \mathrm{cm}^{-1}$
γ	Damping coefficient	$5.6 \times 10^{12}\,s^{-1}$
μ_e	Electron mobility	$1,200\ \mathrm{cm}V^{-1}s^{-1}$

TABLE 16.2 Parameters of the Amplifier Structure

Symbol	Quantity	Value
$\omega_p/2\pi$	Plasma frequency	3 THz
$\omega_{op}/2\pi$	Operating frequency	2 THz
α_T	Transmission loss	$10^2 - 10^3\ \mathrm{cm}^{-1}$
N	Electron concentration	$3 \times 10^{16}\ \mathrm{cm}^{-3}$
V_w	Waveguide volume	$1 \times 1 \times 1 \times \mu\mathrm{m}^3$
P_{in}	Incident optical power	5 dBm

TABLE 16.3 Achievable Performance of a Terahertz Amplifier Structure: Transmission Loss

Symbol	Quantity	0	$10^2\ \mathrm{cm}^{-1}$	$10^3\ \mathrm{cm}^{-1}$
l_p(nm)	Plasma frequency	48.7	48.6	48.4
g_{l_p}(dB)	Operating frequency	0.6	0.48	0.39
g_{l_w}(dB)	Transmission loss	12.3	10.2	8

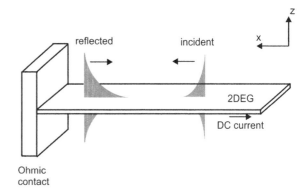

FIGURE 16.13 A perspective view of plasmons propagating along two-dimensional electron channel with a dc current flow [34].

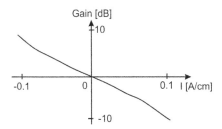

FIGURE 16.14 Calculated gain of plasmons incident on an ohmic contact. They are amplified when incident on contact against the direction of the dc current [34].

16.7 Experimental Search

When taking into account the different methods in exciting SPs, different gain materials/mediums are to be considered. Plasmonic waveguides are capable to confine light lower than the typical diffraction limit, but this confinement increases the propagation loss. Between dielectric and metal interfaces SPPs suffer propagation losses at visible wavelengths. Metallic structures dependent on plasma waves rely on the decrease of optical loss when transmitting EM energy and realizing negative refraction. Although there are ways to optimize the geometry and shape to reduce losses, the optical losses fundamentally come from the physical metals. The only way to solve this problem is to introduce optical gain; dyes, QDs, Erbium (Er) ions, and semiconductors have recently been explored. Gain coefficients are depicted in Table 16.4.

QD nanostructures with a radius ranging from 1 to 10 nm, allow for the emission wavelength to be tunable by controlling the QD material and size. Dyes typically emit in the visible and low near-infrared spectral range (< 1 μm); but since at above 1 μm polymer hosts suffer from absorption by vibrations of carbon–hydrogen (C–H) bonds, it poses a challenge. Er ion doped polymers are capable of producing emissions within 1,310 nm and 1,550 nm ranges. Semiconductors are seen to be a top contender of materials for losses greater than $1,000$ cm^{-1}.

16.8 Plasmonic-Based Lasing in Nanostructures

If there is a significant gain that overcomes losses and an external feedback mechanism is provided, there will be a possibility to generate coherent sources of SSPs. Such devices are known as SSP lasers (also known as spasers) [11]. Spaser devices can be compared to lasers in optics. Spasers, like

TABLE 16.4 Gain Coefficient Approximations for Various Media

Medium	Gain Coefficient (cm^{-1})
AlGaInAs QW	1,800
PbS QD	200
QD-doped PMMA ridges	143
Polymer strip doped with QDs	160
CYTOP and rhodamine dispersed in PMMA	420
PFS in MDMO-PPV	8

lasers, go through a gain medium for amplification, and a cavity where for spasers, SPs reside. To achieve stimulated emissions of SPs, SP modes sustained by the resonator structure must overlap, and have an emission transition that overlaps with the SP modes. Pumping spaser is optically simple, which would allow chromophores, such as nanocrystals or electron–hole excitation of a bulk semiconductor, be a competent gain medium for spasers.

16.8.1 Spasing Process

To begin the spasing process (stimulated emissions), the radiation from light pump excites an electron–hole pair in the gain medium (chromophores). The electron–hole pair then relaxes into an exciton state; with the chromophores coupled to the resonator, the energy is transferred to the SP mode. From this, local fields are created, which then establishes a process that consists of the local field, further exciting the gain medium. This creates somewhat of a feedback loop. If this loop creates a sufficient amount and the SP mode quality factor (Q) is high enough, then spasing will occur with that SP mode. The resonant lifetime of SP can be characterized by their quality factor. The Q of the supported plasmonic mode plays a role in the amplification of local fields generated by SPs. When Q is high, the mode has low loss and is capable of drawing energy from the gain medium for the longest period of time. In the interest of optimizing the spasing process, using a chromophore with string dipole transition and tuning the feedback can be done. Tuning the feedback can be achieved by ensuring that a large enough number of interactions between the gain medium and SP modes are sustained by the resonator.

16.8.2 Spaser Biological Application

Plasmonic lasers over the last several years have spiked interest in the field of cellular biology to study individual biomolecules and cells. For biomedical applications, spaser probes should be bright, water soluble, and safe. As a molecular probe, spasers are conjugated with a low-molecular-weight folic acid to target the folate receptor, which is more prominent on the surface of human cancer cells rather than normal cells. Galanzha et al. [35] explored the difference between the bright field image of a cancer cell with a single spaser versus one with multiple spasers. The cell tested with multiple spasers result in a higher image contrast and hot spots at various laser energies above the spasing threshold. It is noted that the number increases as the laser energy and incubation time increases. To verify the number of spasers within the cell membrane, the following techniques are used: transmission electron microscopy (TEM), inductively coupled plasma mass spectroscopy (ICP-MS), scanning TEM (STEM), energy-dispersive X-ray spectroscopy (EDX) with STEM imaging, photothermal (PT) microscopy, and photoacoustic (PA) flow cytometry.

The cytoxicity of the spasers must be considered in human biological applications. Galanzha et al. estimate that different concentrations and incubation durations for cancer and endothelial cells before and after irradiation yield a low toxicity with a viability of 92–96 similar to the safety of uranine. When incubating with spasers for laser irradiation, there is lower cell viability with an increased spaser concentration with no changes with a similar cell but with no spasers. Upon further experimentation and analysis at various laser energies and multiple spaser concentrations, it uncovered that cell damage was caused by the nanoinduced vapor nanobubbles around the overheated spasers (especially those in a cluster). These nanobubbles would lead to mechanical damage of the cell membrane, eventually leading to cell defragmentation. The nanobubbles created could inadvertently help lead to theranostic agents integrating optical diagnosis and PT-based cell killing by only using a few laser pulses.

16.9 Applications of SPP

16.9.1 Imaging and Spectroscopy

There are many known applications and benefits of SPPs. Specifically, due to spaser technological advancements, near-field scanning optical microscopy (NSOM) typically used for imaging in the nanoscale, is capable to be enhanced. It currently works by utilizing laser beams on pointed tips to create evanescent fields. A disadvantage of using lasers is that it transfers energy throughout, bringing unwanted heat that may cause distortions and noise in the measurements. Spasers can generate these evanescent fields using dark mode operations (which do not couple to the far zone optical field), without generating the unwanted energy transfer (causing heat). As well, since pumping is relatively simple, and the cavity has resonant characteristics, spasers are able to create evanescent fields much more intense than any laser is currently capable of. To implement this idea into NSOM, the approach would be to use a sharp metal probe to enhance optical fields in its area to locally excite the sample being observed. This application is known as tip-enhanced near-field scanning optical microscopy (TENSOM). It can be noted that, due to the dark mode that spasers bring, there is no scattered background from the laser, thus improving the signal-to-noise ratio (SNR). The metal tip must be polarized parallel to the length to utilize the tip enhancement yielding in field enhancement.

16.9.2 Sensors

SP excitation has began to find numerous sensory applications; this is due to the distinct resonance condition associated with the process. SP resonance (SPR) has emerged as a useful process in detecting (in real-time) biomolecular interactions in various fields of study, such as electrochemistry, chemical/vapor detection, spectroscopy, and life sciences. SPR technology is widely found as a transducer for various biosensors. Sensors utilize the information given by the changes in SPR angles, such as the relationship found between time and refractive index changes to attain binding amounts and association/disassociation constants.

16.9.3 Nanolithography

With increasing developments in utilizing SPs, applications in nanolithography have been explored. High-density nanolithography is demonstrated and explored [36]. Here, the potential of using SPs to control light in the subwavelength regime and transmission through subwavelength hole arrays on an opaque metal film display possible benefits in lithography applications. Since SP waves are capable of transmitting through the subwavelength plasmonic masks at UV wavelengths (10–400 nm), patterning nanoscale features using the conventional UV light sources are possible.

Srituravanich et al. conclude that periodic hole array mask utilizes near UV light to excite SPs on metal substrate to increase transmission through subwavelength periodic apertures with a much shorter wavelength compared with the conventional excitation light wavelength used. At 365 nm wavelength, a sub-100 nm dot array pattern on a 170 nm period can be established. This opens the possibility of lithography utilizing apertures much smaller than the exposing wavelength. By changing the hole array period and spacer thickness, the possibility of higher resolution, density, and improved transmission can be applied to optical lithography.

16.9.4 Devices Based on Metal–Dielectric–Metal Stub Resonators

Metal–dielectric–metal (MDM) waveguides devices are reviewed and elaborated by Huang et al. [37]. MDM waveguides have peaked interest due to their ability to support modes with deep subwavelength scales and high group velocity over a vast range of frequencies ranging from DC to visible spectrum. Other waveguide structures to be noted are metallic nanowires, V-shape grooves, and metallic nanoparticle arrays. MDM waveguides provide a possible interface between conventional optics and subwavelength electronic/optoelectrical devices. Currently, many experiments utilize MDM waveguides by using electron beam lithography (EBL) and patterned using a focused ion beam (FIB). Nanoplasmonic waveguides with active materials have been seen in photochromic molecules and CdSe quantum dots. This technology is useful in enhancing and improving integrated photonic devices, such as tunable filters, optical switches, drop filters, reflectors, and utilizing impedance matching elements. MDM plasmonic devices waveguide-cavity system is fabricated by a side-coupling stub resonator consisting of a MDM waveguide of finite length to a MDM waveguide.

Due to the difficulty in achieving active control of optical signals in nanoscale plasmonic devices, there have been an increased interest in actively controlled plasmonic devices, such as switches and modulators. Various proposals have been made to actively control light in the nanoscale, such as changing the refractive index through thermal inducements, direct ultrafast optical excitation of the metal materials, and incorporating multitudes of different gain mediums.

Slowing down light in plasmonic waveguides provide enhanced light–matter interactions, which ultimately leads to performance enhancements of nanoscale plasmonic devices. Although the benefits come at the cost of group velocity tunability, more specifically, once the operating wavelength and modal size are fixed, the group velocity tunability is no longer tunable.

MDM plasmonic waveguides can act as an alternative to previous proposals mentioned for active control of plasmonic devices. This works by altering and tuning the absorption coefficient of the device. It has been fabricated and explored experimentally through optical excitation of photochromic molecules and CdSe quantum dots.

16.9.5 Nanoscale THz and Optical Devices

Essentially, active and passive plasmonic devices are similar to and provide the same role as photonic devices, such as emitters and detectors. SPP devices are not yet developed to such a stage, see review by Dragoman [38]. Many are in the theoretical phase and some are being tested experimentally. We provide a short summary below.

The first of these devices is a modified resonant optical antenna. This optical antenna is enhanced by integrating itself on the facet of a commercially available semiconducting laser diode, where it is known as a plasmonic laser antenna. It comprised a coupled pair of triangularly shaped gold nanoparticles/nanorods separated by 20 nm of space. Field enhancements in the near-field region is produced by the gold nanorods, which is a result of SPP excitation and the reduced plasmon dephasing rate.

Direct SPP monolithic SPP detectors have been demonstrated using an organic p-n diode. This organic p-n diode is fabricated from a heterostructure between two metal electrodes (Ag), where one of these are a plasmonic waveguide. SPPs are excited when the SPP wave is absorbed by the organic structure, which then dissociate into electrons and holes. Currently, the SPP and electric current propagate within the same medium (Ag electrode). The external quantum efficiency is very low ($\approx 10^{-3}\%$); constant research is being done to improve this drawback. However, by using InGaN light emitters (based on quantum wells), SPPs enhance the emission of EM radiation yielding high quantum efficiencies. The light emitters are covered in a 10 nm metal film of either Ag or Al. This change allows for the luminescence to be integrated 17 times greater for Ag films and 8 times better for Al films when compared with emitters without metallic films. This ultimately leads to an increase of up to seven times of the internal quantum efficiency. The mechanism responsible for this increase is the resonant interaction of the electron–hole pairs in the quantum wells and SPP. During the recombination process, SPPs are generated which feeds back and increases the recombination rate.

Due to surface imperfections and roughness, the SPPs are transformed into light, which leads to more applications as a bright light emitting diode.

References

1. D.K. Gramotnev and S.I. Bozhevolnyi. Plasmonics beyond the diffraction limit. *Nature Photonics*, 4:83–91, 2010.
2. M.A. Noginov. Metamaterials with optical gain. In M.A. Noginov and V.A. Podolskiy, editors, *Tutorials in Metamaterials*, pp. 129–161. CRC Press, New York, 2012.
3. S.A. Maier. *Plasmonics: Fundamentals and Applications.* Springer, New York, 2007.
4. M. P. Nezhad, K. Tetz, and Y. Fainman. Gain assisted propagation of surface plasmon polaritons on planar metallic waveguides. *Optics Express*, 12:4072–4079, 2004.
5. VA Markel and AK Sarychev. Propagation of surface plasmons in ordered and disordered chains of metal nanospheres. *Physical Review B*, 75:085426, 2007.
6. A.V. Zayatsa, I.I. Smolyaninov, and A.A. Maradudinc. Nano-optics of surface plasmon polaritons. *Elsevier Physics Reports*, 408:131–314, 2005.
7. L.S. Slaughter, B.P. Khanal, L. Vigderman, E.R. Zubarev W.-S. Chang, B.A. Willingham and S. Link. Low absorption losses of strongly coupled surface plasmons in nanoparticle assemblies. *PNAS*, 108:198791719884, 2011.
8. B.H. Nguyen and V.H. Nguyen. Dispersion and attenuation of surface plasmon polariton at metaldielectric interface. *Advances in Natural Sciences: Nanoscience and Nanotechnology*, 5:035002, 2014.
9. M.Z Alam, J. Meier, J.S Aitchison, and M. Mojahedi. Gain assisted surface plasmon polariton in quantum wells structures. *Optics Express*, 15:176–182, 2007.
10. D.Y. Fedyanin, A.V. Krasavin, A.V. Arsenin, and A.V. Zayats. Surface plasmon polariton amplification upon electrical injection in highly integrated plasmonic circuits. *Nano Letters*, 12:2459–2463, 2012.
11. M.I. Stockman. Spasers explained. *Nature Photonics*, 2:328–329, 2008.
12. D. Li and C.Z. Ning. All-semiconductor active plasmonic system in mid-infrared wavelengths. *Optics Express*, 19:14594–14603, 2011.
13. I. De Leon and P. Berini. Theory of surface plasmon-polariton amplification in planar structures incorporating dipolar gain media. *Rapid Communications*, 78:161401, 2008.
14. M.C. Gather, K. Meerholz, N. Danz, and K. Leosson. Net optical gain in a plasmonic waveguide embedded in a fluorescent polymer. *Nature Photonics*, 4:457–461, 2010.
15. I.P. Radko, M.G. Nielsen, O. Albrektsen, and S.I. Bozhevolnyi. Stimulated emission of surface plasmon

polaritons by lead-sulphide quantum dots at near infra-red wavelengths. *Optics Express*, 18:18633–18641, 2010.

16. V. Lozovskii, S. Schrader, and A. Tsykhonya. Possibility of surface plasmon-polaritons amplification by direct current in two-interface systems with 2d bragg structure on the surface. *Optics Communications*, 282:3257–3265, 2009.

17. M.A. Noginov. Compensation of surface plasmon loss by gain in dielectric medium. *Journal of Nanophotonics*, 2:021855, 2008.

18. S. Kena-Cohen, P. Stavrinou, D. Bradley, and S. Maier. Confined surface plasmon-polariton amplifiers. *Nano Letters*, 13:1323–1329, 2013.

19. F.A. Bass, S.I. Khankina, and V.M. Yakovenko. The low frequency properties of a semiconducting plasma situated in a constant electric field. *Soviet Physics JETP*, 23:70–75, 1966.

20. O. Sydoruk. Amplification and generation of terahertz plasmons in gated two-dimensional channels: Modal analysis. *Journal of Applied Physics*, 115:204507, 2014.

21. M. Cada and J. Pistora. Plasmon dispersion at an interface between a dielectric and a conducting medium with moving electrons. *The IEEE Journal of Quantum Electronics*, 52:7200107, 2016.

22. W.C. Hahn and G.F. Metcalf. Velocity-modulated tubes. *Proceedings of the IRE*, 27:106–116, 1939.

23. S. Ramo. Space charge and field waves in an electron beam. *Physical Review*, 56:276–283, 1939.

24. M. Sumi. Travelling-wave amplification by drifting carriers in semiconductors. *Applied Physics Letters*, 9:251–253, 1966.

25. A. Yariv and D.R. Armstrong. Traveling wave oscillations in the optical region: A theoretical examination. *Journal of Applied Physics*, 44:1664, 1973.

26. A. Gover. Wave interactions in periodic structures and periodic dielectric waveguides. PhD thesis, CalTech, Physics, 1976.

27. M. Cada and J. Pistora. Optical plasmons in semiconductors. The Institute of Photonic Sciences, Barcelona, Spain, 2008.

28. M. Cada and J. Pistora. Design of a new terahertz nanowavegude amplifier. *IEEE Photonics Journal*, 9:2200905, 2017.

29. O. Sydoruk, E. Shamonina, and L. Solymar. Solid-state traveling-wave amplifiers and oscillators in the thz range: Effect of electron collisions. *The European Physical Journal D*, 59:233–240, 2010.

30. O. Sydoruk, E. Shamonina, V. Kalinin, and L. Solymar. Terahertz instability of surface optical-phonon polaritons that interact with surface plasmon polaritons in the presence of electron drift. *Physics of Plasmas*, 17:102103, 2010.

31. O. Sydoruk, R.R.A. Syms, and L. Solymar. Distributed gain in plasmonic reflectors and its use for terahertz generation. *Optics Express*, 20:19618–19627, 2012.

32. O. Sydoruk, R.R.A. Syms, and L. Solymar. Amplifying mirrors for terahertz plasmons. *Journal of Applied Physics*, 112:104512, 2012.

33. O. Sydoruk. Drifting plasmons in open two-dimensional channels: modal analysis. *Journal of Physics D: Applied Physics*, 46:135103, 2013.

34. O. Sydoruk. Amplification of drifting semiconductor plasmons and effects of carrier collisions and diffusion. *Journal of Physics D: Applied Physics*, 46:345101, 2013.

35. E. Galanzha, R. Weingold, D. Nedosekin, and et al. Spaser as a biological probe. *Nature Communications*, 10:1038, 2017.

36. W. Srituravanich, N. Fang, C. Sun, Q. Luo, and Z. Xiang. Plasmonic nanolithography. *Nano Letters*, 4:1085–1088, 2004.

37. L. Yang Y. Huang, C. Min and G. Veronis. Nanoscale plasmonic devices based on metal-dielectric-metal stub resonators. *International Journal of Optics*, 2012.

38. M. Dragoman and D. Dragoman. Plasmonics: Applications to nanoscale terahertz and optical devices. *Progress in Quantum Electronics*, 32:1–41, 2008.

39. M. Cada and J. Pistora. Design of a new terahertz nanowaveguide amplifier. *IEEE Photonics Journal*, 9:2200905, 2017.

Magneto-plasmonics in Purely Ferromagnetic Subwavelength Arrays

17.1 Introduction: Surface Plasmons and Magneto-optics **17**-1
17.2 Properties of Plasmons **17**-2
  Surface Plasmon Polaritons • Excitation and Coupling of SPPs
17.3 Magneto-Optical Kerr Effect **17**-5
  Introductory Remarks • Polar Magneto-Optical Kerr Effect • Enhancement of
  P-MOKE • SPP Enhancement of Polar Kerr Rotation
17.4 Summary **17**-10
Acknowledgments **17**-10
References **17**-10

S. D. Pappas
Technische Universitat Kaiserslautern

E. Th. Papaioannou
Technische Universitat Kaiserslautern
Martin-Luther Universität Halle Wittenberg

17.1 Introduction: Surface Plasmons and Magneto-optics

Plasmonics is a subfield of nanophotonics that deals with the light-matter interaction at metal/dielectric interfaces (Barnes, Dereux, & Ebbesen, 2003; Maier, 2007; Raether, 1988). Systematic investigations of surface plasmons (SPs) started in the early 1970s (Economou, 1969), and the field has experienced a renaissance during the last 20 years, arising from the progress in nanofabrication of particles and nanopatterning of surfaces (Baryakhtar, Demidenko, & Lozovski, 2013; Ctistis, Patoka, Wang, Kempa, & Giersig, 2007; Ebbesen, Lezec, Ghaemi, Thio, & Wolff, 1998). The ground-breaking progress in understanding the underlying physics has opened up the road for utilization of new types of applications, ranging from sensing (Anker et al., 2008; Homola, 2008) to microscopy and telecommunications (Gjonaj et al., 2011; Pendry, 2000; van Oosten, Spasenović, & Kuipers, 2010).

A recent evolution in the plasmonic research is the field of magneto-plasmonics. The main goal of the field is to combine magneto-optical effects with plasmonic resonances. Magneto-optical effects deal with the change of polarization state of light after reflection (Kerr effect) (Kerr, 1877) or transmission (Faraday effect) (Faraday, 1846), which is dependent on the magnetization of the material and geometrical configuration. The presence of SPs in hybrid structures composed of noble metals and magnetic materials can strongly modify the magneto-optic response. Alternatively, the use of magnetic materials in plasmonic structures offers the possibility to influence plasmonic resonances with magnetization and an external magnetic field.

The first studies on the interplay between plasmonic and magneto-optical effects were conducted on metals and semiconductors containing free charge carriers (Brion, Wallis, Hartstein, & Burstein, 1972; Chiu & Quinn, 1972). High static magnetic fields were used to modify the surface magneto-plasmon polaritons, giving rise to surface modes that are not reciprocal, or in other words, the propagation with wavevector $+\vec{k}$ is different from the propagation at the reversed direction $-\vec{k}$, leading to a surface wave of different frequency. Subsequently, an interest for utilizing combined plasmonic and ferromagnetic materials developed, allowing the alteration of the optical response using moderate or low magnetic fields. This approach has generated substantial experimental and theoretical efforts in recent years (Armelles et al., 2009; Armelles, Cebollada, García-Martín, & Gonzalez, 2013; Belotelov, Bykov, Doskolovich, Kalish, & Zvezdin, 2009; Caballero, García-Martín, & Cuevas, 2015; Chetvertukhin, Grunin, Dolgova, Inoue, & Fedyanin, 2013; Deshpande et al., 2012; Inoue, Arai, Fujii, & Abe, 1999; Kostylev et al., 2013; Liu et al., 2009; Papaioannou, Meyer, & Hillebrands, 2014; Pimenov & Kurin, 2012; Sapozhnikov et al., 2010; Torrado et al., 2010; Zvezdin & Belotelov, 2004), partially inspired by new possible applications as, for example, in the case of the use of a magnetic field to modify the dispersion relation of SP polaritons (SPPs) in a structure composed of trilayers of Au/Co/Au (Temnov et al., 2010). Recently, the magneto-plasmonic research has also investigated to what extent pure ferromagnetic patterned films (Fang et al., 2015; Luong, Ai, Zhao, & Nguyen, 2018; Melander et al., 2012; Papaioannou et al., 2010, 2011, 2017; Razdolski et al., 2013; Rollinger et al., 2016; Torrado et al., 2010) and nanoparticles (Bonanni et al., 2011; Kataja et al., 2015;

Maccaferri et al., 2013; Rubio-Roy et al., 2012) can host SPs and can increase the magneto-optic response.

The technological perspectives of magneto-plasmonic structures are large: the potential capability to trap light locally (near-field enhancement) in magneto-plasmonic nanostructures (Inoue, Levy, & Baryshev, 2013; Maksymov, 2016; Rollinger et al., 2016) opens new routes in the research of magnetism, optomagnetism, ultrafast all-optical switching (Bossini, Belotelov, Zvezdin, Kalish, & Kimel, 2016; Lambert et al., 2014), and in the field of biosensing (Maccaferri et al., 2015). This Chapter explains and analyzes about magneto-plasmonic interaction. It is organized in the following way: (i) In the first place, the properties of SPPs will be briefly introduced. (ii) Next, the basic concepts of the magneto-optical effect will be presented and the macroscopic formalism will be provided. (iii) Eventually, the magneto-plasmonic interaction in ferromagnetic antidot arrays, which support SPP excitations, will be elucidated by combining near- and far-field results.

17.2 Properties of Plasmons

17.2.1 Surface Plasmon Polaritons

Free electrons in metals can be treated as a high-density electron liquid and their interactions with optical electromagnetic fields can be rigidly described within the classical framework of Maxwell's equations. The collective motion of free electrons in all three directions and their high density, results in negligible spacing between the electron energy levels when compared with their thermal kinetic energy $\sim k_B T$. Therefore, unless if we treat cases of metallic structures with gap distances that enter the nanometer and subnanometer scales (quantum regime) (Zhu et al., 2016), the optical properties of metals can be understood and modeled in terms of classical theory, without the need of a quantum mechanical approach (Maier, 2007). The electromagnetic interactions, as described by Maxwell's equations, are symmetric with respect to the electric and magnetic fields. However, there is a large asymmetry for nanophotonic properties, such as the excitation of SPPs, where the presence of surfaces reduces the symmetry of the material.

SPs can be generally categorized into two big subgroups: the localized SPs (LSPs) that are present in metallic nanoparticles, and the propagating surface plasmons. In this chapter, we focus on the propagating case and its interaction with magnetic materials. This type of SPs is a propagating electromagnetic surface mode (see upper half of Figure 17.1a). SPs can be generated when light interacts with the free conduction electrons confined at metal/dielectric interfaces, while the associated electromagnetic field decays exponentially with the distance from the interface (see lower half of Figure 17.1a) (Maier, 2007). The generated SP, in this case, is the result of strong coupling of the incident radiation to the surface mode, and is therefore often called SPP. SPP is confined at and propagating along

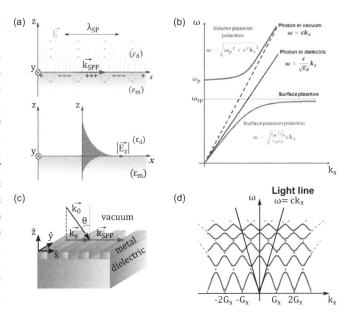

FIGURE 17.1 (a) The schematic visualizes the distribution of the associated confined electric field of an SPP close to the metal/dielectric interface (upper half). The schematic on the lower half is showing the magnitude of the electric field, both in the dielectric and the metal, as a function of the distance z. (b) Generic plot of the dispersion relation of plasmon polaritons corresponds to case (a). The black-dashed line is the dispersion curve of a free space electromagnetic wave, whereas the bold-black line corresponds to an electromagnetic wave propagating through the dielectric medium with dielectric constant ε_d. (c) The sketch depicts the mechanism of an SPP generation via the grating coupling technique. (d) Dispersion relation for SPPs on a grating (Raether, 1988). The dispersion lines corresponding to the free space electromagnetic wave are also shown in the same plot (Light lines).

the planar metal/dielectric interface, while being evanescent perpendicular to the interface.

The generic plot in Figure 17.1b shows the dispersion curves of plasmon polaritons along with their corresponding mathematical expression. The definition of axes for the metal/dielectric interface with relative dielectric constants ε_m and ε_d, respectively, is given in Figure 17.1a. The metal/dielectric interface is defined by the plane $z = 0$, and the plasmon propagates along the x direction ($\vec{k}_{\text{SPP}}$ is aligned along the x direction for simplicity). The linear dispersion curves of free space photons (photon in vacuum) as well as of photons traveling in the dielectric ε_d (photon in dielectric), the so-called "light lines," are also shown in the same plot. The bold-dark gray line in Figure 17.1b shows the dispersion curve corresponding to SPPs (Raether, 1988), which can be calculated by solving Maxwell's equations, considering the boundary conditions that apply for the metal/dielectric interface. As it can be observed, at very small k_x values, the dispersion curve approaches asymptotically the light line in the dielectric medium. At large k_x values, the frequency ω approaches the asymptotic limit ω_{sp}, and the group velocity $(d\omega/dk)$ becomes zero, namely the collective oscillations of the electron plasma near the surface are nonpropagating. The asymptotic frequency limit ω_{sp} is called SP frequency and is given by: $\omega_{sp} = \omega_p \sqrt{1/(1 + \varepsilon_d)}$, where ω_p is the plasma frequency of the corresponding metal (Raether, 1988). It is worth mentioning that, in the frequency region $0 < \omega < \omega_{sp}$, the plasmon wavevector k_x is real, denoting that SPP is a propagating mode along the x direction. On the other hand, k_z is imaginary in this frequency region, denoting that the mode is evanescent in the z direction. If we wish now to examine the frequency region $\omega_{sp} < \omega < \omega_p$ for the case of an ideal nonlossy metal–dielectric interface, we will conclude that both k_x and k_z become imaginary, and therefore, no propagating modes exist in this spectral gap.

For incident electromagnetic waves with frequencies $\omega > \omega_p$, the metal is transparent and the waves are transmitted through it. In this frequency region both k_x and k_z are real, denoting a "radiative mode." The resulting polariton is called volume plasmon polariton or bulk plasmon polariton, and its dispersion curve is shown in Figure 17.1b with a bold-dark gray line (Fox, 2010). At the high frequency limit, the volume plasmon curve follows asymptotically the light line in vacuum (black-dashed line). As it can be observed, the dispersion curve of volume plasmon polaritons lies always above the light line, and therefore, they can be accessible with an incident light source from free space. Volume plasmon polaritons are radiative, nonconstrained modes. They cannot be guided with a waveguide structure, and therefore, their study is not of particular interest for our work. On the other hand, SPPs are modes bounded to and confined at the dielectric/metal interface (Figure 17.1a). The latter implies that SPPs can be guided by the physical constrain of a waveguide structure, and therefore, their study is of great technological and fundamental interest (Fang & Sun, 2015). In the next sections, we will be focusing on the mechanism of SPP generation and their effect on the magneto-optic properties of ferromagnetic materials.

17.2.2 Excitation and Coupling of SPPs

The dispersion curve of SPPs lies always under the light dispersion curve of photons in the dielectric ε_d (Figure 17.1b). For this reason, an SPP can never be directly excited by incident photons on the metallic surface. Practically, the momentum of an incident photon must be enhanced, in order for its wavevector to be matched with the wavevector of the SPP. Various techniques have been invented to make this coupling viable: SPPs can be excited either by evanescent field coupling (Kretschmann & Raether, 1968; Otto, 1968) or through a corrugated surface, such as a diffraction grating (Hibbins, 1999).

The mechanism of the grating coupling, on which the topic of our study is based, is depicted schematically in Figure 17.1c. The grating modifies the in-plane component of the wavevector of the incident light $\vec{k}_o$, by adding (or subtracting) an integer multiple i of the grating wavevector $\vec{G}_x$. An SPP is excited on the grating when the momentum matching condition

$$\vec{k}_{spp} = \vec{k}_x \pm i\vec{G}_x, \tag{17.1}$$

is fulfilled. In Eq. 17.1 $\vec{k}_{spp}$ is the wavevector of the SPP on a continuous interface, $\vec{k}_x$ is the photon wavevector projection on x axis, and $\vec{G}_x = (2\pi/\alpha)\hat{x}$ is the grating wavevector along the x direction with periodicity α. It has been known since a long time ago that the reflectivity of metallic ruled gratings drops significantly when one of the diffracted orders propagates parallel to the surface of the grating. This effect, which historically is called "Wood's anomaly" (Wood, 1902, 1912, 1935), is now known to be caused by the excitation of SPPs in the metal. Assuming a single frequency incident radiation, Eq. 17.1 is satisfied by the values of angle of incidence θ, for which the ith diffracted order becomes parallel to the surface of the grating. The diffracted radiation then becomes evanescent and couples resonantly with an SPP propagating mode. The electromagnetic field of the SPPs excites the electron-hole pairs at the Fermi level, which upon de-excitation produces phonons and causes heating. Due to the creation of this extra mechanism of energy dissipation, the reflectivity drops significantly.

If we treat conceptually the grating grooves as periodic scattering centers, which allow the wavevector of the propagating SPPs to be modified by integer multiples of $\vec{G}_x$, then with the aid of Eq. 17.1 we can obtain the dispersion curves of the excited SPPs on the grating. This is demonstrated in Figure 17.1d, in the extended Brillouin zone (Raether, 1988). The dark gray-dashed lines correspond to the SPP dispersion curve, modified by $\pm iG_x$. The parts of the scattered dispersion curves that fall in the area between the light lines (transparent-gray area) can be now coupled to the incident radiation. In the same dispersion plot, the effect of the degenerate states, which are formed at the points where the dispersion curves are crossing ($k_x = \pm iG_x/2$), is also demonstrated. At these points, standing waves arise, an effect well known in wave physics, which produces small bandgaps between the allowed energy bands (Kitson, Barnes, & Sambles, 1996). The resulting dispersion curves are presented with dark gray-bold lines.

The results presented in this Chapter, are based on a twofold study of (i) the coupling of an external electromagnetic wave with SPPs on ferromagnetic two-dimensional (2D) lattices, as well as (ii) the modification of the magneto-plasmonic behavior of the initially continuous ferromagnetic material caused by the presence of SPPs. The created patterned layers are composed of a nanostructured 2D hexagonal lattice of holes with diameter $d = 275$ nm and pitch size $a = 470$ nm (see Figure 17.2a). Details about the fabrication of samples have been analyzed elsewhere (Ctistis et al., 2009; Melander et al., 2012). The grating coupling that satisfies the so-called momentum matching condition in the case of a 2D lattice is becoming

$$\vec{k}_{spp} = \vec{k} \pm i\vec{G}_x \pm j\vec{G}_y, \tag{17.2}$$

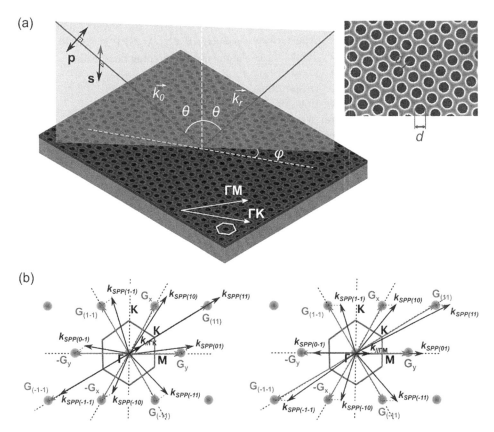

FIGURE 17.2 (a) Left: A schematic that depicts the used geometry for the excitation of SPP modes, the definition of the directions, and the angle of incidence θ. As ϕ is defined the angle between the plane of incidence and the high symmetry axis Γ-K of the reciprocal 2D lattice (nearest neighbor direction noted in the real space). Right: Example of a scanning electron micrograph of a hexagonal array of holes in a Ni film of 100 nm thickness. The relevant dimensions are the diameter d of the holes and the pitch size a. Here, $d = 275$ nm and $a = 470$ nm. Very good quality and long-range order up to 1×1 cm^2 are obtained. (b) Reciprocal lattice and the first Brillouin zone of the 2D hexagonal lattice. Different wavevectors $\vec{k}_{\text{spp}(ij)}$ of resonant SPP modes in the lowest frequency subband are shown for incident light wavevectors aligned along the Γ-K (left panel) or Γ-M (right-panel) directions.

where $\vec{k}_{\text{spp}}$ is the SPP wavevector, $\vec{k} = \vec{k}_0 \sin\theta$ is the component of the incident wavevector that lies in the plane of the sample. $\vec{G}_x$ and $\vec{G}_y$ are the basis vectors of the reciprocal hexagonal lattice while i and j are integers, Figure 17.2. SPP resonances can be tuned by different extrinsic (orientation of the pattern, angle of incidence) and intrinsic factors (material, pattern symmetry). In our discussion we keep the intrinsic factors constant by studying a patterned Ni film with a hexagonal lattice symmetry. We make use of extrinsic factors such as the orientation of the pattern and the angle of incidence.

The interplay between these two aforementioned factors is visualized in Figure 17.2. By aligning the plane of incidence of the incoming p-polarized light with one of the two main symmetry axis of the hexagonal pattern, SPP modes can be excited. If we limit the discussion to the first Brillouin zone and the SPP modes that belong to the lowest subband ($|i| \leq 1$, $|j| \leq 1$), six SPP modes are present for each symmetry direction Γ-K and Γ-M, as is indicated in Figure 17.2b. However, due to symmetry, these six modes are degenerated, while, depending on the refractive index of the material used, some of them are not present at all.

The SPP modes that satisfy Eq. 17.2 are defined each time based on geometrical considerations (by the pattern pitch size) and the dielectric constant of the materials at the interfaces. Factors, like the hole diameter, the thickness, as well as interference effects and structural properties of the interfaces are typically not taken into account. These parameters, though, can modify the resonances that Eq. 17.2 predicts. However, Eq. 17.2 successfully serves as a guideline for the expected energy and angular dependence of SPPs.

In Figure 17.3a, the analytically calculated SPP dispersion curves are indicatively shown for the case of a Ni hexagonal lattice, when the incident light wavevector is aligned along the Γ-M direction. The dispersion curves have been calculated by solving Eq. 17.2 and they correspond to the $\vec{k}_{\text{spp}(i,j)}$, which are graphically illustrated in the right panel of Figure 17.2b. The dispersion of the refractive index of the metal as well as of the substrate was taken into account for the calculations. In the structure of a real plasmonic crystal, there are two interfaces that can support SPP modes: the metal/vacuum and the metal/substrate interface. In Figure 17.3a, the dispersion graphs that correspond to the Ni/vacuum interface are drawn with gray-bold

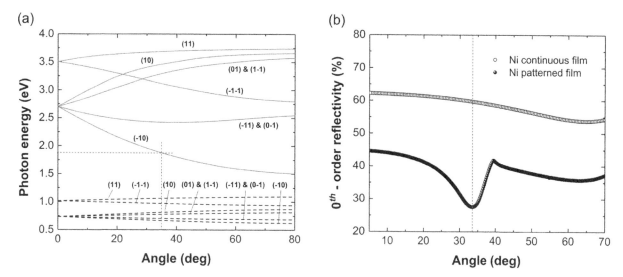

FIGURE 17.3 (a) Dispersion curves corresponding to a Ni hexagonal lattice with $\alpha = 470$ nm, calculated analytically by solving the Eq. 17.2. The gray-bold lines correspond to the metal/substrate (i,j) SPP modes, whereas the black-dashed lines correspond to the metal/vacuum modes. (b) Reflectivity with respect to the angle of incidence, recorded for a Ni patterned film grown on Si substrate, with relevant dimensions $d = 275$ nm (hole diameter) and $\alpha = 470$ nm (pitch size). p-polarized light with wavelength of $\lambda = 660$ nm (1.88 eV) was used. The plane of incidence was aligned along the nearest neighbor direction of the real lattice (Γ-K direction of the reciprocal lattice). The vertical black-dashed line indicates the position of the plasmon resonance, which is well predicted by the dispersion curve calculations in (a) (cross point of the gray-dashed lines). For comparison sake, the reflectivity of a Ni continuous film is presented in the same plot.

lines, whereas those corresponding to the Ni/Si interface are drawn with black-dashed lines. In this way, we can visualize how the different resonant SPP modes evolve in energy as a function of the angle of incidence (or the incident in-plane momentum k). By measuring at a specific energy or angle we can move along the dispersion lines and depict the shape and width of individual plasmon resonances. In Figure 17.3b, an example of a reflectivity measurement is shown as a function of the angle of incidence θ, recorded for a Ni patterned film with structure Au(2 nm)/Ni(100 nm)/Ti(2 nm)/Si substrate(111). The plane of incidence was aligned along the nearest neighbor direction (Γ-K), and p-polarized light with wavelength of $\lambda = 660$ nm (1.88 eV) was used. The reflectivity trough at $\theta = 33.5°$ (Wood's anomaly) is attributed to the excitation of the (-10) SPP mode at the vacuum/Ni interface, as it is indicated with the red-dash lines in Figure 17.3a. We should emphasize that the observed plasmonic resonances are recorded in a magnetic material like Ni. A question that immediately arises from the aforementioned is, could these SPP modes be correlated to the magneto-optical properties of the periodically perforated magnetic film? One way to study the effect of SPP modes to the magneto-optic behavior of the material is to perform Kerr-effect spectroscopy around the spectral vicinity of the plasmonic resonance condition, where the interaction of magnetism with light is at its maximum. In the following sections we address the issue of the modification of the magneto-optic behavior of a ferromagnetic metal by the presence of SPPs, which is explored with a series of different experimental arrangements.

17.3 Magneto-Optical Kerr Effect

17.3.1 Introductory Remarks

The magneto-optical Kerr effect (MOKE) refers to the change of the polarization state of the linearly polarized incident light when the latter is reflected from the surface of a magnetic material. For the appearance of the magneto-optical effect, two requirements have to be simultaneously fulfilled: (i) spin polarization (or generally magnetization) and (ii) Spin–orbit coupling. The spin polarization causes the cancelation of the degeneration of spin–orbit-coupled states, so that different complex refractive indexes appear for right-circular polarized light (RCP) $\tilde{n}_+(\omega)$ and left-circular polarized light (LCP) $\tilde{n}_-(\omega)$, respectively, where ω is the angular frequency of light. The complex refractive index of a material is a unit-less frequency-dependent physical quantity. It can be written as the sum of a real part (refractive index) $n(\omega)$ and an imaginary part (absorptive index) $\kappa(\omega)$:

$$\tilde{n}(\omega) = n(\omega) + i\kappa(\omega) \qquad (17.3)$$

This quantity is generally a tensor of second order, but it reduces to a scalar quantity for isotropic systems.

The MOKE is characterized by a frequency-dependent change of the polarization state of light which is reflected by a medium. Upon reflection, the linearly polarized light is elliptically polarized: the main oscillation axis of the electric field of the reflected light is tilted at an angle θ_K, while at the same time, ellipticity appears at η_K. This is graphically illustrated in the right panel of Figure 17.4. The Kerr ellipticity is due to unequal amplitudes between the incident and reflected amplitudes of the electric field, while the Kerr rotation appears as a consequence of the different (not equal) phases of incident and reflected amplitudes of the electric field. The two factors can be combined in the complex Kerr rotation:

$$\tilde{\Theta}(\omega) = \theta_K(\omega) + i\eta_K(\omega) \tag{17.4}$$

The experimental quantities θ_K and η_K, as they have been defined in Eq. 17.4, are related to the corresponding Fresnel reflection coefficients (r_{pp}, r_{ss}, r_{ps}, r_{sp}). Since the MOKE appears when a p-polarized component (or s-) is present in the reflection from incident s-polarized light (or p-), the complex Kerr rotation $\tilde{\Theta}$ can be written with the help of Fresnel reflection coefficients, for example, for incident s-polarized light, in the form of (Heinrich & Bland, 1994):

$$\tilde{\Theta}(\omega) = \frac{r_{ps}}{r_{ss}} \tag{17.5}$$

$\tilde{\Theta}(\omega)$ can be also expressed in terms of the diagonal and off-diagonal elements of the optical conductivity and dielectric tensor; the two tensors are correlated to each other ($\varepsilon(\omega) = 1 + (4\pi/\omega)\sigma(\omega)$). The appearance of off-diagonal elements has thereby been described with macroscopic as well as microscopic models (see for example, reviews on magneto-optics (Oppeneer, 2001; Reim & Schoenes, 1990; Schoenes, 1991)).

The Kerr effect exists in three main different possible geometries: polar, longitudinal, and transverse, depending on how the magnetization vector is aligned with respect to the plane of incidence of the incoming light. Of these, the polar Kerr effect (P-MOKE), for which the direction of the macroscopic magnetization of the ferromagnetic material and the propagation direction of the linearly polarized incident light beam are perpendicular to the plane of the surface, is by far the largest one. Consequently, this geometry is the most interesting one as far as technological applications are concerned.

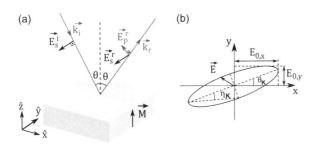

FIGURE 17.4 (a) An explanatory schematic of MOKE. In the polar magneto-optic Kerr effect (P-MOKE) the magnetization lies along the z-axis, i.e. perpendicular to the sample surface, as it is shown in the schematic. (b) The reflected light from a magnetic sample is elliptically polarized with the main axis of oscillation of the electric vector to be tilted at an angle θ_K. The angle η_K expresses the ellipticity of the reflected light.

17.3.2 Polar Magneto-Optical Kerr Effect

In the classical macroscopic framework, P-MOKE can be explained in terms of localized electrons that are excited by the electric field of the incident light and are subjected to a Lorentz force. In the polar geometry at normal incidence with the application of an external field (or presence of a magnetization) along the z-axis (see Figure 17.4a), the symmetry is broken in the xy-plane due to the magnetization: a Hall current appears on the xy-plane (optical conductivity element σ_{xy}) due to the magnetic Lorentz force. This renders the off-diagonal elements of the complex optical conductivity (and dielectric) tensor (σ_{yx}, σ_{xy}) of a magnetized isotropic medium nonzero. Due to symmetry considerations $\sigma_{yx} = -\sigma_{xy}$, the related conductivity tensor for the geometry of Figure 17.4 can be written as

$$\tilde{\sigma} = \begin{pmatrix} \tilde{\sigma}_{xx} & \tilde{\sigma}_{xy} & 0 \\ -\tilde{\sigma}_{xy} & \tilde{\sigma}_{xx} & 0 \\ 0 & 0 & \tilde{\sigma}_{zz} \end{pmatrix} \tag{17.6}$$

By solving Maxwell's equations with the help of adequate boundary conditions for the conductivity given by Eq. 17.6, we can relate Fresnel's reflection matrix with the magneto-optical quantities. In the quantum mechanical model, the calculation of the complex optical conductivity $\tilde{\sigma}_{xy}$ is related to the transitions between magnetically quantized states. For ferromagnetic materials the introduction of spin–orbit coupling is used to explain the strong MO response (Argyres, 1955). Quantitative calculations of $\tilde{\sigma}_{xy}$ for inter- and intraband transitions were first developed by Erskine and Stern (1973). In our case, where we are interested in the polar Kerr rotation, the relation between $\tilde{\Theta}$, the optical conductivity, and complex index of refraction obtains the following form (Fumagalli & Munekata, 1996)

$$\tilde{\Theta} = \frac{r_{ps}}{r_{ss}} = \frac{4\pi i}{\omega} \frac{\tilde{\sigma}_{xy}}{\tilde{n}(1 - \tilde{n}^2)} \tag{17.7}$$

17.3.3 Enhancement of P-MOKE

Equation 17.7 shows that the complex Kerr rotation $\tilde{\Theta}$ is not only dependent on $\tilde{\sigma}_{xy}$ but also on $\tilde{n}$. This means that $\tilde{\Theta}$ is influenced by both factors: the intrinsic magneto-optically active electronic transitions ($\tilde{\sigma}_{xy}$) and the extrinsic reflectivity (refractive index n and extinction k). A proper adjustment of $\tilde{n}$ can therefore lead to optical enhancement effects of $\tilde{\Theta}$. In particular, three main effects can optically enhance $\tilde{\Theta}$ without increasing the intrinsic Kerr effect in the MO material: the interface, the interference, and plasma edge effects (Fumagalli, 1997).

When the magneto-optic active material is not facing the vacuum but has an interface with another material (like, for example, a transparent and nonabsorbing glass substrate), the Kerr effect can be enhanced. The magnitude of the enhancement depends linearly on the value of the index of refraction of the substrate (see Eq. 17.7 where $\tilde{n}_{rel} = \tilde{n}/n_{sub}$). Enhancement of the Kerr effect can also be obtained if destructive interference of light occurs, as it reflects from the front and the back surface of an overlayer on top of the MO active material. This is possible if the thickness of the overlayer fulfills the antireflection condition. This factor has been widely used to optimize the performance of MO data storage disks (Mansuripur, Connell, & Goodman, 1982), while in the last years, it has been applied to increase the MOKE signal contrast in nanopatterned magnetic structures (Holiday & Gibson, 2006). Another amplification mechanism is the plasma edge effect that arises from the excitation of volume plasmon polaritons close to ω_p (Figure 17.1), which shows a minimum in the reflectivity spectrum. As the reflectivity reaches zero, the Kerr rotation is enhanced without the real MO activity to be on a resonance. Such enhancements have been observed in rare-earth chalcogenides (Schoenes, Hüsser, Reim, Kaldis, & Wachter, 1985).

A new type of P-MOKE enhancement has been revealed in recent years with the development of nanostructured hybrid magneto-plasmonic crystals: the LSPs and/or the SPP assisted enhancement. In contrast to the aforementioned pure optical enhancement effects, the SP mechanism includes two factors that contribute to the increase of $\tilde{\Theta}$: (i) the intrinsic magneto-optically active electronic transitions and (ii) the extrinsic reflectivity. By using either LSPs excited in magnetic wires or nanoparticles (Bonanni et al., 2011) or SPP resonances excited in hybrid structures, such as Au/Co or Au/iron garnets, and furthermore in pure patterned magnetic films of Fe (Papaioannou et al., 2010), Co (Ctistis et al., 2009), and Ni (Belotelov et al., 2011; Martin-Becerra et al., 2010; Melander et al., 2012; Papaioannou et al., 2011), the magneto-optic enhancement has been observed. Localized plasmon excitation modes induce an increase of intensity of the electromagnetic field inside the ferromagnetic materials, which leads to a higher polarization conversion. Propagating modes also influence the electromagnetic field distribution inside the magnetic material, leading to an intrinsic type of enhancement, and, at the same time, can be coupled to the lattice periodicity (so-called Bragg plasmons), giving rise to an extrinsic type of amplification. At plasmonic resonances, one observes that the reflectivity (transmission) exhibits minima (maxima), and a MO enhancement appears at the corresponding energies. In the following sections, we will elucidate the enhancement mechanism due to SPP excitation in magneto-plasmonic structures made of ferromagnetic patterned films.

17.3.4 SPP Enhancement of Polar Kerr Rotation

SPP Enhancement of Polar Kerr Rotation in the Far-Field

To study the physical mechanism that drives the magneto-plasmonic interaction in the far-field we present as an example the optical and magneto-optical properties of a Ni antidot film of 100 nm thickness. The hexagonal antidot pattern comprises holes with diameter $d = 275$ nm and a center-to-center distance of $a = 470$ nm.

The reflectivity spectra, shown in Figure 17.5, were recorded with different polarization states of light for the Ni antidot sample and compared to a reference continuous film of exactly the same composition and thickness. The reflectivity measurements were performed at a small angle of incidence $\theta = 6.5°$ and without the application of any external magnetic field. While keeping $\phi = 0°$, the polarization direction was initially aligned horizontally along the Γ-K direction (Figure 17.2). The light polarization was then rotated 90°, namely was aligned along the Γ-M direction. Figure 17.5 shows the absolute reflectivity curves for incident light polarized along the Γ-K and Γ-M direction for the continuous and patterned samples. Minima in the reflectivity of the patterned sample are observed for light polarized along both the Γ-K and Γ-M direction at specific energies. The minima of reflectivity are a result of the resonant coupling of light to SPP excitations at the metal/air interface. The condition for resonant excitation of SPPs is given by the momentum matching between SPP, incident light, and lattice geometry according to Eq. 17.2.

Figure 17.6 (top graph) shows the corresponding magneto-optic spectra of the Ni hole array sample and the reference film for the two polarization states. The polar Kerr experiment was performed at $\theta = 4°$ with the plane of incidence at $\phi = 0$ and at the magnetic saturation of samples. The magneto-optic spectrum of the Ni reference film exhibits two broad features around $\hbar\omega = 1.5$ eV and around $\hbar\omega = 3$ eV for both polarization states. This behavior has been associated in Ni with the d-p transitions of electrons at the top majority spin band and at the bottom of

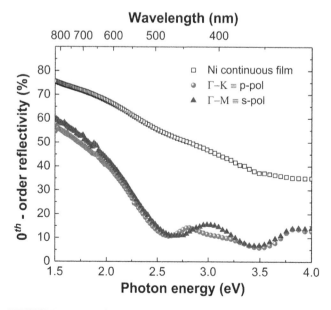

FIGURE 17.5 Absolute reflectivity curves as a function of energy for the antidot Ni-based film. Measurements were performed with incident light polarized along the nearest neighbor (Γ-K) and next-nearest neighbor (Γ-M) direction.

the minority spin band (Buschow, Engen, & R. Jongebreur, 1983). These characteristics are changed in the case of the patterned film as Figure 17.6 reveals. In the low-energy regime, below $\hbar\omega = 2.4$ eV, the Kerr rotation for the patterned films is smaller than in the continuous film, which is expected, if we consider that we have a smaller amount of magneto-optically active material in the hole array. At photon energies above $\hbar\omega = 2.4$ eV, a large enhancement of the polar Kerr rotation θ_K is observed with respect to the continuous film for incident light polarized along both Γ–K and Γ–M directions. The amplification of the Kerr rotation is visible for photon energies up to a photon energy of $\hbar\omega = 3.5$ eV. Furthermore, at larger photon energies, the Kerr rotation changes sign but still remains higher in

absolute values than that of the continuous Ni film for both polarizations.

As already discussed, the complex Kerr rotation is a relative quantity, and it depends on the ratio of polarization conversion (r_{ps}) to the extrinsic reflectivity (r_{ss})(Armelles et al., 2013; Fumagalli & Munekata, 1996; Papaioannou et al., 2014). Accordingly, the experimentally measured real part (θ_K) of the complex Kerr rotation not only partially represents the polarization conversion, but also depends on reflectivity. To reveal this dependence we have normalized the measured Kerr rotation values to the reflectivity. In particular, we introduce the parameter of figure of merit (FOM) that is defined here as the product of Kerr rotation θ_K and the square root of reflectivity $\sqrt{R}$. FOM is a well-established parameter (Zvezdin & Kotov, 1997), and it is usually defined as the product of complex Kerr rotation and the square root of reflectivity. It is used to normalize the measured magneto-optical activity with respect to the excitation intensity (or in other words, to the reflectivity), with the aim to extract the magnitude of the pure magneto-optical contribution, (r_{ps}). In Figure 17.6 (bottom graph), we present the result of such a normalization process with the help of FOM. To obtain our FOM parameter, we used the experimental data from Figure 17.5 and Figure 17.6 (top graph). The presented FOM values for both polarization states in Figure 17.6 (bottom graph) affirms that polarization conversion is enhanced.

To summarize, reflectivity and magneto-optic signal are strongly interconnected in magneto-plasmonic structures in the far-field. The strong enhancement of the Kerr effect is visible at energies where the reflectivity is decreasing due to the fact of surface-plasmon excitation. The increase of Kerr values is always supported by surface-plasmon resonances at specific energies. Moreover, the FOM parameter shows that in the presence of SPPs the pure magneto-optical polarization conversion is enhanced.

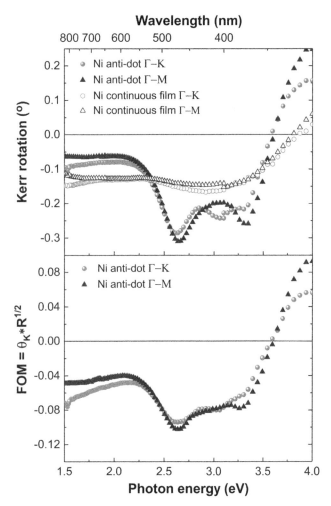

FIGURE 17.6 (Top graph) Polar Kerr rotation spectra of the Ni antidot (filled symbols) and a Ni continuous film (open symbols) are used as a reference (Rollinger et al., 2016). Measurements are performed for 0°- and 90°-polarized light aligned along the Γ-K and Γ-M direction, respectively, for both samples at a magnetic saturation of $B = 1$ T. (Bottom graph) Magneto-optic figure of merit (FOM) calculated for both polarization states reveals the enhancement of the pure magneto-optical conversion. (Modified with permission from Rollinger et al., 2016. Copyright 2016 American Chemical Society.)

SPP Enhancement of Magneto-Optical Effect in the Near-Field

We focus now on the correlation of magneto-optics and SPPs in the near-field. To provide direct evidence for the excitation of SPPs on magneto-plasmonic structures, we present data recorded by Photoemission Electron Microscopy (PEEM). PEEM is a powerful technique for imaging the photoelectron distribution of the sample on a nanometer local scale (Aeschlimann & Zacharias, 2010; Cinchetti et al., 2005; Vogelgesang & Dmitriev, 2010).

For the definitions of directions and angles used below, the reader is referred again to Figure 17.2. For the PEEM experiments, $\phi = 0$ and the angle of incidence of the laser light impinging on the sample surface was $\theta = 4°$ relative to the surface normal, which is equivalent to the one used for the polar Kerr rotation measurements presented in Figure 17.5. This will allow us to directly correlate the near-field obtained with the PEEM experiment with the MOKE signal of the sample in the far-field. In Figure 17.7 we present PEEM

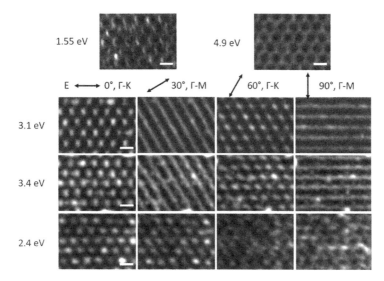

FIGURE 17.7 PEEM images of the Ni-based antidot film, recorded at different photon energies (Rollinger et al., 2016). The scale bar for all images is 500 nm. (First row) Images of the patterned structure by using the excitation with linearly polarized laser light of $\hbar\omega = 1.55$ eV as well as with a mercury discharge lamp at $\hbar\omega = 4.9$ eV. (Second row) Photoemission patterns excited at $\hbar\omega = 3.1$ eV. Spot-like emission is observed from the region in between the holes for incident light aligned along the Γ-K direction. Incident light polarized along the Γ-M direction causes emission patterns that form lines perpendicular to the E-field vector direction. (Third row) Photoemission pattern excited at $\hbar\omega = 3.4$ eV is similar to the patterns observed for $\hbar\omega = 3.1$ eV. (Fourth row) Photoemission pattern excited at $\hbar\omega = 2.4$ eV. Weak spot-like emission as well as homogeneous emission is observed, which does not follow the symmetry of the pattern while rotating the polarization plane of the incident light. (Reprinted with permission from Rollinger et al., 2016. Copyright 2016 American Chemical Society.)

images recorded at different excitation energies. The excitation of SPPs due to the coupling of the incident light with the periodic lattice will cause an enhancement of the electric near-field that can be directly measured with PEEM. The top two images of Figure 17.7 are taken at a photon energy of $\hbar\omega = 1.55$ eV and $\hbar\omega = 4.9$ eV. At these photon energies no SPP excitation is expected; therefore, we take these images as reference, since they only reveal the topography of the sample. The next three rows of Figure 17.7 show the images of the detected electron intensity for different polarization directions, i.e. aligned along the Γ-K direction (0° and 60°) and along the Γ-M direction (30° and 90°). They are recorded at photon energies of $\hbar\omega = 3.1$ eV and 3.4 eV, i.e. close to the onset of SPP resonances and at a photon energy of $\hbar\omega = 2.4$ eV.

At $\hbar\omega = 3.1$ eV PEEM images show specific emission patterns for different polarizations: either spot-like or continuous lines. In particular, when the incident E-field is aligned along the direction of the nearest neighbors (Γ-K direction), PEEM images reveal dot-like emission from the sample surface. The spots are formed between the holes, and they are also arranged in a hexagonal pattern. This behavior is repeated for every 60° of rotation of the polarization state of light, thus every time, the polarization is oriented along the Γ-K direction of the hexagonal hole lattice. When the E-field is rotated by 30° or 90°, Γ-M direction, strong changes in the PEEM images are observed. In this case, bright continuous lines are formed perpendicular to the polarization direction of light. Again the lines appear on the Ni and not in

the holes. The pattern is also repeated for every 60° of rotation of the polarization of the light, mirroring the symmetry of the structure for the Γ-M direction. At $\hbar\omega = 3.4$ eV photon energy the PEEM images exhibit a similar behavior: hexagonal bright spot formation for incident light polarized along the Γ-K direction and continuous lines for incident light polarization along the Γ-M direction.

The results for a photon energy of $\hbar\omega = 2.4$ eV are obtained at the outer edge of an SPP resonance (see Figure 17.5). Spot-like emission is observed from different areas of the surface for 0° and 30° rotation of the polarization of the incident light, while a more homogeneous emission is seen for 60° and 90° of rotation. The small changes in the photoemission patterns are not as distinct as before and furthermore do not mirror the rotational symmetry of the lattice of 60°. A distinct line formation in the photoemission pattern is not discernible here for any angle of polarization.

The dramatic changes in the near-field distribution recorded by PEEM can be correlated to the magneto-optic Kerr response of the structure. PEEM images recorded at $\hbar\omega = 3.1$ eV and $\hbar\omega = 3.4$ eV have revealed the distribution of the near-field either as stripe lines or as bright spots depending on the polarization state. This spatial distribution of the electric near-field recorded by PEEM influences the magneto-optic response. We see that the electric fields are strongly intensified for light polarized either along the Γ-K or Γ-M direction at a spectral region between 2.5 and 3.5 eV. Inside this spectral region, we also observed a strong magneto-optic enhancement, Figure 17.6. This is a clear indication that the intensification of the electric field generated by the spatial distribution of the SPs leads to an increase of the pure magneto-optic contribution. In particular r_{sp} can be written as (Armelles et al., 2013):

$$|r_{sp}| \propto \langle E_p E_s \rangle d |\varepsilon_{mo}| \qquad (17.8)$$

where d is the thickness of the film, ε_{mo} its magneto-optical constant, and $\langle E_p E_s \rangle$ is the mean value of the product of both components of the field inside the MO layer. The enhancement of both components of the electric field, as PEEM images reveal, increases the pure magneto-optic contribution of the Kerr rotation. The enhancement of the electric field is also present for energies above $\hbar\omega = 3.5$ eV, leading to higher positive values of Kerr rotation.

In summary, the correlation of the near- and far-field of the magneto-optic enhancement is elucidated. The Kerr spectra and the FOM diagram in Figure 17.6 affirm that the polarization conversion is enhanced. Furthermore, Figure 17.7 associates the magneto-optical conversion with the near-field intensification of the electric field in the form of spots or lines for both polarizations of the incident light.

17.4 Summary

Magneto-plasmonic lattices, produced upon patterning of purely ferromagnetic metal films, can exhibit large enhancement of the magneto-optic response close to their SPP resonances, when compared with the corresponding continuous ferromagnetic films. In the first part of the chapter, this effect is demonstrated by using periodically patterned Ni hexagonal antidot lattices, along with an introduction on excitation of SPPs by external electromagnetic radiation. In order to again insight into the mechanism of the coupling between plasmonic and magneto-optical properties, we combine in the next section magneto-optical spectroscopy with PEEM. PEEM is a technique that allows imaging capability of the near-field. PEEM images clearly demonstrate that the use of light at specific photon energies that correspond to the excitation of SPPs leads to polarization-dependent distributions of the near-field in the form of spots when excited with light polarized along the Γ-K direction or stripes when excited with the incident electric field vector aligned along the Γ-M direction. The distribution of the electric near-field results in a significant increase of the field intensity in the spectral region above 2.5 eV and thereby defines the enhancement of the Kerr rotation. The field intensification also drives the enhancement of the polarization conversion, which is the intrinsic magneto-optic activity. By combining the aforementioned techniques, we introduce the concept of engineering the magneto-optic response of magneto-plasmonic metastructures: The patterned films can be seen as magneto-plasmonic metastructures, the emergent magneto-optic behavior of which can be tuned by properly tailoring intrinsic or extrinsic factors. The knowledge of the fundamental properties of the magneto-plasmonic interaction can lead to the design of functional optical and magneto-optical devices.

Acknowledgments

We acknowledge the Deutsche Forschungsgemeinschaft (DFG) collaborative research center SFB TRR 173: SPIN+X Project B07 and the Carl Zeiss Foundation. We greatly acknowledge all our colleagues who have contributed in the analysis and characterization of the magneto-plasmonic arrays: Markus Rollinger (PEEM experiments) and Emil Melander (MOKE measurements). We acknowledge Prof. Martin Aeschlimann, Assoc. Prof. Vassilios Kapaklis and Dr Antonio García-Martín for the discussions.

References

Aeschlimann, M., & Zacharias, H. (2010). Time-resolved two-photon photoemission on surfaces and nanoparticles. In *Nanotechnology*. Weinheim: Wiley-VCH Verlag GmbH & Co. KGaA.

Anker, J. N., Hall, W. P., Lyandres, O., Shah, N. C., Zhao, J., & van Duyne, R. P. (2008). Biosensing with plasmonic nanosensors. *Nature Mater.*, *7*(6), 442–453.

Argyres, P. N. (1955). Theory of the Faraday and Kerr effects in ferromagnetics. *Phys. Rev.*, *97*, 334–345.

Armelles, G., Cebollada, A., García-Martín, A., García-Martín, J. M., González, M. U., González-Díaz, J. B., … Torrado, J. F. (2009). Magnetoplasmonic nanostructures: Systems supporting both plasmonic and magnetic properties. *J. Opt. A: Pure Appl. Opt.*, *11*(11), 114023.

Armelles, G., Cebollada, A., García-Martín, A., & Gonzalez, M. U. (2013). Magnetoplasmonics: Combining magnetic and plasmonic functionalities. *Adv. Opt. Mater.*, *1*(1), 10–35.

Barnes, W. L., Dereux, A., & Ebbesen, T. W. (2003). Surface plasmon subwavelength optics. *Nature*, *424*, 824–830.

Baryakhtar, I., Demidenko, Y., & Lozovski, V. (2013). Interaction between localized-on-nanoparticles plasmon polaritons and surface plasmon polaritons. *J. Opt. Soc. Am. B*, *30*(4), 1022–1026.

Belotelov, V. I., Akimov, I. A., Pohl, M., Kotov, V. A., Kasture, S., Vengurlekar, A. S., … Bayer, M. (2011). Enhanced magneto-optical effects in magnetoplasmonic crystals. *Nat. Nanotech.*, *4*, 1–7.

Belotelov, V. I., Bykov, D. A., Doskolovich, L. L., Kalish, A. N., & Zvezdin, A. K. (2009). Extraordinary transmission and giant magneto-optical transverse Kerr effect in plasmonic nanostructured films. *J. Opt. Soc. Am. B*, *26*(8), 1594–1598.

Bonanni, V., Bonetti, S., Pakizeh, T., Pirzadeh, Z., Chen, J., Nogués, J., … Dmitriev, A. (2011). Designer magnetoplasmonics with nickel nanoferromagnets. *Nano Lett.*, *11*(12), 5333–5338.

Bossini, D., Belotelov, V. I., Zvezdin, A. K., Kalish, A. N., & Kimel, A. V. (2016). Magnetoplasmonics and femtosecond optomagnetism at the nanoscale. *ACS Photonics*, *3*(8), 1385–1400.

Brion, J. J., Wallis, R. F., Hartstein, A., & Burstein, E. (1972). Theory of surface magne-toplasmons in semiconductors. *Phys. Rev. Lett.*, *28*(22), 1455–1458.

Buschow, K. H. J., Engen, P., & R.Jongebreur. (1983). Magneto-optical properties of metallic ferromagnetic materials. *J. Magn. Magn. Mater.*, *38*(1), 1–22.

Caballero, B., García-Martín, A., & Cuevas, J. C. (2015). Faraday effect in hybrid magneto-plasmonic photonic crystals. *Opt. Express*, *23*(17), 22238–22249.

Chetvertukhin, A. V., Grunin, A. A., Dolgova, T. V., Inoue, M., & Fedyanin, A. A. (2013). Transversal magneto-optical Kerr effect in two-dimensional nickel magneto-plasmonic crystals. *J. Appl. Phys.*, *113*(17), 17A942.

Chiu, K. W., & Quinn, J. J. (1972). Magnetoplasma surface waves in metals. *Phys. Rev. B*, *5*(12), 4707–4709.

Cinchetti, M., Gloskovskii, A., Nepjiko, S. A., Schönhense, G., Rochholz, H., & Kreiter, M. (2005). Photoemission electron microscopy as a tool for the investigation of optical near fields. *Phys. Rev. Lette.*, *95*(4), 047601.

Ctistis, G., Papaioannou, E., Patoka, P., Gutek, J., Fumagalli, P., & Giersig, M. (2009). Optical and magnetic properties of hexagonal arrays of subwavelength holes in optically thin cobalt films. *Nano Lett.*, *9*(1), 1–6.

Ctistis, G., Patoka, P., Wang, X., Kempa, K., & Giersig, M. (2007). Optical transmission through hexagonal arrays of subwavelength holes in thin metal films. *Nano Lett.*, *7*(9), 2926–2930.

Deshpande, N. G., Hwang, J. S., Kim, K. W., Rhee, J. Y., Kim, Y. H., Chen, L. Y., & Lee, Y. P. (2012). Enhancement of the magneto-optical properties in 2-dimensional bilayered magnetic anti-dot lattice. *Appl. Phys. Lett.*, *100*(22), 222403.

Ebbesen, T. W., Lezec, H. J., Ghaemi, H. F., Thio, T., & Wolff, P. A. (1998). Extraordinary optical transmission through sub-wavelength hole arrays. *Nature*, *391*, 667–669.

Economou, E. N. (1969). Surface plasmons in thin films. *Phys. Rev.*, *182*(2), 539–554.

Erskine, J. L., & Stern, E. A. (1973) Magneto-optic Kerr effect in ni, co, and fe. *Phys. Rev. Lett.*, *30*, 1329–1332.

Fang, H., Caballero, B., Akinoglu, E. M., Papaioannou, E. T., García-Martín, A., Cuevas, J. C., ... Fumagalli, P. (2015). Observation of a hole-size-dependent energy shift of the surface-plasmon resonance in Ni antidot thin films. *Appl. Phys. Lett.*, *106*(15), 153104.

Fang, Y., & Sun, M. (2015). Nanoplasmonic waveguides: Towards applications in integrated nanophotonic circuits. *Light Sci. Appl.*, *4*, e294.

Faraday, M. (1846). Experimental researches in electricity. *Phil. Trans. R. Soc. Lond.*, *136*, 1–20.

Fox, M. (2010). *Optical Properties of Solids*. Oxford University Press Inc., New York.

Fumagalli, P. (1997). Habilitation thesis: Magneto-optic spectroscopy of magnetic thin films: Influence of structural properties on magnetic ordering, exchange, and coercivity. Fakultät für Mathematik, Informatik und Naturwissenschaften der Rheinisch-Westfälischen Technischen Hochschule Aachen.

Fumagalli, P., & Munekata, H. (1996). Magneto-optic properties and ferromagnetism of (In,Mn)As/(In,Al)As/(Ga,Al)Sb heterostructures. *Phys. Rev. B*, *53*, 15045–15053.

Gjonaj, B., Aulbach, J., Johnson, P. M., Mosk, A. P., Kuipers, L., & Lagendijk, A. (2011). Active spatial control of plasmonic fields. *Nat. Photonics*, *5*, 360–363.

Heinrich, B., & Bland, J. (Eds.) (1994). *Magneto-Optical Effects in Ultrathin Magnetic Structures*. Springer-Verlag, Berlin, Heidelberg.

Hibbins, A. P. (1999). *Grating coupling of surface plasmon polaritons at visible and microwave frequencies* (Unpublished doctoral dissertation). University of Exeters.

Holiday, L. F., & Gibson, U. J. (2006). Improved longitudinal magneto-optic kerr effect signal contrast from nano-magnets with dielectric coatings. *Opt. Express*, *14*(26), 13007–13013.

Homola, J. (2008). Surface plasmon resonance sensors for detection of chemical and biological species. *Chem. Rev.*, *108*, 462–493.

Inoue, M., Arai, K., Fujii, T., & Abe, M. (1999). One-dimensional magnetophotonic crystals. *J. Appl. Phys.*, *85*(8), 5768–5770.

Inoue, M., Levy, M., & Baryshev, A. V. (2013). *Magnetophotonics from Theory to Applications*. Springer Series in Materials Science, Springer-Verlag, Berlin Heidelberg.

Kataja, M., Hakala, T. K., Julku, A., Huttunen, M. J., van Dijken, S., & Torma, P. (2015). Surface lattice resonances and magneto-optical response in magnetic nanoparticle arrays. *Nat Commun*, *6*, 7072.

Kerr, J. (1877). On rotation of the plane of the polarization by reflection from the pole of a magnet. *Phil. Mag.*, *3*, 321.

Kitson, S. C., Barnes, W. L., & Sambles, J. R. (1996). Full photonic band gap for surface modes in the visible. *Phys. Rev. Lett.*, *77*, 2670–2673.

Kostylev, N., Maksymov, I. S., Adeyeye, A. O., Samarin, S., Kostylev, M., & Williams, J. F. (2013). Plasmon-assisted high reflectivity and strong magneto-optical Kerr effect in permalloy gratings. *Appl. Phys. Lett.*, *102*, 121907.

Kretschmann, E., & Raether, H. (1968). Radiative decay of non radiative surface plasmons excited by light. *Zeitschrift für Naturforschung A*, *23*(a), 2135–2136.

Lambert, C.-H., Mangin, S., Varaprasad, B. S. D. C. S., Takahashi, Y. K., Hehn, M., Cinchetti, M., ... Fullerton, E. E. (2014). All-optical control of ferromagnetic thin films and nanostructures. *Science*, *345*(6202), 1337–1340.

Liu, Z., Shi, L., Shi, Z., Liu, X. H., Zi, J., Zhou, S. M., ... Xia, Y. J. (2009). Magneto-optical kerr effect in perpendicularly magnetized Co/Pt films on two-dimensional colloidal crystals. *Appl. Phys. Lett.*, *95*(3).

Luong, H. M., Ai, B., Zhao, Y., & Nguyen, T. D. (2018). Weak enhanced resonant Faraday rotation in pure cobalt plasmonic lattices: Thickness dependent Faraday rotation studies. *J. Magn. Magn. Mater.*, *468*, 79–84.

Maccaferri, N., Berger, A., Bonetti, S., Bonanni, V., Kataja, M., Qin, Q. H., ... Vavassori, P. (2013). Surface lattice resonances and magneto-optical response in magnetic nanoparticle arrays. *Phys. Rev. Lett.*, *111*, 167401.

Maccaferri, N., Gregorczyk, E. K., Thales, V. A. G. O., Kataja, M., van Dijken, S., Pirzadeh, Z., ... Vavassori, P. (2015). Ultrasensitive and label-free molecular-level detection enabled by light phase control in magnetoplasmonic nanoantennas. *Nat. Commun, 6*, 6150.

Maier, S. A. (2007). *Plasmonics: Fundamentals and Applications.* Springer, Berlin.

Maksymov, I. S. (2016). Magneto-plasmonic nanoantennas: Basics and applications. *Rev. Phys.*, *1* (Supplement C), 36–51.

Mansuripur, M., Connell, G. A. N., & Goodman, J. W. (1982). Signal and noise in magneto-optical readout. *J. Appl. Phys.*, *53*(6), 4485–4494.

Martin-Becerra, D., Gonzalez-Diaz, J. B., Temnov, V. V., Cebollada, A., Armelles, G., Thomay, T., ... Gonzalez, M. U. (2010). Enhancement of the magnetic modulation of surface plasmon polaritons in Au/Co/Au films. *Appl. Phys. Lett.*, *97*(18), 183114.

Melander, E., Östman, E., Keller, J., Schmidt, J., Papaioannou, E. T., Kapaklis, V., ... Hjörvarsson, B. (2012). Influence of the magnetic field on the plasmonic properties of transparent Ni anti-dot arrays. *Appl. Phys. Lett.*, *101*(6), 063107.

Oppeneer, P. M. (2001). Magneto-optical Kerr spectra. In K. H. J. Buschow (Ed.), *Handbook of Magnetic Materials, vol. 13* (Vol. 13, pp. 229–422). Elsevier Science B.V., Amsterdam.

Otto, A. (1968) Excitation of nonradiative surface plasma waves in silver by the method of frustrated total reflection. *Zeitschrift für Physik A Hadrons and Nuclei*, *216*(4), 398–410.

Papaioannou, E. T., Fang, H., Caballero, B., Akinoglu, E. M., Giersig, M., García-Martín, A., & Fumagalli, P. (2017) Role of interactions in the magneto-plasmonic response at the geometrical threshold of surface continuity. *Opt. Express*, *25*(26), 32792–32799.

Papaioannou, E. T., Kapaklis, V., Melander, E., Hjörvarsson, B., Pappas, S. D., Patoka, P., ... Ctistis, G. (2011) Surface plasmons and magneto-optic activity in hexagonal ni anti-dot arrays. *Opt. Express*, *19*(24), 23867–23877.

Papaioannou, E. T., Kapaklis, V., Patoka, P., Giersig, M., Fumagalli, P., Garcia-Martin, A., ... Ctistis, G. (2010) Magneto-optic enhancement and magnetic properties in Fe antidot films with hexagonal symmetry. *Phys. Rev. B*, *81*(5), 054424.

Papaioannou, E. T., Meyer, T., & Hillebrands, B. (2014). Magneto-optical enhancement in Co/Au patterned nanostructures. *J. Surf. Interfaces Mater.*, *2*, 40–45.

Pendry, J. B. (2000). Negative refraction makes a perfect lens. *Phys. Rev. Lett.*, *85*, 3966–3969.

Pimenov, I., & Kurin, V. (2012) Theory of magneto-optical effects in nanostructured ferromagnetics. *J. Opt. Soc. Am. B*, *29*(7), 1815–1821.

Raether, H. (1988). *Surface Plasmons on Smooth and Rough Surfaces and on Gratings* (Vol. 111). Springer, Berlin.

Razdolski, I., Gheorghe, D. G., Melander, E., Hjörvarsson, B., Patoka, P., Kimel, A. V., ... Rasing, T. (2013) Nonlocal nonlinear magneto-optical response of a magnetoplasmonic crystal. *Phys. Rev. B*, *88*, 075436.

Reim, W., & Schoenes, J. (1990). Magneto-optical spectroscopy of f-electron systems. In E. P. Wohlfarth & K. H. J. Buschow (Eds.), *Ferromagnetic Materials, a Handbook on the Properties of Magnetically Ordered Substances* (Vol. 5, p. 133). Elsevier, North-Holland, Amsterdam.

Rollinger, M., Thielen, P., Melander, E., Östman, E., Kapaklis, V., Obry, B., ... Papaioannou, E. T. (2016). Light localization and magneto-optic enhancement in ni antidot arrays. *Nano Lett.*, *16*(4), 2432–2438.

Rubio-Roy, M., Vlasin, O., Pascu, O., Caicedo, J. M., Schmidt, M., Goi, A. R., ... Herranz, G. (2012). Magneto-optical enhancement by plasmon excitations in nanoparticle/metal structures. *Langmuir*, *28*(24), 9010–9020.

Sapozhnikov, M. V., Gusev, S. A., Rogov, V. V., Ermolaeva, O. L., Troitskii, B. B., Khokhlova, L. V., & Smirnov, D. A. (2010). Magnetic and optical properties of nanocorrugated co films. *Appl. Phys. Lett.*, *96*(12).

Schoenes, J. (1991). In P. H. R. W. Kahn & E. J. Kramer (Eds.), (Vol. 3). Springer, Berlin.

Schoenes, J., Hüsser, O., Reim, W., Kaldis, E., & Wachter, P. (1985). Magneto-optics of semiconducting, intermediate valent and metallic Tm-chalcogenides. *J. Magn. Magn. Mater.*, *47&48*, 481–484.

Temnov, V. V., Armelles, G., Woggon, U., Guzatov, D., Cebollada, A., García-Martín, A., ... Bratschitsch, R. (2010). Active magneto-plasmonics in hybrid metal-ferromagnet structures. *Nat. Photonics*, *4*, 107–111.

Torrado, J. F., Papaioannou, E. T., Ctistis, G., Patoka, P., Giersig, M., Armelles, G., & García-Martín, A. (2010). Plasmon induced modification of the transverse magneto-optical response in Fe antidot arrays. *Phys. Status Solidi (RRL)*, *4*(10), 271–273.

van Oosten, D., Spasenović, M., & Kuipers, L. (2010). Nanohole chains for directional and localized surface plasmon excitation. *Nano Lett.*, *10*(1), 286–290.

Vogelgesang, R., & Dmitriev, A. (2010). Real-space imaging of nanoplasmonic resonances. *Analyst*, *135*(6), 1175–1181.

Wood, R. W. (1902). Xlii. On a remarkable case of uneven distribution of light in a diffraction grating spectrum. *Philos. Mag.*, *4*(21), 396–402.

Wood, R. W. (1912). Xxvii. Diffraction gratings with controlled groove form and abnormal distribution of intensity. *Philos. Mag.*, *23*(134), 310–317.

Wood, R. W. (1935). Anomalous Diffraction gratings. *Phys. Rev.*, *48*, 928–936.

Zhu, W., Esteban, R., Borisov, A. G., Baumberg, J. J., Nordlander, P., Lezec, H. J., ... Crozier, K. B. (2016) Quantum mechanical effects in plasmonic structures with subnanometre gaps. *Nat. Commun.*, 7, 11495.

Zvezdin, A., & Kotov, V. A. (1997). *Modern Magnetooptics and Magnetooptical Materials*. Taylor & Francis Group, New York.

Zvezdin, A. K., & Belotelov, V. I. (2004). Magnetooptical properties of two dimensional photonic crystals. *Eur. Phys. J. B*, *37*(4), 479–487.

Cathodoluminescence of Nanoplasmonics

18.1 Introduction... **18**-1
18.2 Fundamentals **18**-2
 Surface Plasmon Polariton • Localized Surface Plasmon
18.3 Cathodoluminescence **18**-12
 Instrumentation of STEM-CL • CL Measurement • Imaging Mechanism of
 Photon Map
18.4 Application to LSP-Related Phenomena................................ **18**-17
 Thin Triangular Nanoprism • Nanoparticle
18.5 Application to SPP-Related Phenomena............................... **18**-24
 SPP–Light Conversion by Surface Step • One-Dimensional Plasmonic Crystal
 • Two-Dimensional Plasmonic Crystals
18.6 Conclusion **18**-33
References **18**-33

Naoki Yamamoto and
Takumi Sannomiya
Tokyo Institute of Technology

18.1 Introduction

Electrons in materials cause group oscillation as a density wave, which can be quantized and called "plasmon." At the surface of materials, a similar charge oscillation mode is called "surface plasmon: SP." Surface plasmons can be categorized into "localized surface plasmon: LSP," which is localized at the surface of metallic nanoparticles and "surface plasmon polariton: SPP," which propagates along the surface of metal or the interface between metal and dielectric. Both LSP and SPP accommodate electromagnetic wave at the vicinity of the metal surface. SPP was first predicted theoretically by Ritchie (1957) and has been experimentally proved by light absorption and electron energy loss spectroscopy (EELS) followed by a number of other studies (Raether, 1988). SPPs propagating on a purely flat surface cannot interact with photons, or free-space light, because of momentum mismatch. By incorporating a prism or grating, an interexchange between SPPs and photons becomes possible.

The breakthrough of research on SPs was the discovery of extraordinary light transmission in an array of holes with a subwavelength diameter in a metallic film discovered by Ebbesen et al. (1998). Since then various studies have been conducted and spotlighted. The unique feature of SP is the strong field localization and enhancement of optical properties. For instance, high-sensitivity biosensors for organic and biomolecules have been developed and already close to practical use, which utilizes locally enhanced electromagnetic field at the vicinity of the metallic nanoparticles or metallic nanostructures (Homola, 2008). Because SP can confine the electromagnetic field at nanoscales below the diffraction

limit of light, waveguide applications in optical integrated circuits have also been proposed. Other applications include enhancement of energy conversion, such as high-efficiency light emitting diode (LED) or solar cells using metallic nanostructures. Thus, SP-related technologies taking advantage of its integration possibility and fast optical response have been rapidly expanding, which is now called as "plasmonics" (Barnes et al., 2004; Maier, 2007; Brongersma and Kik, 2007).

For the interexchange of SPP and photon, a structured metal surface with a periodicity close to or below the wavelength of light has been used (Heitmann, 1977). On a periodically structured surface, the dispersion relation of SPP resembles that of electrons in a crystal, which can be folded into a Brillouin zone, letting the SPP dispersion inside the light line, giving also a bandgap. SPPs with the energy in this bandgap cannot propagate on such a periodic metal surface. Instead, the density of state of SPPs at the band edge energy becomes large, increasing the probability of conversion to photons. Such a metal surface with a periodic structure is called "plasmonic crystals" and considered as an important building block to design SPP mirror, waveguide, cavity, and related sensor applications.

To realize SPP-photon conversion or SPP optical element using plasmonic crystals, proper control and tuning of the band edge energy and spatial distribution of the mode density are required. However, purely optical means do not have enough spatial resolution to observe such optical modes and cannot provide satisfactory information. Instead of photons, accelerated electrons (electron beam) can excite SPPs and have been used to study dispersion relation of SPP (Yamamoto, 2012). More recently, cathodoluminescence (CL) techniques based on scanning electron microscopy (SEM) or scanning transmission electron microscopy (STEM) have been applied to study, e.g., LSP radiation from particles or holes, SPP propagation, and dispersion relation of SPP on plasmonic crystals (Yamamoto, 2016).

In this chapter, we focus on the STEM-CL technique. First we introduce the basic theory of SP, then describe the measurement technique of STEM-CL, and finally show examples of LSP and SPP measurement and analysis.

18.2 Fundamentals

18.2.1 Surface Plasmon Polariton

Electromagnetic Field at a Plane Interface of Metal and Dielectric

Here we consider the induced surface charge wave accompanied by electromagnetic field localized at the interface between dielectric (medium 1) and metal (medium 2) (Figure 18.1). First of all, the electromagnetic field in each medium is deduced from Maxwell's equations, and the electromagnetic field in both media is coupled by the boundary condition at the interface. If the permittivity of the medium is spatially uniform and there is no space charge or current, the electromagnetic field in each medium is expressed by the following Maxwell's equations.

$$\nabla \cdot \boldsymbol{D} = 0 \tag{18.1a}$$

$$\nabla \cdot \boldsymbol{B} = 0 \tag{18.1b}$$

$$\nabla \times \boldsymbol{E} = -\frac{\partial \boldsymbol{B}}{\partial t} \tag{18.1c}$$

$$\nabla \times \boldsymbol{H} = \frac{\partial \boldsymbol{D}}{\partial t} \tag{18.1d}$$

Here, we consider a propagating wave in the x direction parallel to the interface. The expression of this wave includes a factor $\exp i(k_x x - \omega t)$, where k_x is the x component of

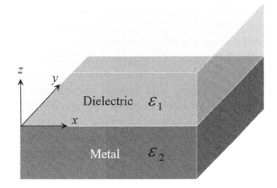

z
y
x
Dielectric ε_1
Metal ε_2

FIGURE 18.1 Boundary between metal and dielectric.

the wave vector and ω is the frequency. By applying the conditions $\frac{\partial}{\partial t} = -i\omega$, $\frac{\partial}{\partial x} = ik_x$, and $\partial/\partial y = 0$ based on the uniformity in the y direction, Eqs. (18.1c) and (18.1d) become as follows.

$$\frac{\partial E_z}{\partial y} - \frac{\partial E_y}{\partial z} = i\omega\mu_0 H_x \rightarrow -\frac{\partial E_y}{\partial z} = i\omega\mu_0 H_x \tag{18.2a}$$

$$\frac{\partial E_x}{\partial z} - \frac{\partial E_z}{\partial x} = i\omega\mu_0 H_y \rightarrow \frac{\partial E_x}{\partial z} - ik_x E_z = i\omega\mu_0 H_y \tag{18.2b}$$

$$\frac{\partial E_y}{\partial x} - \frac{\partial E_x}{\partial y} = i\omega\mu_0 H_z \rightarrow ik_x E_y = i\omega\mu_0 H_z \tag{18.2c}$$

$$\frac{\partial H_z}{\partial y} - \frac{\partial H_y}{\partial z} = -i\omega\varepsilon_0\varepsilon E_x \rightarrow -\frac{\partial H_y}{\partial z} = -i\omega\varepsilon_0\varepsilon E_x \tag{18.2d}$$

$$\frac{\partial H_x}{\partial z} - \frac{\partial H_z}{\partial x} = -i\omega\varepsilon_0\varepsilon E_y \rightarrow \frac{\partial H_x}{\partial z} - ik_x H_z = -i\omega\varepsilon_0\varepsilon E_y \tag{18.2e}$$

$$\frac{\partial H_y}{\partial x} - \frac{\partial H}{\partial y} = -i\omega\varepsilon_0\varepsilon E_z \rightarrow ik_x H_y = -i\omega\varepsilon_0\varepsilon E_z, \tag{18.2f}$$

where ε_0 and μ_0 are the permittivity and permeability in vacuum, ε is the relative dielectric constant of the medium, and the relative magnetic permeability equals 1, assuming a nonmagnetic medium. The expressions (18.2a) to (18.2f) are divided into a set of $\{H_y, E_x, E_z\}$ and $\{E_y, H_x, H_z\}$. The electromagnetic wave represented by the former combination is a transverse magnetic field (TM) mode, and the latter is called a transverse electric field (TE) mode. Since no TE mode solution exists for a guided SPP wave on a planer interface, we only treat TM modes. From Eqs. (18.2d) and (18.2f), the electric field components can be expressed by the magnetic field component H_y as

$$E_x = -\frac{i}{\omega\varepsilon_0\varepsilon}\frac{\partial H_y}{\partial z} \tag{18.3}$$

$$E_z = -\frac{k_x}{\omega\varepsilon_0\varepsilon}H_y \tag{18.4}$$

Substitution of these into Eq. (18.2b) yields an equation to solve for H_y,

$$\frac{\partial^2 H_y}{\partial z^2} - \left(k_x^2 - \varepsilon\frac{\omega^2}{c^2}\right)H_y = 0. \tag{18.5}$$

Here we define

$$\kappa^2 = k_x^2 - \varepsilon\frac{\omega^2}{c^2}, \tag{18.6}$$

and assume the following form as the solution

$$H_y(\boldsymbol{r}, t) = F(z)\exp[i(k_x x - \omega t)]. \tag{18.7}$$

For the surface normal component ($F(z)$), the general solution is

$$F(z) = C^+\exp(\kappa z) + C^-\exp(-\kappa z). \tag{18.8}$$

Since the two exponential terms correspond to the upward and downward waves, we should select one physically acceptable solution that properly represents the attenuation of the wave from the interface. When κ^2 is real and negative, κ becomes a pure imaginary number representing electromagnetic waves propagating in the z direction like light in free space. When κ^2 is real and positive, the electromagnetic field extending in the z direction suffers pure exponential attenuation. Therefore, as an ideal surface propagating wave localized at the interface, we can consider only the case that κ^2 is positive (in reality, κ^2 is complex with a negative real component).

Dispersion Relation of SPP

Here we again consider the configuration in which the upper half space ($z > 0$) is occupied by a dielectric (medium 1) with a dielectric constant ε_1, and the lower half space ($z < 0$) is occupied by a metal (medium 2) with a dielectric constant $\varepsilon_2(\omega)$ (Figure 18.1). From the results of Eqs. (18.7) and (18.8), the y component of the magnetic field in each medium is expressed using the amplitudes $A1$ and $A2$ as

$$H_y^{(1)}(\boldsymbol{r}, t) = A1 \exp\left[i\left(k_x x - \omega t\right)\right]\exp\left(-\kappa_1 z\right)(z > 0) \tag{18.9a}$$

$$H_y^{(2)}(\boldsymbol{r}, t) = A2 \exp\left[i\left(k_x x - \omega t\right)\right]\exp(\kappa_2 z)(z < 0), \tag{18.9b}$$

where

$$\kappa_1^2 = k_x^2 - \varepsilon_1 \frac{\omega^2}{c^2} \tag{18.10a}$$

$$\kappa_2^2 = k_x^2 - \varepsilon_2 \frac{\omega^2}{c^2}. \tag{18.10b}$$

To match the phase at the interface between the waves of medium 1 and 2, the wave number k_x should be common in both media (Snell's law). From Eqs. (18.3) and (18.4), the electric field components of the TM mode in each medium are written as

$$E_x^{(1)}(\boldsymbol{r}, t) = i\frac{\kappa_1}{\omega\varepsilon_0\varepsilon_1}H_y^{(1)}(\boldsymbol{r}, t) \tag{18.11a}$$

$$E_z^{(1)}(\boldsymbol{r}, t) = -\frac{k_x}{\omega\varepsilon_0\varepsilon_1}H_y^{(1)}(\boldsymbol{r}, t) \tag{18.11b}$$

$$E_x^{(2)}(\boldsymbol{r}, t) = -i\frac{\kappa_2}{\omega\varepsilon_0\varepsilon_2}H_y^{(2)}(\boldsymbol{r}, t) \tag{18.12a}$$

$$E_z^{(2)}(\boldsymbol{r}, t) = -\frac{k_x}{\omega\varepsilon_0\varepsilon_2}H_y^{(2)}(\boldsymbol{r}, t) \tag{18.12b}$$

Equations (18.9), (18.11), and (18.12) represent electromagnetic waves localized near the interface with an exponential decay in the z direction and propagating in the x direction (evanescent waves).

The boundary conditions of the electromagnetic field at the interface at $z = 0$ are expressed as

$$\boldsymbol{E}^{(1)} \times \boldsymbol{n} = \boldsymbol{E}^{(2)} \times \boldsymbol{n}\ (E_x^{(1)} = E_x^{(2)}) \tag{18.13}$$

$$\boldsymbol{D}^{(1)} \cdot \boldsymbol{n} = \boldsymbol{D}^{(2)} \cdot \boldsymbol{n}\ (\varepsilon_1 E_z^{(1)} = \varepsilon_2 E_z^{(2)}) \tag{18.14}$$

where $\boldsymbol{n}$ is a unit vector in the z direction normal to the boundary. Using Eq. (18.13) the ratio of the field amplitude $\frac{\kappa_1}{\varepsilon_1}A_1 = -\frac{\kappa_2}{\varepsilon_2}A_2$ is derived. Also from Eqs. (18.11) and (18.12), we obtain

$$k_x E_x^{(1)} + i\kappa_1 E_z^{(1)} = 0\ k_x E_x^{(2)} - i\kappa_2 E_z^{(2)} = 0, \tag{18.15}$$

Combining these with Eq. (18.13), we obtain the following equation at $z = 0$

$$\kappa_1 E_z^{(1)} + \kappa_2 E_z^{(2)} = 0. \tag{18.16}$$

Using Eq. (18.14), this can be changed to

$$\varepsilon_1\kappa_2 + \varepsilon_2\kappa_1 = 0 \tag{18.17}$$

Substituting Eqs. (18.10a) and (18.10b) to Eq. (18.17), the following equation is derived.

$$k_x = \frac{\omega}{c}\sqrt{\frac{\varepsilon_1\varepsilon_2}{\varepsilon_1 + \varepsilon_2}} \tag{18.18}$$

This is an equation expressing the dispersion relation of SPP at the planar interface. As an example, let us consider a Drude model ($\varepsilon_2(\omega) = 1 - (\omega_p/\omega)^2$) for metal (medium 2) and vacuum for dielectric (medium 1, $\varepsilon_1 = 1$). The imaginary term of the dielectric constant of the metal is ignored for simplicity. Figure 18.2a shows the dispersion curve of

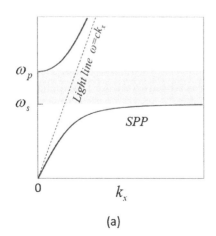

(a)

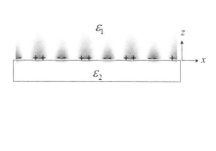

(b)

FIGURE 18.2 (a) Dispersion curve of SPP on the metal surface and (b) electric field distribution (E_z component).

SPP generated on the metal surface. When $\varepsilon_1 + \varepsilon_2 = 0$, k_x becomes infinite from Eq. (18.18). This happens at the frequency that satisfies the equation $\varepsilon_2(\omega_s) = -\varepsilon_1$, and ω_s is called the SP frequency. When using the Drude model, $\omega_s = \omega_p/\sqrt{2}$.

The dispersion curve of SPP lies in the frequency region of $0 < \omega < \omega_s$, where $\varepsilon_1 + \varepsilon_2$ and ε_2 are negative at the same time, and is outside the light dispersion line ($k_x = \omega/c$). The dispersion curve of SPP gradually approaches the dispersion line of light when ω is small and approaches ω_s as k_x increases. On the other hand, since k_x is an imaginary number in the region of $\omega_s < \omega < \omega_p$, the wave of Eq. (18.9) attenuates and cannot propagate in the x direction. For $\omega > \omega_p$, the dispersion line lies inside the light line, and from Eq. (18.10), both κ_1 and κ_2 become purely imaginary, so it represents freely propagating light in a medium, except that the mode frequency stays at the plasma frequency ω_p in the small k_x region (Brewster mode).

In the SPP region of $0 < \omega < \omega_s$, where the real part of ε_2 is negative, the following equations are derived by substituting Eq. (18.18) into (18.10a) and (18.10b)

$$\kappa_1 = \sqrt{\frac{\varepsilon_1}{-\varepsilon_2}} k_x, \quad \kappa_2 = \sqrt{\frac{-\varepsilon_2}{\varepsilon_1}} k_x \qquad (18.19)$$

By further substituting them into Eq. (18.15) at position $z = 0$,

$$E_z^{(1)} = i\frac{k_x}{\kappa_1} E_x^{(1)} = i\sqrt{\frac{-\varepsilon_2}{\varepsilon_1}} E_x^{(1)},$$

$$E_z^{(2)} = -i\frac{k_x}{\kappa_2} E_x^{(2)} = -i\sqrt{\frac{\varepsilon_1}{-\varepsilon_2}} E_x^{(2)} \qquad (18.20)$$

are obtained. $E_z^{(1)}$ and $E_z^{(2)}$ have opposite signs, which shows that the electric field is created by polarized charges at the boundary (Figure 18.2b). The surface charge density induced by SPP can be expressed as follows, using the modified relation from (18.1a) $\nabla \cdot \boldsymbol{D} = \nabla \cdot (\varepsilon_0 \boldsymbol{E} + \boldsymbol{P}) = \nabla \cdot (\varepsilon_0 \boldsymbol{E}) - \sigma_{\text{SPP}} = 0$

$$\sigma_{\text{SPP}}(x, y, t) = \varepsilon_0 \left(1 - \frac{\varepsilon_1}{\varepsilon_2}\right) E_z^{(1)}(x, y, 0, t) \qquad (18.21)$$

Dielectric Function of Metals

The frequency of light is very high; for example, the frequency in the visible light range is about 10^{15} Hz. The dielectric constant representing the response of electrons in a substance to such an oscillating electric field of light has a characteristic difference between a dielectric and a metal. In dielectric materials, electrons bound to atoms and molecules in the solid are displaced from the equilibrium position by the oscillating electric field to create electric dipoles, which generates macroscopic electric polarization $\boldsymbol{P}$. The dimensionless relative dielectric constant ε defined by $\boldsymbol{D} = \varepsilon\varepsilon_0\boldsymbol{E} = \varepsilon_0\boldsymbol{E} + \boldsymbol{P}$ is derived from a simple mechanical model (Kittel, 2004). When only one type of electric dipole with eigenfrequency ω_b associated with a bound electron is primarily induced, the dielectric constant depends on the frequency ω of light

$$\varepsilon_b(\omega) = \varepsilon(\infty) + \frac{(\varepsilon(0) - \varepsilon(\infty))\omega_b^2}{\omega_b^2 - \omega^2 - i\omega/\tau} \text{(Lorentz model)} \qquad (18.22)$$

Here, $\varepsilon(0)$ and $\varepsilon(\infty)$ are values when the frequency is 0 and ∞, and τ is the relaxation time of the oscillation. In addition, ω_b is defined by $\omega_b \equiv N_b\,e^2/m\,\varepsilon_0$, where N_b is the density of the bound electrons, e is the elementary charge, and m is the electron mass. Figure 18.3a shows the real and imaginary parts of the dielectric constant of Eq. (18.22) as a function of frequency. The imaginary part represents the magnitude of absorption of light by material and has the shape of a Lorentz function having a peak in the vicinity of eigenfrequency ω_b. Since the eigenfrequencies of many dielectrics are higher than the optical regions, the real part of the dielectric constant is approximately constant in the optical region and the imaginary part is small, so the solid becomes transparent to visible light.

In the case of metal, free electrons in metals respond to the oscillating electric field of light. Its dielectric constant is expressed as

$$\varepsilon_f(\omega) = 1 - \frac{\omega_p^2}{\omega(\omega + i/\tau)}, \text{(Drude model)} \qquad (18.23)$$

where ω_p is called plasma frequency and is defined by $\omega_p \equiv N_f\,e^2/m\varepsilon_0$ using density N_f of free electrons. This equation corresponds to the equation in which $\omega_b = 0$ in Eq. (18.22). Figure 18.3b shows the dielectric constant of Eq. (18.23). The real part of the dielectric function becomes negative in the range of $\omega < \omega_p$ and approaches 1 with an increase in ω in the range of $\omega > \omega_p$. The imaginary part is always positive, but increases as ω approaches zero.

In an actual metal, the dielectric constant includes the contribution of ε_f of free electrons and ε_b of bound electrons, and it varies complicatedly with respect to frequency in the optical region. As an example, Figure 18.3c–e shows the dielectric functions of metals frequently used in plasmonics. Aluminum (Al) shows a dependency that follows the Drude model well in the ultraviolet region, where the energy corresponding to the plasma frequency ($\hbar\omega_p$: bulk plasmon energy) is 15.8 eV, $\hbar/\tau = 0.6$ eV. Silver (Ag) nicely follows the Drude model in the low-energy region with a 5s electron where $\hbar\omega_p$ of the Eq. (18.23) is at 9.2 eV, but $\hbar\omega_p$ actually drops to 3.78 eV due to the influence of 4d electrons. Since the imaginary part of the dielectric constant is small, meaning that the dielectric loss is small in the optical region, silver is advantageous for the application of SPP. Gold (Au) has a 6s free electron with $\hbar\omega_p$ at 8.45 eV, but the influence of ε_b due to 5d electrons extends to the visible light region, and the dielectric constant shows a complicated change with respect to energy at 2 eV or more (Olmon et al., 2012).

Decay Length of SPP

Since the dielectric constant of a metal is generally complex, the propagation constant k_x in Eq. (18.18) becomes a complex number. When expressing the dispersion relation of SPP, the real part of the right side of Eq. (18.18)

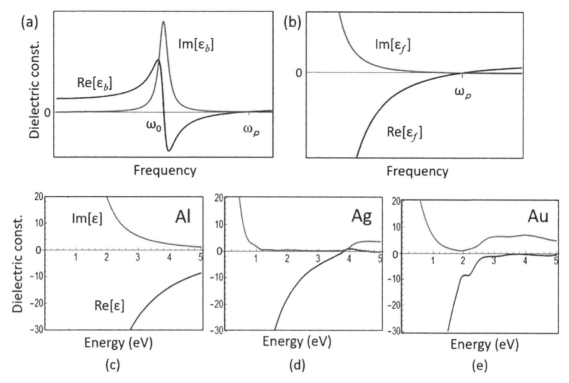

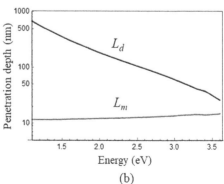

FIGURE 18.3   Dielectric function of materials. (a) Lorentz model (dielectric), (b) Drude model (metal), (c) aluminum, (d) silver, (e) gold, and (c–e) are from the data by Palik (1998).

should be considered. The imaginary part of the propagation constant k_x corresponds to the attenuation as the SPP travels. The SPP propagation length can be defined as

$$L_{\mathrm{SPP}} \equiv \frac{1}{2Im\,[k_x]}. \qquad (18.24)$$

Since the dielectric constant of a metal changes depending on energy, the propagation length also varies with energy. Figure 18.4a shows the energy dependence of the propagation length of SPP on the surface of Ag. The propagation length increases as the energy decreases, meaning that SPP with a longer wavelength can propagate longer distances. This corresponds to the fact that the penetration depth of the SPP electric field into the metal causing the dielectric loss varies with energy. From Eq. (18.9a) and (18.9b), the

penetration depths L_d and L_m of the SPP electric field into the dielectric and metal sides (z direction) are defined by

$$L_d \equiv \frac{1}{2\mathrm{Re}[\kappa_1]} \qquad (18.25a)$$

$$L_m \equiv \frac{1}{2\mathrm{Re}[\kappa_2]}, \qquad (18.25b)$$

where κ_1 and κ_2 are derived from Eqs. (18.10a), (18.10b), and (18.18).

$$\kappa_1^2 = -\left(\frac{\omega}{c}\right)^2 \frac{\varepsilon_1^2}{\varepsilon_1 + \varepsilon_2}, \kappa_2^2 = -\left(\frac{\omega}{c}\right)^2 \frac{\varepsilon_2^2}{\varepsilon_1 + \varepsilon_2} \qquad (18.26)$$

Figure 18.4b shows the result of calculating the energy dependence of the penetration depths L_d and L_m from the

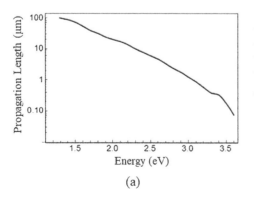

FIGURE 18.4   (a) Propagation length of SPP on silver surface and (b) penetration depth to vacuum and metal sides (L_d: vacuum side, L_m: silver side) calculated using the dielectric constant data of silver by Palik (1998).

Ag surface using Eqs. (18.25a) and (18.25b). The penetration depth to the Ag side is small in the order of 10 nm, whereas that to the vacuum side is large. As the energy increases, the penetration depth on the Ag side increases, correspondingly, the propagation length of SPP becomes shorter.

SPP Excitation by High-Energy Electrons

The SPP on the metal surface can be excited directly by the incidence of fast electrons. When high-energy electrons are incident on a substance, they lose energy by various excitation processes and cause inelastic scattering. Important excitation processes in the optical region include interband transitions of electrons in the solid and plasmon excitation, as well as characteristic photon emission processes for fast electrons such as Cherenkov radiation, transition radiation, and Smith-Purcell radiation (García de Abajo, 2010; Yamamoto et al., 1996, 2001b, 2015). EELS is a method to investigate the excited state in a substance using inelastically scattered electrons, and SPP has also been studied by this method. The inelastic scattering probability of electrons incident on a flat thin film from the normal direction was theoretically derived by Kröger (1968). The probability $P(k_x, \hbar\omega)$ that electrons lose energy of $\hbar\omega$ per unit length along the beam path and receive momentum change of k_x parallel to the surface is defined by using the energy loss W:

$$-\frac{dW}{dz} = \int dk_x \, 2\pi k_x \int d\omega \, \hbar\omega \, P(k_x, \hbar\omega) \qquad (18.27)$$

where $\hbar$ is a Planck's constant divided by 2π. We assume that a thin film of thickness d and dielectric constant ε_2 is surrounded by the medium with the dielectric constant ε_1, and the electron with velocity v is incident from the normal direction to the thin film. The inelastic scattering probability in this film was theoretically given by Kröger using the Gauss unit system as follows.

$$P(k_x, \hbar\omega) = \frac{e^2}{\pi^2 \hbar^2 v^2} \, \mathrm{Im} \left[\frac{1 - \varepsilon_2 \beta^2}{\varepsilon_2 \phi^2} d - \frac{2k_x^2 (\varepsilon_2 - \varepsilon_1)^2}{\phi_0^4 \phi^4} \right.$$
$$\left\{ \frac{\phi_{01}^4}{\varepsilon_1 \varepsilon_2} \left(\frac{\sin^2\left(\frac{\omega d}{2v}\right)}{L^+} + \frac{\cos^2\left(\frac{\omega d}{2v}\right)}{L^-} \right) + \beta^2 \frac{\kappa_1}{\varepsilon_1} \frac{\omega}{v} \phi_{01}^2 \right.$$
$$\times \left(\frac{1}{L^+} - \frac{1}{L^-} \right) \sin\frac{\omega d}{v} - \beta^4 \left(\frac{\omega}{v} \right)^2 \kappa_1 \kappa_2$$
$$\times \left. \left. \left(\frac{\cos^2\left(\frac{\omega d}{2v}\right) \tanh\left(\frac{\kappa_2 d}{2}\right)}{L^+} + \frac{\sin^2\left(\frac{\omega d}{2v}\right) \coth\left(\frac{\kappa_2 d}{2}\right)}{L^-} \right) \right\} \right]$$
$$(18.28)$$

Here $\beta = v/c$. κ_1 and κ_2 are already defined by Eq. (18.10). Other symbols are given as

$$L^+ = \kappa_1 \varepsilon_2 + \kappa_2 \varepsilon_1 \tanh\left(\frac{\kappa_2 d}{2}\right) \qquad (18.29a)$$

$$L^- = \kappa_1 \varepsilon_2 + \kappa_2 \varepsilon_1 \coth\left(\frac{\kappa_2 d}{2}\right) \qquad (18.29b)$$

$$\phi^2 = \kappa_2^2 + \left(\frac{\omega}{v}\right)^2, \quad \phi_0^2 = \kappa_1^2 + \left(\frac{\omega}{v}\right)^2,$$
$$\phi_{01}^2 = \kappa_1^2 + \left(\frac{\omega}{v}\right)^2 - (\varepsilon_1 + \varepsilon_2)\left(\frac{\omega}{v}\right)^2. \qquad (18.30)$$

The first term in the parenthesis in Eq. (18.28) is proportional to the film thickness and represents excitation in the bulk, such as bulk plasmon or Cherenkov radiation. The second term represents excitation at the surface and includes excitation of SPP and transition radiation.

In the limit where the film thickness d is infinite, the second term of Eq. (18.28) is simplified to

$$P(k_x, \hbar\omega) = \frac{e^2}{\pi^2 \hbar^2 v^2}$$
$$\times \mathrm{Im}\left[\frac{2k_x^2 (\varepsilon_2 - \varepsilon_1)^2}{\phi_0^4 \phi^4 L} \left\{ \frac{\phi_{01}^4}{\varepsilon_1 \varepsilon_2} - \beta^4 \left(\frac{2\pi}{\lambda\beta}\right)^2 \kappa_1 \kappa_2 \right\} \right],$$
$$(18.31)$$

where

$$L = \kappa_1 \varepsilon_2 + \kappa_2 \varepsilon_1. \qquad (18.32)$$

This represents the energy loss probability per unit length along the beam path when a single interface between the two media is present.

The energy loss probability due to surface excitation is shown in Figure 18.5a, which was calculated using Eq. (18.31) for an electron with an accelerating voltage of 80 kV ($\beta = 0.5$), where medium 1 is vacuum ($\varepsilon_1 = 1$) and medium 2 is silver. We use data by Palik (1998) for the dielectric constant of silver, which is shown in Figure 18.2d, where a strong contrast is seen along the dispersion line of SPP outside the light line (broken line). The intensity represents the magnitude of SPP excitation by the incident electron. This contrast can be understood from the fact that L in the expression (18.31) becomes 0 along the dispersion line where the dispersion relation of SPP of Eq. (18.17) is satisfied. As the energy becomes close to the SP energy $\hbar\omega_s = 3.68$ eV, the dispersion line goes away from the light line as k_x increases and the inclination approaches horizontal. The intensity distribution inside the light line is due to transition radiation.

Figure 18.5b shows the energy loss probability calculated using Eq. (18.28) for a silver film having a thickness of 50 nm under the same condition. It can be seen that the dispersion line of SPP is divided into two. The dispersion relation of each is represented by $L^+ = 0$ and $L^- = 0$ in Eq. (18.29), the one with the higher energy is called Long-Range (LR) mode and the lower one is called the Short-Range (SR) mode. In the LR-SPP mode, the electric field component in the propagation direction (E_x component) is weak in the thin film; therefore, the energy loss due to propagation is small and can propagate for long distances (Figure 18.6a). The relative excitation intensity of the two modes depends on the energy of incident electrons. At 80 keV, it is about the same, but at lower electron acceleration energy, the LR-SPP mode is more strongly excited, and at higher energy, the SR-SPP

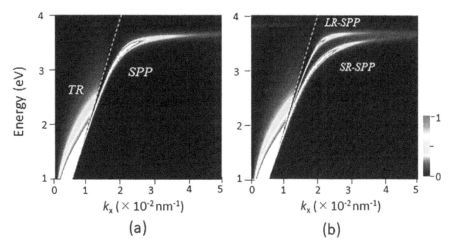

FIGURE 18.5 Electron energy loss probability on silver surface, (a) bulk silver surface, and (b) silver thin film (film thickness 50 nm). Electron energy is 80 keV.

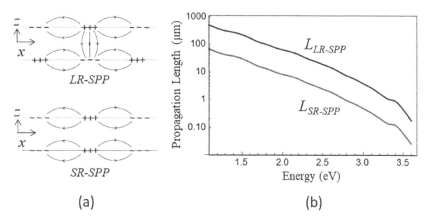

FIGURE 18.6 (a) Electric field distributions of LR-SPP and SR-SPP in silver thin film and (b) their propagation lengths (film thickness 50 nm).

mode is strongly excited. As the film thickness increases, the two dispersion lines approach each other, and when it exceeds 100 nm, it becomes almost the same as the dispersion line of the SPP on the bulk surface. Figure 18.6b shows the energy dependence of the propagation length of LR-SPP and SR-SPP in a silver thin film with a film thickness of 50 nm. The propagation lengths of both are different by one order of magnitude in the wide energy range, and the difference increases as the film thickness decreases.

18.2.2 Localized Surface Plasmon

Multipole Mode of Spherical Nanoparticle

LSPs excited in metal nanostructures are quantized vibration modes of surface charges accompanied with an electromagnetic field localized on the surface. Since the electric field of incident light is enhanced near the surface, and the resonance energy is sensitive to the dielectric properties of the medium surrounding the surface, LSP is applied to plasmonic devices such as sensors (Sannomiya, 2011), surface enhanced Raman spectroscopy (Schlücker, 2014),

and energy conversion (Clavero, 2014). Nanoparticle is the simplest representative of nanostructures, and some examples by CL measurement are shown in Section 18.4.

Light scattering by a metal sphere was rigorously treated by the classical electromagnetic theory based on Maxwell's equations (Mie, 1908). In Mie's theory, it was shown that there is an electromagnetic oscillation mode around the sphere. Figure 18.7 shows a schematic diagram of multipole modes. The multipole modes of the LSP in the metal sphere can be obtained by solving the Laplace equation under the spherical boundary condition if neglecting the retardation effect. When the diameter of the metal sphere is sufficiently smaller than the wavelength of light, the dipole mode dominates. Since the incident electric field $\boldsymbol{E}_0$ of the light can be regarded as being substantially uniform within the nanoparticle, the dipole moment excited in the metallic sphere by the electric field is expressed as

$$\boldsymbol{p} = 4\pi\varepsilon_0\varepsilon_\mathrm{d}a^3\frac{\varepsilon - \varepsilon_d}{\varepsilon + 2\varepsilon_d}\boldsymbol{E}_0, \qquad (18.33)$$

where a is the radius of the particle, ε_0 is the permittivity in vacuum, ε and ε_d are the dielectric constants of the metal

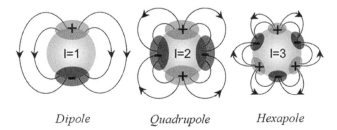

FIGURE 18.7 Illustration of multipole modes in a spherical nanoparticle of metal.

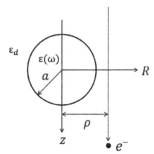

FIGURE 18.8 A schematic diagram of the geometry of nanoparticle and incident electron.

and the dielectric around the particle, respectively. The resonance condition of the dipole moment with respect to the incident electric field is written as

$$\varepsilon + 2\varepsilon_d = 0. \tag{18.34}$$

When the dielectric constant of the metal is expressed as a function of frequency, the resonance frequency of the ℓ-th order multipole mode is determined from the following equation.

$$\varepsilon(\omega_\ell) = -\frac{\ell+1}{\ell}\varepsilon_d \tag{18.35}$$

The dielectric constant of a metal is generally complex, and its real part takes a negative value at the frequency of the optical region. As shown in the dielectric function of silver in Figure 18.3d, silver is a rare substance with a plasma frequency (3.78 eV) in the optical region. The resonance energy of dipole mode ($\ell = 1$) of silver spherical nanoparticles is 3.50 eV in vacuum from Eq. (18.34) and the resonance energy of quadrupole mode ($\ell = 2$) is 3.67 eV. Eqs. (18.34) and (18.35) are valid when the diameter of the fine particle is sufficiently smaller than the wavelength of light, giving a limit value when the diameter approaches zero.

The resonance energy of the LSP in actual metal nanoparticles varies depending on the size of the nanoparticles. It was reported that the retardation effect becomes strong when the size of the nanoparticle increases, and the resonance energies of the multipole modes shift to the low-energy side (Yamamoto et al., 2001a).

In the relativistic treatment involving the retardation effect, EELS and radiation loss due to interaction between a metal nanoparticle and high-energy electron were theoretically treated by García de Abajo (1999). The radiation loss probability for photon energy $E = \hbar\omega$ takes the following form in SI unit when the electron passes at a distance of b from the sphere center, as shown in Figure 18.8.

$$\Gamma^{\text{rad}}(E) = \frac{\alpha}{E}\sum_{\ell=1}^{\infty}\sum_{m=-\ell}^{\ell}K_m^2\left(\frac{\omega b}{v\gamma}\right)\left[C_{\ell m}^M\left|t_\ell^M\right|^2 + C_{\ell m}^E\left|t_\ell^E\right|^2\right] \tag{18.36}$$

This formula represents the emission spectrum obtained by integrating the radiation intensity at all solid angles. Here α is the fine structure constant, v is the electron velocity, γ is the Lorentz factor ($\gamma = 1/\sqrt{1-\beta^2}$), K_m is the modified Bessel function of m-th order, and t_ℓ^M and t_ℓ^E are the

scattering-matrix elements for magnetic and electric modes in the Mie theory. The coefficients $C_{\ell m}^M$ and $C_{\ell m}^E$ depend exclusively on β, and (ℓ, m) is a subscript specifying the SP mode excited on the spherical surface; $\ell = 1$ indicates the dipole mode, and $\ell = 2$ indicates the quadrupole mode. The first term in the parenthesis represents the contribution of the magnetic multipole mode, and the second term represents that of the electric multipole mode. When the particle size is sufficiently smaller than the wavelength of light, the second term is dominant, so for simplicity, only the electric multipole mode is considered here. The peak in the emission spectrum of the particle appears mainly at the resonance energy position that maximizes $\left|t_\ell^E\right|^2$.

Radiation of Multipole Mode Excited by Fast Electrons

The scalar function that gives the electric multipole field generated by an electron incident on the position $\boldsymbol{R} = (\rho, \varphi_0)$ is represented by the following series expansion.

$$\psi^{E,\text{ext}}(\boldsymbol{r}, \boldsymbol{R}) = \sum_{\ell=1}^{\infty}\sum_{m=-\ell}^{\ell}i^\ell j_\ell(kr)Y_{\ell m}(\Omega)\psi_{\ell m}^{E,\text{ext}}(\boldsymbol{R}) \tag{18.37}$$

The electric field created by the incident electron induces surface charges on the metal nanoparticle, and the surface charge oscillation further causes the emission of electromagnetic waves. The scalar function of the radiation field is obtained by using the scattering matrix element in the Mie theory.

$$\psi_{\ell m}^{E,\text{ind}}(\boldsymbol{R}) = t_\ell^E \psi_{\ell m}^{E,\text{ext}}(\boldsymbol{R}) \tag{18.38}$$

The scattering matrix element t_ℓ^E is expressed using the ℓ-th-order spherical Bessel function $j_\ell(x)$ and its differential $j_\ell'(x)$.

$$t_\ell^E = \frac{-\varepsilon_d j_\ell(\rho_0)[\rho_1 j_\ell(\rho_1)]' + \varepsilon[\rho_0 j_\ell(\rho_0)]' j_\ell(\rho_1)}{\varepsilon_d h_\ell^{(+)}(\rho_0)[\rho_1 j_\ell(\rho_1)]' - \varepsilon\left[\rho_0 h_\ell^{(+)}(\rho_0)\right]' j_\ell(\rho_1)} \tag{18.39}$$

Here $\rho_0 = \sqrt{\varepsilon_d}\omega a/c$, $\rho_1 = \sqrt{\varepsilon}\omega a/c$, ε_d is the dielectric constant of the surrounding medium ($\varepsilon_d = 1$ in vacuum), and $\varepsilon(\omega)$ is the dielectric constant of the metal. $h_l^{(+)}$ is expressed by $h_l^{(+)} = ih_l^{(1)}$ using the first kind spherical Hankel function $h_l^{(1)}$. Therefore t_ℓ^E depends on the radius a of the sphere and the dielectric constant, and does not depend on the electron velocity or position.

The energy of the peak appearing in the emission spectrum corresponds to the resonance energy that maximizes $\left|t_\ell^E\right|^2$ in the ℓ-th-order multipole mode and maximizes the scattering matrix element t_ℓ^E in Eq. (18.39). It is determined from the following condition.

$$\varepsilon_d h_\ell^{(+)}(\rho_0) \frac{d\left[\rho_1 j_\ell(\rho_1)\right]}{d\rho_1} - \varepsilon \frac{d\left[\rho_0 h_\ell^{(+)}(\rho_0)\right]}{d\rho_0} j_\ell(\rho_1) = 0 \tag{18.40}$$

This equation explains that the emission peak of the multipole mode shifts to the lower energy side as the radius of the sphere increases. However, when the mode peak becomes broad as the particle size increases, the emission peak energy shifts slightly toward the lower energy side than the resonance energy, because the factor of $K_m(\frac{\omega\rho}{v\gamma})$, which is explained later, is applied.

The coefficient $\psi_{\ell m}^{E,\text{ext}}(\boldsymbol{R})$ of the electric multipole field generated by the incident electron is obtained in the series expansion of the scalar function (García de Abajo, 1999).

$$\psi_{\ell m}^{E,\text{ext}}(\boldsymbol{R}) = \frac{-2\pi i^{1-\ell} k}{c\gamma} \frac{B_{\ell m}}{\ell(\ell+1)} K_m\left(\frac{\omega\rho}{v\gamma}\right) e^{-im\varphi_0 - i\omega z_0/v} \tag{18.41}$$

It is expressed using atomic units according to the paper. In this equation, $\boldsymbol{R} = (\rho, \varphi_0, z_0)$ is taken as the reference point on the electron path, but we take $z_0 = 0$ and ignore z dependence for simplicity. Then $\boldsymbol{R}$ is specified only by the parameter (ρ, φ_0) in the two-dimensional (2D) plane at $z_0 = 0$. Coefficient $B_{\ell m}$ depends only on electron velocity v. The induced field function $\psi^{E,\text{ind}}(\boldsymbol{r}, \boldsymbol{R})$ of the radiation field can be written as follows by substituting $\psi_{\ell m}^{E,\text{ind}}$ instead of $\psi_{\ell m}^{E,\text{ext}}$ in Eq. (18.37) using an outgoing wave.

$$\psi^{E,\text{ind}}(\boldsymbol{r}, \boldsymbol{R}) = \sum_{\ell=1}^{\infty} \sum_{m=-\ell}^{\ell} i^\ell h_\ell^{(+)}(kr) Y_{\ell m}(\Omega) \psi_{\ell m}^{E,\text{ind}}(\boldsymbol{R})$$

$$= \sum_{\ell=1}^{\infty} \sum_{m=-\ell}^{\ell} i^\ell h_\ell^{(+)}(kr) Y_{\ell m}(\Omega) t_\ell^E \psi_{\ell m}^{E,\text{ext}}(\boldsymbol{R}) \tag{18.42}$$

This is rewritten using Eq. (18.41) as

$$\psi^{E,\text{ind}}(\boldsymbol{r}, \boldsymbol{R}) = \frac{-2\pi i k}{c\gamma} \sum_{\ell=1}^{\infty} \frac{1}{\ell(\ell+1)} t_\ell^E h_\ell^{(+)}(kr)$$

$$\sum_{m=-\ell}^{\ell} B_{\ell m} K_m\left(\frac{\omega\rho}{v\gamma}\right) Y_{\ell m}(\Omega) e^{-im\varphi_0}, \tag{18.43}$$

where $\Omega = (\theta, \varphi)$ represents the angle of the emission direction of photons. Considering each term in Eq. (18.42) into two parts, i.e., $\psi_{\ell m}^{E,\text{ind}}(\boldsymbol{R})$ and $h_\ell^{(+)}(kr)Y_{\ell m}(\Omega)$. The former

part represents the excitation amplitude of the (ℓ, m) multipole mode by the incident electron, and the remaining part can be considered as a radiated electromagnetic field created by that mode, because $h_\ell^{(+)}(kr)$ represents a spherical wave sufficiently far from the metal sphere.

The radiation electromagnetic field emitted by the induced charge oscillation without magnetic multipoles is approximated as

$$\boldsymbol{H} \approx -\boldsymbol{L}\psi^E \tag{18.44}$$

$$\boldsymbol{E} \approx -\frac{i}{k}\nabla \times \boldsymbol{L}\psi^E = Z_0 \boldsymbol{H} \times \boldsymbol{n}, \tag{18.45}$$

where $\boldsymbol{L}$ is a vector operator defined by $\boldsymbol{L} \equiv -i\boldsymbol{r} \times \nabla$, $\boldsymbol{n}$ is a radial unit vector, and Z_0 is a vacuum impedance. We now introduce a vector spherical harmonics function $\boldsymbol{X}_{\ell m}(\Omega)$ defined as

$$\boldsymbol{X}_{\ell m}(\theta, \varphi) = \frac{1}{\sqrt{\ell(\ell+1)}} \boldsymbol{L} Y_{\ell m}(\theta, \varphi). \tag{18.46}$$

By inserting Eq. (18.43) into (18.44) and using $\boldsymbol{X}_{\ell m}(\Omega)$ as the angular component of the magnetic field $\boldsymbol{H}$, the magnetic and electric field of the (ℓ, m) mode can be expressed as

$$\boldsymbol{H}_{\ell m}(\boldsymbol{r}) \propto h_\ell^{(+)}(kr) \boldsymbol{X}_{\ell m}(\Omega) \tag{18.47}$$

$$\boldsymbol{E}_{\ell m}(\boldsymbol{r}) \approx Z_0 \boldsymbol{H}_{\ell m} \times \boldsymbol{n}. \tag{18.48}$$

Here the excitation components are omitted. The radiation energy flow of the (ℓ, m) mode is obtained by taking the time average of the Poynting vector at a sufficiently far distance (by omitting the radial component):

$$\langle \boldsymbol{S}(\boldsymbol{r}) \rangle = \frac{1}{2} \text{Re}\left[\boldsymbol{E}_{\ell m}(\boldsymbol{r}) \times \boldsymbol{H}_{\ell m}(\boldsymbol{r})\right]$$

$$= \frac{Z_0}{2} |\boldsymbol{H}_{\ell m}(\boldsymbol{r})|^2 \boldsymbol{n} \propto \frac{Z_0}{2(kr)^2} |\boldsymbol{X}_{\ell m}(\Omega)|^2 \boldsymbol{n} \tag{18.49}$$

The radiation intensity per unit solid angle $\Omega = (\theta, \varphi)$ should be proportional to the time average Poynting vector and therefore also to $|\boldsymbol{X}_{\ell m}(\Omega)|^2$:

$$I_{\ell m}(\theta, \varphi) = r^2 \langle \boldsymbol{S}(\boldsymbol{r}) \rangle \cdot \boldsymbol{n} \propto \frac{Z_0}{2k^2} |\boldsymbol{X}_{\ell m}(\Omega)|^2. \tag{18.50}$$

Because the (ℓ, m) mode is degenerate for m, the total radiation intensity is the squared magnitude of the sum of the radiation of the degenerate modes multiplied by the excitation amplitude $\psi_{\ell m}^{E,\text{ind}}(\boldsymbol{R})$. The radiation intensity of the ℓth mode reads as

$$I_\ell(\theta, \varphi, \boldsymbol{R}) \propto \left|t_\ell^E\right|^2 \left|\sum_{m=-\ell}^{\ell} B_{\ell m} K_m\left(\frac{\omega\rho}{v\gamma}\right) \boldsymbol{X}_{\ell m}(\theta, \varphi) e^{-im\varphi_0}\right|^2. \tag{18.51}$$

For the expression of the eigenmodes, we use the following conversion.

$$\boldsymbol{X}_{\ell m}^+ = \frac{1}{\sqrt{2}}(\boldsymbol{X}_{\ell m} + \boldsymbol{X}_{\ell-m}) \tag{18.52a}$$

$$\boldsymbol{X}_{\ell m}^- = \frac{1}{\sqrt{2i}}(\boldsymbol{X}_{\ell m} - \boldsymbol{X}_{\ell-m}) \tag{18.52b}$$

As the relation $\boldsymbol{X}_{\ell-m} = \pm \boldsymbol{X}_{\ell m}^*$ (+ sign for $m =$ odd, - sign for $m =$ even) holds, both $\boldsymbol{X}_{\ell m}^+$ and $\boldsymbol{X}_{\ell m}^-$ are real vectors for $m =$ odd, and both imaginary vectors for $m =$ even. By using this relation, Eq. (18.51) can be rewritten as

$$I_{\text{non-pol}}(\theta, \varphi, \boldsymbol{R}) \propto \left| \sum_{\substack{m=-\ell \\ m=\text{even}}}^{\ell} C_{\ell m}(\boldsymbol{R}) \left\{ \boldsymbol{X}_{\ell m}^+(\theta, \varphi) \right.\right.$$

$$\left. \cos m\varphi_0 + \boldsymbol{X}_{\ell m}^-(\theta, \varphi) \sin m\varphi_0 \right\} + i \sum_{\substack{m=-\ell \\ m=\text{odd}}}^{\ell} C_{\ell m}(\boldsymbol{R})$$

$$\left. \left\{ \boldsymbol{X}_{\ell m}^-(\theta, \varphi) \cos m\varphi_0 - \boldsymbol{X}_{\ell m}^+(\theta, \varphi) \sin m\varphi_0 \right\} \right|^2. \quad (18.53)$$

The first summation term is for even m, and the second for odd m. For $m = 0$, $\boldsymbol{X}_{\ell 0}^- = 0$ holds for all ℓ. Also the following expression and relations are used.

$$C_{\ell m}(\boldsymbol{R}) = \sqrt{2} t_\ell^E B_{\ell m} K_m \left(\frac{\omega \rho}{v \gamma} \right) \quad (18.54)$$

$$K_m(\xi) = K_{-m}(\xi) \quad (18.55)$$

$$B_{\ell-m} = -B_{\ell m}(\text{for } m = \text{odd}), B_{\ell-m} = B_{\ell m}(\text{for } m = \text{even}) \quad (18.56)$$

$C_{\ell m}(\boldsymbol{R})$ can be considered as a coefficient expressing the excitation amplitude of each mode. Equation (18.53) shows the radiation intensity for nonpolarized light. To decompose it to the polarized light, a scalar product of the polarization unit vector $\boldsymbol{e}$ and the electric field vector should be calculated. We here note that $\boldsymbol{X}$ expresses the magnetic field and that electric field is given by the vector product of the magnetic field and the unit vector $\boldsymbol{n}$ in the radiation direction. The polarized radiation intensity is now given as

$$I_{e-\text{pol}}(\theta, \varphi, \boldsymbol{R}) \propto \left| \sum_{\substack{m=-\ell \\ m=\text{even}}}^{\ell} C_{\ell m}(\boldsymbol{R}) \left\{ \boldsymbol{X}_{\ell m}^+(\theta, \varphi) \right.\right.$$

$$\left. \cos m\varphi_0 + \boldsymbol{X}_{\ell m}^-(\theta, \varphi) \sin m\varphi_0 \right\} \times \boldsymbol{n} \cdot \boldsymbol{e} + i \sum_{\substack{m=-\ell \\ m=\text{odd}}}^{\ell}$$

$$\left. C_{\ell m}(\boldsymbol{R}) \left\{ \boldsymbol{X}_{\ell m}^-(\theta, \varphi) \cos m\varphi_0 - \boldsymbol{X}_{\ell m}^+(\theta, \varphi) \sin m\varphi_0 \right\} \times \boldsymbol{n} \cdot \boldsymbol{e} \right|^2$$

$$(18.57)$$

Angular Distribution of Multipole Mode Radiation

The multipole radiation is described in detail by Jackson (1975). Here as examples, we show angular radiation field distribution of dipole and quadrupole modes with polarization. This result will be used in Section 18.4 for the analysis of photon mapping.

Dipole Mode ($\ell = 1$)

The dipole consists of three degenerate modes, namely $m = 1, 0, -1$. We first derive $\boldsymbol{X}_{1m}$ from $Y_{1m}(\theta, \varphi)$, and then calculate the radiation intensities. $Y_{1m}(\theta, \varphi)$ reads

$$Y_{10}(\theta, \varphi) = \sqrt{\frac{3}{4\pi}} \cos \theta \quad (18.58a)$$

$$Y_{\_}(1 \pm 1)(\theta, \varphi) = \sqrt{(3/8\pi)} \sin \theta e^{\pm i\varphi} \quad (18.58b)$$

Applying Eq. (18.46), we obtain

$$\boldsymbol{X}_{10} = \sqrt{\frac{3}{8\pi}} i \sin \theta \boldsymbol{e}_\varphi = \boldsymbol{X}_{10}^+ \quad (18.59a)$$

$$\boldsymbol{X}_{11} = \sqrt{\frac{3}{16\pi}} e^{i\varphi} \left\{ \boldsymbol{e}_\theta + i \cos \theta \boldsymbol{e}_\varphi \right\} \quad (18.59b)$$

$$\boldsymbol{X}_{1-1} = \boldsymbol{X}_{11}^* = \sqrt{\frac{3}{16\pi}} e^{-i\varphi} \left\{ \boldsymbol{e}_\theta - i \cos \theta \boldsymbol{e}_\varphi \right\}. \quad (18.59c)$$

The conversion of the basis function of Eq. (18.52) gives

$$\boldsymbol{X}_{11}^+ = \sqrt{\frac{3}{8\pi}} \left\{ \cos \varphi \boldsymbol{e}_\theta - \cos \theta \sin \varphi \boldsymbol{e}_\varphi \right\} \quad (18.60a)$$

$$\boldsymbol{X}_{11}^- = \sqrt{\frac{3}{8\pi}} \left\{ \sin \varphi \boldsymbol{e}_\theta + \cos \theta \cos \varphi \boldsymbol{e}_\varphi \right\}. \quad (18.60b)$$

Both $\boldsymbol{X}_{11}^+$ and $\boldsymbol{X}_{11}^-$ are real vectors. $\boldsymbol{e}_\theta$ and $\boldsymbol{e}_\varphi$ are polarization unit vectors and are perpendicular to the radiation direction (vector $\boldsymbol{n}$).

The resultant angular distributions of electric field from $\boldsymbol{X}_{10}^+$, $\boldsymbol{X}_{11}^+$ and $\boldsymbol{X}_{11}^-$ are expressed as

$$\boldsymbol{E}_{10}^+ \propto \boldsymbol{X}_{10}^+ \times \boldsymbol{n} \propto i \sin \theta \boldsymbol{e}_\theta \quad (18.61a)$$

$$\boldsymbol{E}_{11}^+ \propto \boldsymbol{X}_{11}^+ \times \boldsymbol{n} \propto -\cos \theta \sin \varphi \boldsymbol{e}_\theta - \cos \varphi \boldsymbol{e}_\varphi \quad (18.61b)$$

$$\boldsymbol{E}_{11}^- \propto \boldsymbol{X}_{11}^- \times \boldsymbol{n} \propto \cos \theta \cos \varphi \boldsymbol{e}_\theta - \sin \varphi \boldsymbol{e}_\varphi. \quad (18.61c)$$

Figure 18.9 illustrates the calculated angular distribution of the radiation intensity $\left| \boldsymbol{E}_{10}^+ \right|^2$, $\left| \boldsymbol{E}_{11}^+ \right|^2$ and $\left| \boldsymbol{E}_{11}^- \right|^2$.

This result clearly shows that $\boldsymbol{X}_{10}^+$, $\boldsymbol{X}_{11}^+$, and $\boldsymbol{X}_{11}^-$ correspond to the electric dipoles oscillating along the z, y, and x directions, respectively. Here polarization has not been taken into account. To decompose into $\boldsymbol{e}_\theta$ polarization, the polarized intensity is proportional to the square of the $\boldsymbol{e}_\theta$ component of the electric field, leading to

$$I_\theta \left(\boldsymbol{X}_{10}^+ \right) \propto \sin^2 \theta \quad (18.62a)$$

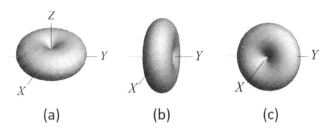

FIGURE 18.9 Angular distribution of radiation by dipole modes. (a) $\left|\boldsymbol{E}_{10}^{+}\right|^{2}$, (b) $\left|\boldsymbol{E}_{11}^{+}\right|^{2}$, and (c) $\left|\boldsymbol{E}_{11}^{-}\right|^{2}$.

$$I_{\theta}\left(\boldsymbol{X}_{11}^{+}\right) \propto \cos^{2}\theta \sin^{2}\varphi \qquad (18.62b)$$

$$I_{\theta}\left(\boldsymbol{X}_{11}^{-}\right) \propto \cos^{2}\theta \cos^{2}\varphi \qquad (18.62c)$$

In the same manner, $\boldsymbol{e}_{\varphi}$–polarized components are

$$I_{\varphi}\left(\boldsymbol{X}_{10}^{+}\right) \propto 0 \qquad (18.63a)$$

$$I_{\varphi}\left(\boldsymbol{X}_{11}^{+}\right) \propto \cos^{2}\varphi \qquad (18.63b)$$

$$I_{\varphi}\left(\boldsymbol{X}_{11}^{-}\right) \propto \sin^{2}\varphi \qquad (18.63c)$$

The charge distribution of the three dipoles are illustrated on the upper left in Figure 18.10, and their angular radiation distributions for nonpolarized, θ–polarized, and φ-polarized light on the upper right in Figure 18.10. As will be discussed in Section 18.4.2, in the CL measurement, these modes are simultaneously excited by the incident electron beam, and the measurement to separately detect each degenerate mode is generally difficult.

Quadrupole Mode ($\ell = 2$)

The quadrupole can be treated as for dipoles and is classified into five degenerate modes with different m. The spherical harmonics for quadrupoles are

$$Y_{20}\left(\theta,\varphi\right) = \sqrt{\frac{5}{16\pi}}(3\cos^{2}\theta - 1) \qquad (18.64a)$$

$$Y_{2\pm 1}\left(\theta,\varphi\right) = \mp\sqrt{\frac{15}{8\pi}}\sin\theta\cos\theta e^{\pm i\varphi} \qquad (18.64b)$$

$$Y_{2\pm 2}\left(\theta,\varphi\right) = \sqrt{\frac{15}{32\pi}}\sin^{2}\theta e^{\pm 2i\varphi} \qquad (18.64c)$$

The corresponding spherical vector harmonics can be derived from Eq. (18.46) as

$$\boldsymbol{X}_{20} = \sqrt{\frac{15}{8\pi}}i\sin\theta\cos\theta\boldsymbol{e}_{\varphi} = \sqrt{2}\boldsymbol{X}_{20}^{+} \qquad (18.65a)$$

$$\boldsymbol{X}_{2\pm 1} = \sqrt{\frac{5}{16\pi}}e^{\pm i\varphi}\left\{\cos\theta\boldsymbol{e}_{\theta} \pm i\cos 2\theta\boldsymbol{e}_{\varphi}\right\} \qquad (18.65b)$$

$$\boldsymbol{X}_{2\pm 2} = \mp\sqrt{\frac{5}{16\pi}}e^{\pm i2\varphi}\left\{\sin\theta\boldsymbol{e}_{\theta} \pm i\sin\theta\cos\theta\boldsymbol{e}_{\varphi}\right\} \qquad (18.65c)$$

The conversion of the basis function of Eq. (18.52) gives

$$\boldsymbol{X}_{20}^{+} = \frac{1}{\sqrt{2}}\boldsymbol{X}_{20} = \sqrt{\frac{15}{16\pi}}i\sin\theta\cos\theta\boldsymbol{e}_{\varphi} \qquad (18.66a)$$

$$\boldsymbol{X}_{21}^{+} = \sqrt{\frac{5}{8\pi}}\left\{\cos\theta\cos\varphi\boldsymbol{e}_{\theta} - \cos 2\theta\sin\varphi\boldsymbol{e}_{\varphi}\right\} \qquad (18.66b)$$

$$\boldsymbol{X}_{21}^{-} = \sqrt{\frac{5}{8\pi}}\left\{\cos\theta\sin\varphi\boldsymbol{e}_{\theta} + \cos 2\theta\cos\varphi\boldsymbol{e}_{\varphi}\right\} \qquad (18.66c)$$

$$\boldsymbol{X}_{22}^{+} = -i\sqrt{\frac{5}{8\pi}}\left\{\sin\theta\sin 2\varphi\boldsymbol{e}_{\theta} + \sin\theta\cos\theta\cos 2\varphi\boldsymbol{e}_{\varphi}\right\} \qquad (18.66d)$$

$$\boldsymbol{X}_{22}^{-} = -i\sqrt{\frac{5}{8\pi}}\left\{\sin\theta\cos 2\varphi\boldsymbol{e}_{\theta} - \sin\theta\cos\theta\sin 2\varphi\boldsymbol{e}_{\varphi}\right\} \qquad (18.66e)$$

The electric field distribution can then be calculated using $\boldsymbol{E}_{2m}^{\pm} \propto \boldsymbol{X}_{2m}^{\pm} \times \boldsymbol{n}$

$$\boldsymbol{E}_{20}^{+} \propto i\sqrt{3}\sin\theta\cos\theta\boldsymbol{e}_{\theta} \qquad (18.67a)$$

$$\boldsymbol{E}_{21}^{+} \propto -(\cos 2\theta\sin\varphi\boldsymbol{e}_{\theta} + \cos\theta\cos\varphi\boldsymbol{e}_{\varphi}) \qquad (18.67b)$$

$$\boldsymbol{E}_{21}^{-} \propto \cos 2\theta\cos\varphi\boldsymbol{e}_{\theta} - \cos\theta\sin\varphi\boldsymbol{e}_{\varphi} \qquad (18.67c)$$

$$\boldsymbol{E}_{22}^{+} \propto -i(\sin\theta\cos\theta\cos 2\varphi\boldsymbol{e}_{\theta} - \sin\theta\sin 2\varphi\boldsymbol{e}_{\varphi}) \qquad (18.67d)$$

$$\boldsymbol{E}_{22}^{-} \propto -i(\sin\theta cos\theta\sin 2\varphi\boldsymbol{e}_{\theta} + \sin\theta\cos 2\varphi\boldsymbol{e}_{\varphi}) \qquad (18.67e)$$

The polarization-decomposed expressions are

$$I_{\theta}\left(\boldsymbol{X}_{20}^{+}\right) \propto 3\sin^{2}\theta\cos^{2}\theta, I_{\varphi}\left(\boldsymbol{X}_{20}^{+}\right) = 0 \qquad (18.68a)$$

$$I_{\theta}\left(\boldsymbol{X}_{21}^{+}\right) \propto \cos^{2}2\theta\sin^{2}\varphi, I_{\varphi}\left(\boldsymbol{X}_{21}^{+}\right) \propto \cos^{2}\theta\cos^{2}\varphi \qquad (18.68b)$$

$$I_{\theta}\left(\boldsymbol{X}_{21}^{-}\right) \propto \cos^{2}2\theta\cos^{2}\varphi, I_{\varphi}\left(\boldsymbol{X}_{21}^{-}\right) \propto \cos^{2}\theta\sin^{2}\varphi \qquad (18.68c)$$

$$I_{\theta}\left(\boldsymbol{X}_{22}^{+}\right) \propto \sin^{2}\theta\cos^{2}\theta\cos^{2}2\varphi, I_{\varphi}\left(\boldsymbol{X}_{22}^{+}\right) \propto \sin^{2}\theta\sin^{2}2\varphi \qquad (18.68d)$$

$$I_{\theta}\left(\boldsymbol{X}_{22}^{-}\right) \propto \sin^{2}\theta\cos^{2}\theta\sin^{2}2\varphi, I_{\varphi}\left(\boldsymbol{X}_{22}^{-}\right) \propto \sin^{2}\theta\cos^{2}2\varphi \qquad (18.68e)$$

The nonpolarized radiation intensity distribution corresponds to simple addition of θ- and φ- polarization components. The quadrupole charge distribution as well as the corresponding radiation distribution of non-, θ-, and φ- polarizations are shown in Figure 18.10. The (2 0) mode has the rotationally symmetric distribution around z axis. (2 1) modes are split into two configurations with charge oscillation in four directions on the y-z plane or x-z plane. The charge distributions of two (2 2) modes lie mainly on the x-y plane, and they are oriented by 45° rotation around the z axis from each other, and their radiation distribution also holds the same symmetry.

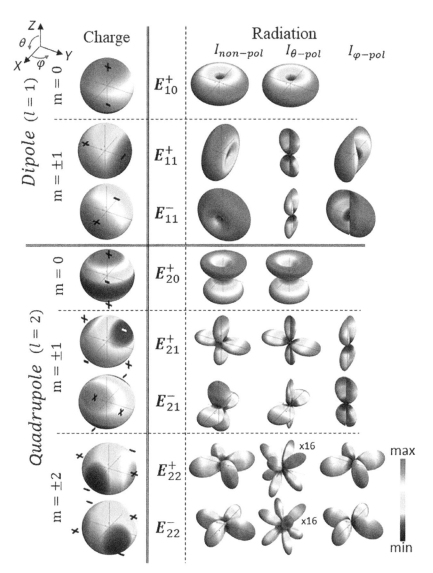

FIGURE 18.10 Angular distribution of radiation by dipole and quadrupole modes with and without polarizations. The charge distribution of each mode is shown on the leftmost column.

18.3 Cathodoluminescence

18.3.1 Instrumentation of STEM-CL

The CL measurement system that we introduce here is based on a STEM equipped with a light detection system (Figure 18.11a). In this section, we discuss 1) the STEM configuration for light collection from the sample and 2) light detection system for polarimetry as well as angle-resolved spectroscopy. Details of the methodology are also found in our previous publication (Yamamoto, 2012, 2016).

STEM

A wide-gap objective lens pole-piece has been adopted in the base STEM (JEM-2100F) system. A parabolic mirror with a height of 8 mm is inserted in the pole-piece gap of 9 mm. The sample at the focal point is surrounded by the parabolic mirror above and below so that the emitted light

towards the electron-incidence and transmission directions can be both detected. We use 80 kV acceleration voltage to avoid sample damages, although the STEM is capable up to 200 kV acceleration.

The use of the wide-gap pole-piece inevitably sacrifices the spatial resolution of the electron microscopy. In addition, the high probe current of the order of 1 nA for sufficient CL signals further increases the probe size, compared with the high-resolution STEM, where the probe current is typically in the order of 10 pA. To overcome these problems we adopted a field emission gun (FEG) and a Cs corrector, which realizes a spatial resolution below 1 nm even at 1 nA probe current and 80 kV acceleration.

Figure 18.11b shows the configuration of the parabolic mirror and the sample holder, which are located in the pole-piece gap. On one side of the mirror there is a 10 mm × 1.5 mm hole to allow the sample holder reaching the focal position of the mirror. A mirror holder to support the parabolic

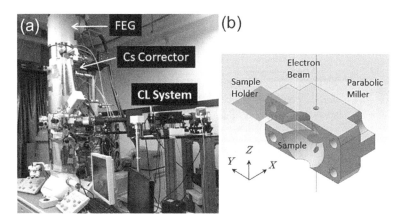

FIGURE 18.11 (a) Photograph of a STEM-CL system and (b) illustration of a parabolic mirror and sample holder.

mirror can be adjusted in X, Y, and Z directions by a micrometer.

CL Light Detection System

Schematic illustrations of the light detection system is shown in Figure 18.12. The collimated light by the parabolic mirror is transferred out of the STEM column and travels through a lens (L1) and a polarizer (P) illuminating the pinhole mask set on an X-Y stage. Only the light going through the pinhole mask reaches the second lens (L2) and is focused on the spectrometer slit, finally being detected by the electron multiplying charge coupled device (EM-CCD) spectrometer (Andor DU970N). After the X-Y stage for the pinhole mask, a three-way selector mirror is placed, which lets the light either straight to the abovementioned spectrometer, flip to a photomultiplier tube (PMT: GaAs photocathode), or flip in the opposite direction to a fiber-plate-coupled CCD camera (Hamamatsu Photonics C9000). The lens L1 forms a same-size image of the parabolic mirror on the X-Y mask stage, which is further imaged on the CCD camera by the lens L3 so that one can observe the emission distribution to the mirror as well as the position of the pinhole mask. PMT has much higher sensitivity compared with the spectrometer and is used to perform panchromatic CL mapping within a very short time.

18.3.2 CL Measurement

The parabolic mirror collimates the emitted light from the sample towards the light detection setup out of the STEM column. This parabolic mirror plane can be mathematically described using the XYZ coordinate fixed on the mirror,

$$Y^2 + Z^2 = 4p(p - X) \tag{18.69}$$

where p is the focal length of the parabola. The used mirror has $p = 1.5$ mm and a total size of 10 mm width and 8 mm height. The parabolic mirror has two holes at $2p = 3$ mm above and below the sample position for the electron beam to pass through. The electron beam travels along the Z axis, hitting the sample at the focal position. The solid angle of

the mirror covering the sample exceeds 60% of the whole space solid angle of 4π. The actual detection solid angle is reduced by the presence of the holes (for the electron beam and sample) and by shadowing from the sample support or sample itself.

The position on the mirror plane can be expressed, using the polar coordinates defined in Figure 18.13a, as follows:

$$X = 2p \frac{\sin\theta\cos\varphi}{1 + \sin\theta\cos\varphi} \tag{18.70a}$$

$$Y = 2p \frac{\sin\theta\sin\varphi}{1 + \sin\theta\cos\varphi} \tag{18.70b}$$

$$Z = 2p \frac{\cos\theta}{1 + \sin\theta\cos\varphi}. \tag{18.70c}$$

Therefore, $Y = 0$ at $\varphi = 0°$ leads to the relation along the Z axis,

$$\sin\theta = \frac{4p^2 - Z^2}{4p^2 + Z^2}. \tag{18.71}$$

Since the parabolic mirror is imaged on the pinhole mask plane in the same size, the pinhole mask position can be expressed using the coordinate on the Y-Z plane.

This CL system allows selection of the energy (wavelength), polarization, and emission angle of photons. Combined with the spatial control of the electron beam position, the following three measurement modes are available.

Angle-Resolved Spectral Pattern

While spotting or scanning the electron beam in a certain area of interest, the CL spectrum is acquired as the pinhole mask position (detection angle) is changed. The plotted 2D spectrum pattern as a function of pinhole position or detection angle is called angle-resolved spectral (ARS) pattern. There are two possible manners of the pinhole scan.

Vertical Scan

As shown in Figure 18.14a, the spectrum is acquired as the pinhole mask is moved along the Z axis. This corresponds to the mask scan along the line of $\varphi = 0°$ in Figure 18.13b. We here take an example of a one-dimensional plasmonic crystal (1D PlC) with a periodic structure along the x axis,

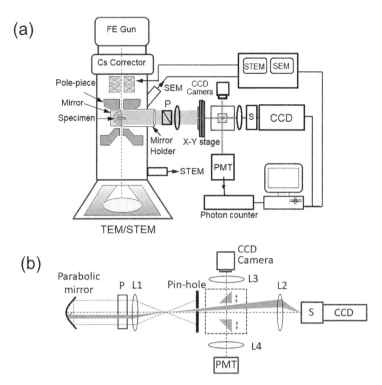

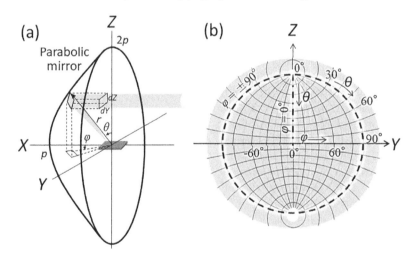

FIGURE 18.12   (a) Schematics of STEM-CL system and (b) light path of the CL light detection.

FIGURE 18.13   (a) The coordinate at the parabolic mirror and (b) the projected polar angles on the Y-Z plane for positioning the pinhole mask.

as shown in Figure 18.14b. The sample normal (z axis) is set along the Z axis, and the x axis is along the X axis, where the coordinate xyz is fixed on the sample. The electron beam is scanned on the sample, and the emitted light from SPP excited by the electron beam is detected. By moving the pinhole mask from $Z = 0$ to $2p$ and acquiring the spectrum, an ARS pattern shown in Figure 18.14c has been obtained. Conversion of the horizontal axis from Z to θ gives a truly angle-resolved pattern with θ ranging from 0° to 90°. The emission angle θ is related to the in-plane wave vector of SPP as the following equation.

$$|k_x| = |k|\sin\theta = \frac{E_{ph}}{\hbar c}\sin\theta \qquad (18.72)$$

where E_{ph} is the energy of detected photons. The ARS pattern can then be further converted to the dispersion pattern, with the horizontal axis representing k_x, as shown in Figure 18.14d. The obtained peak lines well match the calculated dispersion band lines (light gray lines) of SPP propagating along the x axis (Suzuki and Yamamoto, 2009; Yamamoto, 2012).

Horizontal Scan

In the above mentioned measurement, the signal around the Γ point in the reciprocal lattice ($k_x = 0$), which corresponds to the surface normal, is missing due to the presence of the top hole for the electron beam. In order to

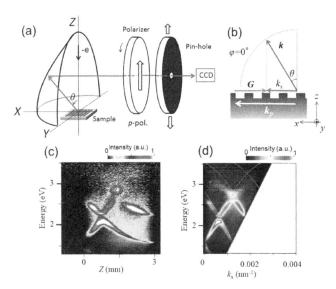

FIGURE 18.14 (a) A schematic illustration of ARS measurement (vertical scan). (b) SPP-photon conversion in a 1D PlC. (c) ARS pattern and (d) dispersion pattern of the 1D PlC.

obtain the signal around the Γ point, one can tilt the sample so that the surface normal is inclined towards the X axis (Figure 18.15a). The polar coordinate angles fixed to the sample, namely polar and azimuthal angles θ and φ, should be shifted on the projection map as shown in Figure 18.15b. In this condition, the horizontal line passing the pole (line A-B) approximately corresponds to a pure θ scan with φ fixed at 90°, when the scan range of θ is small. This way, ARS patterns can be obtained by horizontally moving the pinhole mask. With this sample tilt method, dispersion in any direction including the Γ point (surface normal) becomes available. Here we show an example of ARS patterns of a 2D plasmonic crystal (2D-PlC) with a square lattice. Figure 18.15c shows the calculated dispersion relation, and Figure

18.15d shows the ARS patterns with p- and s-polarizations, where p- and s- polarizations correspond to the polarizer along the Y and Z axes in Figure 18.15a. These ARS patterns clearly show that the different dispersion lines can be selectively visualized by changing the polarization.

Beam Scan Spectral Image

In this mode, the spectra are obtained with a fixed pinhole mask position while linearly scanning the electron beam, and are plotted as a function of the beam position resulting in a 2D plot. This plot is called beam scan spectral (BSS) image, which shows the spatial variation of the spectrum. BSS image examples of a 2D plasmonic crystal are shown in Figure 18.16a,b, which are obtained with the pinhole mask fixed at the Γ point (surface normal) without a polarizer. Panel (a) shows the schematic illustration of the 2D plasmonic crystal and the scanned lines L1 and L2 to obtain the BSS images in panel (b). The antinodes of the standing wave for A-, B-, and E-type eigenmodes at different energies appear at different spatial positions.

Angle Resolved Monochromatic CL Image (Photon Map)

By 2D electron beam scan with a fixed pinhole mask position, a spectral mapping of a 3D data volume is obtained as described in Figure 18.16c. A 2D image showing the intensity distribution at the chosen photon energy is called photon map. Examples of photon maps are shown in Figure 18.16d, which are obtained in the same condition as in Figure 18.16b, where the pinhole mask was fixed at Γ point and the emission was detected without polarizer. Photon maps enable selective visualization of a standing wave of the eigenmode at different energy levels. Degenerate modes can also be separately visualized by properly selecting the polarization.

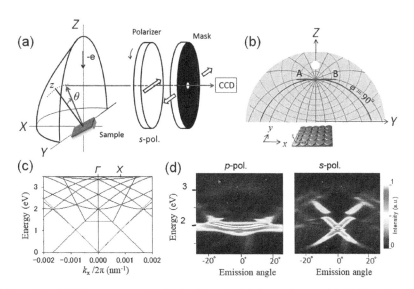

FIGURE 18.15 (a) Schematics of ARS measurement with horizontal pinhole mask scan. (b) Y-Z projection of the polar coordinates with sample tilt. (c) Dispersion relation of a 2D-PlC and (d) ARS patterns obtained with p- and s-polarizations. These ARS patterns correspond to the dispersion in the trapezoid area in panel (c).

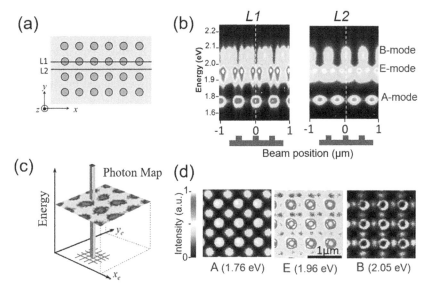

FIGURE 18.16 (a) Illustration of the electron beam scan lines on a 2D-PlC and (b) the corresponding BSS images. (c) 3D data volume of the photon map and (d) monochromatic 2D photon maps of the 2D-PlC.

18.3.3 Imaging Mechanism of Photon Map

For the CL measurement of the lattice defect or impurity in semiconductors, the spatial resolution of the CL method is determined by the spread of the emission center in the sample at a fixed beam position, which is generally determined by three factors, namely the electron beam size, generation area of electron–hole pairs, and diffusion length of the carriers (Yamamoto, 2010). However, for plasmonic structures, the emission area is as large as the SPP propagation length, except LSPs like highly localized gap plasmons. Still, photon maps show fine contrasts in the scale range of 10 nm. To understand this fact, the CL measurement process should be understood by categorizing the process into the SP excitation process by electron beam and emission process by excited SPs. The excitation of SPs by accelerated electrons appears in EELS spectrum together with other excitation processes like dielectric losses. The process of CL has been theoretically treated as radiation losses by García de Abajo and Kociak (2008), and it was shown that the CL intensity corresponds to the electromagnetic local density of state (EMLDOS) for disk structures. Since then, the relation between photon maps and EELS maps have been investigated both theoretically and experimentally (Myroshnychenko, 2008; García de Abajo, 2010; Kociak and García de Abajo, 2012; Kociak and Stéphan, 2014; Losquin and Kociak, 2015). Here we introduce the basic mechanism how the photon maps are generated, and then apply the theory to the analysis of experimental results in Section 18.4 and later.

Flat Surface

Excitation probability of SPs on a structured metal surface by a fast electron depends on the position of the electron beam $\boldsymbol{R} = (X, \mathrm{Y})$ and is proportional to the EMLDOS

projected along the electron beam path or zEMLDOS. zEMLDOS can be calculated as:

$$E_z\left(X, Y, q_z\right) = \int E_z\left(X, Y, Z\right) e^{iq_z z} dZ \ \left(q_z = \omega/v\right) \ (18.73)$$

zEMLDOS is proportional to the square modulus of Fourier-transformed Z-component of the electric field induced by the incident electron itself, i.e., $\left|E_z\left(X, Y, q_z\right)\right|^2$. When the distribution of the electric field is sufficiently smaller than $v/\omega = \lambda/2\pi$, Eq. (18.73) can be approximated as a simple Z-projection of the electric field. For metal surfaces with flat structures, such as terrace shapes, surface charges of the excited SP modes form an electric field perpendicular to the surface except close vicinity of steps. Therefore, the surface charge distribution $\psi(x, y)$ induced by SP modes is proportional to the perpendicular component of the electric field $E_z\left(x, y, z\right)|_{z=0}$. For SPP, the electric field can be expressed as $E_z\left(x, y, z\right) = E_z\left(x, y\right) e^{-i\kappa z}$, which further leads to the relation $\left|E_z\left(X, Y, q_z\right)\right|^2 \propto \left|E_z\left(X, Y\right)\right|^2$. Thus the excitation probability by an electron beam incident at the position $\boldsymbol{R} = (X, Y)$ is approximately proportional to the charge distribution $\left|\psi\left(X, Y\right)\right|^2$.

A photon map in the CL measurement is a 2D plot of the detected emission intensity synchronized with the electron beam scan. By selecting polarization, emission angle, and energy, certain SP modes can be visualized. In general, the detected emission intensity at the electron beam position $\boldsymbol{R}$ is given by the product of the excitation probability of a certain mode at this position and radiation probability of the excited mode. When the SP mode is not degenerate, the excited surface charge distribution is simply expressed as

$$\psi\left(\boldsymbol{r}, \boldsymbol{R}\right) = c_n(\boldsymbol{R})\psi_n(\boldsymbol{r}). \qquad (18.74)$$

The coefficient $c_n(\boldsymbol{R})$ represents the excitation amplitude of the nth mode, and $\psi_n(\boldsymbol{r})$ is the eigenwave function of this mode. Here, we omit the time-dependent harmonic oscillator term $e^{-i\omega_n t}$. The square of the absolute value of the excitation amplitude $c_n(\boldsymbol{R})$ gives the excitation probability of the mode. Since the excitation probability is proportional to $|\boldsymbol{E}_z(\boldsymbol{R})|^2$ for a flat surface, according to the previous discussion, the proportionality $|c_n(\boldsymbol{R})|^2 \propto |\psi_n(\boldsymbol{R})|^2$ can be derived. When multiple modes are degenerate, superposition of all the excited modes should be considered, leading to the following sum of amplitude of all the modes.

$$\psi(\boldsymbol{r}, \boldsymbol{R}) = \sum_n c_n(\boldsymbol{R})\psi_n(\boldsymbol{r}) \qquad (18.75)$$

We now derive the expression of the angular-dependent emission amplitude for each mode to calculate the total CL signal, which is expressed by the product of the excitation and emission amplitudes. With a detection solid angle Ω and unit direction vector $\boldsymbol{n}$ towards the detector, the electric field of the emitted light from the nth mode, defined as $\boldsymbol{E}_n(\Omega)$, is perpendicular to the vector $\boldsymbol{n}$, and is related to the corresponding magnetic field as $\boldsymbol{E}_n(\Omega) = Z_0 \boldsymbol{H}_n(\Omega) \times \boldsymbol{n}$, where Z_0 is the impedance of vacuum. Emission intensity of each mode at a certain angle is given by Poynting vector per solid angle:

$$\frac{dP_n}{d\Omega} = \frac{1}{2} Re\left[\boldsymbol{E}_n^*(\Omega) \times \boldsymbol{H}_n(\Omega) \cdot \boldsymbol{n}\right] \qquad (18.76)$$

Therefore its magnitude is proportional to $|\boldsymbol{E}_n(\Omega)|^2$. From this, the emission intensity of a nondegenerate mode reads

$$I(\Omega, \boldsymbol{R}) = |c_n(\boldsymbol{R})|^2 \frac{dP_n}{d\Omega} \propto |c_n(\boldsymbol{R})\boldsymbol{E}_n(\Omega)|^2. \qquad (18.77)$$

And, with degeneracy,

$$I(\Omega, \boldsymbol{R}) \propto \left|\sum_n c_n(\boldsymbol{R})\boldsymbol{E}_n(\Omega)\right|^2 \qquad (18.78)$$

The emission intensity with polarization in the $\boldsymbol{e}$ direction can be obtained by taking a product with electric field vector, where $\boldsymbol{e}$ is also perpendicular to $\boldsymbol{n}$. Considering the interference of emission of degenerate modes, the total emission intensity is expressed as a square of the absolute value of the sum:

$$I(\Omega, \boldsymbol{R}) \propto \left|\sum_n c_n(\boldsymbol{R})\left[\boldsymbol{e} \cdot \boldsymbol{E}_n(\Omega)\right]\right|^2 \qquad (18.79)$$

Three-Dimensional Structure

For metallic structures with three dimensionally varying surfaces, it is not easy to express the zEMLDOS as a function of the electron beam position. For a spherical particle, the simplest 3D structure has been already discussed in Section 18.2.2, where we expressed the angular-dependent emission intensity without much argument about the excitation amplitude. While the excitation amplitude for a flat structure reflects the surface charge distribution of the eigenmode, the excitation amplitude of a sphere can be expressed

as $\psi_{\ell m}^{E,\text{ind}}(\boldsymbol{R}) = t_\ell^E \psi_{\ell m}^{E,\text{ext}}(\boldsymbol{R})$ for (ℓ, m) multipole mode. The position of the incident electron $\boldsymbol{R} = (\rho, \varphi_0)$ is now given by the polar coordinate on the X-Y plane perpendicular to the electron beam direction (Z axis). The coordinate system is the same as previous sections, with the X axis towards the parabolic mirror. The azimuthal angle φ is measured from the X axis.

For a metal sphere with a radius a, the angular distribution of the emission is expressed as a function of the emission angle $\Omega = (\theta, \varphi)$ and electron beam position $\boldsymbol{R}$, as derived in Section 18.2.2, which holds in the range $\rho > a$. Therefore, the photon map intensity distribution at the detection angle $\Omega_D = (\theta_D, \varphi_D)$ is given by

$$I^{\text{PM}}(\boldsymbol{R}, \Omega_D) = I(\theta_D, \varphi_D, \boldsymbol{R}). \qquad (18.80)$$

If the detection solid angle is fairly large, the intensity should be integrated over a detector solid angle.

$$I^{\text{PM}}(R, \Omega_D) = \int_{\Omega_D} I(\theta, \varphi, \boldsymbol{R}) d\Omega \qquad (18.81)$$

This formulation will be used in Section 18.4.2 to explain the experimentally observed photon maps of silver nanoparticles with selected polarization and detection angles.

18.4 Application to LSP-Related Phenomena

18.4.1 Thin Triangular Nanoprism

SP resonance arises in a nanoscale triangular prism as in a spherical metal particle. Shape of a triangular prism has the following symmetries as shown in Figure 18.17; threefold rotational symmetry axis at the central point P of the triangle normal to the surface (C_3) and the mirror plane perpendicular to the surface passing through both the central point and each corner (σ_v). These symmetric operations create a point group of C_{3v}, including the identity operation E. The index table of this point group is shown in Table 18.1.

The basis function of the plasmon oscillation of a resonant mode with a frequency ω is expressed as

$$\psi(\boldsymbol{r}, t) = \varphi_1(\boldsymbol{r})\exp(-i\omega t). \qquad (18.82)$$

$\varphi_1(\boldsymbol{r})$ can be a function which expresses surface charge distribution on a plasmonic structure. Due to the threefold

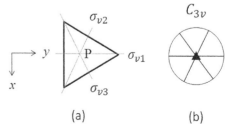

FIGURE 18.17 Symmetry of triangular prism (a) and its point group (b).

TABLE 18.1   Character Table for C_{3v} Point Group

C_{3v}	E	$2C_3$	$3\sigma_v$	Eigenfunction
$A1$	1	1	1	ψ_{A1}
$A2$	1	1	-1	
E	2	-1	0	$\psi_{E(1)}, \psi_{E(2)}$
Γ_{red}	3	0	1	$A1 \oplus E$

rotational symmetry, the other basis functions are deduced as $\varphi_2(\boldsymbol{r}) = C_3^1\varphi_1$ and $\varphi_3(\boldsymbol{r}) = C_3^2\varphi_1$. Since the temporal part is common, it is omitted. The representation based on the three basis functions is {3, 0, 1} for the three classes of symmetry operations (Table 18.1, bottom row). This representation is reducible, and can be decomposed into two irreducible representations, which are expressed as their direct sum,

$$\Gamma_{\text{red}} = A1 \oplus E \qquad (18.83)$$

Hence, the three basic plasmon oscillations should form two eigenstates with two different energies. $A1$ is a 1D representation with total symmetry and E is a 2D representation associated with doubly degenerate states. The eigenfunctions of these irreducible representations can be derived by applying the projection operator to one basis function using the characters in Table 18.1 and are expressed as

$$\psi_{A1} = \varphi_1 + \varphi_2 + \varphi_3, \psi_{E(1)} = 2\varphi_1 - \varphi_2 - \varphi_3, \psi_{E(2)} = \varphi_2 - \varphi_3.$$
$$(18.84)$$

Here the normalization factors are omitted.

If we select the basis functions of $\varphi_1 \sim \varphi_3$ to represent in-plane dipole-like surface charge oscillation (Figure 18.18a), the eigenfunctions of $A1$ and E modes are drawn as shown in Figure 18.18b.

A SiO_2 sputtered film was deposited with a thickness of 200 nm on a silicon substrate, and gold triangular nanoprisms were fabricated on the SiO_2 film by electron beam lithography. Several peaks related to a dipole-like

mode and quadrupole-like mode appeared in the CL spectrum, and the corresponding photon maps were measured (Myroshnychenko et al., 2018). Figure 18.18c shows the photon maps of the energy of the dipole-like mode measured with various polarizations. The prism is 226 nm on one side and 50 nm in height, and one side is set parallel to the parabolic mirror axis (X axis). A circular pinhole mask with a diameter of 6 mm was used with its center fixed on the X axis. This means that light intensity is integrated for half of the total solid angle on the upper side of the sample. Three-fold symmetric pattern appears for nonpolarized light, but the s-polarized photon map represents a pattern corresponding to $E(1)$ mode and the p-polarized photon map represents a pattern corresponding to $E(2)$ mode. This means that, by selecting the polarization direction, the energy-degenerate modes can be separately visualized by the photon map.

Here we explain the interpretation of polarized photon map of energy-degenerate mode using group theory. The surface charge distribution excited on the triangular prism, when the electron beam enters the position $\boldsymbol{R}$, is expressed as a linear combination of each eigenmode,

$A1$ mode: $\psi_{A1}(\boldsymbol{r}, \boldsymbol{R}) = c_0(\boldsymbol{R})\psi_{A1}(\boldsymbol{r})$ (18.85a)

E mode: $\psi_E(\boldsymbol{r}, \boldsymbol{R}) = c_1(\boldsymbol{R})\psi_{E(1)}(\boldsymbol{r}) + c_2(\boldsymbol{R})\psi_{E(2)}(\boldsymbol{r}),$
$(18.85b)$

where the energies of the $A1$ and the doubly degenerate E modes are generally different, so the eigenfunctions are expressed separately. The coefficient $c_n(\boldsymbol{R})$ represents the excitation amplitude of n-th mode, and is related to the eigenfunction of the irreducible expression as $|c_n(\boldsymbol{R})|^2 \propto |\psi_n(\boldsymbol{R})|^2$ (see Section 18.3.3.1). We focus only on E mode here.

Here, we explain the emission intensity in the surface normal direction due to the excited charge oscillation that depends on the polarization direction of light. The emission in this direction is considered to be due to the dipole radiation caused by in-plane electric dipole oscillation. Assuming that $\boldsymbol{p}(\boldsymbol{R})$ is the electric dipole moment in the plane due to the surface charge distribution excited in a triangular prism, the emission intensity should be proportional to $|\boldsymbol{p}(\boldsymbol{R})|^2$, which gives a photon map as a function of $\boldsymbol{R}$. The dipole moment $\boldsymbol{p}(\boldsymbol{R})$ is obtained by integrating the product of $\boldsymbol{r}$ and the charge distribution $\sigma(\boldsymbol{r})$ over the prism.

$$\boldsymbol{p}(\boldsymbol{R}) \equiv \int \boldsymbol{r}\sigma(\boldsymbol{r})\,d\boldsymbol{r} \propto \int \boldsymbol{r}\psi(\boldsymbol{r}, \boldsymbol{R})\,d\boldsymbol{r} \qquad (18.86)$$

The electric dipole moment for each eigenfunction is given by

$$\boldsymbol{p}_{E1}(\boldsymbol{R}) \propto \int \boldsymbol{r}\psi_{E(1)}(\boldsymbol{r}, \boldsymbol{R})\,d\boldsymbol{r} = \begin{pmatrix} 0 \\ p_y \end{pmatrix},$$
$$\boldsymbol{p}_{E2}(\boldsymbol{R}) \propto \int \boldsymbol{r}\psi_{E(2)}(\boldsymbol{r}, \boldsymbol{R})\,d\boldsymbol{r} = \begin{pmatrix} p_x \\ 0 \end{pmatrix}, \qquad (18.87)$$

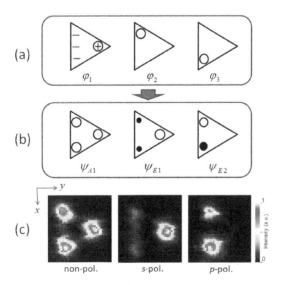

FIGURE 18.18   (a) Basis functions representing charge oscillation of dipole mode on the prism surface, (b) eigenfunctions of the eigenmodes, and (c) photon maps taken with different polarizations.

where

$$p_x \equiv \int x\psi_{E(2)}\left(\boldsymbol{r}\right)d\boldsymbol{r}, p_y \equiv \int y\psi_{E(1)}\left(\boldsymbol{r}\right)d\boldsymbol{r}. \quad (18.88)$$

From these results, it is understood that the $E(1)$ mode contributes to the polarization only in the y direction and the $E(2)$ mode only contributes to the polarization in the x direction.

The photon map of the light polarized in the $\boldsymbol{e}$ direction is written as

$$I_{E\parallel e}(\boldsymbol{R}) \propto |\boldsymbol{e} \cdot \boldsymbol{p}(\boldsymbol{R})|^2 = |c_1(\boldsymbol{R})(\boldsymbol{e} \cdot \boldsymbol{p}_{E1}) + c_2(\boldsymbol{R})(\boldsymbol{e} \cdot \boldsymbol{p}_{E2})|^2 . \quad (18.89)$$

The photon map using polarized light in the x direction, $\boldsymbol{e} = (1,0)$ is written as

$$I_{E\parallel x}(\boldsymbol{R}) \propto |c_2(\boldsymbol{R})(\boldsymbol{e} \cdot \boldsymbol{p}_{E2})|^2 \propto |\psi_{E(2)}(\boldsymbol{R})|^2 p_x^2. \quad (18.90)$$

And the photon map using polarized light in the y direction, $\boldsymbol{e} = (0,1)$ is written as

$$I_{E\parallel y}(\boldsymbol{R}) \propto |c_1(\boldsymbol{R})(\boldsymbol{e} \cdot \boldsymbol{p}_{E1})|^2 \propto |\psi_{E(1)}(\boldsymbol{R})|^2 p_y^2. \quad (18.91)$$

Therefore, the photon map using the polarization in the x direction reveals the square modulus of the eigenfunction of the $E(2)$ mode, while the photon map using the polarized light in the y direction reveals the square modulus of the $E(1)$ mode eigenfunction. Then the nonpolarized photon map is expressed as

$$I_{E\parallel e}(\boldsymbol{R}) \propto \int_0^{2\pi} |c_1(\boldsymbol{R})(\boldsymbol{e} \cdot \boldsymbol{p}_{E1}) + c_2(\boldsymbol{R})(\boldsymbol{e} \cdot \boldsymbol{p}_{E2})|^2 d\theta \quad (18.92)$$

$$\propto |\psi_{E(1)}(\boldsymbol{R})|^2 p_y^2 + |\psi_{E(2)}(\boldsymbol{R})|^2 p_x^2 \propto I_{E\parallel x}(\boldsymbol{R}) + I_{E\parallel y}(\boldsymbol{R}).$$

This result indicates that a nonpolarized photon map is obtained by taking the sum of the intensities of the photon maps in which the polarization is in the x and y directions.

18.4.2 Nanoparticle

Photon Map, BSS Image, and ARS Pattern of NPs

In this subsection, we show application examples of CL measurement on metallic nanoparticles and compare them with the previously discussed theory of CL image formation. In the CL system we present here, one can selectively extract certain information from properly set detection energy, polarization, and detection angle parameters. In other words, meaningful information cannot be separately obtained without such parameter selection. Below we show a concrete example of CL measurement on silver nanoparticles with such parameter selection, which tells the importance of the polarization- and angle-selected CL measurement.

Figure 18.19 shows CL measurement results of a 140 nm silver particle supported on a carbon film (Figure 18.19a). The measurement was conducted using a pinhole mask with a diameter of 6 mm set at the X direction (Figure 18.19b). In this setup the detection solid angle is about 2π sr, which means the detection signal is integrated over this hemisphere.

Figure 18.19c is the acquired emission spectrum that clearly exhibits a dipole (D) and quadrupole (Q) peaks. Figure 18.19d is a panchromatic photon map (without energy selection) acquired using PMT at nonpolarization, p-polarization, and s-polarization. Nonpolarized and p-polarized photon maps show intensity spread over the whole particle, while the s-polarized map shows locally strong signals only at the left and right edges. Figure 18.19e,f show energy-selected photon maps at the dipole

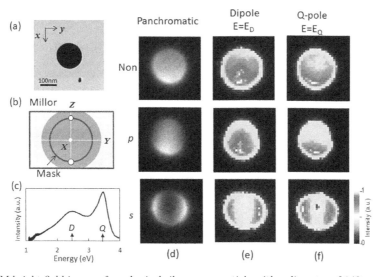

FIGURE 18.19　(a) STEM bright field image of a spherical silver nanoparticle with a diameter of 140 nm. (b) Illustration of the 6 mm "pinhole" mask (solid circle) set at the center of the mirror projected on the Y-Z plane. (c) Emission spectrum of the nanoparticle. (d–f) Photon maps of the nanoparticle with nonpolarization, p-polarization, and s-polarization. (d) Panchromatic photon map, (e) photon map at the dipole energy, and (f) photon map at the quadrupole energy.

and quadrupole energies. The *p*-polarized photon maps have asymmetric intensity distribution in the top-bottom directions, which is due to asymmetric emission depending on the electron beam position with respect to the parabolic mirror direction. The nonpolarized map is basically the sum of *p*- and *s*-polarized maps, and the panchromatic map is the sum of dipole and quadrupole mode photon maps.

Figure 18.20 shows BSS images acquired with the same condition as Figure 18.19. Figure 18.20a shows the STEM image of a 150 nm silver particle, and Figure 18.20b shows the schematics of the mask and pinhole configuration. Figure 18.20c,d shows *p*- and *s*- polarized BSS images with the linear electron beam scan along the line A-B in the STEM image. For the *p*-polarization, the dipole radiation is strongest around the center of the particle, while it is strongest at the edges for *s*-polarization. They correspond to electric dipole oscillations, respectively, along *z* and *y* axes, which we discuss more in detail in the next section. The quadrupole BSS image exhibits similar intensity distribution features with additional intensity maximum at the edge also for the *p*-polarization.

Figures 18.20e,f show the ARS patterns with *p*-polarization obtained with the electron beam fixed at the edge of the particle (position C) and at the center (position D). This ARS measurement has been conducted using a 1-mm pinhole scanned along the *Z* axis, as shown in Figure 18.14 in Section 18.3. In Figure 18.20e with the edge excitation, the dipole radiation is distributed mostly around the surface-normal direction while the quadrupole radiation is distributed more at the higher angles. In Figure 18.20f with the center excitation, stronger dipole radiation is observed especially at the high angles. With

the edge excitation (position C), in-plane electric dipole is dominantly excited and with the center excitation (position D), out-of-plane dipole is excited.

For the measurement of Figure 18.20e,f, the pinhole was scanned with the fixed electron beam position. Inversely, the electron beam can be scanned two-dimensionally with the fixed pinhole position, to produce a photon map of selected emission at a certain angle. Figure 18.21 shows photon maps of dipole and quadrupole modes with the pinhole position at $\theta = 0°, 45°$, and $80°$. The left columns are of *p*-polarization, and the right columns *s*-polarization. The dipole mode photon maps with *p*-polarization exhibits bright contrasts at the upper and lower edges at the emission angle $\theta = 0°$, and at the angle $\theta = 80°$ bright contrasts at the center, which corresponds well to the ARS patterns of Figure 18.20e,f. At $\theta = 0°$, the quadrupole pattern looks similar to the dipole, but with the maximum intensities slightly shifted towards the center. The *s*-polarized photon maps of the quadrupole mode start to show four poles with a fourfold symmetry as the emission angle exceeds $45°$. The photon maps with *s*-polarization rotate to φ direction when the pinhole mask is scanned to the *Y* direction (Thollar et al., 2018).

Formation Mechanism of Photon Map

Photon Map of Dipole Mode

We first consider photon map formation of $\ell = 1$ dipole modes. According to Eq. (18.57), the angular distribution of the emission intensity is expressed for the incident electron beam position $\boldsymbol{R} = (\rho, \varphi_0)$ and the polarization vector $\boldsymbol{e}$:

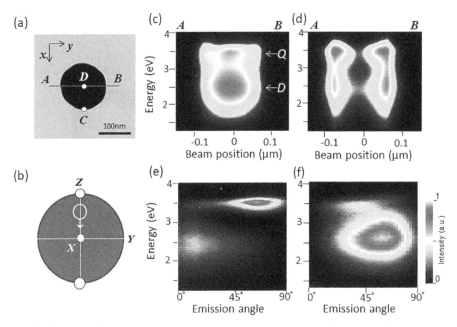

FIGURE 18.20 (a) STEM bright field image of the observed spherical silver particle (diameter 150 nm). (b) Pinhole mask configuration. A 6-mm pinhole for BSS (dark solid circle) and a 1-mm pinhole for ARS (white circle) have been used (c,d). BSS images acquired with the electron beam scan along the A-B line in panel (a), with (c) *p*-polarization and (d) *s*-polarization (e,f). ARS patterns with *p* polarization, with the electron beam fixed (e) at the edge of the particle (position C), and (f) at the center (position D).

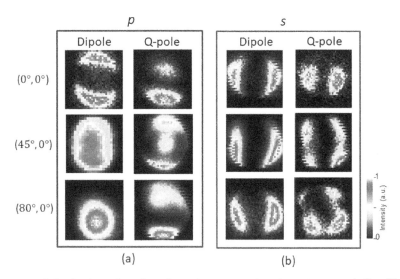

FIGURE 18.21 Photon maps of the dipole and quadrupole modes measured at the emission angle $\theta = 0°, 45°, 80°$. The (a) and (b) columns correspond to the p- and s-polarization, respectively. The scan area size is 200 nm $\times$ 200 nm.

$$I_{e-\text{pol}}(\theta, \varphi, \boldsymbol{R}) \propto \left| \sum_{\substack{m=-\ell \\ m=\text{even}}}^{\ell} C_{\ell m}(\boldsymbol{R}) \left\{ \boldsymbol{E}_{\ell m}^{+}(\theta, \varphi) \right. \right.$$

$$\left. \cos m\varphi_0 + \boldsymbol{E}_{\ell m}^{-}(\theta, \varphi) \sin m\varphi_0 \right\} \cdot \boldsymbol{e} + i \sum_{\substack{m=-\ell \\ m=\text{odd}}}^{\ell} C_{\ell m}(\boldsymbol{R})$$

$$\left. \left\{ \boldsymbol{E}_{\ell m}^{-}(\theta, \varphi) \cos m\varphi_0 - \boldsymbol{E}_{\ell m}^{+}(\theta, \varphi) \sin m\varphi_0 \right\} \cdot \boldsymbol{e} \right|^2 \quad (18.93)$$

The $\ell = 1$ term can be extracted for the detection angle $\Omega_D = (\theta_D, \varphi_D)$

$$I_{e-\text{pol}}^{\text{PM}}(\rho, \varphi_0, \Omega_D) = I_{e-\text{pol}}(\theta_D, \varphi_D, \boldsymbol{R})$$
$$\propto \left| C_{10}(\rho) \boldsymbol{E}_{10}^{+}(\Omega_D) \cdot \boldsymbol{e} + i C_{11}(\rho) \left\{ \boldsymbol{E}_{11}^{-}(\Omega_D) \cos \varphi_0 \right. \right.$$
$$\left. \left. - \boldsymbol{E}_{11}^{+}(\Omega_D) \sin \varphi_0 \right\} \cdot \boldsymbol{e} \right|^2 \quad (18.94)$$

$\boldsymbol{E}_{10}^{+}$, $\boldsymbol{E}_{11}^{+}$ and $\boldsymbol{E}_{11}^{-}$ are found in Section 18.2.2.3. The $\boldsymbol{e}$-polarized photon map is given as

$$I_{e-\text{pol}}^{\text{PM}}(\rho, \varphi_0, \Omega_D) \propto$$
$$\times \left| \begin{array}{l} \{C_{10}(\rho) \sin \theta + C_{11}(\rho) \cos \theta \cos(\varphi_0 - \varphi_D)\} \boldsymbol{e}_{\theta} \cdot \boldsymbol{e} \\ + C_{11}(\rho) \sin(\varphi_0 - \varphi_D) \boldsymbol{e}_{\varphi} \cdot \boldsymbol{e} \end{array} \right|^2 \quad (18.95)$$

Here the coefficients are defined as follows:

$$C_{10}(\rho) = \sqrt{2} t_1^E B_{10} K_0\left(\frac{\omega\rho}{v\gamma}\right) \quad (18.96a)$$

$$C_{11}(\rho) = \sqrt{2} t_1^E B_{11} K_1\left(\frac{\omega\rho}{v\gamma}\right) \quad (\rho > a) \quad (18.96b)$$

$$B_{10} = -\sqrt{\frac{3}{2\pi}} \frac{1}{\beta^2 \gamma}, B_{11} = -B_{1-1} = -\sqrt{\frac{6}{\pi}} \frac{i}{\beta^2 \gamma} \quad (18.97)$$

B_{10} is real, and B_{11} is pure imaginary, which leads to real $C_{10}(\rho)$ and imaginary $C_{11}(\rho)$. This indicates that the out-of-plane electric dipole $(\boldsymbol{p}_z)$ and in-plane electric dipoles $(\boldsymbol{p}_x, \boldsymbol{p}_y)$ have a $\pi/2$ relative phase shift when excited by an electron beam.

The modified Bessel functions K_0 and K_1 in Eq. (18.96) have exponential decay by distance ρ for $\rho > a$. The function shapes of $C_{10}(\rho)$ and $C_{11}(\rho)$ inside the particle $\rho < a$ can be qualitatively presumed from the symmetry of the dipole charge distribution (Figure 18.10 left). The $(1,0)$ mode is the dipole along the z axis, and the excitation amplitude becomes maximum with the electron beam at the center of the particle $(\rho = 0)$. Therefore, $C_{10}(b)$ should have maximum at $\rho \cong 0$ and monotonously decreases as b increases. In the same manner, the $(1, \pm 1)$ modes are in-plane dipoles along x and y axes, and should have the maximum excitation amplitude when the electron beam passes at the edge $(\rho = a)$. Therefore the function $C_{11}(\rho)$ is expected to be zero at $\rho = 0$, and has its maximum at $\rho \cong a$. We approximated the function shape of $C_{10}(\rho)$ and $C_{11}(\rho)$ by Gaussian.

From Eq. (18.95), photon maps for the dipole mode can be reproduced. At the detection angle $\Omega_D = (\theta_D, \varphi_D)$, the p-polarized photon map signal can be approximated as the intensity with θ-polarization, which is therefore expressed as

$$I_{p-\text{pol}}^{\text{PM}}(\rho, \varphi_0, \Omega_D) \propto |C_{10}(\rho) \sin \theta_D$$
$$- C_{11}(\rho) \cos \theta_D \cos(\varphi_0 - \varphi_D)|^2$$
$$= |C_{10}(\rho)|^2 \sin^2 \theta_D + |C_{11}(\rho)|^2 \cos^2 \theta_D \cos^2(\varphi_0 - \varphi_D). \quad (18.98)$$

In the same manner, s-polarized signal can be approximated as the φ-polarized intensity:

$$I_{s-\text{pol}}^{\text{PM}}(\rho, \varphi_0, \Omega_D) \propto |C_{11}(\rho)|^2 \sin^2(\varphi_0 - \varphi_D) \quad (18.99)$$

Nonpolarized signal is a sum of these two.

$$I_{\text{non-pol}}^{\text{PM}}(\rho, \varphi_0, \Omega_D) = I_{\theta-\text{pol}}^{\text{PM}}(\rho, \varphi_0, \Omega_D)$$
$$+ I_{\varphi-\text{pol}}^{\text{PM}}(\rho, \varphi_0, \Omega_D) \propto |C_{10}(\rho)|^2 \sin^2\theta_D$$
$$+ |C_{11}(\rho)|^2 \{\cos^2\theta_D \cos^2(\varphi_0 - \varphi_D) + \sin^2(\varphi_0 - \varphi_D)\}$$
$$(18.100)$$

In Figure 18.22, reproduced photon maps using Eqs. (18.98) and (18.99) are displayed for different detection angles and polarizations.

For the emission in the X-Z plane with p-polarization, only $\boldsymbol{p}_x$ and $\boldsymbol{p}_z$ dipoles form the photon map without contribution from $\boldsymbol{p}_y$ dipole due to the polarization mismatch. For the upward emission to the Z direction, only $\boldsymbol{p}_x$ dipole contributes to the signal showing strong contrasts at the facing two edges along the x axis in the photon map. For the "sideward" emission along the X direction, only $\boldsymbol{p}_z$ dipole contributes with strong contrasts at the center in the photon map. This axis symmetric contrast does not change with the detection angle scan of φ_D around the Z axis at fixed $\theta_D = 90°$.

For s-polarization, the photon map pattern of the dipole depends only on the detection angle component φ_D and not on θ_D, according to Eq. (18.99). As observed in the experiment (Thollar et al., 2018), the dipole contrast at the two edges rotating as φ_D is scanned. Nonpolarization signal is a sum of p- and s- polarization intensities. Therefore a donut shape pattern $|C_{11}(\rho)|^2$ should be obtained when upward emission $(\theta_D, \varphi_D) = (0°, 0°)$ is detected without polarization.

Photon Map of Quadrupole (Q-Pole) Mode

The emission intensity of $\boldsymbol{e}$-polarized light giving the photon map of Q-pole mode with $\ell = 2$ is expressed as

$$I_{e-\text{pol}}^{\text{PM}}(\rho, \varphi_0, \Omega_D) \propto |C_{20}(\rho)\,\boldsymbol{e}\cdot\boldsymbol{E}_{20}^+(\Omega_D) + iC_{21}(\rho)\,\boldsymbol{e}\cdot$$
$$\{\boldsymbol{E}_{21}^-(\Omega_D)\cos\varphi_0 - \boldsymbol{E}_{21}^+(\Omega_D)\sin\varphi_0\} + C_{22}(\rho)\,\boldsymbol{e}\cdot$$
$$\{\boldsymbol{E}_{22}^+(\Omega_D)\cos 2\varphi_0 + \boldsymbol{E}_{22}^-(\Omega_D)\sin 2\varphi_0\}|^2. \quad (18.101)$$

Each term on the right-hand side corresponds to emission in the five modes with different m. The coefficient representing

the excitation amplitude of each mode is written as

$$C_{2m}(\rho) = \sqrt{2}t_2^E B_{2m} K_m\left(\frac{\omega\rho}{v\gamma}\right) \quad (\rho > a). \quad (18.102)$$

Specifically, B_{2m} is expressed as

$$B_{20} = -i10\sqrt{\frac{3}{\pi}}\frac{1}{\beta^3\gamma}, \, B_{21} = -B_{2-1} = \sqrt{\frac{15}{2\pi}}\left(\frac{1}{\beta^3\gamma^2} + \frac{3}{\beta^2}\right),$$
$$B_{22} = B_{2-2} = i2\sqrt{\frac{15}{2\pi}}\frac{1}{\beta^3\gamma}. \quad (18.103)$$

Since B_{20} and B_{22} are pure imaginary and B_{21} is real, $C_{20}(\rho)$ and $C_{22}(\rho)$ are pure imaginary and $C_{21}(\rho)$ is real. Using the expression of $\boldsymbol{E}_{2m}^{\pm}(\Omega_D)$ derived in Section 18.2.2, Eq. (18.101) can be written as

$$I_{p-\text{pol}}^{\text{PM}}(\rho, \varphi_0, \Omega_D) \propto$$
$$\times \left| \begin{array}{c} \sqrt{3/2}C_{20}(\rho)\sin\theta_D\cos\theta_D + C_{21}(\rho)\cos 2\theta_D \\ \cos(\varphi_0 - \varphi_D) - C_{22}(\rho)\sin\theta_D\cos\theta_D\cos 2(\varphi_0 - \varphi_D) \end{array} \right|^2$$
$$(18.104\text{a})$$

$$I_{s-\text{pol}}^{\text{PM}}(\rho, \varphi_0, \Omega_D) \propto |C_{21}(\rho)\cos\theta_D\cos(\varphi_0 - \varphi_D)$$
$$- C_{22}(\rho)\sin\theta_D\sin 2(\varphi_0 - \varphi_D)|^2. \quad (18.104\text{b})$$

Thus we can calculate the intensity distribution of the photon map of p- and s-polarized lights in an arbitrary detection direction. In Eq. (18.104a), considering the expression of C_{2m} and B_{2m}, the second term corresponding to the emission of the $(2,1)$ mode is a real number, whereas the first and third terms are multiplied by a factor of $(-i)$ with respect to the second term. This indicates that the excited $(2,0)$ mode and the $(2,2)$ mode oscillate in the same phase, while the $(2,1)$ mode oscillate with a phase delay by $\pi/2$ with respect to them.

A functional form of $C_{2m}(\rho)$ is given by Eq. (18.102) for $\rho > a$, which decreases with radial coordinate ρ by a modified Bessel function. However, since the function form for $\rho < a$ is unknown, the functional form of $C_{2m}(\rho)$ must be estimated from the surface charge

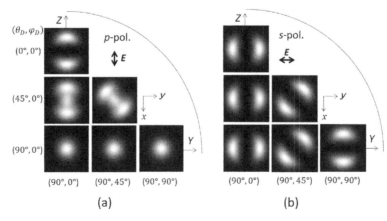

FIGURE 18.22 Calculated photon maps with different polarization and detection angles. (a) p-polarization. (b) s-polarization. The axes coordinate x, y lie on the photon map plane while the axes Y, Z are fixed at the parabolic mirror geometry (see Figure 18.13).

distribution of the Q-pole mode as in the case of the dipole mode. The ratio among $C_{20}(\rho)$, $C_{22}(\rho)$, and $C_{21}(\rho)$ at $\rho = a$ is given as $|C_{20}(\rho)| : |C_{21}(\rho)| : |C_{21}(\rho)| \approx 4.5K_0(\rho) : 3K_1(\rho) : 2K_2(\rho)$. For example, with an acceleration voltage of 80 kV and the particle radius of 50 nm, $|C_{20}(a)| : |C_{21}(a)| : |C_{21}(a)| \approx 1 : 1.42.5$. From the surface charge distribution of each mode shown in Figure 18.10, $C_{20}(\rho)$ intuitively becomes maximum at the center of the nanoparticle, and it is considered that it once attenuates with increasing ρ and becomes maximum at the edge of the nanoparticle. On the other hand, $C_{21}(\rho)$ is close to 0 at the center of the nanoparticle, then maximized between the center and edge of the nanoparticle ($\rho \approx a/\sqrt{2}$), and is expected to decrease as further away from it. Also, $C_{22}(\rho)$ is 0 at the center of the nanoparticle, then maximized near the edge ($\rho \approx a$), and is expected to decrease as further away from it. Figure 18.23 shows the approximate functional form of these $C_{2m}(\rho)$ expressed using the Gaussian function. Figure 18.24a and (b) shows the calculated photon maps of p- and s-polarized lights using Eq. (18.103) for various detection directions using this functional form.

In the case of p-polarized light, emission due to the Q-pole oscillation of X_{21}^+ and X_{22}^- does not contribute to the emission in the X-Z plane due to the difference in polarization direction, and only emissions due to X_{20}^+, X_{21}^+ and X_{22}^+ enter in Eq. (18.104a). In the surface normal direction ($\Omega_D = (0°, 0°)$ in the Z direction) and in the X direction

($\Omega_D = (90°, 0°)$), only emission by X_{21}^- contributes to intensity.

$$I_{p-\text{pol}}^{\text{PM}}(\rho, \varphi_0, \Omega_D) \propto |C_{21}(\rho)|^2 \cos^2\varphi_0 \qquad (18.105)$$

And the bright pair contrast appears on the edges of the particle sphere in the x direction. This selection of the detection angle is a condition for selectively observing the photon map of the (2,1) mode. Therefore, the functional form of $C_{21}(\rho)$ can be determined from the experimental photon map. In the $\theta_D = 45°$ direction, the detection intensity is written as

$$I_{p-\text{pol}}^{\text{PM}}(\rho, \varphi_0, \Omega_D) \propto \left| \sqrt{3}C_{20}(\rho) - C_{21}(\rho)\cos 2\varphi_0 \right|^2, \qquad (18.106)$$

so the functional form of $C_{20}(\rho)$ can be obtained by comparison with the observed photon map measured in this direction using the form of $C_{21}(\rho)$. As the detection direction is changed from X to the Y direction ($\Omega_D = (90°, 90°)$) while keeping $\theta_D = 90°$, the photon map is rotated around the center together with φ_D (see Figure 18.24a).

In the case of s-polarized light, the Q-pole oscillations of X_{21}^+ and X_{22}^- contribute to the emission in the X-Z plane. If the detection direction is the Z-direction, only the emission of X_{21}^+ contributes to the photon map, and bright pair contrast appears at the edge of the sphere in the y direction. When the detection direction is set to the X direction, only the emission of X_{22}^- contributes to the photon map, and the fourfold symmetric contrast appears having a strong intensity in the diagonal direction between the x and the y axis. When changing the detection direction from the X to the Y direction, the emission intensity is given by

$$I_{s-\text{pol}}^{\text{PM}}(\rho, \varphi_0, \Omega_D) \propto |C_{22}(\rho)\sin 2(\varphi_0 - \varphi_D)|^2. \qquad (18.107)$$

Therefore, the fourfold symmetric pattern rotates around the center as φ_D changes, which was confirmed by the angle-resolved CL measurement (Thollar et al., 2018). It is seen in Figure 18.24b that, for $(\theta_D, \varphi_D) = (90°, 45°)$, the fourfold symmetric pattern rotates by 45°, which shows a photon map contributed only by the X_{22}^+.

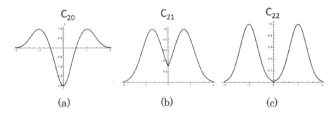

C_{20} C_{21} C_{22}

(a)   (b)   (c)

FIGURE 18.23 Excitation coefficients estimated from comparison with experimental patterns. (a) $C_{20}(\rho)$, (b) $C_{21}(\rho)$, (c) $C_{22}(\rho)$.

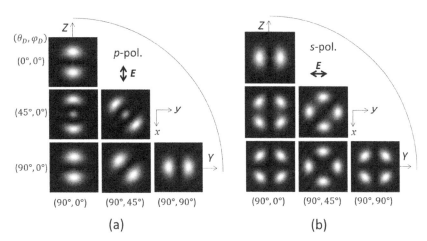

(a)   (b)

FIGURE 18.24 Theoretical polarized photon maps of Q-pole mode for various detection directions. (a) p-polarization and (b) s-polarization.

Size Dependence

The resonance energy of multipole mode of silver nanoparticles varies with particle size (Yamamoto et al., 2001a, 2011). Figure 18.25 shows the BSS images of silver nanoparticles of various sizes supported by the carbon film. Figure 18.25a,b shows BSS images of *p*- and *s*-polarized lights, respectively, measured under the same condition as in Figure 18.20c,d. It can be seen that the peaks of dipole and Q-pole mode shift to the lower energy side as the size increases.

Figure 18.26a shows the emission spectra measured from silver nanoparticles of various sizes on a carbon film. The *s*-polarized emission spectra were acquired with the electron beam incident near the edge of the nanoparticle under the same condition as Figure 18.25b. Several peaks corresponding to the multipole modes appear in the spectrum, and it is seen that the peak shifts to the low-energy side as the particle diameter increases. Figure 18.26b compares the observed peak energies with the theoretical ones as a function of particle diameter. The contrast distribution of the background represents the theoretical contour calculated based on Eq. (18.36), and the plot (solid circle) represents

the experimental peak energy from nanoparticles on a carbon thin film. The experimental values of the resonance energy are seen to be in good agreement with the theoretical background. Open circles in Figure 18.26b show the experimental peak energies from nanoparticles on a silver substrate. There also appear peaks due to the gap plasmon resonance in the lower energy side (Yamamoto et al., 2011).

18.5 Application to SPP-Related Phenomena

18.5.1 SPP–Light Conversion by Surface Step

When an electron incident on a metal surface excites an SPP, the SPP propagates as a spherical wave around the incident point to the surroundings and is converted into light by protuberance on the surface. The CL measurement of the SPP-light conversion by the surface step was carried out using a step of silver surface prepared by electron beam lithography. From the analysis of the results, it is shown

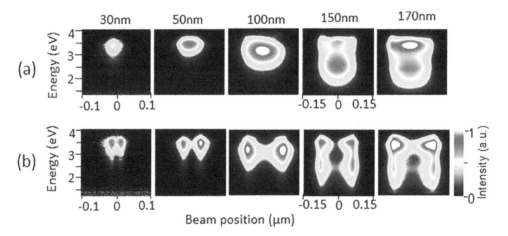

FIGURE 18.25   BSS images of silver nanoparticles of different sizes. (a) *p*-polarized and (b) *s*-polarized images.

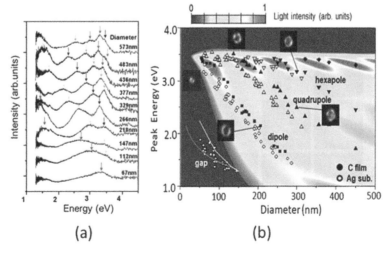

FIGURE 18.26   (a) Emission spectra from nanoparticles of various sizes and (b) size dependence of emission peak energy. Solid and open circles are plots of emission peak energies of silver nanoparticles on a carbon thin film and a silver substrate, respectively.

that the dispersion relation of SPP can be derived. Detailed analysis is described in the previous paper (Yamamoto and Suzuki, 2008; Yamamoto, 2012).

Figure 18.27a shows the arrangement of the experiment. A BSS image of p-polarized light was obtained by fixing a pinhole mask with a diameter of 0.5 mm to an appropriate emission angle θ ($\varphi = 0°$) in the xz plane and scanning the electron beam along the line AB (Figure 18.27b). Interference fringes with different periods appear on the left and right terraces of the center step in the BSS image. This contrast can be explained by interference between light converted from SPP in step and transition radiation caused by electron incidence. Assuming now that the incident position of the electron is taken as the origin, the x axis is perpendicular to the step, and the y axis is parallel, and the amplitude of each emission is expressed as

$$\psi_{\mathrm{TR}} = A(\theta)\frac{1}{R}e^{i(kR - \omega t)} \qquad (18.108)$$

$$\psi_{\mathrm{SPPR}} = \frac{1}{R}e^{i(kR-\omega t)}\int_{-\infty}^{\infty} E\frac{\exp(ik_p r)}{\sqrt{r}}S(\theta)e^{i\delta}e^{-i\boldsymbol{k}\cdot\boldsymbol{r}}dy, \qquad (18.109)$$

where E is the excitation amplitude of the SPP, $S(\theta)e^{i\delta}$ is the SPP-light conversion efficiency per unit length of the step, R is the distance from the origin to the light detection position, and $\boldsymbol{r}$ is the position vector on the step. The expression of the amplitude $A(\theta)$ of the transition radiation was given by Ginzburg and Frank (1946) and Ter-Mikaelian (1972). Since the integral of Eq. (18.109) contributes only from the SPP propagating approximately in the x direction (Yamamoto, 2012), the emission intensity is expressed by the following form,

$$I(x,\theta) = |\psi_{\mathrm{TR}} + \psi_{\mathrm{SPPR}}|^2 = C_1 + C_2\frac{1}{x} + C_3\frac{1}{\sqrt{x}}\cos(Kx+\delta). \qquad (18.110)$$

Here, K is wave number defined by

$$K = k_p \pm k\sin\theta. \qquad (18.111)$$

The double sign depends on whether the electron incident position is on the right or the left side of the step, and k_p and k are the wave numbers of SPP and light, respectively. The period corresponding to the wave number K represents the period of the interference fringes in the BSS image of Figure 18.27b. Since K and δ are determined by fitting the intensity profile of the interference fringe using Eq. (18.110), and k_p and θ can be obtained for each energy using Eq. (18.111). Figure 18.27c,d shows the energy dependence of the SPP wave number thus obtained and the phase shift in the SPP-light conversion.

18.5.2 One-Dimensional Plasmonic Crystal

SPP-Light Conversion by Periodic Structure

When SPP propagates through a periodic structure on a metal surface, the SPP is converted to light by the surface structure. In Figure 18.28, the phase difference between the light emitted by the scatterer at the origin O and the light emitted from the scatterer at the position R, which is separated by a vector $\boldsymbol{r}$, is given by

$$\omega\Delta t - \boldsymbol{k}\cdot\boldsymbol{r} = \boldsymbol{k_p}\cdot\boldsymbol{r} - \boldsymbol{k}\cdot\boldsymbol{r} = (\boldsymbol{k_p} - \boldsymbol{k})\cdot\boldsymbol{r}. \qquad (18.112)$$

Here, Δt is the time the SPP wave front propagates from O to R, and ω is the frequency common to both SPP and light.

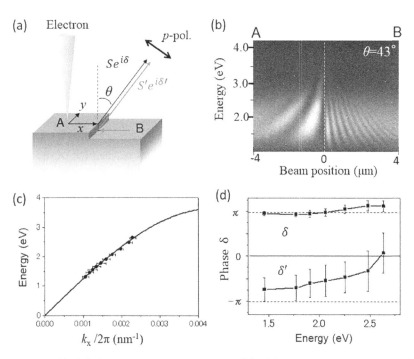

FIGURE 18.27 (a) Illustration of SPP-light conversion by surface step. (b) BSS image taken along the line AB, (c) the SPP dispersion curve on the silver surface, and (d) the phase shift by the SPP-light conversion.

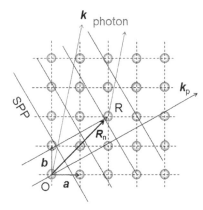

FIGURE 18.28 Conversion from SPP to light by periodic surface structure.

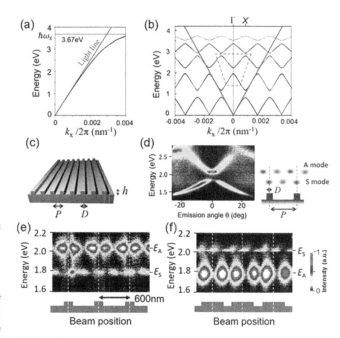

FIGURE 18.29 Dispersion relation of SPP on (a) a flat silver surface and (b) 1D-periodic structure ($P = 600$ nm). (c) Illustration of a 1D PlC. (d) ARS pattern from 1D-PlC (corresponding to the area surrounded by the gray dotted line in (b). BSS images are shown for 1D-PlCs of (e) Diameter to Period ratio D/P = 1/4, (f) D/P = 3/4.

k_p and k are the wave vectors of SPP and the emitted light, respectively. Let r be a lattice vector R_n at the n-th lattice point, and let the amplitude at which SPP is converted to light by one scatterer be $f(k_p, k)$, then, the amplitude of the optical emission converted by all the scatterers is expressed as

$$\psi(k_p, k) = \psi_0 \sum_n f(k_p, k) \exp[i(k_p - k) \cdot R_n]. \quad (18.113)$$

Here ψ_0 is the amplitude of the incident SPP. Because R_n is an in-plane vector $k \cdot R_n = k_\parallel \cdot R_n$ holds with $k_\parallel$ as the in-plane component of k. Therefore, the intensity of the emitted light is

$$I_{ph} = |\psi(k_p, k)|^2 = |\psi_0|^2 \cdot |f(k_p, k)|^2 L(k_p - k_\parallel). \quad (18.114)$$

Here f and L correspond to the atomic scattering factor and Laue function appearing in diffraction theory (Kittel, 2004). From the above, when the SPP propagates on the surface periodic structure, we obtain the following condition under which the emitted light interferes.

$$k_p - k_\parallel = G \quad (18.115a)$$

$$E_{SPP} = E_{ph} \quad (18.115b)$$

Here, G is the reciprocal lattice vector of the surface periodic structure.

One-Dimensional Plasmonic Crystal

Here we treat a 1D periodic structure with the basic translational vector of a and use the corresponding basic reciprocal lattice vector a^* to represent the reciprocal lattice vector as $G = na^*$, where n is an integer. As for an emission in the surface normal direction, $k_\parallel = 0$ (Γ point), the condition of (18.115a) for SPP-optical conversion holds for the plane wave SPP of the wave vector of $k_p = G$. Figure 18.29b shows the dispersion relation of a 1D-PlC in the a^* direction, which is created by shifting the dispersion curve of SPP on a flat silver surface (Figure 18.29a) by the reciprocal lattice vectors (empty lattice approximation). An SPP

wave having a wave vector k_p at a crossing position of the dispersion line (Brillouin zone boundary) is Bragg reflected due to the periodic structure. The reflected wave with the wave vector k_p' that satisfy $k_p' - k_p = G$ is simultaneously generated, and a standing wave is formed as a superposition of them. An SPP in a 1D-PlC is generally represented by a Bloch wave.

$$\psi_k^n(r) = \exp[ik \cdot r] u(r) = \sum_G C_G^n \exp[i(k + G) \cdot r] \quad (18.116)$$

The lowest order standing wave at the Γ point ($k = 0$) is the two plane waves propagating in the opposite direction having the wave vector $k_p = \pm a^*$ ($G = 2\pi/P$) with the same energy. That is, the SPP of the lowest energy state at the Γ point is mainly represented as a standing wave composed of these two plane waves, ignoring the higher order term.

$$\psi(x) \cong C_G \exp\left(2\pi i \frac{x}{P}\right) + C_{-G} \exp\left(-2\pi i \frac{x}{P}\right) \quad (18.117)$$

Here, the function represents the surface charge density, that is, it represents the electric field component perpendicular to the surface. Taking the coordinate origin at the center of the ridge of the 1D periodic structure, the surface structure has twofold rotational symmetry around the axis passing through the center and perpendicular to the surface. Translation of wave function by two-fold rotational symmetric operation C_2 is written as

$$C_2 \psi = \kappa \psi. \quad (18.118)$$

Since the wave function returns to the original for 360° rotation, $\kappa^2 = 1$ is hold. Thus we obtain $\kappa = \pm 1$. The eigenfunctions corresponding to the two eigenvalues are deduced as

$$\psi^S(x) \cong \sqrt{2}\cos\left(2\pi\frac{x}{P}\right), \psi^A(x) \cong \sqrt{2}\sin\left(2\pi\frac{x}{P}\right) \tag{18.119}$$

where subscripts S and A represent symmetric and antisymmetric standing waves, respectively, relative to the center of the ridge. Since the energy of the electromagnetic field formed by the two standing waves is different, an energy gap is opened to the dispersion line that intersects at the Γ point.

Barnes et al. (1996) showed that the magnitude of the bandgap at the Brillouin zone boundary of the SPP band structure depends on the Fourier series at the corresponding wave number of the surface shape function. They analytically derived the energy of SPP at the Γ point for the surface periodic structure with only two sinusoidal components. The band edge energy at the upper and lower bandgap of the Γ point is expressed as

$$E_{A,S} = E_0$$
$$\times \left[\{1 - (k_p h_2)^2\} \pm 2(k_p h_2)\frac{1}{\sqrt{-\varepsilon_0 \varepsilon}}\left\{1 - \frac{7}{2}(k_p h_2)^2\right\}\right]^{\frac{1}{2}} \tag{18.120}$$

where ε_0 and ε are the relative dielectric constant of vacuum and metal, respectively, and k_p is equal to $2\pi/P$, with P being the period of the surface structure. E_0 is the energy of SPP with the wave number $2\pi/P$, which is determined from the dispersion relation of SPP (Figure 18.29a), and $E_0 = 2.0$ eV when $P = 600$ nm. h_2 is the amplitude of the sinusoidal component of the wave number $2k_p$, that is, the second-order Fourier coefficient of the surface shape function. The n-th order Fourier coefficient of the surface shape function of the rectangular cross section (width D, height h) is expressed by the following equation.

$$h_n = \frac{h}{n\pi}\sin\left(n\pi\frac{D}{P}\right) \tag{18.121}$$

The double sign on the right side of Eq. (18.120) corresponds to the energy of the symmetric mode (-: S mode) and the antisymmetric mode (+: A mode) of the SPP standing wave, but the magnitude of $E_{A,S}$ depends on the sign of h_2.

We used a metal surface structure in which ridges having a rectangular cross section are periodically arranged as a 1D-PlC (Figure 18.29c). The period P was fixed at 600 nm, and the height of the ridge was fixed at 70 nm. 1D-PlCs with various ridge widths D were fabricated on an InP substrate by electron beam lithography, and silver was deposited to a thickness of 200 nm thereon. Figure 18.29d shows a representative ARS pattern from the 1D-PlC obtained by setting the sample and pinhole mask as shown in Figure 18.15. Figure 18.29e,f shows BSS images measured by scanning the electron beam in the direction crossing the ridge and using emission in the surface normal direction. These figures reveal

SPP standing waves in 1D-PlCs with D/P of 1/4 and 3/4, respectively, and the magnitude of energy in the symmetric mode and antisymmetric mode is reversed in the two plasmonic crystals. The symmetric mode shows a sharp peak because of the small radiation loss. When the height of the ridge becomes lower, the peak becomes sharper. Honda and Yamamoto (2014) reported that the nonradiative symmetric mode shows a high Q value of ~198 in the 1D-PlC with $D/P = 3/4$ with a height of 30 nm.

Figure 18.30a shows the change of the emission spectrum taken in the surface normal direction (Γ point) when D/P is changed. Figure 18.30b shows the result of plotting the peak energy versus D/P. The gray and white points of the plot indicate the peak energies of the symmetric and antisymmetric modes of the standing wave, respectively. The bandgap becomes maximum when D/P is near 1/4 and 3/4, and it is seen that the two bands cross at around 1/2. The background intensity distribution in Figure 18.30b represents the D/P dependence of the absorption spectrum of light calculated by the rigorous coupled-wave analysis (RCWA) method.

The white solid line and dotted line are the energies of symmetric and antisymmetric modes calculated using Eq. (18.120) derived by Barnes et al. (1996). Compared with the experimental result, the higher energy peak is relatively coincident with the theoretical one because it is close to the upper limit given by the photon energy of the same wave number. However, the lower peak energy is much lower than the calculated one by Eq. (18.120). This is considered to be due to the higher order terms included in the rectangular shape function. The result of the RCWA method is close to the experimental one, but the deviation to a slightly higher energy of the experimental peak is considered to be due to the fact that the step edge is rounded and deviates from the rectangular shape in the actual sample (Watanabe et al., 2014).

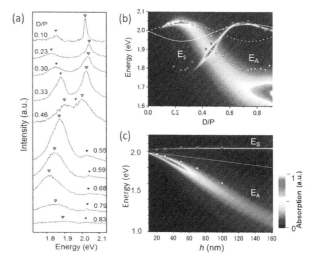

FIGURE 18.30 (a) Emission spectra from 1D-PlCs with various D/P, (b) D/P dependence of band edge energies (ridge height 70 nm), and (c) ridge height dependence of the band edge energies (D/P = 3/4).

Here we explain the difference in the SPP-light conversion efficiency between the symmetric and antisymmetric SPP standing waves by taking the emission of the Γ point as an example. Figure 18.31 depicts the first and second frequency components of the surface shape function of the (a) $D/P = 1/4$ and (b) $D/P = 3/4$ plasmonic crystal. Although the bandgap of Eq. (18.120) depends only on the second-order component (h_2), the first-order component (h_1) is necessary for light emission. The charge distributions of the two modes of the Γ point are shown on the lower side of Figure 18.31. The electric field vector of light emitted in the surface normal direction is parallel to the surface. It is obvious from the figure that the charge distribution that generates the electric field component parallel to the surface is in the antisymmetric mode, and it can be seen that this mode can be strongly coupled with light irrespective of the value of D/P. However, due to the difference in the charge distribution in the second-order periodic structure, the energy of the two modes becomes opposite in the case of $D/P = 1/4$ and $3/4$ (Figure 18.30b). That is, a mode in which the charge distribution concentrates on the peak positions of the second-order structure always appears on the low-energy side.

SPP Cavity

We describe the fundamental properties of SPP confined in 1D cavity composed of two walls placed at distance L on the metal surface. Assume that SPP is reflected by the amplitude $\text{Re}^{i\phi}$ at each wall of the cavity. When the SPP plane wave of the wave number k_p expressed by $\psi = e^{i(k_p x - \omega t)}$ exists in the cavity, the internal wave undergoing multiple reflection by the side wall at $x = 0$ and $x = L$ is written as

$$\psi = e^{i(k_p x - \omega t)} + \text{Re}^{i\phi} e^{i\{k_p(L-x)+k_p L - \omega t\}} + \left(\text{Re}^{i\phi}\right)^2$$
$$\times e^{i\{k_p x + 2k_p L - \omega t\}} + \cdots = \left\{ e^{ik_p x} + \text{Re}^{i\phi} e^{i\{k_p(L-x)+k_p L\}} \right\}$$
$$\times \left\{ 1 + R^2 e^{i(2k_p L + 2\phi)} + \cdots \right\} e^{-i\omega t}. \qquad (18.122)$$

If we approximate $R \approx 1$, this is rewritten as

$$\psi \approx 2\cos\left\{ (k_p (x - L) - \frac{\phi}{2} \right\}$$
$$\times \frac{1}{1 - R^2 e^{i(2k_p L + 2\phi)}} e^{i\left(k_p L + \frac{\phi}{2}\right)} e^{-i\omega t}. \qquad (18.123)$$

The condition that maximizes this amplitude is expressed as

$$k_p L + \phi = n\pi, \qquad (18.124)$$

where n is an integer. Under this cavity condition, the SPP forms a standing wave in the cavity, and the standing wave has a following form,

$$\psi_{\text{SPP}} \propto \cos\left\{ \left(k_p \left(x - \frac{L}{2}\right) - \frac{k_p L + \phi}{2} \right\}$$
$$= \cos\left\{ k_p \left(x - \frac{L}{2}\right) - \frac{n\pi}{2} \right\}. \qquad (18.125)$$

When n is an even number, this function is symmetric with respect to the center of the cavity, and when n is an odd number, it is antisymmetric. The amplitude at the wall is $\pm\cos\left(\frac{\phi}{2}\right)$. Therefore a standing wave has an open end at the wall when the phase shift ϕ is 0, and has a fixed end when ϕ is π.

Figure 18.32 shows the relationship between wave number and phase ϕ of SPP satisfying the cavity condition. With the cavity width L being constant, there are two solutions of SPP standing waves, symmetrical one (broken line) with respect to the center of the cavity and antisymmetric one (solid line) for a certain wave number (or energy). As ϕ changes, the end shape of the SPP standing wave changes between the open and fixed ends, but the shape of the center does not change. Which solutions actually appear depends on the nature of SPP reflection at the wall of the cavity. If the wave numbers of the two SPP waves are the same, which shape of the SPP standing wave appears depends on the phase ϕ.

Cavity of a Step Pair

The result of angle-resolved measurement of 1D cavity sandwiched between two steps on a metal surface is shown in

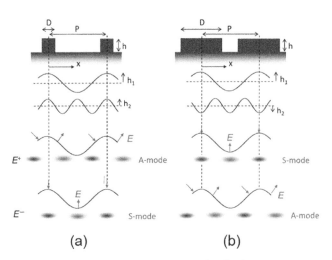

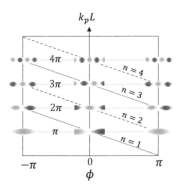

FIGURE 18.31 First-order and second-order frequency components and surface charge distribution of the two eigenmodes for (a) $D/P = 1/4$ and (b) $D/P = 3/4$ surface shape functions.

FIGURE 18.32 Relationship between k_p and phase ϕ satisfying cavity condition. A standing wave in the cavity has a solution of a symmetrical form for $n =$ even (broken line) and an antisymmetric form for $n =$ odd (solid line).

Figure 18.33. A sectional view of the cavity is schematically shown on each figure. The width of the groove or ridge is 300 nm, and the height (depth) is 50 nm. As shown in Figure 18.33a, the sample was slightly tilted around the Y axis, and the pinhole mask was moved horizontally from A to B, measuring the ARS pattern using the s polarized light. This is the same arrangement as in Figure 18.15, where the circle at the center of line A-B corresponds to the surface normal direction. The pinhole mask moves substantially along the $\varphi = 90°$ line along which the emission angle θ changes with φ fixed at 90°.

Figure 18.33b shows the ARS pattern measured while scanning the electron beam near the center in the groove-type cavity sandwiched between Up-steps. A symmetrical standing wave pattern with $n = 2$ is generated. This is because the emission intensity in the surface normal direction is 0. Since the energy is 2.0 eV, i.e., the wavelength of SPP of that energy is 600 nm, the phase shift of SPP reflection at the Up-step is determined to be $\phi = \pi$ from the cavity condition.

Figure 18.33c is a similar ARS pattern for the ridge-type cavity sandwiched between Down-steps. Although emission intensity is weak, emission due to an antisymmetrical standing wave of $n = 1$ appears at $E = 2.0$ eV. Therefore,

in the ridge-type cavity, the phase shift of SPP reflection at the Down-step is determined to be $\phi = 0$. That is, it can be seen that Up-step works as fixed end of SPP wave and Down-step works as an open end.

Cavity of Groove and Ridge Pair

Ridge-Type Cavity with a Pair of Grooves. Figure 18.34 shows the results of angle-resolved measurement of a ridge-type cavity sandwiched between rectangular grooves with width $w = 150$ nm. Figure 18.34a,b shows the ARS pattern and BSS image of the cavity with the inner width $d = 300$ nm measured under the same conditions as in Figure 18.33, respectively, and Figure 18.34c shows a sectional view of the cavity. In the measurement of the BSS image, a pinhole mask with a diameter of 1 mm was set at the position corresponding to the surface normal direction (circle in the middle of the line AB in Figure 18.33a), so only an antisymmetric standing wave of odd n is visualized.

Standing waves with energy of 2 eV (SPP wavelength of 600 nm) show strong contrast. This is considered because the cavity satisfies the condition that the outer Up-step acts as the fixed end ($\phi = \pi$) and the inner Down-step simultaneously acts as the open end ($\phi = 0$), which causes a high reflection efficiency. From this result, a ridge-type cavity

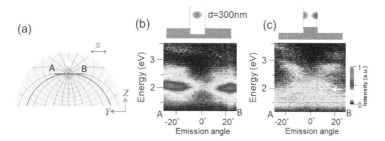

FIGURE 18.33 (a) Pinhole mask movement for ARS measurement, and ARS patterns from (b) groove-type and (c) ridge-type cavities.

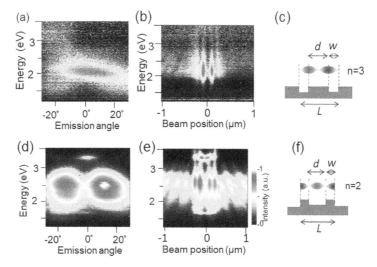

FIGURE 18.34 ARS measurement of the cavities of groove and ridge pair on the silver surface. (a) ARS pattern of ridge-type cavity composed of two grooves, (b) BSS image, and (c) a schematic view of the cavity section. (d) ARS pattern of groove-type cavity composed of two ridges, (e) BSS image, and (f) a schematic view of the cavity section. Inner width $d = 300$ nm, groove or ridge width $w = 150$ nm, and depth (height) $h = 50$ nm.

formed by a pair of grooves can realize high efficiency when the cavity width L ($= d + 2\,w$) and the groove width w satisfy the following conditions at the same time,

$$(1)\ L = \frac{(n-1)}{2}\lambda_p,\ (2)\ w = \frac{(2m+1)}{4}\lambda_p, \qquad (18.126)$$

where λ_p is the wavelength of the SPP determined by the SPP dispersion relation from the wavelength of the light. When n is odd and L is a multiple of λ_p, an antisymmetric SPP standing wave is excited in the cavity. Such a cavity has high emission efficiency because the SPP standing wave forms an electric dipole on the central terrace. As for the cavity in Figure 18.34c, $L = \lambda_p (n = 3)$ and $w = \lambda_p/4$ for $\lambda_p = 600\mathrm{nm}$, so the strong emission appears at the energy of 2 eV. In this cavity, it is expected that the antisymmetric SPP standing wave can be efficiently excited by the incidence of 2 eV light from the surface normal direction. However, as the cavity width increases, many emission peaks of the higher order cavity modes appear in the ARS pattern, making it difficult to signify enhancement conditions such as in Eq. (18.126).

Groove-Type Cavity with a Pair of Ridges. The property of a groove-type cavity sandwiched between two rectangular ridges is different from that of a ridge-type cavity. Figure 18.34d,e shows the ARS pattern and the BSS image from the cavity composed of two ridges with ridge width $w = 150$ nm and height $h = 50$ nm. In this groove-type cavity, if assuming that the outer Down-step is the open end ($\phi = 0$), the conditions of the phase matching should be written as

$$(1)\ L = \frac{n}{2}\lambda_p,\ (2)\ w = \frac{(2m+1)}{4}\lambda_p. \qquad (18.127)$$

Figure 18.34f depicts the SPP standing wave of $\lambda_p = 600$ nm as expected from the above equations. However, although the ARS pattern reflects a symmetric mode, the energy of the measured SPP standing wave is much higher than the 2 eV expected from the above equations. This indicates that the assumption of the phase shift at the outer Down-step ($\phi = 0$) is not valid.

The SPP reflection by the rectangular ridge was studied by Ozawa and Yamamoto (2018), which depends on rectangular shape (width and height) and SPP energy. They calculated the SPP reflectivity of a single rectangular ridge based on the Green tensor method and deduced the optimum conditions of the groove-type cavity.

Cavity in 1D PlC. In the case where there is a locally different structure in the 1D-PlC, a localized SPP mode is formed. The bandgap at the Γ point of the 1D-PlC opens largely when the ratio D/P of terrace width D and period P is 1/4 and 3/4. As shown in Figure 18.35, when changing the lattice spacing at one position in the perfect periodic structure from the original period P to $3/2P$, the antisymmetric mode (D/P = 1/4) or a symmetric mode (D/P = 3/4) standing wave is formed. The energy of this groove-type cavity mode shifts from the band edge on the high-energy side toward the band edge on the low-energy side as the cavity width increases. The property of this cavity was investigated by CL method using silver 1D-PlC (Honda and

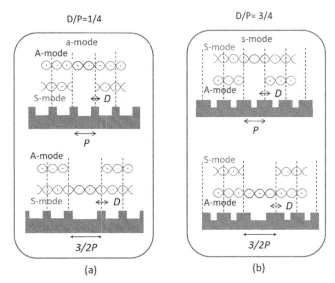

FIGURE 18.35 A schematic diagram of an SPP standing wave formed in the groove-type cavity of 1D-PlC. (a) $D/P = 1/4$ and (b) $D/P = 3/4$.

Yamamoto, 2013). It was found that, inside the cavity, the wave number of the SPP standing wave follows the dispersion relation of the SPP propagating on the plane, and the phase shift of the SPP reflection at the cavity wall appearing in the cavity condition (Eq. 18.124) is approximately expressed as

$$\phi\left(E^{\mathrm{cav}}\right) = \pi\frac{E^{\mathrm{cav}} - E^-}{E^+ - E^-}, \qquad (18.128)$$

where E^{cav} is the cavity mode energy.

When the cavity width is further increased from $3/2P$, cavity modes of the next order enter into the bandgap in order. Figure 18.36 shows the ARS pattern by the cavity mode appearing within the bandgap of the Γ point when changing the cavity width using a 1D-PlC of silver with a period of 600 nm. The left side is the groove-type and on the right side is the ridge-type cavity, respectively, each of which shows patterns of $D/P = 1/4$ and $D/P = 3/4$. It should be noted that the groove width (or ridge width) d in the cavity is different from the cavity width $d + D$ defined in Figure 18.35. It can be seen that the symmetry of the standing wave is different between groove-type and ridge-type cavities of the same d. Also, as d increases by P/2 (= 300 nm), the symmetry of the standing wave changes alternately.

18.5.3 Two-Dimensional Plasmonic Crystals

Here, we deal with the SPP mode of a 2D-PlC in which cylindrical pillars are arranged in a square lattice (Takeuchi and Yamamoto, 2011; Yamamoto and Saito, 2014). The reciprocal lattice of the square lattice is shown in Figure 18.37a. The dispersion plane of the SPP propagating on the 2D-PlC is obtained by superimposing the dispersion plane of

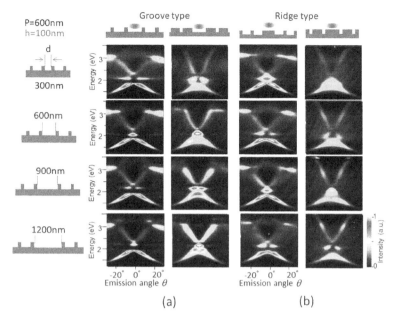

FIGURE 18.36 ARS patterns from cavities in 1D-PlC. (a) Groove type and (b) ridge type, with $D/P = 1/4$ on the left and $D/P = 3/4$ on the right, respectively.

the SPP propagating in the plane by the reciprocal lattice vectors. Figure 18.37b shows the band dispersion curves along the X and M directions from the Γ point ($\boldsymbol{k}_{\mathrm{p}} = 0$).

The eigenmodes of an SPP at the Γ point in the SQ-PlC are deduced using group theory. The $\boldsymbol{k}$ group at the Γ point corresponds to the C_{4v} point group (Table 18.2). The wave vector for the band structure at the Γ point (Figure 18.37b) is $\boldsymbol{k}_{\mathrm{p}} = \boldsymbol{a}^* = (2\pi/P, 0, 0)$. This particular wave vector can be transformed into other equivalent wave vectors via operations involving the C_{4v} point group. These transformations yield a set of basic wave vectors, $\boldsymbol{g}_1 = \boldsymbol{a}^*$, $\boldsymbol{g}_2 = \boldsymbol{b}^*$, $\boldsymbol{g}_3 = -\boldsymbol{a}^*$, $\boldsymbol{g}_4 = -\boldsymbol{b}^*$. The eigenfunctions of the band edge states are constructed by a linear combination of basis functions, $\varphi_j(\boldsymbol{r}) = e^{i\boldsymbol{g}_j \cdot \boldsymbol{r}}$ ($j = 1 \sim 4$). The representation based on the four basis functions is $\{4,0,0,2,0\}$ for the five classes of C_{4v} (Table 18.1, bottom row). This representation is reducible and can be decomposed into three irreducible representations, which are expressed as their direct sum,

$$\Gamma_{\mathrm{red}} = A_1 \oplus B_1 \oplus E. \qquad (18.129)$$

Hence, the four bands should form three eigenstates with three different energies at the Γ point.

The eigenfunctions of these irreducible representations can be derived by applying the projection operator to one of the basis functions using the characters in Table 18.2 and are expressed as

$$\psi_{A1}(x, y) = \cos 2\pi x + \cos 2\pi y, \qquad (18.130a)$$

$$\psi_{B1}(x, y) = \cos 2\pi x - \cos 2\pi y \qquad (18.130b)$$

$$\psi_{E(1)}(x, y) = \sin 2\pi x \qquad (18.130c)$$

$$\psi_{E(2)}(x, y) = \sin 2\pi y \qquad (18.130d)$$

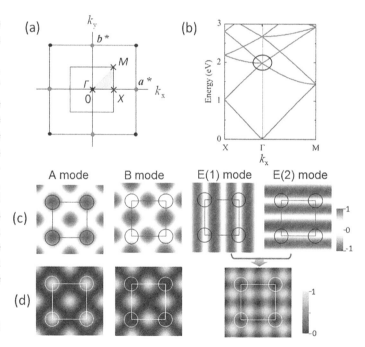

FIGURE 18.37 (a) Reciprocal lattice of square lattice and (b) energy dispersion of the 2D-PlC (empty lattice approximation). (c) Spatial distribution of the electric field of the eigenmodes at the Γ point and (d) field strength distribution.

where the normalization factors are omitted. Because the E mode in Eqs. (18.130c) and (18.130d) is energetically doubly degenerate, it can be expressed as a set of other eigenfunctions orthogonally transformed (Takeuchi and Yamamoto, 2011). These functions represent the surface charge distribution of each mode, that is, the surface normal component of the electric field, $\psi_n(\boldsymbol{r}) \propto E_z^n(\boldsymbol{r})$. The distribution of

TABLE 18.2   Character Table for C_{4v} Point Group

C_{4v}	E	$2C_4$	C_2	$2\sigma_v$	$2\sigma_d$	Eigenfunction
$A1$	1	1	1	1	1	ψ_{A1}
$A2$	1	1	1	-1	-1	
$B1$	1	-1	1	1	-1	ψ_{B1}
$B2$	1	-1	1	-1	1	
E	2	0	-2	0	0	$\psi_{E(1)}, \psi_{A(1)}$
Γ_{red}	4	0	0	2	0	$A_1 \oplus B_1 \oplus E$

$E_z^n(\boldsymbol{r})$ in each mode is shown in Figure 18.37c. As described in Section 18.3.3, the photon map reflects the time average of the field strength, $\left\langle \left| \mathrm{Re}[\psi_n(\boldsymbol{r})\exp(-i\omega t)]\right|^2 \right\rangle_{\mathrm{t}}$. Calculating this for each mode, we obtain

$A1$ mode:$(\cos 2\pi x + \cos 2\pi y)^2$,

$B1$ mode:$(\cos 2\pi x - \cos 2\pi y)^2$,

$E(1)$ mode:$\sin^2 2\pi x$,

$E(2)$ mode:$\sin^2 2\pi y$.

Figure 18.37d shows the spatial distribution of the eigenmodes, where the field strength of the E mode is given by the sum of the $E(1)$ and $E(2)$ modes, assuming that the two modes are equally excited.

Energetically degenerate modes like E mode can be visualized individually using specific polarized light for photon map measurement. Figure 18.38a shows a panchromatic photon map of a 2D-PlC. Cylindrical pillars with diameter $D = 500$ nm and height $h = 100$ nm are arranged in a square lattice with a period $P = 600$ nm covered with silver, having a thickness of 200 nm. The band dispersion curves at the Γ point of this 2D-PlC is schematically shown in Figure 18.38b deduced from the ARS pattern (Yamamoto and Saito, 2014). Doubly degenerate E mode is at the lowest energy position (1.867 eV). In the arrangement shown in Figure 18.15, a pinhole mask with a diameter of 1 mm was

set at a position corresponding to the surface normal direction, and monochromatic photon maps of E mode measured with nonpolarized light and s polarized light are shown in Figure 18.38c,d, respectively.

The photon map of Figure 18.38c shows a pattern similar to Figure 18.38e, which is a pattern derived from the group theory. On the other hand, the s-polarized photon map in Figure 18.38d reveals the pattern corresponding to the $E(1)$ mode shown in Figure 18.37c. As is clear from the electric field distribution in Figure 18.37c, the electric dipoles formed in the cylindrical pillar are in the horizontal direction in the $E(1)$ mode and in the vertical direction in the $E(2)$ mode. Therefore only the electric field of the $E(1)$ mode can contribute to the emission of s-polarized light. Figure 18.38f shows the surface electric field strength calculated by RCWA method when collimated s-polarized light is incident from the surface normal direction. This result well reproduces the photon map of Figure 18.38d.

The band edge energy of each mode at the Γ point of 2D-PlC consisting of a periodic square array of cylindrical pillars depends on the cylindrical diameter D. Figure 18.39 shows the D dependence of the band edge energy of each mode at the Γ point. Figure 18.39a depicts the position of the line scanned by the electron beam when measuring the BSS image with the arrangement shown in Figure 18.15. Figure 18.39b,c shows nonpolarized BSS images obtained by scanning the electron beam along the lines L1 and L2, respectively. The D dependence of the band edge energy of each mode obtained from this image is plotted on the graph of Figure 18.39d. It turns out that the energy of the E mode is higher than that of the A mode when D is small and it is reversed when D is large. This can be understood from the analogy of symmetric and antisymmetric modes in 1D-PlC. The second-order Fourier coefficient corresponding to Eq. (18.121) is as follows in the case of a cylinder

$$h_2 = \pi h \left(\frac{D}{2P}\right)^2 \frac{2J_1(X)}{X}. \tag{18.131}$$

Here, J_1 is the Bessel function of the first kind and $X = 2\pi D/P$. This function is 0 when $X = 3.832$, i.e., $D = 0.61 P = 366$ nm. Even in Figure 18.39d, the energy intersection between A mode and E mode occurs in the vicinity of this value. On the other hand, since the electric field of the B mode is distributed in a flat area around the cylinder, as shown in Figure 18.37c, its energy does not change significantly depending on D.

The band structure of 2D-PlC in which circular holes are periodically arranged on a square lattice instead of a cylindrical pillar was also investigated by the angle-resolved CL measurement (Yamamoto and Saito, 2014). In the case of a circular hole, Fourier coefficient h in Eq. (18.131) becomes negative so that the magnitudes of the band edge energy of the A and E bands are opposite to those in the case of a cylindrical pillar. When one row is removed from the 2D-PlC square lattice, a 1D cavity mode is formed between the A and B bands at the Γ point and also between B and E bands. It was shown that their cavity energy positions can

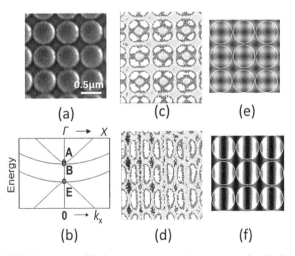

FIGURE 18.38   (a) A panchromatic photon map of cylindrical pillars arranged in square lattice and (b) band structure at Γ point (period $P = 600$ nm, cylindrical diameter $D = 500$ nm, height $h = 100$ nm). Monochromatic photon maps of E mode; (c) nonpolarized light and (d) s-polarized light. Photon maps calculated from theory: (e) group theory and (f) RCWA method.

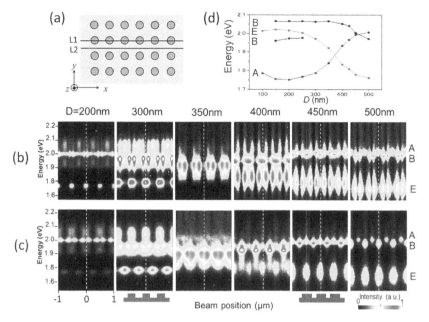

FIGURE 18.39 (a) A schematic diagram of 2D-PlC with a square lattice (period $P = 600$ nm, cylindrical height $h = 100$ nm), nonpolarized BSS images of PlCs of various diameter D taken with the electron beam scanning along the lines (b) L1 and (c) L2. (d) Diameter dependence of band edge energy of each mode at Γ point.

be controlled within the bandgap by changing the interval between the neighboring rows (Saito and Yamamoto, 2015a). Furthermore, it was presented that 2D cavity can be formed by introducing a 2D heterostructure in 2D-PlC with tetragonal lattice (Saito et al., 2015). 2D-PlC with a tetragonal lattice has been applied to plasmonic laser (Beijnum et al., 2013), chemical sensors (Anker et al., 2008), photodetectors (Senanayake et al., 2011), and plasmonic solar cells (Atwater and Polman, 2010) because of the high density of state at the band edge and cavity state.

The SPP band structure at the Γ point of 2D-PlC in which cylindrical pillars and circular holes are arranged in a hexagonal lattice is examined by angle-resolved CL (Saito and Yamamoto, 2015b). In the hexagonal plasmonic structure, there is an energy band in which the bandgaps at points M and K are overlapped and form a full gap in all directions (Kitson et al., 1996). Therefore, it has been an interesting application for SPP waveguide (Bozhevolnyi et al., 2001) and Bragg mirror (Ditlbacher et al., 2002). By introducing a linear defect into 2D-PlC with a hexagonal lattice and being coupled with a 1D-PlC, the property of nonradiative SPP waveguide mode was investigated by angle-resolved CL method (Saito et al., 2017).

BSS image, and the photon map by selecting the emission angle (wave number) due to angle-resolved measurement in addition to the selection of polarization of light and wavelength (energy). From the ARS pattern, the dispersion pattern of SPP in the plasmonic structure is visualized, and the band structure in the reciprocal space is obtained. In the BSS image and photon map, the standing wave pattern of the eigenmode of SPP is visualized, and the mode can be identified. In particular, degenerate modes can be individually visualized by choosing emission direction and polarization of light. The application of the STEM-CL method to the representative nanostructures in plasmonics, such as metal nanoparticles and plasmonic crystals, has been demonstrated, and the reader can understand how to use this method and what kind of information can be obtained. Although we mainly dealt with basic and simple plasmonic structures here, this method has been applied to other nanostructures of various shapes such as rods (Yamamoto et al., 2006; Gomez-Medina et al., 2008) and nanoholes (Sannomiya et al., 2016). In future, it is expected that high-resolution STEM-CL system for the characterization of SPs in nanoscale structures and nanomaterials of more complicated plasmonic devices will be utilized.

18.6 Conclusion

In this chapter, we introduce the high-resolution CL method combined with STEM as a useful technique for evaluating nanomaterials in the field of rapidly developing plasmonics. The electron beam is focused down to about 1 nm at any position on the sample and can generate a point source of SP. In the CL system, it is possible to acquire the ARS pattern,

References

Anker, JN, Hall, WP, Lyandres, O, Shah, NC, Zhao, J, and Van Duyne, RP, 2008, *Nat. Mater.* **7**(6), 442–453.

Atwater, HA, and Polman, A, 2010, *Nat. Mater.* **9**(3), 205–213.

Barnes, WL, Murray,WA, Dintinger, J, Devaux, E, and Ebbesen, TW, 2004, *Phys. Rev. Lett.* **92**(10), 107401.

Barnes, WL, Preist, TW, Kitson, SC, and Samble, JR, 1996, *Phys. Rev. B* **54**(9), 6227–6244.

Beijnum, F, Veldhoven, PJ, Geluk, EJ, Dood, MJA, Hooft, GW, and Exter1, MP, 2013, *Phys. Rev. Lett.* **110**, 206802.

Bozhevolnyi, SI, Erland, J, Leosson, K, Skovgaard, PMW, and Hvam, JM, 2001, *Phys. Rev. Lett.* **86**(14), 3008–3011.

Brongersma, ML, and Kik, PG (eds.), 2007, *Surface Plasmon Nanophotonics*, Springer, Dordrecht.

Clavero, C, 2014, *Nat. Photonics* **8**, 95–103.

Ditlbacher, H., Krenn, J. R., Schider, G., Leitner, A., and Aussenegg, F. R., 2002, *Appl. Phys. Lett.* **81**(10), 1762–1764.

Ebbesen, TW, Lecec, HJ, Chaemi, HF, Thio, T, and Wolf, PA, 1998, *Nature* **391**, 667–669.

García de Abajo, FJ, 1999, *Phys. Rev. B* **59**, 3095–3107.

García de Abajo, FJ, 2010, *Rev. Mod. Phys.* **82**, 209–275.

García de Abajo, FJ, and Kociak, M, 2008, *Phy. Rev. Lett.* **100**, 106804.

Ginzburg, VL, Frank, IM, 1946, *JETP (USSR)* **16**, 15; 1946, *J. Phys.* **9**, 353.

Gomez-Medina, R, Yamamoto, N, Nakano, M, and Garcia de Abajo, FJ, 2008, *New J. Phys.* **10**, 105009 (1–13).

Heitmann, D, 1977, *J. Phys. C.* **10**, 397.

Homola, J, 2008, *Chem. Rev.* **108**, 462–493.

Honda, M, and Yamamoto, N, 2014, *Appl. Phys. Lett.* **104**, 081112.

Honda, M, and Yamamoto, N, 2013, *Opt. Express* **21**, 11973–11983.

Jackson, J.D. 1975. *Classical Electrodynamics*, Wiley, New York.

Kitson, SC, Barnes, WL, and Sambles, JR, 1996, *Phys. Rev. Lett.* **77**, 2670–2673.

Kittel, C, 2004, *Introduction to Solid State Physics*, 8th edn, Wiley, New York.

Kociak, M, and García de Abajo, FJ, 2012, *MRS Bull.* **37**, 39–46.

Kociak, M, and Stéphan, O, 2014, *Chem. Soc. Rev.* **43**, 3865–3883.

Kröger, E, 1968, *Z. Phys.* **216**, 115–135.

Losquin, A, and Kociak, M, 2015, *ACS Photonics* **2**, 1619–1627.

Maier, SA, 2007, *Plasmonics: Fundamentals and Applications*, Springer, New York.

Mie, G, 1908, *Ann. Phys. (Leipzig)* **330**, 377–445.

Myroshnychenko, V, Rodríguez-Fernández, J, Pastoriza-Santos, I, Funston, AM, Novo, C, Mulvaney, P, Liz-Marzán, LM, García de Abajo, FJ, 2008, *Chem. Soc. Rev.* **37**, 1792–1805.

Myroshnychenko, V, Nishio, N, García de Abajo, FJ, Förstner, J, and Yamamoto, N, 2018, *ACS Nano* **12**, 8436–8446.

Olmon, RL, Slovick, B, Johnson, TW, Shelton, D, Oh, S-H, Boreman, GD, and Raschke, MB, 2012, *Phys. Rev. B* **86**, 235147.

Ozawa, K, and Yamamoto, N, 2018, *Phys. Rev. B* **100**, 075406.

Palik, ED (ed.), 1998, *Handbook of Optical Constants of Solids*, Academic Press, San Diego, CA.

Raether, H, 1988, *Surface Plasmons on Smooth and Rough Surfaces and on Gratings*, Springer-Verlag, Berlin.

Ritchie, RH, 1957, *Phys. Rev.* **106**, 874.

Saito, H, and Yamamoto, N, 2015a, *Nano Lett.* **15**, 5764–5769.

Saito, H, and Yamamoto, N, 2015b, *Opt. Express* **23**, 2524–2540.

Saito, H, Mizuma, M, and Yamamoto, N, 2015, *Nano Lett.* **15**, 6789–6793.

Saito, H, Yamamoto, N, and Sannomiya, T, 2017, *ACS Photonics* **4**, 1361–1370.

Sannomiya, T, 2011, *Trends Biotechnol.* **29**, 343–351.

Sannomiya, T, Saito, H, Junesch, J, and Yamamoto, N, 2016, *Light Sci. Appl.* **5**, e16146.

Schlücker, S, 2014, *Angew. Chem. Int. Ed.* **53**, 4756–4795.

Senanayake, P, Hung, C-H, Shapiro, J, Lin, A, Liang, B, Williams, BS, and Huffaker, DL, 2011, *Nano Lett.* **11**(12), 5279–5283.

Suzuki, T, and Yamamoto, N, 2009, *Opt. Express* **17**, 23664–23671.

Takeuchi, K, and Yamamoto, N, 2011, *Opt. Express* **19**, 12365–12374.

Ter-Mikaelian, ML, 1972, *High-Energy Electromagnetic Processes in Condensed Media*, Wiley, New York.

Thollar, Z, Wadell, C, Matsukata, T, Yamamoto, N, and Sannomiya, T, 2018, *ACS Photonics* 5, 2555-2560.

Watanabe, H, Honda, M, and Yamamoto, N, 2014, *Opt. Express* **22**, 5155–5165.

Yamamoto, N, 2010, Cathodoluminescence of nanomaterials, In: *Handbook of Nanophysics; Nanoelectronics and Nanophotonics*, Ed. Sattler, KD, Ch. 21, Taylor & Francis Publisher and CRC Press, Boca Raton, FL, pp. 1–25.

Yamamoto, N, 2012, Cathodoluminescence of surface plasmon induced light emission, In: *The Transmission Electron Microscope*, Ed. Khan, M, Ch. 15, InTech, Croatia, pp. 1–24.

Yamamoto, N, 2016, *Microscopy* **65**(4), 282–295.

Yamamoto, N, and Saito, H, 2014, *Opt. Express* **22**, 29761–29777.

Yamamoto, N, and Suzuki, T, 2008, *Appl. Phys. Lett.* **93**, 093114.

Yamamoto, N, Araya, K, and García de Abajo, FJ, 2001a, *Phys. Rev.B* **64**, 205419.

Yamamoto, N, Araya, K, Toda, A, and Sugiyama, H, 2001b, *Surf. Interface Anal.* **31**, 79–86.

Yamamoto N, García de Abajo, FJ, and Myroshnychenko, V, 2015, *Phys. Rev. B* **91**, 125144.

Yamamoto, N, Nakano, M, and Suzuki, T, 2006, *Surf. Interface Anal.* **38**, 1725–1730.

Yamamoto, N, Ohtani, S, Garca de Abajo, FJ, 2011, *Nano Lett.* **11**, 91–95.

Yamamoto, N, Sugiyama, H, and Toda, A, 1996, *Proc. Roy. Soc. London A* **452**, 2279–2301.

19

Biosensing under Surface Plasmon Resonance Conditions

19.1 Introduction... 19-1
19.2 Biosensing Concept in Analytical Science 19-2
19.3 Physics of SPR 19-2
   Excitation of SPP • Utilizing the SPR Phenomena in Sensors • Simple Model of an SPR Sensor with a Linear Standard Curve • Technology Overview
19.4 Interfacial Design... 19-7
   Sensing Architectures for Transducer-Based Systems: General Requirements • PT of SPR Sensors: Optimization of the Metal Layer • Selection of the Optimal Interfacial Arrangement: Major Factors Determining the Interaction of a Macromolecule with a Functionalized Surface
19.5 Research Methodology: SPR-Specific Analytical Workflow 19-9
   Determination of the Presence of the Analyte in the Sample: Direct Measurements • Quantitative Determination of Analytes, Especially of a Small Size: Competition Assay Formats • Protein–Protein Interaction in the Solution: Capture of an Analyte–Receptor Complex with Standard Additions of the Receptor (CARSAR) Approach • SPR Analysis of "3D" Objects: Density Variations in the Fixed Architectures Approach (DViFA) • Validation Procedure: Checking for Surface-Induced Analyte Transitions in Adsorbed Biofilms • Analyte-Specific Contrast Enhancement
19.6 SPR-Specific Data Mining 19-13
   Measurement Uncertainty in SPR-Based Analytical Studies • Sizing the SPR Response: Optical Quantification of the Adsorbate Concentration • Formalization of the Evolving Interfacial Landscape: Comprehensive Kinetic Analysis
19.7 Concluding Remarks.. 19-18
Acknowledgment 19-18
References 19-18

Boris Snopok
National Academy of Sciences of Ukraine

19.1 Introduction

The surface wave, which first appeared more than a century ago as a "nonphysical" solution of Maxwell's equations, is considered today as a kind of "panacea" for bridging the nano and macroworlds in nanophotonics and, especially, plasmonics. Despite the advances made in the understanding of surface waves and their widespread use in various elements and systems of modern optoelectronics, an ever-increasing number of investigators from various branches of science are showing interest in "surface-captured light." This is mostly due to the discovered potential control of the localization and direction of propagation of electromagnetic excitations at the nanoscale, which creates new opportunities for local photoconversions in chemistry, subwave cartography, chemical and biological sensors, spatial movement, and directed heat transfer at the nanolevel, active photonic elements with new information carriers, etc. (Brongersma and Kik 2007).

One of the most successful applications of the "surface-captured light" is optical biosensors based on surface plasmon resonance (SPR) phenomena (Raether 1985, Mukhopadhyay 2005, Singh 2016, Snopok 2012, Long and Jing 2014, Narayanaswamy and Wolfbeis 2004). Depending on the shape and structure of the nano-object that limits the electronic "plasma" of the material (noble metals, etc.), various effects are possible—in particular, local accumulation of electromagnetic energy (local SPR, l-SPR) or generation of hybrid excitations "gliding" along the surface in the form of captured light (propagating SPR, p-SPR), etc.

In analytical applications utilizing label-free technology to monitor the macromolecular interactions, most widely used are the systems based on "surface-captured light" under p-SPR conditions (hereinafter referred to as SPR). The generation of an evanescent wave at the interface of a dielectric and electronic plasma of noble metals involves partial transfer of the optical wave energy to a subsystem of surface plasmons.

Generated under these conditions hybrid surface plasmon polariton (SPP) state is localized close to and propagated along the interface of two phases (Schasfoort and Tudos 2008, Barnes et al. 2003).

SPR systems offer a set of advantages for label-free recognition of biomolecular associations. These systems have broad applications in many fields, but find their main use in biosensing, materials screening, and drug discovery (Mariani and Minunni 2014, Fritzsche and Popp 2012, Mol and Fisher 2010). Among commercially available SPR-based systems, in the global market for SPR instrumentation prominent participants are Bio-Rad Labor Inc., Horiba Ltd., Biosensing Instrument, Reichert Technologies, GE Healthcare, and some other companies. In line with a report of Future Market Insights, the worldwide market for SPR-related technologies is likely to grow to US$1.11 billion by the end of 2025. This means that an increasing number of scholars will be involved in the activities associated with SPR technologies, and they need to be professionally trained to work in this field. Both the content and presentation style of this chapter is focused on acquainting a wide range of stakeholders and interested students with the concept and practical implementation of SPR-based biosensors.

19.2 Biosensing Concept in Analytical Science

A distinguishing feature of present-day trends is the need for quick biochemical analysis, which becomes progressively more and more sophisticated. The most common amongst modern commercially available control systems are the sensors of the "physical" type, which determine intramolecular markers of the target analyte: specific absorption band, mass or mass/charge ratio, etc. (Snopok and Kruglenko 2002). An alternative approach is based on chemical sensors, which allow to obtaining quantitative or semiquantitative analyte-specific information, thanks to a natural or synthetic recognition element, which is in direct contact with a physical transducer (PT) sensitive to the changes in the "amount of matter" on its surface (Thevenot et al. 1999, Schasfoort et al. 2017). This approach significantly broadens the spectrum of possible analytes and conditions in which they can be determined.

It is a common "feeling" that the limitations related to chemical sensor performance are specific to the sensitive layer. This is a keystone in the context of chemical sensors, since the recognition "efficiency" of the receptor sites on a certain transducer is the source of sensor functionality. It seems evident that chemists are presently in a strong position to have a significant impact on future developments of the materials for sensitive layers, but to date, exclusive selectivity profiles are only possible to find within biological systems. It is the main reason for the design and development of biosensors—the chemical sensors with recognition centers of biological origin. A typical biosensor transforms information on the process of a biochemical interaction at the recognition layer into a signal that is convenient for further processing and must be presented in a form that can be processed by a human user. The latter is essential, since final decision-making depends on the results obtained by the analysis and interpretation of collected data. Thus, designing a biosensor implies solving various problems specific for physical engineering, biotechnology-related surface science, and analytical informatics to integrate into the final instrument:

✓ a PT that provides high sensitivity to the changes within the recognition layer and does not influence the process of interaction between the receptor and the analyte;

✓ a sensing layer with fixed reception centers that are responsible for the selectivity of the interaction with the given molecules in the solution;

✓ an analytical workflow, including related calibrating procedures, raw data processing algorithms, and protocols, which should display information on the process of the intermolecular interaction "on your screen" in an acceptable form.

This chapter is dedicated only to the more conventional SPR-based sensors as a typical representative of optical biosensors used for molecular interaction analysis. The presentation of the material is constructed in such a way that the basic concepts could easily be adapted to different types of affine sensors for which "the bus" (the analyte) will someday necessarily arrive at its "stop" (the receptor). And although such systems do not require the presence of special labels for the designated analyte (label free technology), they still have to build a broad grid of "bus stations" at which they are expected to arrive. A distinctive feature of this chapter in comparison to other publications on this issue (Homola 2006, Klimov 2012, Schasfoort and Tudos 2008, Long and Jing 2014, Marks et al. 2007, Howell 2017, Rasooly and Prickril 2017) is a generalized discussion of the issues related to SPR biosensors and an attempt to give a physical interpretation of the accompanying phenomena in the form of simple but adequate models.

19.3 Physics of SPR

19.3.1 Excitation of SPP

To explain the numerous effects arising from SPR phenomenon, various theoretical models have been developed in terms of classical electrodynamics (local response approximation, theory of optical antenna theory and contours, etc.) and quantum theory (various modifications of the Drude model, density functional theory), and new hydrodynamic models have been proposed. However, to analyze the processes utilized in classical SPR transducers, it is sufficient to use classical dispersion theory (Ashcroft and Mermin 1976, Sommerfeld and Bethe 1933). According to the free electron plasma model, the metal is represented as a gas of free noninteracting electrons moving in relation to a positively charged crystal lattice. In plasma, the energy of the electric field is drawn from the kinetic energy of the thermal motion of the particles. Since plasma is a statistical electroneutral assembly of charged particles in the presence of long-range electrostatic interactions, any changes in the positions of the electrons in relation to the fixed ions of the crystal lattice give rise to recurrent forces of electrostatic nature. Therefore, charge oscillations in plasma are due to correlated electrostatic–kinetic effects: any concentration gradient leads to the appearance of restorative forces, and, naturally, results in free electron movements, which again induce the concentration gradient and so forth. Finally, if oscillations of the self-coordinated movement of electrons driven by electrostatic forces arise in the bulk, these oscillations have plasma (Langmuir) frequency ω_{pl}, the resonance mode of free-electron plasma in the absence of external fields.

The positions of the electrons in the plasma of metals can be changed as a result of an action of an external field—for example, under the illumination of light in a region of frequencies close to the plasma frequency. In the case of a monochromatic external field, the displacement of electrons leads to the appearance of a dipole moment of macroscopic polarization. These dynamic processes can be described adequately by the means of classical macroscopic Maxwell equations; a description of the properties of the electronic plasma in this case is conveniently made with the use of the dielectric constant $\varepsilon(\omega)$—the coefficient of proportionality between the vectors of induction in a material and the intensity of the electromagnetic field. In terms of Drude–Sommerfeld theory, the material equation characterizing the individual properties of the free electrons in metal in the region of high frequencies, where quenching is low, is represented by the dielectric constant which depends on ω (Wilson 1953):

$$\varepsilon(\omega) = 1 - \frac{\omega_{\mathrm{pl}}^2}{\omega^2} \tag{19.1}$$

Hereby, $\varepsilon(\omega)$ is a real value, the sign of which depends on ω: $\varepsilon(\omega)$ can be both positive and negative quantities in various frequency ranges. The physical reason for the dependence of the dielectric constant on the frequency is the change in the phase of the currents induced in the metal in relation to the phase of the incident light. If $\omega > \omega_{\mathrm{pl}}$, the electrons aren't responding fast enough to compensate for the external field, and the metal transmits external radiation. If $\omega < \omega_{\mathrm{pl}}$, the dielectric constant is negative, which means that the vector of electromagnetic induction is directed towards the vector of the intensity of the external electromagnetic field; i.e., the plasma tries to redistribute free electrons shielding the bulk. If the mobility and the concentration of the carriers are sufficient to "shield" the effect of the external field, the external electromagnetic wave is reflected from the surface of the plasma (the screening effect).

If changes in the positions of free electrons under external illumination arise near the surface, the potential barrier at the interface restricts their movement from the surface side and reduces the space, within which the movement occurs. This leads to the accumulation of electrons at the interface (surface charge) guided by the external electromagnetic wave as a result of their collisions with the surface (Figure 19.1). The electric fields induced by these spatially modulated and time-dependent surface charges are inseparably linked to their sources (localized). The nonlinear interaction of the evanescent wave and the induced basifixed electromagnetic excitations lead to the formation of SPP—an excitation of dual nature, combining the characteristics of a photon (the electromagnetic wave) and a plasmon (the permitted oscillation modes of an electronic plasma). The latter, in accordance with Bohren and Huffman: "Although a plasmon is made up of electrons, it is not an electron: it is a gang, or collection, of electrons that get together under urging of the long-range Coulomb force and decide

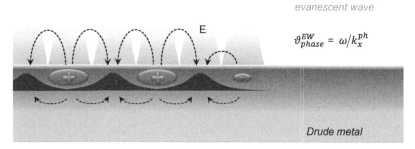

FIGURE 19.1 Illustration of the formation of plasmon–polariton state at the interface of a dielectric and metal with free-electron plasma under the conditions of total internal reflection. The electric field of the propagated (with the phase velocity v_{phase}) evanescent wave (the exponentially vanishing component of the electric field is shown for illustration purposes only) leads to the redistribution of free electrons in the metal: a positive metal surface charge is due to the deficit of electrons in this area, while a negative one is due to their excess. This leads to the formation of spatially modulated surface charge and charge-related electric fields. Under resonance conditions, the evanescent wave and electrical fields related to space-discretized surface charge induced by the evanescent wave form a hybrid polariton state, which "lives" in two mediums simultaneously: the propagated longitudinal plasmon in the metal and the transversal photon in the dielectric.

to act in concert. Hence, for the purpose of discussing their behavior, that may, like an orchestra or choir, be considered as a single entity following the same (Coulombic) conductor" (Bohren and Huffman 1998). As a result of self-consistent processes, the hybrid electromagnetic polyariton (photon–plasmon) excitation is "stuck" to the interface but still "gliding" along the surface in the form of "heavy" light with the wave vector, which is larger than that for the light wave under vacuum.

In the case of naturally (elliptically) polarized light, the surface charges become blurred due to the presence of electrical components in both Y and Z directions perpendicular to the direction of propagation X (Figure 19.2); therefore, it is not possible to create a spatially localized system of charges required for SPP generation. The best conditions for the excitation of a spatially modulated and localized system of charges arise in the case of a linearly polarized (TM (transverse magnetic) or p-polarized) electromagnetic wave, when the oscillations of the electric vector remain unchanged and take place in the XZ plane perpendicular to the surface.

In the most general case of a metal (m)–dielectric (d) interface (Figure 19.2) and TM ($k_y = 0$) polarization of the incident light, it is possible to establish a simple relationship between the vectors k_z and k_x in both media (Yamamoto 2002, Wilson 1953, Ashcroft and Mermin 1976):

$$K_{z,l} = \sqrt{\varepsilon_l(\omega) \cdot (\omega/c) - k_x^2}, \quad l = m, d \quad (19.2)$$

By including the boundary conditions while preserving the continuity of the components of the intensities of the electric fields

$$\varepsilon_m \cdot \vec{E}_{zm} = \varepsilon_d \cdot \vec{E}_{zd} \quad (19.3)$$

it is possible to write the dispersion relation

$$\frac{\varepsilon_m}{k_{zm}} + \frac{\varepsilon_d}{k_{zd}} = 0, \quad (19.4)$$

the fulfillment of which indicates the existence of propagating waves. The frequency dependence of the component of the evanescent wave k_x can be obtained from Eqs. (19.2) and (19.4):

$$k_x = \frac{\omega}{c} \cdot \sqrt{\frac{\varepsilon_m(\omega) \cdot \varepsilon_d(\omega)}{\varepsilon_m(\omega) + \varepsilon_d(\omega)}} \quad (19.5)$$

and, accordingly, the components of the wave vector in the direction of the normal to the surface can be written in the form

$$k_{z,l}^2 = \frac{\varepsilon_l^2}{\varepsilon_m + \varepsilon_d} \cdot \left(\frac{\omega}{c}\right)^2, \quad l = m, d \quad (19.6)$$

We are interested in a solution for the propagating (i.e., k_x is real) and surface-localized (i.e., both k_{zm} and k_{zd} must be imaginary) waves. From the mathematical perspective, this means that the expression under the square root sign in Eq. (19.5) must be positive, while the denominator in Eq. (19.6) must be negative. If we disregard the imaginary parts of the dielectric functions (ε_m and ε_d are real numbers), the condition for the existence of waves localized on the surface but propagating along it can be written in the form

$$\varepsilon_m(\omega) \cdot \varepsilon_d(\omega) < 0 \quad (19.7a)$$

$$\varepsilon_m(\omega) + \varepsilon_d(\omega) < 0 \quad (19.7b)$$

Simultaneous fulfillment of conditions (19.7) is possible in the case where the dielectric constant of one of the media is negative and greater than the second. Such a situation does indeed arise in the case of metals for which the real part of the dielectric constant is negative in the region of $\omega < \omega_{\text{pl}}$. Here the value of the real part of $\varepsilon_m(\omega)$ in the visible region of the spectrum is fairly large (e.g., for gold at 650 nm, it amounts to ~ -10 to 11), whereas the imaginary part is relatively small (~ -1 to 2 under the same conditions), and this gives rise to the longer propagation paths of the surface wave (approximately units to tens of micrometers) (Barnes et al. 2003). Typically used in biosensing, dielectric media (e.g., water or air) are transparent (imaginary part is negligible) with low and positive $\varepsilon_d(\omega)$.

Despite the large number of methods for the experimental determination of SPR conditions, the overwhelming majority of them are based on the analysis of the dependence of the intensity of light reflected by the "plasmon structure" on the angle of incidence or on the wavelength of incident radiation. The SPP hybrid states of the evanescent wave and the electronic plasma of the metal can be excited by p-polarized radiation (the vector E lies in the plane of incidence (XZ), while the vector H is directed along the Y axis (Figure 19.2)), by means of a prism inlet (in the Otto (1968) or the most widely used Kretschmann (1971) configuration (Figure 19.3)). The excitation of such polariton states takes place if the tangential component of the incident p-polarized radiation (k_x^{ph}, the x-component of the wave vector of the photons in the prism (ε_p)) coincides with the wave vector of the SPP k_x^{SPP}:

$$k_x^{\text{ph}} = \sqrt{\varepsilon_p} \cdot \frac{2\pi}{\lambda} \cdot \sin\theta \cong \frac{\omega}{c} \cdot \sqrt{\left(\frac{\varepsilon_m \cdot \varepsilon_d}{\varepsilon_m + \varepsilon_d}\right)} = k_x^{\text{SPP}} \quad (19.8)$$

Here it is expedient to mention once again that SPP states cannot be directly excited by natural light, since the SPP wave vector $k_x^{\text{SPP}} = k_x^{\text{ph}}$ is larger than the wave vector of bulk waves in vacuum (ω/c). This is the reason for using the evanescent wave generated under total internal conditions: at the interface between two media, a tiny fraction of incident energy leaks across the boundary. This energy appears

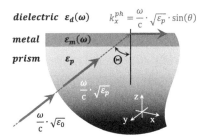

dielectric $\varepsilon_d(\omega)$ $k_x^{ph} = \frac{\omega}{c} \cdot \sqrt{\varepsilon_p} \cdot \sin(\theta)$

metal $\varepsilon_m(\omega)$

prism ε_p Θ

$\frac{\omega}{c} \cdot \sqrt{\varepsilon_p}$

$\frac{\omega}{c} \cdot \sqrt{\varepsilon_0}$

FIGURE 19.2 A schematic representation of an SPR PT in the Kretschmann configuration.

(a)

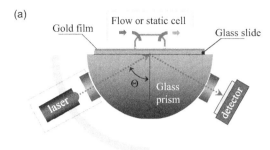

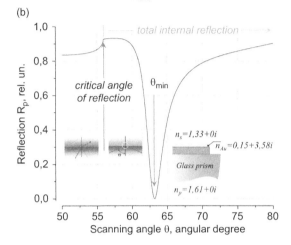

FIGURE 19.3   A typical construction of an SPR spectrometer with angular scanning (a) and the characteristic form of the reflection curve (SPR curve) under SPR conditions (b).

as a wave attached to the surface, which decays exponentially with the distance from the surface.

The use of SPR for the analysis of various processes occurring in the range of units, tens, or hundreds of nanometers from the surface is based on the strong dependence of the resonance conditions (19.8) on the optical parameters of the medium $(\varepsilon_d(\omega))$ in direct contact with the metal $(\varepsilon_m(\omega))$. The most important features of SPR from the analytical standpoint are (i) the capture of light by the surface (2D localization); (ii) the presence of a strong nonuniform electric field at the interface (concentration of the optical field); (iii) the 1D propagation of SPP states along the surface through distances that significantly exceed their wavelength.

The propagation length for the red region of the spectrum in metals such as gold and silver lies in the range of units and tens of microns and depends on the topographical features of the interface and the presence of interfacial structures on it. The penetration depth in a medium with a smaller refractive index lies in the range of hundreds and thousands of nanometers, depending mainly on the wavelength λ of the irradiation used, with the largest value $(\sim\lambda)$ near the critical angle of total internal reflection and decreasing down to a saturation value $\sim\lambda/10$ with the increase of θ.

When k_x is small, the longitudinal component of the electric field of the plasmon is also small, the field is predominantly transverse and the dispersion law is close to linear, typical for light. For large k_x, on the other hand, the contribution of the longitudinal component of the electric field is

large, and SPP approaches bulk plasmon in its properties. In this range, the energy losses of SPP sharply increase due to the heating of conduction electrons.

19.3.2   Utilizing the SPR Phenomena in Sensors

The main idea on which the use of SPR for sensing applications is based is the fact that the analyte, interacting with a selective center attached at the interface, increases the refractive index of the medium $\left(n_d = \sqrt{\varepsilon_d}\right)$ in the region occupied by the evanescent wave, thereby varying the resonance conditions (19.8). By following the changes of the latter with the variation of the SPR angle θ, it is possible to directly monitor the amount of the analyte attached to the receptor (Figure 19.4).

However, the amplitude of the nonuniform electric field under SPR conditions decreases exponentially in the direction normal to the surface. Therefore, the resonance conditions are determined by the overlap integral of the space occupied by the evanescent wave in the dielectric medium with the refractive index of the architecture spatially dependent on the direction normal to the surface $n(z)$ (Figure 19.5a). This approach makes it possible to introduce an effective refractive index, namely (Liedberg et al. 1993, Lukosz 1997)

$$\eta_{\text{eff}} = \frac{2}{l_d} \cdot \int_0^\infty n(z) \cdot |E_z|^2 \, dz \qquad (19.9)$$

where l_d is the characteristic penetration depth of the wave into the dielectric medium above the surface of the metal. For most cases of practical importance, l_d amounts to $0.37 \pm 13\%$ of the emission wavelength λ (Jung et al. 1998).

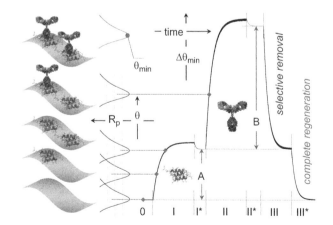

FIGURE 19.4   Illustration of the formation of an SPR kinetic curve during the adsorption of protein "A" on the transducer surface and its following interaction with protein "B": "0" baseline corresponds to the running buffer; "I," "II"—to the additions of samples with proteins "A" and "B," respectively; "I*," "II*" — to the washing by the running buffer in order to remove the weakly bonded components from the surface; "III," "III*" —to the selective removal of protein "B" and complete regeneration, respectively.

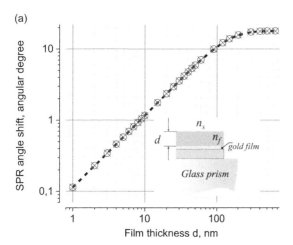

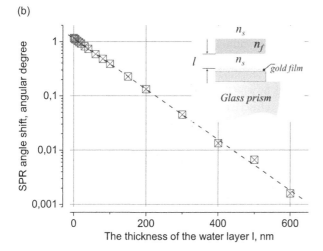

FIGURE 19.5 Dependence of the shift of the SPR angle on (a) the thickness of the adsorbed film d and the refractive index $n_f = 1.46 + 0i$ (in relation to the value for $d = 0$); (b) the thickness l of the interlayer of water between the surface of the transducer and the film of 10 nm thickness ($n_f = 1.46 + 0i$) in relation to the value for $l = \infty$. The data was calculated in the Winspall 3.02 software environment for a model of a semi-infinite prism ($1.61 + 0i$), gold film of thickness 50 nm ($0.15 + 3.58i$), aqueous medium ($n_s = 1.333 + 0i$), and p-polarized laser radiation with 650 nm wavelength.

Equation (19.9) means that with the increase of the distance from the surface, the effective region of intersection of one and the same object with the evanescent wave also changes at various distances from the surface (Figure 19.5b). This in effect means that the resonance conditions for the same macromolecular object situated at various distances from the surface will differ, and the same object will accordingly give rise to a different shift of the SPR curve.

Taking into account the fact that the amplitude of the evanescent wave decreases exponentially in the direction of the normal to the surface Z, Eq. (19.9) for an analyte layer of thickness d with a refractive index n_f in contact with an infinite external medium (bulk) with a refractive index n_b has the following form:

$$\eta_{\text{eff}}(d) = \frac{2}{l_d} \int_0^\infty n(z) \exp\left(-\frac{2z}{l_d}\right) dz$$

$$= \frac{2}{l_d} \left[\int_0^d n_f \exp\left(-\frac{2d}{l_d}\right) dz \right.$$

$$\left. + \int_d^\infty n_b \exp\left(-\frac{2z}{l_d}\right) dz \right] \quad (19.10)$$

and after integration

$$\eta_{\text{eff}}(d) = n_f - \exp\left(-\frac{2d}{l_d}\right) \cdot (n_f - n_b) \quad (19.11)$$

The dependence of η_{eff} on d has a nonlinear character similar to the one presented in Figure 19.5, even if the dependence of l_d on the film thickness d is disregarded. Actually, despite the fairly rough approximation used for the production of Eq. (19.11), the function describes the results from the numerical calculation by using Fresnel equations for the optical model of the system quite well. Taking into consideration the changes of l_d depending on the thickness of the analyte film d (as mentioned in (Yeatman 1996), the main error in Eq. (19.11) comes from the ambiguity in the determination of l_d) can further improve the approximation, but the general trend remains. This fact makes it possible to construct a simple analytical model of an SPR sensor.

19.3.3 Simple Model of an SPR Sensor with a Linear Standard Curve

Quasilinear calibration characteristics for an SPR transducer can be obtained only for film thicknesses on the metal surface not exceeding ~ 50–60 nm ($\sim \lambda/10$) (Figure 19.5a), which restricts direct use of SPR for structures with large thicknesses (Boltovets and Snopok 2009). In this region, for the change in the effective refractive index (Eq. 19.11) after appropriate expansion of the exponential factor into a series, we obtain

$$\Delta\eta_{\text{eff}}(d) = \eta_{\text{eff}}(d) - \eta_{\text{eff}}(0)$$

$$= \left(n_f - \exp\left(-\frac{2d}{l_d}\right) \cdot (n_f - n_b) - n_b \right)$$

$$\xrightarrow[d\to 0]{} \frac{2d}{l_d}(n_f - n_b) \quad (19.12)$$

The monitoring of SPR angle $\theta_{\min}$ (Figure 19.3) is most widely used in analytical applications, since the response of the SPR transducer $\Delta\theta_{\min}$ (the shift of the SPR angle) depends linearly on the value η_{eff}:

$$\Delta\theta_{\min}(d) = m \cdot \Delta\eta_{\text{eff}}(d), \quad m = \frac{\Delta\theta_{\min}(d)}{\Delta\eta_{\text{eff}}(d)} \quad (19.13)$$

where m represents the sensitivity of the SPR transducer—the tangent of the slope (for the linear case) of the dependence of the output signal on a characteristic parameter of the medium.

By substituting the expression (19.12) in Eq. (19.13), it is possible to analytically derive the result obtained by numerical calculations (Snopok et al. 1998), according to which the following expression is true for thin films with $d < \lambda/10$ nm in one and the same medium ($n_b = $ const):

$$\frac{\Delta\theta_{\min}}{d} = \frac{2m}{l_d}(n_f - n_b) \qquad (19.14)$$

Such a narrow range of thicknesses ($d < \lambda/10$) at which Eq. (19.14) holds is due to the fact that the changes in l_d during film growth under these conditions can be disregarded. The value of l_d in this range of d is determined predominantly by the optical parameters of the gold and the external medium and not by the parameters of the adsorbed layer.

The expression (19.14) is suitable for determining the values of $d \cdot (n_f - n_b)$ needed to calculate the amount (surface density) of the analyte, using effective medium approaches (see section 6.2). The results from Fresnel equation based computer simulations (Snopok 2012) confirm that standard solutions with given refractive indices are suitable for the calibration of SPR transducers if such transducers are to be used for the analysis of surface architectures with thickness not exceeding 50 nm.

19.3.4 Technology Overview

Both the excitation conditions of SPP and their subsequent evolution (propagation, thermalization, scattering, transformation into bulk waves, etc.) are widely used to generate an informative signal in different versions of analytical instrumentation. In the huge variety of engineered systems, a few main directions can be highlighted. The first one is aimed at screening or analyzing small quantities of matter and includes the design of multichannel systems (simultaneous analysis of multiple samples), microarrays (simultaneous analysis of many analytes in one sample) (Hardiman 2009), SPR imaging (Scarano et al. 2011) (in particular the subsurface one (Naumenko et al. 2013)) and SPR transducers for Lab-on-a-Chip devices. The second class of devices is based on the so-called "hyphenated" SPR techniques and is represented by a large number of different combinations of SPR detectors plus some other technique—such as quartz crystal microbalance (SPR-QCM), mass spectrometry (SPR-MS), electrochemistry, fluorescence (surface plasmon fluorescence spectroscopy, SPFS), Raman scattering, IR spectroscopy (surface enhanced infrared absorption, SEIRA), etc. (Schasfoort and Tudos 2008).

Whilst typical analytical tasks specify a definite design of the instrument, the more significant difference is related to the physical property being monitored under SPR conditions. Most classical SPR transducers use the Kretschmann configuration based on the prismatic input as the most simple and convenient in daily operation. The differences are only in the scanning method for the resonance conditions monitoring Eq. (19.8); possible options are angular (θ) and wavelength ($\varepsilon_m(\omega)$) mechanical scanning, plus a fan-shaped beam configuration. In a number of cases, other methods of excitation are also used, among which the

most common are the grating coupler and the local input (utilizing tapered metal rods, etc.) due to the possibility of them being used in conjunction with various "hyphenated" SPR techniques. Modifications of the basic configuration of a PT also allow to obtain a number of new or required for a specific task features. For example, if the thickness of the transparent dielectric layer on the metal surface is increased above $\sim\lambda/2$, in such a system, the generation of the waveguide mode is also possible, and this opens up possibilities for high-sensitivity gas analysis using plasmon structures with porous transparent sensing layers (Lysenko et al. 2005). Multilayer structures containing magnetic film are the basis of a whole class of magneto-optic SPR (MO-SPR) sensor (Manera et al. 2014, David et al. 2017).

A promising class of devices utilizes polarization effects observed during the excitation and propagation of SPP. It is generally agreed that these systems have the greatest sensitivity among detectors. The main difficulty in using this approach is that the light phase cannot be measured directly due to high light frequency. Therefore, the phase shift caused by the interaction with the surface plasmon is measured indirectly, using an interference phenomenon. Typical approaches include analysis of the interference pattern, spectral ellipsometry, and heterodyne (Sotnikov 2015); another impressive example is direct visualization of nanoparticles using wide-field SPR microscopy (Nizamov et al. 2016, Nizamov and Mirsky 2018).

A number of effects that are not taken into account in the classical SPR approach are also used to generate an analytical signal. For example, in the case where there are some periodic structures present on the surface (including natural roughness, which can be regarded as a statistical set of periodic lattices or some specific structures of the analyte (Savchenko et al. 2008)) with a period close to the wavelength of the surface waves, SPP scattering (e.g., the transformation of the surface into bulk waves) becomes possible. This phenomenon was used for the development of the symbolic SPR spectrometer that makes it possible to visualize macromolecular processes at the interface using various characteristics of the scattered radiation under SPR conditions as analytical signals (Savchenko et al. 2007).

19.4 Interfacial Design

In order to gain an understanding of the optimal arrangement of surface-coupled receptor centers, there have been extensive studies of various concepts controlling the specific adsorption of a particular analyte and of preventing the nonspecific adsorption of other biological molecules onto the sensing architecture. The aim of all these investigations is to better understand, at the molecular level, the impact of the immediate surroundings on the selective receptor–analyte interactions and to develop optimal schemes to either limit or promote, as necessary, these adhesive reactions.

19.4.1 Sensing Architectures for Transducer-Based Systems: General Requirements

Since intimate contact between biomolecules (receptors) and transducer is essential for the overall biosensor functioning, the preparation of a functional immobilization matrix is a key factor in biosensor fabrication. When choosing the optimal interfacial nanoarchitecture for bioreceptors, it is essential to take into account specific features of sensing elements of transducer-based sensors, namely

✓ immobilization milieu similar to the natural state: conditions on the surface should maintain or even increase receptor stability and functioning;

✓ strong receptor–transducer binding, preferably *via* covalent coupling;

✓ oriented immobilization: to increase the number of receptors, their recognition centers should face outwards from the transducer surface;

✓ spatial arrangement of receptor centers: packing density of receptor centers at the interface should be optimal to provide high binding activity;

✓ negligible nonspecific binding: to achieve high sensitivity and selectivity, nonspecific binding at both the receptor and at support layer (matrix) should be suppressed;

✓ "smooth" controlled functionalization of the transducer: a mechanism of sensitive layer formation should preferably be based on self-limiting/self-assembled approaches;

✓ "physically" inert interfacial architecture: a sensitive layer should permit the transducing mechanism to function adequately;

✓ low-cost technology.

19.4.2 PT of SPR Sensors: Optimization of the Metal Layer

Effective operation of an affinity biosensor depends on two main factors: the recognition ability of the receptor layer as well as the sensitivity and stability of the PT. The optimization of the PT for an SPR biosensor focuses mainly on two directions:

✓ the "physical" one, aiming to provide the most efficient transformation of the changes in the sensitive layer into the output signal; the optimization efficiency is determined by both optical and structural (topological and morphological) properties of the metal film;

✓ the "chemical" one, aiming to create conditions for the oriented and nondestructive immobilization of the reception centers.

"Physical" optimization primarily concerns the structure of the metal layer where plasmon oscillations occur. Typically, the thin gold film in SPR sensors is deposited on the surface

of the glass prism/slide through chromium/titanium adhesive layer. Since both the incident and reflected beams in the Kretschmann geometry cross the metal layer, the influence of both the morphology and topography of the film on the shape of the plasmon resonance curve is very important (Snopok et al. 2001). In fact, absorption and scattering both inside of the film and on the interfaces can disturb the response variations induced by the excitation of plasmon–polariton states. Since the Fresnel equations do not take these effects into account, the accuracy of the SPR curve approximation can decrease. Other typical problems include the time stability of polycrystalline metal films (caused by mechanical strains) and the possible penetration of the solutions inside the film along the boundaries of grains (alkali solutions can dissolve the adhesive Cr layer), etc.

Besides the optimization of sensitivity, stability, and reproducibility of the PT, one also needs to prevent the denaturation of biomolecules on the surface and to prepare the conditions for the formation of organized structures on the rough surface of polycrystalline films. Indeed, uncoated metal films can induce changes in native conformations of adsorbed biomolecules. This is accompanied by both considerable changes in the physicochemical properties of the protein and partial or complete loss of its biological activity (Snopok et al. 1998, Snopok and Kostukevich 2006). The changes in native conformation usually occur due to the breaking of disulfide bonds and the presence of hydrophobic environment on the surface. Thus, if the surface is covered by a hydrophilic layer preventing intramolecular disulfide bond breaking, this may provide a retention of the native conformation for biological molecules.

19.4.3 Selection of the Optimal Interfacial Arrangement: Major Factors Determining the Interaction of a Macromolecule with a Functionalized Surface

In contrast to most homogeneous reactions in solution, the occurrence of heterogeneous reactions at the interface is due to the following characteristic features: the two-dimensionality of the sensitive layer, the interaction of molecules with each other and/or with the matrix, the polyfunctionality, the presence of vibrational and rotational degrees of freedom in the absence of lateral mobility, and the ultralow absolute content of the analyte in the sample. From the standpoint of biosensor applications, this suggests that it is necessary to take into account the composition (the type and state of the selective surface center, the chemical functionality of the matrix), the structure (surface architecture and grafting density), the immobilization procedures (one-center/multicenter, with/without an additional spacer), the stability of the structure during analysis/reuse, and also the technological compatibility with existing and future production processes. Thus, the particular structure of the interfacial architecture on the one hand and the molecular nature of the SPR response shaping recognition process on the other force us to develop technologies for the structuring of the surface within a nanometer range. Major factors determining the interaction efficiency of a macromolecular analyte with the functionalized surface, namely the ones regarding the peculiarities of the spatial organization in interfacial architectures in nanoscale, are listed and illustrated in Table 19.1. Table 19.2 summarizes specific for SPR biosensors conceptual approaches in surface chemistry for the design and development of immobilization matrices utilizing different driven forces for the retention of bioreceptors near the surface. Typical examples of sensing architectures can be found in numerous publications (Altintas et al. 2012, Schasfoort et al. 2017, Snopok et al. 2008), where they demonstrate the diversity of immobilization matrices such as dextran layers, lipid or thiol monolayers, protein A/G

(Figure 19.6a) or biotin–streptavidin systems, etc. A variety of coupling procedures, including relevant activation chemistries starting from the physical entrapment in polymers and insoluble matrices and up to the chemical attachment by crosslinking and covalent binding onto a native-like membranes, are widely described in detail (Rusmini et al. 2007, Wijaya et al. 2011, Wong et al. 2009, Hoa et al. 2007).

19.5 Research Methodology: SPR-Specific Analytical Workflow

Any biochemical method of analysis aims to obtain quantitative information using measurements that should give an

TABLE 19.1 Typical Effects of Nanoenvironment on an Adsorption Process in Heterogeneous Biosensing

Title of Mechanism and Its Typical Spatial Scale (TSS)	Schematic Diagram of Core Processes	Mechanism of the Process and Its Driven Forces
I. *Molecular recognition* of the immobilized ligand L by the selective center of the receptor R determines the presence of an interaction process. TSS is the transverse dimension of the ligand molecule, ca. 1–2 nm in diameter	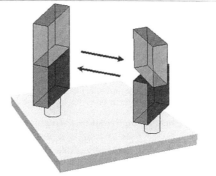 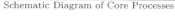$pH < pK_a \rightleftharpoons pH > pK_a$	Intermolecular interactions are determined by the complementarity of the recognition center of the receptor R and the shape of the van der Waals surface, the charge state, and the accessibility of specific groups in the ligand molecule L
II. *Steric blocking* of the ligand–receptor interaction in the contact region may substantially suppress the binding process. TSS is around the effective diameter of the region adjoining the recognition epitope, ~2–5 nm	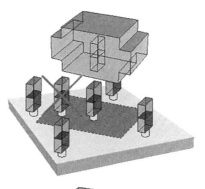	Steric hindrances for the process of receptor–ligand molecular recognition due to the presence of additional ligand molecules in the contact region
III. *Suppression of the R~L (receptor R – ligand L) interaction* in the vicinity of the contact region has little effect on macroscopic parameters due to process IV. TSS is 1.2–1.5 effective receptor diameters, ~5–10 nm	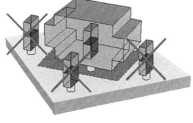	Steric blocking (by the macromolecular receptor) of the immobilized ligand, found in the vicinity of and directly in the region of the receptor surface projection
IV. *Stochastic packing* leads to a decrease in adsorption capacity of the surface down to 60%–70% of the entire surface area. TSS is at least several effective receptor diameters, ~10–20 nm		Stochastic packing of the receptor in a 2D layer on the surface, leading to the formation of voids smaller than the area of the projection of R onto the surface

TABLE 19.2   Sensor Functionality Controlled by Receptor Orientation: Design of Typical Immobilization Matrices Utilized in Biosensors for the Retention of Bioreceptors Near the Surface

Immobilization Matrices	Summarized Characteristics and Average Operating Conditions
Reactionless coating for nondestructive adsorption of macromolecules	• balanced base for *weak undirected* immobilization • kinetically and thermodynamically stable inorganic layer • simple production technology • low cost *Typical example: the formation of $Au_x S_y$ interfacial layer by reactive annealing of gold films in $H_2 S$ atmosphere*
Coating based on nanoscale-specific electrostatic levitation	• "charged" base for *weak oriented* immobilization by site-selective electrostatic interaction • kinetically and thermodynamically stable inorganic layers • simple production technology • low cost *Typical example: $(SCN)^-$-modified gold surface, the position of the molecule in relation to the "charged" surface may be controlled by changing the charge of definite amino acid groups via the change of ambience pH*
Mixed monolayers of compounds with sulfur end-groups	• monolayer base for *oriented* immobilization • thickness in nm range • variable 2D arrangement of active sites in nm scale • thermodynamically stable organic layer • self-assembled production technology • low cost *Typical examples: mixed monolayers of aliphatic thiols with functional end-groups, monolayers based on thio-substituted DNA/RNA, etc.*
Bioinspired interfacial architectures	• native base for *oriented* immobilization • biologically optimized 2D arrangement of active sites • thickness in nm range • typically self-assembled protocols *Typical examples: protein A or protein G-antibody complexes, streptavidin–biotin, etc.*
Coating with a given range of thickness produced by layer-by-layer deposition	• layered base for *oriented* immobilization • precision thickness control • fixed 2D distribution of active sites • kinetically and thermodynamically stable organic/inorganic layers • self-assembled layer-by-layer production technology • low cost *Typical examples: layered structures based on $([Fe(CN)_5 NH_3]^{3-})$ and Cu^{2+}, etc.*
3D sensing architectures	• 3D base for *spatially distributed* immobilization • thermodynamically stable organic layers with low density • thickness in tens to hundreds of nm range • low cost *Typical examples: dextran layers, virion interfacial structures, etc.*

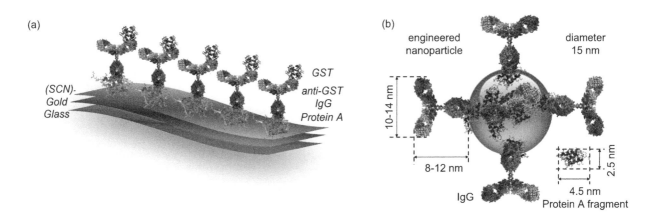

FIGURE 19.6   Scheme of protein A based interfacial architectures on an isothiocyanate $(SCN)^-$-modified SPR transducer (a) (Snopok et al. 2006b) and on magnetic nanoparticles (b) (Kolotilov 2006). Protein A is immobilized on the surface and binds the antibody molecules (anti-Glutathione S-transferase, GST). The antibodies can capture any GST fusion protein (GST on the figure).

adequate correlation between the experimental value and a certain standard. For SPR systems, the correspondence between graduation and measurement may be achieved in a variety of ways: by using the internal standard (direct methods), an extrinsic measure (competitive format), or the so-called method of additions (a combined calibration-measurement procedure based on the consecutive addition of the standard).

19.5.1   Determination of the Presence of the Analyte in the Sample: Direct Measurements

Despite many advantages provided by SPR-based sensors, they remain useless for the quantitative characterization of analytes by direct measurements (when the analyte is in the solution and the receptor is immobilized on the surface and is selective against the analyte). Indeed, the determination of the analyte concentration in the solution is not unambiguous and can lead to controversial results. This is due to the fundamental limitation of methods utilizing surface-based PTs with restricted sensitive area. The quantitative analysis by direct methods may be realized only if the amount of analyte in the sample is low with respect to the number of binding sites on the surface. Otherwise, the response is determined only by the adsorption capacity of the surface and is not dependent on the analyte concentration in the solution. Various kinetic approaches (e.g., the rate analysis within the Henry region) directed at the quantitative analysis also generally suffer from low signal and poor reproducibility induced by the sample injection procedure. Moreover, due to fluctuations of external factors (e.g., temperature), the analyte with a molecular weight below 5 kDa cannot generate an SPR response with a signal-to-noise ratio that is sufficient for a sensitive enough optical readout. An additional problem is that analytes are typically located in complex samples such as blood or saliva, where various high molecular mass components dominate; therefore, for practical applications, unspecific sorption must be taken into account as well. In other words, the main disadvantage of using SPR transducers for the direct quantitative analysis is the fact that it is impossible to control the actual level of surface filling and the composition of the adsorbate while using only the internal measure.

However, if substances A and B in the solution are sufficient to completely cover the surface, the results of direct measurements allow to (i) establish the presence of an interaction between A and B; (ii) determine the stoichiometry of the formed A-B complex (taking into account the ratio of their molecular weights); (iii) investigate the influence of a third molecule on their interaction; (iv) assess the stability and type of dominant interactions in A-B using selective detergents, etc.; (v) find out the mechanism of the interfacial reaction by analyzing the observed kinetic within certain models; and (vi) evaluate the affinity of the reaction. The most successful application of direct measurements is the technology of microarrays (SPR imaging), which allows to reduce the volume of the sample and provides the possibility of a parallel analysis of a large number of different analytes (Schasfoort et al. 2017).

19.5.2 Quantitative Determination of Analytes, Especially of a Small Size: Competition Assay Formats

Competitive methods allow to overcome the limitations of direct methods for surface-type transducers, and therefore are a reasonable alternative (Boltovets et al. 2017). From the analytical point of view, the difference between the competitive methods of analysis and direct measurements is that, instead of internal calibration, some external standard is used at the stage of preliminary incubation of the sample or during the measurement process.

In indirect methods of analysis, two clones of the same molecule—namely the free analyte in the solution and its immobilized analog on the surface, compete for the binding with the macromolecular receptor (an external standard macromolecular receptor (MR), usually high-molecular-weight dedicated proteins—e.g., antibodies, receptors, etc.). For the most commonly used inhibitory-type competitive analysis, a fixed amount of sample is initially mixed with a predetermined amount of MR (which is optimized for the required concentration range of the analyte), and then is incubated for the time required to establish equilibrium binding of the analyte with the MR in the solution. The resulting mixture is analyzed with a sensitive element modified by the analyte analog; the amount of MR that did not react with the analyte in the solution and thus is capable of binding to the surface is determined. The number of these MR that did not react with the analyte depends on the amount of MR that was initially added, the number of binding sites on the surface, and the concentration of the analyte. By varying the concentrations of the analyte, one can obtain the calibration curve for the following quantitative determination of the analyte in the sample. From a physical point of view, this approach allows to deal with the difference in the amount of MR instead of the difference in the amount of the analyte. Taking into account the mechanism of the SPR phenomenon, this results in a formal amplification of the sensor signal by the value that is the ratio of the molecular weight of the MR and the analyte (in the case of monodentate receptors).

According to the principle of traditional competitive analysis in the steady-state mode, the MR concentration on the surface $\Gamma_{\text{MR}\approx\text{analyte-clone}}^{\text{surface}}$ is given by the Morgan−Mercer−Flodin equation or the logistic curve (Snopok et al. 2006a):

$$\Gamma_{\text{MR}\approx\text{analyte-clone}}^{\text{surface}} = \frac{\left(\frac{V}{S}\right) C_{\text{MR}}^{\text{solution}}}{1 + \left(\frac{C_{\text{analyte}}^{\text{solution}}}{\Gamma_{\text{analyte-clone}}^{\text{surface}}} \cdot \frac{V}{S}\right)^p} \qquad (19.15)$$

Here p is the order of reaction with respect to the analyte (an effective quantity that carries information on the mechanism of the processes occurring in the system: the bigger the parameter p value, the greater the extent to which the processes of univalent analyte binding to a polydentate MR take place concurrently); $C_{\text{MR}}^{\text{solution}}$ is the initial concentration of MR in the solution; $C_{\text{analyte}}^{\text{solution}}$ and $\Gamma_{\text{analyte-clone}}^{\text{surface}}$ are the concentrations of the analyte in the solution and of its immobilized analog at the surface, respectively; V and S are the cell volume and the sensitive surface area, respectively. Equation (19.1) also illustrates the linear dependence of the response on the MR concentration and the possibility to optimize both the dynamic range and the minimal detectable value by varying $C_{\text{MR}}^{\text{solution}}$ and $\Gamma_{\text{analyte-clone}}^{\text{surface}}$ (see Snopok et al. 2006a for details).

It should be noted that competitive methods with immobilized analyte analogs and selective bioreceptors make possible a quantitative analysis with high operational efficiency as a result of: the reproducible formation of a sensitive surface with low-molecular-weight organic centers, the long storage period of such sensors without change in their chemical functionality, the possibility of developing analytical assays in which the biological components are in a lyophilized state with retention of their biological activity for a lengthy period of time, the effective restoration of the system to its original state (since the sensor does not contain biological components on the surface), and the possibility of rearranging the analytical system against a new target by replacing the immobilized analyte analogs and the biological component.

19.5.3 Protein–Protein Interaction in the Solution: Capture of an Analyte–Receptor Complex with Standard Additions of the Receptor (CARSAR) Approach

Traditional SPR methods to detect an interaction between certain proteins *in vivo* have limitations because it is difficult to identify a specific signal corresponding to their complexes against the background. Since the purpose of analysis is to establish the fact of the interaction between the wild-type analyte and the MR (usually a purified fusion protein with a specific surface tag is used as a standard), of most interest are the methods of coprecipitation, allowing to separate the analyte from the multicomponent sample by concentrating the analyte and MR complex on the surface. CARSAR helps with this (Snopok et al. 2012). The heart of the approach is a selective adsorption of the analyte–MR complex (this is the investigated reaction) on the functionalized sensor; the surface is able to selectively bind the MR or MR–analyte complex and prevent adsorption of other components of the sample. In this case, calibration and measurements are combined: a certain amount of addition (MR) is introduced into the complex mixture, and the dependence of the SPR response on the concentration of the addition is analyzed. This procedure provides invariability of the sample composition and is applicable to different analytes and samples. The most prominent feature of this approach for analyzing a multicomponent sample is the minimization of the influence of nonspecific sorption on the characteristics of the functional dependence (since repeated measurements with different MR additions take place in the same sample). The CARSAR approach is based on the surface exclusion effects in adsorption processes and is a logical continuation of competition methods with a series-dosing external measure.

19.5.4 SPR Analysis of "3D" Objects: Density Variations in the Fixed Architectures Approach (DViFA)

Conventional SPR studies typically use "2D" interfacial architectures whose thickness is significantly smaller than the penetration depth of an exponentially decaying evanescent wave in a dielectric medium. The use of SPR techniques for the investigation of "bigger" objects (e.g., intact virions, cells, etc.) has some limitations, since their characteristic size exceeds tens of nanometers, which is required to ensure adequacy of the quasi-linear approximation.

In the conventional approach, the magnitude of the SPR response depends on the effective thickness of the analyte layer that is bound to the layer of receptors on the surface. Typical analysis assumes that the density of both layers (receptor and analyte) is uniform, i.e., the variation of the SPR signal is due to the changes of the thickness of the molecular assembly of the interacting molecules in the vertical plane. However, an SPR shift depends not only on the layer thickness but also on the change of the refractive index within the layer. Therefore, variations of the molecular layer density can also affect the response value due to the variations of the refractive index inside the layer ("variative" refraction). One of the possible mechanisms is the process of changing the packing of the objects of different size and shape on the surface (e.g., a virus, a laponite platelets, an antibody, a small molecule, etc.) (Boltovets et al. 2013, Manilo et al. 2017); another well-known example is thick (ca. 100 nm) hydrogels (e.g., dextran and its derivatives with spatially distributed receptor centers) (Löfås et al. 1993). Since in these cases the thickness of the surface layer is fixed due to the design (dextran layers) or the constant form of one of the interacting components (e.g., virion), the SPR shift is a single-valued function of the molecular assembly compactness: in the last case, the compactness of the interfacial structure depends on the ratio of the concentrations of the virus virions and the antibodies against their shell proteins.

19.5.5 Validation Procedure: Checking for Surface-Induced Analyte Transitions in Adsorbed Biofilms

Controlling the irreversible protein adsorption is a major concept for practically every SPR-based sensor system. Despite the fact that the surface itself is a new phase for analyte molecules, all of the abovementioned approaches assume the same molecule structure in both free and adsorbed states; the structure of protein-surface complexes, their affinity, and their thermodynamic characteristics depict a frozen state of the complex. However, biomolecules adsorbed onto the surface can undergo various types of conformational changes. As a result of these changes, new structures of biomolecules will become exposed and their original surface will acquire new characteristics. In particular, on metal surfaces, the proteins tend to adjust their orientation and conformation, resulting in their stronger interaction with the metal and greater irreversibility of adsorption. To this end, it is essential to have an approach that helps to clarify the situation concerning possible surface-induced protein transitions in adsorbed biofilms.

One of the most realistic models of protein adsorption is based on the fact that proteins adsorbed initially in their native state can then either desorb or undergo some conformational transformations (if a neighboring protein cannot block these changes). These transformations lead to an increased surface contact area and a decrease in the number of adsorbed molecules per unit area. The SPR technique can be efficiently used to check the presence of postadsorption transformations utilizing the two-stage model of surface-induced transitions in adsorbed biofilms (Snopok and Kostukevich 2006).

In line with the generally accepted approach, the saturation levels of the sensor response must be constant for the analyte concentration in the solution over the full range of concentrations under investigation. It is a consequence of the restricted area of measurements specific for transducer-based sensors like SPR. However, in a situation where unfolding influences the structure of the biofilm, the composition of the coexisting interfacial forms (native and unfolded) can change with time. At low concentrations, the unfolded form occupies a larger surface that is not limited by neighboring molecules, whereas the native form represents a small fraction of the surface coverage. At high protein concentrations, the adsorption process is accelerated since the rate of adsorption is proportional to the concentration of the protein in the solution. Thus, the surface becomes covered with proteins before a considerable amount of them can unfold. As a result, a higher number of proteins will be able to bind to the surface during the same time interval. The surface forms will be represented in this case by the partly unfolded form (which occupies an area determined mainly by the free surface between randomly adsorbed proteins) and the prevailing native form. Finally, in the case of the presence of surface-induced analyte transitions in adsorbed biofilms, the dependence of the saturation level on the analyte concentration in the solution is nonlinear and follows the expression presented by Snopok and Kostukevich (2006).

19.5.6 Analyte-Specific Contrast Enhancement

In addition to the approaches described above, the SPR response specific for the definite components on the surface can be "enhanced" in different ways (Shalabney and Abdulhalim 2011, Goodrich et al. 2004); the most popular of them is the use of high-molecular-weight secondary antibodies (the so-called "sandwich assay") or functionalized nanoparticles as a "high-contrast developer" (see, for example, Figure 19.6b). In many cases, secondary antibodies allow to enhance the primary antibody binding signal six to eight times. In the case of plasmonic nanoparticles (gold, silver, etc.), signal amplification is the result of both their high mass and, under certain conditions, the resonant coupling of the localized electromagnetic excitations in nanoparticles and the evanescent field of plasmon–polariton states (Lyon et al. 1998).

19.6 SPR-Specific Data Mining

19.6.1 Measurement Uncertainty in SPR-Based Analytical Studies

To be able to apply SPR sensing to the monitoring of interfacial processes, some data processing method is needed to extract quantitative characteristics of the adsorbed layer from raw SPR data (e.g., the dependence of the reflected light on the angle) (see Section 19.1.2). Procedures based on the approximation of an SPR curve by multilayer Fresnel equations are the most precise (within the limitations of the optical system model) but at the same time very time-consuming. In real-time SPR measurements, various procedures based on analytical functions for finding the resonance angle in the SPR curve are applied, for example: polynomial approximation, centroid algorithms, Lorentzian approximation, etc. (Boltovets and Snopok 2009). Most used methods represent the SPR curve "minima hunt" techniques to improve the temporal resolution of the kinetics. Taking into account the fact that the accuracy of the determination of the minimum increases with the number of experimental points used for the calculation and the SPR curve asymmetry near the minimum, the approximation with a third-order polynomial is most useful and accurate.

The second essential point is that the sensitivity of SPR transducers depends on the thickness and spatial organization of the interfacial structures on their surfaces. This is due to the fact that the response of an SPR sensor is calculated by integrating the distance-dependent refractive index outside the metal, weighted by the square of the exponentially decaying electromagnetic field, from zero to infinite distance (see Section 19.3.2). For example, the presence of a protein layer with ca. 5 nm thickness changes the refractometric sensitivity of the SPR transducer by about 3%. Thus, the main sources of uncertainly in SPR measurements arise from both data processing and the assumption of linearity for the internal calibration plot. The estimated refractometric accuracy (ca. 3×10^{-4}) is larger than the value (about 10^{-4}) of the variation of the water refraction coefficient, when temperature changes by 1 K (Kim and Kihm 2007). These uncertainties in SPR measurements should be taken into account when analyzing data obtained with the SPR technique.

19.6.2 Sizing the SPR Response: Optical Quantification of the Adsorbate Concentration

Although the mass of a material is determined mainly by the "total amount of nuclei in the sample," its refractive index is determined by a combination of all its electrons in the nonresonance region of the spectrum. This idea makes it possible to find a connection between the mass of the material and its refraction by using various models. According to the effective medium approach proposed by de Feijter and developed by Ramsden (de Feijter et al. 1978, Ball and

Ramsden 1988), the surface concentration Γ of an analyte bound to the surface, defined in units of mass per area (e.g., g/m^2), can be calculated from experimentally observed quantities, such as the refractive index n_f and the thickness of the adsorbed layer d. Indeed, for any system with linearly dependent parameters, the following is true:

$$\Delta n = \frac{\partial n}{\partial c} \cdot \Delta c \qquad (19.16)$$

where Δn is the change of the refraction index upon the increase Δc (g/m^3) of a certain component volume concentration, and $\partial n / \partial c$ is the refractive index increment. It is assumed that the refractive index increment is constant up to high concentrations specific for analyte's adlayers (e.g., 0.188 cm^3/g is the lower limit of the refractive index increment for proteins). The surface concentration Γ would then be

$$\Gamma = \frac{\Delta m}{S} = \Delta c d = \Delta n d \cdot \left(\frac{\partial n}{\partial c}\right)^{-1} = \Delta \mathrm{SPR} \cdot \xi^{-1} \cdot \left(\frac{\partial n}{\partial c}\right)^{-1} \qquad (19.17)$$

The values of n_f and d can be estimated using an SPR simulation program (e.g., Winspall II, Max Planck Institute for Polymer Research, Mainz), which solves complex Fresnel equations for a user-desired multilayer optical model. Since in practice the analyte concentration in the solution is usually extremely small, it is commonly assumed that the refractive index of the analyte solution and the running buffer are the same. An alternative way is estimating Γ directly from the measured value of $\Delta \mathrm{SPR}$ using the instrument-specific constant ξ. (Snopok et al. 1998)

19.6.3 Formalization of the Evolving Interfacial Landscape: Comprehensive Kinetic Analysis

Keeping in mind the ability of SPR instruments to monitor the formation of surface architectures, it makes sense to assume that if one takes into account the dynamics of the interfacial growth and unifies kinetic manifestations by parametric formalization, then it is possible to further characterize the interaction of the analyte with the sensitive layer. Indeed, for example, the presence of nonspecific sorption can significantly affect the analytical determination result due to the fact that sensory response is only an "evidence" of the formation of some surface complex—hopefully, the desired analyte but, perhaps, something else. Making decisions in such uncertain conditions requires additional information regarding the fact that the sensor response is the result of the "chosen" selective process on the surface. In fact, the kinetics fully reflects the processes at the interface. To identify the mechanisms that are expressed in the macroscopic kinetic behavior, it is necessary to compare experimental data with the predictions of theoretical models that

take into account various aspects of the processes at the interface.

Idealized Kinetic Analysis Based on Deterministic Models

The simplest way of describing adsorption processes is based on phenomenological models with idealized partition processes, where the degree of surface coverage is expressed by some analytic function of time that reflects the mechanism of dominant processes in the system. The first model of such type was the classical theory of adsorption, proposed by Langmuir in 1916 (Table 19.3) (Langmuir 1916). In the case of an energy-uniform bulk medium and surface, the adsorption–desorption process represents a classical two-level system with a high-energy state corresponding to the molecules in the bulk and the low-energy state corresponding to the particles in the adsorbed layer. The presence of macroscopic concentration gradients near the surface (mass transfer) immediately leads to the nonexponential character of the relaxations even at the phenomenological level (Table 19.3). This is due to the fact that the variations in the rate of the interfacial reactions are taken into account in the diffusion model; in the frame of the phenomenological approach, the "efficient" rate constant $(1/\langle \tau^* \rangle_D)$ is a time-dependent parameter (see Table 19.3). In the case of the Langmuir model, this delay and variations are disregarded (the uniform bulk phase), which leads to substantial changes in kinetics.

In classical cases, the above examined adsorption processes that are uniform in the surface plane with their uniformity quite often observed in experiment are no less frequently disrupted. The differences from classical behavior, often observed in practice, lead to a series of kinetic relationships, which is not surprising from this standpoint (Martinez et al. 2009, Minton 2001, Snopok 2014). The simplest case is concentration depletion processes occurring in a closed system with no access to reagents. More complex models take into account the nonuniformity of the analyte flow toward the surface, the differences in the bonding constants for the surface sites, the multivalent bonding, the effects of the geometric surface blockage, the change in the structure of the analyte after bonding, etc. Numerical models can be illustrated by various versions of random or cooperative sequential adsorption (random sequential adsorption, RSA) (Evans 1993), etc. All these models illustrate some specific aspects of the nonuniform adsorption process, when the driving forces of particle motion are modeled by a single dedicated mechanism. An increasing number of processes in a more and more advanced phenomenological model leads to the fact that an initially simple and clear situation turns into an evolving reaction space with a huge number of parameters. The microscopic dynamics of the analyte under these conditions is so complex that it is not possible to describe the features of a single particle trajectory or the kinetic behavior of their ensemble as a whole in terms of some deterministic mathematical formalism.

TABLE 19.3 Typical Theoretical Models of Adsorption Processes in the Presence of Various Sources of Nonuniformity at the Interface and in the Bulk of the Medium Containing the Analyte Molecules

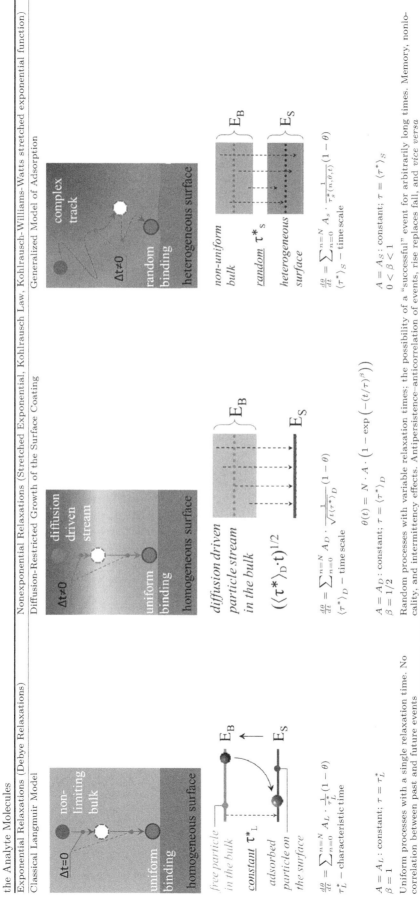

Exponential Relaxations (Debye Relaxations)	Nonexponential Relaxations (Stretched Exponential, Kohlrausch Law, Kohlrausch–Williams–Watts stretched exponential function)	
Classical Langmuir Model	Diffusion-Restricted Growth of the Surface Coating	Generalized Model of Adsorption
Δt=0 / non-limiting bulk / uniform binding / homogeneous surface	Δt≠0 / diffusion driven stream / uniform binding / homogeneous surface	Δt≠0 / complex track / random binding / heterogeneous surface
free particle in the bulk / *constant* τ_L^* / *adsorbed particle on the surface*	*diffusion driven particle stream in the bulk* $(\langle\tau^*\rangle_D \cdot \mathbf{t})^{1/2}$	*non-uniform bulk* / *random* τ_S^* / *heterogeneous surface*
$\frac{d\theta}{dt} = \sum_{n=0}^{n=N} A_L \cdot \frac{1}{\tau_L^*}(1-\theta)$	$\frac{d\theta}{dt} = \sum_{n=0}^{n=N} A_D \cdot \frac{1}{\sqrt{t\langle\tau^*\rangle_D}}(1-\theta)$	$\frac{d\theta}{dt} = \sum_{n=0}^{n=N} A_s \cdot \frac{1}{\tau_s^*(n,\theta,t)}(1-\theta)$
τ_L^* – characteristic time	$\langle\tau^*\rangle_D$ – time scale	$\langle\tau^*\rangle_S$ – time scale
	$\theta(t) = N \cdot A \cdot \left(1 - \exp\left(-(t/\tau)^\beta\right)\right)$	
$A = A_L$: constant; $\tau = \tau_L^*$ $\beta = 1$	$A = A_D$: constant; $\tau = \langle\tau^*\rangle_D$ $\beta = 1/2$	$A = A_S$: constant; $\tau = \langle\tau^*\rangle_S$ $0 < \beta < 1$
Uniform processes with a single relaxation time. No correlation between past and future events		Random processes with variable relaxation times; the possibility of a "successful" event for arbitrarily long times. Memory, nonlocality, and intermittency effects. Antipersistence–anticorrelation of events, rise replaces fall, and *vice versa*

The Stochastic Generalization of Interfacial Binding: Unified Evolution Coding by Stretched Exponential

In accordance with the molecular kinetic theory, an adequate formalism in the description of adsorption processes is to use the background of statistically random processes specific for variegated interfacial reactions. Hereupon, the idealized two-level energy scheme with a single relaxation time τ_L in the Langmuir model is transformed into a two-band model of energy states (Table 19.3). The course of the adsorption process is no longer determined by a simple sum of identical reactions, and the macroscopic kinetics of the adsorption is now calculated by averaging over the whole assembly of different elementary reactions with different rate constants (and, therefore, different activation energies) in accordance with the Arrhenius equation. Indeed, the sensory element is an open nonlinear system that is in a nonequilibrium state with the environment, which causes multivariance and unpredictability of the analyte motion processes at the microscopic level. At the same time, the sensor response is probable (i.e., logical, predictable) and is the result of the formation of spatially organized evolving structures controlled by self-coordinated processes at the bulk-surface interphase. Consideration of such systems is possible only with the use of a synergistic approach, when the microscopic chaotic motion is considered as a factor of evolution, which harmonizes the interconnection of individual components to maintain the system integrity under the influence of external factors.

The generalization of the results of experimental studies, numerical modeling, and phenomenological models suggests that the macroscopic kinetics of such systems in general occurs in accordance with the stretched exponential function (SEF), whose parameters (β and τ) are unique identifiers of the processes at the interface (Table 19.3) (Snopok 2014). The formation of the adsorbed layer from the stream of particles is considered as a cumulative result of rare

"successful" binding acts. Given the fractality of the reaction space (i.e., the space-time modulation within the interphase region), the Gaussian distribution cannot be used. In this case, the sum of individual "successful" binding acts converges not to the normal distribution, but to the Levy distribution (Paul Pierre Levy), which is a stable boundary distribution of the sum of random variables with inverse-power asymptotics (such distributions are usually observed in nature) (Mandelbrot 1982). The nonexponential nature of the relaxation is a reflection of the low probability of single successful analyte adsorption acts (a rare event-dominated phenomenon) and of the multiscale coordination of the interfacial processes in time and space. A set of these processes can be obtained from the corresponding probability density distribution function, which is described by the Levy distribution for a given β value. On the whole, one can say that, at the phenomenological level, nonexponential relaxation is the general form of complex systems evolution, and β is the kinetic marker of a certain combination of elementary processes, which allows it to be distinguished from a series of similar ones (Snopok and Snopok 2018).

Parameters of Nonexponential Kinetics: The Physical Meaning

The value of the parameter τ (which has the dimension of time) determines the scale of the dynamic processes occurring in the system, whereas the value of β determines both the "capture bandwidth" of the elementary processes in the temporal domain and their relative distribution (Figure 19.7). For example, diffusion transfer can take place in different phases; so, the value of β will remain the same, but τ will increase when considering the same process in the gas, liquid, and solid phases.

For complex systems, β determines such a subset of the full set of possible states of the system, relaxation through which corresponds to the principle of least action. This set includes processes with different temporal scales, which are

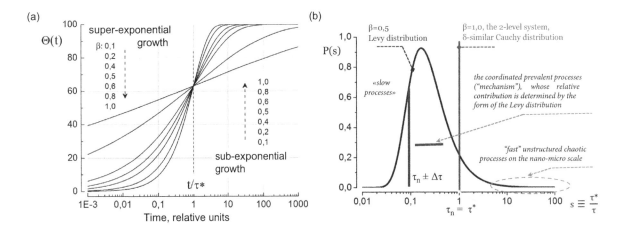

FIGURE 19.7 Typical dependencies of nonexponential relaxation for different values of β (a) and illustration of the difference between functions of the distribution of probabilities density $P(s)$ corresponding to relaxation processes in Langmuir model ($\beta = 1$) and the diffusion-restricted growth of the surface coating ($\beta = 0.5$).

interconnected in the sense that they are coordinated within the system framework while exchanging energy, matter, etc. (Figure 19.7). This means that there are dependencies between these processes, unchanged in a certain range of the spatial scale (spatially invariant), since spatial invariance is directly related to the temporal. In other words, the set of elementary relaxation processes in complex systems is not arbitrary and is defined by a Levy distribution with the corresponding value of β. Therefore, the change of the value of β is a redistribution of the whole set of elementary processes in the system, some of which become active, while others stop.

Indeed, to get to the receptor, the analyte must not only go a long way but also face it with the correct side and approach it with an acceptable speed. Its local trajectory, stops and accelerations, turns and rotations cannot be predicted. The situation changes conceptually when we are dealing with a large ensemble of such identical analytes. In this case, the entire set of particles directed to the receptor forms a statistically definite "image" of sensing architecture in terms of the probability to find the analyte particles located within definite area of its space. As a consequence of this, the biosensor can be considered as an open, distributed nonlinear system. Violation of the principle of superposition during the process of analyte capturing by the sensing surface is caused by selective and coordinated redistribution of energy and matter between system elements by changing the "probability of realization" of the processes occurring between them. At any given moment in time, the system configuration is a certain specific set of simple elements and processes. These configurations of the internal structure of the system are hierarchical with few levels of organizations; definite configuration is determined by the amount of bound analyte and is statically stable ensemble. It is important that

the transition from one configuration to another occurs when the principle of least action is followed, and these configurations themselves do not violate the macroscopic integrity of the entire system. The connection between different levels of organizations, e.g., slow processes of the "upper level" (analyte concentration gradient change) and rapid chaotic motion (which is usually indistinguishable in detail), is made through distributive nonlocal "control" factors ("mechanism"). It is these factors that reflect long-term collective correlations dependent on the generalized structure features (symmetry, flows, etc.) and provide the connection between micro- and macrolevels, resulting in a mutually coordinated coexistence of different system components. Ultimately, in nonequilibrium conditions initiated by the analyte appearance, the independence of system elements ("close order") is replaced by their "synchronous" behavior ("distant order"), aimed at achieving a new state of dynamic equilibrium. The mechanism of such reconstruction of a complex system depends on the analyte concentration, the surface centers selectivity, the values of their adsorption and desorption barriers, the surface topography, and many other factors that take place at the interface of phases.

Application of the Kinetic "Tuning Dial" in Biosensor Engineering and Analytical Science

The methodology of SEF-specific kinetic methods of analysis is an effective tool for both the development and efficient use of biosensors due to the fact that the dynamic relaxation of the response implicitly contains information about all possible system states (Figure 19.8a). This approach opens up the possibility to either limit or promote (as necessary) specific processes at the interface (selective binding, mass transfer, rebinding, competing, polydentate binding, etc.) using the given value of β as a criterion, which

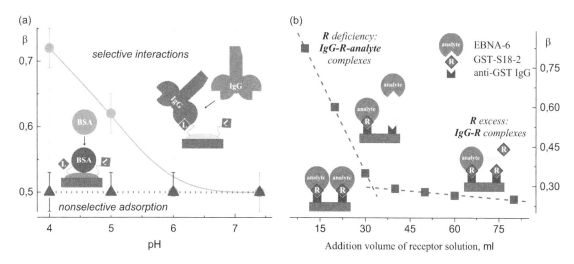

FIGURE 19.8 Illustration of an SEF analysis application for analytical purposes. Identification of the reaction mechanism for direct analysis methods (a): the dependence of the exponent power β on running buffer pH level reveals a diffusion mechanism with $\beta = 0.5$ for nonspecific bovine serum albumin protein (BSA) adsorption, whereas the kinetics of the IgG demonstrates the growth of selective interactions with the immobilized Glycitein L when the pH goes down (Boltovets 2011). Kinetic "visualization" of the change of the interfacial complex composition in CARSAR approach (b) depending on the added volume of receptor solution (Snopok et al. 2012). The values of the exponent power β obtained by kinetic curves approximation using SEF.

allows to purposefully optimize sensitive coatings (Snopok et al. 2006b). Moreover, dynamic methods can significantly improve the procedures of the analytical determination of the analyte when using the value of β as a parameter directly related to the analyte concentration (Snopok and Kruglenko 2015). This approach makes it possible to control interfacial processes (Figure 19.8b) and provides researchers and engineers with an efficient diagnostic tool and purposeful system optimization for achieving the goal (Snopok and Kruglenko 2005), which in turn allows to avoid sorting through endless possible options, to minimize errors and wrong technical decisions and, accordingly, to accelerate the development process and improve its efficiency.

19.7 Concluding Remarks

Further development of modern industrial society implies the constant use of elements and systems to measure the parameters of physical, chemical, or biological processes. By applying SPR phenomenon to the development of measuring systems, it is possible to make cheap, compact, and express biosensors high acceptable at the market. Such systems offer a set of advantages:

i. SPR detectors are sensitive to the change in the refractive index of the medium, the optical "equivalent" of the material mass; this fact is the basis for the "universality" of SPR detectors, whose nonspecific response indicates the "presence" of a substance on the surface;

ii. SPR techniques bridge the technological gap between nanoscale concept and practical instrumentation, since specific chemical functionality (i.e., molecular recognition, energy transfer, chemical conversions, etc.) is realized at the nanoscale, a macroscopic SPR transducer serves as a suitable tool for the spatial confinement of "concentrated" light near the surface within the nanoscale area for the direct monitoring of intramolecular processes;

iii. SPR systems make it possible to analyze molecular processes online by recording the kinetics of interfacial reactions.

Finally, the widespread use of nanoprobes in biofriendly sensing under SPR conditions opens up new possibilities for the detection and identification of potent xenobiotics by utilizing novel PTs (superstrong focusing and localization of light waves), advanced technologies (high-performance screening, subwave cartography), small signal recording methods (surface enhanced Raman spectroscopy, fluorescence, etc.), pinpoint chemistry (activation of catalytic and photosensitive reactions), mechanics (spatial movement of nanoobjects), local heat technology (local heating and directed heat transfer at the nanolevel), nanoelectronics (information carriers with a data signal transmission rate close to the velocity of light are capable of being propagated along nanostructures), etc. The possibility of converting light, electric, or chemical data flows into plasmon excitations and vice versa opens up paths for the direct creation of detection, transformation, processing, and data storage devices at the nanolevel, thus securing their mutual transformation according to the tasks at hand. Successful development of plasmonics in close collaboration with other branches of science will make it possible to finally put into practice the potential discovered in the nanoworld, for the realization of fundamentally new high-performance sensors for the smooth and continuous development of the industrial society.

Acknowledgment

The author expresses his sincere gratitude to Oleksiy Snopok and Tetiana Snopok for assistance in the preparation of the present manuscript material.

References

Altintas, Z., Uludag, Y., Gurbuz, Y. et al. 2012. Development of surface chemistry for SPR based sensors for the detection of proteins and DNA molecules. *Analytica Chimica Acta* 712: 138–44.

Ashcroft, N. W., Mermin, N. D. 1976. *Solid State Physics.* New York: Holt, Rinehart and Winston.

Ball, V., Ramsden, J. 1998. Buffer dependence of refractive index increments of protein solutions. *Biopolymers* 46: 489–92.

Barnes, W. L., Dereus, A., Ebbesen, T. W. 2003. Surface plasmon subwavelength optics. *Nature* 424: 824–930.

Bohren, C. F., Huffman, D. R. 1998. *Absorption and Scattering of Light by Small Particles.* New York: Wiley.

Boltovets, P., Shinkaruk, S., Bennetau-Pelissero, C. et al. 2011. The effect of low pH on the glycitein-BSA conjugate interaction with specific antiserum: competitive inhibition study using surface plasmon resonance technique. *Talanta* 84: 867–73.

Boltovets, P. M., Snopok, B. A. 2009. Measurement uncertainty in analytical studies based on surface plasmon resonance. *Talanta* 80: 466–72.

Boltovets, P. M., Polischuk, O. M., Kovalenko, O. G. et al. 2013. A simple SPR-based method for the quantification of the effect of potential virus inhibitors. *The Analyst* 138: 480–86.

Boltovets, P., Shinkaruk, S., Vellutini, L. et al. 2017. Self-tuning interfacial architecture for Estradiol detection by surface plasmon resonance biosensor. *Biosensors and Bioelectronics* 90: 91–5.

Brongersma, M. L., Kik, P. G. (Eds) 2007. *Surface Plasmon Nanophotonics.* Netherlands: Springer.

David, S., Polonschii, C., Gheorghiu, M. et al. 2017. Biosensing based on magneto-optical surface plasmon resonance. In: *Biosensors and Biodetection. Methods in Molecular Biology*, eds. A. Rasooly, B. Prickril, vol. 1571. New York: Humana Press, 10920–10928.

de Feijter, J. A., Benjamins, J., Veer, F. A. 1978. Ellipsometry as a tool to study the adsorptiom behaviour of polymers at the air-water interface. *Biopolymers* 17: 1759.

Evans, J. W. 1993. Random and cooperative sequential adsorption. *Reviews of Modern Physics* 65: 1281–329.

Fritzsche, W., Popp, J. (Eds) 2012. *Optical Nano- and Microsystems for Bioanalytics.* Heidelberg: Springer.

Goodrich, T. T., Lee, H. J., Corn, R. M. 2004. Direct detection of genomic DNA by enzymatically amplified SPR imaging measurements of RNA microarrays. *Journal of the American Chemical Society* 126: 4086–87.

Hardiman, G. (Ed) 2009. *Microarray Innovations: Technology and Experimentation.* Boca Raton, FL: CRC Press.

Hoa, X. D., Kirk, A. G., Tabrizian, M. 2007. Towards integrated and sensitive surface plasmon resonance biosensors: A review of recent progress. *Biosensors and Bioelectronics* 23 (2): 151–60.

Homola, J. 2006 (Ed). *Surface Plasmon Resonance Based Sensors.* Springer Series on Chemical Sensors and Biosensors, Series 4. Berlin: Springer Science.

Howell, D. (Ed) 2017. *Surface Plasmon Resonance (SPR): Advances in Research and Applications.* Series: Nanotechnology Science and Technology. New York: Nova Science.

Jung, L. S., Campbell, C. T., Chinowsky, T. M. et al. 1998. Quantitative interpretation of the response of surface plasmon resonance sensors to adsorbed films. *Langmuir* 14 (19): 5636–48.

Kim, I. T., Kihm, K. D. 2007. Full-field and real-time surface plasmon resonance imaging thermometry. *Optics Letters* 32: 3456–58.

Klimov, V.V. 2012. *Nanoplasmonic.* Moscow: Phizmatlit.

Kolotilov, S.V., Boltovets, P.N., Snopok, et al. 2006. Nanosized magnetic composite for extraction of γ-Immunoglobilins from biological media. *Theoretical and Experimental Chemistry* 42: 204–09.

Kretschmann, E. 1971. Die Bestimmung op-tischer Konstanten von Metallen durch An-regung von Oberflächenplasmaschwingungen. *Zeitschrift für Physik* 241: 313–24.

Langmuir, I. 1916. The constitution and fundamental properties of solids and liquids (part 1). *Journal of the American Chemical Society* 38: 2221–95.

Liedberg, B., Lundstrem, I., Stenberg, E. 1993. Principles of biosensing with an extended coupling matrix and surface plasmon resonance. *Sensors and Actuators B* 11: 63—72.

Löfås, S., Johnsson, B., Tegendal, K. et al. 1993. Dextran modified gold surfaces for surface plasmon resonance sensors: Immunoreactivity of immobilized antibodies and antibody-surface interaction studies. *Colloids and Surfaces B: Biointerfaces* 1 (2): 83–9.

Long, Y.-T., Jing, C. 2014. *Localized Surface Plasmon Resonance Based Nanobiosensors.* Berlin-Heidelberg: Springer-Verlag.

Lukosz, W. 1997. Integrated-optical and surface-plasmon sensors for direct affinity sensing. Part II: Anisotropy of adsorbed or bound protein adlayers. *Biosensors and Bioelectronics* 12: 175–84.

Lyon, L. A., Musick, M. D., Natan, M. J. 1998. Colloidal Au-enhanced surface plasmon resonance immunosensing. *Analytical Chemistry* 70: 5177–83.

Lysenko, S. I., Kaganovich, E. B., Kizyak, I. M. et al. 2005. Multiparametric chemical sensor based on nanocrystalline silicon waveguide. *Sensor Letters* 3 (2): 117–25.

Mandelbrot, B. B. 1982. *The Fractal Geometry of Nature.* San Francisco: Freeman.

Manera, M. G., Ferreiro-Vila, E., Garcia-Martin, J. M. et al. 2014. Enhanced antibody recognition with a magneto-optic surface plasmon resonance (MO-SPR) sensor. *Biosensors and Bioelectronics* 58: 114–20.

Manilo, M., Boltovets, P., Snopok, B. et al. 2017. Anomalous interfacial architecture in laponite aqueous suspensions on a gold surface. *Colloid and Surfaces A: Physiochemical and Engineering Aspects*: 520: 883-91.

Mariani, S., Minunni, M. 2014. Surface plasmon resonance applications in clinical analysis. *Analytical and Bioanalytical Chemistry* 406: 2303–23.

Marks, R. S., Lowe, C. R., Cullen, D. C. et al. (Eds) 2007. *Handbook of Biosensors and Biochips*, vol 2. New York: John Wiley & Sons, Inc.

Martinez, A. S., González, R. S., Espíndola, A. L. 2009. Generalized exponential function and discrete growth models. *Physica A* 388: 2922–30.

Minton, A. P. 2001. The influence of macromolecular crowding and macromolecular confinement on biochemical reactions in physiological media. *Journal of Biological Chemistry* 276: 10577–80.

Mol, N.J., Fisher, M. J. E. (Eds) 2010. *Surface Plasmon Resonance: Methods and Protocols.* Totowa, NJ: Humana Press.

Mukhopadhyay, R. 2005. Surface plasmon resonance instruments diversify. *Analytical Chemistry* 77: 313A–7A.

Narayanaswamy, R., Wolfbeis, O.S. (Eds) 2004. *Optical Sensors.* Heidelberg: Springer.

Naumenko, D., Snitka, V., Serviene, E. et al. 2013. In-vivo characterization of protein uptake by yeast cell envelope: Single cell AFM imaging and μ-tip-enhanced Raman scattering study. *The Analyst* 138 (18): 5371–83.

Nizamov, S., Mirsky, V. M. 2018. Wide-field surface plasmon resonance microscopy for in-situ characterization of nanoparticle suspensions. In: *In-situ Characterization Techniques for Nanomaterials*, ed. C. S. S. R. Kumar. Heidelberg: Springer, 61–105.

Nizamov, S., Scherbahn, V., Mirsky, V. M. 2016. Detection and quantification of single engineered nanoparticles in complex samples using template matching in wide-field surface plasmon microscopy. *Analytical Chemistry* 88 (20): 10206–14.

Otto, A. 1968. Excitation of nonradiative surface plasma waves in silver by the method of frustratied total reflection. *Zeitschrift für Physik* 216: 398.

Raether, H. 1985. *Surface Plasmons on Smooth and Rough Surfaces and on Gratings.* Berlin: Springer-Verlad.

Rasooly, A., Prickril, B. (Eds) 2017. *Biosensors and Biodetection. Methods and Protocols Volume 1: Optical-Based Detectors.* New York: Humana Press.

Rusmini, F., Zhong, Z., Feijen J. 2007. Protein immobilization strategies for protein biochips. *Biomacromolecules* 8 (6): 1775–89.

Savchenko, A., Kashuba, E., Kashuba, V. et al. 2007. A novel imaging technique for the screening of protein–protein interactions using scattered light under surface plasmon resonance conditions. *Analytical Chemistry* 79: 1349–55.

Savchenko, A., Kashuba, E., Kashuba, V. et al. 2008. Imaging of plasmid DNA microarrays by scattering light under surface plasmon resonance conditions. *Sensor Letters* 6: 705–13.

Scarano, S., Scuffi, C., Mascini, M. et al. 2011. Surface plasmon resonance imaging-based sensing for anti-bovine immunoglobulins detection in human milk and serum. *Anal Chim* Acta 707: 178–83.

Schasfoort, R. B. M. (Ed) 2017. *Handbook of Surface Plasmon Resonance*, 2nd edition. Cambridge: Royal Society of Chemistry.

Schasfoort, R. B. M., Tudos, A. J. 2008. *Handbook of Surface Plasmon Resonance.* Cambridge: Royal Society of Chemistry.

Shalabney, A., Abdulhalim, I. 2011. Sensitivity-enhancement methods for surface plasmon sensors. *Laser and Photonics Reviews* 5: 571–606.

Singh, P. 2016. SPR biosensors: Historical perspectives and current challenges. *Sensors and Actuators B: Chemical* 229: 110–30.

Snopok, B. A. 2012. Theory and practical use of surface plasmon resonance for analytical purposes (review). *Theoretical and Experimental Chemistry* 48 (N5): 265–84.

Snopok, B. A. 2014. Nonexponential kinetics of surface chemical reactions (review). *Theoretical and Experimental Chemistry* 50 (2): 67–95.

Snopok, B. A., Kostukevich, E. V. 2006. Kinetic studies of protein–surface interactions: A two-stage model of surface-induced protein transitions in adsorbed biofilms. *Analytical Biochemistry* 348 (2): 222–31.

Snopok, B. A., Kruglenko, I. V. 2002. Multisensor systems for chemical analysis: State-of-the-art in electronic nose technology and new trends in machine olfaction. *Thin Solid Films* 418: 21–41.

Snopok, B. A., Kruglenko, I. V. 2005. Nonexponential relaxations in sensor arrays: Forecasting strategy for electronic nose performance. *Sensors and Actuators B: Chemical* 106 (1): 101–13.

Snopok, B., Kruglenko, I. 2015. Analyte induced water adsorbability in gas phase biosensors: The influence of ethinylestradiol on the water binding protein capacity. *Analyst* 140: 3225–32.

Snopok, B. A., Snopok, O. B. 2018. Information processing in chemical sensing: Unified evolution coding by stretched exponential. In: *Detection of CBRN-Nanostructured Materials*, ed. J. Bonca, S. Kruchinin,

Chapter 18. Netherlands: Springer, Dordrecht, 233–243.

Snopok, B. A., Boltovets, P. N., Rowell, F. J. 2008. Interface functional architecture using mixed thiol monolayers: Effect of the composition and spatial organization on the immobilization of low-molecular-weight ligands. *Theoretical and Experimental Chemistry* 44 (N3): 165–171.

Snopok, B. A., Boltovets, P. N., Rowell, F. J. 2006a. Simple analytical model of biosensor competition analysis for detection of low molecular weight analytes. *Theoretical and Experimental Chemistry* 42 (2): 106–12.

Snopok, B. A., Darekar, S., Kashuba, E. V. 2012. Analysis of protein–protein interactions in a complex environment: Capture of an analyte–receptor complex with standard additions of the receptor (CARSAR) approach. *Analyst* 137: 3767–72.

Snopok, B. A., Kostyukevich, K. V., Lysenko, S. I. et al. 2001. Optical biosensors based on the surface plasmon resonance phenomenon: Optimization of the metal layer parameters. *Semiconductor Physics, Quantum Electronics and Optoelectronics* 4 (1): 56–69.

Snopok, B. A., Kostyukevych, K. V., Rengevych, O. V. et al. 1998. A biosensor approach to probe the structure and function of the adsorbed proteins: Fibrinogen at the gold surface. *Semiconductor Physics, Quantum Electronics and Optoelectronics (SPQEO)* 1 (1): 121–34.

Snopok, B. A., Yurchenko, M., Szekely, L. et al. 2006b. SPR based immuno-capture approach for in vitro analysis of protein complex formation: Mapping of MRS18-2 binding site on retinoblastoma protein. *Analytical and Bioanalytical Chemistry* 386: 2063–73.

Sommerfeld A., Bethe H. (1933). Elektronentheorie der Metalle. In: *Aufbau Der Zusammenhängenden Materie. Handbuch der Physik*, vol 24/2. Springer, Berlin, Heidelberg.

Sotnikov, D. V., Zherdev, A. V., Dzantiev, B. B. 2015. REVIEW: Detection of intermolecular interactions based on surface plasmon resonance. *Biochemistry Moscow* 80: 1820.

Thevenot, D., Toth, K., Durst, R. et al. 1999. Electrochemical biosensors: Recommended definitions and classification. *Pure and Applied Chemistry* 71 (12): 2333.

Wilson, A. H. 1953. *The Theory of Metals.* Cambridge: Cambridge University Press.

Wijaya, E., Lenaerts, C., Maricot, S. et al. 2011. Surface plasmon resonance-based biosensors: From the development of different SPR structures to novel surface functionalization strategies. *Current Opinion in Solid State and Materials Science* 15 (5): 208–24.

Wong, L. S., Khan, F., Micklefield, J. 2009. Selective covalent protein tmmobilization: Strategies and applications. *Chemical Reviews* 109 (9): 4025–53.

Yamamoto, M. 2002. Surface Plasmon Resonance (SPR) theory: Tutorial. *Review of Polarography* 48 (3): 209–37.

Yeatman, E. M. 1996. Resolution and sensitivity in surface Plasmon microscopy and sensing. *Biosensors and Bioelectronics* 11: 635–49.

Plasmonic Optical Antenna and Its Enhancement to Infrared Photodetectors

Lin Li
Neda Mojaverian
University of Massachusetts Lowell

Guiru Gu
Stonehill College

Xuejun Lu
University of Massachusetts Lowell

20.1 Introduction... 20-1
20.2 A Brief Review of the Basic Concepts of Conventional RF Antennas... 20-1
 $\vec{E}$ and $\vec{H}$ Fields of a Half-Wave Dipole Antenna • Directivity and Gain of the Half-Wave Dipole Antenna • Electromagnetism Reciprocity and Antenna Receiving Patterns
20.3 Surface Plasmonic Waves and Optical Antennas 20-4
 Surface Plasmonic Waves • Wave Vector of SPR and Scaling Factor • Difference between RF Antennas and Optical Antennas
20.4 Optical Antenna Enhanced QDIPs 20-6
 Transformation of the Far Field to the Near Field in a Pointed Dipole Optical Antenna • Polarization-Dependent Enhancement in QDIP
20.5 Chapter Summary.. 20-10
References 20-12

20.1 Introduction

Since their invention by Heinrich Hertz in 1886, antennas have been extensively utilized in transmitting and receiving electromagnetic (EM) waves in the radio frequency and microwave spectral regimes (referred to as RF antenna henceforth). Antennas can collect a large area of free-propagating EM radiation in these spectrum regimes and covert them to localized electric current or, conversely, radiate EM waves into free space. Antennas can significantly enhance the transmission efficiency and receiving sensitivity of EM waves. They can also allow EM waves to be emitted and received with specific radiation patterns and directions (i.e. gains). It is a critical enabling technology for virtually all wireless communication devices, such as satellites, cell phones, televisions, and radars. As the counterparts of antennas in the optical regimes, optical antennas[1−11] have shown great promises in improving the transmission and receiving performance in a broad range of applications, including near-field microscopy and spectroscopy,[4,12] biological sensing,[13−15] solar cell energy harvesting,[2,13,16,17] and infrared (IR) detection.[6,18−20]

Similar to the RF antennas, optical antennas have also been demonstrated to be effective components in mediating the free-space propagation EM waves and localized fields[1−11,18] with engineerable transmission and receiving patterns.[18,19,21] Since the wavelength is much shorter in the optical spectral regimes, the dimensions of optical antennas are much smaller than that of conventional RF antennas. The dimensions of optical antennas can be scaled down with

effective wavelength scaling factors.[22] Despite the similarity, however, due to the difference in signal collection schemes, optical antennas interact with emitters, detectors, or other devices through their near fields. By engineering the near fields, one can not only improve the receiving and transmission efficiency but also provide strong enhancement with desired receiving patterns.

This chapter will cover the introduction of optical antennas, their comparison with RF antennas, the surface plasmonic resonance (SPR) excitation in optical antennas, the near-field profiles and enhancement, directional gains, and far-field detection and patterns. The applications of the plasmonic optical antennas in IR photodetection will also be discussed.

20.2 A Brief Review of the Basic Concepts of Conventional RF Antennas

Antenna theories have been thoroughly presented in numerous textbooks.[23−25] For the readers' convenience, this section gives a brief review of the basic concepts of RF antennas, with more focus on receiving antennas and electromagnetism reciprocity.

20.2.1 $\vec{E}$ and $\vec{H}$ Fields of a Half-Wave Dipole Antenna

Figure 20.1 shows a schematic of a dipole antenna. When the length of each arm of the antenna L equals $\lambda/4$ and the total

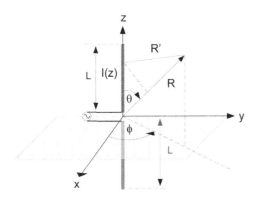

FIGURE 20.1 A schematic of a dipole antenna. The current distribution varies along the antenna arm with the current at the tip end of the antenna equals zero due to the open circuit at the tip end.

length of antenna is $\lambda/2$, the antenna is the well-known half-wavelength dipole antenna. The current distribution can be written as

$$I\left(z\right)=I_0\cos(kz),\qquad(20.1)$$

where I_0 is the peak current and $k=\frac{2\pi}{\lambda}$ is the propagation constant on the antenna.

The magnetic vector potential $\vec{A}$ can be expressed as

$$\vec{A}\left(R\right)=\hat{z}\frac{\mu_0}{4\pi}\int_{-L}^{L}\frac{I\left(z\right)e^{-ikR'}}{R'}dz,\qquad(20.2)$$

where the integral is from $-L$ to L. When R is large compared with z, $R'\approx R-z\cos\ \theta$. The magnetic vector potential $\vec{A}$ can be approximately written as

$$\vec{A}\left(R\right)\approx\hat{z}\frac{\mu_0 I_0}{4\pi}\frac{e^{-ikR}}{R}\int_{-L}^{L}\cos\left(kz\right)e^{ikz\cos\theta}dz$$

$$=\hat{z}\frac{\mu_0 I_0}{4\pi}\frac{e^{-ikR}}{R}\int_{-L}^{L}[\cos\left(kz\right)\cos\left(kz\cos\theta\right)$$
$$+\ i\cos\left(kz\right)\sin(kz\cos\theta)]\,dz$$

$$=\hat{z}\frac{\mu_0 I_0}{4\pi}\frac{e^{-ikR}}{R}\int_{-L}^{L}\frac{1}{2}\left[\cos\left(kz(1+\cos\theta)\right)\right.$$
$$\left.+\cos\left(kz(1-\cos\theta)\right)\right]dz$$

$$=\hat{z}\frac{\mu_0 I_0}{4\pi}\frac{e^{-ikR}}{kR}\left[\frac{\sin\left(kL(1+\cos\theta)\right)}{1+\cos\theta}\right.$$
$$\left.+\frac{\sin\left(kL(1-\cos\theta)\right)}{1-\cos\theta}\right],\qquad(20.3)$$

In the spherical coordinate, the $\vec{A}$ components are

$$A_R=\hat{R}\frac{\mu_0 I_0}{4\pi}\frac{e^{-ikR}}{kR}$$
$$\times\cos\theta\left[\frac{\sin\left(kL(1+\cos\theta)\right)}{1+\cos\theta}+\frac{\sin\left(kL(1-\cos\theta)\right)}{1-\cos\theta}\right],$$
$$=\hat{R}\frac{\mu_0 I_0}{4\pi}\frac{e^{-ikR}}{kR}$$
$$\times\cos\theta\left[\frac{2\sin\left(kL\right)\cos\left(kL\cos\theta\right)+2\cos\theta\cos\left(kL\right)\sin\left(kL\cos\theta\right)}{\sin^2\theta}\right],$$

$$=\hat{R}\frac{\mu_0 I_0}{4\pi}\frac{e^{-ikR}}{kR}\cos\theta\left[\frac{2\cos\left(\frac{\pi}{2}\cos\theta\right)}{\sin^2\theta}\right],\qquad(20.4)$$

$$A_\theta=\hat{\theta}\frac{-\mu_0 I_0}{4\pi}\frac{e^{-ikR}}{kR}\left[\frac{2\cos\left(\frac{\pi}{2}\cos\theta\right)}{\sin\theta}\right],\qquad(20.5)$$

$$A_\varphi=0,\qquad(20.6)$$

$$\vec{H}=\frac{1}{\mu_0}\nabla\times\vec{A}$$

$$=\hat{\varphi}\frac{1}{R}\left\{\frac{iI_0}{4\pi}\left[\frac{2\cos\left(\frac{\pi}{2}\cos\theta\right)}{\sin\theta}\right]e^{-ikR}\right.$$
$$\left.-\frac{I_0}{4\pi}\frac{e^{-ikR}}{kR}\frac{\partial}{\partial\theta}\left[\cos\theta\left[\frac{2\cos\left(\frac{\pi}{2}\cos\theta\right)}{\sin^2\theta}\right]\right]\right\}$$

$$=\hat{\varphi}\frac{I_0}{4\pi}\frac{e^{-ikR}}{kR^2}\left[\frac{2\cos\left(\frac{\pi}{2}\cos\theta\right)}{\sin\theta}\right](ikR+1)$$
$$-\hat{\varphi}\frac{I_0}{4\pi}\frac{e^{-ikR}}{kR^2}\cos\theta\left\{\frac{\partial}{\partial\theta}\left[\left[\frac{2\cos\left(\frac{\pi}{2}\cos\theta\right)}{\sin^2\theta}\right]\right]\right\}$$

$$=\hat{\varphi}\frac{I_0}{4\pi}\frac{e^{-ikR}}{kR^2}\left[\frac{2\cos\left(\frac{\pi}{2}\cos\theta\right)}{\sin\theta}\right](ikR+1)$$
$$-\hat{\varphi}\frac{I_0}{4\pi}\frac{e^{-ikR}}{kR^2}\cos\theta\left\{\frac{\pi\sin\left(\frac{\pi}{2}\cos\theta\right)}{\sin\theta}-\frac{2\cos\left(\frac{\pi}{2}\cos\theta\right)\cos\theta}{3\sin^3\theta}\right\}$$
$$(20.7)$$

Only considering θ close to $\pi/2$, the magnetic field $\vec{H}$ can be approximately expressed as

$$\vec{H}\approx\hat{\varphi}\frac{I_0}{4\pi}\frac{e^{-ikR}}{kR^2}\left[\frac{2\cos\left(\frac{\pi}{2}\cos\theta\right)}{\sin\theta}\right](ikR+1),$$
$$=\hat{\varphi}\frac{jI_0 k}{4\pi}e^{-ikR}\left[\frac{2\cos\left(\frac{\pi}{2}\cos\theta\right)}{\sin\theta}\right]\left(\frac{1}{kR}-\frac{j}{(kR)^2}\right)$$
$$(20.8)$$

The electric field $\vec{E}$ is thus

$$\vec{E}=\frac{1}{j\omega\epsilon_0}\nabla\times\vec{H}$$

$$E_R=\hat{R}\frac{1}{j\omega\epsilon_0}\frac{I_0}{4\pi}\frac{e^{-ikR}}{kR^3\sin\theta}(ikR+1)\frac{\partial}{\partial\theta}\left[2\cos\left(\frac{\pi}{2}\cos\theta\right)\right],$$
$$=\hat{R}\frac{1}{j\omega\epsilon_0}\frac{I_0}{4\pi}\frac{e^{-ikR}}{kR^3\sin\theta}(ikR+1)\left[\pi\sin\theta\sin\left(\frac{\pi}{2}\cos\theta\right)\right],$$
$$=\hat{R}\eta_0\frac{I_0 k}{4}e^{-ikR}\left(\frac{1}{(kR)^2}-\frac{j}{(kR)^3}\right)\left[\sin\left(\frac{\pi}{2}\cos\theta\right)\right],$$
$$(20.9)$$

where $\eta_0=\sqrt{\frac{\mu_0}{\epsilon_0}}$ is the impedance of the vacuum.

$$E_\theta=\hat{\theta}\frac{-1}{j\omega\epsilon_0}\frac{I_0}{4\pi}\frac{1}{kR}\left[\frac{2\cos\left(\frac{\pi}{2}\cos\theta\right)}{\sin\theta}\right]\frac{\partial}{\partial R}\left[(ikR+1)\frac{e^{-ikR}}{R}\right],$$
$$=\hat{\theta}\frac{-1}{j\omega\epsilon_0}\frac{I_0}{4\pi}\frac{1}{kR}\left[\frac{2\cos\left(\frac{\pi}{2}\cos\theta\right)}{\sin\theta}\right]\frac{e^{-ikR}}{R}$$

$$\times \left[(ik) + (ikR+1)\,(-ik) - (ikR+1)\,\frac{1}{R} \right],$$

$$= \hat{\theta}\frac{-1}{j\omega\epsilon_0}\frac{I_0}{4\pi}\frac{e^{-ikR}}{kR^2}\left[\frac{2\cos\left(\frac{\pi}{2}\cos\theta\right)}{\sin\theta}\right]\left[k^2R - ik - \frac{1}{R}\right],$$

$$= \hat{\theta}\frac{-1}{j\omega\epsilon_0}\frac{I_0}{4\pi}\frac{e^{-ikR}}{k}\left[\frac{2\cos\left(\frac{\pi}{2}\cos\theta\right)}{\sin\theta}\right]\left[\frac{k^2}{R} - \frac{ik}{R^2} - \frac{1}{R^3}\right],$$

$$= \hat{\theta}\frac{jk\eta_0 I_0}{4\pi}e^{-ikR}\left[\frac{2\cos\left(\frac{\pi}{2}\cos\theta\right)}{\sin\theta}\right]\left[\frac{1}{(kR)} - \frac{i}{(kR)^2} - \frac{1}{(kR)^3}\right],$$

$$(20.10)$$

In the far field $(R \gg \lambda)$, the electric field $\vec{E}$ and the magnetic field $\vec{H}$ can be written as

$$\vec{E} = \hat{\theta}\frac{j\eta_0 I_0}{4\pi}\frac{e^{-ikR}}{R}\left[\frac{2\cos\left(\frac{\pi}{2}\cos\theta\right)}{\sin\theta}\right],$$

$$\vec{H} = \hat{\varphi}\frac{jI_0}{4\pi}\frac{e^{-ikR}}{R}\left[\frac{2\cos\left(\frac{\pi}{2}\cos\theta\right)}{\sin\theta}\right], \qquad (20.11)$$

20.2.2 Directivity and Gain of the Half-Wave Dipole Antenna

The EM energy flux, i.e. the Poynting vector $\vec{S}$ can be expressed as

$$\vec{S} = \vec{E} \times \vec{H}^* = \hat{R}\left(\frac{I_0}{4\pi}\right)^2\frac{\eta_0}{R^2}\left[\frac{2\cos\left(\frac{\pi}{2}\cos\theta\right)}{\sin\theta}\right]^2,$$

$$= \hat{R}\left(\frac{I_0}{4\pi}\right)^2\frac{\eta_0}{R^2}F\left(\theta\right), \qquad (20.12)$$

where $F\left(\theta\right) = \left[\frac{2\cos\left(\frac{\pi}{2}\cos\theta\right)}{\sin\theta}\right]^2$ is the far-field radiation pattern. The directivity D and gain G of the antenna can be written as

$$D = \frac{4\pi}{\int_0^{2\pi}d\varphi\int_0^\pi\left[\frac{2\cos\left(\frac{\pi}{2}\cos\theta\right)}{\sin\theta}\right]^2\sin\theta d\theta}, \qquad (20.13)$$

$$G = \xi D = \frac{P_{\text{rad}}}{P_{\text{total}}}, \qquad (20.14)$$

where ξ is antenna efficiency, and P_{rad} and P_{total} are the radiation and total powers, respectively. P_{rad} and P_{total} can be written as

$$P_{\text{rad}} = \oiint\vec{S}\cdot d\vec{S}_R = \left(\frac{I_0}{4\pi}\right)^2\eta_0 2\pi\int_0^\pi\left[\frac{2\cos\left(\frac{\pi}{2}\cos\theta\right)}{\sin\theta}\right]^2$$

$$\times\sin\theta d\theta, \qquad (20.15)$$

$$P_{\text{total}} = P_{\text{rad}} + P_{\text{loss}},$$

$$= P_{\text{rad}} + \frac{1}{2}R_{\text{loss}}I_0^2, \qquad (20.16)$$

where R_{loss} is the loss resistance of the antenna.

20.2.3 Electromagnetism Reciprocity and Antenna Receiving Patterns

Assume two E-fields $\vec{E}_1$, $\vec{E}_2$. Their corresponding magnetic fields and current densities are $\vec{H}_1$, $\vec{H}_2$, and $\vec{J}_1$, $\vec{J}_2$, respectively. The $\nabla \cdot \left(\vec{E}_1 \times \vec{H}_2 - \vec{E}_2 \times \vec{H}_1\right)$ can be written as

$$\nabla \cdot \left(\vec{E}_1 \times \vec{H}_2 - \vec{E}_2 \times \vec{H}_1\right) = \vec{H}_2 \cdot \left(\nabla \times \vec{E}_1\right)$$

$$- \vec{E}_1 \cdot \left(\nabla \times \vec{H}_2\right) - \vec{H}_1 \cdot \left(\nabla \times \vec{E}_2\right) + \vec{E}_2 \cdot \left(\nabla \times \vec{H}_1\right)$$

$$= -\vec{H}_2 \cdot j\omega\mu_0\vec{H}_1 - \vec{E}_1 \cdot \vec{J}_2 - \vec{E}_1 \cdot j\omega\epsilon_0\vec{E}_2 + \vec{H}_1 \cdot j\omega\mu_0\vec{H}_2$$

$$+ \vec{E}_2 \cdot \vec{J}_1 + \vec{E}_2 \cdot j\omega\epsilon_0\vec{E}_1$$

$$= -\vec{E}_1 \cdot \vec{J}_2 + \vec{E}_2 \cdot \vec{J}_1, \qquad (20.17)$$

Equation (20.12) can be expressed as

$$\oiint\left(\vec{E}_1 \times \vec{H}_2 - \vec{E}_2 \times \vec{H}_1\right) \cdot d\vec{S} = \iiint\left(\vec{E}_2 \cdot \vec{J}_1 - \vec{E}_1 \cdot \vec{J}_2\right)dv, \qquad (20.18)$$

when the surface is infinitely far away, the surface integral at the left side of Eq. (20.18) equals zero. Equation (20.18) can thus be written as

$$\iiint\left(\vec{E}_2 \cdot \vec{J}_1 - \vec{E}_1 \cdot \vec{J}_2\right)dv = 0, \qquad (20.19)$$

Equation (20.19) indicates that the coupling between $\vec{E}_1$ and $\vec{J}_2$ is the same as between $\vec{E}_2$ and $\vec{J}_1$.

To understand the reciprocity of the transmission and receiving antenna, one can define $\vec{E}_1$ and $\vec{J}_1$ for Antenna #1, and $\vec{E}_2$, $\vec{J}_2$ for Antenna #2. Equation (20.19) indicates that the E-filed of Antenna #1 coupled to the current of Antenna #2 (i.e. $\vec{E}_1 \cdot \vec{J}_2$) is the same as that of Antenna #2 coupled to Antenna #1, i.e. the transmitting and receiving roles of the two antennas are exchangeable. The antenna transmission patterns can be measured by rotating the receiving antenna (Antenna #2) around the transmission antenna (Antenna #1) as shown in Figure 20.2. When Antenna #1 is the transmitter and Antenna #2 is the receiver, the received signals

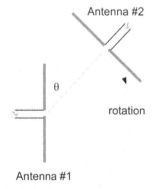

FIGURE 20.2 Measurement of antenna transmission and receiving patterns.

(i.e. $\vec{E}_1 \cdot \vec{J}_2$) along the rotation angle θ are the transmission patterns. Due to reciprocity, it will be the same as those of Antenna #1 as the receiver and Antenna #2 as the transmitter. In the latter configuration, the antenna transmission patterns are the same as the receiving patterns due to reciprocity.

20.3 Surface Plasmonic Waves and Optical Antennas

20.3.1 Surface Plasmonic Waves

Ritchie predicted in 1957[26] that coherent electron oscillation at a metal surface can generate surface EM waves (also called surface plasmon polaritons, SPPs). The current continuous equation can be expressed:

$$e\frac{\partial n}{\partial t} = \nabla \cdot \vec{J} \qquad (20.20)$$

where $\vec{J}$ is the current density, n is the electron density, and e is the charge of an electron, $e = -1.6 \times 10^{-19}(C)$. The current density $\vec{J}$ is related to the velocity of the electrons v by

$$\vec{J} = env = en\frac{dx}{dt}, \qquad (20.21)$$

where x is the position of the electrons. Taking a partial derivative on Eq. (20.20) with regard to time t, one gets

$$e\frac{\partial^2 n}{\partial t^2} = \nabla \cdot \frac{\partial \vec{J}}{\partial t} = \nabla \cdot \left(en\frac{d^2 x}{dt^2} \right),$$
$$= \frac{en}{m}\nabla \cdot (ma), \qquad (20.22)$$

where a is acceleration, m is the mass of an electron, and $ma = F = e\vec{E}$ is the electric force on the electron. Combining Eq. (20.22) with Gauss's Law, one gets

$$e\frac{\partial^2 n}{\partial t^2} = \frac{en}{m}\frac{ne^2}{\varepsilon_0}, \qquad (20.23)$$

In the frequency domain,

$$-en\omega_p^2 = \frac{en}{m}\frac{ne^2}{\varepsilon_0}, \qquad (20.24)$$

where ω_p is plasma frequency. $\omega_p = \sqrt{\frac{e^2 n}{m\varepsilon_0}}$. Equation (20.24) indicates that the electron concentration fluctuates at the plasma frequency ω_p.

Taking partial derivative on Eq. (20.21) with regard to time t, one gets

$$\frac{\partial \vec{J}}{\partial t} = en\frac{d^2 x}{dt^2} = \frac{en}{m}\left(e\vec{E} + \Gamma mv \right) = \varepsilon_0 \omega_p^2 \vec{E} + \Gamma \vec{J}, \quad (20.25)$$

where the Γmv term represents the dumping force, which is proportional to the velocity v.

In the frequency domain,

$$(j\omega - \Gamma)\vec{J} = \varepsilon_0 \omega_p^2 \vec{E}, \qquad (20.26)$$

The conductivity σ is thus

$$\sigma = \frac{\varepsilon_0 \omega_p^2}{(j\omega - \Gamma)}, \qquad (20.27)$$

The complex permittivity ε_c is thus

$$\varepsilon_c = \varepsilon_\infty + \frac{\sigma}{j\omega} = \varepsilon_0 \left(\varepsilon_{\infty,r} - \frac{\omega_p^2}{\omega^2 + j\omega\Gamma} \right), \qquad (20.28)$$

where $\varepsilon_{\infty,r}$ is the relative permittivity at a high frequency well beyond the plasma frequency ω_p.

Equation (20.28) is the well-known Drude model.[27] Equation (20.28) indicates the permittivity variation at different frequencies. When the frequency ω is much smaller than the plasmonic frequency ω_0, the metal has a large negative permittivity ε_c, whereas when the frequency ω is much larger than the plasmonic frequency ω_0, the metal behaves like regular dielectric materials. Ordal, etc., listed the complex permittivity ε_c of various metals at different frequencies.[28]

20.3.2 Wave Vector of SPR and Scaling Factor

The analysis of plasmonic waves at a dielectric/metal interface was given by Raether.[29] For the readers' convenient reference, a brief summary is presented as the following:

Figure 20.3 shows the incidence of a transmagnetic (TM) plane wave on a dielectric/metal interface with the H-field in the y direction. The H fields of the incident wave (i.e. $\vec{H}_{1i}$), reflection wave (i.e. $\vec{H}_{1r}$), and the transmission wave (i.e. $\vec{H}_{2t}$) are marked on the figure.

The E-fields are related to the magnetic fields by

$$\vec{E} = \frac{1}{j\omega\varepsilon_c}\nabla \times \vec{H} \qquad (20.29)$$

From Equation (20.29), the E_x can be expressed as

$$\vec{E}_{1i} = \hat{x}\frac{-jk_{1z}}{j\omega\varepsilon_1}H_{1i}e^{-jk_{1x}x-jk_{1z}z} \qquad (20.30)$$

$$\vec{E}_{1r} = \hat{x}\frac{jk_{1z}}{j\omega\varepsilon_1}H_{1r}e^{-jk_{1x}x+jk_{1z}z} \qquad (20.31)$$

$$\vec{E}_{2t} = \hat{x}\frac{-jk_{2z}}{j\omega\varepsilon_2}H_{2t}e^{-jk_{2x}x-jk_{2z}z} \qquad (20.32)$$

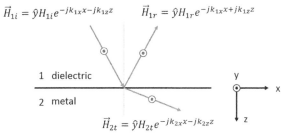

FIGURE 20.3 Scheme of TM wave incidence at the dielectric and metal interface.

where k_{1x}, k_{1z} are the x and z components of the propagation constant k_1, respectively. Similarly, k_{2x}, k_{2z} are the x and z components of the propagation constant k_2, respectively.

$$k_{1x}^2 + k_{1z}^2 = k_1^2 = \omega^2 \varepsilon_1 \mu_1 \qquad (20.33)$$

$$k_{2x}^2 + k_{2z}^2 = k_2^2 = \omega^2 \varepsilon_{2c} \mu_2 \qquad (20.34)$$

where $\mu_1 \approx \mu_2 \approx \mu_0$ are the permeability of the material 1, 2 and vacuum, respectively. The boundary conditions are H_y and E_x continuous at the interface (i.e. $z = 0$). Applying the boundary conditions and assuming the reflection is zero, one gets

$$k_{1x} = k_{2x} \qquad (20.35)$$

$$\frac{k_{1z}}{\varepsilon_1} = \frac{k_{2z}}{\varepsilon_{2c}} \qquad (20.36)$$

Combining Eqs. (20.34)−(20.36), one gets

$$k_{2x}^2 + k_{2z}^2 = \frac{\varepsilon_{2c}^2}{\varepsilon_1^2} k_{1z}^2 + k_{2x}^2 = \omega^2 \varepsilon_{2c} \mu_0 \qquad (20.37)$$

$$k_{1x}^2 = \omega^2 \mu_0 \frac{\varepsilon_{2c} \varepsilon_1}{(\varepsilon_1 + \varepsilon_{2c})} \qquad (20.38)$$

$$k_x = k_0 \mathrm{Re}\left[\sqrt{\frac{\varepsilon_{2cr} \varepsilon_{1r}}{(\varepsilon_{1r} + \varepsilon_{2cr})}} \right] + j k_0 \mathrm{Im}\left[\sqrt{\frac{\varepsilon_{2cr} \varepsilon_{1r}}{(\varepsilon_{1r} + \varepsilon_{2cr})}} \right] \qquad (20.39)$$

where ε_{1r} and ε_{2cr} are the relative permittivity, and $\mathrm{Re}(.)$ and $\mathrm{Im}(.)$ are taking the real and imaginary parts, respectively.

$$k_{2z} = k_0 \mathrm{Re}\left[\sqrt{\frac{\varepsilon_{2cr} \varepsilon_{2cr}}{(\varepsilon_{1r} + \varepsilon_{2cr})}} \right] + j k_0 \mathrm{Im}\left[\sqrt{\frac{\varepsilon_{2cr} \varepsilon_{2cr}}{(\varepsilon_{1r} + \varepsilon_{2cr})}} \right] \qquad (20.40)$$

The plasmonic wave propagates in x and z directions with loss described by the imaginary parts of the corresponding directions, i.e. $k_{\mathrm{spp}} = k_x$. Using gold as an example, the complex permittivity ε_{2cr} of gold at different wavelengths are listed in the following table[28] together with the real ($k_{x,re}$) and imaginary ($k_{x,im}$) parts of k_{spp}.

Table 20.1 shows that, at low frequency, SPR travels at the same wave vector as in air with low loss, whereas at high frequency, the wave vector and the loss of the SPR waves increase. Since both the RF and the IR frequencies are far from ω_p, the SPR waves in these frequency regimes behave in a similar way to the wavelength scaling factor.[22] Therefore, for frequencies much lower than ω_p, antenna theories developed for the RF frequencies can be applied to optical antennas with a simple scaling factor determined by the square root of the relative permittivity of the dielectric material, i.e. $1/\sqrt{\varepsilon_{1r}}$.

TABLE 20.1 Complex Permittivity ε_{2cr} of Gold at Different Wavelengths

λ (µm)	$\varepsilon_{2cr,re}$	$\varepsilon_{2cr,im}$	$k_{x,re}$	$k_{x,im}$
32.0	$-3.69\mathrm{E}+04$	$2.54\mathrm{E}+04$	$k_0\sqrt{\varepsilon_{1r}}$	0
10.0	$-5.05\mathrm{E}+03$	$1.09\mathrm{E}+03$	$k_0\sqrt{\varepsilon_{1r}}$	0
3.0	$-4.75\mathrm{E}+02$	$3.07\mathrm{E}+01$	$k_0\sqrt{\varepsilon_{1r}}$	0
0.8	-23.4	1.55	$1.02\,k_0\sqrt{\varepsilon_{1r}}$	$0.0015\,k_0\sqrt{\varepsilon_{1r}}$
0.5	-2.68	3.09	$1.07\,k_0\sqrt{\varepsilon_{1r}}$	$0.12\,k_0\sqrt{\varepsilon_{1r}}$

On the other hand, for the frequencies comparable to ω_p, the scaling factor can be analyzed as follows:

For ideal lossless perfect electrical conductors, the relative complex permittivity ε_{cr} is

$$\varepsilon_{cr} = \varepsilon_{\infty,r} - \frac{\omega_p^2}{\omega^2} = \varepsilon_{\infty,r} - \frac{\omega_p^2}{\omega^2} = \varepsilon_{\infty,r} - \frac{\lambda^2}{\lambda_p^2} \qquad (20.41)$$

where λ and λ_p are the wavelength in vacuum for the frequencies ω and ω_p, respectively.

$$k_x = k_0 \mathrm{Re}\left[\sqrt{\frac{\varepsilon_{2cr} \varepsilon_{1r}}{(\varepsilon_{1r} + \varepsilon_{2cr})}} \right] + j k_0 \mathrm{Im}\left[\sqrt{\frac{\varepsilon_{2cr} \varepsilon_{1r}}{(\varepsilon_{1r} + \varepsilon_{2cr})}} \right] \qquad (20.42)$$

$$= k_0 \mathrm{Re}\left[\sqrt{\frac{\varepsilon_{1r}\left(\varepsilon_{\infty,r} - \frac{\lambda^2}{\lambda_p^2}\right)}{\left(\varepsilon_{1r} + \varepsilon_{\infty,r} - \frac{\lambda^2}{\lambda_p^2}\right)}} \right] + j k_0 \mathrm{Im}\left[\sqrt{\frac{\varepsilon_{1r}\left(\varepsilon_{\infty,r} - \frac{\lambda^2}{\lambda_p^2}\right)}{\left(\varepsilon_{1r} + \varepsilon_{\infty,r} - \frac{\lambda^2}{\lambda_p^2}\right)}} \right] \qquad (20.43)$$

when $\omega \ll \omega_p$, i.e. $\lambda \gg \lambda_p$, and $\left|\varepsilon_{\infty,r} - \frac{\lambda^2}{\lambda_p^2}\right| \gg \varepsilon_{1r}$, the wave vector of the SPR waves k_x is thus

$$k_x = k_0 \mathrm{Re}\left[\sqrt{\frac{\varepsilon_{1r}}{\left(\frac{\varepsilon_{1r}}{\varepsilon_{\infty,r} - \frac{\lambda^2}{\lambda_p^2}} + 1\right)}} \right] + j k_0 \mathrm{Im}\left[\sqrt{\frac{\varepsilon_{1r}\left(\varepsilon_{\infty,r} - \frac{\lambda^2}{\lambda_p^2}\right)}{\left(\varepsilon_{1r} + \varepsilon_{\infty,r} - \frac{\lambda^2}{\lambda_p^2}\right)}} \right] \qquad (20.44)$$

$$= k_0 \sqrt{\varepsilon_{1r}} \mathrm{Re}\left[\frac{1}{\sqrt{\left(\frac{\varepsilon_{1r}}{\varepsilon_{\infty,r} - \frac{\lambda^2}{\lambda_p^2}} + 1\right)}} \right] + j k_0 \mathrm{Im}\left[\sqrt{\frac{\varepsilon_{1r}\left(\varepsilon_{\infty,r} - \frac{\lambda^2}{\lambda_p^2}\right)}{\left(\varepsilon_{1r} + \varepsilon_{\infty,r} - \frac{\lambda^2}{\lambda_p^2}\right)}} \right] \qquad (20.45)$$

$$\approx k_0 \sqrt{\varepsilon_{1r}} \left(1 - \frac{1}{2}\frac{\varepsilon_{1r}}{\left(\varepsilon_{\infty,r} - \frac{\lambda^2}{\lambda_p^2}\right)}\right) \approx k_0 \sqrt{\varepsilon_{1r}}\left(1 + \frac{\varepsilon_{1r}}{2}\frac{\lambda_p^2}{\lambda^2}\right) \qquad (20.46)$$

As ω increases and $\left|\varepsilon_{\infty,r} - \frac{\lambda^2}{\lambda_p^2}\right| \approx \varepsilon_{1r}$, the wave vector of the SPR waves k_x can be expressed as

$$k_x = k_0 \mathrm{Re}\left[\sqrt{\frac{\varepsilon_{1r}\left(\varepsilon_{\infty,r} - \frac{\lambda^2}{\lambda_p^2}\right)}{\left(\varepsilon_{1r} + \varepsilon_{\infty,r} - \frac{\lambda^2}{\lambda_p^2}\right)}} \right]$$

$$+ jk_0 \text{Im} \left[\sqrt{\frac{\varepsilon_{1r} \left(\varepsilon_{\infty,r} - \frac{\lambda^2}{\lambda_p^2} \right)}{\left(\varepsilon_{1r} + \varepsilon_{\infty,r} - \frac{\lambda^2}{\lambda_p^2} \right)}} \right] \qquad (20.47)$$

$$k_x = k_0 \sqrt{\varepsilon_{1r}} \sqrt{\frac{1}{\left(1 + \frac{\varepsilon_{\infty,r} - \frac{\lambda^2}{\lambda_p^2}}{\varepsilon_{1r}} \right)}} \qquad (20.48)$$

The scaling factor is $\sqrt{1 + \frac{\varepsilon_{\infty,r} - \frac{\lambda^2}{\lambda_p^2}}{\varepsilon_{1r}}}$, which is very small. As ω increases and $\left| \varepsilon_{\infty,r} - \frac{\lambda^2}{\lambda_p^2} \right| \ll \varepsilon_{1r}$, the wave vector of the SPR waves k_x can be expressed as

$$k_x = k_0 \text{Re} \left[\sqrt{\frac{\varepsilon_{1r} \left(\varepsilon_{\infty,r} - \frac{\lambda^2}{\lambda_p^2} \right)}{\left(\varepsilon_{1r} + \varepsilon_{\infty,r} - \frac{\lambda^2}{\lambda_p^2} \right)}} \right]$$

$$+ jk_0 \text{Im} \left[\sqrt{\frac{\varepsilon_{1r} \left(\varepsilon_{\infty,r} - \frac{\lambda^2}{\lambda_p^2} \right)}{\left(\varepsilon_{1r} + \varepsilon_{\infty,r} - \frac{\lambda^2}{\lambda_p^2} \right)}} \right] \qquad (20.49)$$

$$= jk_0 \left[\sqrt{\frac{\lambda^2}{\lambda_p^2} - \varepsilon_{\infty,r}} \right] \qquad (20.50)$$

The SPR wave becomes a pure lossy wave.

20.3.3 Difference between RF Antennas and Optical Antennas

As discussed in the previous section, in the IR spectral regime, since $\omega \ll \omega_p$ and $k_x \cong k_0 \sqrt{\varepsilon_{1r}}$, the SPR waves behave similar to those in the RF antennas with a sample scaling factor $1/\sqrt{\varepsilon_{1r}}$. Despite the similarity, however, there is still a major difference between the optical and RF antenna. In RF antennas, though the RF signal feeding (transmitting antennas) and collection (receiving antennas) are the feeding centers, whereas for optical antennas, the signal collection is mainly through the near fields of the EM waves. Therefore, for optical antennas, the main focus is to maximize the near fields to achieve the highest performance enhancement. In the following section, a couple of optical antenna enhanced quantum dot IR photodetectors (QDIPs) will be presented together with their near-field analysis and polarization and receiving pattern measurements.

20.4 Optical Antenna Enhanced QDIPs

20.4.1 Transformation of the Far Field to the Near Field in a Pointed Dipole Optical Antenna

Similar to antennas in the RF and microwave spectral regimes, optical antennas have been demonstrated to be effective components in mediating free-space propagation EM waves and localized fields.[1-11,18] Such mediation

between the free-space propagation and localized fields is also referred to as the transformation between far fields and near fields. In this chapter, we analyze the transformation between the far field and near field of a pointed dipole optical antenna. The surface current distribution profiles were simulated at different wavelengths and polarizations. The induced surface current shows plasmonic resonance with the optical antenna structure. The surface current and the E-field are correlated by the current continuous equation. Based on the EM surface equivalence theorem (SET) introduced by Schekunoff,[30] the far field can be uniquely defined by the electric surface current density $\overrightarrow{J}_s$ and magnetic surface current density $\overrightarrow{M}_s$ and vice versa. The $\overrightarrow{J}_s$ and $\overrightarrow{M}_s$ are defined by the EM field boundary conditions:

$$\overrightarrow{M}_s = \hat{n} \times \left(\overrightarrow{E}_{\text{in}} - \overrightarrow{E}_{\text{out}} \right), \qquad (20.51)$$

$$\overrightarrow{J}_s = -\hat{n} \times \left(\overrightarrow{H}_{\text{in}} - \overrightarrow{H}_{\text{out}} \right), \qquad (20.52)$$

where the unit vector $\hat{n}$ indicates the normal to the surface direction and points to the outside surface, and $\overrightarrow{E}_{\text{in}}$, $\overrightarrow{E}_{\text{out}}$, $\overrightarrow{H}_{\text{in}}$, $\overrightarrow{H}_{\text{out}}$ are the E-field and magnetic field inside and outside the surface, respectively. The magnetic current density $\overrightarrow{M}_s$ is not physically realizable.[31] It is mainly introduced for mathematical convenience in SET[30] to count for the tangential E-field discontinuity when setting a zero E-field inside the surface. In this paper, we investigate the E-fields in all the regions and the E-field continuity holds. Therefore, the $\overrightarrow{M}_s = 0$ throughout the analysis in this chapter.

Figure 20.4a shows the three-dimensional (3D) light incident scheme on the pointed dipole optical antenna. The gap between the two antenna pairs and the width of the antennas are labeled as g and w, respectively. The lengths of the bar and the pointed parts are labeled as L_d and L_b, respectively. The light is the TM polarized with the magnetic field $\overrightarrow{H}$ aligned in the y direction. The light incident angle is θ. Figure 20.4b shows the cross section of the light incident scheme.

The thickness of the antenna is labeled as t. The TM wave is also known as the p-wave with the $\overrightarrow{E}$ parallel to the incident plane as shown in Figure 20.4b. Under proper excitation conditions,[29,32,33] SPR waves[29,34] can be generated by the light incident on metallic optical antennas.[19,22] For the TM polarized light, (i.e. the p-wave) shown in Figure 20.4, the generated SPR travels along the x direction, which leads to the $\overrightarrow{J}_s$ defined by Eq. (20.52). The resonant condition can be written as

$$k_{\text{sp}} L_d = \pi, \qquad (20.53)$$

where k_{sp} is the wave vector of the surface plasmonic wave. k_{sp} can be written as[29]

$$k_{\text{sp}} = \frac{2\pi}{\lambda_0} \text{Re} \left\{ \left[\frac{\varepsilon_m \varepsilon_d}{\varepsilon_m + \varepsilon_d} \right]^{1/2} \right\}, \qquad (20.54)$$

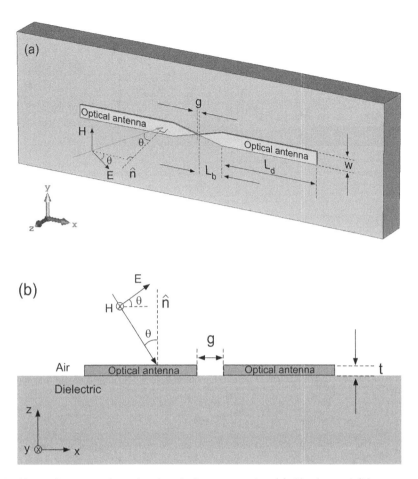

FIGURE 20.4 Light incident schemes on the pointed optical antenna pairs: (a) 3D view and (b) cross section view. The light is a TM wave (p wave) with an incidence angle of θ.

where λ_0 is the free-space wavelength, ε_m and ε_d are the relative permittivity of the metal and GaAs, respectively. The dielectric constant of GaAs in the long-wave IR region is 10.98.[35] The dielectric constant of Au is calculated to be $-3412 + i1735$ at 10.6 μm.[26] For the optical antenna length $L_d = 1.6$ μm, the calculated plasmonic resonant wavelength is $\lambda_{\mathrm{sp}} = 10.6$ μm.

For the TE polarized light (i.e. the s-wave) where the E-field is along the y direction, the resonant condition is

$$k_{\mathrm{sp}}W = \pi, \qquad (20.55)$$

Equation (20.55) gives a plasmonic resonant wavelength of $\lambda_{\mathrm{sp}} = 0.92$ μm, which is out of the detecting spectrum of QDIP. For this reason, we only consider the p-waves in this paper.

Figure 20.5a–d show the simulated surface current distributions on the optical antenna for the p-waves under the surface-normal incidence (i.e. $\theta = 0$) at the wavelengths of 8.0, 10.0, 10.6, and 12.0 μm, respectively. The simulation was performed using the CST's Microwave Studio® with the open boundary conditions. The E-field magnitude was set to be 1 V/m in each simulation. The surface current is the highest at the plasmonic resonant wavelength of $\lambda_{\mathrm{sp}} = 10.6$ μm, which agrees with Eq. (20.53).

The surface currents follow the sinusoidal-type current distribution in the dipole optical antenna.[31] The sinusoidal current distribution is associated with the full wavelength dipole antenna receiving pattern.[18,31] The surface current can be modeled using the transmission line (TL) theory,[31,36] with open circuit loads at both ends of the antenna pair and a capacitor load at the gap. The near fields at the gap can be obtained from the charge accumulation at the tip ends. The tips with opposite charges form a dipole radiator. Its near field can be expressed as[37]

$$\vec{E} = \frac{1}{j\omega\varepsilon\varepsilon_0 R^3}\left[3\hat{R}\left(\hat{R}\cdot\vec{p}\right) - \vec{p}\right], \qquad (20.56)$$

where $\vec{p} = g\int \vec{J}\cdot dy$ is the dipole moment with a small gap.

Figure 20.6a–c illustrate the top views of the simulated E_x components of the near fields for the gaps of 20, 50 and 100 nm, respectively.

Figure 20.7a–c illustrate the cross-section views of the simulated E_z components of the near fields for the various gaps.

The E_z components are quite similar at the different gaps, whereas the E_x components can be modified by varying the gaps. The integrated currents (i.e. $I = \int Jdy$) are $I = 7.7 \times 10^{-9}$ A for the optical antennas with different gaps.

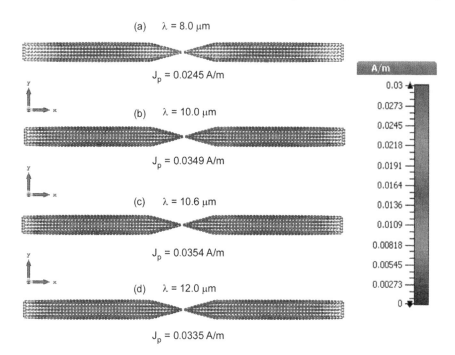

FIGURE 20.5 Simulated surface current distributions for p-waves under surface normal incidence at different incident wavelengths: (a) 8.0 μm, (b) 10.0 μm, (c) 10.6 μm, and (d) 12.0 μm. The surface current is the strongest at the plasmonic resonant wavelength of 10.6 μm.

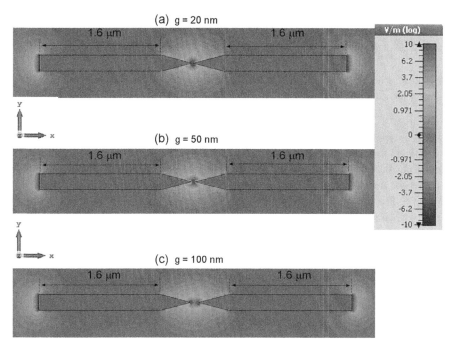

FIGURE 20.6 Top views of the simulated E_x components of optical antenna pairs with different gaps: (a) 20 nm, (b) 50 nm, and (c) 100 nm.

The integrated E-field magnitudes over the whole surface $\iint |E|\,ds$ are also the same for these optical antennas, $\iint |E|\,ds = 8.8 \times 10^{-12}$ Vm. The current continuous relation can be written as

$$I = \int J\,dy = j\omega\varepsilon_{r,\text{eff}}\varepsilon_0 \iint |E|\,ds, \qquad (20.57)$$

where ε_0 and $\varepsilon_{r,\text{eff}}$ are the permittivity of vacuum and the effective relative permittivity, respectively. The current continuous relation Eq. (20.57) holds with an effective $\varepsilon_{r,\text{eff}} = 1.1$ at the air/dielectric interface. This indicates that the optical antenna collects the far-field light incidence and converts it to the near field though the induced surface current in the optical antenna. In other words, the

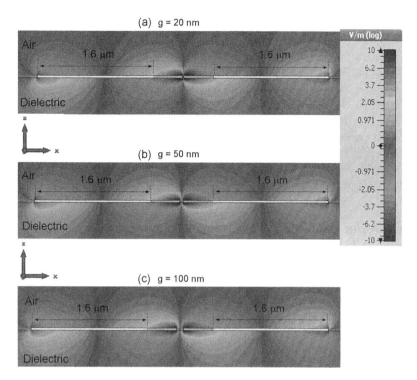

FIGURE 20.7 Cross-section views of the simulated E_z components of optical antenna pairs with different gaps: (a) 20 nm, (b) 50 nm, and (c) 100 nm.

transformation between the far field and the near field is achieved through the induced surface current in the pointed optical antenna. Note that the charges are mainly accumulated at the tips of the antenna pair. There is still a small amount of charge on the edges of the bow-tie parts of the antennas, which leads to the deviation from the ideal dipole–radiator type of E-fields. Figure 20.8a–c shows the top views of the simulated E_x components of antenna pairs with various pointed angles of $\varphi = 15°$, $\varphi = 45°$, and $\varphi = 60°$, respectively. The dipole parts of the antennas were kept the same in the simulation.

From Figure 20.8, the E-field profile of the sharp pointed angle closely resembles the near field of a dipole radiator, whereas that of a large angle antenna shifts to a parallel

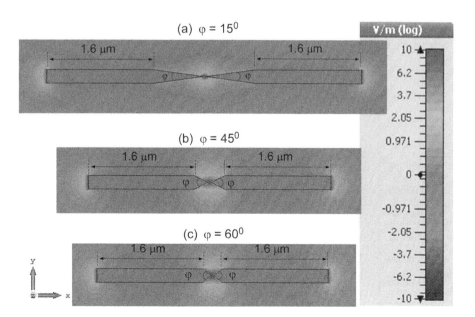

FIGURE 20.8 Top view of the simulated E_x components of optical antenna pairs with various pointed angles: (a) $\varphi = 15°$, (b) $\varphi = 45°$ nm, and (c) $\varphi = 60°$. The scale bar is the same as that in Figure 20.7. The near-field profiles vary with the pointed angles, but the integrated E-field magnitudes $\iint \vec{E} \cdot \vec{ds}$ are similar.

plate capacitor type of E-field. The integrated E-field magnitudes $\iint |E|\, ds$ are 8.8×10^{-12} V m, 8.8×10^{-12} V m, and 8.6×10^{-12} V m for $\varphi = 15°$, $\varphi = 45°$, and $\varphi = 60°$, respectively. Even though the near-field profiles vary with the pointed angles, the integrated E-fields are quite similar, indicating that the dipole part is the light collection part, whereas the bow-tie part changes the near-field profiles. The surface current and the E-field are correlated and follow the current continuous relation, which indicates that the transformation of the far-field to the near-field is through the surface current generated in the optical antenna.

20.4.2 Polarization-Dependent Enhancement in QDIP

The excitation of surface plasmonic waves in optical antennas can not only change the near-field distribution but also greatly enhance the near fields at the plasmonic resonant wavelengths. The strongly enhanced near fields can increase the interaction with a QDIP[38] close to the optical antenna, leading to the performance enhancement in the QDIP.

As discussed earlier, the pointed dipole optical antenna can only collect longwave infrared (LWIR) light incidence with the TM polarization (the p-wave). A general linear polarization can be decomposed into the TM polarization and TE polarization. For a linear polarization with an α angle with the TM-polarization direction (referred to the polarization angle henceforth), since only the TM polarization can generate surface current for LWIR light incidence, the surface current is expected to show a $\cos\alpha$ dependence on the polarization angle α. Figure 20.9 shows the simulated surface currents vs. $\cos\alpha$ plot for different polarization angles. A linear relationship is obtained, which verifies the $\cos\alpha$ dependence. Such polarization dependence further verifies that the effectiveness of the optical antenna in transforming far fields to the near fields through currents in the optical antenna.

The polarization dependence of the optical antenna enhanced QDIP was measured using a Bruker Tensor 27 Fourier Transform Infrared (FTIR) Spectrometer. An IR wire-grid polarizer with a maximum transmission of 50% in the LWIR spectral regime was used in front of the FTIR source to provide the polarized LWIR light source. The polarization angle (α) was tuned by rotating the IR polarizer. The 0° polarization angle corresponds to the transmagnetic (TM) polarized light, whereas the 90° polarization refers to the transelectric (TE) polarized light.

Figure 20.10a–f shows the measured photocurrent spectrum of the optical antenna enhanced QDIP compared with a reference QDIP without the optical antenna structures at different polarization angles $\alpha = 0°$, 30°, 45°, 60°, 75°, and 90°. The optical antenna enhanced QDIP shows a strong dependence on the incident field polarization, while the reference QDIP shows very little or no dependency on the polarization angle. The photocurrent enhancement ratio reaches a maxima for the TM polarized light (i.e. $\alpha = 0°$). When the polarization angle increases, the photocurrent enhancement ratio decreases to its minima at the TE polarized light ($\alpha = 90°$).

The enhancement ratio is defined as the ratio of the photocurrent of the optical antenna enhanced QDIP over that of the reference QDIP. Figure 20.11 shows the photocurrent enhancement ratio at different polarization angles (α). The black diamonds are the measurement data and the dashed lines are the $|\cos(\alpha)|$ curve fitting. The measurement data follows the $\cos(\alpha)$ function. This well agrees with the polarization dependence simulation presented in Figure 20.11 and the surface current distribution analysis in the pointed dipole optical antenna.

20.5 Chapter Summary

This chapter reviews the basic concepts of conventional RF antennas and compares them with optical antennas. When the frequency ω is much lower than the plasma frequency

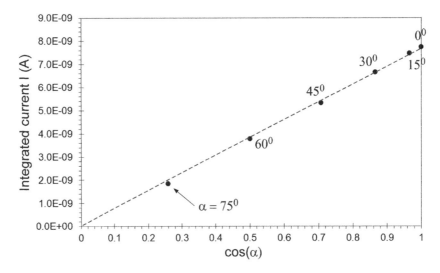

FIGURE 20.9 Simulated surface currents vs. $\cos\alpha$ plot. A linear dependence of the integrated surface current I on $\cos\alpha$ is verified.

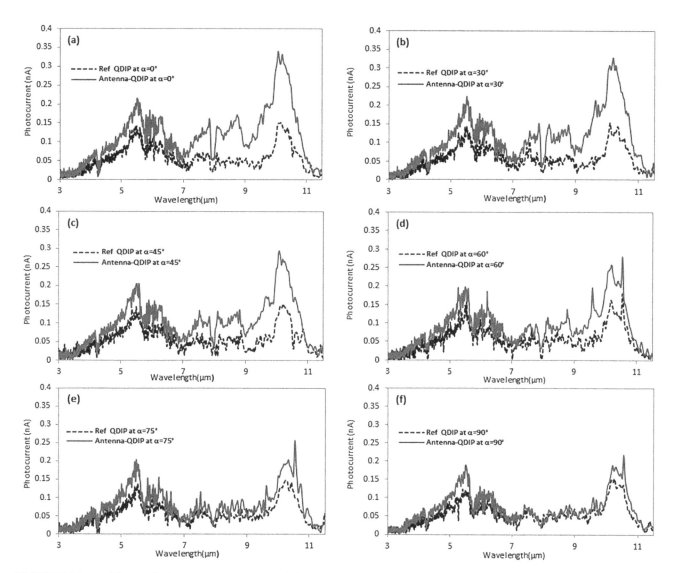

FIGURE 20.10 Measured photocurrent spectrum of the antenna-coupled QDIP compared with the reference QDIP for different polarization angles: (a) $\alpha = 0°$, (b) $\alpha = 30°$, (c) $\alpha = 45°$, (d) $\alpha = 60°$, (e) $\alpha = 75°$, and (f) $\alpha = 90°$.

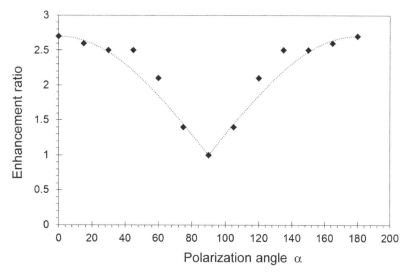

FIGURE 20.11 Photocurrent enhancement ratios of the antenna-coupled QDIP compared at different polarization angles. The diamonds indicate the measurement data and the dashed line is the $\cos(\alpha)$ function fitting.

ω_p, optical antennas have similar properties as the conventional RF antennas with a simple scaling factor. The scaling factors are discussed at different frequency regimes. Despite the similarity, however, optical antennas have a different signal collection scheme than conventional RF antennas. In RF antenna, electrical current is collected and delivered to the load via TLs, whereas for optical antennas, the antenna and photodetector interaction is through the near fields. The near fields are analyzed for a pointed dipole optical antenna and compared with the photodetector enhancement. Similar to RF antennas, optical antennas are expected to play important roles in a broad range of applications, including near-field microscopy and spectroscopy,[4,12] biological sensing,[13−15] solar cell energy harvesting,[2,13,16,17] and IR detection.[6,18−20]

References

1. L. Novotny and N. van Hulst, "Antennas for light," *Nature Photonics* **5** (2), 83–90 (2011).
2. P. Bharadwaj, B. Deutsch and L. Novotny, "Optical Antennas," *Advances in Optics and Photonics* **1** (3), 438–483 (2009).
3. A. E. Krasnok, I. S. Maksymov, A. I. Denisyuk, P. A. Belov, A. E. Miroshnichenko, C. R. Simovski and Y. S. Kivshar, "Optical nanoantennas," *Physics-Uspekhi* **56** (6), 539 (2013).
4. T. Kalkbrenner, U. Håkanson, A. Schädle, S. Burger, C. Henkel and V. Sandoghdar, "Optical Microscopy via Spectral Modifications of a Nanoantenna," *Physical Review Letters* **95** (20), 200801 (2005).
5. V. Giannini, A. I. Fernández-Domínguez, S. C. Heck and S. A. Maier, "Plasmonic Nanoantennas: Fundamentals and Their Use in Controlling the Radiative Properties of Nanoemitters," *Chemical Reviews* **111** (6), 3888–3912 (2011).
6. G. Gu, J. Vaillancourt and X. Lu, "Analysis of near-field components of a plasmonic optical antenna and their contribution to quantum dot infrared photodetector enhancement," *Optics Express* **22** (21), 24970–24976 (2014).
7. N. Yu, P. Genevet, M. A. Kats, F. Aieta, J.-P. Tetienne, F. Capasso and Z. Gaburro, "Light Propagation with Phase Discontinuities: Generalized Laws of Reflection and Refraction," *Science* **334** (6054), 333–337 (2011).
8. G. Rui, D. C. Abeysinghe, R. L. Nelson and Q. Zhan, "Demonstration of beam steering via dipole-coupled plasmonic spiral antenna," *Scientific Reports* **3** (2237), (2013).
9. T. Kosako, Y. Kadoya and H. F. Hofmann, "Directional control of light by a nano-optical Yagi-Uda antenna," *Nature Photonics* **4** (5), 312–315 (2010).
10. T. H. Taminiau, F. D. Stefani and N. F. van Hulst, "Enhanced directional excitation and emission

11. M. Ramezani, A. Casadei, G. Grzela, F. Matteini, G. Tütüncüoglu, D. Rüffer, A. Fontcuberta i Morral and J. Gómez Rivas, "Hybrid Semiconductor Nanowire–Metallic Yagi-Uda Antennas," *Nano Letters* **15** (8), 4889–4895 (2015).
12. L. Novotny and S. J. Stranick, "Near-field optical microscopy and spectroscopy with pointed probes*," *Annual Review of Physical Chemistry* **57**, 303–331 (2006).
13. J. A. Schuller, E. S. Barnard, W. Cai, Y. C. Jun, J. S. White and M. L. Brongersma, "Plasmonics for extreme light concentration and manipulation," *Nature Materials* **9** (3), 193–204 (2010).
14. T. S. van Zanten, M. J. Lopez-Bosque and M. F. Garcia-Parajo, "Imaging Individual Proteins and Nanodomains on Intact Cell Membranes with a Probe-Based Optical Antenna," *Small* **6** (2), 270–275 (2010).
15. C. Höppener and L. Novotny, "Antenna-based optical imaging of single Ca2+ transmembrane proteins in liquids," *Nano Letters* **8** (2), 642–646 (2008).
16. H. A. Atwater and A. Polman, "Plasmonics for improved photovoltaic devices," *Nature Materials* **9** (3), 205–213 (2010).
17. E. Yu, D. Derkacs, P. Matheu and D. Schaadt, "Plasmonic nanoparticle scattering for enhanced performance of photovoltaic and photodetector devices," *Presented at the NanoScience + Engineering*, 2008: International Society for Optics and Photonics, pp. 70331V-70331V-9.
18. M. Neda, G. Guiru and L. Xuejun, "A plasmonic dipole optical antenna coupled quantum dot infrared photodetector," *Journal of Physics D: Applied Physics* **48** (47), 475102 (2015).
19. T. Kemsri, G. Gu, Y. Zhang, X. Lan, H. Zhang, J. Tice and X. Lu, "Angular-dependent photodetection enhancement by a metallic circular disk optical antenna," *AIP Advances* **7** (2), 025013 (2017).
20. S. Schwarz and B. Ulrich, "Antenna-coupled infrared detectors," *Journal of Applied Physics* **48** (5), 1870–1873 (1977).
21. L. Xuejun, M. Neda, L. Lin, G. Guiru, Z. Yingjie and K. Thitikorn, "A back side configured pointed dipole plasmonic optical antenna array enhanced quantum dot infrared photodetector," *Semiconductor Science and Technology* **32** (12), 125017 (2017).
22. L. Novotny, "Effective Wavelength Scaling for Optical Antennas," *Physical Review Letters* **98** (26), 266802 (2007).
23. C. A. Balanis, *Antenna Theory: Analysis and Design*, 4th edition. John Wiley & Sons, Inc., Hoboken, New Jersey (2016).
24. H. J. Visser, *Antenna Theory and Applications*. John Wiley & Sons Ltd, Chichester (2012).

25. R. S. Elliott, *Antenna Theory and Design.* Wiley, New York (2003).
26. R. H. Ritchie, "Plasma Losses by Fast Electrons in Thin Films," *Physical Review* **106** (5), 874–881 (1957).
27. P. Drude, "Zur Elektronentheorie der Metalle," *Annalen der Physik* **306** (3), 566–613 (1900).
28. M. A. Ordal, L. L. Long, R. J. Bell, S. E. Bell, R. R. Bell, R. W. Alexander and C. A. Ward, "Optical properties of the metals Al, Co, Cu, Au, Fe, Pb, Ni, Pd, Pt, Ag, Ti, and W in the infrared and far infrared," *Applied Optics* **22** (7), 1099–1119 (1983).
29. H. Raether, *Surface Plasmons on Smooth Surfaces.* Springer-Verlag, Berlin Heidelberg (1988).
30. S. A. Schelkunoff, "Some Equivalence Theorems of Electromagnetics and Their Application to Radiation Problems," *Bell System Technical Journal* **15** (1), 92–112 (1936).
31. C. A. Balanis, *Antenna Theory Analysis and Design,* 3rd edition. John Wiley & Son, Hoboken, NJ (2005).
32. H. Ghaemi, T. Thio, D. E. A. Grupp, T. W. Ebbesen and H. Lezec, "Surface plasmons enhance optical transmission through subwavelength holes," *Physical Review B* **58** (11), 6779 (1998).
33. K. A. Willets and R. P. Van Duyne, "Localized surface plasmon resonance spectroscopy and sensing," *Annual Review of Physical Chemistry* **58**, 267–297 (2007).
34. W. L. Barnes, A. Dereux and T. W. Ebbesen, "Surface plasmon subwavelength optics," *Nature* **424** (6950), 824–830 (2003).
35. E. Palik, *Handbook of Optical Constants of Solids.* Academic Press, Cambridge, Massachusetts (1998).
36. Q.-Y. Wen, Y.-S. Xie, H.-W. Zhang, Q.-H. Yang, Y.-X. Li and Y.-L. Liu, "Transmission line model and fields analysis of metamaterial absorber in the terahertz band," *Optics Express* **17** (22), 20256–20265 (2009).
37. J. D. Jackson, *Classical Electrodynamics,* 3rd edition. John Wiley & Sons, New York (1998).
38. Y. Zhang, J. Vaillancourt, G. Gu, W. Guo and X. Lu, "Quantum selection rule dependent plasmonic enhancement in quantum dot infrared photodetectors," *Journal of Applied Physics* **119** (19), 193103 (2016).

Index

A

ABS, *see* Andreev bound states (ABS)
Absorption processes
    excitonic absorption, 3-9–3-11
    fundamental absorption, 3-8–3-9, 3-10
Absorption rate
    1PF, 2-3
    2PF, 2-9
Actin filaments molecular order, 2-7
Adsorption process, 19-12, 19-13
    in heterogeneous biosensing, 19-9
    in kinetic analysis, 19-14, 19-16
    in various sources of nonuniformity, 19-15
AEAPTMS, *see* N-(2-Aminoethyl)-3-aminopropyl-
        trimethoxysilane (AEAPTMS)
AFM, *see* Atomic force microscopy (AFM)
Aggregation kinetics, 14-7
ALD, *see* Atomic layer deposition (ALD)
Al-InAs-Al junction, 13-32–13-33
Amplification of surface plasmons, 16-1–16-2
    compensating losses, 16-4–16-5
        dipolar gain media incorporation, 16-5, 16-6
        electrical injection, 16-7, 16-8
        fluorescent polymer, 16-6, 16-7
        lead-sulphide quantum dots, 16-6, 16-7
        quantum wells, 16-5
        semiconductor structure, 16-5, 16-6
        SPP propagation, 16-8, 16-9
        thin organic gain medium, 16-8, 16-9
        2D Bragg structure, 16-8
    in DC electric field
        Cada's group analysis, 16-9–16-10
        Sydoruk's group analysis, 16-10–16-11
    mathematical formulation, 16-2–16-3
    observation, 16-8
    plasmonic-based lasing, 16-11
        spaser biological application, 16-11–16-12
        spasing process, 16-11
    plasmonic materials, losses in
        Ohmic loss, 16-4
        scattering, 16-4
    surface plasmon polariton (*see* Surface plasmon
        polariton (SPP))
Analytical models, SET, 12-7–12-8
Analytical science, 19-2
    biosensing concept, 19-2
    kinetic analysis in, 19-17–19-18
Analytical workflow, SPR, 19-9
    competition assay formats, 19-11–19-12
    contrast enhancement, 19-13
    determination of analyte, 19-10–19-11
    DViFA approach, 19-12
    protein–protein interaction, 19-12

    quantitative determination of analyte,
        19-11–19-12
    validation procedure, 19-12–19-13
Andreev bound states (ABS), 13-16–13-18, 13-34–13-35
Andreev reflections, in ballistic S-N-S JJs,
    13-16–13-18, 13-36
Angle resolved monochromatic image, 18-15, 18-16
Angle-resolved spectral (ARS) pattern, 18-13
    for groove and ridge type cavity, 18-29–18-31
    in horizontal scan, 18-14, 18-15
    p-polarization, 18-20
    for ridge type cavity, 18-29
    in vertical scan, 18-13, 18-14
Angular distribution of multipole mode LPP,
    18-10–18-12
Angular radiation intensity distribution
    diffractometry patterns, SiO_2, 5-11
    FWHM variations, 5-7
    measurement of X-ray characteristic radiation,
        5-6–5-7
    slit width, 5-7–5-8
    spatial intensity distribution of $MoK_{\alpha\beta}$ radiation
        beam, 5-12
    waveguide-resonators, 5-8
Anisotropic plasmonic nanoparticles, 14-15–14-17
ANNs, *see* Artificial neural networks (ANNs)
Antenna/lightning rod effect, 7-6
Antenna transmitting and receiving patterns, 20-3–20-4
Aqueous dispersions, 14-3
ARS pattern, *see* Angle-resolved spectral (ARS) pattern
Artificial neural networks (ANNs), 4-11
Asymmetric double-junction structure, 15-4
Atomic force microscopy (AFM), 7-13, 8-5
Atomic layer deposition (ALD), 13-30
Atomic switches, 11-3
Attenuated total internal reflection (ATIR), 5-3
Au–Ag
    head-body nanosnowmen, 7-17
    nanospheres, 7-20
    plasmonic nanomushrooms, 7-18
Au-nanobridged nanogap particles (Au-NNPs),
    7-19, 7-21
Au nanostructures, 7-25
Au-PS Janus particles, 6-10

B

Back-gate sweep, 13-20, 13-21
Back-scattered detection method (BSD), 6-9–6-10
Band edge energy, 18-27, 18-32, 18-33
Bandgap widening, 10-3, 10-4
Band profile, quantum well, 10-3
Band-to-band transitions, 3-12
Band topology, 13-9

Barycenter method, 6-3
Beam scan spectral (BSS), 18-15, 18-16, 18-20,
  18-24, 18-25
Berry curvature, 13-6
Bessel beams, 6-12
Bessel function, 13-13, 18-8, 18-21, 18-22, 18-32
Beyond CMOS, 9-2, 9-24, 9-25
Binary model dispersions, 14-3
Bioinspired computing, 11-7
Biomolecules, 14-7
Biosensing, 19-1–19-2
  advantages of, 19-2, 19-18
  analytical science, 19-2
  analytical workflow
    competition assay formats, 19-11–19-12
    contrast enhancement, 19-13
    determination of analyte, 19-10–19-11
    DViFA approach, 19-12
    protein–protein interaction, 19-12
    surface-induced analyte transitions, 19-12–19-13
  data mining, 19-13–19-18
    interfacial landscape formalization, 19-14–19-18
    kinetic analysis, 19-14–19-18
    measurement uncertainty, 19-13
    optical quantification, of adsorbate
      concentration, 19-13–19-14
  excitation of SPP, 19-2–19-5
  interfacial design, 19-7
    optimal interfacial arrangement, 19-8–19-9
    physical transducer of SPR, 19-8
    transducer-based systems, 19-8
  with linear standard curve, 19-6–19-7
  physical transducer of, 19-8
  for sensing applications, 19-5–19-6
  spectrometer with angular scanning, 19-4, 19-5
  technology overview, 19-7
Biosensors, 19-2, 19-8
Birth–death process, single-electron device, 12-4
Bloch function, 3-1–3-2
Bloch wave, 18-26
Blueshift, 10-2
Bottom-emitting OLEDs, 1-11–1-12
Bound exciton emission, 3-12
Bound state energy, 13-17
Brillouin zone, band structure, 13-6
BSD, *see* Back-scattered detection method (BSD)
BSS, *see* Beam scan spectral (BSS)

C

Cada's group analysis, 16-9–16-10
Camera, 6-4
CAMSET, *see* Circuit Analysis Model of Single
  Electron Tunneling (CAMSET)
CARSAR approach, *see* Complex with Standard
  Additions of the Receptor (CARSAR)
  approach
Cathodoluminescence (CL), of nanoplasmonics

instrumentation of STEM-CL, 18-12–18-14
  light detection system, 18-13, 18-14
  localized surface plasmon, 18-1
    multipole mode, of spherical nanoparticle,
      18-7–18-12
    nanoparticle, photon map, 18-19–18-24
    thin triangular nanoprism, 18-17–18-18
  measurements, 18-13–18-16
    angle resolved monochromatic image, 18-15,
      18-16
    angle-resolved spectral pattern, 18-13,
      18-14, 18-15
    beam scan spectral image, 18-15, 18-16
  overview of, 18-1–18-2
  photon map imaging, 18-16
    flat surface, 18-16–18-17
    at three-dimensional structure, 18-17
  surface plasmon polariton, 18-1
    cavity, 18-28–18-30
    decay length of, 18-4–18-6
    dielectric function of metals, 18-4
    dispersion relation of, 18-3–18-4, 18-26, 18-33
    electromagnetic field at plane interface,
      18-1–18-3
    excitation by high-energy electrons, 18-6–18-7
    light conversion, 18-24–18-25, 18-28
    one-dimensional plasmonic crystal, 18-25–18-28
    two-dimensional plasmonic crystal, 18-30–18-33
Cavity, 18-28–18-30
  of groove and ridge pair, 18-29–18-30, 18-31
  of step pair, 18-28–18-29
CB, *see* Conduction band (CB)
CCD, *see* Charge coupled device (CCD)
CdTe band structure, 3-3
Cell membrane molecular order imaging, 2-6–2-7
Cetyltrimethylammonium bromide (CTAB), 14-7–14-9
Cetyltrimethylammonium chloride (CTAC), 14-5, 14-6
Charge carriers, 13-3, 13-10, 13-12
Charge coupled device (CCD), 6-4
Charge oscillation, 18-1, 18-8, 18-9, 18-11, 18-18
Charge pumping, 13-15–13-16, 13-26–13-28
Charge relaxation time, 13-24
Charge stability diagram
  GDQD, 13-26
  SQD and DQD, 13-30–13-31
Charge transport mechanism, 12-2, 12-6
Chemical optimization, 19-8
Chemical sensors, 19-2
Child–Langmuir's threehalves- power law, 1-6
Chiral configuration, 14-3
Chiralityin graphene, 13-3
CiC, *see* Cube-in-cube (CiC) nanoparticles
Circuit Analysis Model of Single Electron Tunneling
  (CAMSET), 12-7
Circuit elements, 11-1
CL, *see* Cathodoluminescence (CL)
Classical electromagnetic model, 15-3, 15-4

Classical SPR approach, 19-7

CMGT, *see* Confinement-modulated gap transistor (CMGT)

CMOS, *see* Complementary metal oxide semiconductor (CMOS)

Colloidal self-assembly, 14-1–14-3

Colloidal synthesis, 14-14, 14-15

Compensating losses, 16-4–16-5
  dipolar gain media incorporation, 16-5, 16-6
  electrical injection, 16-7, 16-8
  fluorescent polymer, 16-6, 16-7
  lead-sulphide quantum dots, 16-6, 16-7
  quantum wells, 16-5
  semiconductor structure, 16-5, 16-6
  thin organic gain medium, 16-8, 16-9
  2D Bragg structure, 16-8

Competition assay formats, 19-11–19-12

Complementary metal oxide semiconductor (CMOS)
  beyond, 9-2, 9-25
  dimensional scaling rules, 9-6
  memory scaling, 9-13–9-19
  mixing nanowires, 9-24
  scaling of, 9-1–9-3
  technology, 9-3
  transistors, 9-19, 9-20

Complex permittivity, 20-4

Complex with Standard Additions of the Receptor (CARSAR) approach, 19-12

Conductance of O_2, 13-18–13-20

Conduction band (CB), 3-2, 10-3

Conductive tip atomic force microscopy (CAFM), 4-7–4-8

Conductivity, 20-4

Confinement-modulated gap transistor (CMGT), 10-9–10-11

Constant interaction model, 13-10, 13-13

Continuum memristance, 11-2, 11-4

Contrast enhancement, 19-13

Conventional electronic components, 12-8–12-9

Conventional nanolithography, 1-14

Conventional optical tweezers, 6-5

Conventional plasmonic materials, 16-1

Conventional RF antennas, 20-1
  electromagnetism reciprocity and antenna receiving patterns, 20-3–20-4
  half-wave dipole antenna, 20-1–20-3

Convergence level, 15-6

Cooper pairs, 13-16, 13-17, 13-33, 13-38

Copper (Cu) interconnects, 9-23

Core–petal nanoparticles (CPNs), 7-19

Coulomb blockade, 12-9, 13-10, 13-11, 13-19
  and magnetotransport, 13-25–13-26
  at zero field, 13-20–13-22

Coulomb blockade plasmonic switch, 15-1–15-2, 15-4
  nanodevice preparation, 15-4–15-5
  numerical calculations of, 15-6–15-7
  quantum-corrected model
    overview of, 15-3–15-4
    Wentzel–Kramers–Brillouin approximation, 15-2–15-3
    single-electron transistor, 15-1, 15-2

Coulomb diamond, 13-11, 13-12, 13-19, 13-21, 13-22, 13-28–13-30

Coulombic repulsion, 1-2, 1-5, 1-7, 1-12

Coulomb oscillations, 12-2–12-3

Coulomb peaks, 13-22, 13-23, 13-28

CPNs, *see* Core–petal nanoparticles (CPNs)

Cross-capacitance coupling, 13-13–13-15

Cross-coupling effect, 13-14

CST's Microwave Studio®, 20-7

CTAB, *see* Cetyltrimethylammonium bromide (CTAB)

CTAC, *see* Cetyltrimethylammonium chloride (CTAC)

Cube-in-cube (CiC) nanoparticles, 7-29–7-31

Curve fitting parameters, 12-9–12-10

D

Dark-field (DF) imaging, 6-9

Data mining
  interfacial landscape formalization, 19-14–19-18
  kinetic analysis, 19-14–19-18
  measurement uncertainty, 19-13
  optical quantification, of adsorbate concentration, 19-13–19-14

DBTBT, *see* Direct band-to-band tunneling (DBTBT)

DC electric field, SP in
  Cada's group analysis, 16-9–16-10
  Sydoruk's group analysis, 16-10–16-11

DDA, *see* Discrete dipole approximation method (DDA)

DE, *see* Droplet epitaxy (DE)

Dealloyed intra-nanogap particles (DIPs), 7-21

Decay length of SPP, 18-4–18-6

Degenerate bands
  $k \cdot p$ method, 3-3–3-5
  tight-binding method, 3-5

Dennard's scaling, 9-6

Density functional theory (DFT), 3-5, 10-7

Density Variations in the Fixed Architectures Approach (DViFA) approach, 19-12

Depletion interaction, 14-3

Depletion processes, 19-14

Derjarguin, Landau, Verwey, and Overbeek (DLVO) model, 14-2

Detection intensity, 18-23

Detection polarization state, 2-8

Device capacitance, 1-11

Device fabrication
  QD-OLED with edge injection, 1-12–1-13

DF imaging, *see* Dark-field (DF) imaging

DFT, *see* Density functional theory (DFT)

DIBL, *see* Drain-induced barrier lowering (DIBL)

Dielectric constant, 18-4, 18-8, 18-9, 19-3, 19-4

Dielectric function, 7-4–7-5, 18-4, 18-5

Dielectric-metal particle, 6-10–6-11

Dielectric particles, 6-8

Dielectrophoresis, 14-5

Diffusion current, 1-10, 1-11

Dipolar gain media incorporation, 16-5, 16-6

Dipole antenna, half-wave, 20-1–20-3

Dipole mode, photon map, 18-20–18-22

DIPs, *see* Dealloyed intra-nanogap particles (DIPs)

Direct band-to-band tunneling (DBTBT), 9-13

Directed self-assembly, molecular interactions, 14-7, 14-8, 14-9

Discrete dipole approximation method (DDA), 6-7

Dispersion relation of SPP, 16-3, 18-3–18-4, 18-26, 18-33

Displacement measurement, 6-3

DLVO model, *see* Derjarguin, Landau, Verwey, and Overbeek (DLVO) model

DNA

    gold nanogap particles, 7-21

    hybridization, 14-7

Dolmen-like plasmonic nanostructure, 7-29

Domain wall memory (DWM), 9-17

Double quantum dot (DQD), 13-13–13-16

    electrostatic network model for, 13-13

    MoS_2 nanoribbon, 13-30–13-32

    SiO_2/Si substrates, 13-25

    stability diagram with coupling regimes, 13-14–13-15

Drain-induced barrier lowering (DIBL), 9-6, 9-8, 9-9

DRAMs, *see* Dynamic random access memories (DRAMs)

Drift current, 1-10, 1-11

Droplet epitaxy (DE), 8-1

Drude model, 15-3, 15-6, 18-3–18-5, 20-4

Drude–Sommerfeld theory, 19-3

DViFA approach, *see* Density Variations in the Fixed Architectures Approach (DViFA) approach

DWM, *see* Domain wall memory (DWM)

Dynamic random access memories (DRAMs), 9-3, 9-13–9-14

E

ECM, *see* Electrochemical metallization (ECM)

Edge conduction, 13-38–13-40

Edge emission, QD-OLED, 1-13, 1-14

Edge injection, QD-OLED, 1-12–1-13

EDX, *see* Energy dispersive X-ray spectroscopy (EDX)

Effective medium approach, 19-13–19-14

Effective refractive index, 19-6

Electrical field- induced bandgap, 13-29

Electrical gating, 13-29

Electrical injection, 16-7, 16-8

Electric multipole field, 18-8, 18-9

Electrochemical metallization (ECM), 11-3

Electrochemical potential of single QD, 13-10, 13-11

Electromagnetic local density of state (EMLDOS), 18-16

Electromagnetic scattering theory, 6-7–6-8

Electromagnetic (EM) waves, 20-1

Electromagnetism reciprocity, 20-3–20-4

Electron-beam lithography, 14-12

Electron diffraction patterns, 12-2

Electron energy loss spectroscopy (EELS), 18-6, 18-8, 18-16

Electron–hole crossover, 13-13, 13-22

Electronic band structure, semiconductor nano structures

    Bloch function, 3-1–3-2

    degenerate bands

        $k \cdot p$ method, 3-3–3-5

        tight-binding method, 3-5

    low-dimensional nanostructures

        metallurgical junctions, 3-6–3-7

        metal semiconductor junctions, 3-6

        type I quantum well, 3-7

        type II quantum well, 3-7

    SL, 3-7–3-8

    tight-binding method, 3-3

    two-band model, 3-2

Electron microscopy imaging, 2-2

Electron multiplying charge coupled device (EM-CCD) spectrometer, 18-13

Electrostatic network model, 13-13

Emission intensity, 18-17, 18-18, 18-23, 18-25, 18-26

Emission processes, semiconductor nano structures

    band-to-band transitions, 3-12

    bound exciton emission, 3-12

    PL excitation, 3-12–3-13

Emission rate, 2-3

Emission spectrum, 18-8, 18-9, 18-19, 18-24, 18-27

EMLDOS, *see* Electromagnetic local density of state (EMLDOS)

EM wave equations, 7-2

Energy band diagram, 15-4

Energy dispersion, 13-3, 13-6, 18-31

Energy dispersive X-ray spectroscopy (EDX), 8-5

Energy gap

    graphene nanoribbon, 13-19, 13-20, 13-30

    graphene quantum dots, 13-29

    S-N-S junction, 13-34

Energy harvesting, 9-25–9-27

Energy loss probability, 18-6, 18-7

Enhanced spectroscopies

    direct photoluminescence, 7-9–7-10

    molecular plasmonics, 7-5–7-9

EQE, *see* External quantum efficiency (EQE)

Equivalent circuit components, 12-11

Excitation process, 16-7, 18-6–18-7

    grating coupling, 16-3

    optical beams focus, 16-4

    prism coupling/attenuated total internal reection, 16-3

Excitonic absorption, 3-9–3-11

External field-directed assembly, 14-5–14-7

External quantum efficiency (EQE), 1-9

F

Fabrication
  quantum-dot organic light-emitting diode, 1-16
Fabrication of Janus particle, 6-11
Fabry–Prot cavity, 13-34
Far-field coupled PCMs
  plasmonics, 4-8–4-9
    reconfigurable metasurfaces, 4-7, 4-8
    reflective displays, 4-7–4-8
FDSOI channels, 9-7, 9-8
FDTD, *see* Finite-difference time domain (FDTD)
FEM, *see* Finite element method (FEM)
Fermi energy, 13-16, 13-20, 13-21, 13-30, 13-35, 13-38
Fermi level pinning, 10-9
Fermi-level shift, 1-3, 1-9
Ferromagnetic two-dimensional (2D) lattices, 17-3–17-4
Fictitious conductive material, 15-3, 15-4
Field-effect transistor (FET), 10-1, 11-6
Field-emission scanning electron microscopy
    (FESEM), 8-5–8-6
Field-induced interactions, 14-5
Field-programmable gate arrays, 11-6
Figure of merit (FOM), 17-8, 17-10
Filament formation, 11-3–11-4
FinFET, 9-1, 9-7, 9-8, 9-10
Finite-difference time domain (FDTD), 6-7, 7-11
Finite element method (FEM), 7-15
Flash memories, 9-15
Floating gate flash memories, 9-14, 9-16–9-19
Flory–Krigbaum theory, 14-2
Fluorescence
  definition, 2-3
  intensity, 2-3
  labeling, 2-7
  microscopy, 2-3
    absorption and emission, 2-3
    1PF microscopy, 2-3–2-5
    polarized fluorescence, 2-4, 2-5–2-6
    retrieved molecular order, interpretation, 2-5
Fluorescence intensity, 7-7
Fluorescent polymer, 16-6, 16-7
FM growth mode, *see* Frank–van der Merwe (FM)
    growth mode
Fock–Darwin diagram, 13-12, 13-22, 13-31, 13-32
Forward-scattered detection method (FSD), 6-9–6-10
Frank–van der Merwe (FM) growth mode, 8-4
Free electron gas model, 7-4
Free electron plasma model, 19-3
Fresnel equations, 19-6–19-8, 19-13, 19-14
FSD, *see* Forward-scattered detection method (FSD)
Full-width at half-maximum (FWHM), 8-5
Fundamental absorption, 3-8–3-9
FWHM, *see* Full-width at half-maximum (FWHM)

G

GaAs, dielectric constant of, 20-7
Gallium nitride (GaN); *see also* Nanoflowers (NFs)

characteristics of, 8-4–8-6
  growth of, 8-2–8-6
  photodetection devices, 8-6–8-15
GaN, *see* Gallium nitride (GaN)
Gate-depleted area, 13-10
Gatemon qubit, 13-32–13-33
Gaussian laser beam, 6-7
Gauss's Law, 20-4
Gauss unit system, 18-6
GDQDs, *see* Graphene double quantum dots (GDQDs)
Geometrical optics method, 6-6–6-7
GNR, *see* Graphene nanoribbon (GNR)
Gold nanogap particles (Au-NNPs), *see* Au-nanobridged
    nanogap particles (Au-NNPs)
Gold nanoparticles, 6-9
Gold nanorod aggregation, 14-7, 14-9
Gold NPs with DNA strands, 14-7, 14-9
Gold–silver core–shell nanodumbbells (GSNDs),
    7-13–7-15
Gold triangle assembly, 14-5, 14-6
GOS, *see* Graphene-oxide-semiconductor (GOS)
GQDs, *see* Graphene quantum dots (GQDs)
Gradient force, 6-1–6-2
Graphene, 13-1
  based JJs, 13-33–13-35
  Dirac nature of, 13-3, 13-4, 13-22
  single-electron transport in, 13-18–13-20
    charge pumping, 13-26–13-28
    charge relaxation time, 13-24
    Coulomb blockade, 13-20–13-22, 13-25–13-26
    DQDs on SiO_2/Si substrates, 13-25
    electron–hole crossover, 13-22
    GQDs on hBN substrate, 13-28–13-29
    magnetotransport, 13-25–13-26
    spin states, in-plane magnetic field, 13-22–13-23
    SQDs on SiO_2/Si substrates, 13-20
  2D materials, 13-2–13-5
Graphene double quantum dots (GDQDs),
    13-25–13-26
  charge pumping, 13-26–13-28
  charge stability diagram of, 13-26
Graphene nanoribbon (GNR), 13-18, 13-27
  conductance of O_2, 13-18–13-20
  energy gap, 13-19, 13-20
Graphene-oxide-semiconductor (GOS), 1-3–1-4
  I–V characteristic of, 1-6, 1-7
  photocurrent response, 1-6, 1-7, 1-8
  spectral dependence, 1-9–1-11
  UV-enhanced photodetection in, 1-4
  2DEG emission, 1-5–1-6
    transport of photocarrier, 1-8–1-9
  with void channel, 1-6, 1-7, 1-8, 1-11
Graphene quantum dots (GQDs), 13-1, 13-2, 13-12
  on hBN substrate, 13-28–13-29
Graphene single quantum dots (GSQDs), *see* Graphene,
    single-electron transport in
Graphene work function, 1-3–1-4

Grating coupling, 16-3
Groove type cavity, 18-29–18-30, 18-31
Growth kinetics, 8-4
GSNDs, *see* Gold–silver core–shell
        nanodumbbells (GSNDs)

H

Hadley and Lientschnig method, 12-5
Half-wave dipole antenna, 20-1–20-3
Hall effect, 13-6, 13-7
Hamiltonian, 13-3, 13-5, 13-7, 13-12, 13-30–13-31
Hard disk drives (HDD), 9-13
HBN, *see* Hexagonal boron nitride (hBN)
Helmholtz equation, 7-2
HEMT, *see* High-electron-mobility transistor (HEMT)
Heterogeneous biosensing, 19-9
Hexagonal boron nitride (hBN), 13-2–13-5
    GQDs on, 13-28–13-29
High-contrast developer, 19-13
High electron mobility transistors (HEMT), 1-2, 3-7
High numerical aperture (NA) objective lens, 6-1
High resistance state (HRS), 11-2–11-3
High-resolution X-ray diffraction (HRXRD), 8-4–8-5
Hodgkin–Huxley neuronal model, 11-5
Holographic optical tweezers (HOTs), 6-5
Hopfield neural network, 11-6
HRS, *see* High resistance state (HRS)
HRXRD, *see* High-resolution X-ray
        diffraction (HRXRD)
Hybridization model, 7-4
Hydrodynamic drag method, 6-3
Hydrophobic interactions, 14-7, 14-8, 14-10
Hyphenated SPR techniques, 19-7
HZG-4 diffractometer, 5-4–5-5

I

Image-charge potential, 1-1, 1-7, 1-8
Image-potential states, 1-7
Immobilization matrices, 19-10
Indium-tin-oxide (ITO), 1-11, 1-12, 1-13
Inelastic scattering probability, 18-6
Infinite chain limit, 14-8
Infrared (IR)-emitting fluorophores, 7-1
Inhibitory-type competitive analysis, 19-11
Integrate-and-fire neuronal model, 11-5
Integrated photonics and PCMs, 4-1, 4-2
Interconnection capacitance, 12-9, 12-10
Interdot coupling energy, 13-15, 13-25
Interfacial design, 19-7
    optimal interfacial arrangement, 19-8–19-9
    physical transducer of SPR, 19-8
    sensing architectures, 19-8
Interfacial landscape formalization, 19-14–19-18
Inter-nanogap-based plasmonic particles, 7-12–7-15
International Technology Roadmap for Semiconductors
        (ITRS), 9-1, 9-5
Internet of Everything (IoE), 9-3

Internet of Things (IoT), 9-3
Intra-nanogap-based plasmonic particles, 7-19–7-22
ITO, *see* Indium-tin-oxide (ITO)
ITRS, *see* International Technology Roadmap for
        Semiconductors (ITRS)

J

Janus particles, *see* Dielectric-metal particle
Josephson junctions (JJs), 13-32–13-33, 13-37
    graphene-based, 13-33–13-35
        TMD-based, 13-35–13-38
        2D material-based, 13-32–13-33
Joule heating, 11-3, 11-4

K

Kerr-effect spectroscopy, 17-5
Kinetic analysis, 19-14
    adsorption process in, 19-14, 19-16
    in biosensor engineering, 19-17–19-18
    on deterministic models, 19-14
    nonexponential kinetics, parameters of, 19-16–19-17
    stochastic generalization of interfacial
        binding, 19-16
Kretschmann configuration, 16-3, 16-8, 19-4, 19-8

L

Laguerre-Gaussian beams, 6-12
Landau Level (LL) energy, 13-3–13-4, 13-12, 13-13
Langmuir–Blodgett technique, 14-7
Langmuir model, 19-14, 19-16
Large particles, 6-6–6-7
Laser-scanning-assisted dark-field (DF), 7-15
Laser system, 6-2
Lead-sulphide quantum dots, 16-6, 16-7
Leaky capacitor, 12-2
LER, *see* Line edge roughness (LER)
Levy distribution, 19-16, 19-17
Light conversion, SPP, 18-24–18-25, 18-28
Light detection system, 18-13, 18-14
Light energy and momentum, 6-1
Light microscopy, 2-1
Linear dispersion curves, free space photons, 17-2
Line edge roughness (LER), 9-7
Line optical tweezers (LOT), 6-10
Lipid membranes, 2-2
Lithography, 7-29
LL energy, *see* Landau Level (LL) energy
Localized surface plasmon (LSP), 7-2–7-3, 17-2, 18-1
    multipole modes, 18-7–18-8
        angular distribution of, 18-10–18-12
        radiation of, 18-8–18-10
    photon map, 18-19–18-24
        dipole mode, 18-20–18-22
        quadrupole mode, 18-22–18-23
    size dependence, 18-24
    thin triangular nanoprism, 18-17–18-18
Localized surface plasmon resonance (LSPR), 7-11, 14-1

Logistic curve, 19-11

Long-junction limit, ABS, 13-17

Long-range (LR) mode, 18-6, 18-7

Long-range SPP (LRSPP), 16-6, 16-7

Long-term memory, 11-5, 11-6

Long-term plasticity (LTP), 11-5

Lorentz function, 18-4, 18-5

Lorentz symmetry, 13-37

LOT, *see* Line optical tweezers (LOT)

Low-dimensional nanostructures

    metallurgical junctions, 3-6–3-7

    metal semiconductor junctions, 3-6

    type I quantum well, 3-7

    type II quantum well, 3-7

Low resistance state (LRS), 11-2, 11-3

Low-voltage emission, 2DEG, 1-2–1-3; *see also*
        Two-dimensional electron gas (2DEG)

    edge emission from, 1-13, 1-14

    with edge injection, 1-12–1-13

    fabrication and characterization, 1-16

    graphene-oxide-semiconductor, 1-3–1-4

        I–V characteristic of, 1-6, 1-7

        photocurrent response, 1-6, 1-7, 1-8

        spectral dependence, 1-9–1-11

        UV-enhanced photodetection in, 1-4

        with void channel, 1-6, 1-7, 1-8, 1-11

    nanoparticle-patterned etch mask, 1-14–1-15

    nanoscale quantum-dot organic light-emitting
        diode, 1-11–1-12

LR mode, *see* Long-range (LR) mode

LRS, *see* Low resistance state (LRS)

LRSPP, *see* Long-range SPP (LRSPP)

LSP, *see* Localized surface plasmon (LSP)

LSPR, *see* Localized surface plasmon resonance (LSPR)

LTP, *see* Long-term plasticity (LTP)

M

Macromolecular receptor (MR), 19-11, 19-12

Macroscopic kinetic behavior, 19-14, 19-16

Magnetic field

    electron–hole crossover in perpendicular, 13-22

    spin states in-plane, 13-22–13-23

Magneto-optical effects, 17-1

Magneto-optical Kerr effect (MOKE)

    characterization, 17-5–17-6

    P-MOKE enhancement, 17-6–17-7

    possible geometries, 17-6

    requirements, 17-5

Magneto-optic SPR (MO-SPR) sensor, 19-7

Magneto-plasmonics, 17-2

Magnetotransport, 13-25–13-26

Majorana bound states (MBS), 13-38

Majorana fermions (MFs), 13-38

Malachite green isothiocyanate (MGITC), 7-11

MAR, *see* Multiple Andreev reflections (MAR)

Marangoni flow suppression, 14-3, 14-5

Markov property, 12-4

Master equation of single-electron tunneling,
        12-4–12-5, 12-7

Mathematical model, 12-1–12-2

Maxwell electromagnetic tensor, 6-7

Maxwell's equation, 7-2, 16-2, 18-2, 18-7

MBE, *see* Molecular beam epitaxy (MBE)

MBS, *see* Majorana bound states (MBS)

Mean Squared Error (MSE), 12-8

MEF, *see* Metal-enhanced fluorescence (MEF)

Memoryless property, 12-4

Memory scaling

    floating gate devices, 9-15–9-19

    mass storage, 3D integration, 9-15–9-16

    nonvolatile memories, 9-13–9-17

    SRAM *vs.* DRAM, 9-13–9-14

    zero intrinsic variability, 9-19–9-23

Memory window, 11-5

Memristance/memory-dependent conductivity,
        11-2–11-4, 11-7

Memristor, 11-1–11-2

    types of, 9-17

Mesoscopic particles, 6-7–6-8

Metal–dielectric–metal stub resonators, 16-12–16-13

Metal-enhanced fluorescence (MEF), 7-1, 7-6–7-7

Metal-induced gap states (MIGS), 10-12

Metal nanostructures, 7-9–7-10

Metal–organic chemical vapor deposition (MOCVD), 8-1

Metal–organic vapor phase epitaxy (MOVPE), 8-1

Metal-oxide semiconductor (MOS), 1-1, 12-1, 12-6

    emission of 2DEG, 1-5–1-6

Metal oxide semiconductor field effect transistors
        (MOSFETs), 1-1, 1-2, 10-1

    microelectronics, 9-1–9-4, 9-10–9-12, 9-19–9-21

Metal particles, 6-8–6-9

Metamolecules, 14-12

MFs, *see* Majorana fermions (MFs)

MGITC, *see* Malachite green isothiocyanate (MGITC)

MIB model, 12-8

Micron-sized plasmonic supercrystals, 14-15

Micro-objects, 6-8–6-11

Mid-infrared (MIR), 16-5

Mie theory, 6-6, 7-3–7-4, 18-8

MIR, *see* Mid-infrared (MIR)

MOCVD, *see* Metal–organic chemical vapor deposition
        (MOCVD)

Modulated ellipsometry/reflectance difference
        spectroscopy, 3-11

Modulation techniques, semiconductor nano structures,
        3-11–3-12

MOKE, *see* Magneto-optical Kerr effect (MOKE)

Molecular beam epitaxy (MBE), 8-1–8-6

Molecular order, 2-1–2-3

    actin stress fibers, 2-7–2-8

    lipid membranes, 2-7

Molecular plasmonics, 7-5–7-9

Molecular recognition, 19-9

Monitoring devices, 6-3

Monodisperse colloidal synthesis, 14-1–14-2
Monomaterial electronic devices
    CMGT, 10-9–10-11
    Schottky-type junctions, 10-8–10-9
Monte Carlo method, 12-5–12-6
Moore's law, 12-1, 13-1
    and scaling of CMOS, 9-1–9-3
    3D into zero power systems, 9-24–9-27
More Moore, 9-24–9-27
More than Moore, 9-24–9-27
Morgan−Mercer−Flodin equation, 19-11
MOS, *see* Metal-oxide semiconductor (MOS)
MOSFETs, *see* Metal oxide semiconductor field effect
    transistors (MOSFETs)
MoS_2 nanoribbon
    AFM image of, 13-30
    Fock–Darwin spectrum of, 13-32
    gate-defined, 13-31
MO-SPR sensor, *see* Magneto-optic SPR (MO-SPR)
    sensor
MOVPE, *see* Metal−organic vapor phase epitaxy
    (MOVPE)
MR, *see* Macromolecular receptor (MR)
MSE, *see* Mean Squared Error (MSE)
Multiple Andreev reflections (MAR), 13-36
Multiple scattering regime, 16-4
Multipole mode LPP
    angular distribution of, 18-10–18-12
    radiation of, 18-8–18-10
    spherical nanoparticle, 18-7–18-8

N

NA, *see* Numerical aperture (NA)
N-(2-Aminoethyl)-3-aminopropyl-trimethoxysilane
    (AEAPTMS), 1-15
NAND flash memories, 9-3, 9-4
Nanocrevice-based plasmonic particles, 7-15–7-19
Nanodevice(s)
    definition of, 13-41
    preparation
        asymmetric double-junction structure, 15-4
        transmission measurement setup, 15-4–15-5
Nanoelectromechanical systems (NEMS), 9-24
Nanoelectronics
    CMOS, 9-1–9-3
    market drivers, innovations, and trends, 9-3–9-5
    memory scaling
        floating gate devices, 9-15–9-19
        mass storage, 3D integration, 9-15–9-16
        nonvolatile memories, 9-13–9-17
        SRAM *vs.* DRAM, 9-13–9-14
        zero intrinsic variability, 9-19–9-23
    More Moore, 9-24–9-27
    More than Moore, 9-24–9-27
    MOSFET, 9-6–9-12
    zero intrinsic variability, 9-19–9-23

zero power systems, 9-24–9-27
Nanoflowers (NFs); *see also* Gallium nitride (GaN)
    characteristics of, 8-4–8-6
    growth of, 8-2–8-6
    nitride-based, 8-2–8-3
    photodetection devices, 8-6–8-15
Nanogap-enhanced Raman spectroscopy (NERS), 7-6
Nanohole fabrication
    nanoparticle-patterned etch mask for, 1-14–1-15
Nanolithography, 16-12
Nanoobjects, 6-8–6-11
Nanoparticle-patterned etch mask
    for nanohole fabrication, 1-14–1-15
Nanoparticles (NPs), 11-5–11-6, 14-1
    photon map, 18-19–18-24
    plasmonic properties of, 14-1
Nanopatterned materials
    behaviour of materials, 10-2
    semmeals, 10-3–10-4
Nanophotonic waveguides
    loss performance, 4-3
    optical resonators, 4-3–4-4
    ridge-type, 4-2
Nanoplasmonic structures, 16-1
Nano-QD-OLED fabrication, 1-16
Nanoscale films, 11-5, 11-6
Nanoscale memristance, 11-2–11-4
Nanoscale quantum-dot organic light-emitting diode,
    1-11–1-12
    edge emission from, 1-13, 1-14
    with edge injection, 1-12–1-13
    fabrication and characterization, 1-16
    nanoparticle-patterned etch mask, 1-14–1-15
Nanoscale THz and optical devices, 16-13
Nanosheets, 9-12
Nanostructures (NSs), 8-1
Nanowires, 9-10, 9-11, 9-12
Near-infrared (IR) laser, 6-2
NEMS, *see* Nanoelectromechanical systems (NEMS)
NERS, *see* Nanogap-enhanced Raman spectroscopy
    (NERS)
Neuromorphic computing, 4-10–4-11, 11-4–11-5
Neuromorphic metal-oxide memristors, 11-7
Neuromorphic nanoelectronics, 11-1–11-2
    aim of, 11-2
    nanoparticles, 11-5–11-6
    nanoscale memristance, 11-2–11-4
    networks of, 11-6–11-7
    neuromorphic computing, 11-4–11-5
    thin films and 2D materials, 11-5, 11-6
Nitride-based nanoflowers (NFs), 8-2–8-3
Nonexponential kinetics, parameters of, 19-16–19-18
Nonvolatile memories (NVMs), 9-4, 9-5, 9-13–9-17
Normalized loss, plasmonic resonance, 15-5–15-7
Novel optical tweezers, 6-4–6-6
NSs, *see* Nanostructures (NSs)

n-type dopants, 10-1
Nucleation mechanism, 8-4
Numerical aperture (NA), 6-1
NVMs, *see* Nonvolatile memories (NVMs)
NW, *see* Quantum wires/nanowires (NW)

O

Objective lens, 6-2–6-3
Ohmic loss, 16-4
OLED, *see* Organic Light-Emitting Diode (OLED)
One-dimensional nanoparticle assembly, 14-8, 14-9, 14-11, 14-12
One dimensional plasmonic crystal, SPP, 18-25–18-30
One-photon fluorescence (1PF), 2-3–2-5
One-photon photoluminescence (PL), 7-9–7-10
One transistor one resistor (1T1R) memory, 9-16, 9-18
1T'-transition metal dichalcogenides, 13-7–13-10
    band structures of, 13-8
    quantum spin hall edge states in, 13-38–13-41
Optical antennas, RF antennas *vs.,* 20-6
Optical beams focus, 16-4
Optical force measurement, 6-4
Optical gain, 16-8
Optical imaging, 2-1
Optical microscopy, 2-1
Optical trap stiffness calibration, 6-3
    hydrodynamic drag method, 6-3
    power spectrum method, 6-3–6-4
Optical tweezers
    application of, 6-1–6-2
    biology, 6-11–6-12
    biosamples, 6-9
    measurement, 6-2–6-6
    operation, 6-2–6-6
    physical principle of, 6-6–6-8
    physics, 6-12
    principle of, 6-1–6-2
    setup, 6-2–6-3, 6-5
Optical vortex beams, 6-12
Optimal interfacial arrangement, 19-8–19-9
Optothermal method, 14-5, 14-6
Organic Light-Emitting Diode (OLED), 1-11–1-12
    circular-window-pattern, 1-14
    quantum-dots in, 1-11 (*see also* Quantum-Dot Organic Light-Emitting Diode (QD-OLED))
    ring trench pattern, 1-13, 1-14
Organic memristors, 11-7
Orthodox theory of single-electron tunneling, 12-3, 12-7, 12-8
Oscillatory properties, 12-2
Otto configuration, 16-3, 16-8

P

Pair-pulsed depression (PPD), 11-5
Pair-pulsed facilitation (PPF), 11-5

PAMBE, *see* Plasma-assisted molecular beam epitaxy (PAMBE)
Panchromatic photon map, 18-32
Particle surface functionalizations, 14-7
Patterning, 9-23–9-24
Pauli matrices, 13-5
PEEM, *see* Photoemission electron microscopy (PEEM)
Permanent-oriented corrugations, 14-12
Permittivity
    of metal, 15-3, 15-4
    tunneling region, 15-6
Phalloidin-Alexa Fluor 488, 2-8
Phase-change materials (PCMs)
    different materials, 4-4
    electrical properties, 4-2
    far-field coupled
        plasmonics, 4-8–4-9
        reconfigurable metasurfaces, 4-7
        reflective displays, 4-7–4-8
    multilevel memory cell, 4-9
    neuromorphic computing, 4-10–4-11
    optical properties, 4-1–4-2
    optical switching and routing, 4-9
    photonic arithmetic processing, 4-9–4-10
    switching mechanism, 4-4–4-5
    usage, 4-1
    waveguide-coupled
        experimental techniques, 4-5
        networks, 4-6–4-7
        switching events, 4-5–4-6
        transmission measurement, 4-5
Phase-change photonics
    far-field coupled PCMs
        plasmonics, 4-8–4-9
        reconfigurable metasurfaces, 4-7, 4-8
        reflective displays, 4-7–4-8
    waveguide-coupled PCMs
        experimental techniques, 4-5
        networks, 4-6–4-7
        switching events, 4-5–4-6
        transmission measurement, 4-5
Photocarrier 2DEG
    emission and transport of, 1-8–1-9
    multiplication effect, 1-10, 1-11
Photoconductor gain effect, 1-10
Photocurrent response
    of graphene-oxide-semiconductor, 1-6, 1-7, 1-8
    spectral dependence of, 1-9–1-11
Photocurrent-saturation voltage, 1-9
Photodetection
    applications, 8-13
    background of, 8-6–8-8
    performance evaluation parameters, 8-8–8-10
    self-driven operation, 8-11–8-13
    UV illumination, 8-10–8-11
Photodetector-based position detection, 6-4

Photoemission electron microscopy (PEEM), 17-8–17-10
Photogeneration rate, 1-9
Photoluminescence (PL), 7-9–7-10
    Au nanostructures, 7-25
    CiC nanoparticles, 7-31
    excitation, 3-12–3-13
    gold nanobipyramids, 7-26
    gold nanorods, 7-26
    metal enhancement, 7-22–7-32
    metal nanostructures, 7-9–7-10, 7-27
    plasmonic coupling effect, 7-30
Photon map, 18-19–18-24
    angle resolved monochromatic image, 18-15, 18-16
    dipole mode, 18-20–18-22
    imaging mechanism, 18-16
      flat surface, 18-16–18-17
      three-dimensional structure, 18-17
    quadrupole mode, 18-22–18-23
    size dependence of, 18-24
Photothermal therapy, 7-11
Physical optimization, 19-8
Physical transducer (PT), 19-2, 19-4, 19-7, 19-8
Pinhole mask configuration, 18-20, 18-29, 18-33
PL, *see* Photoluminescence (PL)
Planar X-ray waveguides (PXW)
    angular radiation intensity distribution
      energy spectrum modification, 5-9, 5-10
      slit width, 5-8
    reflectors, 5-2–5-5
Plasma-assisted molecular beam epitaxy (PAMBE), 8-1, 8-6
Plasma enhanced chemical vapor deposition (PECVD), 8-1
Plasmon hybridization model, 7-4
Plasmonic-based lasing in nanostructures, 16-11
    spaser biological application, 16-11–16-12
    spasing process, 16-11
Plasmonic coupling effect, 7-29–7-32
Plasmonic crystal (PlC)
    defintion of, 18-1
    one-dimensional, 18-13, 18-15, 18-25–18-30
    two-dimensional, 18-15, 18-16, 18-30–18-33
Plasmonic effects, 7-5–7-10
Plasmonic intra-nanogap particles, 7-22
Plasmonic materials, losses in
    Ohmic loss, 16-4
    scattering, 16-4
Plasmonic metal nanostructures
    dielectric function, 7-4–7-5
    EM wave equations, 7-2
    localized surface plasmons, 7-2–7-3
    Maxwell's equation, 7-2
    Mie theory, 7-3–7-4
    plasmon hybridization model, 7-4
Plasmonic molecules, 14-3

Plasmonic nanocrevice, 7-17
Plasmonic nanoparticles, 7-12, 7-13, 14-1
    anisotropic physical properties, 14-15–14-17
    colloidal self-assembly, 14-1–14-3
    directed self-assembly, 14-7, 14-8, 14-9
    external field-directed assembly, 14-5–14-7
    with magnetic properties, 14-5
    one-dimensional nanoparticle assembly, 14-8, 14-9, 14-11, 14-12
    optical properties of, 14-11
    self-assembled structures, 14-7, 14-8, 14-9, 14-10
    solvent evaporation self-assembly, 14-3
    structural diversity of, 14-1, 14-2
    template-directed assembly, 14-3–14-5
    three-dimensional nanoparticle assembly, 14-13, 14-14, 14-15, 14-16–14-17
    two-dimensional nanoparticle assembly, 14-12, 14-14, 14-15
    by wrinkle-assisted assembly, 14-11–14-13
Plasmonic oligomer meta molecules, 14-14
Plasmonic resonance, 15-5, 15-6
Plasmonics, 15-2, 15-4, 17-1, 18-1; *see also* Coulomb blockade plasmonic switch
Plasmonic waveguiding, 14-11
Plasmon oscillation, 18-17, 18-18
Plasmon–polariton state, formation of, 19-3
Plasmon structure, 19-4
PlC, *see* Plasmonic crystal (PlC)
P-MOKE, *see* Polar magneto-optic Kerr effect (P-MOKE)
Poisson equation, 1-3
Poisson processes, 12-6
Polarizability, 7-3
Polarization, 2-1
Polarization-dependent enhancement, 20-10, 20-11
Polarization-resolved SHG microscopy, 2-11
Polarized fluorescence microscopy, 2-5–2-6
Polarized nonlinear microscopy
    second harmonic generation
      polarization, 2-13–2-14
      scattering process, 2-11–2-12
      subdiffraction vectorial imaging, 2-12–2-13
      tissue imaging, collagen, 2-12
    2PF, 2-9–2-10
Polarized radiation intensity, 18-10, 18-11
Polar magneto-optic Kerr effect (P-MOKE), 17-6
    enhancement, 17-6–17-7
    enhancement by SPPs
      far field, 17-7–17-8
      near field, 17-8–17-10
Poly(vinylpyrrolidone) (PVP)
    Ag, 7-16–7-17
    Au, 7-16–7-17
Poly(9-vinylcarbazole) (PVK), 1-12
Polydimethylsiloxane (PDMS), 14-5, 14-12, 14-13
Polymer-on-polymer stamping, 14-3

Polystyrene-coated gold NPs, 14-7, 14-10

Poole–Frenkel emission, 11-4

Position-sensitive detector (PSD), 6-4

Power spectrum method, 6-3–6-4

PPF, *see* Pair-pulsed facilitation (PPF)

Prism coupling/attenuated total internal reflection, 16-3

Probability of realization, 19-17

Protein–protein interaction, 19-12

Proximity effect, 13-16, 13-33, 13-34

Pruvost model, 12-8

PSD, *see* Position-sensitive detector (PSD)

Pseudospin degree of freedom, 13-3

PT, *see* Physical transducer (PT)

p-type dopants, 10-1

Pulse gating techniques, 13-24, 13-41

PVP, *see* Poly(vinylpyrrolidone) (PVP)

PVK, *see* Poly(9-vinylcarbazole) (PVK)

PXW, *see* Planar X-ray waveguides (PXW)

PXW resonators (PXWR)

  emergent beam divergence, 5-9–5-10

  improving beam characteristics, 5-30–5-33

  nonparallelism evaluation, 5-23–5-24

  problems

    angular divergence reduction, 5-25–5-30

    integral intensity, 5-24–5-25

  radiation beam, 5-8–5-9

  standing wave propagation, Bragg PXWR, 5-22

  x-ray flux propagation, wave guide resonance, 5-20

Pysical type sensors, 19-2

Q

QD, *see* Quadrant photodiode (QD)

QDIPs, *see* Quantum dot IR photodetectors (QDIPs)

QD-OLED, *see* Quantum-Dot Organic Light-Emitting Diode (QD-OLED)

QDs, *see* Quantum dots (QDs)

QSH, *see* Quantum spin Hall (QSH)

Quadrant photodiode (QD), 6-4

Quadrupole mode, photon map, 18-22–18-23

Quantitative analysis, 19-10

Quantitative determination of analyte, 19-11–19-12

Quantum bits (qubits), 13-1

Quantum computing, 13-1

Quantum confinement, 11-1, 13-12, 13-18, 13-20

  application, 10-6

  electron and hole pocket, 10-7

  increase in bandgap, 10-5–10-6

  mechanical aspects, 10-4

  quantum size, 10-4–10-5

  subband formation, 10-6–10-7

Quantum-corrected model (QCM)

  experimental findings, in Coulomb blockade

    nanodevice preparation, 15-4–15-5

    numerical calculations, 15-6–15-7

  overview of, 15-3–15-4

Wentzel–Kramers–Brillouin approximation, 15-2–15-3

Quantum dot IR photodetectors (QDIPs)

  optical antennas, 20-6–20-10

  polarization-dependent enhancement in, 20-10, 20-11

Quantum-Dot Organic Light-Emitting Diode (QD-OLED), 1-11–1-12

  edge emission from, 1-13, 1-14

  with edge injection, 1-12–1-13

  fabrication and characterization, 1-16

  nanoparticle-patterned etch mask, 1-14–1-15

Quantum dots (QDs), 13-1, 13-10; *see also* Double quantum dot (DQD); Single quantum dot (SQD)

  Coulomb diamond, 13-11, 13-12, 13-19, 13-21, 13-22, 13-28–13-30

  in GaAs/AlGaAs 2DEG system, 13-10, 13-12

  Raman spectra, 3-16–3-17

  v-shape feature, 13-12, 13-13, 13-19

  WSe$_2$, 13-30

  Zeeman splitting, 13-22, 13-23

Quantum efficiency, 1-9–1-10

Quantum Hall effect (QHE), 13-3, 13-4

Quantum resistance, 12-2

Quantum-size/confinement, 10-2

Quantum spin Hall (QSH), 13-9, 13-10, 13-38–13-41

Quantum transport

  Andreev reflections, 13-16–13-18

  double quantum dot, 13-13–13-16

  single QD, 13-10–13-13

Quantum tunneling, Wentzel-Kramers-Brillouin approximation, 15-2–15-3

Quantum wells, 16-5

Quantum wires/nanowires (NW), 10-2–10-3

  crystallographic orientation, 10-9

  electron energy gap, 3-3

  mechanism for synthesis, 10-3

  Raman spectrum, 3-15, 3-15–3-16

  surface chemistry on SM NWs, 10-11–10-13

Quasilinear calibration, 19-6

R

Radiation intensity, 18-9

Radiation intensity distribution, 5-4–5-5

Radiation of multipole mode LPP, 18-8–18-10

Radiation pressure, 6-1

Raman efficiency, 3-13

Raman enhancement, 7-10–7-11

  inter-nanogap-based plasmonic particles, 7-12–7-15

  intra-nanogap-based plasmonic particles, 7-19–7-22

  nanocrevice-based plasmonic particles, 7-15–7-19

  single particles, 7-11–7-12

Raman scattering

  Feynman diagram, 3-13

Raman scattering (*cont.*)
    mechanism, 3-13–3-14
    NERS, 7-6
    phonons in low-dimensional structures, 3-15–3-17
    selection rules, 3-14–3-15
    SERS, 7-5–7-6
    TERS, 7-6
Raman spectroscopy, 7-1
Random dopant fluctuations (RDF), 9-6, 9-7
Rayleigh distance, 6-3
Rayleigh scattering theory, 6-6
Ray-optics model, 6-7
RC time, *see* Resistance-capacitance (RC) time
RCWA, *see* Rigorous coupled-wave analysis (RCWA)
RDF, *see* Random dopant fluctuations (RDF)
Reactive-ion etching (RIE), 1-6, 1-14, 1-15, 1-16,
    13-29–13-30
Refractive index, 19-5, 19-6, 19-14
Resistance-capacitance (RC) time, 1-11
Resistive switching, 11-2–11-4
Resonant Raman scattering (RRS), 3-13
Retardation effects, 14-11
RF antennas *vs.* optical antennas, 20-6
RIE, *see* Reactive-ion etching (RIE)
Rigorous coupled-wave analysis (RCWA),
    18-27, 18-32
Ring deposition mechanism, 14-3
RRAMs, 9-18–9-19
Rutherford backscattering (RBS), 5-3

S

SAM, *see* Self-assembled monolayer (SAM)
Scanning transmission electron microscopy (STEM),
    18-12–18-13
Scanning tunneling microscopy (STM), 13-4, 13-5
Scattering force, 6-1–6-2
Scattering process, 16-4
Schottky barrier, 13-35
Schottky-barrier diode, 16-4, 16-7
Schottky–Mott rule, 10-9
Schottky-type junctions, 10-8–10-9
Schrödinger equation, 15-2
SCM, *see* Storage class memories (SCM)
Second harmonic generation (SHG)
    polarization, 2-13–2-14
    scattering process, 2-11–2-12
    subdiffraction vectorial imaging, 2-12–2-13
    tissue imaging, 2-12
SE devices, *see* Single-electron (SE) devices
SEF, *see* Stretched exponential function (SEF)
SEIRA, *see* Surface-enhanced IR absorption (SEIRA)
Self-assembled monolayer (SAM), 15-2, 15-4–15-6
Self-assembled structures, 14-7, 14-8, 14-9, 14-10
Semiconductor nano structures
    electronic band structure
        Bloch function, 3-1–3-2

        degenerate bands, 3-3–3-5
        metallurgical junctions, 3-6–3-7
        metal semiconductor junctions, 3-6
        SL, 3-7–3-8
        tight-binding method, 3-3
        two-band model, 3-2
        type I quantum well, 3-7
        type II quantum well, 3-7
    emission processes
        band-to-band transitions, 3-12
        bound exciton emission, 3-12
        PL excitation, 3-12–3-13
    excitonic absorption, 3-9–3-11
    fundamental absorption, 3-8–3-9
    modulation techniques, 3-11–3-12
    Raman scattering
        Feynman diagram, 3-13
        mechanism, 3-13–3-14
        phonons in low-dimensional structures,
            3-15–3-17
        selection rules, 3-14–3-15
Semiconductor structure, 16-5, 16-6
Semidigital loops, 11-4
Semimetals
    electronic structure, 10-3
    quantum confinement effect
        application, 10-6
        electron and hole pocket, 10-7, 10-8
        increase in bandgap, 10-5–10-6
        mechanical aspects, 10-4
        quantum size, 10-4–10-5
        subband formation, 10-6–10-7
        surface effects, subband formation, 10-7–10-8
SERS, *see* Surface-enhanced Raman scattering (SERS)
S-graphene-S junction, electrical characterization,
    13-33–13-34
SHG, *see* Second harmonic generation (SHG)
Shockley equation, 12-8–12-9
Short-channel effects, 10-2
Short-junction limit, 13-17
Short-range (SR) mode, 18-6, 18-7
Short-term memory, 11-4
Short-term plasticity (STP), 11-4
Sideward emission, 18-22
SILC, *see* Stress-induced leakage current (SILC)
Silicon NWs (SiNWs), 10-3, 10-5
Silicon photodiodes, 1-4
Simulated Program with Integrated Circuits Emphasis
    (SPICE), 12-9, 12-10
SIMulation Of Nanostructures (SIMON),
    12-5–12-6, 12-8
Simulation softwares, 12-3–12-4
Single-electron (SE) devices
    overview of, 12-1–12-2
    simulation methods, 12-3–12-4
        Monte Carlo method, 12-5–12-6

single-electron tunneling, 12-4–12-5
single-electron transistor, 12-2–12-3, 12-6–12-7
analytical models, 12-7–12-8
macromodels, 12-8–12-11
tunneling and tunnel junctions, 12-2
Single-electron transistors (SETs), 12-1–12-3
Coulomb blockade, 15-1, 15-2
proposed models, 12-6–12-7
analytical models, 12-7–12-8
macromodels, 12-8–12-11
Single-electron transport
in graphene, 13-18–13-20
charge pumping, 13-26–13-28
charge relaxation time, 13-24
Coulomb blockade, 13-20–13-22, 13-25–13-26
DQDs on SiO_2/Si substrates, 13-25
electron–hole crossover, 13-22
GQDs on hBN substrate, 13-28–13-29
magnetotransport, 13-25–13-26
spin states, in-plane magnetic field, 13-22–13-23
SQDs on SiO_2/Si substrates, 13-20
in 2H-TMDs, 13-29–13-32
Single-electron tunneling, 12-4–12-5
master equation of, 12-4–12-5, 12-7
orthodox theory of, 12-3, 12-7, 12-8
Single nanoparticles, 7-24–7-28
Single-particle optical spectroscopy, 14-9
Single particles, 7-11–7-12
Single quantum dot (SQD), 13-10–13-13, 13-20
Single scattering regime, 16-4
Size dependence, 18-24
SK growth mode, *see* Stranski–Kranstanov (SK) growth
mode
Skin depth, 16-2
Slitless collimator, 5-2
SLM, *see* Spatial light modulators (SLM)
Small particles, 6-6
S-N-S junction, Andreev reflections in, 13-16–13-18,
13-35
SOC, *see* Strong spin–orbit coupling (SOC)
Soft corona, 14-1–14-3
Solvent evaporation, self-assembly of NP, 14-3
SP, *see* Surface plasmon (SP)
Spaser biological application, 16-11–16-12
Spasing process, 16-11
Spatial distribution, 18-31, 18-32
Spatial light modulators (SLM), 6-5
Spectral dependence, GOS, 1-9–1-11
Spectrometer with angular scanning, 19-4, 19-5
Spectroscopic ellipsometry, 3-11
Spherical harmonics function, 18-9
SPICE, *see* Simulated Program with Integrated Circuits
Emphasis (SPICE)
Spike timing-dependent plasticity (STDP), 11-4–11-6
Spin–momentum locking, 13-9
Spin–orbit coupling, 17-5

Spin polarization, 17-5
Spin states, in-plane magnetic field,
13-22–13-23
SPP, *see* Surface plasmon polariton (SPP)
SPR, *see* Surface plasmon resonance (SPR)
SQUID-like S-G-S junction, 13-34–13-35
SRAM, *see* Static random access memories (SRAM)
SR mode, *see* Short-range (SR) mode
Stability diagram, 13-14–13-15
Static random access memories (SRAM), 9-4, 9-13
STDP, *see* Spike timing-dependent plasticity (STDP)
STEM, *see* Scanning transmission electron microscopy
(STEM)
STEM-CL, instrumentation of, 18-12–18-13, 18-14
Steric blocking, 19-9
STM, *see* Scanning tunneling microscopy (STM)
Stochastic description, single-electron device, 12-4
Stochastic generalization of interfacial binding, 19-16
Stochastic packing, 19-9
Stochatic optical reconstruction microscopy
(dSTORM), 2-8
Storage class memories (SCM), 9-4
STP, *see* Short-term plasticity (STP)
Stranski–Kranstanov (SK) growth mode, 8-4
Stress-induced leakage current (SILC), 9-4
Stretched exponential function (SEF), 19-16, 19-17
Strong spin–orbit coupling (SOC), 13-2, 13-5, 13-7
Strurrock criteria, 16-8
Subdiffraction vectorial imaging, metal nanostructures,
2-12–2-13
Subthreshold region, 10-10
Subthreshold swing (SS), 10-10
Super-resolution fluorescence polarized imaging,
2-7–2-8
Suppression of receptor–ligand, 19-9
Surface-captured light, 19-1
Surface charge density, 18-26
Surface chemistry
definition, 10-8
electrostatic effects, 10-11–10-13
Surface-enhanced IR absorption (SEIRA), 7-7–7-8,
7-8–7-9
Surface-enhanced Raman scattering (SERS), 7-1,
7-5–7-6, 7-13–7-15, 14-12
Surface equivalence theorem (SET), 20-6
Surface-induced analyte transitions, 19-12–19-13
Surface plasmon (SP), 17-1, 18-1, 18-16–18-17
compensating losses, 16-4–16-5
dipolar gain media incorporation, 16-5, 16-6
electrical injection, 16-7, 16-8
fluorescent polymer, 16-6, 16-7
lead-sulphide quantum dots, 16-6, 16-7
quantum wells, 16-5
semiconductor structure, 16-5, 16-6
SPP propagation, 16-8, 16-9
thin organic gain medium, 16-8, 16-9

Surface plasmon (SP) (*cont.*)
  2D Bragg structure, 16-8
  in DC electric field
    Cada's group analysis, 16-9–16-10
    Sydoruk's group analysis, 16-10–16-11
  mathematical formulation, 16-2–16-3
  overview of, 16-1–16-2
  plasmonic-based lasing in nanostructures, 16-11
    spaser biological application, 16-11–16-12
    spasing process, 16-11
  plasmonic materials, losses in
    Ohmic loss, 16-4
    scattering process, 16-4
Surface plasmonic waves, 20-4
Surface plasmon polariton (SPP), 6-5, 16-1, 17-1, 18-1,
    19-2, 19-4, 19-7
  applications of
    imaging and spectroscopy, 16-12
    metal–dielectric–metal stub resonators,
      16-12–16-13
    nanolithography, 16-12
    nanoscale THz and optical devices, 16-13
  categorization, 17-2
  cavity, 18-28–18-30
  coupling (2D) lattices, 17-3–17-4
  decay length of, 18-4–18-6
  dielectric function of metals, 18-4
  dispersion curves, 17-2, 17-3
    Ni hexagonal lattice, 17-4–17-5
  dispersion relation of, 16-3, 18-3–18-4, 18-26, 18-33
  electromagnetic field at plane interface, 18-2–18-3
  enhancement of polar Kerr rotation, 17-7–17-8
  excitation, 16-7, 17-3
    grating coupling, 16-3
    by high-energy electrons, 18-6–18-7
    optical beams focus, 16-4
    prism coupling/attenuated total internal
      reflection, 16-3
  grating coupling technique, 17-3
  light conversion, 18-24–18-25, 18-28
  one-dimensional plasmonic crystal, 18-25–18-30
  propagation, 16-2–16-4
  propagation length, 18-5, 18-6
  scattering process, 16-4
  SP frequency, 17-3
  two-dimensional plasmonic crystal, 18-30–18-33
Surface plasmon resonance (SPR), 20-1
  advantages of, 19-2, 19-18
  analysis of 3D objects, 19-12
  analytical workflow, 19-9, 19-10, 19-11–19-13
  based biosensors (*see* Biosensing)
  based systems in global market, 19-2
  data mining, 19-13–19-18
  description of, 19-1–19-2
  excitation of SPP, 19-2–19-5
  with linear standard curve, 19-6–19-7
  physical transducer of, 19-2, 19-4, 19-8

scaling factor, 20-6
for sensing applications, 19-5–19-6
spectrometer with angular scanning, 19-4, 19-5
technology overview, 19-7
wave vector of, 19-4, 20-4–20-6
Sydoruk's group analysis, 16-10–16-11
Symmetry order decomposition, 2-2
Synaptic weight, 11-2, 11-4, 11-5
Synergistic approach, 19-16

T
Tansverse electric field (TE), 18-2
TASA, *see* Template-assisted self-assembly (TASA)
TE-like mode, dielectric waveguides, 4-2
TEM, *see* Transmission electron microscopy (TEM)
Template-assisted self-assembly (TASA), 14-4–14-5,
    14-14, 14-15
Template-directed assembly, plasmonic nanoparticles,
    14-3–14-5
TER, *see* Total external reflection (TER)
Terminal velocity, 1-6
TERS, *see* Tip-enhanced Raman scattering (TERS)
Testing X-ray beams, 5-1
  polycapillary optics systems, 5-2
  radiation intensity distribution, 5-4–5-5
TFET, *see* Tunnel field effect transistor (TFET)
Thermophoresis, 14-5
Thin film(s), 10-3
  batteries, 9-27
  neuromorphic nanoelectronics, 11-5, 11-6
Three-dimensional materials, 9-9–9-10
Three-dimensional nanoparticle assembly, 14-13, 14-14,
    14-15, 14-16–14-17
TI, *see* Topological insulators (TI)
Tight-binding method, 3-3
Time-reversal invariant momenta (TRIM), 13-9
Time-reversal symmetry (TRS), 13-6, 13-9, 13-32, 13-38
Tip-enhanced Raman scattering (TERS), 3-16, 7-6
Tip-induced deformation, 13-29
Tissue imaging, 2-12
TL theory, *see* Transmission line (TL) theory
TM, *see* Transverse magnetic field (TM)
T-matrix method, 6-7
TMDs, *see* Transition metal dichalcogenides (TMDs)
TM-like mode, dielectric waveguides, 4-2
Top-gate geometry, 13-20
Topological insulators (TI), 13-2
Topological phase, 13-7, 13-9, 13-10
Total external reflection (TER), 5-18
Traditional competitive analysis, 19-11
Transducer-based sensors, 19-8, 19-13
Transistor
  doping profile, 10-1
  miniaturization of designs, challenges, 10-2
  types, 10-1
Transition metal dichalcogenides (TMDs),
    10-3–10-4, 13-1

atomistic structures, 13-8

1T′ phase, 13-7–13-10

2H phase, 13-2, 13-5–13-7

    single-electron transport in, 13-29–13-32

Transmagnetic (TM) plane wave, 20-4, 20-6

Transmission electron microscopy (TEM), 7-23

Transmission line (TL) theory, 20-7

Transmission probability, 15-3

Transmon qubit, 13-32, 13-33

Transparency window, 4-3

Transverse magnetic field (TM), 18-2, 18-3

Trapped objects, 6-5

Trigonal prismatic coordination, 13-6

TRIM, *see* Time-reversal invariant momenta (TRIM)

Triple point, 13-16

TRS, *see* Time-reversal symmetry (TRS)

TSS, *see* Typical Spatial Scale (TSS)

Tunnel field effect transistor (TFET), 9-12–9-13

Tunneling and tunnel junctions, 12-2

    master equation of single-electron, 12-4–12-5

Tunneling probability, 15-3, 15-4, 15-6

Two-band model, 3-2

2D Bragg structure, 16-8

Two-dimensional electron gas (2DEG) system, 1-1, 13-3, 13-4, 13-10

    emission and transport, 1-8–1-9

    graphene/SiO_2/Si structure, 1-3–1-4

    low-voltage emission of, 1-2–1-3

    overview of, 1-1–1-2

2D gold metamolecules, 14-14

Two-dimensional hole gas (2DHG), 1-1

2D materials

    graphene and hBN, 13-2–13-5

    Josephson junctions, 13-32–13-33

        graphene-based, 13-33–13-35

        TMD-based, 13-35–13-38

    nanowires and nanosheets, 9-10–9-12

    neuromorphic nanoelectronics, 11-5

    1T′-transition metal dichalcogenides, 13-7–13-10

        quantum spin hall edge states in, 13-38–13-41

    overview of, 13-1–13-2

    quantum transport

        Andreev reflections, 13-16–13-18

        double quantum dot, 13-13–13-16

        single QD, 13-10–13-13

    single-electron transport, in graphene, 13-18–13-20

        charge pumping, 13-26–13-28

        charge relaxation time, 13-24

        Coulomb blockade, 13-20–13-22, 13-25–13-26

        DQDs on SiO_2/Si substrates, 13-25

        electron–hole crossover, 13-22

        GQDs on hBN substrate, 13-28–13-29

        magnetotransport, 13-25–13-26

        spin states, 13-22–13-23

        SQDs on SiO_2/Si substrates, 13-20

    single-electron transport, in 2H-TMDs, 13-29–13-32

2H-transition metal dichalcogenides, 13-5–13-7

    single-electron transport in, 13-29–13-32

Two-dimensional nanoparticle assembly, 14-12, 14-14, 14-15

Two-dimensional plasmonic crystal, 18-30–18-33

2H-transition metal dichalcogenides (2H-TMDS), 13-2

    single-electron transport in, 13-29–13-32

Two-photon fluorescence (2PF), 2-9–2-10

Typical Spatial Scale (TSS), 19-9

U

Uchida model, 12-7, 12-8

Ultraviolet (UV) illumination, 8-10–8-11

UV-enhanced photodetection, 1-4

V

Valence band (VB), 3-2, 10-3

Valence change mechanism (VCM), 11-3

Valley index, 13-5–13-7, 13-12

van der Waals forces, 14-2

Vapor–liquid–solid (VLS), 8-1

Vapor–solid (VS), 8-1

VCM, *see* Valence change mechanism (VCM)

Very Large Scale Integration (VLSI), 11-4, 11-6, 11-7, 12-3

Video-based position detection, 6-4

VLS, *see* Vapor–liquid–solid (VLS)

VLSI, *see* Very Large Scale Integration (VLSI)

Volatile memories, 9-4

Volmer–Weber (VW) growth mode, 8-4

Voltage scaling, 9-12–9-13

Volume plasmon polariton, 17-3

von Neumann bottleneck, 11-2

VS, *see* Vapor–solid (VS)

V-shape feature, 13-12, 13-13, 13-19

VW growth mode, *see* Volmer–Weber (VW) growth mode

W

Wave function, 15-2–15-3, 15-6

Waveguide-coupled PCMs

    experimental techniques, 4-5

    networks, 4-6–4-7

    switching events, 4-5–4-6

    transmission measurement, 4-5

Waveguide-resonance mechanism, 5-2

    nonzero rest mass, 5-33–5-34

Wavevector, SPP, 16-3

Wentzel–Kramers–Brillouin (WKB) approximation, 15-2–15-3, 15-6

Weyl semimetals, 13-36–13-38

Widespread technique, 14-4, 14-5

Winner-takes-all neuronal model, 11-5

WKB approximation, *see* Wentzel–Kramers–Brillouin (WKB) approximation

Wood's anomaly, 17-3

Wulff shapes, 14-14

X

X-ray beam reflection, 5-14–5-16
X-ray flux total reflection, 5-16–5-19
X-ray flux waveguide, 5-19–5-24
X-ray photon fluxes, 5-1
X-ray standing wave (XSW), 5-1
    intensity distribution function, 5-20
    interference field, 5-16, 5-18
X-ray tracing code, 5-2

Y

Young modulus, 14-14

Z

Zeeman splitting, 13-22, 13-23
Zero-dimensional structures, 10-2
Zero intrinsic variability, 9-19–9-23
Zigzag nanoribbons, 13-18

Printed and bound by CPI Group (UK) Ltd, Croydon, CR0 4YY

17/10/2024

01775672-0020